Selected Topics in Gravity, Field Theory and Quantum Mechanics

Selected Topics in Gravity, Field Theory and Quantum Mechanics

Editors

Steven Duplij
Michael L. Walker

MDPI • Basel • Beijing • Wuhan • Barcelona • Belgrade • Manchester • Tokyo • Cluj • Tianjin

Editors

Steven Duplij
University of Münster
Germany

Michael L. Walker
University of New South
Wales
Australia

Editorial Office
MDPI
St. Alban-Anlage 66
4052 Basel, Switzerland

This is a reprint of articles from the Special Issue published online in the open access journal *Universe* (ISSN 2218-1997) (available at: https://www.mdpi.com/journal/universe/special_issues/GFTQM).

For citation purposes, cite each article independently as indicated on the article page online and as indicated below:

LastName, A.A.; LastName, B.B.; LastName, C.C. Article Title. *Journal Name* **Year**, *Volume Number*, Page Range.

ISBN 978-3-0365-5905-6 (Hbk)
ISBN 978-3-0365-5906-3 (PDF)

Contents

About the Editors

Steven Duplij

Steven Duplij (Stepan Douplii) is a theoretical and mathematical physicist from the University of Munster, Germany. He was born in Chernyshevsk–Zabaikalsky, Russia, and studied at Kharkiv University, Ukraine, where he gained his PhD in 1983. While working at Kharkiv, he received the title Doctor of Physical and Mathematical Sciences by Habilitation in 1999.

Dr Duplij is the editor-compiler of 'Concise Encyclopedia of Supersymmetry' (2005, Springer), and is the author of more than a hundred scientific publications and several books. He is listed in the World Directory Of Mathematicians, Marques Who Is Who In America, the Encyclopedia of Modern Ukraine, the Academic Genealogy of Theoretical Physicists and the Mathematics Genealogy Project. His scientific directions include supersymmetry and quantum groups, advanced algebraic structures, gravity and nonlinear electrodynamics, constrained systems and quantum computing.

Michael L. Walker

Dr Michael Walker completed his Ph.D. at the Australian National University studying chiral symmetry breaking in supersymmetric QED. He then went to study the monopole condensate in the QCD vacuum using the Cho-Dun-Ge decomposition which he adapted to Higgsless symmetry breaking and supersymmetry breaking. His more recent work uses the Clairaut-based formalism to study the implications for the particle interpretation in second quantization, both in QCD and gravity.

His work also includes the application of machine learning to drug design, computer modelling of hormone transport in plants and epidemiological modelling. Currently, he is working on a model of spacetime in which time and relativity emerge spontaneously from four dimensional Euclidean space. His academic affiliation is with the Kirby Institute at the University of New South Wales.

Editorial

Editorial: Selected Topics in Gravity, Field Theory and Quantum Mechanics

Michael L. Walker [1] and Steven Duplij [2,*]

1 Kirby Institute, University of New South Wales, Kensington, NSW 3010, Australia
2 Center for Information Technology (WWU IT), Universität Münster, Röntgenstrasse 7-13, D-48149 Münster, Germany
* Correspondence: douplii@uni-muenster.de

"Selected topics in Gravity, Field Theory and Quantum Mechanics" is for physicists wanting a fresh perspective into quantum gravity. Its content therefore does not include refinements of established approaches but rather brings new methods and approaches to various aspects of the problem. Our expectation that this will lead to further insight is supported by some papers having been cited already [1–5].

The first four contributions bring new, or at least unconventional, mathematical tools to describe the Hamiltonian dynamics of either conformable manifolds or non-trivial background curvature, with consequences for second quantization, spacetime dynamics and the constants of motion. The opening article by the editors [6] uses the Clairaut-based generalisation of the Hamiltonian formalism to study the effects of a non-trivial ground state in a gauged Lorentz symmetry theory on second quantisation. The Clairaut formalism alters the Poisson bracket to rigorously incorporate degrees of freedom which are not dynamic in the usual sense. In a similar vein, Hounnkonnou et al. consider a Poisson algebra whose bracket is based on a conformable differential and construct, among other things, Hamiltonian vector fields and other related objects on conformable Poisson-Schwarzchild and FLRW manifolds [7]. The paper by Znojil [1] addresses the issues of using the Wheeler-de Witt equation to describe the quantum evolution of the cosmos near the big bang singularity. The problem of solutions being "void of a physical meaning" is addressed by replacing the (non-Hermitian) Schroedinger picture with the corresponding Dirac interaction picture. A highly detailed review of quantum current algebra symmetry representations in integrable Hamiltonian systems from both a geometric and analytical perspective is provided by Prykarpatski [8].

The next three papers focus on quantum mechanics. Krivoruchenko [9] presents a logical construction of the linear vector nature of the quantum state, and by extension linear superposition, from the basic principles of quantum statics, number theoretic basis of physics and quantum covariance. The following paper [2] generalises Huygens-Fresnel superposition to massive particles and non-linear field theories using Kirchhoff's integral theorem. Zooming in from quantum mechanics to quantum gravity [3] shows that the non-Abelian component of the dynamic algebra is essential to general covariance. We have also included detailed analyses of the polyadic and ternary algebraic properties of quantum mechanics. One of the editors (S. Duplij) generalised the algebra of the direct product [4] in quantum mechanics with implications for the particle content of any elementary particle model. Also exploring the generalised algebraic properties of quantum mechanics, Bruce [5] reviews the construction of semiheaps and their operators on a Hilbert space and explores how symmetries in a quantum induce homomorphisms between semiheaps and ternary algebras. The final paper [10] is a review covering topics which intersect with the other papers in this collection.

Citation: Walker, M.L.; Duplij, S. Editorial: Selected Topics in Gravity, Field Theory and Quantum Mechanics. *Universe* **2022**, *8*, 572. https://doi.org/10.3390/universe8110572

Received: 24 October 2022
Accepted: 24 October 2022
Published: 30 October 2022

Funding: This research received no external funding.

Conflicts of Interest: The authors declare no conflict of interest.

References

1. Znojil, M. Wheeler-DeWitt equation and the applicability of crypto-Hermitian interaction representation in quantum cosmology. *Universe* **2022**, *8*, 385. [CrossRef]
2. Brezhnev, Y.V. Linear superposition as a core theorem of quantum empiricism. *Universe* **2022**, *8*, 217. [CrossRef]
3. Bojowald, M. Abelianized Structures in Spherically Symmetric Hypersurface Deformations. *Universe* **2022**, *8*, 184. [CrossRef]
4. Duplij, S. Polyadic analogs of direct product. *Universe* **2022**, *8*, 230. [CrossRef]
5. Bruce, A.J. Semiheaps and Ternary Algebras in Quantum Mechanics Revisited. *Universe* **2022**, *8*, 56. [CrossRef]
6. Walker, M.L.; Duplij, S. Gauge gravity vacuum in constraintless Clairaut-type formalism. *Universe* **2022**, *8*, 176. [CrossRef]
7. Hounkonnou, M.N.; Landalidji, M.J.; Mitrovic, M. Einstein field equation, recursion operators, Noether and master symmetries in conformable Poisson manifolds. *Universe* **2022**, *8*, 247. [CrossRef]
8. Prykarpatski, A.K. Quantum Current Algebra in Action: Linearization, Integrability of Classical and Factorization of Quantum Nonlinear Dynamical Systems. *Universe* **2022**, *8*, 288. [CrossRef]
9. Krivoruchenko, M.I. Superposition Principle and Kirchhoff's Integral Theorem. *Universe* **2022**, *8*, 315. [CrossRef]
10. Obukhov, V.V. Maxwell's Equations in Homogeneous Spaces for Admissible Electromagnetic Fields. *Universe* **2022**, *8*, 245. [CrossRef]

MDPI

Article

Gauge Gravity Vacuum in Constraintless Clairaut-Type Formalism

Michael L. Walker [1] and Steven Duplij [2,*]

[1] Kirby Institute, University of New South Wales, Kensington, NSW 2033, Australia; m.walker@aip.org.au
[2] Center for Information Technology (WWU IT), Universität Münster, Röntgenstrasse 7-13, D-48149 Münster, Germany
* Correspondence: douplii@uni-muenster.de

Abstract: The gauged Lorentz theory with torsion has been argued to have an effective theory whose non-trivial background is responsible for background gravitational curvature if torsion is treated as a quantum-mechanical variable against a background of constant curvature. We use the CDG decomposition to argue that such a background can be found without including torsion. Adapting our previously published Clairaut-based treatment of QCD, we go on to study the implications for second quantisation.

Keywords: gravity; Clairaut equation; Cho–Duan–Ge decomposition; constraintless formalism

Citation: Walker, M.L.; Duplij, S. Gauge Gravity Vacuum in Constraintless Clairaut-Type Formalism. *Universe* **2022**, *8*, 176. https://doi.org/10.3390/universe8030176

Academic Editors: Fabbri Luca and Maxim Chernodub

Received: 10 January 2022
Accepted: 9 March 2022
Published: 10 March 2022

1. Introduction

The dual superconductor model of QCD confinement requires the vacuum to contain a condensate of (chromo) magnetic monopoles. This led several authors to consider embedded, usually Abelian, subgroups within gauge groups. The early focus was on the $U(1)$ subgroup of $SU(2)$, with analyses by Savvidy [1], Nielsen and Olesen [2] and t'Hooft [3] considering the maximal Abelian gauge in which the Abelian subgroup is assumed to lie along the internal e_3 axis. While they did find a magnetic condensate to be a lower energy state than the perturbative vacuum, their analyses blatantly violated gauge covariance and offered no evidence that the chromomagnetic background was due to monopoles. There was also considerable controversy regarding the stability of such a vacuum. These issues were resolved by the Cho–Duan–Ge (CDG) decomposition [4,5], which introduces an internal vector to covariantly allow a subgroup embedding within a theory's gauge group to vary throughout spacetime. Analyses based on this approach confirmed this magnetic background [1,3] and careful consideration of renormalisation and causality [6–9] finally resolved such a condensate to be stable through several independent arguments.

It is common for analyses of QCD based on the CDG decomposition to assume the monopole condensate comprising the vacuum to provide a slow-moving vacuum background to the quantum degrees of freedom (DOFs) [6]. This was the basis of a novel approach to Einstein–Cartan gravity, in which contorsion (or torsion) is the quantised dynamic degree of freedom confined by a slow-moving classical background gravitational curvature et al. [10–12]. Their work was based on the Lorentz gauge field theory initially put forward by Utiyam, Kibble and Sciama [13–15] for which it has long been known that the non-compact nature of the Lorentz group led to the theory not being positive semi-definite. They dealt with this by performing their initial analyses in Euclidean space, transforming the Lorentz gauge group to $SO(4) \simeq SU(2) \times SU(2)$, until later work found the theory to be well defined with propagators for its canonical DOFs [16].

The theory considered in this paper is also a Lorentz gauge theory quadratic in curvature except that we set contorsion to be zero. Instead of including contorsion, we consider the Abelian decomposition of the Lorentz gauge field, whose details and consequences would be obscured by the complexities of handling contorsion properly. Because we also deal with the non-compact nature of the Lorentz group by working with $SU(2) \times SU(2)$ in Euclidean space [16], we can draw on a considerable body of literature concerning the Abelian decomposition of $SU(2)$ Yang–Mills theory and find that an interesting structure emerges without the introduction of contorsion. Additionally, like Pak et al. [16], we take our DOFs to be those of the Lorentz gauge fields instead of the metric and/or vierbein. To avoid third-order derivatives from entering the equations of motion (EOMs), our theory does not include localised translation symmetry (for which vierbein are required), despite it being accepted that spacetime respects the full Poincaré symmetry group. We restrict ourselves to the subgroup in this work to avoid complications and so that we can find conventional propagators for the gauge bosons with a Lagrangian quadratic in gravitational curvature. We remain mindful, however, that this is a reduced symmetry group of gravitational dynamics rendering our model to be either low-energy effective or perhaps even just a toy.

One of the more confusing mathematical subtleties of the CDG decomposition was the number of canonical degrees of freedom. Shabanov argued that an additional gauge-fixing condition is needed to remove a supposed "two extra degrees" [17] introduced by the internal unit vector field to covariantly describe the embedded subgroup(s). Bae, Cho and Kimm later clarified that this internal vector did not introduce two degrees of freedom requiring to be fixed but non-canonical DOFs without EOMs [18], while the proposed constraint was merely a consistency condition. The interested reader is referred to [6,19–21] for further details (see, also [22,23]). Cho et al. [24] approached the issue with Dirac quantisation using second-order restraints. In an earlier paper [25], however, the authors present a new approach to rigorously elucidate the dynamic DOFs from the topological. It is based on the Clairaut-type formulation, proposed by one of the authors (SD) [26,27], in a constraintless generalisation of the standard Hamiltonian formalism to include Hessians with zero determinant. It provides a rigorous treatment of the non-physical DOFs in the derivation of EOMs and the quantum commutation relations. In this paper, we apply our Clairaut approach to the gauged Lorentz group [28,29] theory with a Lagrangian quadratic in curvature.

A review of the CDG decomposition is given in Section 2, beginning with an introduction in the context of QCD before illustrating its application to $SU(2) \times SU(2)$. In Section 3, we illustrate the reduction of our theory to two copies of two-colour QCD and use one-loop results from the latter to inform us about the former. Section 4 gives a brief overview of the Clairaut–Hamiltonian formalism and uses it to study the quantisation of this theory, sorting canonical dynamic DOFs from DOFs describing the embedding of important subgroups and finding deviations from canonical second quantisation even for dynamic fields. We consider the one-loop effective dynamics in Section 5, discussing the effective particle spectrum in Section 5.1 and the possible emergence of the Einstein–Hilbert (EH) term in Section 5.2. Our final discussion is in Section 6.

2. A Review of the Covariant Abelian Decomposition of Lorentz Gauge Theory

2.1. The CDG Decomposition in $SU(2)$ QCD

2.1.1. Formalism

Abelian dominance has played a major role in our understanding of the QCD vacuum, facilitating the demonstration of a monopole condensate. That a magnetic condensate suitable for colour confinement can have lower energy than the perturbative vacuum has been known since the 1970s [1–3], but in early work the internal direction supporting the magnetic background could not be specified in a covariant manner and nor was there support for the magnetic condensate being due to monopoles. The apparent existence of destabilising tachyon modes was also an issue for some time [2,8,30]. These issues were rectified by the introduction of the CDG decomposition, which specifies the internal direction of the Abelian subgroup in a gauge covariant manner, allowing the internal direction to vary arbitrarily throughout spacetime.

The application of the CDG decomposition in N-colour ($SU(N)$) QCD is as follows: The Lie group $SU(N)$ has N^2-1 generators $\lambda^{(a)}$ ($a=1,\dots N^2-1$), of which $N-1$ are Abelian generators $\Lambda^{(i)}$ ($i=1,\dots N-1$).

The gauge transformed Abelian directions (Cartan generators) are denoted as

$$\hat{n}_i(x) = U(x)^{\dagger}\Lambda^{(i)}U(x). \tag{1}$$

Gluon fluctuations in the $\hat{n}_i$ directions are described by $c_\mu^{(i)}$, where μ is the Minkowski index. There is a covariant derivative which leaves the $\hat{n}_i$ invariant,

$$\hat{D}_\mu\hat{n}_i(x) \equiv (\partial_\mu + g\vec{V}_\mu(x)\times)\hat{n}_i(x) = 0, \tag{2}$$

where $\vec{V}_\mu(x)$ is of the form

$$\vec{V}_\mu(x) = c_\mu^{(i)}(x)\hat{n}_i(x) + \vec{C}_\mu(x), \quad \vec{C}_\mu(x) = g^{-1}\partial_\mu\hat{n}_i(x)\times\hat{n}_i(x). \tag{3}$$

The vector notation refers to the internal space, and summation is implied over $i=1,\dots N-1$. For later convenience, we define

$$F_{\mu\nu}^{(i)}(x) = \partial_\mu c_\nu^{(i)}(x) - \partial_\nu c_\mu^{(i)}(x), \tag{4}$$

$$\vec{H}_{\mu\nu}(x) = \partial_\mu\vec{C}_\nu(x) - \partial_\nu\vec{C}_\mu(x) + g\vec{C}_\mu(x)\times\vec{C}_\nu(x) = \partial_\mu\hat{n}_i(x)\times\partial_\nu\hat{n}_i(x), \tag{5}$$

$$H_{\mu\nu}^{(i)}(x) = \vec{H}_{\mu\nu}(x)\cdot\hat{n}_i(x), \tag{6}$$

$$\vec{F}_{\mu\nu}^{(i)}(x) = F_{\mu\nu}^{(i)}(x)\hat{n}_i(x) + \vec{H}_{\mu\nu}(x). \tag{7}$$

The second last term in Equation (5) follows from the definition in Equation (3). Its being a cross-product is significant as it prevents μ,ν from having the same value. The Lagrangian contains the square of this value, namely

$$H_{\mu\nu}^{(i)}(x)H_{(i)}^{\mu\nu}(x) = (\partial_\mu\hat{n}_i(x)\times\partial_\nu\hat{n}_i(x))\cdot(\partial^\mu\hat{n}_i(x)\times\partial^\nu\hat{n}_i(x)), \tag{8}$$

The form of Equation (3) might suggest the possibility of third or higher time derivatives in a quadratic Lagrangian, but we have now seen that the specific form of the Cho connection does not allow this.

The dynamical components of the gluon field in the off-diagonal directions of the internal space vectors are denoted by $\vec{X}_\mu(x)$, so if $\vec{A}_\mu(x)$ is the gluon field then

$$\vec{A}_\mu(x) = \vec{V}_\mu(x) + \vec{X}_\mu(x) = c_\mu^{(i)}(x)\hat{n}_i(x) + \vec{C}_\mu(x) + \vec{X}_\mu(x), \tag{9}$$

where

$$\vec{X}_\mu(x)\perp\hat{n}_i(x), \ \forall 1\le i<N, \quad \vec{D}_\mu = \partial_\mu + g\vec{A}_\mu(x). \tag{10}$$

The Lagrangian density is still

$$\mathcal{L}_{gauge}(x) = -\frac{1}{4}\vec{R}_{\mu\nu}(x)\cdot\vec{R}^{\mu\nu}(x), \tag{11}$$

where the field strength tensor of QCD expressed in terms of the CDG decomposition is

$$\vec{R}_{\mu\nu}(x) = \vec{F}_{\mu\nu}(x) + (\hat{D}_\mu\vec{X}_\nu(x) - \hat{D}_\nu\vec{X}_\mu(x)) + g\vec{X}_\mu(x)\times\vec{X}_\nu(x). \tag{12}$$

Gauge transformations are effected with a gauge parameter $\vec{\alpha}(x)$. Under a gauge transformation δ with $SU(2)$ parameter $\vec{\alpha}(x)$

$$\begin{aligned}\delta\hat{V}(x) &= \hat{D}_\mu\vec{\alpha}(x)\\ \delta c_\mu(x) &= (\partial_\mu\vec{\alpha}(x)\cdot\hat{n}(x)),\\ \delta\hat{n}(x) &= \hat{n}(x)\times\vec{\alpha}(x),\\ \delta\vec{C}_\mu(x) &= (\partial_\mu\vec{\alpha}(x))_{\perp\hat{n}} + g\vec{C}_\mu(x)\times\vec{\alpha}(x),\\ \delta\vec{X}_\mu(x) &= g\,\vec{X}_\mu(x)\times\vec{\alpha}(x).\end{aligned} \tag{13}$$

The form of the transform for $\vec{X}_\mu$ is the same as that for a coloured source, so that these components are sometimes described as "valence". This gauge transformation tell us two interesting things. The first is that the Abelian component c_μ combined with the Cho connection $\vec{C}_\mu$ is enough to represent the full Lorentz symmetry even without the valence components $\vec{X}_\mu$ Cho et al. [28,29] described as the "restricted" theory. The second is that the valence components transform like a source transforms. There is a corresponding situation in $N = 2$ Yang–Mills theory where the valence gluons are interpreted as colour sources. The importance of this observation is that we shall later discuss the possibility of mass generation for the valence gluons and this form for the gauge transformation leaves such mass terms covariant. We note however that a bare mass for $\vec{X}_\mu$ cannot be inserted artificially without spoiling renormalisability.

2.1.2. The Degrees of Freedom in the CDG Decomposition

Henceforth, we restrict ourselves to the $SU(2)$ theory, for which there is only one $\hat{n}$, and neglect the (i) indices.

The unit vector $\hat{n}$ posseses two DOFs and so its inclusion in the gluon field together with the Abelian component c_μ and the valence gluons $\vec{X}_\mu$ raises questions about the DOF of the decomposed gluon, with one paper [17] advocating the gauge condition

$$\hat{D}_\mu\vec{X}_\mu(x) = 0, \tag{14}$$

to remove two apparent extra degrees of freedom. The matter was sorted by Bae et al. [18], who demonstrated that the DOFs of $\hat{n}$ were not canonical but topological, indicating the embedding of the Abelian subgroup in the gauge group. The canonical DOFs are carried by the components c_μ, $\vec{X}_\mu$ and Equation (14) is a consistency condition expected of valence gluons. Kondo et al. [31] considered a stronger condition guaranteed not to be unaffected by Gribov copys.

The topological nature of $\hat{n}$ has significance beyond making the canonical DOFs add up correctly. As is well known, monopole configurations in gauge theories are topological configurations corresponding to the embedding of an Abelian subgroup. The other important consequence is that $\hat{n}$ does not have a canonical EOM from the Euler–Lagrange equation.

We took an alternative approach to this issue by applying a new method for finding the effects of degenerate variables called the Clairaut formalism. We further assumed that, as a unit vector, its dynamics were best described by angular variables.

2.2. *CDG Decomposition of $SU(2)\times SU(2)$ in Euclidean Space*

As is well known [10,13–16], the non-compact nature of the Lorentz group causes Lorentz gauge theories to be non-positive semi-definite. In fact, our attempts to apply the CDG decomposition to the Lorentz gauge field strength tensor in Minkowski space led to negative kinetic energy terms for some of the gauge fields (not shown). As demonstrated by Pak et al. [10,16], this can be avoided by Wick rotating the theory to Euclidean space and then either considering effective theories or finding a way to rotate back later without spoiling the quantum theory.

This procedure also rotates the internal Lorentz group to $SO(4)$ which is locally isomorphic to $SU(2)_R \times SU(2)_L$, corresponding to the right- and left-handed groups generated by

$$_{\pm}\hat{e}_l \equiv \frac{1}{\sqrt{2}}(J_l \pm iK_l), \tag{15}$$

where J_l, K_l are the rotation and boost operators, respectively, and $_{\pm}\hat{e}_l$ is used to represent the corresponding direction in the internal space of the corresponding group. The two $SU(2)$ subgroups in our gauge theory, though separate, are not independent but are built from the same rotation and boost operators, albeit in combinations of opposite chirality. It follows that their respective Abelian directions must correspond, but represent operators of different chirality. We denote them $\hat{n}_R, \hat{n}_L$, respectively, using these suffices for other field objects also when appropriate, including $_R\hat{e}_l$, $_L\hat{e}_l$, and apply previously published analyses [1,3,6–8] to each symmetry group.

We apply the CDG decomposition to $SU(2)_R \times SU(2)_L$ gauge group. Their Abelian components we denote $_Rc_\mu$ and $_Lc_\mu$, respectively, and the valence components we denote as $_R\vec{X}_\mu$ and $_L\vec{X}_\mu$, respectively. For each chirality $\chi \in \{R, L\}$, we have the Cho connection

$$_\chi C_\mu(x) = g^{-1}\partial_\mu\, {}_\chi\hat{n}(x) \times {}_\chi\hat{n}(x), \tag{16}$$

and monopole field strength

$$\begin{aligned} {}_\chi\vec{H}_{\mu\nu} \equiv \partial_\mu {}_\chi\vec{C}_\nu(x) - \partial_\nu {}_\chi\vec{C}_\mu(x) + g\, {}_\chi\vec{C}_\mu(x) \times {}_\chi\vec{C}_\nu(x) = \partial_\mu\, \hat{n}_\chi(x) \times \partial_\nu\, \hat{n}_\chi(x) \\ \equiv {}_\chi H_{\mu\nu}(x)\, \hat{n}_\chi(x). \end{aligned} \tag{17}$$

Similarly defining field strengths $_\chi\vec{F}(x), {}_\chi\vec{R}_{\mu\nu}(x)$, we see from the direct product structure of the group that the Lagrangian is simply

$$\mathcal{L}^E_{gauge}(x) = \frac{1}{4} \sum_{\chi\, \in \{R,L\}} {}_\chi\vec{R}_{\mu\nu}(x) \cdot {}_\chi\vec{R}_{\mu\nu}(x). \tag{18}$$

3. The Vacuum of $SU(2)_R \times SU(2)_L$

Since the component $SU(2)$ symmetry groups have generators mutually orthogonal in the internal space, their contributions to the ground state may be calculated independently and summed. Furthermore, their identical fundamental dynamics imply that $_\chi H_{\mu\nu}$ is independent of χ when we are not considering an internal vector and may be replaced with $H_{\mu\nu}$, which we do henceforth.

It is sufficient to calculate to one loop to find a non-zero monopole condensate in the effective action of $SU(2)$ Yang–Mills theory. The authors of [6–8] have shown this by a variety of methods. Useful material on this theory at one-loop order can also be found in references [32–34].

Calculating the relevant one-loop Feynman diagrams in Feynman gauge with dimensional regularisation [7,8], we have

$$\Delta S_{eff} = -\frac{11g^2}{96} \sum_{\chi=R,L} \int d^4p\, {}_\chi\vec{F}_{\mu\nu}(p)\, {}_\chi\vec{F}_{\mu\nu}(-p) \left(\frac{2}{\epsilon} - \gamma - \ln\left(\frac{p^2}{\mu^2}\right)\right). \tag{19}$$

An imaginary part is generated by the $\ln \frac{p^2}{\mu^2}$ term only when the momentum p is timelike, leading to the well-known result [7,8,35] that it is the electric backgrounds are unstable but magnetic ones are not. Using this information, we then have the effective potential

$$V = \frac{H^2}{g^2}\left[1 + \frac{11g^2}{24}\left(\ln\frac{\sqrt{H^2}}{\mu^2} - c\right)\right] \tag{20}$$

It should be remembered that this close parallel with the corresponding $N = 2$ calculation does not hold beyond one loop because then there are diagrams including fields from both $SU(2)$ subgroups.

Defining the running coupling $\bar{g}$ by [7,8]

$$\left.\frac{\partial^2 V}{\partial H^2}\right|_{\sqrt{H^2}=\bar{\mu}^2} = \frac{1}{\bar{g}^2}, \tag{21}$$

leads to a non-trivial local minimum at

$$\langle H \rangle = \bar{\mu}^2 \exp\Big(-\frac{24\pi^2}{11\bar{g}^2} + 1 \Big). \tag{22}$$

The specific value of H^2 is less important than knowing it has a strictly positive value lying in two orthogonal directions in the $SU(2)_R \times SU(2)_L$ internal space.

4. Application of Clairaut Formalism to the Rotation-Boost Decomposition of the Gravitational Connection

4.1. A Review of the Hamiltonian-Clairaut Formalism

Here, we review the main ideas and formulae of the Clairaut-type formalism for singular theories [26,36,37]. Let us consider a singular Lagrangian $L(q^A, v^A) = L^{\deg}(q^A, v^A)$, $A = 1, \ldots n$, which is a function of $2n$ variables (n generalised coordinates q^A and n velocities $v^A = \dot{q}^A = dq^A/dt$) on the configuration space TM, where M is a smooth manifold, for which the Hessian's determinant is zero. Therefore, the rank of the Hessian matrix $W_{AB} = \frac{\partial^2 L(q^A, v^A)}{\partial v^B \partial v^C}$ is $r < n$, and we suppose that r is constant. We can rearrange the indices of W_{AB} in such a way that a non-singular minor of rank r appears in the upper left corner. Then, we represent the index A as follows: if $A = 1, \ldots, r$, we replace A with i (the "regular" or "canonical" index), and, if $A = r+1, \ldots, n$ we replace A with α (the "degenerate" or "non-canonical" index). Obviously, $\det W_{ij} \neq 0$, and rank $W_{ij} = r$. Thus any set of variables labelled by a single index splits as a disjoint union of two subsets. We call those subsets regular (having Latin indices) and degenerate (having Greek indices). Canonical DOFs are obviously described by the former of these subsets while other DOFs can be placed in the second if their contribution to the Wronskian vanishes. As was shown in [26,36], the "physical" Hamiltonian can be presented in the form

$$H_{phys}\left(q^A, p_i\right) = \sum_{i=1}^{r} p_i V^i\left(q^A, p_i, v^\alpha\right) + \sum_{\alpha=r+1}^{n} B_\alpha\left(q^A, p_i\right) v^\alpha - L\left(q^A, V^i\left(q^A, p_i, v^\alpha\right), v^\alpha\right), \tag{23}$$

where the functions

$$B_\alpha\left(q^A, p_i\right) \stackrel{def}{=} \left.\frac{\partial L(q^A, v^A)}{\partial v^\alpha}\right|_{v^i = V^i\left(q^A, p_i, v^\alpha\right)} \tag{24}$$

are independent of the unresolved velocities v^α since rank $W_{AB} = r$. Additionally, the r.h.s. of (23) does not depend on the degenerate velocities v^α

$$\frac{\partial H_{phys}}{\partial v^\alpha} = 0, \tag{25}$$

which justifies the term "physical". The Hamilton–Clairaut system which describes any singular Lagrangian classical system (satisfying the second-order Lagrange equations) has the form

$$\frac{dq^i}{dt} = \left\{q^i, H_{phys}\right\}_{phys} - \sum_{\beta=r+1}^{n} \left\{q^i, B_\beta\right\}_{phys} \frac{dq^\beta}{dt}, \quad i = 1, \dots r \tag{26}$$

$$\frac{dp_i}{dt} = \left\{p_i, H_{phys}\right\}_{phys} - \sum_{\beta=r+1}^{n} \left\{p_i, B_\beta\right\}_{phys} \frac{dq^\beta}{dt}, \quad i = 1, \dots r \tag{27}$$

$$\sum_{\beta=r+1}^{n} \left[\frac{\partial B_\beta}{\partial q^\alpha} - \frac{\partial B_\alpha}{\partial q^\beta} + \left\{B_\alpha, B_\beta\right\}_{phys}\right] \frac{dq^\beta}{dt} = \frac{\partial H_{phys}}{\partial q^\alpha} + \left\{B_\alpha, H_{phys}\right\}_{phys}, \quad \alpha = r+1, \dots, n \tag{28}$$

where the "physical" Poisson bracket (in regular variables q^i, p_i) is

$$\{X, Y\}_{phys} = \sum_{i=1}^{n-r} \left(\frac{\partial X}{\partial q^i}\frac{\partial Y}{\partial p_i} - \frac{\partial Y}{\partial q^i}\frac{\partial X}{\partial p_i}\right). \tag{29}$$

Whether the variables $B_\alpha\left(q^A, p_i\right)$ have a non-trivial effect on the time evolution and commutation relations is equivalent to whether or not the so-called "q^α-field strength"

$$\mathcal{F}_{\alpha\beta} = \frac{\partial B_\beta}{\partial q^\alpha} - \frac{\partial B_\alpha}{\partial q^\beta} + \left\{B_\alpha, B_\beta\right\}_{phys} \tag{30}$$

is non-zero. The reader is referred to [26,27,36] for more details.

4.2. The Contribution of the Clairaut Formalism

4.2.1. q^α Curvature

Substituting in this notation, the angles ϕ, θ are seen, in parallel with our previously published analysis [25], to be degenerate DOFs with unresolved velocities. Indeed, their contribution to both Lagrangian and Hamiltonian vanishes when their derivatives vanish.

We use the CDG decomposition in which the embedding of a dominant direction $U(1)$ is denoted by $\hat{n}_\chi$ which, from the discussion in Section 2.2, is expressed by,

$$\hat{n}_\chi(x) \equiv \cos\theta(x)\sin\phi(x)\,{}_\chi\hat{e}_1 + \sin\theta(x)\sin\phi(x)\,{}_\chi\hat{e}_2 + \cos\phi(x)\,{}_\chi\hat{e}_3. \tag{31}$$

We note that the angles are ϕ, θ are independent of χ for the reasons discussed after Equation (15) and need not be labelled. The following will prove useful:

$$\begin{aligned} \sin\phi(x)\,{}_\chi\hat{n}_\theta(x) &\equiv \int dy^4 \frac{d\hat{n}(x)}{d\theta(y)} = \sin\phi(x)\left(-\sin\theta(x)\,{}_\chi\hat{e}_1 + \cos\theta(x)\,{}_\chi\hat{e}_2\right), \\ {}_\chi\hat{n}_\phi(x) &\equiv \int dy^4 \frac{d\hat{n}_\chi(x)}{d\phi(y)} = \cos\theta(x)\cos\phi(x)\,{}_\chi\hat{e}_1 \\ &\quad + \sin\theta(x)\cos\phi(x)\,{}_\chi\hat{e}_2 - \sin\phi(x)\,{}_\chi\hat{e}_3. \end{aligned} \tag{32}$$

For later convenience, we note that

$$\begin{aligned} {}_\chi\hat{n}_{\phi\phi}(x) = -{}_\chi\hat{n}(x),\ {}_\chi\hat{n}_{\theta\theta}(x) &= -\sin\phi\,{}_\chi\hat{n}(x) - \cos\phi(x)\,{}_\chi\hat{n}_\phi(x)\,, \\ {}_\chi\hat{n}_{\theta\phi}(x) = 0,\ \ {}_\chi\hat{n}_{\phi\theta}(x) &= \cos\phi(x)\,{}_\chi\hat{n}_\theta(x), \end{aligned} \tag{33}$$

and that the vectors ${}_\chi\hat{n} = {}_\chi\hat{n}_\phi \times {}_\chi\hat{n}_\theta$ form an orthonormal basis of the internal space. Substituting the above into the Cho connection in Equation (3) gives

$$
\begin{aligned}
g\,{}_\chi\vec{C}_\mu(x) &= (\cos\theta(x)\cos\phi(x)\sin\phi(x)\partial_\mu\theta(x) + \sin\theta(x)\partial\phi(x))\,{}_\chi\hat{e}_1 \\
&\quad + (\sin\theta(x)\cos\phi(x)\sin\phi(x)\partial_\mu\theta(x) - \cos\theta(x)\partial\phi(x))\,{}_\chi\hat{e}_2 - \sin^2\phi(x)\partial_\mu\theta(x)\,{}_\chi\hat{e}_3 \\
&= \sin\phi(x)\,\partial_\mu\theta(x)\,{}_\chi\hat{n}_\phi(x) - \partial_\mu\phi(x)\,{}_\chi\hat{n}_\theta(x)
\end{aligned}
\tag{34}
$$

from which, it follows that

$$
g^2\,{}_\chi\vec{C}_\mu(x) \times {}_\chi\vec{C}_\nu(x) = \sin\phi(x)(\partial_\mu\phi(x)\partial_\nu\theta(x) - \partial_\nu\phi(x)\partial_\mu\theta(x))\hat{n}_\chi(x), \tag{35}
$$

where we again see that higher-order time derivatives are thwarted.

Since their Lagrangian terms do not fit the form of a canonical DOFs we consider them instead to be degenerate, having no canonical DOFs of their own but manifesting through their alteration of the EOMs of the dynamic variables. Finding these alterations first requires the Clairaut-related quantities

$$
\begin{aligned}
B_\phi(x) &= \int dy^3 \frac{\delta\mathcal{L}}{{}_x\partial_0\phi(x)} \\
&= \sum_{\chi=R,L} \int dy^3 \int dy_0\,\delta(x_0-y_0)\Big(\sin\phi(y)_y\partial_\mu\theta(y)\hat{n}_\chi(y) \\
&\qquad + {}_\chi\hat{n}_\theta(y)\times{}_\chi\vec{X}_\mu(y)\Big)\cdot{}_\chi\vec{R}_{0\mu}(y)\,\delta^3(\vec{x}-\vec{y}) \\
&= \sum_{\chi=R,L}\Big(\sin\phi(x)\,\partial_\mu\theta(x)\hat{n}_\chi(x) + {}_\chi\hat{n}_\theta(x)\times{}_\chi\vec{X}_\mu(x)\Big)\cdot{}_\chi\vec{R}_{0\mu}(x),
\end{aligned}
\tag{36}
$$

$$
\begin{aligned}
B_\theta(x) &= \int dy^3 \frac{\delta\mathcal{L}}{{}_x\partial_0\theta(x)} \\
&= -\sum_{\chi=R,L} \int dy^3 \int dy_0\delta(x_0-y_0)\sin\phi(y)\Big({}_y\partial_\mu\phi(y)\,\hat{n}_\chi(y) \\
&\qquad + \sin\phi(y)\,{}_\chi\hat{n}_\phi(y)\times{}_\chi\vec{X}_\mu(y)\Big)\cdot{}_\chi\vec{R}_{0\mu}(y)\,\delta^3(\vec{x}-\vec{y}) \\
&= -\sum_{\chi=R,L}\sin\phi(x)\Big(\partial_\mu\phi(x)\,\hat{n}_\chi(x) + {}_\chi\hat{n}_\phi(x)\times{}_\chi\vec{X}_\mu(x)\Big)\cdot{}_\chi\vec{R}_{0\mu}(x).
\end{aligned}
\tag{37}
$$

$$
\frac{\delta B_\phi(x)}{\delta\theta(y)} = \sum_{\chi=R,L}\Big(\sin\phi(x)\,{}_\chi\hat{n}_{\theta\theta}(x)\times{}_\chi\vec{X}_\mu\cdot{}_\chi\vec{R}_{0\mu}(x) - {}_\chi T_\phi(x)\Big)\delta^4(x-y), \tag{38}
$$

$$
\begin{aligned}
\frac{\delta B_\theta(x)}{\delta\phi(y)} &= -\sum_{\chi=R,L}\Big(\cos\phi(x)\Big(\partial_\mu\phi(x)\,\hat{n}_\chi(x) + {}_\chi\hat{n}_\phi(x)\times{}_\chi\vec{X}_\mu(x)\Big)\cdot\Big({}_\chi\vec{R}_{0\mu}(x) + {}_\chi\vec{H}_{0\mu}(x)\Big) \\
&\qquad + {}_\chi T_\theta(x)\Big)\delta^4(x-y),
\end{aligned}
\tag{39}
$$

where

$$
{}_\chi T_\phi(x) = \partial_k\Big[\sin\phi(x)\,\hat{n}_\chi\cdot{}_\chi\vec{R}_{0k}(x) - \Big(\sin\phi(x)\partial_k\theta(x) + {}_\chi\hat{n}_\theta(x)\times{}_\chi\vec{X}_k\cdot\hat{n}_\chi\Big)\partial_0\phi(x)\Big], \tag{40}
$$

$$
{}_\chi T_\theta(x) = -\partial_k\Big[\sin\phi(x)\Big(\hat{n}_\chi\cdot{}_\chi\vec{R}_{0k}(x) + \Big(\partial_k\phi(x) + {}_\chi\hat{n}_\phi(x)\times{}_\chi\vec{X}_k\cdot\hat{n}_\chi\Big)\partial_0\theta(x)\Big)\Big], \tag{41}
$$

are the surface terms arising from derivatives $\frac{\delta(\partial\theta)}{\delta\theta}$, $\frac{\delta(\partial\phi)}{\delta\phi}$ and the latin index k is used to indicate that only spacial indices are summed over.

This yields the q^α-curvature

$$\begin{aligned}\mathcal{F}_{\theta\phi}(x) &= \int dy^4 \Big(\frac{\delta B_\theta(x)}{\delta\phi(y)} - \frac{\delta B_\phi(x)}{\delta\theta(y)}\Big)\delta^4(x-y) + \{B_\phi(x), B_\theta(x)\}_{phys} \\ &= -\sum_{\chi=R,L} \cos\phi(x)\Big(\partial_\mu\phi(x)\,\hat{n}_\chi(x) + {}_\chi\hat{n}_\phi(x)\times{}_\chi\vec{X}_\mu(x)\Big)\cdot\Big({}_\chi\vec{R}_{0\mu}(x) + {}_\chi\vec{H}_{0\mu}(x)\Big) \\ &\quad - \sum_{\chi=R,L} \sin\phi(x)\,{}_\chi\hat{n}_{\theta\theta}(x)\times{}_\chi\vec{X}_\mu(x)\cdot{}_\chi\vec{R}_{\mu 0}(x) + \sum_{\chi=R,L}\Big({}_\chi T_\phi(x) - {}_\chi T_\theta(x)\Big). \end{aligned} \quad (42)$$

where we have used that the bracket $\{B_\phi, B_\theta\}_{phys}$ vanishes because B_ϕ and B_θ share the same dependence on the dynamic DOFs and their derivatives.

In earlier work on the Clairaut formalism [26,36], this was called the q^α-field strength, but we call it q^α-curvature in quantum field theory applications to avoid confusion.

This non-zero $\mathcal{F}^{\theta\phi}$ is necessary, and usually sufficient, to indicate a non-dynamic contribution to the conventional Euler–Lagrange EOMs. More significant is a corresponding alteration of the quantum commutators, with repurcussions for canonical quantisation and the particle number.

4.2.2. Corrections to the Equations of Motion

Generalising Equations (7.1,7.3,7.5) in [26] (see also the discussion around Equation (23)

$$\partial_0 q(x) = \{q(x), H_{phys}\}_{new} = \frac{\delta H_{phys}}{\delta p(x)} - \int dy^4 \sum_{\alpha=\phi,\theta} \frac{\delta B_\alpha(y)}{\delta p(x)}\partial_0\alpha(y), \quad (43)$$

the derivative of the Abelian component, complete with corrections from the monopole background is

$$\partial_0\,{}_\chi c_\sigma(x) = \frac{\delta H_{phys}}{\delta\,{}_\chi\Pi^\sigma(x)} - \int dy^4 \sum_{\alpha=\phi,\theta} \frac{\delta B_\alpha(y)}{\delta\,{}_\chi\Pi^\sigma(x)}\partial_0\alpha(y). \quad (44)$$

The effect of the second term is to remove the monopole contribution to $\frac{\delta H_{phys}}{\delta_\chi\Pi^\sigma}$. To see this, consider that, by construction, the monopole contribution to the Lagrangian and Hamiltonian is dependent on the time derivatives of θ, ϕ, so the monopole component of $\frac{\delta H_{phys}}{\delta_\chi\Pi^\sigma}$ is

$$\begin{aligned}\frac{\delta}{\delta\,{}_\chi\Pi_\sigma(x)} H_{phys}|_{\theta\phi} &= \frac{\delta}{\delta\,{}_\chi\Pi_\sigma(x)}\Big(\frac{\delta H_{phys}}{\delta\partial_0\theta(x)}\partial_0\theta(x) + \frac{\delta H_{phys}}{\delta\partial_0\phi(x)}\partial_0\phi(x)\Big) \\ &= \frac{\delta}{\delta\,{}_\chi\Pi_\sigma(x)}\Big(\frac{\delta L_{phys}}{\delta\partial_0\theta(x)}\partial_0\theta(x) + \frac{\delta L_{phys}}{\delta\partial_0\phi(x)}\partial_0\phi(x)\Big) \\ &= \frac{\delta}{\delta\,{}_\chi\Pi_\sigma(x)}\Big(B_\theta(x)\partial_0\theta(x) + B_\phi(x)\partial_0\phi(x)\Big), \end{aligned} \quad (45)$$

which is a consistency condition for Equation (44). This confirms the necessity of treating the monopole as a non-dynamic field.

We now observe that

$$\frac{\delta B_\theta(x)}{\delta\,{}_\chi c_\sigma(y)} = \frac{\delta B_\phi(x)}{\delta\,{}_\chi c_\sigma(y)} = 0, \quad (46)$$

from which it follows that the EOMs of ${}_\chi c_\sigma$ receives no correction. However, its $\{,\}_{phys}$ contribution, corresponding to the terms in the conventional EOM for the Abelian component, already contains a contribution from the monopole field strength.

Repeating the above steps for the valence gluons ${}_\chi\vec{X}_\mu$, assuming $\sigma \neq 0$ and combining

$$\hat{D}_0\,{}_\chi\vec{\Pi}_\sigma(x) = \frac{\delta H}{\delta\,{}_\chi\vec{X}_\sigma(x)} - \int dy^4 \sum_{\alpha=\phi,\theta} \frac{\delta B_\alpha(y)}{\delta\,{}_\chi\vec{X}_\sigma(x)}\partial_0\alpha(y). \tag{47}$$

with

$$\frac{\delta B_\phi(y)}{\delta\,{}_\chi\vec{X}_\sigma(x)} = -\Big(\Big(\sin\phi(y)_y\partial_\sigma\theta(y)\hat{n}_\chi(y) + {}_\chi\hat{n}_\theta(y)\times{}_\chi\vec{X}_\sigma(y)\Big)\times{}_\chi\vec{X}_0(y)$$
$$- {}_\chi\hat{n}_\phi(y)\hat{n}_\chi\cdot{}_\chi\vec{R}_{0\sigma}(y)\Big)\delta^4(x-y), \tag{48}$$

$$\frac{\delta B_\theta(y)}{\delta\,{}_\chi\vec{X}^\sigma(x)} = \Big(\Big(\partial_\sigma\phi(y)\hat{n}(y) + \sin\phi(y)\,\hat{n}_\phi(y)\times{}_\chi\vec{X}_\sigma(y)\Big)\times{}_\chi\vec{X}_0(y)$$
$$- \sin\phi(y)\,{}_\chi\hat{n}_\theta(y)\hat{n}_\chi(y)\cdot{}_\chi\vec{R}_{0\sigma}(y)\Big)\delta^4(x-y), \tag{49}$$

gives

$$\begin{aligned}\hat{D}_0\,{}_\chi\vec{\Pi}_\sigma(x) =& \frac{\delta H}{\delta\,{}_\chi\vec{X}_\sigma(x)} - \frac{1}{2}\Big(\Big(\sin\phi(x)(\partial_\sigma\phi(x)\partial_0\theta(x) - \partial_\sigma\theta(x)\partial_0\phi(x)\Big)\hat{n}_\chi(x) \\ &+ \Big(\sin\phi(x)\,{}_\chi\hat{n}_\phi(x)\partial_0\theta(x) - {}_\chi\hat{n}_\theta(x)\partial_0\phi(x)\Big)\times{}_\chi\vec{X}_\sigma(x)\Big)\times{}_\chi\vec{X}_0(x) \\ =& \frac{\delta H}{\delta\,{}_\chi\vec{X}_\sigma(x)} - \frac{1}{2}g^2\Big({}_\chi\vec{C}_\sigma(x)\times{}_\chi\vec{C}_0(x) + {}_\chi\vec{C}_0(x)\times{}_\chi\vec{X}_\sigma(x)\Big)\times{}_\chi\vec{X}_0(x).\end{aligned} \tag{50}$$

This is the converse situation of the Abelian gluon, where their derivatives ${}_\chi\vec{X}_\sigma$ is uncorrected while their EOM receives a correction which cancels the monopole's electric contribution to $\{\hat{D}_0\,{}_\chi\vec{X}_\sigma, H_{phys}\}_{phys}$. This is required by the conservation of topological current, but a further implication is that the monopole background, even if assumed to be present, does not contribute to the EOMs of motion and therefore makes no impact at the classical level. Note that this is strictly limited to the monopole field and the effects of backgrounds due to the dynamic fields are not affected. Monopole field contributions are not cancelled from quantum corrections however, although calculating loop effects is beyond the scope of this paper.

4.2.3. Corrections to the Commutation Relations

Corrections to the classical Poisson bracket correspond to corrections to the equal-time commutators in the quantum regime. We shall see corrections for commutators with fields of different $SU(2)_\chi$ representations even though there were no such crossover terms in the effective potential calculation.

Denoting conventional commutators as $[,]_{phys}$ and the corrected ones as $[,]_{new}$, for $\mu,\nu \neq 0$, we have

$$\begin{aligned}[{}_\chi c_\mu(x), {}_{\tilde{\chi}} c_\nu(z)]_{new} =& [{}_\chi c_\mu(x), {}_{\tilde{\chi}} c_\nu(z)]_{phys} \\ &- \int dy^4\Big(\frac{\delta B_\theta(y)}{\delta\,{}_\chi\Pi_\mu(x)}\mathcal{F}^{-1}_{\theta\phi}(z)\frac{\delta B_\phi(y)}{\delta\,{}_\chi\Pi_\nu(z)} - \frac{\delta B_\phi(y)}{\delta\,{}_{\tilde{\chi}}\Pi_\mu(x)}\mathcal{F}^{-1}_{\phi\theta}(z)\frac{\delta B_\theta(y)}{\delta\,{}_\chi\Pi_\nu(z)}\Big)\delta^4(x-z) \\ =& [{}_\chi c_\mu(x), {}_{\tilde{\chi}} c_\nu(z)]_{phys} \\ &- \sin\phi(x)\sin\phi(z)(\partial_\mu\phi(x)\partial_\nu\theta(z) - \partial_\nu\phi(z)\partial_\mu\theta(x))\mathcal{F}^{-1}_{\theta\phi}(z)\delta^4(x-z).\end{aligned} \tag{51}$$

The second term on the final line, after integration over d^4z, clearly becomes

$$H_{\mu\nu}(x)\sin\phi(x)\mathcal{F}^{-1}_{\theta\phi}(x), \tag{52}$$

indicating the role of the monopole condensate in the correction. By contrast, the commutation relations

$$\begin{aligned}[{}_\chi c_\mu(x),\, {}_{\tilde\chi}\Pi_\nu(z)]_{new} &= [{}_\chi c_\mu(x),\, {}_{\tilde\chi}\Pi_\nu(z)]_{phys},\\ [{}_\chi \Pi_\mu(x),\, {}_{\tilde\chi}\Pi_\nu(z)]_{new} &= [{}_\chi \Pi_\mu(x),\, {}_{\tilde\chi}\Pi_\nu(z)]_{phys},\end{aligned} \tag{53}$$

are unchanged. Nonetheless, the deviation from the canonical commutation shown in Equation (51) is inconsistent with the particle creation/annihilation operator formalism of conventional second quantisation, so that particle number is no longer well defined for the ${}_\chi c_\mu$ fields.

Repeating for the valence part,

$$[{}_\chi \Pi^a_\mu(x), {}_{\tilde\chi}\Pi^b_\nu(z)]_{new} \tag{54}$$

$$\begin{aligned}=&[{}_\chi \Pi^a_\mu(x),\, {}_{\tilde\chi}\Pi^b_\nu(z)]_{phys} - \int dy^4 \Big(\frac{\delta B_\theta(y)}{\delta\, {}_\chi X^a_\mu(x)} \frac{\delta B_\phi(y)}{\delta\, {}_{\tilde\chi} X^b_\nu(z)} - \frac{\delta B_\phi(y)}{\delta X^a_\mu(x)} \frac{\delta B_\theta(y)}{\delta\, {}_{\tilde\chi} X^b_\nu(z)} \Big) \mathcal{F}^{-1}_{\theta\phi}(z)\\ =&[{}_\chi \Pi^a_\mu(x),\, {}_{\tilde\chi}Pi^b_\nu(z)]_{phys} + \Big(\sin\phi(z) n^a_\phi(x) n^b_\theta(z)\, {}_\chi\vec{R}_{0\mu}(x)\cdot\hat{n}_\chi(x)\, {}_{\tilde\chi}\vec{R}_{0\nu}(z)\cdot\hat{n}_{\tilde\chi}(z)\\ &- \sin\phi(x) n^a_\theta(x) n^b_\phi(z)\, {}_\chi\vec{R}_{0\mu}(z)\cdot\hat{n}_\chi(z)\, {}_{\tilde\chi}\vec{R}_{0\nu}(x)\cdot\hat{n}_{\tilde\chi}(x)\Big) \times \mathcal{F}^{-1}_{\theta\phi}(z)\,\delta^4(x-z),\end{aligned} \tag{55}$$

where the second term on the final line, integrates over d^4z to become

$$(n^a_\phi(x) n^b_\theta(x) - n^a_\theta(x) n^b_\phi(x)) \sin\phi(x)\, {}_\chi\vec{R}^{0\mu}(x)\cdot\hat{n}_\chi(x)\, {}_{\tilde\chi}\vec{R}^{0\nu}(x)\cdot\hat{n}_{\tilde\chi}(x)\mathcal{F}^{-1}_{\theta\phi}(x), \tag{56}$$

while other relevant commutators are unchanged

$$\begin{aligned}[{}_\chi X^a_\mu(x),\, {}_{\tilde\chi}\Pi^b_\nu(z)]_{new} &= [{}_\chi X^a_\mu(x),\, {}_{\tilde\chi}\Pi^b_\nu(z)]_{phys},\\ [{}_\chi X^a_\mu(x),\, {}_{\tilde\chi}X^b_\nu(z)]_{new} &= [{}_\chi X^a_\mu(x),\, {}_{\tilde\chi}X^b_\nu(z)]_{phys}.\end{aligned} \tag{57}$$

5. Effective Action

5.1. Particle Number and the Monopole Background

It is textbook knowledge that gravitational curvature spoils canonical quantisation, but our approach gives a detailed mechanism. It also provides some narrowly defined circumstances under which it may be salvaged. For monopole background ${}_\chi\vec{H}_{\mu\nu}$ the form of Equation (51) indicates that they would arise for ${}_\chi c_\sigma$ polarised along either of the μ,ν directions. The only way to avoid this is if ${}_\chi c_\sigma$ is polarised in the direction of the monopole field strength, requiring that the Abelian component of the connection propagate at a right angle to the monopole field strength. However, the form of the monopole field strength requires that a non-vanishing field must have a varying orientation in space, since it is proportional to the derivatives of the angles ϕ,θ. So even if the Abelian gauge component is propagating at a right angle to the monopole field strength with its polarisation in the direction of the field strength, in general this could not be assumed to continue as the orientation of the monopole field strength varied. However, if the variation were gradual over space in comparison to the wavelength of ${}_\chi c_\mu$, then it might continue to propagate while adjusting to the required orientations in a manner analogous to photon polarisation being rotated by successive, closely oriented, polarising filters. On the other hand, if the wavelength of ${}_\chi c_\mu$ is significant compared to the length scale of the field variation, then such a mechanism could not act and the particle's energy would be either absorbed or deflected by the condensate, effectively suppressing the longer wavelengths and providing a measure of the background curvature.

One important observation is that the background field is (Lorentz) magnetic, so that at any point in spacetime a reference frame exists where the monopole field and its associated potential lie entirely along the spatial directions.

The particle inconsistent contribution from Equation (54) only occurs in the presence of a background electric component of the monopole field strength, vanishing when the polarisation of ${}_\chi\vec{X}_\mu$ is orthogonal to the electric component of the background field. This restricts the polarisation for a transversally polarised field whose direction of propagation is not in the direction of this electric component, but not otherwise. Of course, the electric component of the background monopole field can always be removed by a suitable Lorentz transformation, but this still leaves the particle interpretation frame-dependent.

Some authors have argued that the valence gluons in two-colour QCD gain an effective mass term [20,21] via their quartic interaction with the non-trivial monopole condensate. A similar mechanism could apply to the valence components of this theory. Consider the following quartic term from Equations (11) and (12),

$$\begin{aligned}&\frac{g^2}{4}({}_\chi\vec{C}_\mu(x)\times{}_\chi\vec{X}_\nu(x))\cdot({}_\chi\vec{C}^\mu(x)\times{}_\chi\vec{X}^\nu(x))\\&=\frac{g^2}{4}({}_\chi\vec{C}_\mu(x)\cdot{}_\chi\vec{C}^\mu(x)\,{}_\chi\vec{X}_\nu(x)\cdot{}_\chi\vec{X}^\nu(x)-{}_\chi\vec{C}_\mu(x)\cdot{}_\chi\vec{X}^\mu(x)\,{}_\chi\vec{X}_\mu(x)\cdot{}_\chi\vec{C}^\mu(x)).\end{aligned}\tag{58}$$

Remembering that the Lorentz monopole fields ${}_\chi\vec{C}_\mu$ have non-zero condensates yields the terms

$$\frac{g^2}{4}\langle{}_\chi\vec{C}_\mu(x)\cdot{}_\chi\vec{C}^\mu(x)\rangle\,{}_\chi\vec{X}_\nu(x)\cdot{}_\chi\vec{X}^\nu(x),\tag{59}$$

so that the monopole condensate is seen to generate a mass term for the valence component. Such a mass term is covariant under the gauge transformation because, as shown in the discussion of Equation (13), the valence components transform as sources although explicitly adding a mass term for these fields would spoil renormalisability. In this case the valence components could also be longitudinally polarised. With longitudinal polarisation the only restriction is that the direction of propagation be orthogonal to the background electric component of the monopole field strength. The valence component might therefore enjoy a limited particle interpretation under a range of circumstances.

We observe that the two monopole field strengths ${}_R\vec{H}_{\mu\nu}$, ${}_L\vec{H}_{\mu\nu}$ sum to give a net field strength lying purely along the rotation directions in the internal space. Exactly how this affects the observed dynamics of the theory, or even if it does, is unclear. We were unable to find a linear combination of the gauge fields to separate rotation and boost generators which was equivalent to the original theory. If there is an effect, then a reasonable scenario is that the coupling to linear momentum would dominate that to rotational momentum at large distances, as determined by the length scale of the condensate.

5.2. The Hilbert–Einstein Term

Kim and Pak [10] considered the effects of a torsion condensate. They found the resulting background field strength, if constant, spontaneously generated an EH term if the curvature tensor is expanded around it (see the discussion of Equation (45) in their paper [10]). EH terms have been shown to stabilise theories with higher-order derivatives by rendering the propagator poles gauge invariant [38,39] and Kim and Pak suggest that this may stabilise their theory also. Since our background is attributable to an Abelian background field, we expect the effective theory to have an Abelianised EH term, similar to that derived by Cho et al. [28,29] when applying the CDG decomposition to the Levi-Civita tensor. Such details must await further work, but we are encouraged to believe that the theory may be Wick rotated back to Lorentz space for a positive semi-definite effective theory. Not only do all quantum fields have kinetic terms with the correct sign, but the Lagrangian's lowest-order derivative terms come from an emergent term sometimes added to rectify the non-semi-positive definiteness.

6. Discussion

We have applied the CDG decomposition to a Lorentz gauge theory and confirmed that it has a monopole condensate at one loop. Using the Clairaut formalism, we have found how the monopole background modifies the canonical EOMs for the physical DOFs. Lorentz gauge theory has the problem of being non-positive semi-definite, which can be handled by adding a EH term. We did not add such a term but instead postponed the problem by Wick rotating the theory into Euclidean space, where the Lorentz gauge group becomes locally isomorphic to $SU(2)_R \times SU(2)_L$. We found the spontaneous generation of a vacuum condensate which others have argued [10,16] leads to an effective Hilbert–Einstein term.

The CDG decomposition introduces an internal unit vector to indicate the local internal direction of the Abelian subgroup of the gauged symmetry group. However, the unit vector used to specify this subgroup does not form a canonical EOM and is degenerate. If we expand it in terms of its angular dependence, since its information content is purely directional, then those angles are also degenerate and we do not derive canonical EOMs for them. They do however add additional terms with important consequences for the theory's physics. They may not be ignored therefore, but require appropriate theoretical tools to analyse them. The authors addressed these issues in a previous analysis of QCD. The purpose of this paper was to do so for a theory relevant to gravity. The main advantages of working in a gauged Lorentz theory for us is that the gauge fields have quadratic kinetic terms well suited to our Clairaut-based approach in addition to the opportunity to apply analyses and even results from $SU(2)$ Yang–Mills theories.

We have not considered the effects of matter fields in the fundamental representation. We do note in passing that differences in this part of the spectrum must lead to variations in the magnitude for the monopole condensate, so the differences in their matter spectra suggest that this theory has significantly different infrared behaviour from that of $SU(2)$ QCD.

We also observe that the net monopole condensate lies in a direction of a rotation generator. We have not been able to derive corresponding canonical DOFs to reflect this, so the physical significance of this observation, if any, remains obscure.

We have left the inclusion of translation symmetry to subsequent work. A full gravitational theory must of course include the full Poincaré symmetry group, but we submit that our Lorentz-only theory makes a sufficiently good approximation to indicate some relevant phenomenology.

Author Contributions: Conceptualization, M.L.W. and S.D.; methodology, S.D.; software, S.D.; validation, M.L.W. and S.D.; formal analysis, M.L.W.; resources, M.L.W.; writing—original draft preparation, M.L.W.; writing—review and editing, S.D.; visualization, S.D.; supervision, S.D.; project administration, S.D. All authors have read and agreed to the published version of the manuscript.

Funding: This research received no external funding.

Data Availability Statement: Not applicable.

Conflicts of Interest: The authors declare no conflict of interest.

References

1. Savvidy, G.K. Infrared instability of the vacuum state of gauge theories and asymptotic freedom. *Phys. Lett.* **1977**, *B71*, 133. [CrossRef]
2. Nielsen, N.K.; Olesen, P. An unstable Yang-Mills field mode. *Nucl. Phys.* **1978**, *B144*, 376. [CrossRef]
3. 't Hooft, G. Topology of the Gauge Condition and New Confinement Phases in Nonabelian Gauge Theories. *Nucl. Phys.* **1981**, *B190*, 455. [CrossRef]
4. Cho, Y.M. Colored monopoles. *Phys. Rev. Lett.* **1980**, *44*, 1115–1118. [CrossRef]
5. Duan, Y.S.; Ge, M.L. $SU(2)$ gauge theory and electrodynamics of N moving magnetic monopoles. *Sci. Sinica* **1979**, *11*, 1072–1075.
6. Cho, Y.M.; Pak, D.G. Monopole condensation in $SU(2)$ QCD. *Phys. Rev.* **2002**, *D65*, 074027.
7. Cho, Y.M.; Walker, M.L. Stability of monopole condensation in $SU(2)$ QCD. *Mod. Phys. Lett.* **2004**, *A19*, 2707–2716. [CrossRef]
8. Cho, Y.M.; Walker, M.L.; Pak, D.G. Monopole condensation and confinement of color in $SU(2)$ QCD. *JHEP* **2004**, *05*, 073. [CrossRef]
9. Kay, D.; Kumar, A.; Parthasarathy, R. Savvidy vacuum in $SU(2)$ Yang-Mills theory. *Mod. Phys. Lett.* **2005**, *A20*, 1655–1662. [CrossRef]
10. Kim, S.W.; Pak, D.G. Torsion as a dynamic degree of freedom of quantum gravity. *Class. Quantum Gravity* **2008**, *25*, 065011. [CrossRef]

11. Pak, D. Confinement, vacuum structure: from QCD to quantum gravity. *Nucl. Phys. A* **2010**, *844*, 115c–119c. [CrossRef]
12. Cho, Y.M.; Pak, D.G.; Park, B.S. A Minimal model of Lorentz gauge gravity with dynamical torsion. *Int. J. Mod. Phys. A* **2010**, *25*, 2867–2882. [CrossRef]
13. Utiyama, R. Invariant theoretical interpretation of interaction. *Phys. Rev.* **1956**, *101*, 1597. [CrossRef]
14. Kibble, T. Lorentz invariance and the gravitational field. *J. Math. Phys.* **1961**, *2*, 212–221. [CrossRef]
15. Sciama, D. The physical structure of general relativity. *Rev. Mod. Phys.* **1964**, *36*, 463. [CrossRef]
16. Pak, D.; Kim, Y.; Tsukioka, T. Lorentz gauge theory as a model of emergent gravity. *Phys. Rev. D* **2012**, *85*, 084006. [CrossRef]
17. Shabanov, S.V. Yang–Mills theory as an Abelian theory without gauge fixing. *Phys. Lett.* **1999**, *B463*, 263–272. [CrossRef]
18. Bae, W.S.; Cho, Y.M.; Kimm, S.W. Qcd versus skyrme-faddeev theory. *Phys. Rev.* **2002**, *D65*, 025005.
19. Shabanov, S.V. An effective action for monopoles and knot solitons in Yang-Mills theory. *Phys. Lett.* **1999**, *B458*, 322–330. [CrossRef]
20. Kondo, K.I.; Murakami, T.; Shinohara, T. BRST symmetry of $SU(2)$ Yang-Mills theory in Cho-Faddeev-Niemi decomposition. *Eur. Phys. J.* **2005**, *C42*, 475–481. [CrossRef]
21. Kondo, K.I. Gauge-invariant gluon mass, infrared Abelian dominance and stability of magnetic vacuum. *Phys. Rev.* **2006**, *D74*, 125003. [CrossRef]
22. Lavrov, P.M.; Merzlikin, B.S. Legendre transformations and Clairaut-type equations. *Phys. Lett.* **2016**, *B756*, 188–193. [CrossRef]
23. Ren, J.; Wang, H.; Wang, Z.; Qu, F. The Wu-Yang potential of Magnetic Skyrmion from $SU(2)$ Flat Connection. *Sci. China Phys. Mech. Astron.* **2019**, *62*, 950021, [CrossRef]
24. Cho, Y.; Hong, S.T.; Kim, J.; Park, Y.J. Dirac quantization of restricted QCD. *Mod. Phys. Lett. A* **2007**, *22*, 2799–2813. [CrossRef]
25. Walker, M.L.; Duplij, S. Cho-Duan-Ge decomposition of QCD in the constraintless Clairaut-type formalism. *Phys. Rev. D* **2015**, *91*, 064022. [CrossRef]
26. Duplij, S. Generalized duality, Hamiltonian formalism and new brackets. *J. Math. Phys. Anal. Geom.* **2014**, *10*, 189–220. [CrossRef]
27. Duplij, S. Formulation of singular theories in a partial Hamiltonian formalism using a new bracket and multi-time dynamics. *Int. J. Geom. Methods Mod. Phys.* **2015**, *12*, 1550001. [CrossRef]
28. Cho, Y.; Oh, S.; Kim, S. Abelian dominance in Einstein's theory. *Class. Quantum Gravity* **2012**, *29*, 205007. [CrossRef]
29. Cho, Y.M.; Oh, S.H.; Park, B.S. Abelian decomposition of Einstein's theory: Restricted gravity. *Grav. Cosm.* **2015**, *21*, 257–269. [CrossRef]
30. Schanbacher, V. Gluon propagator and effective lagrangian in QCD. *Phys. Rev.* **1982**, *D26*, 489. [CrossRef]
31. Kondo, K.I.; Murakami, T.; Shinohara, T. Yang–Mills theory constructed from Cho-Faddeev-Niemi decomposition. *Prog. Theor. Phys.* **2006**, *115*, 201–216. [CrossRef]
32. Itzykson, C.; Zuber, J.B. *Quantum Field Theory*; McGraw-Hill: New York, NY, USA, 2012.
33. Peskin, M.; Schroeder, D. *An Introduction to Quantum Field Theory*; Addison-Wesley: Reading, MA, USA, 1995.
34. Weinberg, S. *The Quantum Theory of Fields*; Cambridge University Press: Cambridge, UK, 1996; Volumes 1–3.
35. Schwinger, J. On gauge invariance and vacuum polarization. *Phys. Rev.* **1951**, *82*, 664. [CrossRef]
36. Duplij, S. A new Hamiltonian formalism for singular Lagrangian theories. *J. Kharkov Univ. Ser. Nuclei Part. Fields* **2011**, *969*, 34–39.
37. Duplij, S. Constraintless Hamiltonian formalism and new brackets. In *Exotic Algebraic and Geometric Structures in Theoretical Physics*; Duplij, S., Ed.; Nova Publishers: New York, NY, USA, 2018; pp. 197–230.
38. Antoniadis, I.; Tomboulis, E. Gauge invariance and unitarity in higher-derivative quantum gravity. *Phys. Rev. D* **1986**, *33*, 2756. [CrossRef] [PubMed]
39. Johnston, D. Sedentary ghost poles in higher derivative gravity. *Nucl. Phys. B* **1988**, *297*, 721–732. [CrossRef]

MDPI

Article

Wheeler–DeWitt Equation and the Applicability of Crypto-Hermitian Interaction Representation in Quantum Cosmology

Miloslav Znojil [1,2]

[1] Department of Physics, Faculty of Science, University of Hradec Králové, Rokitanského 62, 500 03 Hradec Králové, Czech Republic; znojil@ujf.cas.cz

[2] Nuclear Physics Institute, The Czech Academy of Sciences, Hlavní 130, 250 68 Řež, Czech Republic

Abstract: In the broader methodical framework of the quantization of gravity, the crypto-Hermitian (or non-Hermitian) version of Dirac's interaction picture is considered. The formalism is briefly outlined and shown to be well suited for an innovative treatment of certain cosmological models. In particular, it is demonstrated that the Wheeler–DeWitt equation could be a promising candidate for the description of the evolution of the quantized Universe near its initial Big Bang singularity.

Keywords: quantum gravity and the problem of the Big Bang; hidden Hermitian formulations of quantum mechanics; stationary Wheeler–DeWitt system; physical Hilbert space metric; non-stationary Wheeler–DeWitt system

Citation: Znojil, M. Wheeler–DeWitt Equation and the Applicability of Crypto-Hermitian Interaction Representation in Quantum Cosmology. *Universe* **2022**, *7*, 385. https://doi.org/10.3390/universe8070385

Academic Editors: Steven Duplij and Michael L. Walker

Received: 12 June 2022
Accepted: 18 July 2022
Published: 20 July 2022

1. Introduction

The concept of the wave function ψ of the Universe (introduced, 55 years ago, as a solution of the Einstein–Schrödinger *alias* Wheeler–DeWitt (WDW) equation [1,2]) is contradictory. On the positive side, this concept played a key role during the development of the canonical quantization of gravity [3]. These efforts climaxed in the recent comparatively satisfactory and constructive formulation of the so-called loop quantum gravity (LQG, [4–6]). At the same time, Mostafazadeh pointed out, in his review of the recent progress in quantum theory [7], that the solutions ψ themselves remain "void of a physical meaning", without "finding an appropriate inner product on the space of solutions of the WDW equation" (see p. 1291 in review [7]). In *loc. cit.*, Mostafazadeh also emphasized that "the lack of a satisfactory solution to this problem has been one of the major obstacles in transforming canonical quantum gravity and quantum cosmology into genuine physical theories". Precisely, this obstacle is to be addressed and discussed in what follows.

In the cited review, we can further read that "in ... quantum cosmology ... the relevant field equations ... are second order differential equations in a time variable ... [which] have the ... general form

$$\frac{d^2}{dt^2}\psi(t) + D(t)\,\psi(t) = 0 \tag{1}$$

where t denotes a dimensionless time variable, $\psi : \mathbb{R} \to \mathcal{L}$ is a function taking values in some separable Hilbert space $\mathcal{L}$, and $D : \mathcal{L} \to \mathcal{L}$ is a positive-definite operator that may depend on t". Treating the latter variable as "a fictitious evolution parameter in quantum cosmology" (see p. 1292 in [7]), the same author later adds that "the cases in which D is t-dependent (that arises in quantum cosmological models) require a more careful examination". In this sense, we are prepared to discuss some of the open questions and subtleties of the theory.

In *loc. cit.*, Mostafazadeh redirected interested readers to his earlier study [8]. In a series of our subsequent unpublished comments on this topic [9] (which were later finalized and summarized in papers [10,11]), we showed that an appropriate means of "dealing with these cases" is, *simultaneously* , less complicated and more complicated than it seems—less complicated in the sense that some of the technical obstacles have later been found surmountable, and more complicated because it appeared necessary to amend the overall quantum-theoretical framework and to replace the non-Hermitian Schrödinger picture (NSP) interpretation of the evolution of ψ (as presented, basically, in [7] or [8]) by the more involved formalism called the interaction picture (IP) (or Dirac's representation; see a comprehensive review of its non-Hermitian form (NIP) in [12]).

In what follows, we intend to outline the implementation of the NIP approach in the WDW case. The key purpose of our paper is to provide an explicit explanation of the connection between several challenging and open physical questions (a typical one concerns the quantum Big Bang problem, as formulated in Section 2) and the most recent progress in the hidden unitary version of quantum mechanics (the basic features of this theoretical innovation are reviewed). Our main message (viz, the detailed description of the theory and of its application to the WDW equation) will finally be outlined in Section 3 (devoted to a specific schematic toy model of the quantum geometry of the Universe), in Section 4 (on the fully fledged NIP formalism), and in Section 5 (in which the mechanism of transition to the Big Bang singularity will be given its ultimate model-independent construction recipe form). Our results will be discussed in Section 6 and summarized in Section 7.

2. Challenge: Quantum Big Bang Problem

At present, it is widely believed that up to the "youngest age" of the Universe (i.e., for times $t > t_1$ with $t_1 \approx 10^{32}$ s), the evolution (i.e., slow expansion) of the Universe is more or less safely controlled by the classical theoretical cosmology. In contrast, in the interval of times (t_0, t_1) (where $t_0 = 0$ denotes the hypothetical time of the Big Bang), we still lack a fully consistent and rigorous quantum theory behind the early history of the Universe [3].

2.1. *Could the Degeneracy Survive Quantization? Yes, It Could*

In our present study, we felt strongly motivated by the deep relevance of the understanding of the evolution of the Universe near its Big Bang origin, i.e., in a genuine quantum dynamical regime. In this regime, the theoretically most ambitious LQG formalism still seems to lead to at least some contradictory results. In one of the LQG predictions [13], for example, the Big Bang singularity (compatible with the classical Einstein theory of gravity) has been found to be smeared out by the quantization. In the series of papers [14–19] or in Section 8 in [5]), for example, it is claimed that the Big Bang singularity of classical theory must necessarily be replaced by a regularized "Big Bounce" mechanism. In contrast, more recently, Wang with Stankiewicz [20] came forward with the opposite conclusion, claiming that, within the scale-invariant LQG framework, "the quantized Big Bang is *not* replaced by a Big Bounce".

At first sight, the latter claim appears suspicious. In Rovelli's words, the quantization-related "absence of singularities" is in fact "what one would expect from a quantum theory of gravity" (see p. 297 in [5]). An elementary support of such an intuitive expectation can be provided by the following schematic observable

$$\Lambda(t) = \begin{bmatrix} 0 & -1+it & 0 & 0 \\ -1-it & 0 & -1+it & 0 \\ 0 & -1-it & 0 & -1+it \\ 0 & 0 & -1-it & 0 \end{bmatrix} \tag{2}$$

and by the inspection of its spectrum (see Figure 1). As long as the matrix is Hermitian, its spectrum must be real. Moreover, in the generic case (i.e., unless we impose a symmetry upon the matrix), the spectrum *must* remain non-degenerate. This is the reason that the levels avoid the crossing (which would simulate the regularized Big Bounce). In our example, the proof of the phenomenon is elementary: up to a small vicinity of the "Big Bang time" $t^{(BB)} = 0$, the matrix as well as its spectrum are dominated by their asymptotic components, which are strictly linear in t. One might even suspect that the eigenvalues could cross due to an accidental symmetry emerging at $t = 0$, but such a symmetry is manifestly broken by the t-independent component of the model.

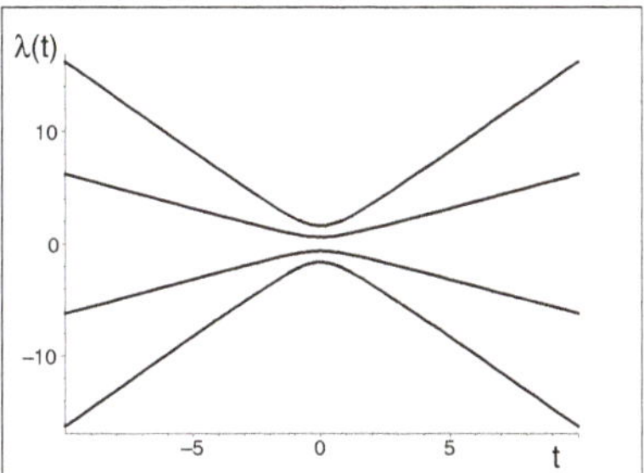

Figure 1. Eigenvalues of matrix (2) (avoided-crossing phenomenon).

We intend to show that, against all expectations, the latter argument is not foolproof. Admitting that it need not *necessarily* lead to the wrong conclusions, we will only show that Wang's and Stankiewicz's alternative scenario [20] may equally well be supported by an equally elementary toy model. The essence of such a claim is that the Hermiticity property (cf. relation $\Lambda = \Lambda^{\dagger}$ satisfied by our toy model matrix (2), with the superscript † marking the matrix transposition plus complex conjugation) depends on a mathematically motivated *a priori* specification of the inner product in our physical Hilbert space of states [21].

A deeper abstract foundation of our "constructive scepticism" concerning the genericity of the Big Bounce may be found in the literature on quantum mechanics using non-Hermitian operators [7,22–24]. In this sense, the common requirement of the self-adjointness of the operators of observables $\Lambda(t)$ can be weakened and replaced by the condition of their Hermitizability *alias* quasi-Hermiticity [22]. In many non-Hermitian models, indeed, the Hermiticity may be restored by the mere *ad hoc* amendment of the inner product [25].

In our present paper, we will narrow the scope of the discussion to the Big Bang and to the WDW equations. Simultaneously, we will broaden the theoretical framework, emphasizing that, in the genuine Big Bang spatial-degeneracy context, it is necessary to replace the most common NSP mathematics with its perceivably more complicated NIP amendment. In a preparatory step, let us now return to the toy model (2) and let us Wick-rotate the time $t \to -it$ and shift the origin, $t \to t-1$. The resulting new matrix

$$Q(t) = \begin{bmatrix} 0 & -2+t & 0 & 0 \\ -t & 0 & -2+t & 0 \\ 0 & -t & 0 & -2+t \\ 0 & 0 & -t & 0 \end{bmatrix} \tag{3}$$

is simply a hidden Hermitian (i.e., via an amendment of the inner product, Hermitizable) candidate for a toy model observable representing, in the context of quantum cosmology, say, a potentially measurable discrete spatial grid [26–30].

In essence, the latter example indicates that the Big-Bang-type singularities *need not* necessarily be smeared out by the quantization. Indeed, at the not too large values of the positive time parameter $t > 0$, the spectrum of our manifestly non-Hermitian model (3) may be shown to be real and non-degenerate. This is illustrated in Figure 2. At $t = 0$, the spectrum becomes degenerate and the matrix itself ceases to be diagonalizable.

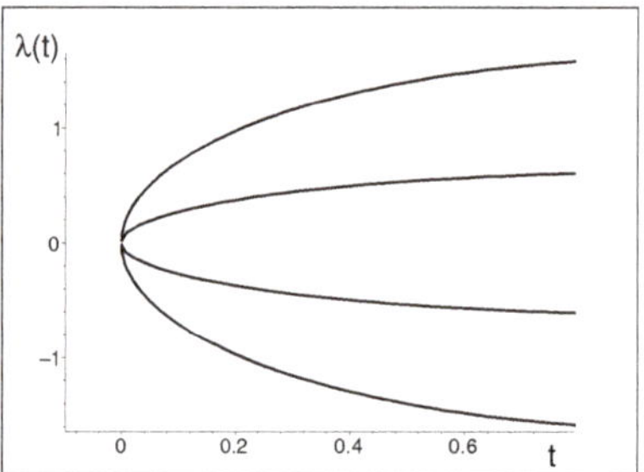

Figure 2. The reality of the spectrum of non-Hermitian matrix (3) at not too large $t \geq 0$.

The latter simulation of the Big Bang singularity is called the exceptional point (EP) in mathematics [31,32]. In the complementary context of physics, the spatial-grid interpretation of the time-dependent eigenvalues $\lambda_n(t)$ as sampled in Figure 2 enables us to speak about the "inflation period" of the history of the related hypothetical and highly schematic (i.e., four-point) quantized Universe immediately after its birth. Naturally, the corresponding internally consistent quantum theory must be reformulated accordingly [22].

2.2. Stationary Theory (Non-Hermitian Schrödinger Picture, NSP)

In the pedagogically oriented and compact review of the history of quantum mechanics [33], the authors emphasized that there exists no universal version of quantum theory and that "no formulation produces a royal road to quantum mechanics". This explains the incessant emergence of new versions of the theory, including its recent "non-selfadjoint-operator" formulations [23].

Incidentally, the "non-selfadjoint-operator" characteristics of these theories could be misleading. As we have already indicated, the mathematical concept of non-selfadjointness (or, in the shorthand terminology used by physicists, of non-Hermiticity) is ambiguous, covering, in various branches of physics, *both* the generators of the unitary evolution [22] *and* of the non-unitary evolution [34]. It is necessary to emphasize that only the former (i.e., unitarity-compatible) meaning of the word "non-Hermiticity" will be considered and taken into account in what follows.

The disambiguation in fact deserves an early mention because the difference is often less clear in applications. Moreover, the formulation of the *hphysical* background of the problems happens to suffer from ambiguities. The details will be discussed later (see, first of all, the introduction to the problem, as given in Appendix A). Now, let us only repeat that the questions that we intend to discuss have their origin in the field of quantum gravity [3]. In this broad context, our attention will be paid, first of all, to the possible role played by the WDW equation and to the questions of physics near the Big Bang (BB) singularity.

2.3. Stationary Wheeler–DeWitt Equation

In the stationary case, the WDW problem becomes formally equivalent to the Klein–Gordon (KG) problem known in the relativistic quantum mechanics [8]. In their simplest versions, both of these problems may be characterized, in suitable units, by the linear differential equation. Thus, in the KG case (where the suitable units are $\hbar = c = 1$ and where one omits, for the sake of simplicity, the electromagnetic field), we have, for example,

$$\left(\frac{\partial^2}{\partial t^2} + D^{(KG)}\right)\psi^{(KG)}(\vec{x},t) = 0\,, \qquad D^{(KG)} = -\triangle + m^2\,. \tag{4}$$

The kinetic energy is represented here by the elementary Laplacian $\triangle$, and the dynamics can be maximally reduced to the mere scalar mass term, which may be made position-dependent, $m^2 = m^2(\vec{x})$.

In the simplest non-stationary WDW model, the analogue of the mass term would be a time-dependent function (cf., e.g., Section 3.5 of review [7] for further references). The KG–WDW analogy enables us to use the same mathematical tools. The relevant literature is fairly extensive but, for our present purposes, it is sufficient to cite the paper by Feshbach and Villars ([35], cf. also Ref. [36]) in which the change of variables

$$\psi^{(WDW)}(\vec{x},t) \quad \rightarrow \quad \langle \vec{x}|\psi^{(FV)}(t)\rangle = \left(\begin{array}{c} \mathrm{i}\partial_t \psi^{(WDW)}(\vec{x},t) \\ \psi^{(WDW)}(\vec{x},t) \end{array} \right) \tag{5}$$

was shown to lead to a replacement of the hyperbolic partial differential Equation (4) by the Schrödinger-like parabolic equation for the two-component wave function (5),

$$\mathrm{i}\frac{\partial}{\partial t}\,|\psi^{(FV)}(t)\rangle = G_{(FV)}(t)\,|\psi^{(FV)}(t)\rangle\,. \tag{6}$$

This equation can be interpreted as controlling the unitary evolution of the system via the generator *alias* FV Hamiltonian

$$G_{(FV)}(t) = \left(\begin{array}{cc} 0 & D(t) \\ I & 0 \end{array} \right). \tag{7}$$

Such an operator is, in the FV Hilbert space

$$\mathcal{H}^{(FV)} = \mathcal{L}^2(\mathbb{R}^3) \bigoplus \mathcal{L}^2(\mathbb{R}^3)$$

manifestly non-Hermitian, $G_{(FV)} \neq G^{\dagger}_{(FV)}$. Pauli with Weisskopf [37] noticed that the same operator can in fact be treated as selfadjoint with respect to another, indefinite inner product,

$$\langle \psi_1|\psi_2\rangle \quad \rightarrow \quad (\psi_1,\psi_2)_{(Krein)} = \langle \psi_1|\mathcal{P}_{(FV)}|\psi_2\rangle\,. \tag{8}$$

i.e., that it is selfadjoint in another, *ad hoc* Krein space. In the modern terminology, one would say that this operator is non-Hermitian but $\mathcal{PT}$-symmetric [38].

Decisive progress achieved under the stationarity assumption $G_{(FV)} \neq G_{(FV)}(t)$ (or, more precisely, after its generalized form, called the quasi-stationarity assumption) is due to Mostafazadeh. In his papers [8,39], he imagined that the FV pseudometric $\mathcal{P}$ could be replaced by the positive definite metric $\Theta_{(stationary)}$, converting the Krein-space physics (in which, during evolution, the usual norm is not conserved) into the fully standard and norm-conserving Hilbert-space physics. In essence, only a straightforward change in the inner product was needed,

$$(\psi_1,\psi_2)_{(Krein)} \quad \rightarrow \quad (\psi_1,\psi_2)_{(Mostafazadeh)} = \langle \psi_1|\Theta_{(stationary)}\,|\psi_2\rangle\,. \tag{9}$$

This opened the way towards a consistent picture of unitary physics in which the stationary Hamiltonian $G_{(FV)} = H_{(FV)}$ controls the NSP quantum evolution, which is, with respect to the amended inner product (9), unitary.

After either the KG or the WDW interpretation of Equation (4) in the stationary case, the Hilbert-space metrics in (9) can be given a formal block-diagonal-operator structure

$$\Theta_{(stationary)} = \left(\begin{array}{cc} 1/\sqrt{D} & 0 \\ 0 & \sqrt{D} \end{array} \right). \tag{10}$$

This leads to the first quantization of both of these systems.

3. Fine-Tuned Nature of the Quantum Big Bang

The conventional mental operation called "quantization of the classical theory" does really very naturally lead to the conclusion that the singularity is "smeared out" near $t \approx 0$

due to "quantum effects" [40,41] (see also the four-by-four Hermitian matrix (2)). In our text, we pointed out that the support of such a regularization hypothesis is only unavoidable in the conventional "textbook" quantum mechanics. In a more general, hiddenly Hermitian theory, such an assumption is artificial and unfounded (cf. Appendix A or toy model (3)). Once one overcomes the mental barrier, one reveals that the inner product may start playing the central descriptive role.

3.1. The N-Grid-Point Toy Model of Kinematics

In the literature, the manifestly non-Hermitian but Hermitizable Wheeler–DeWitt equation has only been considered in the stationary (or, better, quasi-stationary) mathematical NSP regime (cf. [7] or Section 2.2 above). In Section 4, we will turn attention to the conceptual necessity of keeping the WDW-related Hilbert space time-dependent. In the overall context of the canonical quantization of gravity, we have to be prepared to address, therefore, a number of purely technical questions and tasks.

In the first one, the point-like Big Bang must be made compatible with a consequent theoretical unitary evolution scenario. Thus, we have to complement the abstract argumentation of Section 2 with a detailed description of a suitable concrete toy model. In the model, the measurable values of the spatial grid points (say, the necessarily real and time-dependent values $q_j(t)$ with $j = 1, 2, \dots N$) will have to be assumed obtainable, in principle at least, as eigenvalues of a suitable non-Hermitian geometry-representing "effective kinematical input" operator (say, $Q^{(N)}(t)$).

Secondly, we have to keep in mind that, in a way indicated by our four-by-four matrix (3), we may assume that the general N by N matrix $Q^{(N)}(t)$ will still be real and tridiagonal. Indeed, in a way explained in [42], the reality and tridiagonality is an important merit of *any* candidate for an observable because it enables one to construct the metric algebraically, in a recurrent manner. In this sense, we may recall the existing results in linear algebra [43] and choose the one-parametric family of our N by N toy model "effective kinematics" as follows:

$$Q^{(N)}(z) = \left[\begin{array}{cccccc} -i\,(N-1)z & -\sqrt{N-1} & 0 & 0 & \dots & 0 \\ -\sqrt{N-1} & -i\,(N-3)z & -\sqrt{2(N-2)} & 0 & \ddots & \vdots \\ 0 & -\sqrt{2(N-2)} & -i\,(N-5)z & \ddots & \ddots & 0 \\ 0 & 0 & -\sqrt{3(N-3)} & \ddots & -\sqrt{2(N-2)} & 0 \\ \vdots & \ddots & \ddots & \ddots & i\,(N-3)z & -\sqrt{N-1} \\ 0 & \dots & 0 & 0 & -\sqrt{N-1} & i\,(N-1)z \end{array} \right] . \qquad (11)$$

The non-triviality of this matrix and the arbitrariness of its dimension N in combination with its non-numerical tractability [44] will enable us to show how the requirement of the existence of the quantum Big Bang singularity becomes supported by a consistent reconstruction of the related physical time-dependent Hilbert-space metric. As long as $z = z(t)$ can be any suitable function of time, we may restrict our considerations to the interval of $z \in (-1, 1)$ in the interior of which the grid-point-coordinate spectrum of $Q^{(N)}(z)$ remains non-degenerate, real, and discrete, and at the boundaries of which one can visualize the realization of the Big Bang. Thus, after the simplest choice of $z(t) = -1 + t$, we obtain an immediate $N-$level analogue of the graphical evolution pattern of Equation (2), where we had $N = 4$.

One of the main constraints imposed upon our toy model "geometry operator" (11) is its compatibility with the unitarity of the quantum evolution, i.e., with the existence of the Hilbert-space metric. Naturally, the process of the evolution of the corresponding schematic Universe will have to start at the Big Bang single-point-degeneracy singularity, which is such that

$$\lim_{t\to 0^+} q_n(z(t)) = q_1(z(0)), \quad n = 1,2,\ldots,N \tag{12}$$

On the technical level, one can really speak about a challenge because even the purely formal construction of a highly schematic "Big-Banging" model of the quantum Universe must remain compatible with the basic theoretical requirement of compatibility between the kinematical input information (12) and the dynamical input information as represented by the WDW Hamiltonian operator. The details will be discussed below. For the time being, let us only assume that with the kinematical spatial-grid input (12) adapted to *any* phenomenological requirements, the dynamics of the WDW-related Universe will remain reflected by a suitable non-stationary form of the operator D in its form entering the non-stationary analogue of the stationary (or, if you wish, adiabatic) form (10) of the WDW Hilbert-space metric.

3.2. *The Fine-Tuned Nature of the Hilbert-Space Metric* $\Theta(z)$

One of our most important WDW-related model-building tasks can be seen in the generalization of the qualitative and consistent picture of the quantum Big Bang singularity as mediated by its $N = 4$ grid-point realization via Equation (3) above (cf. also Figure 2). In such a project, we encounter the two main technical obstacles. The first one lies in the necessity of the guarantee of the existence of the metric Θ at all times $t > 0$ (i.e., in our model, at all of the sufficiently small positive times) *up to the very Big Bang birth-of-the-Universe EP limit* $t \to 0^+$. In our toy model, due to its exact solvability [43], such a guarantee will have an exact, non-numerical form.

The means of circumventing the second technical obstacle (viz., the necessity of a guarantee that the Hilbert-space metric remains, at all of the relevant times, non-singular and positive definite) is equally difficult to find. In our model, we shall see that, for the model in question, this goal can be achieved by non-numerical means as well.

The respective solutions of both of the above-mentioned problems are closely interrelated. Their essence can be identified with the necessity of the coexistence of the singularity in the grid with the singularity-free nature of the metric $\Theta(z)$. The most universal approach to this problem has been promoted by Scholtz et al. [22], who proposed to use the *complete* information about the set of the observables $\Lambda_1(z), \Lambda_2(z), \ldots$. Such an "extreme" model-building strategy yielded a unique physical metric $\Theta^{(N)}(z)$. In principle, its applicability is strongly N-dependent of course. Thus, our methodical considerations will only concern the systems with the smallest dimensions.

3.2.1. The Eligible Hilbert-Space Metrics at $N = 2$

At $N = 2$, the grid-point operator (11) reads

$$Q^{(2)}(z) = \begin{bmatrix} -iz & -1 \\ -1 & iz \end{bmatrix}, \quad z \in (0,1). \tag{13}$$

with the four real parameters a, c, d and $\chi \in (0, 2\pi)$, with, for the sake of definiteness, positive $z \in (0,1)$, and with the general ansatz

$$\Theta^{(ansatz)}(a,c,d,\chi) = \begin{bmatrix} a & c\,e^{-i\chi} \\ c\,e^{i\chi} & d \end{bmatrix} \tag{14}$$

For the Hilbert-space metric, the condition of quasi-Hermiticitiy degenerates to the two elementary relations,

$$d = a = z^{-1}\,c\,\sin\chi\,.$$

and without any loss of generality, we may set $c = z$ and evaluate the eigenvalues of matrix of Equation (14),

$$\lambda_{\pm} = \sin\chi \pm z\,.$$

Thus, this matrix may be declared acceptable as a metric if and only if it is positive definite, i.e., if and only if

$$\sin\chi > z\,. \tag{15}$$

This relation clearly indicates that near the EP limit $z \to 1$, the range of variability of the admissible parameter χ (numbering the admissible Hilbert-space metrics) becomes extremely narrow. Moreover, whenever the dynamics-controlling parameter z moves closer to the EP singularity, the interval quickly shrinks so that our choice of the metric must be, in the Big Bang vicinity, very precisely "fine-tuned".

Equation (15) becomes further simplified when we reparametrize the strength of the non-Hermiticity $z = \sin\beta$ in terms of the new variable $\beta \in (0, \pi/2)$. Now, the Hermitian limit corresponds to $\beta = 0$ while the singular EP (or, if you wish, Big Bang or Big Crunch) extreme is reached at $\beta = \pi/2$. Ultimately, formula

$$\Theta(\beta,\chi) = \left[\begin{array}{cc} \sin\chi & e^{-i\chi}\sin\beta \\ e^{i\chi}\sin\beta & \sin\chi \end{array}\right], \qquad \chi \in (\beta, \pi - \beta) \tag{16}$$

defines, up to an inessential overall factor, all of the eligible correct metric operators at $N = 2$.

3.2.2. $N = 3$ and the Requirement of Positivity

Once we move to the next geometry operator (11) with $N = 3$, the general ansatz for the metric may be reduced to a six-parametric Hermitian matrix

$$\Theta = \left[\begin{array}{ccc} a & be^{i\phi} & ce^{i\chi} \\ be^{-i\phi} & f & be^{i\phi} \\ ce^{-i\chi} & be^{-i\phi} & a \end{array}\right]. \tag{17}$$

This reveals that the construction of the metric remains a purely routine linear-algebraic problem. At the same time, the weakness of the construction is found to lie in the less easy determination of the domain of parameters for which the metric operator Θ remains positive definite. Although the domain of positivity of the metric is still implicitly defined by the $N = 3$ secular determinant and by the relation

$$\lambda^3 + (-f - 2a)\lambda^2 + \left(-2b^2 - c^2 + a^2 + 2fa\right)\lambda + c^2 f - fa^2 + 2ab^2 - 2b^2 c\cos(2\phi - \chi) = 0$$

the $N = 3$ analogue of the $N = 2$ Equation (16) would be complicated for an explicit display. The task still remains non-numerical because the secular polynomial remains linear in the parameters b^2, f, and/or $\cos(2\phi - \chi)$. This still allows the determination of the range of the admissible parameters to be straightforward. A typical example of such a determination is provided by Figure 3.

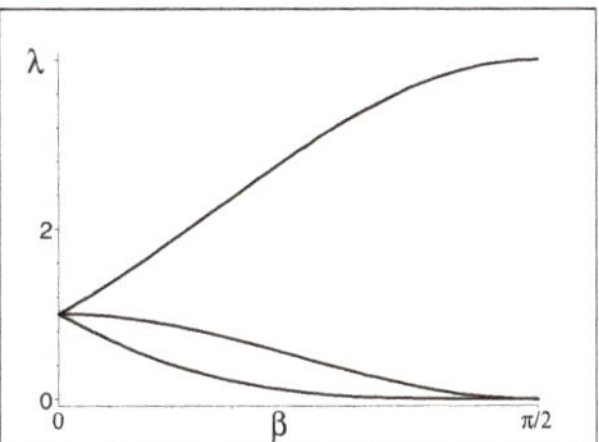

Figure 3. Eigenvalues of our singularity-free metric (17) as functions of one of the parameters at $N = 3$.

In this illustrative picture, we see that the eigenvalues of the metric remain real and non-degenerate in a large interval of one of the dynamical $N = 3$ parameters β. In a small vicinity of the singular EP/BB limit $\beta \to \pi/2$, we may deduce that the rank of the metric becomes approximately equal to one. The picture even shows the confluence of the eigenvalues of the metric in the trivial-metric Hermitian-system limit $\Theta \to I$, i.e., very far from the EP/BB dynamical regime.

Naturally, the technical difficulties will grow with the dimension. At the larger N, the construction has to be given an alternative, purely graphical form. This strategy has been used in paper [45], where it has been shown that the use of the graphical method remains feasible even for the higher-order secular polynomials (cf. Figures 16 and 17 in *loc. cit.*). Nonetheless, one has to expect that at the truly large matrix dimensions, the construction becomes purely numerical.

3.3. *Candidates for the Other Observables*

For any given non-Hermitian grid-point operator $Q^{(N)}(z)$ with the real and nondegenerate spectrum $\{q_n(z)\}$, one can construct the arbitrarily normalized eigenvectors,

$$Q^{(N)}(z)\,|n(z)\rangle = q_n(z)\,|n(z)\rangle\,, \quad n = 1,2,\ldots,N\,. \tag{18}$$

For the same spectrum, the arbitrarily normalized double-bra-marked left eigenvectors may be also defined as the standard right eigenvectors of a Hermitian conjugate operator,

$$\left[Q^{(N)}(z)\right]^{\dagger} |n(z)\rangle\!\rangle = q_n(z)\,|n(z)\rangle\!\rangle\,, \quad n = 1,2,\ldots,N\,. \tag{19}$$

It is easy to deduce that $\langle\!\langle m|n\rangle = 0$ for $m \neq n$. In the generic case, the overlaps $\langle\!\langle m|m\rangle$ will be real and non-vanishing. Whenever N is finite, the resulting biorthogonal basis can be used in a generalized spectral representation of the operator

$$Q^{(N)}(z) = \sum_{n=0}^{N-1} |n(z)\rangle \frac{q_n(z)}{\langle\!\langle n(z)|n(z)\rangle} \langle\!\langle n(z)|\,. \tag{20}$$

We may conclude that the general (though not necessarily invertible or positive definite) N-parametric Hilbert-space metric can be then defined by formula

$$\Theta^{(N)} = \sum_{n=0}^{N-1} |n\rangle\!\rangle\,\kappa_n\,\langle\!\langle n|\,. \tag{21}$$

The parameters κ_n must be all real. The acceptability of the matrix in the role of the physical Hilbert-space metric (i.e., the necessary invertibility and positivity properties) is then guaranteed if and only if $0 < \kappa_n < \infty$ at all n [46]. In such a setting, one can easily use an analogous generalized spectral representation to define also any other operator of an acceptable quantum observable.

4. Mathematics: Non-Hermitian Interaction Picture (NIP)

Naturally, the (quasi-) stationarity restriction becomes, in the WDW case, hardly acceptable, especially if one tries to deal with the quantum dynamics near a singularity such as the Big Bang. In such a case, a much deeper modification of the formalism of the non-Hermitian quantum theory is needed.

After one decides to relax the assumptions of stationarity, an increase in the complexity of the system of equations is partially compensated by the clarification of several conceptual problems. In this sense, our main methodical recommendation is that, in analogy with the Hermitian interaction picture of textbooks, one still keeps in mind the necessity of the description of the dynamics in terms of both the operators and wave functions. In other words, it is necessary to avoid several existing and widespread misunderstandings that can be found in the current literature. Paradoxically, the root of these misunderstandings may be seen in an insufficiently careful use of the terminology (see, e.g., the explanatory "Rosetta-stone-like" Table 1 in [12]). Indeed, once we replace a stationary NSP model by its non-stationary IP and/or NIP alternative and extension, the concept of quantum Hamiltonian ceases to be unique and adequate.

4.1. *Non-Stationary Quantum Systems*

In the non-stationary quantum theory, the use of the time-dependent metric is known to lead to the loss of the unitarity of the evolution or to the loss of the observability of the NSP Hamiltonian [7]. In fact [9], the puzzle is artificial and purely terminological. The problem disappears when one employs the non-Hermitian version of Dirac's interaction picture (NIP, [12]).

4.1.1. Evolution Law for the NIP Ket Vectors

In the non-stationary non-Hermitian cases, there is no need for the observability of the generator of the evolution of the ket vectors [47–50]. Easily, the stationary version of the Dyson map (A3) can be replaced by its time-dependent generalization

$$|\psi(t)\succ = \Omega(t)\,|\psi(t)\rangle \in \mathcal{H}^{(T)}\,, \qquad |\psi(t)\rangle \in \mathcal{H}^{(F)}\,. \tag{22}$$

In $\mathcal{H}^{(F)}$, similarly, Schrödinger Equation (A4) acquires the form

$$\mathrm{i}\frac{\partial}{\partial t}\,|\psi(t)\rangle = G(t)\,|\psi(t)\rangle \tag{23}$$

in which the generator is only one of the two unobservable components of the observable instantaneous-energy operator

$$H(t) = G(t) + \Sigma(t)\,. \tag{24}$$

Only the sum will be called Hamiltonian in what follows. The other component of the Hamiltonian can be defined directly in terms of the Dyson map,

$$\Sigma(t) = \mathrm{i}\Omega^{-1}(t)\,\dot{\Omega}(t)\,, \quad \dot{\Omega}(t) = \frac{d}{dt}\,\Omega(t) \tag{25}$$

(see [10–12] for details).

In the unitary evolution case, the observable version of the non-Hermitian but Hermitizable Hamiltonian (24) is connected with its selfadjoint partner by formula

$$H(t) = \Omega^{(-1)}(t)\,\mathfrak{h}_{(NSP)}(t)\,\Omega(t)\,. \tag{26}$$

In $\mathcal{H}^{(F)}$, operator (26) has the property of quasi-Hermiticity,

$$H^\dagger(t)\,\Theta(t) = \Theta(t)\,H(t)\,, \quad \Theta(t) = \Omega^\dagger(t)\,\Omega(t)\,. \tag{27}$$

In an internally consistent theory of a unitary (or hidden unitary) quantum system, the Hamiltonian still has to have the real and discrete spectrum representing the instantaneous (but still observable) bound-state energies.

It is unfortunate that, in the literature, only too many people assign the name of a Hamiltonian also to *both* of the other operators $G(t)$ and $\Sigma(t)$, neither of which represents an observable quantity [9,11,49]. We prefer calling operator $G(t)$ a "generator" (which does not represent an observable, while still controlling and generating the evolution of the IP/NIP wave functions). In parallel, we would also propose calling operator $\Sigma(t)$, say, a "Coriolis force".

4.1.2. Evolution Law for the NIP Bra Vectors

Equation (22) has a dual-space alternative

$$|\psi(t)\succ = \left[\Omega^{\dagger}(t)\right]^{-1}|\psi_{\Theta}(t)\rangle \in \mathcal{H}^{(T)}, \qquad |\psi_{\Theta}(t)\rangle \equiv \Theta(t)\,|\psi(t)\rangle \in \mathcal{H}^{(F)}. \tag{28}$$

This enables us to treat the new states $|\psi_{\Theta}(t)\rangle \equiv \Theta(t)\,|\psi(t)\rangle$ as solutions of another Schrödinger equation in $\mathcal{H}^{(F)}$ [10,11],

$$\mathrm{i}\frac{\partial}{\partial t}\,|\psi_{\Theta}(t)\rangle = G^{\dagger}(t)\,|\psi_{\Theta}(t)\rangle\,. \tag{29}$$

The process of the solution of the two Schrödinger equations is maximally economical. The key merit of this recipe (see also more commentaries in [12]) is that it circumvents the necessity of the technically much more complicated direct construction of the metric as used, e.g., in papers [49,51–54].

The present version of the process must be initiated by the specification of the respective states $|\psi(t)\rangle$ and $|\psi_{\Theta}(t)\rangle$ at $t = t_i = 0$. Thus, Equations (23) and (29) have to be complemented by the specification of the initial values represented by the kets $|\psi(t_i)\rangle$ and $|\psi_{\Theta}(t_i)\rangle$. Naturally, such values must obey constraints (22) and (28) at $t = t_i = 0$. This, in turn, is closely connected with the experiment and with the preparation of the system in question.

4.2. *Non-Hermitian Operators in Interaction Picture*

It is well known that even in the conventional Hermitian version of IP, the Coriolis-force operators obey the Heisenberg-type equations. These equations control the evolution of *every* relevant operator of an observable.

In the non-Hermitian NIP formalism, the role of $\Sigma(t)$ is analogous. In both of the IP and NIP cases, the ultimate goal of the theory lies in the derivation of the predictions of the results of measurements. In our present version of the recipe, this merely requires the evaluation of the overlaps

$$\langle\psi_{\Theta}(t_f)\,|Q(t_f)\,|\psi(t_f)\rangle\,. \tag{30}$$

Due to the identity

$$\mathrm{i}\,\frac{\partial}{\partial t}\,\Theta(t) = \Theta(t)\,\Sigma(t) - \Sigma^{\dagger}(T)\,\Theta(t) \tag{31}$$

or due to its alternative version (cf. Equation (27)),

$$\mathrm{i}\,\frac{\partial}{\partial t}\,\Theta(t) = G^{\dagger}(t)\,\Theta(t) - \Theta(t)\,G(t) \tag{32}$$

the NIP formalism is internally consistent, indeed. At the same time, one has to keep in mind that the operators of the IP or NIP observables are manifestly time-dependent and that their time dependence is not arbitrary.

4.2.1. Evolution Law for the Density Matrices

In the non-Hermitian but unitary pure-state quantum systems of our present interest, the state is defined by *a pair* of the ket vectors, i.e., by the projectors

$$\pi_{\psi,\Theta}(t) = |\psi(t)\rangle \frac{1}{\langle\psi_\Theta(t)\,|\psi(t)\rangle} \langle\psi_\Theta(t)|\,. \tag{33}$$

Alternatively, one can speak about the non-Hermitian density matrix

$$\widehat{\varrho}(t) = \sum_k |\psi^{(k)}(t)\rangle \frac{p_k}{\langle\psi_\Theta^{(k)}(t)\,|\psi^{(k)}(t)\rangle} \langle\psi_\Theta^{(k)}(t)|\,, \qquad \sum_k p_k = 1\,. \tag{34}$$

Due to Equations (23) and (29), this operator has to obey the specific evolution equation

$$\mathrm{i}\,\partial_t\,\widehat{\varrho}(t) = G(t)\,\widehat{\varrho}(t) - \widehat{\varrho}(t)\,G(t) \tag{35}$$

which opens the way towards the formulation of quantum statistics in the non-Hermitian Liouvillean picture [12].

4.2.2. The Evolution of Observables

The requirement

$$Q^\dagger(t)\,\Theta(t) = \Theta(t)\,Q(t) \tag{36}$$

guarantees the observability status of any operator $Q(t)$. This relation is equivalent, due to Equation (A7), to the NSP Hermiticity of $\mathfrak{q}(t)$ in $\mathcal{H}^{(T)}$ since

$$Q(t) = \Omega^{(-1)}(t)\,\mathfrak{q}_{(NSP)}(t)\,\Omega(t) \tag{37}$$

The Heisenberg-type evolution equation follows:

$$\mathrm{i}\,\frac{\partial}{\partial t}\,Q(t) = Q(t)\,\Sigma(t) - \Sigma(t)\,Q(t) + K(t)\,, \qquad K(t) = \Omega^{(-1)}(t)\,\mathrm{i}\,\dot{\mathfrak{q}}_{(NSP)}(t)\,\Omega(t)\,. \tag{38}$$

It is recommendable to assume that the partial derivatives $\dot{\mathfrak{q}}_{(NSP)}(t)$ vanish so that the related operator $K(t)$ would be vanishing as well, making the process of the solution of Equation (38) more user-friendly.

Given the generator $G(t)$, the choice of the Coriolis force $\Sigma(t)$ is far from arbitrary. First of all, it is constrained by the experiment-related initial state vectors. Secondly, it must be compatible with its relation (24) to the initial instant energy $H^{(NIP)}(t_i)$ and to its evolution law

$$\mathrm{i}\,\frac{\partial}{\partial t}\,H^{(NIP)}(t) = H^{(NIP)}(t)\,\Sigma^{(NIP)}(t) - \Sigma^{(NIP)}(t)\,H^{(NIP)}(t) + K^{(NIP)}(t) \tag{39}$$

or, equivalently,

$$\mathrm{i}\,\frac{\partial}{\partial t}\,H^{(NIP)}(t) = G^{(NIP)}(t)\,H^{(NIP)}(t) - H^{(NIP)}(t)\,G^{(NIP)}(t) + K^{(NIP)}(t)\,. \tag{40}$$

Next, one will also frequently decide to accept the important simplification obtained for the vanishing NSP-time-derivative operators

$$K^{(NIP)}(t) = \Omega^{(-1)}(t)\,\mathrm{i}\,\dot{\mathfrak{h}}_{(NSP)}(t)\,\Omega(t)\,.$$

As long as $\Sigma^{(NIP)}(t) = H^{(NIP)}(t) - G^{(NIP)}(t)$, there remains no freedom left. In particular, as long as we have the definition

$$\mathrm{i}\frac{\partial}{\partial t}\,\Omega^{(NIP)}(t)\rangle = \Omega^{(NIP)}(t)\,\Sigma^{(NIP)}(t)\,, \tag{41}$$

the only ambiguity of $\Omega^{(NIP)}(t)$ is contained in its initial-value specification.

5. The Construction of Non-Stationary WDW Universe Admitting Big Bang

In our present study of the applicability of the NIP approach to the various models in cosmology, we felt particularly interested in a guarantee of the Big Bang degeneracy property

$$\lim_{t\to 0^-} q_j(t) = 0\,, \quad j = 0,1,\ldots \tag{42}$$

which, in the formal context of quantum mechanics, prescribes and restricts the behavior of certain time-dependent eigenvalues $q_j(t)$ of a suitable operator characterizing the spatial geometry (or at least the size) of the Universe, sampled, say, by $Q(t)$ of Equation (37), or of Equation (3), with the spectrum as sampled in Figure 2. For this purpose, let us now return to some less general, simplified WDW models.

5.1. The Evolution of the WDW Ket Vectors

Even in the non-stationary cases, many KG and WDW models remain formally equivalent. For this reason, let us now return to Equation (4), replaced by its non-stationary generalization

$$\left(\frac{\partial^2}{\partial t^2} + D(t)\right)\psi^{(WDW)}(\vec{x},t) = 0\,, \qquad D(t) = -\triangle + m^2(\vec{x},t)\,. \tag{43}$$

Using the same amendment of the wave functions as before,

$$\langle \vec{x}|\psi^{(NIP)}(t)\rangle = \left(\begin{array}{c} \mathrm{i}\partial_t\psi^{(WDW)}(\vec{x},t) \\ \psi^{(WDW)}(\vec{x},t) \end{array}\right) \tag{44}$$

we are able to replace Equation (43) by an analogue of Equation (23), i.e., by the correct NIP Schrödinger equation

$$\mathrm{i}\frac{\partial}{\partial t}|\psi^{(NIP)}(t)\rangle = \left(\begin{array}{cc} 0 & D(t) \\ I & 0 \end{array}\right)|\psi^{(NIP)}(t)\rangle\,. \tag{45}$$

Here, the spectrum of the WDW generator $G_{(NIP)}(t)$ need not be real of course (see, for example, an elementary illustrative example as given in [11]).

5.2. The Evolution of the WDW Bra Vectors

It is obvious that the time dependence of the metric $\Theta(t)$ may be highly sensitive to its initial value at $t = t_i$ [49,51,52,55]. Unfortunately, the direct analysis of this dependence via the solution of Equation (32) is complicated. For this reason, we recommended, in [12], to follow the guidance of papers [47,55] and to circumvent the solution of the auxiliary *operator* evolution Equation (32) (which was characterized, in [49], as the "time-dependent quasi-Hermiticity relation") and to solve the second Schrödinger equation (for the mere bra *vectors*) instead.

This leads to the implementation of the NIP recipe with the evolution of

$$|\psi_\Theta^{(NIP)}\rangle = \Theta(t)\,|\psi^{(NIP)}\rangle \tag{46}$$

controlled by Schrödinger Equation (29),

$$\mathrm{i}\frac{\partial}{\partial t}|\psi_\Theta^{(NIP)}(t)\rangle = \left(\begin{array}{cc} 0 & I \\ D^*(t) & 0 \end{array}\right)|\psi_\Theta^{(NIP)}(t)\rangle\,. \tag{47}$$

Here, it is necessary to emphasize that once we identified the NIP generator $G(t)$ with the WDW generator in Equations (45) and (47), we made, in effect, a certain highly nontrivial decision. It has two aspects. In the phenomenological context, such a decision implies that the WDW generator *does not* represent an observable. We believe that there are many reasons for such a preference, especially in the context of the possible quantization of gravity, because, in such a context, the WDW eigenstates are usually treated as a means of specification of the Hilbert space, rather than as the observable states that would be directly connected with the energy [5].

5.3. Reconstruction of the Metric $\Theta(t)$ from the Generator $G(t)$

In the NIP framework, it is sufficient to admit that only the sum (24) of the generator $G(t)$ and of the Coriolis force $\Sigma(t)$ (of a purely kinematical origin) can be interpreted as the observable Hamiltonian. In such a non-stationary NIP scenario, several open questions emerge and have to be resolved, of course.

5.3.1. Big Bang Rendered Possible by the Time Dependence of the Metric

Let us now accept the model-building strategy in which one is given the kinematical input operator $G(t)$. Then, the general non-Hermitian interaction picture can be declared exceptional because only this picture is in fact a candidate for a realization of the quantum Big-Bang-like phase transitions via a unitary evolution process [43,56,57]. Naturally, the details of such a realization remain nontrivial even when we restrict our attention to the Wheeler–DeWitt form of the most elementary differential-operator generators $G(t)$ and to the Big-Bang-like quantum phase transitions. Nevertheless, what we achieve is that we avoid and eliminate the danger of the Big-Bounce smearing after quantization. In Hermitian theory, this smearing is unavoidable, caused by an effective level repulsion, as sampled in Figure 1 above. In the quasi-Hermitian NIP context, the Big-Bang-related exceptional-point degeneracy is rendered possible via the "fine-tuning" of the metric: a few non-numerical, exactly solvable simulations of such a fine-tuning may be found described, e.g., in [56].

A complementary word of warning has been formulated in our brief methodical note [47]. We revealed there that in the Heisenberg picture (HP), the Big Bang degeneracy cannot be realized at all. Indeed, the underlying constant choice of vanishing $G^{(HP)}(t) = 0$ has been shown to imply the stationarity of the HP metric, $\Theta^{(HP)} \neq \Theta^{(HP)}(t)$ (recall Equation (32) for the quick proof). The HP form of Equation (24) implies that we have $\Sigma_{(HP)}(t) = H_{(HP)}(t)$ so that only the solution of the Heisenberg Equations (38) is needed. The only advantage of using the HP simplification is that both of the underlying Schrödinger equations remain trivial. Nevertheless, as long as the realization of the Big Bang degeneracy necessarily requires that the Hilbert-space metric $\Theta(t)$ has to vary with time, the use of the NIP formalism with nontrivial $G^{(NIP)}(t)$ is unavoidable.

Unfortunately, no help has been reached in an extended Heisenberg picture (EHP, [55]). In a slightly amended formalism, we proposed the use of a constant-operator choice of a *non-vanishing* generator $G_{(EHP)}(t) = G_{(EHP)}(0) \neq 0$. We found that the EHP formalism can already describe the evolution equivalent to the one generated by the manifestly time-dependent selfadjoint quantum Hamiltonians $\mathfrak{h}(t)$ (cf. Abstract of Ref. [55] or a rediscovery of this possibility in [54]). Nonetheless, the description of the phase transitions (such as the Big Bang) remains beyond the capacity of the amended EHP approach. The fully fledged NIP is needed.

5.3.2. The Detailed WDW NIP Recipe

In [12], we outlined some of the details of the constructive treatment of the quantum phase transitions. We pointed out that our "dynamical input" knowledge of the non-observable Hamiltonian $G(t)$ enables us to solve the pair of our Schrödinger Equations (23) and (29) at any initial conditions. In this sense, every initial N-plet

$$|\psi_1(0)\rangle\,, |\psi_2(0)\rangle\,, \ldots\,, |\psi_N(0)\rangle \tag{48}$$

and

$$|\psi_{1,\Theta}(0)\rangle\,, |\psi_{2,\Theta}(0)\rangle\,, \ldots\,, |\psi_{N,\Theta}(0)\rangle \tag{49}$$

chosen at $t = 0$ can be used to construct the time-dependent N-plets of the kets

$$|\psi_1(t)\rangle\,,|\psi_2(t)\rangle\,,\ldots\,,|\psi_N(t)\rangle$$

and

$$|\psi_{1,\Theta}(t)\rangle\,,|\psi_{2,\Theta}(t)\rangle\,,\ldots\,,|\psi_{N,\Theta}(t)\rangle\,.$$

Under an elementary working hypothesis of a finite-, N-dimensional Hilbert space, the additional initial bi-orthonormality assumption

$$\langle\psi_{m,\Theta}(0)\,|\psi_n(0)\rangle = \delta_{m,n}\,,\quad m,n = 1,2,\ldots,N \tag{50}$$

and the completeness

$$\sum_{n=1}^{N}|\psi_n(0)\rangle\langle\psi_{n,\Theta}(0)| = I \tag{51}$$

become immediately extended to all times $t > 0$,

$$\sum_{n=1}^{N}|\psi_n(t)\rangle\langle\psi_{n,\Theta}(t)| = I\,,\qquad \langle\psi_{1,\Theta}(t)\,|\psi_n(t)\rangle = \delta_{m,n}\,,\quad m,n = 1,2,\ldots,N\,. \tag{52}$$

Moreover, the time-dependent metric operator $\Theta(t)$ acquires the standard representation in $\mathcal{H}^{(F)}$,

$$\Theta(t) = \sum_{n=1}^{N}|\psi_{n,\Theta}(t)\rangle\,\langle\psi_{n,\Theta}(t)|\,. \tag{53}$$

This means that the choice of a suitable generator $G(t)$ and of the two suitable initial vector sets (48) and (49) with properties (50) and (51) does not leave too much space for the further requirements concerning the dynamics.

Fortunately, we come to the conclusion that the space left by the NIP formalism is still sufficient for our present purposes. Indeed, in our construction, we started from the assumption of the knowledge of a preselected WDW form of the generator $G(t)$. Such specific "kinematical-like" input information is still not in conflict with the Big Bang dynamics. Indeed, such a version of the general NIP formalism still admits the use of the formal spectral representation of the observables. In this sense, there exist the two most important operators of our present interest. The first one is the "dynamical", observable-energy-representing operator $H(t)$ called Hamiltonian. In its spectral representation of the form

$$H(t) = \sum_{n=1}^{N}|\psi_n(t)\rangle E_n(t)\langle\psi_{n,\Theta}(t)| \tag{54}$$

the choice of the energy eigenvalues $E_n(t)$ remains unrestricted.

In the climax of the story, an entirely analogous expansion should be finally introduced in order to define the complementary, "kinematical", background-representing operator of a suitable "geometry" or "spatial-grid" operator (37). In its analogous spectral representation

$$Q(t) = \sum_{n=1}^{N}|\psi_n(t)\rangle q_n(t)\langle\psi_{n,\Theta}(t)| \tag{55}$$

we are free to require the validity of the Big Bang constraint (42) imposed upon all of its spatial-background-representing eigenvalues $q_n(t)$.

6. Discussion

The non-Hermitian innovation of the NSP framework opened, in [8], the way towards a deeper understanding of the KG- and WDW-like quantum systems in stationary approximation. Later, the birth of the more sophisticated non-Hermitian version of Dirac's interaction picture [10] seemed to be, initially, merely an artificial mathematical exercise.

Nobody seemed to be willing to admit that the NIP formalism might find application in quantum gravity. The main reason was that in the most advanced version of quantum gravity (i.e., in the canonical LQG approach), virtually all the results seemed to indicate that the classical Big Bang singularity has to be replaced by its quantized Big Bounce alternative.

Even among mathematicians, it has been firmly believed that the quantization must necessarily smear out the singularities of the classical Einstein's general relativity [3]. In this sense, before any return to the quantum Big Bang hypothesis, it was necessary to wait for a renewal of its support in the realistic LQG context [20]. Naturally, the problem is technically complicated. In this sense, the present methodical support of the latter hypothesis is also merely schematic and incomplete. In its framework, we had to leave many important phenomenological requirements aside. Let us now mention some of them in the form of brief comments.

6.1. The Background Independence Requirement

In Isham's foreword preceding Thiemann's comprehensive 2007 monograph on canonical quantum gravity [3], the Hamiltonian constraint $\hat{H}\psi = 0$ *alias* "the famous Wheeler–DeWitt equation" is characterized as "arguably one of the most elegant equations in theoretical physics, and certainly one of the most mathematically ill-defined". In the introductory part of the book itself, one reads, indeed, that the sufficiently rigorous specification of a suitable Hilbert space in which the Wheeler–DeWitt operator $\hat{H}$ would be defined represents one of the most important unresolved theoretical challenges.

The latter Hilbert-space problem may be found thoroughly discussed in Section 9.2 of Mostafazadeh's 2010 study [7]. Even the authors of the LQG study admit that such an approach does not yet provide a fully consistent description of the physical reality. Nonetheless, their approach addresses, successfully, the necessary background independence of the theory [5]. In some sense, such a requirement should be incorporated into any theory that pretends to be "fundamental" rather than merely "effective".

From the perspective of our present approach based on the drastically simplified WDW equation, the constructions that would be background-independent were found feasible. In some sense, such a requirement can be perceived as lying in the very center of the NIP approach, in which, admittedly, one starts from the knowledge of the explicit WDW form of the operator $G(t)$, but in which the theory admits the introduction of an "observable background". Although our present spectral representation definition (55) of such an *independent* kinematical background may appear rather abstract, a more specific example may be sought, say, in [58], where a nontrivial coordinate/background has constructively been obtained in an elementary dynamical model.

In our considerations, the role of a geometric background has been played by the "dynamical input" operator $Q(t)$ sampled by a matrix in Equation (3), with the spectrum $q_n(t)$ simulating the "observable" spatial grid points and guaranteeing the existence of the Big Bang singularity at $t = 0$ (cf. Equation (42) or Figure 2). In discussion, one only has to emphasize the mathematical subtlety of the correspondence between the hidden Hermiticity of $Q(t)$ and the fine-tuned nature of the corresponding Hilbert-space metric $\Theta(t)$, which guarantees the unitarity of the system (i.e., of the evolution of the Universe from the very beginning of its observability and existence).

In the latter considerations, the truly drastic simplifications of the picture seem still absolutely necessary at present, skipping, typically, the Lorentz-covariance requirements and working with the models in which the time is a parameter and in which, for methodical reasons, the "Universe" is one-dimensional and discretized via a finite mesh of the time-dependent grid points $q_j(t)$, $j = 1, 2, \ldots, N$. In such a "Universe", only the degeneracy $q_j(t) \to 0$ in the classical physics Big Bang limit $t \to 0^-$ is asked for.

6.2. Problems with Terminology

The conventional belief that the avoided crossings of the eigenvalues are generic is equivalent to the (usually, only tacit) assumption of the time independence of the physical inner-product metric Θ. In the opposite direction, once we replace the conventional selfadjoint grid matrix $\mathfrak{q}(t)$ with its isospectral but merely hidden Hermitian partner $Q(t)$, we discover the existence of a new freedom in the formalism as carried by the Dyson map $\Omega(t)$. As a consequence, the existence of the singular Big Bang grid-point limit (42) is rendered possible.

In the language of mathematics, the innovation lies in the enhancement of the flexibility of the dynamical laws. One arrives at the less usual, non-Hermitian NIP formulation of quantum mechanics. In its framework, the unitary and closed quantum systems may be defined via structures using more than one inner product, i.e., strictly speaking, more than one Hilbert space.

One of the most welcome consequences is an enhancement of the flexibility, while, on the other hand, one may find terminological misunderstandings. In the phenomenologically oriented literature, several similar terms denote more or less the same theory. Thus, in different papers, one encounters, e.g., a reference to the quasi-Hermitian quantum theory [22,59], to the pseudo-Hermitian quantum theory [7], to the non-Hermitian but $\mathcal{PT}$-symmetric quantum theory (usually also with $\mathcal{P}$ = parity and with $\mathcal{T}$ = time reversal [38]), to the three-Hilbert-space quantum theory [11], or to the crypto-Hermitian quantum theory [60], etc.

6.3. The Danger of an Over-Determination of the Dynamical Input

All of the latter approaches lead to a perceivable gain in flexibility of the realistic models of various quantum systems. This is to be countered merely by the necessity of keeping trace of the more sophisticated forms of Hermitian conjugations. One can conclude that the subject is still relevant. On a model-independent level of discussion, it is worth adding that the consistency of the dynamical input need not in fact be easily guaranteed. In review [22], for example, the authors stressed that in the over-determined cases, the necessary Hilbert-space metric (and, hence, the theory itself) need not exist at all. In [61], such a non-existence has been shown to occur even in some fairly popular realistic models. An abstract analysis of such an unpleasant possibility was presented in [62]. Only recently, more encouraging results were obtained in [25], offering a certain systematic guide to the construction of the mutually compatible non-Hermitian observables.

Once we restrict attention to the applicability of the NIP approach in cosmology, encouragement may be sought in the progress and simplifications of the canonical quantization [3,5]. The latter two reviews of the state of the art differ by the language, with the former one being more mathematically oriented. Nonetheless, both of these monographs share the traditional philosophy interpreting the quantum theory as a result of a modification of its classical predecessor. In our final remark, we would like to point out that one could also try to weaken our dependence on the classical-physics-based intuition by treating, as primary, the tentative quantum hypotheses in a way defended, e.g., by Brody and Hughston [63].

7. Summary

The core of our present message is that the consistency of the quantum-mechanical interpretation of the non-stationary WDW systems requires that the Schrödinger equation ceases to be perceived as offering a complete picture of the evolution. In this sense, it is necessary to add a parallel and fully fledged description of the evolution of the operators of observables using the Heisenberg-like evolution equations. In the natural physical quantum-gravity context, the unitarity of the WDW-controlled evolution can be then guaranteed. The apparently non-unitary evolution of the left and right wave functions (controlled by the respective two Schrödinger-type equations) is precisely compensated by the apparent non-unitarity of the evolution of the operators representing the observables (controlled, in parallel, by non-Hermitian Heisenberg-type equations).

Having accepted such a philosophy, our present paper can be read as a more or less purely methodical return to the question of whether, in the framework of quantum cosmology, the birth of our Universe should be perceived as a point-like Big Bang or as a smeared Big Bounce. In essence, we have presented here a few arguments supporting our persuasion that, in the purely theoretical NIP framework, such a question remains, at present, open.

Funding: This research received no external funding

Conflicts of Interest: The author declares no conflicts of interest.

Appendix A. Two Hilbert Spaces in Quantum Mechanics

In the conventional quantum mechanics of textbooks [21], the predictions of the results of experiments have their mathematical background, in NSP , in the evaluation of matrix elements

$$\prec \psi^{(NSP)}(t_f)\,|\mathfrak{q}_{(NSP)}(t_f)\,|\psi^{(NSP)}(t_f)\succ \ . \tag{A1}$$

The symbol $\mathfrak{q}_{(NSP)}$ denotes here a selfadjoint operator of the observable in question; usually, this operator is time-independent, $\mathfrak{q}_{(NSP)} \neq \mathfrak{q}_{(NSP)}(t)$. All information about the evolution of the system in time is carried, in the pure state regime, by a ket vector element $|\psi^{(NSP)}(t)\succ$ of a Hilbert space of states $\mathcal{H}^{(textbook)}$. This state is assumed prepared at $t_{initial} = 0$ and measured at $t = t_{final} = t_f$. Prediction (A1) is probabilistic and contains only the NSP wave-ket solutions $|\psi^{(NSP)}(t)\succ$ of Schrödinger equation

$$\mathrm{i}\frac{\partial}{\partial t}\,|\psi(t)\succ = \mathfrak{h}_{(NSP)}\,|\psi(t)\succ\ , \qquad |\psi(t)\succ\, \in \mathcal{H}^{(textbook)}\,. \tag{A2}$$

Due to the Stone theorem, the evolution is unitary if and only if the Hamiltonian is selfadjoint in $\mathcal{H}^{(textbook)}$, $\mathfrak{h}_{(NSP)} = \mathfrak{h}^{\dagger}_{(NSP)}$ [64].

One of many efficient simplifications of the practical solution of Equation (A2) is due to Dyson [65]. He revealed that, in many cases, one has to work with a technically unfriendly Hamiltonian, which can be perceivably simplified via a suitable isospectral preconditioning $\mathfrak{h}_{(NSP)} \to H_{(NSP)} = \Omega^{-1}\,\mathfrak{h}_{(NSP)}\,\Omega$. This is formally equivalent to the transformation of the ket vector wave functions,

$$|\psi_n^{(textbook)}\succ\, = \Omega\,|\psi_n^{(auxiliary)}\rangle\,, \quad n = 0, 1, \ldots\,. \tag{A3}$$

Operator Ω has to be n-independent and stationary ($\Omega \neq \Omega(t)$). Dyson also recommended to make the choice of Ω non-unitary ($\Omega^{\dagger}\Omega = \Theta \neq I$). In analogy with the so-called coupled-cluster method based on a similar idea [66], one may also treat the simpler partner of the Hilbert space $\mathcal{H}^{(textbook)}$ as formally different, denoted by a different dedicated symbol, say, $\mathcal{H}^{(friendlier)}$.

Schrödinger Equation (A2) becomes replaced, in the majority of applications of the Dyson-recommended and Ω-mediated change of space $\mathcal{H}^{(textbook)} \to \mathcal{H}^{(friendlier)}$, by a friendlier equation

$$\mathrm{i}\frac{\partial}{\partial t}\,|\psi^{(Dyson)}(t)\rangle = H_{(Dyson)}\,|\psi^{(Dyson)}(t)\rangle\,, \qquad |\psi^{(Dyson)}(t)\rangle \in \mathcal{H}^{(friendlier)}\,. \tag{A4}$$

The transformed Hamiltonian is de-Hermitized since $H = \Omega^{-1}\,\mathfrak{h}\,\Omega \neq H^{\dagger}$ in $\mathcal{H}^{(friendlier)}$. In the early review [22] of the procedure, a change in the philosophy has been proposed, resulting in a reformulation of the textbook NSP approach called, in the spirit of the mathematician's terminology [59], quasi-Hermitian quantum mechanics. In this framework, the model-building process has to start directly from Equation (A4) and from a guarantee of the user-friendliness of the preconditioned Hamiltonian H. Whenever necessary, one may, after all, re-Hermitize the model, say, via a reconstruction of Ω [7].

The non-unitarity of the map Ω implies, for the manifestly auxiliary Hilbert space $\mathcal{H}^{(friendlier)}$, the loss of its physical-space status. Fortunately, it appeared sufficient to amend the inner product and to convert $\mathcal{H}^{(friendlier)}$ into a fully acceptable and physical Hilbert space $\mathcal{H}^{(standard)}$. By construction, the latter space has to be unitary equivalent to $\mathcal{H}^{(textbook)}$, with the most straightforward method being the reconstruction of the so-called metric $\Theta = \Omega^\dagger \Omega$. The mathematical details can be found in reviews [22] and [7]. The essence of the trick is that the correct space $\mathcal{H}^{(standard)}$ can be represented via the mere amendment of the bra vectors in $\mathcal{H}^{(friendlier)}$,

$$\langle\psi| \rightarrow \langle\psi|\Theta \equiv \langle\psi_\Theta| \quad \text{for} \quad \mathcal{H}^{(friendlier)} \rightarrow \mathcal{H}^{(standard)} . \tag{A5}$$

In the terminology of functional analysis, the definition of the dual *alias* bra vector space of the linear functionals is merely amended and transferred back, from $\mathcal{H}^{(standard)}$ to $\mathcal{H}^{(friendlier)}$, via formula $\mathcal{V}' \rightarrow \mathcal{V}'_\Theta$. In other words, one simply converts the conventional, unphysical bra-ket inner product $\langle\psi|\chi\rangle$ into its physical alternative,

$$\langle\psi|\chi\rangle \rightarrow \langle\psi|\Theta|\chi\rangle \equiv \langle\psi_\Theta|\chi\rangle . \tag{A6}$$

In light of this relation, it is possible to perform all calculations in $\mathcal{H}^{(friendlier)}$. Nonetheless, in practice, the redundancy of the introduction of the manifestly unphysical Hilbert space $\mathcal{H}^{(friendlier)}$ must be well motivated. The expense must be more than compensated by the simplification of the evaluation of the experimental predictions. Moreover, the loss of the direct connection with $\mathcal{H}^{(textbook)}$ has to be taken into account because, in this space, we usually define the operators of observables using the principle of correspondence [21].

One can often pull at least some of the necessary operators from $\mathcal{H}^{(textbook)}$ up to the auxiliary Hilbert space $\mathcal{H}^{(friendlier)}$ (see, e.g., [58]), e.g., whenever one knows the Dyson map, one can define the necessary operators in $\mathcal{H}^{(friendlier)}$ using formula

$$Q_{(Dyson)} = \Omega^{(-1)}_{(Dyson)}\, \mathfrak{q}_{(NSP)}\, \Omega_{(Dyson)} \neq Q^\dagger_{(Dyson)} . \tag{A7}$$

The experiment-predicting NSP equation (A1) then acquires the upgraded form,

$$\prec\psi^{(NSP)}(t_f)\,|\mathfrak{q}_{(NSP)}(t_f)\,|\psi^{(NSP)}(t_f)\succ = \langle\psi_\Theta^{(Dyson)}(t_f)\,|Q_{(Dyson)}(t_f)\,|\psi^{(Dyson)}(t_f)\rangle \tag{A8}$$

in which one can use, at worst, merely some reasonably precise approximate form of the physical Hilbert-space metric $\Theta = \Omega^\dagger \Omega$ in Equation (A5) (cf. [22,67]).

References

1. DeWitt, B.S. Quantum Theory of Gravity. I. The Canonical Theory. *Phys. Rev.* **1967**, *160*, 1113–1148. [CrossRef]
2. Hamber, H.W.; Williams, R.M. Discrete Wheeler-DeWitt Equation. *Phys. Rev.* **2011**, *84*, 104033. [CrossRef]
3. Thiemann, T. *Introduction to Modern Canonical Quantum General Relativity;* Cambridge University Press: Cambridge, UK, 2007.
4. Rovelli, C.; Smolin, L. Loop space representation of quantum general relativity. *Nucl. Phys. B* **1990**, *331*, 80–152. [CrossRef]
5. Rovelli, C. *Quantum Gravity;* Cambridge University Press: Cambridge, UK, 2004.
6. Ashtekar, A.; Lewandowski, J. Background independent quantum gravity: A status report. *Class. Quantum Grav.* **2004**, *21*, R53–R152. [CrossRef]
7. Mostafazadeh, A. Pseudo-Hermitian Representation of Quantum Mechanics. *Int. J. Geom. Meth. Mod. Phys.* **2010**, *7*, 1191–1306. [CrossRef]
8. Mostafazadeh, A. Quantum mechanics of Klein-Gordon-type fields and quantum cosmology. *Ann. Phys.* **2004**, *309*, 1–48. [CrossRef]
9. Znojil, M. Which operator generates time evolution in Quantum Theory? *arXiv* **2007**, arXiv:0711.0535.
10. Znojil, M. Time-dependent version of cryptohermitian quantum theory. *Phys. Rev. D* **2008**, *78*, 085003. [CrossRef]
11. Znojil, M. Three-Hilbert-space formulation of Quantum Mechanics. *Symm. Integ. Geom. Meth. Appl. SIGMA* **2009**, *5*, 001. [CrossRef]
12. Znojil, M. Non-Hermitian interaction representation and its use in relativistic quantum mechanics. *Ann. Phys.* **2017**, *385*, 162–179. [CrossRef]
13. Bojowald, M. Absence of a singularity in loop quantum cosmology. *Phys. Rev. Lett.* **2001**, *86*, 5227–5230. [CrossRef] [PubMed]
14. Ashtekar, A.; Pawlowski, T.; Singh, P. Quantum nature of the big bang: Improved dynamics. *Phys. Rev. D* **2006**, *74*, 084003. [CrossRef]
15. Bojowald, M. Quantum nature of cosmological bounces. *Gen. Rel. Grav.* **2008**, *40*, 2659–2683. [CrossRef]

16. Ashtekar, A.; Corichi, A.; Singh, P. Robustness of key features of loop quantum cosmology. *Phys. Rev. D* **2008**, *77*, 024046. [CrossRef]
17. Malkiewicz, P.; Piechocki, W. Turning Big Bang into Big Bounce: II. Quantum dynamics. *Class. Quant. Gravity* **2010**, *27*, 225018. [CrossRef]
18. Bojowald, M.; Paily, G.M. A no-singularity scenario in loop quantum gravity. *Class. Quant. Gravity* **2012**, *29*, 242002. [CrossRef]
19. Yang, J.-S.; Zhang, C.; Ma, Y.-G. Loop quantum cosmology from an alternative Hamiltonian. *Phys. Rev. D* **2019**, *100*, 064026. [CrossRef]
20. Wang, C.; Stankiewicz, M. Quantization of time and the big bang via scale-invariant loop gravity. *Phys. Lett. B* **2020**, *800*, 135106. [CrossRef]
21. Messiah, A. *Quantum Mechanics*; North Holland: Amsterdam, The Netherlands, 1961.
22. Scholtz, F.G.; Geyer, H.B.; Hahne, F.J.W. Quasi-Hermitian Operators in Quantum Mechanics and the Variational Principle. *Ann. Phys.* **1992**, *213*, 74–101. [CrossRef]
23. Bagarello, F.; Gazeau, J.-P.; Szafraniec, F.; Znojil, M. (Eds.) *Non-Selfadjoint Operators in Quantum Physics: Mathematical Aspects*; Wiley: Hoboken, NJ, USA, 2015.
24. Bender, C.M. *PT Symmetry in Quantum and Classical Physics*; World Scientific: Singapore, 2018.
25. Znojil. M. Feasibility and method of multi-step Hermitization of crypto-Hermitian quantum Hamiltonians. *Eur. Phys. J. Plus* **2022**, *137*, 335. [CrossRef]
26. Rovelli, C.; Smolin, L. Discreteness of area and volume in quantum gravity. *Nucl. Phys. B* **1995**, *442*, 593–619. [CrossRef]
27. Thiemann, T. A length operator for canonical quantum gravity. *J. Math. Phys.* **1998**, *39*, 3372–3392. [CrossRef]
28. Znojil, M. Quantum Big Bang without fine-tuning in a toy-model. *J. Phys. Conf. Ser.* **2012**, *343*, 012136. [CrossRef]
29. Znojil, M. Quantization of Big Bang in crypto-Hermitian Heisenberg picture, In *Non-Hermitian Hamiltonians Quantum Physics*; Bagarello, F., Passante, R., Trapani, C., Eds.; Springer: Cham, Switzerland, 2016; Volume 184, pp. 383–399.
30. Brody, D.C.; Hughston, L.P. Quantum measurement of space-time events. *J. Phys. A Math. Theor.* **2021**, *54*, 235304. [CrossRef]
31. Kato, T. *Perturbation Theory for Linear Operators*; Springer: Berlin/Heidelberg, Germany, 1966.
32. Znojil, M. Parity-time symmetry and the toy models of gain-loss dynamics near the real Kato's exceptional points. *Symmetry* **2016**, *8*, 52. [CrossRef]
33. Styer, D.F.; Balkin, M.S.; Becker, K.M.; Burns, M.R.; Dudley, C.E.; Forth, S.T.; Gaumer, J.S.; Kramer, M.A.; Oertel, D.C.; Park, L.H.; et al. Nine formulations of quantum mechanics. *Am. J. Phys.* **2002**, *70*, 288 – 297. [CrossRef]
34. Moiseyev, N. *Non-Hermitian Quantum Mechanics*; Cambridge University Press: Cambridge, UK, 2011.
35. Feshbach, H.; Villars, F. Elementary relativistic wave mechanics of spin 0 and spin 1/2 particles. *Rev. Mod. Phys.* **1958**, *30*, 24–45. [CrossRef]
36. Znojil, M. Relativistic supersymmetric quantum mechanics based on Klein-Gordon equation. *J. Phys. A Math. Gen.* **2004**, *37*, 9557–9571. [CrossRef]
37. Pauli, W.; Weisskopf, V. Uber die Quantisierung der skalaren relativistischen Wellengleichung. *Helv. Phys. Acta* **1934**, *7*, 709–731.
38. Bender, C.M. Making sense of non-Hermitian Hamiltonians. *Rep. Prog. Phys.* **2007**, *70*, 947–1118. [CrossRef]
39. Mostafazadeh, A. Hilbert space structures on the solution space of Klein-Gordon type evolution equations. *Class. Quant. Grav.* **2003**, *20*, 155–171. [CrossRef]
40. Gielen, S.; Turok, N. Perfect Quantum Cosmological Bounce. *Phys. Rev. Lett.* **2016**, *117*, 021301. [CrossRef]
41. Ashtekar, A.; Bianchi, E. A short review of loop quantum gravity. *Rep. Prog. Phys.* **2021**, *84*, 042001. [CrossRef] [PubMed]
42. Znojil, M. Quantum inner-product metrics via recurrent solution of Dieudonne equation. *J. Phys. A Math. Theor.* **2012**, *45*, 085302. [CrossRef]
43. Znojil, M. Tridiagonal PT-symmetric N by N Hamiltonians and a fine-tuning of their observability domains in the strongly non-Hermitian regime. *J. Phys. A Math. Theor.* **2007**, *40*, 13131–13148. [CrossRef]
44. Znojil, M. Maximal couplings in PT-symmetric chain-models with the real spectrum of energies. *J. Phys. A Math. Theor.* **2007**, *40*, 4863–4875. [CrossRef]
45. Znojil, M. N-site-lattice analogues of $V(x) = ix^3$. *Ann. Phys.* **2012**, *327*, 893–913. [CrossRef]
46. Znojil, M. On the role of the normalization factors κ_n and of the pseudo-metric P in crypto-Hermitian quantum models. *Symm. Integ. Geom. Meth. Appl. SIGMA* **2008**, *4*, 001. [CrossRef]
47. Znojil, M. Non-Hermitian Heisenberg representation. *Phys. Lett. A* **2015**, *379*, 2013–2017. [CrossRef]
48. Miao, Y.-G.; Xu, Z.-M. Investigation of non-Hermitian Hamiltonians in the Heisenberg Picture. *Phys. Lett. A* **2016**, *380*, 1805–1810. [CrossRef]
49. Fring, A.; Moussa, M.H.Y. Unitary quantum evolution for time-dependent quasi-Hermitian systems with non-observable Hamiltonians. *Phys. Rev. A* **2016**, *93*, 042114. [CrossRef]
50. Luiz, F.S.; Pontes, M.A.; Moussa, M.H.Y. Unitarity of the time-evolution and observability of non-Hermitian Hamiltonians for time-dependent Dyson maps. *arXiv* **2016**, arXiv:1611.08286.
51. Gong, J.-B.; Wang, Q.-H. Time-dependent PT-symmetric quantum mechanics. *J. Phys. A Math. Theor.* **2013**, *46*, 485302. [CrossRef]
52. Bíla, H. Non-Hermitian Operators in Quantum Physics. Ph.D. Thesis, Charles University, Prague, Czech Republic, 2008.
53. Bíla, H. Adiabatic time-dependent metrics in PT-symmetric quantum theories. *arXiv* **2009**, arXiv:0902.0474.

54. Fring, A.; Frith, T. Exact analytical solutions for time-dependent Hermitian Hamiltonian systems from static unobservable non-Hermitian Hamiltonians. *Phys. Rev. A* **2017**, *95*, 010102(R). [CrossRef]
55. Znojil, M. Crypto-unitary forms of quantum evolution operators. *Int. J. Theor. Phys.* **2013**, *52*, 2038. [CrossRef]
56. Znojil, M. Passage through exceptional point: Case study. *Proc. Roy. Soc. A Math. Phys. Eng. Sci.* **2020**, *476*, 20190831. [CrossRef]
57. Znojil, M. Horizons of stability. *J. Phys. A Math. Theor.* **2008**, *41*, 44027. [CrossRef]
58. Mostafazadeh, A.; Batal, A. Physical Aspects of Pseudo-Hermitian and PT-Symmetric Quantum Mechanics. *J. Phys. A Math. Gen.* **2004**, *37*, 11645–11679. [CrossRef]
59. Dieudonne, J. Quasi-Hermitian operators. In *Proceedings of the International Symposium on Linear Spaces* ; Pergamon: Oxford, UK, 1961; pp. 115–122.
60. Smilga, A.V. Cryptogauge symmetry and cryptoghosts for crypto-Hermitian Hamiltonians. *J. Phys. A Math. Theor.* **2008**, *41*, 244026. [CrossRef]
61. Znojil, M.; Semorádová, I.; Růžička, F.; Moulla, H.; Leghrib, I. Problem of the coexistence of several non-Hermitian observables in PT-symmetric quantum mechanics. *Phys. Rev. A* **2017**, *95*, 042122. [CrossRef]
62. Krejčiřík, D.; Lotoreichik, V.; Znojil, M. The minimally anisotropic metric operator in quasi-hermitian quantum mechanics. *Proc. Roy. Soc. A Math. Phys. Eng. Sci.* **2018**, *474*, 20180264. [CrossRef] [PubMed]
63. Brody, D.C.; Hughston, L.P. Geometric quantum mechanics. *J. Geom. Phys.* **2001**, *38*, 19–53. [CrossRef]
64. Stone, M.H. On one-parameter unitary groups in Hilbert Space. *Ann. Math.* **1932**, *33*, 643–648. [CrossRef]
65. Dyson, F.J. General Theory of Spin-Wave Interactions. *Phys. Rev.* **1956**, *102*, 1217. [CrossRef]
66. Bishop, R.F.; Znojil, M. The coupled-cluster approach to quantum many-body problem in a three-Hilbert-space reinterpretation. *Acta Polytech.* **2014**, *54*, 85–92. [CrossRef]
67. Znojil, M. The cryptohermitian smeared-coordinate representation of wave functions. *Phys. Lett. A* **2011**, *375*, 3176–3183. [CrossRef]

 MDPI

Review

Quantum Current Algebra in Action: Linearization, Integrability of Classical and Factorization of Quantum Nonlinear Dynamical Systems

Anatolij K. Prykarpatski

Department of Computer Science and Telecomunication, Cracow University of Technology, 31-155 Cracow, Poland; pryk.anat@cybergal.com

Abstract: This review is devoted to the universal algebraic and geometric properties of the non-relativistic quantum current algebra symmetry and to their representations subject to applications in describing geometrical and analytical properties of quantum and classical integrable Hamiltonian systems of theoretical and mathematical physics. The Fock space, the non-relativistic quantum current algebra symmetry and its cyclic representations on separable Hilbert spaces are reviewed and described in detail. The unitary current algebra family of operators and generating functional equations are described. A generating functional method to constructing irreducible current algebra representations is reviewed, and the ergodicity of the corresponding representation Hilbert space measure is mentioned. The algebraic properties of the so called coherent states are also reviewed, generated by cyclic representations of the Heisenberg algebra on Hilbert spaces. Unbelievable and impressive applications of coherent states to the theory of nonlinear dynamical systems on Hilbert spaces are described, along with their linearization and integrability. Moreover, we present a further development of these results within the modern Lie-algebraic approach to nonlinear dynamical systems on Poissonian functional manifolds, which proved to be both unexpected and important for the classification of integrable Hamiltonian flows on Hilbert spaces. The quantum current Lie algebra symmetry properties and their functional representations, interpreted as a universal algebraic structure of symmetries of completely integrable nonlinear dynamical systems of theoretical and mathematical physics on functional manifolds, are analyzed in detail. Based on the current algebra symmetry structure and their functional representations, an effective integrability criterion is formulated for a wide class of completely integrable Hamiltonian systems on functional manifolds. The related algebraic structure of the Poissonian operators and an effective algorithm of their analytical construction are described. The current algebra representations in separable Hilbert spaces and the factorized structure of quantum integrable many-particle Hamiltonian systems are reviewed. The related current algebra-based Hamiltonian reconstruction of the many-particle oscillatory and Calogero–Moser–Sutherland quantum models are reviewed and discussed in detail. The related quasi-classical quantum current algebra density representations and the collective variable approach in equilibrium statistical physics are reviewed. In addition, the classical Wigner type current algebra representation and its application to non-equilibrium classical statistical mechanics are described, and the construction of the Lie–Poisson structure on the phase space of the infinite hierarchy of distribution functions is presented. The related Boltzmann–Bogolubov type kinetic equation for the generating functional of many-particle distribution functions is constructed, and the invariant reduction scheme, compatible with imposed correlation functions constraints, is suggested and analyzed in detail. We also review current algebra functional representations and their geometric structure subject to the analytical description of quasi-stationary hydrodynamic flows and their magneto-hydrodynamic generalizations. A unified geometric description of the ideal idiabatic liquid dynamics is presented, and its Hamiltonian structure is analyzed. A special chapter of the review is devoted to recent results on the description of modified current Lie algebra symmetries on torus and their Lie-algebraic structures, related to integrable so-called heavenly type spatially many-dimensional dynamical systems on functional manifolds.

Citation: Prykarpatski, A.K. Quantum Current Algebra in Action: Linearization, Integrability of Classical and Factorization of Quantum Nonlinear Dynamical Systems. *Universe* **2022**, *8*, 288. https://doi.org/10.3390/universe8050288

Academic Editors: Steven Duplij and Michael L. Walker

Received: 2 March 2022
Accepted: 5 May 2022
Published: 20 May 2022

Keywords: diffeomorphism group; current algebra symmetry; current Lie algebra representation; fock space; generating functional; distribution functions; Lie–Poisson structure; coherent states; Lie-Poisson action; Hilbert space linearization; hamiltonian systems; symmetry reduction; integrability; idiabatic states; factorization; heavenly type dynamical systems; integrable dynamical systems; dirac reduction; hydrodynamic flows; entropy; vortex flows

PACS: 11.10.Ef; 11.15.Kc; 11.10.-z; 11.15.-q; 11.10.Wx; 05.30.-d

1. Introduction

It is an old classical result that the nonrelativistic quantum current algebra realizes [1–3] a representation of the Lie algebra $\mathcal{G}$, related to the semidirect product $G := \mathrm{Diff}(\mathbb{R}^m) \ltimes F(\mathbb{R}^m;\mathbb{R})$ of the topological diffeomorphism group $\mathrm{Diff}(\mathbb{R}^m)$ of the real space $\mathbb{R}^m$ and the space $F(\mathbb{R}^m;\mathbb{R})$ of smooth Schwartz type functions on it. As it was later shown by G. Goldin, with collaborators [4–7], in fact, all nonrelativistic quantum many-particle Hamiltonian systems allow the equivalent representation by means of the current algebra operators and their realization on some specially constructed generalized Hilbert spaces with cyclic vector structure, strongly depending on the groundstate vectors of the corresponding Hamiltonian operators. The detailed analysis of this representation [8–10] made it possible to reveal a deep connection of the specially factorized operator structure of Hamiltonian operators and the quantum complete integrability of the corresponding Heisenberg type operator dynamical systems. Moreover, studying vector field representations of the quantum current algebra, related to the semidirect product $\mathrm{Diff}(\mathbb{S}^1) \ltimes F(\mathbb{S}^1;\mathbb{R})$ of the topological diffeomorphism group $\mathrm{Diff}(\mathbb{S}^1)$ of the circle $\mathbb{S}^1$ and the space $F(\mathbb{S}^2;\mathbb{R})$ of the smooth periodic functions on it, their isomorphism was stated [11–13] with all completely known and up to date Lax type integrable classical dynamical systems on spatially one-dimensional functional manifolds. Closely related algebraic aspects of representation theory of the canonical creation–annihilation operators, both on the Fock space and on related cyclic Hilbert spaces, gave rise to the construction of the effective linearizing scheme of any smooth dynamical system on functional Hilbert space. As the main analytical trick of these schemes is based on the coherent vector representation of the canonical creation–annihilation operators on the Fock space, we describe their unbelievable and impressive applications to the theory of nonlinear dynamical systems on Hilbert spaces, their linearization and integrability, previously initiated in [14,15] and continued in [16]. We briefly review the coherent vector representations of the Bargmann–Segal space $\mathcal{H}_k$ of complex holomorphic functions on $\mathbb{C}^k$, and describe a general approach to constructing coherent states and their applications both to the linearization of nonlinear dynamical systems on Hilbert spaces, and to describing their complete integrability. The latter is developed using the modern Lie-algebraic approach [11,17–19] to nonlinear dynamical systems on Poissonian functional manifolds, and proved to be both unexpected and important for the classification of integrable Hamiltonian flows on Hilbert spaces.

Other very important applications of the current algebra representations are related both to statistical physics, classical and quantum, and to hydrodynamics. The quantum current algebra quasi-classical representations made it possible to analytically describe [20–22] the so-called collective variable approach in equilibrium statistical physics and calculate the main thermodynamical quantities at finite temperatures. The related quantum current algebra quasi-classical Wigner type representations proved to be effective in describing the kinetic theory [23,24] of many-particle systems and calculating both the corresponding evolution equations for the infinite hierarchy of many-particle distribution functions, and developing a new approach to their dynamically compatible splitting, based on the well known Dirac type reduction of Poissonian systems on functional submanifolds.

A very rich geometric structure of liquid flow in a domain $\Omega \subset \mathbb{R}^3$ and its properties can be deeply described by means of the corresponding diffeomorphism group $\mathrm{Diff}(\Omega)$

and its semi-direct products with different functional spaces on the domain $\Omega \subset \mathbb{R}^3$. It is well known that the same physical system is often described using different sets of variables, related with their different physical interpretation. It was observed [25–33] that the corresponding mathematical structures used for describing the analytical properties of hydrodynamical systems are canonically related to each other. Simultaneously, mathematical properties, against a background of their analytical description, make it possible to study additional important parameters [34–50] of different hydrodynamic and magnetohydrodynamic systems. Amongst these, we will mention integral invariants, describing such internal fluid motion peculiarities as vortices, topological singularities [51] and other different instability states, strongly depending [52,53] on imposed isentropic fluid motion constraints. Being interested in their general properties and mathematical structures which are responsible for their existence and behavior, we present [54] a detailed differential geometrical approach to thermodynamically investigating quasi-stationary isentropic fluid motions, paying more attention to the analytical argumentation of tricks and techniques used during the presentation. Amongst the systems analyzed here, we mention the Hamiltonian analysis and adiabatic magneto-hydrodynamic superfluid motion, as well constructing a modified current Lie algebra and describing magneto-hydrodynamic invariants and their geometry. In particular, we studied a modified current Lie algebra symmetry on torus, its Lie-algebraic structure and related integrable heavenly type dynamical systems, describing the quasi-conformal metrics of Riemannian spaces in general relativity.

2. The Fock Space, Non-Relativistic Quantum Current Algebra and Its Cyclic Representations

2.1. The Fock Space Representation

Let a Hilbert space Φ_F possess the standard canonical Fock space structure [5,11,55–60], that is

$$\Phi_F = \oplus_{n\in\mathbb{Z}_+} \Phi_{(s)}^{\otimes n}, \tag{1}$$

where subspaces $\Phi_{(s)}^{\otimes n}$, $n \in \mathbb{Z}_+$, are the symmetrized tensor products of the Hilbert space $H \simeq L_2^{(s)}(\mathbb{R}^m;\mathbb{C}^k)$. If a vector $\varphi := (\varphi_0, \varphi_1, ..., \varphi_n, ...) \in \Phi_F$, its norm

$$\|\varphi\|_\Phi := \left(\sum_{n\in\mathbb{Z}_+} \|\varphi_n\|_n^2\right)^{1/2}, \tag{2}$$

where $\varphi_n \in \Phi_{(s)}^{\otimes n} \simeq L_2^{(s)}((\mathbb{R}^m)^{\otimes n};\mathbb{C})$ and $\| ... \|_n$ is the corresponding norm in $\Phi_{(s)}^{\otimes n}$ for all $n \in \mathbb{Z}_+$. Note here that concerning the rigging structure (18), there holds the corresponding rigging for the Hilbert spaces $\Phi_{(s)}^{\otimes n}$, $n \in \mathbb{Z}_+$, that is

$$\mathcal{D}_{(s)}^n \subset \Phi_{(s),+}^{\otimes n} \subset \Phi_{(s)}^{\otimes n} \subset \Phi_{(s),-}^{\otimes n} \tag{3}$$

with some suitably chosen dense and separable topological spaces of symmetric functions $\mathcal{D}_{(s)}^n$, $n \in \mathbb{Z}_+$. Concerning expansion (1), we obtain by means of projective and inductive limits [55,57,61,62] the quasi-nucleus rigging of the Fock space Φ in the form (18).

Consider now any basis vector $|(\alpha)_n) \in \Phi_{(s)}^{\otimes n}$, $n \in \mathbb{N}$, which can be written [56,57,63–65] in the following canonical Dirac ket-form:

$$|(\alpha)_n) := |\alpha_1, \alpha_2, ..., \alpha_n), \tag{4}$$

where, by definition,

$$|\alpha_1, \alpha_2, ..., \alpha_n) := \frac{1}{\sqrt{n!}} \sum_{\sigma\in S_n} |\alpha_{\sigma(1)}) \otimes |\alpha_{\sigma(2)}) ... |\alpha_{\sigma(n)}) \tag{5}$$

and vectors $|\alpha_j) \in H_+, \Phi_{(s)}^{\otimes 1} \simeq H, j,k \in \mathbb{N}$, are bi-orthogonal to each other, that is $(\alpha_k|\alpha_j)_H = \delta_{k,j}$ for any $k,j \in \mathbb{N}$. The corresponding scalar product of base vectors as (5) is given as follows:

$$\begin{aligned} &((\beta)_n|(\alpha)_n) := (\beta_n, \beta_{n-1}, ..., \beta_2, \beta_1|\alpha_1, \alpha_2, ..., \alpha_{n-1}, \alpha_n) \\ &= \sum_{\sigma \in S_n} (\beta_1|\alpha_{\sigma(1)})_H ... (\beta_n|\alpha_{\sigma(n)})_H := per\{(\beta_i|\alpha_j)_H\}_{i,j=\overline{1,n}}, \end{aligned} \tag{6}$$

where "*per*" denotes the permanence of the matrix and $(\cdot|\cdot)$ is the corresponding scalar product in the Hilbert space H. Based now on the representation (4), one can define an operator $a^+(\alpha) : \Phi_{(s)}^{\otimes n} \longrightarrow \Phi_{(s)}^{\otimes(n+1)}$ for any $|\alpha) \in H_-$ as follows:

$$a^+(\alpha)|\alpha_1, \alpha_2, ..., \alpha_n) := |\alpha, \alpha_1, \alpha_2, ..., \alpha_n), \tag{7}$$

which is called the "*creation*" operator in the Fock space Φ_F. The adjoint operator $a(\beta) := (a^+(\beta))^* : \Phi_{(s)}^{\otimes(n+1)} \longrightarrow \Phi_{(s)}^{\otimes n}$ with respect to the Fock space Φ_F (1) for any $|\beta) \in H_-$, called the "*annihilation*" operator, acts as follows:

$$a(\beta)|\alpha_1, \alpha_2, ..., \alpha_{n+1}) := \sum_{\sigma \in S_n} (\beta|\alpha_j)|\alpha_1, \alpha_2, ..., \alpha_{j\;1}, \hat{\alpha}_j, \alpha_{j+1}, ..., \alpha_{n+1}), \tag{8}$$

where the hat "$\hat{\ }$" over a vector denotes that it should be omitted from the sequence.

It is easy to check that the commutator relationship

$$[a(\alpha), a^+(\beta)] = (\alpha|\beta)_H \tag{9}$$

holds for any vectors $|\alpha) \in H$ and $|\beta) \in H$. Expression (9), owing to the rigging structure (18), can be naturally extended to the general case, when vectors $|\alpha)$ and $|\beta) \in H_-$, conserving its form. In particular, taking $|\alpha) := |\alpha(y)) = \left\{\frac{1}{\sqrt{2\pi}} e^{i\langle y|x\rangle}\right\}^k \in H_- := L_{2,-}(\mathbb{R}^m; \mathbb{C}^k)$ for any $y \in \mathbb{E}^m$, one easily gets from (9) that

$$[a_i(x), a_j^+(y)] = \delta_{ij}\delta(x-y) \tag{10}$$

for any $i,j = \overline{1,k}$, where we put, by definition, $\langle\cdot|\cdot\rangle$ the usual scalar product in the m-dimensional Euclidean space $\mathbb{E}^m := (\mathbb{R}^m; \langle\cdot|\cdot\rangle)$, $a_j^+(y) := a_j^+(y(x))$ and $a_j(y) := a_j(y(x)), j = \overline{1,k}$, for all $x, y \in \mathbb{R}^m$ and denoted by $\delta(\cdot)$ the classical Dirac delta-function.

The construction above makes it possible to observe easily that there exists the unique vacuum vector $|0) \in \Phi_{(s)}^{\otimes 1}$, such that for any $x \in \mathbb{R}^m$

$$a_j(x)|0) = 0 \tag{11}$$

for all $j \in \overline{1,k}$, and the set of vectors

$$\left(\prod_{j=1}^{k}\prod_{i=1}^{n_j}\left(a_j^+\right)(x_j^{(i)})\right)|0) \in \Phi_{(s)}^{\otimes n} \tag{12}$$

is total in $\Phi_{(s)}^{\otimes n}$, that is, their linear integral hull over the functional spaces $\Phi_{(s)}^{\otimes n}$ is dense in the Hilbert space $\Phi_{(s)}^{\otimes n}$ for every $n = \sum_{j=1}^{k} n_j \in \mathbb{N}$. This means that for any vector $\varphi \in \Phi_F$, the following canonical representation

$$\varphi = \sum_{n=\sum_{j=1}^{k} n_j \in \mathbb{Z}_+}^{\oplus} \int_{(\mathbb{R}^m)^n} \varphi^{(n)}_{n_1 n_2 \ldots n_s}(x_1^{(1)}, x_1^{(2)}, \ldots, x_1^{(n_1)}; x_2^{(1)}, x_2^{(2)}, \ldots, x_2^{(n_2)}; \ldots \tag{13}$$

$$; x_k^{(1)}, x_k^{(2)}, \ldots, x_k^{(n_m)}) \prod_{j=1}^{k} \frac{1}{\sqrt{n_j!}} \prod_{s=1}^{n_j} a_j^+(x_j^{(s)})|0)$$

holds with the Fourier type coefficients $\varphi^{(n)}_{n_1 n_2 \ldots n_s} \in \Phi_{(s)}^{\otimes n}$ for all $n = \sum_{j=1}^{k} n_j \in \mathbb{Z}_+$. The latter is naturally endowed with the Gelfand type quasi-nucleus rigging, dual to

$$H_+ \subset H \subset H_-, \tag{14}$$

making it possible to construct a quasi-nucleous rigging of the dual Fock space $\Phi_F := \oplus_{n \in \mathbb{Z}_+} \Phi_{(s)}^{\otimes n}$. Thereby, the chain (14) generates the dual Fock space quasi-nucleolus rigging

$$\mathcal{D} \subset \Phi_{F,+} \subset \Phi_F \subset \Phi_{F,-} \subset \mathcal{D}' \tag{15}$$

with respect to the Fock space Φ_F, easily following from (1) and (14).

Construct now the following self-adjoint operator $\rho(x) : \Phi_F \to \Phi_F$ as

$$\rho(x) := \langle a^+(x) | a(x) \rangle, \tag{16}$$

called the density operator at a point $x \in \mathbb{R}^m$, satisfying the commutation properties:

$$\begin{aligned} [\rho(x), \rho(y)] &= 0, \\ [\rho(x), a(y)] &= -a(y)\delta(x-y), \\ [\rho(x), a^+(y)] &= a^+(y)\delta(x-y) \end{aligned} \tag{17}$$

for any $x, y \in \mathbb{R}^m$.

Assume now that Φ is a separable Hilbert space, F is a topological real linear space and $\mathcal{A} := \{A(\mathrm{f}) : \mathrm{f} \in F\}$ is a family of commuting self-adjoint operators in Φ (i.e., these operators commute in the sense of their resolutions of the identity) with dense in Φ domain $\mathrm{Dom} A(\mathrm{f}) := D_{A(\mathrm{f})} \subset \Phi, \mathrm{f} \in F$. Consider the corresponding Gelfand rigging [57,61,66] of the Hilbert space Φ, i.e., a chain

$$\mathcal{D} \subset \Phi_+ \subset \Phi \subset \Phi_- \subset \mathcal{D}' \tag{18}$$

in which Φ_+ is a Hilbert space, topologically (densely and continuously) and quasi-nucleus (the inclusion operator $i : \Phi_+ \longrightarrow \Phi$ is of the Hilbert–Schmidt type) embedded into Φ, the space Φ_- is the dual to Φ_+ as the completion of functionals on Φ_+ with respect to the norm $||\mathrm{f}||_- := \sup_{||u||_+=1} |(\mathrm{f}|u)_\Phi|$, $u \in \Phi$, a linear dense in Φ_+ topological space $\mathcal{D} \subseteq \Phi_+$ is such that $\mathcal{D} \subset D_{A(\mathrm{f})} \subset \Phi$ and the mapping $A(\mathrm{f}) : \mathcal{D} \to \Phi_+$ is continuous for any $\mathrm{f} \in F$. Then, the following structural theorem [4,5,16,57,61,62,67–69] about the cyclic representations of the family $\mathcal{A} := \{A(\mathrm{f}) : \mathrm{f} \in F\}$ of commuting self-adjoint operators in the separable Hilbert space Φ holds.

Theorem 1. *Assume that the family of operators $\mathcal{A}$ satisfies the following conditions:*

(a) for $A(\mathrm{f})$, $\mathrm{f} \in F$, the closure of the operator $\overline{A(\mathrm{f})}$ in Φ coincides with $A(\mathrm{f})$ for any $\mathrm{f} \in F$, that is $\overline{A(\mathrm{f})} = A(\mathrm{f})$ on domain $D_{A(\mathrm{f})}$ in Φ;

(b) the Range $A(\mathrm{f}) \subset \Phi$ for any $\mathrm{f} \in F$;

(c) for every $\varphi \in \mathcal{D}$ the mapping $F \ni \mathrm{f} \longrightarrow A(\mathrm{f})|\varphi) \in \Phi_+$ is linear and continuous;

(d) there exists a strong cyclic vector $|\Omega) \in \bigcap_{\mathrm{f} \in F} D_{A(\mathrm{f})}$, such that the set of all vectors $|\Omega)$ and $\prod_{j=1}^{n} A(\mathrm{f}_j)|\Omega)$, $n \in \mathbb{Z}_+$, is total in Φ_+ (i.e., their linear hull is dense in Φ_+).

Then there exists a probability measure μ on $(F', C_\sigma(F'))$, where F' is the dual of F and $C_\sigma(F')$ is the σ-algebra generated by cylinder sets in F' such that, for $\mu-$almost every $\eta \in F'$ there is a generalized common eigenvector $\omega(\eta) \in \Phi_-$ of the family $\mathcal{A}$, corresponding to the common eigenvalue $\eta \in F'$, that is for any $\varphi \in \mathcal{D} \subset \Phi_+$ and $A(\mathrm{f}) \in \mathcal{A}$

$$(\omega(\eta)|A(\mathrm{f})\varphi)_{\Phi_- \times \Phi_+} = \eta(\mathrm{f})(\omega(\eta)|\varphi)_{\Phi_- \times \Phi_+} \tag{19}$$

with $\eta(\mathrm{f}) \in \mathbb{R}$, denoting here the result of the pairing between F and F'.

The mapping

$$\mathcal{D} \ni |\varphi) \longrightarrow (\omega(\eta)|\varphi)_{\Phi_- \times \Phi_+} := \varphi(\eta) \in \mathbb{C} \tag{20}$$

for any $\eta \in F'$ can be continuously extended to a unitary surjective operator $\mathcal{F}_\eta : \Phi_+ \longrightarrow L_2^{(\mu)}(F'; \mathbb{C})$, where

$$\mathcal{F}_\eta\, |\varphi) := \eta(\varphi) \tag{21}$$

for any $\eta \in F'$ is a generalized Fourier transform, corresponding to the family $\mathcal{A}$. Moreover, the image of the operator $A(\mathrm{f})$, $\mathrm{f} \in F'$, under the $\mathcal{F}_\eta$- mapping is the operator of multiplication by the function $F' \ni \eta \to \eta(\mathrm{f}) \in \mathbb{R}$.

Now, if to construct the following self-adjoint family $\mathcal{R} := \{\rho(\mathrm{f}) := \int_{\mathbb{R}^m} \rho(x)\mathrm{f}(x)dx : \mathrm{f} \in F\}$ of linear operators in the Hilbert space Φ_μ, where $F := \mathcal{S}(\mathbb{R}^m; \mathbb{R})$ is the Schwartz functional space dense in H, one can derive, making use of Theorem 1, that there exists the generalized Fourier transform (21), such that

$$\Phi_\mu = L_2^{(\mu)}(F'; \mathbb{C}) \simeq \int_{F'}^{\oplus} \Phi_{(\eta)} d\mu(\eta) \tag{22}$$

for some Hilbert space sets $\Phi_{(\eta)}$, $\eta \in F'$, and a suitable measure μ on F', with respect to which the corresponding joint eigenvector $\omega(\eta) \in \Phi_-$ for any $\eta \in F'$ generates the Fourier transformed family $\{\eta(\mathrm{f}) \in \mathbb{R} : \mathrm{f} \in F\}$. Moreover, if $\dim \Phi_\eta = 1$ for all $\eta \in F'$, the Fourier transformed eigenvector $\omega(\eta) := \Omega(\eta) = 1$ for all $\eta \in F'$.

Now we will consider the family of self-adjoint operators $\rho(\mathrm{f}) : \Phi_\mu \to \Phi_\mu, \mathrm{f} \in F$, as generating a unitary family $\mathcal{U} := \{U(\mathrm{f}) : \mathrm{f} \in F\}$, where the operator

$$U(\mathrm{f}) := \exp[i\rho(\mathrm{f})] \tag{23}$$

is unitary, satisfying the abelian commutation condition

$$U(\mathrm{f}_1)U(\mathrm{f}_2) = U(\mathrm{f}_1 + \mathrm{f}_2) \tag{24}$$

for any $\mathrm{f}_1, \mathrm{f}_2 \in F$. Since, in general, the unitary family $\mathcal{U}$ is defined in the Hilbert space Φ_μ, not coinciding, in general with the canonical Fock type space, the important problem of describing its cyclic unitary representation spaces arises, within which the factorization jointly with relationships (17) hold for any $\mathrm{f} \in F$. This problem can be treated using mathematical tools devised both within the representation theory of $\mathbb{C}^*$-algebras [4,5,57,63] and the Gelfand–Vilenkin [66] approach. Below we will describe the main features of the Gelfand–Vilenkin formalism, being much more suitable for the task, providing a reasonably unified framework of constructing the corresponding representations. The next definitions will be used in our construction.

Definition 1. *Let F be a locally convex topological vector space, $F_0 \subset F$ be a finite dimensional subspace of F. Let $F^0 \subseteq F'$ be defined by*

$$F^0 := \{\sigma \in F' : \ \sigma|_{F_0} = 0\}, \tag{25}$$

and called the annihilator of F_0.

The quotient space $F'^0 := F'/F^0$ may be, evidently, identified with $F_0' \subset F'$, the adjoint space of F_0.

Definition 2. *Let $Q \subseteq F'^0$; then the subset*

$$X_{F^0}^{(Q)} := \left\{ \sigma \in F' : \sigma + F^0 \subset Q \right\} \tag{26}$$

is called the cylinder set with the base Q and the generating subspace F^0.

Definition 3. *Let $n = \dim F_0 = \dim F_0' = \dim F'^0$. One says that a cylinder set $X^{(Q)}$ has Borel base, if Q is a Borel set, when regarded as a subset of $\mathbb{R}^m$.*

The family of cylinder sets with Borel base forms an algebra of sets, which is a key stone for defining measurable sets in and the corresponding measures on F'.

Definition 4. *The measurable sets in F' are the elements of the σ-algebra generated by the cylinder sets with Borel base.*

Definition 5. *A cylindrical measure in F' is a non-negative σ-pre-additive function μ defined on the algebra of cylinder sets with a Borel base and satisfying the conditions $0 \leq \mu(X) \leq 1$ for any X, $\mu(F') = 1$ and $\mu\left(\coprod_{j\in\mathbb{N}} X_j\right) = \sum_{j\in\mathbb{N}} \mu(X_j)$, if all sets $X_j \subset F'$, $j \in \mathbb{N}$, have a common generating subspace $F_0 \subset F$.*

Definition 6. *A cylindrical measure μ satisfies the commutativity condition if, and only if, for any bounded continuous function, $\alpha : \mathbb{R}^n \longrightarrow \mathbb{R}$ of $n \in \mathbb{N}$ real variables the function*

$$\alpha[\mathrm{f}_1, \mathrm{f}_2, ..., \mathrm{f}_n] := \int_{F'} \alpha(\eta(\mathrm{f}_1), \eta(\mathrm{f}_2), ..., \eta(\mathrm{f}_n)) d\mu(\eta) \tag{27}$$

is sequentially continuous in $\mathrm{f}_j \in F$, $j = \overline{1,m}$.

Remark 1. *It is known [4,57,66] that in countably normalized spaces, the properties of sequential and ordinary continuity are equivalent.*

Definition 7. *A cylindrical measure μ is countably additive if, and only if, for any cylinder set $X = \coprod_{j\in\mathbb{N}} X_j$, which is the union of countably many mutually disjoints cylinder sets $X_j \subset F'$, $j \in \mathbb{N}$, $\mu(X) = \sum_{j\in\mathbb{N}} \mu(X_j)$.*

The next two standard propositions [4,57,66,70,71], characterizing extensions of the measure μ on $X = \coprod_{j\in\mathbb{N}} X_j$, hold.

Proposition 1. *A countably additive cylindrical measure μ can be extended to a countably additive measure on the σ-algebra, generated by the cylinder sets with a Borel base. Such a measure will also be called a cylindrical measure.*

Proposition 2. *Let F be a nuclear space. Then, any cylindrical measure μ on F', satisfying the continuity condition, is countably additive.*

2.2. Non-Relativistic Quantum Current Algebra and Its Cyclic Representations

Based on the Fock space Φ_F, defined by (18) and generated by the creation–annihilation operators (7) and (8), the current operator $J(x) : \Phi_F \to \Phi_F^m$, $x \in \mathbb{R}^m$, can be easily constructed as follows:

$$J(x) = \frac{1}{2i}[a^+(x)\, \nabla_x a(x) - \nabla_x a^+(x)\, a(x)], \tag{28}$$

satisfying jointly with the density operator $\rho(x) : \Phi_F \to \Phi_F$, $x \in \mathbb{R}^m$, defined by (16), the following quantum current Lie algebra symmetry [4–7,59,68,72] relationships:

$$[J(\mathrm{g}_1), J(\mathrm{g}_2)] = iJ([\mathrm{g}_1,\mathrm{g}_2]), \ [\rho(\mathrm{f}_1), \rho(\mathrm{f}_2)] = 0, \qquad (29)$$
$$[J(\mathrm{g}_1), \rho(\mathrm{f}_1)] = i\rho(\langle \mathrm{g}_1 | \nabla \mathrm{f}_1 \rangle \ ),$$

holding for all $\mathrm{f}_1, \mathrm{f}_1 \in F$ and $\mathrm{g}_1, \mathrm{g}_2 \in F^m$, where we put, by definition,

$$[\mathrm{g}_1,\mathrm{g}_2] := \langle \mathrm{g}_1 | \nabla \rangle \ \mathrm{g}_2 - \langle \mathrm{g}_2 | \nabla \rangle \ \mathrm{g}_1, \qquad (30)$$

being the usual commutator of vector fields $\langle \mathrm{g}_1 | \nabla \rangle$ and $\langle \mathrm{g}_2 | \nabla \rangle$ on the configuration space $\mathbb{R}^m$. It is easy to observe that the current algebra (29) is the Lie algebra $\mathcal{G}$, corresponding to the Banach group $G := \mathrm{Diff}\ (\mathbb{R}^m) \ltimes F$, the semidirect product of the Banach group of diffeomorphisms $\mathrm{Diff}\ (\mathbb{R}^m)$ of the m-dimensional space $\mathbb{R}^m$ and the Abelian group F. As the Lie algebra $\Gamma(\mathbb{R}^m)$ of smooth vector fields on $\mathbb{R}^m$ with the Lie bracket (17) is isomorphic to the Lie algebra $\mathrm{Diff}(\mathbb{R}^m)$ of the Banach diffeomorphism group $\mathrm{Diff}\ (\mathbb{R}^m)$, it is natural to construct the corresponding unitary operators

$$V(\varphi_t^{\mathrm{g}}) := \exp[iJ(\mathrm{g})], \qquad (31)$$

on the Representation Hilbert space Φ_μ, where for any $\mathrm{g} \in F^m$, there holds $d\varphi_t^{\mathrm{g}}/dt = \mathrm{g}(\varphi_t^{\mathrm{g}}), \varphi_t^{\mathrm{g}}(x)|_{t=0} = x \in \mathbb{R}^m$, where $\varphi_t^{\mathrm{g}} \in \mathrm{Diff}(\mathbb{R}^m), t \in \mathbb{R}$. The constructed above exponential currents (23) and (31) constitute together a unitary operator group on the Hilbert space Φ, endowed with the following composition law

$$U(\mathrm{f}_1)U(\mathrm{f}_2) = U(\mathrm{f}_1 + \mathrm{f}_2), V(\varphi_1)V(\varphi_2) = V(\varphi_2 \circ \varphi_1), \qquad (32)$$
$$V(\varphi)U(\mathrm{f}) = U(\mathrm{f} \circ \varphi)V(\varphi)$$

for all $\mathrm{f}_1, \mathrm{f}_2, \mathrm{f} \in F$ and $\varphi, \varphi_2, \varphi \in \mathrm{Diff}(\mathbb{R}^m)$. The operator group (32) is, evidently, isomorphic to the semidirect product group G, which is endowed, respectively, with the natural composition law

$$(\varphi_1, \mathrm{f}_1) \circ (\varphi_2, \mathrm{f}_2) = (\varphi_2 \circ \varphi_1, \mathrm{f}_1 + \mathrm{f}_2 \circ \varphi_1) \qquad (33)$$

for all $\mathrm{f}_1, \mathrm{f}_2 \in F$ and $\varphi_1, \varphi_2 \in \mathrm{Diff}(\mathbb{R}^m)$. Concerning a more adequate mathematical description of the Banach diffeomorphism group $\mathrm{Diff}(\mathbb{R}^m)$, it is useful to consider the subgroup $\mathrm{Diff}_0(\mathbb{R}^m)$ of smooth diffeomorphisms of $\mathbb{R}^m$ with compact supports, which is a topological space with the topology given by a counted family of the metrics $||\varphi_1 - \varphi_2||_n := \max_{|k|=\overline{0,n}} \sup_{x \in \mathbb{R}^m} (1 + |x|^2)^n |\varphi_1^{(k)}(x) - \varphi_2^{(k)}|$ for all $n \in \mathbb{Z}_+$ and $\varphi_1, \varphi_2 \in \mathrm{Diff}_0(\mathbb{R}^m)$. So, the diffeomorphism group $\mathrm{Diff}(\mathbb{R}^m)$ can be defined as the completion of the space $\mathrm{Diff}_0(\mathbb{R}^m)$ with respect to the topology introduced above. This way, the constructed group $\mathrm{Diff}(\mathbb{R}^m)$ is topological, locally linear connected and metrizable with a countable topology basis at each of its points. In particular, the group $\mathrm{Diff}(\mathbb{R}^m)$ contains diffeomorphisms with noncompact supports, yet in the limit $|x| \to \infty, x \in \mathbb{R}^m$, they can be approximated by the identity mapping in $\mathrm{Diff}(\mathbb{R}^m)$. The latter makes it possible to state that for any $\mathrm{g} \in F^m$ the element $\varphi_t^{\mathrm{g}} \in \mathrm{Diff}(\mathbb{R}^m)$ for all $t \in \mathbb{R}$ generates the uniform continuous mapping $F^m \ni \mathrm{g} \to \varphi_t^{\mathrm{g}} \in \mathrm{Diff}(\mathbb{R}^m)$.

Proceeding now to the Banach group of currents $G = \mathrm{Diff}\ (\mathbb{R}^m) \ltimes F$, we have that the separable Hilbert space Φ_μ for every irreducible cyclic representation will be unitary equivalent to the Hilbert space (45), which in many physical applications reduces in the case $\dim \Phi_{(\eta)} = 1$ for all $\eta \in F'$ to the following form:

$$\Phi_\mu \simeq L_2^{(\mu)}(F'; \mathbb{C}), \qquad (34)$$

being the space of square integrable functions with respect to the measure μ on F'.

Assume now that an element $\omega \in \Phi_\mu$ is taken arbitrarily and consider the action of the Banach group of currents G on it:

$$U(\mathrm{f})\omega(\eta) = \exp[i(\eta(\mathrm{f})]\omega(\eta),$$
$$V(\varphi)\omega(\eta) = \chi_\varphi(\eta)\omega(\varphi^*\eta)\left[\frac{d\mu(\varphi^*(\eta))}{d\mu(\eta)}\right]^{1/2}, \tag{35}$$

where, by definition, $\varphi^*\eta(\mathrm{f}) := \eta(\mathrm{f}\circ\varphi)$ for all $\mathrm{f} \in F$, $\frac{d\mu(\varphi^*\eta)}{d\mu(\eta)}$ is the corresponding Radon–Nikodym derivative [19,73] of the measure $\mu \circ \varphi^*$ with respect to the measure μ on F' and $\chi_\varphi(\eta)$ is a complex-valued character of the unit norm, satisfying the relationship

$$\chi_{\varphi_2}(\eta)\chi_{\varphi_1}(\varphi_2^*\eta) = \chi_{\varphi_1\circ\varphi_2}(\eta) \tag{36}$$

for all $\varphi_j \in \mathrm{Diff}(\mathbb{R}^m), j = \overline{1,2}, \eta \in F'$. For the Radon–Nikodym derivative above to exist, the measure μ on F' should be quasi-invariant with respect to the diffeomorphism group $\mathrm{Diff}(\mathbb{T}^n)$, that is, for any measurable set $Q \subset F'$ the condition $\mu(Q) = 0$ if, and only if, $\mu(\varphi^*Q)$ for arbitrary $\varphi \in \mathrm{Diff}(\mathbb{R}^m)$.

In physics applications, the representation (35) is uniquely determined by the measure μ on F', which in the general case has a very complicated [20,72] structure, and its analytic construction is nontrivial. One of the fairly effective approaches to this problem is the quantum method of Bogolubov generating functionals developed in [2,6,7,20,74]. Another approach, which is of considerable interest for the theory of dynamical systems, is based on algebraic methods of constructing self-adjoint functional-operator representations of the original current Lie algebra (29). In particular, the representation (35), corresponding to a quantum-mechanical system of $N \in \mathbb{N}$ identical bose-particles localized at points $x_j \in \mathbb{R}^m$, has a measure μ with supports [20,72] on Dirac delta-functions $\eta := \eta_N \in F'$ of the form:

$$\eta_N(x) = \sum_{j\in\overline{1,N}} \delta(x - x_j) \tag{37}$$

at any $x \in \mathbb{R}^m$ with a measure μ of the form:

$$d\mu(\eta_N) = \Omega_N^*\Omega_N \prod_{j=\overline{1,N}} dx_j\delta(\eta - \eta_N(x)), \tag{38}$$

where $\Omega_N \in \Phi_N \simeq L_2^{(s)}((\mathbb{R}^m)^{\otimes N};\mathbb{C})$ is the corresponding symmetric ground-state wave function of the related quantum Hamiltonian system, satisfying the conditions (49) and (49), reduced on the invariant subspace Φ_N. Moreover, the following general expressions hold: $\Omega(\eta) = 1$ and for any $\omega \in L_2^{(\mu)}(F';\mathbb{C})$

$$\rho(x)\omega(\eta_N) = \sum_{j\in\overline{1,N}} \delta(x - x_j)\omega(\eta_N), \tag{39}$$

$$J(x)\omega(\eta_N) = \frac{1}{2i}\sum_{j\in\overline{1,N}} [\delta(x - x_j)\circ\partial/\partial x_j + \partial/\partial x_j\circ\delta(x - x_j)]\omega(\eta_N),$$

where, by definition, $\omega(\eta_N) \in \Phi_N \simeq L_2^{(s)}((\mathbb{R}^m)^{\otimes N};\mathbb{C})$. As a simple consequence of the actions (39), one derives that

$$U(f)\omega(\eta_N) = \exp[i\sum_{j\in\overline{1,N}} f(x_j)]\omega(\eta_N), \tag{40}$$

$$V(\varphi)\omega(\eta_N) = \omega(\varphi^*\eta_N)\left[\left|\det\left(\frac{\partial\varphi(x)}{\partial x}\right)\right|\right]^{1/2},$$

where we put, for brevity, that the character $\chi_\varphi(\eta_N) = 1$ for all $\varphi \in \mathrm{Diff}(\mathbb{R}^m)$.

2.3. The Generating Functional Equation, Cyclic Current Algebra Representation and Hamiltonian Operator Groundstate

Concerning the Fourier transform of a cylindrical measure μ in F', we will use the following natural definitions.

Definition 8. *Let μ be a cylindrical measure in F'. The Fourier transform of μ is the nonlinear functional*

$$\mathcal{L}(\mathrm{f}) := \int_{F'} \exp[i\eta(\mathrm{f})]d\mu(\eta), \tag{41}$$

coinciding with the characteristic functional of the measure μ.

Definition 9. *The nonlinear functional $\mathcal{L} : F \longrightarrow \mathbb{C}$ on F, defined by (41), is called positive definite, if, and only if, for all $\mathrm{f}_j \in F$ and $\lambda_j \in \mathbb{C}$, $j = \overline{1,n}$, the condition*

$$\sum_{j,k=1}^{n} \bar{\lambda}_j \mathcal{L}(\mathrm{f}_k - \mathrm{f}_j)\lambda_k \geq 0 \tag{42}$$

holds for any $n \in \mathbb{N}$.

The following important proposition, owing to Gelfand and Vilenkin [4,66], Araki [75] and Goldin [1,4], holds.

Proposition 3. *The functional $\mathcal{L} : F \longrightarrow \mathbb{C}$ on F, defined by (41), is the Fourier transform of a cylindrical measure on F' if, and only if, it is positive definite, sequentially continuous and satisfying the condition $\mathcal{L}(0) = 1$. Suppose now that we have a continuous unitary representation of the unitary family $\mathcal{U}$ in a suitable Hilbert space Φ_μ with a cyclic vector $|\Omega) \in \Phi_\mu$. Then we can put*

$$\mathcal{L}(\mathrm{f}) := (\Omega|U(\mathrm{f})|\Omega) \tag{43}$$

for any $\mathrm{f} \in F := \mathcal{S}(\mathbb{R}^n;\mathbb{R})$, being the Schwartz space on $\mathbb{R}^m$, and observe that functional (43) is continuous on F owing to the continuity of the representation. Therefore, this functional is the generalized Fourier transform of a cylindrical measure μ on F':

$$(\Omega|U(\mathrm{f})|\Omega) = \int_{\mathcal{S}'} \exp[i\eta(\mathrm{f})]d\mu(\eta). \tag{44}$$

From the spectral point of view, based on Theorem 1, there is an isomorphism between the Hilbert spaces Φ_μ and $L_2^{(\mu)}(F;\mathbb{C})$, defined by $|\Omega) \longrightarrow \Omega(\eta) = 1$ and $U(\mathrm{f})|\Omega) \longrightarrow \exp[i\eta(\mathrm{f})]$ and next extended by linearity upon the whole Hilbert space Φ. In the non-cyclic case, there exists a finite or countably infinite family of measures $\{\mu_k : k \in \mathbb{Z}_+\}$ on F', with $\Phi_\mu \simeq \oplus_{k\in\mathbb{Z}_+} L_2^{(\mu_k)}(F';\mathbb{C})$ and the unitary operator $U(\mathrm{f}) : \Phi_\mu \longrightarrow \Phi_\mu$ for any $\mathrm{f} \in F$ corresponds in all $L_2^{(\mu_k)}(F';\mathbb{C})$, $k \in \mathbb{Z}_+$, to a multiplication operator on the exponent function $\exp[i\eta(\mathrm{f})]$. This means that there exists a single cylindrical measure μ on F' and a $\mu-$ measurable field of Hilbert spaces $\Phi_{(\eta)}$ on F', such that

$$\Phi_\mu \simeq \int_{F'}^{\oplus} \Phi_{(\eta)} d\mu(\eta), \tag{45}$$

with $\mathrm{U(f)} : \Phi_\mu \longrightarrow \Phi_\mu$, *corresponding* [66] *to the operator of multiplication by* $\exp[i\eta(\mathrm{f})]$ *for any* $\mathrm{f} \in F$ *and* $\eta \in F'$. *Thereby, having constructed the nonlinear functional* (41) *in an exact analytical form, one can retrieve the representation of the unitary family* $\mathcal{U}$ *on the corresponding Hilbert space* Φ_μ, *as follows:* $\Phi_\mu = \oplus_{n \in \mathbb{Z}_+} \Phi_n$, *where*

$$\Phi_n = \prod_{j=\overline{1,n}} \rho(x_j)|\Omega), \tag{46}$$

for all $n \in \mathbb{N}$.

The cyclic vector $|\Omega) \in \Phi_\mu$ can be, in particular, obtained as the ground state vector of some unbounded self-adjoint positive definite Hamiltonian operator $\mathrm{H} : \Phi_\mu \longrightarrow \Phi_\mu$, commuting with the self-adjoint non-negative particle number operator

$$\mathrm{N} := \int_{\mathbb{R}^m} dx \rho(x), \tag{47}$$

that is $[\mathrm{H}, \mathrm{N}] = 0$. Moreover, the conditions

$$\mathrm{H}|\Omega) = 0 \tag{48}$$

and

$$\inf_{\varphi \in D_{\mathrm{H}}} (\varphi|\mathrm{H}|\varphi) = (\Omega|\mathrm{H}|\Omega) = 0 \tag{49}$$

hold for the operator $\mathrm{H} : \Phi_\mu \rightarrow \Phi_\mu$, where D_{H} denotes its domain of definition, dense in Φ_μ. To find the functional (43), which is called the generating Bogolubov type functional for moment distribution functions

$$f_n(x_1, x_2, ..., x_n) := (\Omega| : \rho(x_1)\rho(x_2)...\rho(x_n) : |\Omega), \tag{50}$$

where $x_j \in \mathbb{R}^m$, $j = \overline{1,n}$, and the normal ordering operation: $\cdot$: is defined [4,6,7,55,56,68] as

$$: \rho(x_1)\rho(x_2)...\rho(x_n) : = \prod_{j=1}^{n} \left(\rho(x_j) - \sum_{k=1}^{j-1} \delta(x_j - x_k) \right), \tag{51}$$

it is convenient first to choose the Hamilton operator $\mathrm{H} : \Phi_F \rightarrow \Phi_F$ in the following secondly quantized [4,5,56] representation

$$\mathrm{H} := \frac{1}{2} \int_{\mathbb{R}^m} \langle \nabla_x a^+(x) | \nabla_x a(x) \rangle dx + \mathrm{V}(\rho), \tag{52}$$

on the related Fock space Φ_F, where the sign "∇_x" means the usual gradient operation with respect to $x \in \mathbb{R}^m$ in the Euclidean space $\mathbb{E}^m \simeq (\mathbb{R}^m; \langle \cdot | \cdot \rangle)$. If the energy spectrum density of the Hamiltonian operator (52) on the cyclic representation Hilbert space Φ_μ is bounded from below, in works done by Goldin G.A., Grodnik J., Menikov R. Powers R.T. and Sharp D. [4,5,76] it was stated that this Hamiltonian, modulo the ground state energy eigenvalue, can be algebraically represented on a suitably constructed *current algebra symmetry representation Hilbert space* Φ_μ, as the positive definite gauge type factorized operator

$$\mathrm{H} = \frac{1}{2} \int_{\mathbb{R}^m} \Big\langle (\mathrm{K}^+(x) - \mathrm{A}(x;\rho)) | \rho^{-1}(x) (\mathrm{K}(x) - \mathrm{A}(x;\rho)) \Big\rangle \, dx, \tag{53}$$

satisfying conditions (48) and (49), where $\mathrm{A}(x;\rho) : \Phi_\mu \rightarrow \Phi_\mu^m$, $x \in \mathbb{R}^n$, is some specially constructed [68,77] linear self-adjoint operator, satisfying the condition

$$\mathrm{K}(x)|\Omega) = \mathrm{A}(x;\rho)|\Omega) \tag{54}$$

with the ground state $|\Omega) \in \Phi_\mu$, corresponding to chosen potential operators $\mathrm{V}(\rho) : \Phi_\mu \to \Phi_\mu$. The singular structure of the operator (53) was previously analyzed in detail in [2] where, in part, its well-posedness was showed.

The *"potential"* operator $\mathrm{V}(\rho) : \Phi_\mu \to \Phi_\mu$ is, in general, a polynomial (or analytical) functional of the density operator $\rho(x) : \Phi_\mu \longrightarrow \Phi_\mu$ for any $x \in \mathbb{R}^m$, and the operator $\mathrm{K}(x) : \Phi_\mu \to \Phi_\mu^m$ is defined as

$$\mathrm{K}(x) := \nabla_x \rho(x)/2 + iJ(x), \tag{55}$$

where the self-adjoint "current" operator $J(x) : \Phi_\mu \to \Phi_\mu^m$ can be naturally defined (but non-uniquely) from the continuity equality

$$\partial\rho/\partial t = i[\mathrm{H}, \rho(x)] = -\langle \nabla | J(x) \rangle, \tag{56}$$

holding for all $x \in \mathbb{R}^m$. Such an operator $J(x) : \Phi_\mu \to \Phi_\mu^m$, $x \in \mathbb{R}^m$, can exist owing to the commutation condition $[\mathrm{H}, \mathrm{N}] = 0$, giving rise to the continuity relationship (56), if, additionally, to take into account that supports $\operatorname{supp} \rho$ of the density operator $\rho(x) : \Phi_\mu \to \Phi_\mu$, $x \in \mathbb{R}^m$, can be chosen arbitrarily, owing to the independence of (56) on the potential operator $\mathrm{V}(\rho) : \Phi_\mu \to \Phi_\mu$, but its strict dependence on the corresponding representation (45).

Remark 2. *The self-adjointness of the operator $\mathrm{A}(g;\rho) : \Phi_\mu \to \Phi_\mu$, $g \in F$, can be stated following schemes from works [5,68,72] under the additional existence of such a linear anti-unitary mapping $\mathrm{T} : \Phi_\mu \to \Phi_\mu$ that the following invariance conditions hold:*

$$\mathrm{T}\rho(x)\mathrm{T}^{-1} = \rho(x), \qquad \mathrm{T}\, J(x)\, \mathrm{T}^{-1} = -J(x), \qquad \mathrm{T}|\Omega) = |\Omega) \tag{57}$$

for any $x \in \mathbb{R}^m$. Thereby, owing to conditions (57), the following equalities

$$\mathrm{K}(x)|\Omega) = \mathrm{A}(x;\rho)|\Omega) \tag{58}$$

hold for any $x \in \mathbb{R}^m$, giving rise to the self-adjointness of the operator $\mathrm{A}(g;\rho) : \Phi_\mu \longrightarrow \Phi_\mu$, $g \in F^m$.

It is easy to observe that the time-reversal condition (57) imposes the real value relationship for the real valued ground state $\Omega_N = \overline{\Omega}_N \in \Phi_N \simeq L_2^{(s)}(\mathbb{R}^{m\times N};\mathbb{C})$ of the canonically represented N-particle Hamiltonian $\mathrm{H}_N : \Phi_N \to \Phi_N$ for arbitrary $N \in \mathbb{N}$. Moreover, taking into account the relationship (58), one can easily observe that on the invariant subspace $\Phi_N \subset \Phi_F$, the operator $\mathrm{K}(x) : \Phi_N \longrightarrow \Phi_N$ is representable as

$$K_N(x) = \sum_{j=\overline{1,N}} \delta(x - x_j)\frac{\partial}{\partial x_j}, \tag{59}$$

entailing the following expression for the related operator $A_N(x;\rho) : \Phi_N \to \Phi_N$ on the subspace $\Phi_N \subset \Phi$:

$$A_N(x;\rho) = \sum_{j=\overline{1,N}} \delta(x - x_j)\nabla_{x_j} \ln|\Omega_N(x_1, x_2, ..., x_N)|. \tag{60}$$

The latter makes it possible to derive its secondly quantized [56,57,78,79] expression as

$$\mathrm{A}(x;\rho) = \int_{\mathbb{R}^{m\times N}} dx_2 dx_3 ... dx_N : \rho(x)\rho(x_2)\rho(x_3)...\rho(x_N) : \nabla_x \ln|\Omega_N(x, x_2, ..., x_N)|, \tag{61}$$

which holds for any $x \in \mathbb{R}^m$ and arbitrary $N \in \mathbb{Z}_+$. Being interested in the infinite particle case when $N \to \infty$, the expression (61) can be naturally decomposed [77,79] as

$$\begin{gathered}\mathrm{A}(x;\rho) := \rho(x)\nabla \tfrac{\delta}{\delta\rho(x)}\mathrm{W}(\rho) = \\ = \textstyle\sum_{n\in\mathbb{Z}_+} \frac{1}{n!} \int_{\mathbb{R}^{m\times n}} dy_1 dy_2 ... dy_n : \rho(x)\rho(y_1)\rho(y)\rho(y_3)...\rho(y_n) : \nabla_x W_{n+1}(x; y_1, y_2, ..., y_n),\end{gathered} \tag{62}$$

where the corresponding real-valued coefficients $W_n \in H_2^{(1)}(\mathbb{R}^{m\times n};\mathbb{R})$ should be such functions that the series (62) were convergent in a suitably chosen representation Fock space Φ_F, for which the resulting ground state $\lim_{N\to\infty}\Omega_N \simeq |\Omega) \in \Phi_F$ is necessarily cyclic and normalized.

Based now on the construction above, one easily deduces from expression (55) that the generating Bogolubov type functional (43) obeys for all $x \in \mathbb{R}^m$ the following functional-differential equation:

$$[\nabla_x - i\nabla_x \mathrm{f}]\frac{1}{2i}\frac{\delta\mathcal{L}(\mathrm{f})}{\delta \mathrm{f}(x)} = \mathrm{A}\left(x;\frac{1}{i}\frac{\delta}{\delta \mathrm{f}}\right)\mathcal{L}(\mathrm{f}), \tag{63}$$

whose solutions should satisfy [3,74] the Fourier transform representation (44), and which were, in part, studied in [74]. In particular, a wide class of special so-called Poissonian white noise type solutions to the functional-differential Equation (63) was obtained in [5,61,62,68,71,80] by means of functional-operator methods in the following generalized form:

$$\mathcal{L}(\mathrm{f}) = \exp\left\{2\int_{\mathbb{R}^m} \mathrm{W}\left(\frac{1}{i}\frac{\delta}{\delta \mathrm{f}}\right)dx\right\}\exp\left(\bar{\rho}\int_{\mathbb{R}^m}\{\exp[i\mathrm{f}(x)]-1\}dx\right), \tag{64}$$

where $\bar{\rho} = (\Omega|\rho|\Omega) \in \mathbb{R}_+$ is a suitable Poisson process parameter and the operator $\mathrm{A}(x;\rho) : \Phi_\mu \to \Phi_\mu^m, x \in \mathbb{R}^m$, resulting from the expression (62) for some scalar operator $\mathrm{W}(\rho) : \Phi_\mu \to \Phi_\mu$.

Remark 3. *It is worth remarking here that solutions to Equation (63) realize the suitable physically motivated representations of the abelian Banach subgroup F of the Banach group $G = \mathrm{Diff}(\mathbb{R}^m) \ltimes F$, mentioned above. In the general case of this Banach group G one can also construct [5,6,16,81] a generalized Bogolubov type functional equation, whose solutions give rise to suitable physically motivated representations of the corresponding current Lie algebra $\mathcal{G}$.*

Recalling now the Hamiltonian operator representation (53), one can readily deduce that the following weak representation Hilbert space Φ_μ weak relationship

$$\left(\left\langle \mathrm{A}|\rho^{-1}\mathrm{A}\right\rangle - \left\langle \mathrm{K}^*|\rho^{-1}\mathrm{A}\right\rangle - \left\langle \mathrm{A}|\rho^{-1}\mathrm{K}\right\rangle\right)/2 - \mathrm{V}(\rho) = \epsilon_0, \tag{65}$$

where $\epsilon_0 \in \mathbb{R}$ is the corresponding ground state energy density value. Thus, the main analytical problem is now reduced to constructing the expansion (62) corresponding to a suitable cyclic representation Hilbert space Φ_μ of the quantum current algebra (29), compatible with the Hamiltonian operator structure (52).

Remark 4. *Here we mention that the operator $\mathrm{K}(x) : \Phi_\mu \to \Phi_\mu^m$, $x \in \mathbb{R}^m$, defined by (55), relates to that from the work [4,5,76] via scaling $\mathrm{K}(x) \to \mathrm{K}(x)/2$, $x \in \mathbb{R}^m$.*

2.4. The Hamiltonian Operator Reconstruction and the Cyclic Current Algebra Representation

We will assume that we are given a Banach current group $G = \mathrm{Diff}(\mathbb{R}^m) \ltimes F$ cyclic representation in a Hilbert space Φ_μ with respect to F with a cyclic vector $|\Omega) \in \Phi_+ \subset \Phi_\mu$. Based on the well known Araki reconstruction theorem [5,75] for the canonical Weyl commutation relations, we can first readily obtain from (56) that

$$[\mathrm{H},U(\mathrm{f})] = J(\nabla \mathrm{f})\mathrm{U}(\mathrm{f}) - 1/2\rho(\langle\nabla \mathrm{f}_1|\nabla \mathrm{f}_2\rangle)U(\mathrm{f}), \tag{66}$$

where $U(\mathrm{f}) = \exp[i\rho(\mathrm{f})], \mathrm{f} \in F$, is an element of the unitary family $\mathcal{U}$. The expression (66) makes it possible to calculate the bilinear form

$$\begin{aligned}(U(\mathrm{f}_1)\Omega|\mathrm{H}|U(\mathrm{f}_2)\Omega) = (U(\mathrm{f}_1)\Omega|J(\nabla \mathrm{f}_1)|U(\mathrm{f}_2)\Omega)-\\ -1/2(U(\mathrm{f}_1)\Omega|\rho(\langle\nabla \mathrm{f}_1|\nabla \mathrm{f}_2\rangle)|U(\mathrm{f}_2)\Omega)\end{aligned} \tag{67}$$

for any $f_1, f_2 \in F$. Taking into account the symmetry properties (57), we finally deduce from (67) that for arbitrary functions $f_1, f_2 \in F$

$$(U(f_1)\Omega|H|U(f_2)|\Omega) = 1/2(U(f_1)\Omega|\rho(\langle\nabla f_1|\nabla f_2\rangle)|U(f_2)\Omega). \tag{68}$$

The standard reasonings make it possible to state that the bilinear symmetric form (68) determines on Φ_μ a self-adjoint non-negative definite Hamiltonian operator $H : \Phi_\mu \to \Phi_\mu$, densely defined on the domain $D_H := \underset{f\in F}{span}\{\exp[i\rho(f)]|\Omega) \in \Phi_\mu\}$. Really, for any set of functions $f_j \in F, j = \overline{1,n}$, the following inequalities

$$\sum_{j,k=\overline{1,n}} \bar{s}_j s_k \langle\nabla f_j|\nabla f_k\rangle \geq 0, \quad \sum_{j,k=\overline{1,n}} \bar{s}_j s_k (U(f_j)\Omega|\rho(x)|U(f_k)\Omega) \geq 0 \tag{69}$$

hold for any complex numbers $s_j \in \mathbb{C}, j = \overline{1,n}$, and arbitrary $n \in \mathbb{N}$. Since, for any non-negative definite complex matrices $A, B \in \text{End}\,\mathbb{R}^n$, the matrix $C := \{A_{jk}B_{jk} : j,k = \overline{1,n}\} \in \text{End}\,\mathbb{C}^n$ proves to be non-negative definite [75,82] too, one ensures that the bilinear form (69) is also non-negative definite. Then, as follows from the classical Friedrichs' theorem [69,83–85], there exists a self-adjoint densely defined and non-negative definite operator $H : \Phi_\mu \to \Phi_\mu$.

2.5. Current Algebra Representations, Generating Functional Method and Ergodicity of the Hilbert Space Representation Measure

In view of the importance of the current algebra representations of the Banach group $G = \text{Diff}\,(\mathbb{R}^m) \ltimes F$ for physics applications, we consider their construction by means of the generating functional method [6,7,20,86]. From the very beginning, let us introduce a governing definition in connection with this method.

Definition 10. *A generating functional on a group G is a complex-valued function E on G with the following conditions: (1) $E(1) = 1, 1 \in G$; (2) $E(a_1 \exp(tA) a_2)$ is a continuous function of the parameter $t \in \mathbb{R}$ for all $A \in \mathcal{G}$ and $a_1, a_2 \in G$; (3) the matrix $\left|\left|E(a_k^{-1}a_j)\right|\right|, k, j = \overline{1,N}$, is positive definite for any $N \in \mathbb{N}$;*

The following theorem [75] holds.

Theorem 2. *The function E is a generating functional on G if, and only if, there exists a continuous unitary representation $\pi : G \to \text{Aut}(\Phi_\mu)$ on a separable Hilbert space $(\Phi_\mu; (\cdot|\cdot))$ with a cyclic vector $\Omega \in \Phi_\mu$, such that*

$$E(a) = (\Omega|\pi(a)\Omega) \tag{70}$$

holds for all $a \in G$.

The vector $\Omega \in \Phi_\mu$ is said to be cyclic with respect to the representation $\pi : G \to Aut(\Phi_\mu)$, if the set $\{\pi(a)\Omega : a \in G\}$ is complete in Φ_μ, i.e., is dense in Φ_μ, if taken together with its linear combinations over $\mathbb{C}$. The significance of this theorem is that one can implicitly construct unitary representations of the Banach current group $G = \text{Diff}(\mathbb{R}^m) \ltimes F$ and, thus, the current Lie algebra $\mathcal{G}$ by means of an appropriately defined generating functional on G. This is important, since frequently the latter problem is much simpler than the initial problem.

We now consider the representation $\pi : G \to \text{Aut}(\Phi_\mu)$, restricted to the Abelian subgroup F in the group $G = \text{Diff}\,(\mathbb{R}^m) \ltimes F$ and its corresponding generating functional $\mathcal{L}(f), f \in F$, in the form

$$\mathcal{L}(f) := (\Omega|\exp[i\rho(f)]\Omega) = \int_{F'} d\mu(\eta) \exp[i\eta(f)], \tag{71}$$

where the cyclic vector $\Omega \in \Phi_\mu$ is normalized to unity: $(\Omega|\Omega) = 1$. In many physically interesting cases [6,7,20,86] the expression (71) can be replaced by means of the following equivalent trace-representation:

$$\mathcal{L}(\mathrm{f}) = \mathrm{Tr}(P\exp[i\rho(\mathrm{f})]), \tag{72}$$

where $P : \Phi_\mu \to \Phi_\mu$ is the corresponding so called statistical operator, depending on the Hamiltonian operator $\mathrm{H} : \Phi_\mu \to \Phi_\mu$. The constructed above generating functional (71) should possess the following necessary properties: (1) $\mathcal{L}(\mathrm{f}) = \overline{\mathcal{L}(-\mathrm{f})}$ for all $\mathrm{f} \in F$; (2) $\mathcal{L}(0) = 1$; (3) $|\mathcal{L}(\mathrm{f})| \leq 1$ for all $\mathrm{f} \in F$; (4) $\mathcal{L}(\mathrm{f})$ is a positive definite functional on F : the inequality $\sum_{j,k=\overline{1,N}} \bar{c}_k \mathcal{L}(\mathrm{f}_k - \mathrm{f}_j) c_j \geq 0$ holds for all $c_j \in \mathbb{C}, j = \overline{1,N}$, and arbitrary $N \in \mathbb{N}$. As one can show, a generating functional $\mathcal{L} : F \to \mathbb{C}$, satisfying the properties (1)–(4) always defines [66] a measure μ on F', defining the searched for unitary representation $\pi : G \to \mathrm{Aut}(\Phi_\mu)$ of the Abelian subgroup F of the current Banach group $G = \mathrm{Diff}(\mathbb{R}^m) \ltimes F$. If the measure μ is in addition quasi-invariant and the factors $\chi_\varphi(\eta)$ in (35) are known for all $\varphi \in \mathrm{Diff}(\mathbb{R}^m), \eta \in F'$, the corresponding representation of the current Lie algebra $\mathcal{G}$ is completely determined. Yet, if being interested only by irreducible representations of the current Banach algebra $\mathcal{G}$, it is well known [66] that the corresponding measure μ on F' is ergodic for the diffeomorphism subgroup $\mathrm{Diff}(\mathbb{R}^m)$, that is for any measurable and invariant subset $Q \subset F'$ either $\mu(Q) = 0$, or $\mu(F' \backslash Q) = 0$. Moreover, an arbitrary invariant set is in the general case a nondenumerable union of a family of mutually non-intersecting orbits. Assuming that the orbits containing an invariant subset $Q \subset F'$ are measurable, we obtain that there exist only two possibilities for ergodicity of the cylindrical measure μ on F' : either it is concentrated on one orbit, or each orbit has zero measure, and these two possibilities really occur in applications. For instance, the case when the measure is concentrated on functionals of the form (37) leads to irreducibility of the generating functional representation on the Hilbert space $L_2(\mathbb{R}^{mN}; \mathbb{C})$ for any finite $N \in \mathbb{N}$.

2.6. The Creation–Annihilation Heisenberg Algebra, Its Coherent State Representations and Linearization of Nonlinear Dynamical Systems on Hilbert Spaces

It is well known [87,88] that the representation theory of the quantum current algebra in a separable Hilbert space Φ_μ is very close to the cyclic Hilbert space representations of the canonical creation–annihilation operator Heisenberg algebra $\mathcal{H}$ family $\{a^+(\mathrm{f}), a(\mathrm{f}) : \Phi_F \to \Phi_F : \mathrm{f} \in F\}$, defined on the Fock space Φ_F. The coherent states, being venerable objects in physics, were invented by Schrëdinger [89], as far back as in 1926, in the context of the quantum harmonic oscillator, they seemed to have lapsed into oblivion for some obscure reasons. About thirty-five years later, they were rediscovered, almost simultaneously, by Glauber [90], Klauder [91,92] and Sudarshan [93], in the context of a quantum optical description of coherent light beams emitted by lasers. Since then, coherent states have pervaded nearly all branches of quantum physics—including, of course, quantum optics in the study of lasers, nuclear, atomic and solid state physics, quantum electrodynamics, quantization and dequantization problems and path integrals, to mention just a few. For original references, the reader is referred to the review [88] and reprint volume of Klauder and Skagerstam [94]. In many of these applications, the question naturally poses itself as to whether it might not be possible to find other families of states, sharing some properties of the original or canonical coherent states, emanating from the quantum oscillator and which could possibly be useful to yet other areas of physics.

Already, in 1926, Schrëdinger had tried unsuccessfully to construct coherent states appropriate to the hydrogen atom problem. This was motivated by the quasi-classical character of the canonical coherent states which made them very desirable for studying the quantization of classical dynamical systems, a point which we discuss in some detail below. The key to the generalization of the notion of a coherent state was the observation by Perelomov [95] and independently by Gilmore [96,97], that the construction of the oscillator coherent states could be reformulated as a problem in group representation theory: the canonical coherent states could be obtained by acting on the oscillator ground state with the operators of a unitary representation of the group generated by the creation and annihilation operators, namely the Weyl–Heisenberg group.

The link between the Schrëdinger and the Perelomov approaches is the uniqueness theorem [98,99] of von Neumann for the quantum mechanics of a system with finitely many degrees of freedom. In addition, a unitary representation $T : G \to U(G)$ of a compact symmetry group G in a separable Hilbert space Φ, used for building up the system of canonical coherent states, has the property of square integrability with respect to the left (or right) invariant Haar measure on G. Furthermore, the physical states, associated with the coherent states, are not indexed by elements of G itself, but by points in the coset space G/G^c, where G^c is the Cartan subgroup of G and is isomorphic to the torus.

Since its introduction in 1972, the concept of coherent states was widely exploited [14–16,87,100,101] in many fields of mathematical physics, whose leading idea consisting of considering the translates of a fixed cyclic vector under a group action is as old as the celebrated Gel'fand-Raikov theorem [66] on locally bicompact groups. Their common properties, namely that the related homogeneous space has a complex homogeneous structure, and the corresponding representation Hilbert space can be identified in the coherent state basis with a space of holomorphic functions on the homogeneous space. As it was stated in [87], a homogeneous complex structure is actually present quite generally, and on the basis of the homogeneous complex structure, the related homogeneous manifolds are just the classical phase spaces on which the group acts through canonical transformations. From this point of view, coherent states can be interpreted just as probability wave packets over the classical phase space, that is a well-known result for the harmonic oscillator coherent states. The converse problem, i.e., the construction of irreducible unitary representations of the group, starting from its phase space realization, was considered in [102] and found a definite mathematical setting.

To look at the coherent vector representation problem within the Fock type space, its main idea becomes very transparent and motivative owing to the classical Bargmann–Segal [103] construction. Namely, there is considered the Hilbert space

$$\mathcal{H}_k := \{f \in H(\mathbb{C}^k) : \int_{\mathbb{C}^n} |f(z)|^2 d\mu(z) \tag{73}$$

of holomorphic functions $H(\mathbb{C}^k), k \in \mathbb{N}$, with the scalar product $(f|g) := \int_{\mathbb{C}^k} \overline{f(z)} g(z) d\mu(z)$ for arbitrary $f, g \in \mathcal{H}_k$ with respect to the measure $d\mu(z) = \pi^{-k} \exp(-\langle z|z\rangle) \frac{d\bar{z} \wedge dz}{(2i)^k}$ for $z \in \mathbb{C}^k$. It is easy to observe that the Hilbert space $\mathcal{H}_k$ is the direct sum of the symmetric polynomial subspaces $\mathcal{H}_k^{(s)}, s \in \mathbb{Z}_+$:

$$\mathcal{H}_k = \oplus_{s=0}^{\infty} \mathcal{H}_k^{(s)}, \tag{74}$$

where, by definition,

$$\mathcal{H}_k^{(s)} := \{ \sum_{s=n_1+n_2+\dots n_k} c_{n_1 n_2 \dots n_k} z_1^{n_1} z_2^{n_2} \dots z_k^{n_k} : c_{n_1 n_2 \dots n_k}, z_j \in \mathbb{C}, j = \overline{1,k}\}. \tag{75}$$

Moreover, it is easy to check that the polynomials

$$e_{n_1 n_2 \dots n_k}(z) = \prod_{j=\overline{1,k}} \frac{z_j^{n_j}}{\sqrt{n_j!}} \tag{76}$$

form a complete and orthogonal base in $\mathcal{H}_k$, that is $(e_{n_1 n_2 \dots n_k} | e_{m_1 m_2 \dots m_k}) = \prod_{j=\overline{1,k}} \delta_{n_j m_j}$. The next important observation, made by V. Bargmann, was the point boundedness of any function $f \in \mathcal{H}_k : |f(z)| \leq ||f|| \exp(\langle z|z\rangle/2), z \in \mathbb{C}^n$. Really, for any $f \in \mathcal{H}_k$ there holds the expansion

$$f(z) = \sum_{s\in\mathbb{Z}_+} \sum_{s=n_1+n_2+\dots n_k} (e_{n_1 n_2 \dots n_k}|f) \prod_{j=\overline{1,k}} \frac{z_j^{n_j}}{\sqrt{n_j!}}, \tag{77}$$

from which one easily ensues, owing to the closedness property and Schwartz inequality on $l_2(\mathbb{C})$, that

$$\begin{array}{c} |f(z)| \leq \left(\sum_{s\in\mathbb{Z}_+} \sum_{s=n_1+n_2+\dots n_k} |(e_{n_1 n_2 \dots n_k}|f)|^2\right)^{1/2} \times \\ \times \left(\sum_{s\in\mathbb{Z}_+} \sum_{s=n_1+n_2+\dots n_k} \prod_{j=\overline{1,k}} \frac{|z_j|^{2n_j}}{n_j!}\right)^{1/2} == ||f|| \exp(\langle z|z\rangle/2) \end{array}. \tag{78}$$

The latter makes it possible to define for any $u \in \mathbb{C}^n$ the following dual to (78) bounded functional

$$\hat{u}(f) := f(u), \quad ||\hat{u}|| \leq \exp(\langle u|u\rangle/2) \tag{79}$$

on $\mathcal{H}_k$, whose Riesz representation

$$\hat{u}(f) = (h_u|f) \tag{80}$$

defines the unique element $h_u \in \mathcal{H}_k$, or equivalently

$$f(u) = \int_{\mathbb{C}^k} \overline{h_u(\xi)} f(\xi) \exp(-\langle \xi|\xi\rangle) \frac{d\bar{\xi} \wedge d\xi}{(2i)^k}. \tag{81}$$

Taking into account the orthogonality of the base vectors (76) in $\mathcal{H}_k$, it is easy to calculate that the vector $h_u(\xi) = \exp(\langle u|z\rangle) \in \mathcal{H}_k$, whose norm $||h_u|| = \exp(\langle u|u\rangle/2)$ for any $u \in \mathbb{C}^n$ and which is called the "*coherent vector*". It is worth remarking here that the function representation (81) is well known in the operator theory [104,105] and is called the "reproducing kernel" representation with the kernel $h_u \in \mathcal{H}_k, u \in \mathbb{C}^k$.

The Hilbert space $\mathcal{H}_k$, as the direct sum (74) of symmetrical polynomial subspaces, possesses the Fock space structure, allowing the introduction of the creation operators $a_j^+ : \mathcal{H}_k^{(s)} \to \mathcal{H}_k^{(s+1)}$ for any $j = \overline{1,k}$ and all $s \in \mathbb{Z}_+$ as multiplication operators: for any $f \in \mathcal{H}_k^{(s)}$ $a_j^+ f(z) := z_j f(z)$ for any $j = \overline{1,k}$ and all $s \in \mathbb{Z}_+$. The corresponding adjoint expressions $\left(a_j^+\right)^* := a_j : \mathcal{H}_k^{(s)} \to \mathcal{H}_k^{(s-1)}$ act as $a_j f(z) = \partial/\partial z_j f(z)$ on arbitrary $f \in \mathcal{H}_k^{(s)}$ for any $j = \overline{1,k}$ and all $s \in \mathbb{Z}_+$, where, by definition, $\mathcal{H}_k^{(0)} \simeq \mathbb{C}$. Now one can easily check that the coherent vector $h_u = \exp(\langle u|\cdot\rangle) \in \mathcal{H}_k$ for any $u \in \mathbb{C}^k$ is a common eigenvector of the annihilation operators $a_j : \mathcal{H}_k \to \mathcal{H}_k, j = \overline{1,k}$:

$$a_j h_u(z) = u_j h_u(z) \tag{82}$$

with the eigenvalues $u_j \in \mathbb{C}, j = \overline{1,k}$. It is important also to mention here that the creation–annihilation operators defined above satisfy the canonical commutation relationships:

$$[a_j, a_n] = 0 = [a_j^+, a_n^+], \; [a_j, a_n^+] = \delta_{j,n} \tag{83}$$

for all $j, n = \overline{1,k}$.

The coherent vector representation scheme described above can be respectively generalized to arbitrary symmetric Fock space Φ that will be effectively used in the sections proceeding below. Returning back to the algebraic properties of coherent states, we proceed to describing their unbelievable and impressive applications to theory of nonlinear dynamical systems on Hilbert spaces, their linearization and integrability, previously initiated in [14,15] and continued in [16]. We briefly reviewed the cyclic Hilbert space representations of the quantum Heisenberg algebra and presented a general approach to constructing the coherent states and their applications both to the linearization of nonlinear dynamical systems on Hilbert spaces, and to describing their complete integrability. The latter is developed using the modern Lie-algebraic approach [11,17–19] to nonlinear dynamical systems on Poissonian functional manifolds, and proved to be both unexpected and important for the classification of integrable Hamiltonian flows on Hilbert spaces.

Jointly with the cyclic Hilbert space representations of the Heisenberg algebra $\mathcal{H}$, we briefly reviewed the closely related cyclic Hilbert space density representations [4,6,87,88] of the canonical quantum current algebra $\mathcal{G}$ on the circle $\mathbb{S}^1$, whose vector field representations on smooth spatially one-dimensional functional manifolds coincide exactly with the related symmetry algebra of completely integrable nonlinear Hamiltonian systems on these manifolds. Based on the current algebra symmetry structure and their functional representations, an effective integrability criterion is formulated for a wide class of completely integrable Hamiltonian systems on smooth spatially one-dimensional functional manifolds. The algebraic structure of the Poissonian operators and an effective algorithm of their analytical construction are also described.

2.7. The Canonical Heisenberg Algebra and Its Cyclic Hilbert Space Representations

Let $(\Phi;(\cdot|\cdot))$ be a separable Hilbert space, F be a topological real linear space and $\mathcal{A} := \{A(\mathrm{f}) : \mathrm{f} \in F\}$ a family of commuting self-adjoint operators in Φ (i.e., these operators commute in the sense of their resolutions of the identity) with dense in Φ domain *Dom* $A(\mathrm{f}) := D_{A(\mathrm{f})}, \mathrm{f} \in F$. Consider the Gelfand rigging [57,61,66] of the Hilbert space Φ, i.e., a chain

$$\mathcal{D} \subset \Phi_+ \subset \Phi \subset \Phi_- \subset \mathcal{D}' \tag{84}$$

in which Φ_+ is a Hilbert space, topologically (densely and continuously) and quasi-nucleus (the inclusion operator $i : \Phi_+ \longrightarrow \Phi$ is of the Hilbert-Schmidt type) embedded into Φ, the space Φ_- is the dual to Φ_+ as the completion of functionals on Φ_+ with respect to the norm $||\mathrm{f}||_- := \sup_{||u||_+=1} |(\mathrm{f}|u)_\Phi|,\ u \in \Phi$, a linear dense in Φ_+ topological space $\mathcal{D} \subseteq \Phi_+$ is such that $\mathcal{D} \subset D_{A(\mathrm{f})} \subset \Phi$ and the mapping $A(\mathrm{f}) : \mathcal{D} \to \Phi_+$ is continuous for any $\mathrm{f} \in F$. Then, owing to the structural theorem (1) there exists a cyclic representation of the canonical creation–annihilation Heisenberg operator algebra $\mathfrak{H}$ family $\{a^+(\mathrm{f}), a(\mathrm{f}) : \Phi_\mu \to \Phi_\mu : \mathrm{f} \in F\}$ on the separable Hilbert space Φ_μ, whose generalized Fourier transform is given by the expression

$$\Phi_\mu = L_2^{(\mu)}(F';\mathbb{C}) \simeq \int_{F'}^{\oplus} \Phi_{(\eta)} d\mu(\eta) \tag{85}$$

for some Hilbert space sets Φ_η, $\eta \in F'$, and a suitable measure μ on F', with respect to which the corresponding joint eigenvector $\omega(\eta) \in \Phi_-$ for any $\eta \in F'$ generates the Fourier transformed family $\{\eta(\mathrm{f}) \in \mathbb{R} : \ \mathrm{f} \in F\}$. Moreover, if $\dim \Phi_\eta = 1$ for all $\eta \in F'$, the Fourier transformed eigenvector $\omega(\eta) := \Omega(\eta) = 1$ for all $\eta \in F'$.

Next, we will consider the family of self-adjoint operators $\{P(\mathrm{f}), Q(\mathrm{g}) : \Phi_\eta \to \Phi_\eta : \mathrm{f}, \mathrm{g} \in F\}$, as generating a unitary Heisenberg group

$$\begin{aligned} \mathfrak{H} := \{&\exp(iP(\mathrm{f})), V(\mathrm{g}) = \exp(iQ(\mathrm{g}) : \\ &P := (a^+ + a)/2, Q := i(a - a^+)/2, \mathrm{f}, \mathrm{g} \in F, \} \end{aligned} \tag{86}$$

satisfying the commutation conditions

$$\begin{aligned} &U(\mathrm{f})V(\mathrm{g}) = \exp(-i(f|g))V(\mathrm{g})U(\mathrm{f}), \\ &U(\mathrm{f}\,)U(\mathrm{g}) = U(\mathrm{f}+\mathrm{g}\,), \quad V(\mathrm{f}\,)V(\mathrm{g}) = V(\mathrm{f}+\mathrm{g}\,), \end{aligned} \tag{87}$$

for any $\mathrm{f}, \mathrm{g} \in F$. Since, in general, the unitary Heisenberg group $\mathfrak{H}$ is defined on a representation Hilbert space Φ_μ, not coinciding, in general, with the canonical Fock type space

Φ_F, the important problem of describing its cyclic unitary representation spaces arises, within which the factorization (86) jointly with relationships (87) should hold. Below, we will briefly describe only the main features of the Gelfand–Vilenkin formalism, being much more suitable for the task, providing a reasonably unified framework of constructing the corresponding cyclic representations of the family $\mathcal{A} := \{Q(\mathrm{f}) : \mathrm{f} \in F\}$ of commuting self-adjoint operators in a separable Hilbert space Φ.

Proceeding now to the Heisenberg group $\mathfrak{H}$, the separable Hilbert space Φ_μ for its every irreducible representation will be unitary equivalent to the Hilbert space (45), which in many physical applications reduces in the case $\dim \Phi_{(\eta)} = 1$ for all $\eta \in F'$ to the following form:

$$\Phi_\mu \simeq L_2^{(\mu)}(F'; \mathbb{C}), \tag{88}$$

being the space of square integrable functions with respect to the measure μ on F'.

Assume now that an element $\omega \in \Phi_\mu$ is taken arbitrarily and consider [75] the action of the Heisenberg group $\mathfrak{H}$ on it:

$$\begin{aligned} U(\mathrm{f})\omega(\eta) &= \exp[i(\eta(\mathrm{f})]\omega(\eta), \\ V(\mathrm{g})\omega(\eta) &= \chi_{\mathrm{g}}(\eta)\omega(\eta+\mathrm{g})\left[\frac{d\mu(\eta+\mathrm{g})}{d\mu(\eta)}\right]^{1/2}, \end{aligned} \tag{89}$$

where, by definition, for any $\mathrm{f} \in F$ the expression $\frac{d\mu(\eta+\mathrm{g})}{d\mu(\eta)}$ at $\eta \in F'$ means the corresponding Radon–Nikodym derivative [19,73] of the measure $\mu(\circ + \mathrm{g})$ with respect to the measure μ on F' and $\chi_{\mathrm{g}}(\eta)$ is a complex-valued character of the unit norm, satisfying the relationship

$$\chi_{\mathrm{f}}(\eta)\chi_{\mathrm{g}}(\eta+\mathrm{f}) = \chi_{\mathrm{f}+\mathrm{g}}(\eta) \tag{90}$$

for all $\mathrm{f}, \mathrm{g} \in F \subset H$ and arbitrary $\eta \in F'$. For the Radon–Nikodym derivative above to exist, the measure μ on F' should be quasi-invariant with respect to the shift group elements $\{F' \ni \eta \to \eta + \mathrm{g} \in F'\}$, that is, for any measurable set $Q \subset F'$ the condition $\mu(Q) = 0$ if, and only if, $\mu(Q + \mathrm{g})$ for arbitrary $\mathrm{g} \in F \subset F'$.

Definition 11. *A vector $|u) \in \Phi_\mu$ is called a coherent vector state in the representation Hilbert space Φ_μ with respect to an element $u \in H \simeq L_2(\mathbb{R}^m; \mathbb{R}^k)$, if it satisfies the eigenfunction condition*

$$a_j(x)|u) = u_j(x)|u) \tag{91}$$

for each $j = \overline{1,k}$ and all $x \in \mathbb{R}^m$.

It is easy to check that for any $u \in H$ the coherent ket-vector $|u) \in \Phi_\mu$ exists: really, the following vector expression

$$|u) := \exp[(u|a^+)_H]|\Omega) \tag{92}$$

where $\Omega \in \Phi_+ \subset \Phi_\mu$ is a cyclic vector for the creation–annihilation operator algebra family $\{a^+(\mathrm{f}), a(\mathrm{f}) : \Phi_\mu \to \Phi_\mu : \mathrm{f} \in F\}$ and satisfies the defining condition (91), where the operator $a^+(u) : \Phi_\mu \to \Phi_\mu, u \in H$, action ensues from the determining condition (19): for any $\varphi \in \Phi_\mu$ there exists a unique vector $\omega(\eta_a) \in \Phi_\mu$ for which

$$(\omega(\eta_a)|(a^+(u)\varphi)_\mu = \eta_a(u)\ (\omega(\eta_a)|\varphi)_\mu \tag{93}$$

for all $u \in H$. Moreover, as the Hilbert space $H \subset F'$, the eigenvalue $\eta_a(u) \in \mathbb{R}$ is bounded jointly with the Hilbert space Φ_μ norm

$$\|u\| := (u|u)^{1/2} = \exp(\frac{1}{2}\|u\|_H^2) < \infty, \tag{94}$$

since $u \in H$ and its Hilbert space norm $\|u\|_H$ is *a priori* bounded.

Consider now any function $u \in H$ and observe that the Hilbert spaces embedding mapping

$$\xi : H \ni u \longrightarrow |u) \in \Phi_\mu, \tag{95}$$

defined by means of the coherent vector expression (92), realizes a smooth isomorphism between the Hilbert spaces H and the image $\xi(H) \subset \Phi_\mu$. The inverse mapping $\xi^{-1} : \xi(H) \subset \Phi_\mu \longrightarrow H$ is given by the following exact expression:

$$(u|\eta)_H = (\Omega|a(\eta)|u)/(\Omega|u), \tag{96}$$

holding for any $\eta \in H$. Owing to condition (94), one finds from (96) and the classical Riesz type theorem [85,106] that the corresponding function $u \in H$.

Let now define on the Hilbert space H a nonlinear in general dynamical system (which can, in general, be non-autonomous) in partial derivatives

$$du/dt = K[u], \tag{97}$$

where $t \in \mathbb{R}_+$ is the corresponding evolution parameter, $[u] := (x; u, u_x, u_{xx}, ...,) \in J^{(k)}(\mathbb{R}^m; \mathbb{R}^s)$ belongs to the jet-space $J^{(k)}(\mathbb{R}^m, \mathbb{R}^n)$ of the order $k \in \mathbb{Z}_+$, and, in general, a nonlinear mapping $K : H \longrightarrow H$ is Frechet smooth. Assume also that the corresponding Cauchy problem

$$u|_{t=+0} = u_0 \tag{98}$$

for the nonlinear dynamical system (97) is solvable in the Hilbert space H for any $u_0 \in H$ on an interval $[0, T) \subset \mathbb{R}^1_+$ for some $T > 0$. Thus, there is determined a smooth evolution mapping

$$T_t : H \ni u_0 \longrightarrow u(t|u_0) \in H, \tag{99}$$

for all $t \in [0, T)$. Now, it is natural to consider the following commuting diagram:

$$\begin{array}{ccc} H & \xrightarrow{\xi} & \Phi_\mu \\ T_t \downarrow & & \downarrow \mathrm{T}_t \\ H & \xrightarrow{\xi} & \Phi_\mu, \end{array} \tag{100}$$

where the mapping $\mathrm{T}_t : \Phi_\mu \longrightarrow \Phi_\mu$, $t \in [0, T)$, is defined from the conjugation relationship on the image $\xi(H) \subset \Phi_\mu$ of the mapping (95):

$$\xi \circ T_t = \mathrm{T}_t. \circ \xi \tag{101}$$

Now take coherent vector $|u_0) \in \Phi_\mu$, corresponding to the Cauchy data $u_0 \in H$, and construct the vector

$$|u) := \mathrm{T}_t \cdot |u_0) \in \Phi_\mu \tag{102}$$

for all $t \in [0, T)$. Since the vector (102) is, by construction, coherent, that is

$$a_j(x)|u) := u_j(x, t|u_0)|u) \tag{103}$$

for each $j = \overline{1, k}$, $t \in [0, T)$ and almost all $x \in \mathbb{R}^m$, owing to the smoothness of the mapping $\xi : H \longrightarrow \Phi_\mu$ with respect to the corresponding norms in the Hilbert spaces H and Φ_μ, we derive that the coherent vector (102) is differentiable with respect to the evolution parameter $t \in [0, T)$. Thus, one can easily find [14,15] that

$$\frac{d}{dt}|u) = \mathrm{K}(a^+, a)|u), \tag{104}$$

where

$$|u)|_{t=+0} = |u_0) \tag{105}$$

and an operator mapping $\mathrm{K}(a^+, a) : \Phi_\mu \longrightarrow \Phi_\mu$ is defined by means of the exact analytical expression

$$\mathrm{K}(a^+, a) := (a^+ | K[a])_H. \tag{106}$$

As a result of the consideration above we obtain the following theorem.

Theorem 3. *Any smooth nonlinear dynamical system (97) in Hilbert space H is representable by means of the Hilbert spaces embedding isomorphism $\xi : H \longrightarrow \Phi_\mu$ via the completely linear form (104).*

We now make some comments concerning the solution to the linear Equation (104) under the Cauchy condition (105) in the case of the Fock representation space Φ_F. Since any vector $|\omega) \in \Phi_F$ allows the series representation

$$\begin{aligned} |\omega) = \bigoplus_{n=\sum_{j=1}^k n_j \in \mathbb{Z}_+} & \int_{(\mathbb{R}^m)^n} \omega^{(n)}_{n_1 n_2 \ldots n_s}(x_1^{(1)}, x_1^{(2)}, \ldots, x_1^{(n_1)}; \\ & x_2^{(1)}, x_2^{(2)}, \ldots, x_2^{(n_2)}; \ldots; x_k^{(1)}, x_k^{(2)}, \ldots, x_k^{(n_m)}) \prod_{j=1}^k \left(\frac{1}{\sqrt{n_j!}} \prod_{k=1}^{n_j} dx_k^{(j)} a_j^+(x_j^{(k)}) \right) |\Omega), \end{aligned} \tag{107}$$

where for any $n = \sum_{j=1}^k n_j \in \mathbb{N}$ functions

$$\omega^{(n)}_{n_1 n_2 \ldots n_k} \in \bigotimes_{j=1}^k L_2^{(s)}((\mathbb{R}^m)^{n_j}; \mathbb{C}) \simeq L_2^{(s)}(\mathbb{R}^{mn_1} \times \mathbb{R}^{mn_2} \times \ldots \mathbb{R}^{mn_k}; \mathbb{C}), \tag{108}$$

its Fock space norm is easily calculated as

$$||\omega||^2 = \sum_{n=\sum_{j=1}^k n_j \in \mathbb{N}} ||\omega^{(n)}_{n_1 n_2 \ldots n_k}||_2^2. \tag{109}$$

For the case of the coherent vector $|u) \in \Phi_F$ its norm is easily obtained as $||u|| = \exp(||u||_H^2/2)$, coinciding with the result (94). Moreover, substituting (107) into Equation (104), reduces (104) to an infinite recurrent set of linear evolution equations in partial derivatives on coefficient functions (108). The latter can often be solved [14] step by step analytically in exact form, thereby, making it possible to obtain, owing to representation (96), the exact solution $u \in H$ to the Cauchy problem (98) for our nonlinear dynamical system in partial derivatives (97).

Concerning possible applications of nonlinear dynamical systems like (95) in mathematical physics, it is very important to construct their so called conservation laws or smooth invariant functionals $\gamma : H \longrightarrow \mathbb{R}$ on the Hilbert space H. Making use of the quantum mathematics technique described above, one can suggest an effective algorithm for constructing these conservation laws in exact form.

Indeed, consider a vector $|\gamma) \in \Phi_\mu$, satisfying the linear equation:

$$\frac{\partial}{\partial t}|\gamma) + \mathrm{K}^*(a^+, a)|\gamma) = 0. \tag{110}$$

Then, the following proposition [14,15] holds.

Proposition 4. *The functional*

$$\gamma := (u|\gamma) \tag{111}$$

is a conservation law for dynamical system (95), that is

$$d\gamma/dt|_K = 0 \tag{112}$$

along all orbits of the evolution mapping (99).

It is interesting to reanalyze the dynamical system (104) from the Lie-algebraic point of view [11,19] and represent it as a coadjoint canonical Hamiltonian flow on the corresponding adjoint space to the Hilbert space Φ_μ, considered as a Lie algebra over the field $\mathbb{C}$. To do this, it is necessary to define the related Lie commutator on the Hilbert space Φ_μ : for any vectors $|K_\alpha) := \mathrm{K}_\alpha(a^+,a)^*|\omega) \in \Phi_\mu$ and $|K_\beta) := \mathrm{K}_\beta(a^+,a)^*|\omega) \in \Phi_\mu$, where $\mathrm{K}_\alpha(a^+,a)^*$ and $\mathrm{K}_\beta(a^+,a)^* \in \mathrm{End}\,\Phi_\mu$ are smooth mappings and a central vector $|\omega) \in \Phi_\mu$ is chosen to be fixed, their commutator, defined as

$$[|K_\beta),|K_\alpha)] := [\mathrm{K}_\beta(a^+,a)^*,\mathrm{K}_\alpha(a^+,a)^*|\omega), \tag{113}$$

allows the construction of the related co-adjoint action $\Phi_\mu^* \ni (l| \to ad^*_{|K_\alpha)}(u| \in \Phi_\mu^*$ of a vector $|K_\alpha) \in \Phi_\mu$ on a fixed element $(l| \in \Phi_\mu^*$, where for any vector $|\eta) \in \Phi_\mu$ there holds the following identity:

$$(ad^*_{|K_\alpha)}(l|\eta) = -(l|[K_\alpha),\eta]). \tag{114}$$

The latter makes it possible to define on the adjoint space Φ_μ^* the classical Lie–Poisson bracket for any smooth functionals $\alpha := (l|\alpha)$ and $\beta := (l|\beta) \in \mathcal{D}(\Phi_\mu^*)$:

$$\{\alpha,\beta\} := (l|[\operatorname{grad}\alpha(l),\operatorname{grad}\beta(l)]_\vartheta|\omega) = (l|\operatorname{grad}\alpha(l)\vartheta(a^+)\operatorname{grad}\beta(l)|\omega), \tag{115}$$

where, by definition, $\vartheta(a^+) : \Phi_\mu \to \Phi_\mu$ is some skew-symmetric Poisson operator, the element $(l| := (l(u)| \in \Phi_\mu^*$ for any $u \in H$ is interpreted as the corresponding momentum mapping $H \ni u \xrightarrow{l} (l(u)| := (u| \in \Phi_\mu^*$ for the Poissonian action of the Lie algebra Φ_μ on the Hilbert space H:

$$\Phi_\mu \times H \ni (\operatorname{grad}\gamma(u) \times u) \to \quad \mathrm{K}_\gamma|u) \in \Phi_\mu \tag{116}$$

with $\mathrm{K}_\gamma := (a^+|\,K_\gamma[a])_H \in \mathrm{End}\,\Phi_\mu$, $K_\gamma[a]^* = -\vartheta(a^+)\operatorname{grad}\gamma(l) : \Phi_\mu \to \Phi_\mu$ for arbitrary $\gamma \in \mathcal{D}(\Phi_\mu^*)$. The related action

$$(\operatorname{grad}\gamma(u) \times u) \;= K_\gamma[u] \tag{117}$$

is a Hamiltonian vector field on H, generated by the corresponding Hamiltonian vector field $K_\gamma : H \to H$ on the Hilbert space H commonly with the invariant Hamiltonian function $\gamma = (u|\gamma) \in \mathcal{D}(H)$. Simultaneously, the flow (117) for any $\gamma \in \mathcal{D}(\Phi_\mu^*)$ naturally generates the linear flow on the adjoint space $\Phi_\mu^* \simeq \Phi_\mu$ in the form $^+$

$$ad^*_{\vartheta(a^+)|\gamma)}(u| = (u|\mathrm{K}_\gamma^*. \tag{118}$$

Moreover, one easily checks that the commutator of vector fields $\mathrm{K}_\alpha|u)$ and $\mathrm{K}_\beta|u) \in \Phi_\mu$ equals

$$[\mathrm{K}_\alpha|u),\mathrm{K}_\beta|u)] := [\mathrm{K}_\alpha,\mathrm{K}_\beta]|u) \tag{119}$$

for any smooth conservation laws $\alpha = (u|\alpha)$ and $\beta = (u|\beta) \in \mathcal{D}(H)$ of the dynamical system (104), easily following from the evident conditions $\mathrm{K}^*|\alpha) = 0$ and $\mathrm{K}^*\,|\mu) = 0$. Consider now the following representations of the gradient vectors $\operatorname{grad}\alpha(u) = |\alpha(a^+)|\omega)$ and $\operatorname{grad}\beta(u) = |\beta(a^+)|\omega)$ for some fixed central element $|\omega) \in \Phi_\mu$. Then, the Poisson bracket (115) for any $\alpha,\beta \in \mathcal{D}(H)$ is representable as

$$\begin{aligned}\{\alpha,\beta\} &= -(u|[\mathrm{K}_\alpha^*|\omega),\mathrm{K}_\beta^*|\omega)]_\vartheta) = (u|[\mathrm{K}_\alpha,\mathrm{K}_\beta]^*|\omega) = \\ &= (u|\mathrm{K}^*_{\{\alpha,\beta\}}|\omega) = -(u|\vartheta\operatorname{grad}\{\alpha,\beta\}),\end{aligned} \tag{120}$$

being completely compatible with the Poissonian action of the Lie algebra Φ_μ on the Hilbert space H. The obtained result can be summarized as the following theorem.

Theorem 4. *If the momentum mapping* $H \ni u \xrightarrow{l} (l(u)| := (u| \in \Phi^*_\mu$, *related with the nonlinear dynamical system ((97) on the Hilbert space* H, *is Poissonian, then all its symmetries (117), generated by smooth invariants* $\gamma \in \mathcal{D}(\Phi^*_\mu)$, *are represented as linear Hamiltonian flows (118) on the adjoint Hilbert space* $\Phi^*_\mu \simeq \Phi_\mu$ *with respect to the canonical Lie–Poisson bracket (120).*

The theorem above plays a decisive role in constructing within the suitably modified Adler–Kostant–Souriau [11,19] scheme integrable Hamiltonian flows on the adjoint space Φ^*_μ, equivalent to nonlinear integrable Hamiltonian systems on the functional Hilbert space H.

2.8. Conclusions

Within the scope of this Section we have described the main mathematical preliminaries and properties of the quantum mathematics techniques suitable for analytical studying of the important linearization problem for a wide class of nonlinear dynamical systems in partial derivatives in Hilbert spaces. This problem was analyzed in much detail using the Gelfand–Vilenkin representation theory [66] of infinite dimensional groups and the Goldin–Menikoff–Sharp theory [4,5,74] of generating Bogolubov type functionals, classifying these representations. The related problem of constructing cyclic Hilbert space representations and retrieving their creation–annihilation generating structure still needs a deeper investigation within the approach devised. Here we mention only that some aspects of this problem within the so-called Poissonian White noise analysis which was analyzed in a series of works [55,70,107,108], based on some generalizations of the Delsarte type characters technique. The above-stated theorem about the Hamiltonian structure of symmetries of a nonlinear dynamical system on a Hilbert space and their linearization on a suitably constructed Hilbert space presents, from a practical point of view, a strong interest, if the related results, obtained in [14,15,109,110] and devoted to the application of the Hilbert spaces embedding method to finding conservation laws and the so called recursion operators for the well [17,111] known Korteweg–de Vries type nonlinear dynamical systems, are taken into account. Moreover, a development of these results within the modern Lie-algebraic approach, based on the Adler–Kostant–Symes construction and applied to nonlinear dynamical systems on Poissonian functional manifolds, proves to be both unexpected and important for the classification of integrable Hamiltonian flows on Hilbert spaces, and inspires a hope for new investigations of coherent states and their applications.

3. Quantum Current Lie Algebra as a Universal Algebraic Structure of Symmetries of Completely Integrable Nonlinear Dynamical Systems

3.1. Quantum Lie Algebra of Currents and Its Vector Field Representations

We consider the non-relativistic quantum Lie algebra $\mathcal{G}$ of currents [4,12,112,113] on the torus $\mathbb{T}^n$, realized by means of the density $\rho(\mathrm{f})$ and current $J(\mathrm{g})$ operators on the separable Hilbert subspace Φ_μ:

$$\begin{aligned} &[\rho(\mathrm{f}_1), \rho(\mathrm{f}_2)] = 0, [\rho(\mathrm{f}), J(g)] = J(\langle \mathrm{g}|\nabla \mathrm{f}\rangle), \\ &[J(\mathrm{g}_1), J(\mathrm{g}_2)] = iJ([\mathrm{g}_2, \mathrm{g}_1]), \end{aligned} \tag{121}$$

where $\rho(\mathrm{f}) = \int_{\mathbb{T}^n} \mathrm{f}(x)\rho(x)dx$, $J(\mathrm{g}) = \int_{\mathbb{T}^n} g(x)J(x)dx$ for $\mathrm{f}, \mathrm{f}_j \in F \simeq C^\infty(\mathbb{T}^n; \mathbb{R}), g, g_j \in F^n, j = \overline{1,2}$. Their representation on the Fock space Φ_F is given, respectively, by the following operator expressions: $\rho(x) = a^+(x)a(x)$ and $J(x) = \frac{1}{2i}[a^+(x)\nabla a(x) - \nabla a^+(x)a(x)]$, where $a^+(x)$ is the creation and $a(x)$ is the annihilation operators of bose-particle states at point $x \in \mathbb{T}^n$, satisfying the canonical commutation relationships:

$$\begin{aligned} &[a(x), a(y)] = 0, [a^+(x), a^+(y)], \\ &[a(x), a^+(y)] = \delta(x - y) \end{aligned}$$

for all $x, y \in \mathbb{T}^n$. The current Lie algebra (121) is the infinite-dimensional Lie algebra of the semi-direct product $G := \text{Diff}(\mathbb{T}^n) \ltimes F$ of the Banach Lie group of currents $G := \text{Diff}(\mathbb{T}^n)$ and the abelian functional group F, where $\text{Diff}(\mathbb{T}^n)$ is the topological group of diffeomorphisms [2,20] of the torus $\mathbb{T}^n$. If, to introduce [2,20,22,112,114,115] a family of unitary operators $U(\mathrm{f})$ and $V(\varphi_t^{\mathrm{g}}) : \Phi_\mu \to \Phi_\mu$, acting on a Hilbert space Φ_μ and defined by the formulas

$$U(\mathrm{f}) = \exp[i\rho(\mathrm{f})], V(\varphi_t^{\mathrm{g}}) = \exp[itJ(\mathrm{g})], \tag{122}$$

where $d\varphi_t^{\mathrm{g}}/dt := \mathrm{g}(\varphi_t^{\mathrm{g}}), t \in \mathbb{R}$, $\varphi_t^{\mathrm{g}} \in Diff(\mathbb{T}^n)$ and $\varphi_t^{\mathrm{g}}|_{t=0} = x \in \mathbb{T}^n$, then the following relations

$$\begin{aligned} U(\mathrm{f}_1)U(\mathrm{f}_2) &= U(\mathrm{f}_1 + \mathrm{f}_2), V(\varphi)U(\mathrm{f}) = U(\mathrm{f} \circ \varphi)V(\varphi), \\ V(\varphi_1)V(\varphi_2) &= V(\varphi_2 \circ \varphi_1) \end{aligned} \tag{123}$$

hold for all $\mathrm{f}, \mathrm{f}_j \in F$ and $\varphi, \varphi_j \in Diff(\mathbb{T}^n), j = \overline{1,2}$. As was argued in [2], the various unitary representations of the current group G describe different physical systems and their states, and the study of the set of cyclic unitarily irreducible representations of the Banach Lie group relationships (123) is an extremely important and topical problem in the quantum theory of dynamical systems.

For every irreducible cyclic representation of the unitary current group G on the separable Hilbert space Φ_μ there exists a unitarily equivalent Hilbert space

$$\Phi_\mu \simeq \int_{F'}^{\oplus} d\mu(\eta)\Phi_{(\eta)}, \tag{124}$$

where μ is the measure on the space F' of continuous real linear functionals on F and $\Phi_{(\eta)}$ are complex linear finite-dimensional spaces labeled by the index $\eta \in F'$. In the case when $\dim \Phi_{(\eta)} = 1$, $\Phi_\mu \simeq L_2^{(\mu)}(F'; \mathbb{C})$, the space of complex-valued functions on F', integrable with respect to the measure μ on F'. Moreover, if an element $\omega \in \Phi_\mu$, then for the action of the current group G on this element we have the following representations:

$$\begin{aligned} U(\mathrm{f})\omega(\eta) &= exp[i(\eta(\mathrm{f})]\omega(\eta), \\ V(\varphi)\omega(\eta) &= \chi_\varphi(\eta)\omega(\varphi^*\eta)\left(\frac{d\mu(\varphi^*\eta)}{d\mu(\eta)}\right)^{1/2} \end{aligned} \tag{125}$$

where, by definition, $\varphi^*\eta(\mathrm{f}) := \eta(\mathrm{f} \circ \varphi)$ for all $\mathrm{f} \in F$, $\frac{d\mu(\varphi^*\eta)}{d\mu(\eta)}$ is the corresponding Radon–Nikodym derivative of the measure $\mu \circ \varphi^*$ with respect to the measure μ on F' and $\chi_\varphi(\eta)$ is a complex-valued character of the unit norm, satisfying the relationship

$$\chi_{\varphi_2}(\eta)\chi_{\varphi_1}(\varphi_2^*\eta) = \chi_{\varphi_1 \circ \varphi_2}(\eta) \tag{126}$$

for all $\varphi_j \in Diff(\mathbb{T}^n), j = \overline{1,2}, \eta \in F'$. For the Radon–Nikodym derivative above to exist, the measure μ on F' should be quasi-invariant with respect to the diffeomorphism group $Diff(\mathbb{T}^n)$, that is for any measurable set $Q \subset F'$ the condition $\mu(Q) = 0$ if, and only if, $\mu(\varphi^*Q)$ for arbitrary $\varphi \in Diff(\mathbb{T}^n)$.

In physics applications, the representation (125) is uniquely determined by the measure μ on F', which in the general case has a very complicated [20,72,114] structure, and its analytic construction is nontrivial. One of the fairly effective approaches to this problem is the quantum method of Bogolubov generating functionals developed in [5–7,20]. Another approach, which is of considerable interest for the theory of dynamical systems, is based on algebraic methods of constructing self-adjoint functional-operator representations of the original current Lie algebra (121). We proceed to its description in the case of the current group $G = Diff(\mathbb{S}^1) \ltimes F$, where $F \simeq C^\infty(\mathbb{S}^1; \mathbb{R})$ on the circle $\mathbb{S}^1$, taking into account results in [12,22,116–118].

We now introduce the following basis operators of the Lie current Lie algebra (121) for $n = 1$:

$$\rho_j := \int_{\mathbb{S}^1} exp(ijx + i\varepsilon x)\rho(x)dx, \qquad J_k := \int_{\mathbb{S}^1} exp(ikx)J(x)dx, \tag{127}$$

where $j,k \in \mathbb{Z}$ and $\varepsilon \in \mathbb{R}$ is a parameter. Then from (121) and (127), we find that

$$[\rho_j, \rho_k] = 0, \ \ [J_k, \rho_j] = (j+\varepsilon)\rho_{j+k}, \ \ [J_k, J_j] = (j-k)J_{k+j} \tag{128}$$

for all $j,k \in \mathbb{Z}$, that is the set $\mathcal{G} := \{ \rho_j, J_k : \Phi_\mu \to \Phi_\mu : j,k \in \mathbb{Z}\}$ of operators (128) on the representation Hilbert space Φ_μ, is equivalent to the semidirect product $\mathcal{G}\{J\} \ltimes \mathcal{G}\{\rho\}$ of the Lie subalgebra $\mathcal{G}\{J\} := \{ J_k : \Phi_\mu \to \Phi_\mu : k \in \mathbb{Z}\}$ and the Abelian subalgebra $\mathcal{G}\{\rho\} := \{\rho_j : \Phi_\mu \to \Phi_\mu : j \in \mathbb{Z}\}$ and isomorphic to the current Lie algebra (121) for $n = 1$. It is also worth mentioning [119] that in the case of functional-operator representations, the Lie algebra (128) admits the following central extension by means of the Schwinger cocycle:

$$[\rho_j, \rho_k] = \zeta\rho_{j,-k}, \ [J_k, \rho_j] = (j+\varepsilon)\rho_{j+k}, \ [J_k, J_j] = (j-k)J_{k+j} + \nu k(k^2-1)\delta_{j,-k}, \tag{129}$$

where $j,k \in \mathbb{Z}$ and $\zeta, \nu \in \mathbb{R}$ are the Schwinger parameters. The current Lie algebra (129) is called the generalized Virasoro current algebra [44] and has many applications in modern theoretical physics.

It is easy to show that the current Lie algebra (128) for $\varepsilon = 0$ admits the standard representation in the ring of operators $\mathbb{C}[\lambda, \lambda^{-1}][\partial/\partial\lambda]$, $\lambda \in \mathbb{C}^N$, regarded as a Lie subalgebra of the Lie algebra of rational vector fields on $\mathbb{C}^N, N \in \mathbb{N}$. Namely, if we set

$$\rho_j = \sum_{n=\overline{1,N}} \lambda_n^j, \ \ J_k = \sum_{n=\overline{1,N}} \lambda_n^{k+1}\partial/\partial\lambda_n, \tag{130}$$

for $j,k \in \mathbb{Z}$, then the current Lie algebra relations (128) are satisfied identically. In this case. if we make the restriction $|\lambda_n| = 1, \lambda_n = \exp(i\theta_n), \theta_n \in [0, 2\pi], n = \overline{1,N}$, then for the current algebra operators $\rho(x)$ and $J(x), x \in \mathbb{S}^1$, we obtain the expressions

$$\rho(x) = \sum_{n=\overline{1,N}} \delta(x-\theta_n), \ J(x) = \frac{1}{2i}\sum_{n=\overline{1,N}} [\delta(x-\theta_n)\partial/\partial\theta_n + \partial/\partial\theta_n\delta(x-\theta_n)]. \tag{131}$$

It is readily seen that the operators (131) are N-particle representations of the current Lie algebra (121) on the circle $\mathbb{S}^1$, and that the support of the measure μ on F' in the representation (124) is concentrated on functionals $\eta = \sum_{n=\overline{1,N}} \delta(x-\theta_n)$ and the Hilbert space $L_2^{(\mu)}(F';\mathbb{C}) \simeq L_2^{(s)}(\mathbb{T}^N;\mathbb{C})$, the space symmetric square integrable functions on the torus $\mathbb{T}^N$. In the general case, the current generalized Lie algebra (129) possesses numerous functional-operator representations by means of vector fields on special infinite dimensional manifolds. As will be shown below, these vector fields are defined on these manifold's so-called completely integrable infinite-dimensional Hamiltonian systems, many of which have applications in theoretical and mathematical physics.

On the infinite-dimensional smooth functional manifold $M \subset C^\infty(\mathbb{T}^n;\mathbb{R}^m)$, $n,m \in \mathbb{N}$ are finite, we consider a homogeneous autonomous nonlinear dynamical system

$$u_t = K[u], \tag{132}$$

where $K : M \to T(M)$ is a Frechet-smooth vector field on M, $[u] \in J(\mathbb{T}^n;\mathbb{R}^m)$ denotes a point of a finite order [59,120] at the jet-manifold $J(\mathbb{T}^n;\mathbb{R}^m)$ and $t \in \mathbb{R}$ is the evolution parameter. We assume that the vector field (132) is Hamiltonian, i.e., there exists a skew-symmetric Poissonian [18,59,120,121] operator $\vartheta : T^*(M) \to T(M)$ such that condition

$$L_K\vartheta = 0 \sim \vartheta_t - \vartheta K'^{,*} - K'\vartheta = 0 \tag{133}$$

where L_K denotes the Lie derivative [11,26,59,120,122,123] along the vector field $K : M \to T(M)$, *"prime"* denotes the usual Frechet derivative of a mapping and "*" denotes the adjoint mapping subject to the standard bilinear convolution form $(\cdot|\cdot)$ on the product $T^*(M) \times T(M)$ of the tangent and cotangent spaces over the functional manifold M. If the condition (133) holds, there exists such a smooth Hamiltonian functional $H_\vartheta \in \mathcal{D}(M) \subset C^\infty(M;\mathbb{R})$ that

$$K[u] = -\vartheta \operatorname{grad} H_\vartheta[u] \tag{134}$$

Assume now that the dynamical system (132) possesses one further algebraically independent solution $\eta : T^*(M) \to T(M)$ to the Equation (133), that is $L_K\eta = 0$, which is Poissonian.

Definition 12. *A dynamical system (132) possessing a (ϑ,η)-pair of Poissonian operators is said [11,13,18,59] to be bi-Hamiltonian, if for any $\lambda \in \mathbb{R}$ the pencil $(\vartheta\lambda + \eta) : T^*(M) \to T(M)$ is also Poissonian. The Poissonian (ϑ,η)-pair is called the Magri type compatible.*

Definition 13. *If the Poisson operator $\vartheta : T^*(M) \to T(M)$ is invertible, the operator $\Lambda := \vartheta^{-1}\eta : T^*(M) \to T^*(M)$ is said to be gradient-recursive and satisfies the Noether–Lax equation*

$$L_K\Lambda = 0 \sim \Lambda_t - [\Lambda, K^{',*}] = 0. \tag{135}$$

Similarly, the operator $\Phi := \eta\vartheta^{-1} : T(M) \to T(M)$ is said to be symmetry-recursive and satisfies the Noether–Lax equation

$$L_K\Phi = 0 \sim \Phi_t - [K^{'}, \Lambda] = 0. \tag{136}$$

The inverse operator $\vartheta^{-1} : T(M) \to T^*(M)$ is said to be symplectic, and the operator $\vartheta : T^*(M) \to T(M)$ itself is often called cosymplectic.

Yet, if the inverse operator $\vartheta^{-1} : T(M) \to T^*(M)$ does not exist, the notions of gradient-recursive and symmetry-recursive operators remain the same: $L_K\Lambda = 0$ and $L_K\Phi = 0$, respectively.

Definition 14. *The operator $\Phi : T(M) \to T(M)$ is said to be hereditary-recursive if the bilinear operator*

$$[\Phi', \Phi] : T(M) \times T(M) \to T(M) \tag{137}$$

is symmetric.

It is easy to check that the operator $\Phi = \eta\vartheta^{-1} : T(M) \to T(M)$ is hereditary-recursive [11,59,121] if the Poissonian pair (ϑ,η)-pair is compatible. The Poissonian (ϑ,η)-pair is compatible if, and only if, the operator $\eta\vartheta^{-1}\eta : T^*(M) \to T(M)$ is Poissonian too. Moreover, the operators $\vartheta(\vartheta^{-1}\eta)^n : T^*(M) \to T(M)$ for all $n \in \mathbb{Z}_+$ are Poissonian also.

Definition 15. *A vector field $\alpha : M \to T(M)$ is called a homogeneous symmetry of the dynamical system (132) if $L_K\alpha = 0 \sim [K,\alpha] = 0$. Respectively, a vector field $\tau : M \to T(M)$ is called an inhomogeneous symmetry of the dynamical system (132), $\partial\tau/\partial t + [K,\tau] = 0$.*

It is easy to observe that subsets of homogeneous and inhomogeneous symmetries, respectively, are Lie subalgebras of the symmetry space $\Gamma(M)$. Suppose now that for a consistent bi-Hamiltonian dynamical system (132) there exist two nontrivial homogeneous symmetry $\alpha_0 \in \Gamma(M)$ and homogeneous symmetry $\tau_0 \in \Gamma(M)$, such that

$$\begin{aligned} &L_{\tau_0}\alpha_0 = \varepsilon\alpha_0,\ L_{\alpha_0}\vartheta = 0 = L_{\alpha_0}\eta,\ L_{\tau_0}\vartheta = (\xi - 1/2)\vartheta, \\ &L_{\tau_0}\alpha_0 = \varepsilon\alpha_0, L_{\tau_0}\eta = (\xi + 1/2)\eta,\ \ L_{\tau_0}\Phi = \Phi, \end{aligned} \tag{138}$$

where $\varepsilon, \xi \in \mathbb{R}$ are certain numerical parameters. Having assumed that the symmetry-recursive operator $\Phi : T(M) \to T(M)$ is invertible, one can construct the following subsets $Q\{\alpha\} \subset \Gamma(M)$ and $Q\{\tau\} \subset \Gamma(M)$, where

$$Q\{\alpha\} := \{\alpha_j : \Phi^j \alpha_0 : j \in \mathbb{Z}\}, \quad Q\{\tau\} := \{\tau_j : \Phi^j \tau_0 : j \in \mathbb{Z}\}. \tag{139}$$

The following proposition holds.

Proposition 5. *The semi-direct product $Q := Q\{\tau\} \ltimes Q\{\alpha\}$ is a Lie subalgebra of symmetries of the dynamical system (132) isomorphic to the current Lie algebra (128).*

Proof. The proof is a direct consequence of the relations (138) and (139).

$$u_t = u_{xxx} + uu_x := K[u], \tag{140}$$

□

As a simplest example we consider the classical nonlinear Korteweg–de Vries dynamical system on the functional manifold $M \subset C^\infty(\mathbb{S}^1; \mathbb{R})$, possessing two compatible Poissonian operators

$$\vartheta = \partial, \quad \eta = \partial^3 + (u\partial + \partial u)/3, \tag{141}$$

where $\partial := \partial/\partial x, x \in \mathbb{R}$. Its symmetry-recursive operator equals to the expression

$$\Phi = \eta\vartheta^{-1} = \partial^2 + (u + \partial u \partial^{-1})/3, \tag{142}$$

where $\partial^{-1}(\cdot) := 1/2\left[\int_0^x dx(\cdot) - \int_x^{2\pi} dx(\cdot)\right]$ is the operator of inverse differentiation, $\partial \cdot \partial^{-1} = I$. Then, taking into account the homogeneous symmetry, $\alpha_0 = u_x$ and inhomogeneous $\tau_{-1} = 3/2(1 + tu_x)$ generate, respectively, two subalgebras $Q\{\alpha\} := \{\Phi^j \alpha_0 : j \in \mathbb{Z}\}$ and $Q\{\tau\} := \{\Phi^{j+1}\tau_{-1} : j \in \mathbb{Z}\}$, whose semidirect product $Q = Q\{\tau\} \ltimes Q\{\alpha\}$ is isomorphic to the current Lie algebra $\mathcal{G}$ (128).

3.2. Completely Integrable Hamiltonian Systems and the Current Algebra Symmetry Integrability Criterion

In analyzing the dynamical system (132) above, we assumed for it the existence of the consistent (ϑ, η)-pair of Poissonian operators, with respect to which it is bi-Hamiltonian. However, if the dynamical system (132) is not bi-Hamiltonian but only Hamiltonian and integrable, then obviously the Noether–Lax Equation (133) has only one solution, which is determined up to multiplication by a constant. On the other hand, if the dynamical system (132) is invariant with respect to the universal Banach Lie group symmetry $G = \mathrm{Diff}(\mathbb{T}^n) \ltimes F$, then for the corresponding Lie algebra of symmetries $Q = Q\{\tau\} \ltimes Q\{\alpha\}$, which is isomorphic to the current Lie algebra $\mathcal{G}$, the next conditions should hold:

$$L_\alpha \vartheta = 0, \qquad L_\tau \vartheta = 0 \tag{143}$$

for all $\alpha \in Q\{\alpha\}$ and $\tau \in Q\{\tau\}$. In addition, one easily ensues from (128) the following commutation relationships:

$$\begin{aligned} &(j+\varepsilon)\alpha_{j+1} = [\tau_1, \alpha_j], (j+\varepsilon)\alpha_{j-1} = [\tau_{-1}, \alpha_j], (j+\varepsilon)\alpha_j = [\tau_0, \alpha_j], \\ &(j-1)\tau_{j+1} = [\tau_1, \tau_j], \quad j\tau_j = [\tau_0, \tau_j], \quad (j+1)\tau_{j-1} = [\tau_{-1}, \tau_j]. \end{aligned} \tag{144}$$

Algebraic relationships (144) give rise to the following Lie-algebraic relationships

$$\begin{aligned} &L_{\alpha_j}\vartheta = 0 = L_{\alpha_j}\eta, \, , L_{\tau_j}\Lambda = \Lambda^{j+1}, L_{\alpha_j}\Lambda = 0 = L_{\alpha_j}\Phi, L_{\tau_j}\Phi = \Phi^{j+1}, \\ &L_{\tau_j}\vartheta = (\xi - j - 1/2)\vartheta\Lambda^j, \quad L_{\tau_j}\eta = (\xi - j + 1/2)\eta\Lambda^j, \end{aligned} \tag{145}$$

and show that on the basis of the sl(2) Lie subalgebra $\{\tau_{-1}, \tau_0, \tau_1\}$ jointly with the set of initial homogeneous symmetries $\{\alpha_{-1}, \alpha_1\}$ for $\varepsilon = 0$ or $\{\alpha_0\}$ for $\varepsilon \notin \mathbb{Z}$, as well as the inhomogeneous symmetries $\{\tau_{-2}, \tau_2\}$, one can construct recursively an entire infinite-hierarchy of symmetries $Q\{\tau\} \ltimes Q\{\alpha\}$, which is isomorphic to the current Lie algebra $\mathcal{G}$ (128) by virtue of the construction. In addition, in accordance with the Noether relations (143) there exist two infinite hierarchies of conservation laws to the dynamical system (132), namely, the homogeneous functionals $\gamma_j \in \mathcal{D}(M), j \in \mathbb{Z}$, and the inhomogeneous $\zeta_j \in \mathcal{D}(M), j \in \mathbb{Z}$, satisfying the conditions

$$\begin{gathered}\tau_j = -\vartheta \operatorname{grad} \zeta_j, \;\; \alpha_j = -\vartheta \operatorname{grad} \gamma_j, \;\; \{\gamma_j, \gamma_k\} = 0, \\ (j+\varepsilon)\gamma_{j+k} = (\operatorname{grad} \gamma_j | \tau_k) = \{\gamma_j, \tau_k\}, \\ \partial \tau_k + \{H, \zeta_k\} = 0, \;\; \{H, \gamma_j\} = 0, \;\; \{\zeta_j, \zeta_k\} = (j-k)\zeta_{j+k}\end{gathered} \tag{146}$$

for any $j, k \in \mathbb{Z}$, where, by definition, $K[u] = -\vartheta \operatorname{grad} H$, and $\{\cdot,, \cdot\} := (\operatorname{grad}(\cdot)|\vartheta \operatorname{grad}(\cdot))$ denotes the Poisson bracket on the space of functionals $\mathcal{D}(M)$ on the functional manifold M.

Direct calculations show that the results described above are valid for all the currently known completely integrable nonlinear dynamical systems, including the nonlinear equations of Schrëdinger type [21,59,114,124,125], the Benney—Kaup and Ito equations [114], the Davey—Stewartson and Yajima—Mel'nikov equations [13], and others, defined on infinite-dimensional manifolds, whose symmetry groups are isomorphic to the universal Banach current group $Diff(\mathbb{S}^1) \ltimes F$ on the circle $\mathbb{S}^1$. With regard to *"two- dimensionalized"* integrable dynamical systems of the Kadomtsev–Petviashvily type, it can be asserted that they are closely related [11,13,112,117] to special operator-valued nonlinear integrable dynamical systems, generated by suitably defined iso-spectral Lax type problems [111,126] and which are bi-Hamiltonian with respect to the Poissonian operators on these operator-valued manifolds.

The analysis made above of the correspondence between the universal Lie algebra of currents (128) and the functional Lie algebras of symmetries of integrable infinite-dimensional dynamical systems makes it possible to formulate the following working algorithm as an effective criterion of testing integrability of an arbitrary homogeneous nonlinear dynamical system (132) on the infinite-dimensional manifold M.

Algorithm*: If for the dynamical system $u_t = K[u]$ on the functional manifold M there exists the nontrivial sl(2) Lie subalgebra $\{\tau_{-1}, \tau_0, \tau_1\}$ together with a subset of "initial" inhomogeneous symmetries $\{\tau_{-2}, \tau_2\}$ and homogeneous $\{\alpha_{-1}, \alpha_1 : \varepsilon = 0\}$ symmetries, satisfying the conditions*

$$\begin{gathered}[\tau_0, \tau_2] = 2\tau_2, [\tau_0, \tau_{-2}] = -2\tau_{-2}, [\tau_1, \tau_{-2}] = -3\tau_{-1}, \\ [\alpha_{-1}, \alpha_1] = 0, \;\; [\tau_{-1}, \tau_2] = 3\tau_1, \;\; [\tau_0, \tau_0] = \varepsilon\alpha_0,\end{gathered} \tag{147}$$

then this dynamical system on M possesses an infinite-dimensional Lie algebra of symmetries $Q = Q\{\tau\} \ltimes Q\{\alpha\}$, isomorphic to the current Lie algebra G (128) of the Banach group $\mathrm{Diff}(S^1) \ltimes F$ *on the circle S^1, and if there exists a nontrivial solution of the Noether–Lax equation $L_K\vartheta = 0$, then our dynamical system is an infinite-dimensional completely integrable Hamiltonian flow on the functional manifold M. If at the same time the relations (145) are satisfied, then the dynamical system $u_t = K[u]$ on M is bi-Hamiltonian and possesses an hereditary-recursive operator $\Lambda = \vartheta^{-1}\eta$, where $\eta(\xi - 3/2) = L_{\tau_1}\vartheta$, and, by virtue of the gradient-holonomic algorithm [11,13,59,116], a standard Lax type representation.*

If the conditions (143) are not satisfied, the dynamical system does not possess bi- Hamiltonian structure, but there is an additional infinite-dimensional inhomogeneous hierarchy of conservation laws satisfying the conditions (146).

3.3. *Integrable Systems, Their Symmetry Analysis and Structure of the Poissonian Operators*

Suppose we are given the homogeneous nonlinear dynamical system (132) on the functional manifold M and pose a question of the existence for this dynamical system of a bi-Hamiltonian structure on M and effective methods of determining it in explicit form.

In accordance with the gradient-holonomic algorithm [11,13,59,116] for investigating the integrability of nonlinear dynamical systems, we can successively establish in explicit form the presence for our system (132) of an infinite functionally independent and naturally ordered by means of the parameter $\lambda \in \mathbb{R}$ hierarchy $\gamma_j \in \mathcal{D}(M), j \in \mathbb{Z}_+$, of conservation laws. In addition, by virtue of the homogeneity of the dynamical system (132), it always possesses *a priori* two commuting to each other homogeneous symmetries, which are defined on M by the vector fields d/dx and d/dt. We can also consider the equivalent realization of these vector fields d/dx and d/dt on M as Hamiltonian systems [11,17,18,123] on the infinite-dimensional manifold of jets $J(\mathbb{S}^1;\mathbb{R}^m) \simeq M$ with respect to a symplectic structure $\omega^{(2)} \in \Omega^1(J(\mathbb{S}^1;\mathbb{R}^m)$. We denote by $\alpha^{(1)} \in \Omega^1(J(\mathbb{S}^1;\mathbb{R}^m)$ a Liouville type 1-form, for which $d\alpha^{(1)} = \omega^{(2)}$ and take into account that, by definition, there holds the conditions $i_{d/dx}\omega^{(2)}[u] = -d\gamma[u]$ and $L_{d/dx}\,\omega^{(2)}[u] = 0$, where $\gamma[u] \in \Omega^0(J(\mathbb{S}^1;\mathbb{R}^m)$ denotes the density of the corresponding conservation law $\gamma \in \mathcal{D}(M)$ at point $u \in M$. Based now on the Cartan representation [127] of the Lie derivative $L_{d/dx} = i_{d/dx}d + di_{d/dx}$, one easily obtains the following general relationship: $\gamma[u] = \alpha^{(1)}[u](d/dx) = (\psi[u]|u_x) \bmod(d/dx)$ for some element $\psi \in T^*(M)$ at any point $u \in M$, which should simultaneously satisfy the compatibility condition $d\,L_{d/dt}\psi = 0$ subject to the vector field d/dt on M. The latter gives rise to the analytical expression that is useful in applications, $\vartheta^{-1} = \psi'[u] - \psi'^{,*}[u]$, for the corresponding cosymplectic operator on the manifold M, whose inverse mapping $\vartheta : T^*(M) \to T(M)$ is our searched for Poissonian operator for the dynamical system (132).

3.3.1. Two-Dimensional Korteweg–de Vries Type Hydrodynamic System

We consider an example of a nonlinear bi-Hamiltonian Korteweg–de Vries type hydrodynamic system on "*two-dimensionalized*" smooth functional manifold $M \simeq J(\mathbb{T}^2;\mathbb{R})$ for which the Noether–Lax property (143), mentioned above, is not satisfied. This system [12,21] has the form

$$u_t = u_{xxy} + 2u_x\partial_x^{-1}u_y + 4uu_y := K[u] \tag{148}$$

and possesses two algebraically-independent Poissonian operators

$$\vartheta = \partial_x, \;\; \eta = \partial_x^3 + 2(u\partial_x + \partial_x u) \tag{149}$$

Moreover, it is readily shown that for the dynamical system, (148) allows the representation $K[u] = \Phi u_y$, where $\Phi := \eta\vartheta^{-1} = \partial_x^2 + 2(u + \partial_x u\partial_x^{-1})$ is the corresponding symmetry-recursive operator on M. One can check by direct calculations that the set $\{\tau_{-1}^{(x)} = 1/4(1+tu_x),\ \tau_{-1}^{(y)} = 1/4(1+tu_y)\}$ consists of inhomogeneous symmetries of the dynamical system (148) and the set $\{\alpha_0^{(x)} = u_x, \alpha_0^{(y)} = u_x\}$ consists of homogeneous symmetries. From them, one constructs the following hierarchies of symmetries:

$$Q\{\alpha^{(x)}\} := \{\alpha_j^{(x)} = \Phi^j\alpha_0^{(x)} : j \in \mathbb{Z}\}, Q\{\alpha^{(y)}\} := \{\alpha_j^{(y)} = \Phi^j\alpha_0^{(y)} : j \in \mathbb{Z}\}, \tag{150}$$

$$Q\{\tau^{(x)}\} := \{\tau_j^{(x)} = \Phi^{j+1}\tau_{-1}^{(x)} : j \in \mathbb{Z}\}, Q\{\tau^{(y)}\} := \{\tau_j^{(y)} = \Phi^{j+1}\tau_{-1}^{(y)} : j \in \mathbb{Z}\}$$

The resulting Lie subalgebras $Q^{(x)} := Q\{\tau^{(x)}\} \ltimes Q\{\alpha^{(x)}\}$ and $Q^{(y)} := Q\{\tau^{(y)}\} \ltimes Q\{\alpha^{(y)}\}$ have the following commutation relationships:

$$[\tau_j^{(x)}, \alpha_k^{(y)}] = k\alpha_{j+k}^{(y)},\;\; [\tau_j^{(y)}, \alpha_k^{(x)}] = (k+1/2)\alpha_{j+k}^{(x)}, \tag{151}$$

$$[\tau_j^{(x)}, \tau_k^{(y)}] = k\tau_{j+k}^{(y)} - j\tau_{j+k}^{(x)},\quad [\alpha_j^{(x)}, \alpha_k^{(y)}] = 0$$

4. The Current Algebra Representations and the Factorized Structure of Quantum Integrable Many-Particle Hamiltonian Systems

4.1. The Current Algebra Representation and the Hamiltonian Reconstruction of the Calogero–Moser–Sutherland Quantum Model

The periodic Calogero–Moser–Sutherland quantum bosonic model on the finite interval $[0,l] \simeq \mathbb{R}/\{[0,l]\mathbb{Z}\}$ is governed by the N-particle Hamiltonian

$$H_N := -\sum_{j=\overline{1,N}} \frac{\partial^2}{\partial x_j^2} + \sum_{j\neq k=\overline{1,N}} \frac{\pi^2\beta(\beta-1)}{l^2\sin^2[\frac{\pi}{l}(x_j-x_k)]} \tag{165}$$

on the symmetric Hilbert space $L_2^{(s)}([0,l]^N;\mathbb{C})$, where $N\in\mathbb{Z}_+$ and $\beta\in\mathbb{R}$ is an interaction parameter. As it was stated in very interesting and highly speculative works [132,133], there exists linear differential operators

$$\mathcal{D}_j := \frac{\partial}{\partial x_j} - \frac{\pi\beta}{l}\sum_{k=\overline{1,N},k\neq j} ctg[\frac{\pi}{l}(x_j-x_k)] \tag{166}$$

for $j=\overline{1,N}$, such that the Hamiltonian (165) is factorized as the bounded from below symmetric operator

$$H_N = \sum_{j=\overline{1,N}} \mathcal{D}_j^+\mathcal{D}_j + E_n, \tag{167}$$

where

$$E_N = \frac{1}{3}\left(\frac{\pi\beta}{l}\right)^2 N(N^2-1) \tag{168}$$

is the ground state energy of of the Hamiltonian operator (165), that there exists such a vector $|\Omega_N) \in L_2^{(s)}([0,l]^N;\mathbb{C})$, satisfying for any $N\in\mathbb{Z}_+$ the eigenfunction condition

$$H_N|\Omega_N) = E_N|\Omega_N) \tag{169}$$

and equals

$$|\Omega_N) = \prod_{j<k=\overline{1,N}} \left(\sin[\frac{\pi}{l}(x_j-x_k)]\right)^\beta, \tag{170}$$

coinciding with the corresponding Bethe anzatz representation [134,135] for the groundstate of the quantum Calogero–Moser-Sutherland model (165).

Being additionally interested in proving the quantum integrability of the Calogero–Moser–Sutherland model (165), we will proceed to its second quantized representation [9,10,13,56,57,60,68,135,136] and studying it by means of the density operator representation approach to the current algebra, described above in Section 2 and devised previously in [1,4–7,76,77].

The secondly quantized form of the Calogero–Moser–Sutherland Hamiltonian operator (165) looks as

$$\mathrm{H} = \int_0^l dx\psi_x^+(x)\psi_x(x) + \left(\frac{\pi}{l}\right)^2\beta(\beta-1)\int_0^l dx\int_0^l dy\frac{\psi^+(x)\psi^+(y)\psi(y)\psi(x)}{\sin^2[\frac{\pi}{l}(x-y)]}, \tag{171}$$

acting on the corresponding Fock space $\Phi_F := \oplus_{n\in\mathbb{Z}_+}\Phi_n^{(s)}$, $\Phi_n^{(s)} \simeq L_2^{(s)}([0,l]^n;\mathbb{C}), n\in\mathbb{Z}_+$. To proceed to the current algebra representation of the Hamiltonian operator (171), it would useful to recall the factorized representation (167) and construct preliminarily the following singular Dunkl type [132,133,137,138] symmetrized differential operator

$$D_N(x) := \sum_{j=\overline{1,N}} \delta(x - x_j)\frac{\partial}{\partial x_j} - \\ -\frac{\pi\beta}{2l}\sum_{j\neq k=\overline{1,N}}(\delta(x - x_j)ctg[\tfrac{\pi}{l}(x_j - x_k)] + \delta(x - x_k)ctg[\tfrac{\pi}{l}(x_k - x_j)]) \tag{172}$$

on the Hilbert space $L_2^{(s)}([0,l]^N;\mathbb{C})$, $N \in \mathbb{Z}_+$, parametrized by a running point $x \in \mathbb{R}/\{[0,l]\mathbb{Z}\}$. The corresponding secondly quantized representation of the operator (172) looks as

$$\mathrm{D}(x) = \psi^+(x)\psi_x(x) - \frac{\pi\beta}{l}\int_0^l dy\, ctg[\frac{\pi}{l}(x-y)] : \psi^+(x)\psi^+(y)\psi(y)\psi(x) : \tag{173}$$

for any $x \in \mathbb{R}/[0,l]\mathbb{Z}$, or on the density operator $\rho : \Phi_F \to \Phi_F$ representation form, as

$$\mathrm{D}(x) = \nabla_x\rho(x)/2 + iJ(x) - \\ -\frac{\pi\beta}{2l}\int_0^l dy\, \left[ctg[\tfrac{\pi}{l}(x-y)] : \rho(x)\rho(y) : - ctg[\tfrac{\pi}{l}(y-x)] : \rho(y)\rho(x) :\right], \tag{174}$$

which is equivalently representable in a suitable current algebra symmetry representation Hilbert space Φ, as

$$\mathrm{D}(x) = K(x) - \\ -\frac{\pi\beta}{2l}\int_0^l dy\left[ctg[\tfrac{\pi}{l}(x-y)] : \rho(x)\rho(y) : - ctg[\tfrac{\pi}{l}(y-x)] : \rho(y)\rho(x) :\right]. \tag{175}$$

Now, based on the operator (174), one can formulate [10] the following proposition.

Proposition 7. *The secondly quantized Calogero–Moser–Sutherland Hamiltonian operator (171) in a suitable current algebra symmetry representation Hilbert space Φ is weakly equivalent to the factorized Hamiltonian operator*

$$\hat{\mathrm{H}} = \int_0^l dx \mathrm{D}^+(x)\rho(x)^{-1}\mathrm{D}(x) \tag{176}$$

modulo the ground state energy operator $\mathrm{E} : \Phi \to \Phi$, where

$$\mathrm{E} = \frac{1}{3}\left(\frac{\pi\beta}{l}\right)^2 : \mathrm{N}^3 : + \left(\frac{\pi\beta}{l}\right)^2 : \mathrm{N}^2 :, \tag{177}$$

where, as before,

$$\mathrm{N} := \int_0^l \rho(x)dx \tag{178}$$

is the particle number operator, and satisfies the determining conditions

$$(\mathrm{H} - \mathrm{E})|\Omega) = 0, \ \ \mathrm{D}(x)|\Omega) = 0 \tag{179}$$

on the suitably renormalized groundstate vector $|\Omega) \in \Phi$ for all $x \in \mathbb{R}/[0,l]\mathbb{Z}$. Moreover, for any integer $\mathrm{N} \in \mathbb{Z}_+$ the corresponding projected vector $|\Omega_N) := |\Omega)|_{\Phi_N}$ exactly coincides with the *related Bethe groundstate vector for the N-particle Calogero–Moser–Sutherland model (165) and satisfies the following eigenfunction relationships:*

$$\mathrm{N}|\Omega_N) = N|\Omega_N), \;\; \mathrm{E}|\Omega_N) = \left(\frac{1}{3}\left(\frac{\pi\beta}{l}\right)^2 : \mathrm{N}^3 : + \left(\frac{\pi\beta}{l}\right)^2 : \mathrm{N}^2 : \right)|\Omega_N) = \tag{180}$$

$$= \left[\frac{1}{3}\left(\frac{\pi\beta}{l}\right)^2 (N^3 - 3N^2 + 2N) + \left(\frac{\pi\beta}{l}\right)^2 N(N-1)\right]|\Omega_N) =$$

$$- \left[\frac{1}{3}\left(\frac{\pi\beta}{l}\right)^2 (N^3 - 3N^2 + 2N + 3N^2 - 3N)\right]|\Omega_N) =$$

$$= \left[\frac{1}{3}\left(\frac{\pi\beta}{l}\right)^2 N(N^2 - 1)\right]|\Omega_N) := E_N|\Omega_N),$$

exactly ensuing the result (168).

Remark 5. *When deriving the expression (180), we have used the identities*

$$\begin{aligned} \rho(x)\rho(y) = &\; :\rho(x)\rho(y): + \rho(y)\delta(x-y), \\ \rho(x)\rho(y)\rho(z) = &\; :\rho(x)\rho(y)\rho(z): + :\rho(x)\rho(y): \delta(y-z) + \\ &+ :\rho(y)\rho(z): \delta(z-x) + :\rho(z)\rho(x): \delta(x-y) + :\rho(x)\delta(y-z)\delta(z-x), \end{aligned} \tag{181}$$

which hold [5,55,56,77] for the density operator $\rho : \Phi \to \Phi$ at any points $x, y, z \in \mathbb{R}/\{[0,l]\mathbb{Z}\}$.

Observe now that the operator (173) can be rewritten down in Φ as

$$\mathrm{D}(x) = \mathrm{K}(x) - \mathrm{A}(x), \tag{182}$$

where, by definition,

$$\mathrm{K}(x) := \nabla_x \rho(x)/2 + iJ(x), \;\; \mathrm{A}(x) := \frac{\pi\beta}{l}\int_0^l dy\, ctg[\frac{\pi}{l}(x-y)] : \rho(x)\rho(y) : \tag{183}$$

for all $x \in \mathbb{R}/\{[0,l]\mathbb{Z}\}$. Recalling now the second condition of (179), one can rewrite it equivalently as

$$\mathrm{K}(x)|\Omega) = \mathrm{A}(x)|\Omega) \tag{184}$$

on the renormalized ground state vector $|\Omega) \in \Phi$ for all $x \in \mathbb{R}/\{[0,l]\mathbb{Z}\}$. On the other hand, owing to the expression (176), we obtain the searched for current algebra representation

$$\hat{\mathrm{H}} = \int_0^l dx(\mathrm{K}^+(x) - \mathrm{A}(x))\rho(x)^{-1}(\mathrm{K}(x) - \mathrm{A}(x)) \tag{185}$$

of the Calogero–Moser–Sutherland Hamiltonian operator (165) on the suitably renormalized Hilbert space Φ, as it was already demonstrated in the work [76,77], using the condition (184) in the form (61).

4.2. The Current Algebra Representation and Integrability of the Calogero–Moser–Sutherland Quantum Model

We now briefly discuss the quantum integrability of the Calogero–Moser–Sutherland model (165). Owing to the factorized representation (185), one can easily observe [8–10] that for any integer $p \in \mathbb{Z}_+$, the suitably symmetrized Hamiltonian operator densities $\mathrm{h}(x) := \mathrm{D}^+(x)\rho(x)^{-1}\mathrm{D}(x) : \Phi \to \Phi, x \in \mathbb{R}/\{[0,l]\mathbb{Z}\}$, commute to each other and with the particle number operator $\mathrm{N} : \Phi \to \Phi$, that is

$$[\mathrm{h}(x), \mathrm{h}(y)] = 0, \ [\mathrm{h}(x), \mathrm{N}] = 0 \tag{186}$$

for any $x, y \in \mathbb{R}/[0,l]\mathbb{Z}$. As a result of the commutation property (186), one easily obtains that for any integer $p \in \mathbb{Z}_+$ the symmetric operators

$$\hat{\mathrm{H}}^{(p)} := \int_0^l dx \mathrm{h}(x)^p \tag{187}$$

also commute to each other

$$[\hat{\mathrm{H}}^{(p)}, \hat{\mathrm{H}}^{(q)}] = 0 \tag{188}$$

for all integers $p, q \in \mathbb{Z}_+$, and in particular, commute to the Calogero–Moser–Sutherland Hamiltonian operator (176):

$$[\hat{\mathrm{H}}^{(p)}, \hat{\mathrm{H}}] = 0. \tag{189}$$

Concerning the related N-particle differential expressions for the operators (187), it is enough to calculate their projections on the N-particle Fock subspace $\Phi_N^{(s)} \subset \Phi_F, N \in \mathbb{N}$. Namely, let an arbitrary vector $|\varphi_N) \in \Phi_N^{(s)}$ be representable as

$$|\varphi_N) := \int_{[0,l]^N} \varphi_N(x_1, x_2, ..., x_N) \prod_{j=\overline{1,N}} dx_j \psi^+(x_j)|0) \tag{190}$$

for some coefficient function $\varphi_N \in L_2^{(s)}([0,l]^N; \mathbb{C})$. Then, by definition,

$$\hat{\mathrm{H}}^{(p)}|\varphi_N) := |\varphi_N^{(p)}), \tag{191}$$

where

$$|\varphi_N^{(p)}) = \int_{[0,l]^N} (H_N^{(p)}\varphi_N)(x_1, x_2, ..., x_N) \prod_{j=\overline{1,N}} dx_j \psi^+(x_j)|0) \tag{192}$$

for a given $p \in \mathbb{Z}_+$ any $N \in \mathbb{Z}_+$. In particular, for $p = 2$, when $\hat{\mathrm{H}}^{(2)} + \mathrm{E} = \mathrm{H} : \Phi_F \to \Phi_F$, one easily retrieves the shifted Calogero–Moser–Sutherland Hamiltonian operator (165):

$$H_N^{(2)} = -\sum_{j=\overline{1,N}} \frac{\partial^2}{\partial x_j^2} + \sum_{j \neq k = \overline{1,N}} \frac{\pi^2 \beta(\beta - 1)}{l^2 \sin^2[\frac{\pi}{l}(x_j - x_k)]} - \left(\frac{\pi\beta}{l}\right)^2 \frac{N(N^2 - 1)}{3}. \tag{193}$$

Respectively for higher integers $p > 2$ the resulting N-particle differential operator expressions $H_N^{(p)} : L_2^{(s)}([0,l]^N; \mathbb{C}) \to L_2^{(s)}([0,l]^N; \mathbb{C}), N \in \mathbb{Z}_+$, can be obtained the described above way by means of simple yet cumbersome calculations, and which will prove to be completely equivalent to those calculated previously in good work [132].

Remark 6. *In the thermodynamical limit, when* $\lim_{N\to\infty, l\to\infty} N/\pi l := \bar{\rho} > 0$, *the structural operator* $\mathrm{D}(x) : \Phi \to \Phi, x \in \mathbb{R}/\{[0,l]\mathbb{Z}\}$, *reduces to*

$$\bar{\mathrm{D}}(x) := \lim_{N/l \to \bar{\rho}} \mathrm{D}(x) = \nabla_x \rho(x)/2 + iJ(x) - \beta \int_{\mathbb{R}} dy \frac{:\rho(y)\rho(x):}{x - y}, \tag{194}$$

and, respectively, the operator (165) reduces to

$$\bar{H}_N = -\sum_{j=\overline{1,N}} \frac{\partial^2}{\partial x_j^2} + \beta(\beta - 1) \sum_{j \neq k = \overline{1,N}} \frac{1}{(x_j - x_k)^2} \tag{195}$$

on the Hilbert space $L_2^{(s)}(\mathbb{R}^N;\mathbb{C})$ *for any* $N \in \mathbb{Z}_+$, *whose density operator representation in a suitable Hilbert space* Φ, *respectively, equals*

$$\bar{H} = \int_{\mathbb{R}} dx \left(\bar{D}^+(x)\rho(x)^{-1}\bar{D}(x) + \epsilon_0 \right), \tag{196}$$

where $\epsilon_0 := \lim_{N/l \to \bar{\rho}} \frac{E_N}{l} = \bar{\rho}^3/3$ *denotes the average energy density of the reduced Calogero–Moser–Sutherland Hamiltonian operator (195) as* $N \to \infty$, *exactly coinciding with the before obtained results in [77,133].*

5. The Dual Current Algebra Density Representation and the Factorized Structure of Quantum Integrable Many-Particle Hamiltonian Systems

5.1. The Current Algebra Density Representation

We are now interested in constructing a special density functional representation of the local current algebra (29) on the corresponding representation Hilbert space $\Phi_\mu \simeq \oplus_\rho$ with the cyclic vector $|\Omega) = 1 \in \Phi_\rho$. To do this, let us first consider the *creation* $\psi^+(x)$ and annihilation operators $\psi(x), x \in \mathbb{R}^m$, defined via (56) on the canonical Fock space Φ_F, which can be formally represented as

$$\psi^+(x) = \sqrt{\rho(x)}\exp[-i\vartheta(x)], \psi(x) = \exp[i\vartheta(x)]\sqrt{\rho(x)}, \tag{197}$$

where $\rho(x) : \Phi_F \to \Phi_F$ is our density operator and $\vartheta(x) : \Phi_F \to \Phi_F, x \in \mathbb{R}^m$, is some self-adjoint operator. What is important is the operators $\rho(x)$ and $\vartheta(x) : \Phi_F \to \Phi_F$ realize the canonical [55,56,60,63,85] commutation relationships

$$[\rho(x), \rho(y)] = 0 = [\vartheta(x), \vartheta(y)], \tag{198}$$

$$[\rho(y), \vartheta(x)] = i\delta(x-y)$$

for any $x, y \in \mathbb{R}^m$. Concerning the current operator $J(x) : \Phi_F \to \Phi_F^m, x \in \mathbb{R}^m$, one can easily obtain its equivalent expression

$$J(x) = \rho(x)\nabla\vartheta(x). \tag{199}$$

Based on the canonical relationships (198) one can easily obtain, following [72,85,139], that

$$\vartheta(x) = \frac{1}{i}\frac{\delta}{\delta\rho(x)} + i\sigma[\rho(x)], \tag{200}$$

where $\sigma[\rho(x)] : \Phi_\rho \to \Phi_\rho$ acts on the corresponding Hilbert representation space Φ_ρ and is some function of the density operator $\rho(x) : \Phi_\rho \to \Phi_\rho, x \in \mathbb{R}^m$. Then, respectively, the current operator (199) is representable in Φ_ρ as

$$J(x) = -i\rho(x)\nabla\frac{\delta}{\delta\rho(x)} + \rho(x)\nabla\sigma[\rho(x)]. \tag{201}$$

The functional-operator expression (201) proves to make sense [5,72,75,139] as operators on the Hilbert space Φ_ρ of functional valued complex-functions on the manifold $\mathcal{M}$, coordinated by the density functional parameter $\rho : \Phi_\rho \to \Phi_\rho$ and endowed with the scalar product $(a|b)_{\Phi_\rho} := \int_{\mathcal{M}} \overline{a(\rho)} b(\rho) d\mu(\rho)$ subject to some measure μ on $\mathcal{M}$. To calculate this measure μ on $\mathcal{M}$, we will present an explicit isomorphism between this Hilbert space Φ_ρ and the corresponding Fock space Φ of spinless bosonic particles in $\mathbb{R}^m$. First, we determine the support $supp\ \mu \subset \mathcal{M}$ of the measure μ, having assumed that the manifold

$$\mathcal{M} = \cup_{n \in \mathbb{Z}_+} \mathcal{M}_n, \tag{202}$$

where $\mathcal{M}_n := \{a(\rho) : \rho(x) := \sum_{j=1}^{n} \delta(x - c_j) : a \in C^\infty(F'; End\ \Phi_\rho)\}$, where $c_j \in \mathbb{R}^m, j = \overline{1,n}$, $n \in \mathbb{N}$, are arbitrary vector parameters. The restriction $d\mu_n$ of the measure μ on the submanifold $\mathcal{M}_n$ can be presented [4,5,57,68,71] as

$$d\mu_n = \gamma_n(c_1, c_2, ..., c_n) \prod_{j=\overline{1,n}} dc_j, \tag{203}$$

where functions $\gamma_n : \mathbb{R}^{m\times n} \to \mathbb{R}_+, n \in \mathbb{N}$, should be determined from the condition (201). In accordance with the manifold structure (202), we can decompose the Hilbert space Φ_ρ as

$$\Phi_\rho = \oplus_{n\in\mathbb{N}} \Phi_n, \tag{204}$$

where the space Φ_n depends on the mapping $\sigma : \mathcal{M} \to \mathrm{End}(\Phi_\rho)$ and consists of functionals that are bounded on $\mathcal{M}_n$, in particular, for any $a(\rho) \in \mathcal{M}$ the restrictions $a(\rho)|_{\Phi_n}, n \in \mathbb{N}$, consist of functions of vectors $(c_1, c_2, ..., c_n) \in \mathbb{R}^{m\times n}, n \in \mathbb{N}$, respectively. The scalar product in $\Phi_n, n \in \mathbb{N}$, is suitably defined by means of the expressions (203). Now we can construct the isomorphism between the Hilbert spaces $\Phi_n, n \in \mathbb{N}$, and the corresponding components $\Phi_n, n \in \mathbb{N}$, of the corresponding Fock space Φ, representing spinless bosonic particles in $\mathbb{R}^m$. In the Hilbert space $\Phi_n := \Phi_n^{(\sigma)}, n \in \mathbb{N}$, one can easily calculate the eigenfunctions $\varphi^{(\sigma)}_{p_1,p_2,...,p_n}(\rho) \in \Phi_n^{(\sigma)}$ of the free Hamiltonian

$$\mathrm{H}_0^{(\sigma)} := \frac{1}{2} \int_{\mathbb{R}^m} dx \langle K^+(x) | \rho^{-1}(x) K(x) \rangle \tag{205}$$

with structural

$$K(x) := \frac{1}{2} \nabla \rho(x) + i J^{(\sigma)}(x), \ \ K^+(x) := \frac{1}{2} \nabla \rho(x) - i J^{(\sigma)}(x) \tag{206}$$

and the momentum

$$\mathrm{P}^{(\sigma)} := \int_{\mathbb{R}^m} dx J^{(\sigma)}(x) \tag{207}$$

operators:

$$\mathrm{H}_0^{(\sigma)} \varphi^{(\sigma)}_{p_1,p_2,...,p_n}(\rho) = (\sum_{j=\overline{1,n}} E_j) \varphi^{(\sigma)}_{p_1,p_2,...,p_n}(\rho), \tag{208}$$

$$\mathrm{P}^{(\sigma)} \varphi^{(\sigma)}_{p_1,p_2,...,p_n}(\rho) = (\sum_{j=\overline{1,n}} p_j) \varphi^{(\sigma)}_{p_1,p_2,...,p_n}(\rho),$$

where $p_j \in \mathbb{R}^m$, $j = \overline{1,n}$, are momentums of bose-particles in $\mathbb{R}^m$, the operator $\mathrm{H}_0^{(\sigma)} : \Phi_\rho \to \Phi_\rho$ is given by the expressions (55), (201) and (205) and the operator $\mathrm{P}^{(\sigma)} : \Phi_\rho \to \Phi_\rho$ is given by the expressions (201) and (206), respectively, within which the current operator $J^{(\sigma)}(x) : \Phi_\rho \to \Phi_\rho$ is realized under the condition $\nabla\sigma[\rho(x)] := \sigma\rho(x)^{-1}\nabla\rho(x)$ as

$$J^{(\sigma)}(x) = -i\rho(x)\nabla \frac{\delta}{\delta\rho(x)} + i\sigma\nabla\rho(x) \tag{209}$$

where $\sigma \in \mathbb{R}$ is a fixed real-valued parameter. In this case, the eigenfunctions $\varphi^{(\sigma)}_{p_1,p_2,...,p_n}(\rho) \in \Phi_n^{(\sigma)}, n \in \mathbb{N}$, can be expressed [72,139] as

$$\varphi^{(\sigma)}_{p_1,p_2,...,p_n}(\rho) = \frac{1}{n!} \bar{\varphi}_0^{(\sigma)}(\rho) \left(\prod_{j=\overline{1,n}} B_{p_j}(\rho) \cdot 1 \right), \tag{210}$$

where

$$\bar{\varphi}_0^{(\sigma)}(\rho) := \exp\left[(\sigma - 1/2)\int_{\mathbb{R}^m} dx\rho(x)\ln\rho(x)\right], \tag{211}$$

$$B_{p_j}(\rho) := \int_{\mathbb{R}^m} dx \exp(i\langle p|x\rangle\ )\rho(x)\exp(-\frac{\delta}{\delta\rho(x)}).$$

The corresponding n-particle Fock subspaces $\Phi_n^{(\sigma)}, n \in \mathbb{N}$, can be naturally represented by means of the vectors

$$|\varphi_n^{(\sigma)}) := \frac{1}{\sqrt{n!}}\int_{\mathbb{R}^{m\times n}}\prod_{j=\overline{1,n}} dp_j\ \varphi_n^{(\sigma)}(p_1,p_2,...,p_n)a^+(p_1)a^+(p_2)...a^+(p_n)|0) \tag{212}$$

with functions $\varphi_n^{(\sigma)} \in L_2^{(s)}(\mathbb{R}^{m\times n};\mathbb{C}), n \in \mathbb{N}$, where

$$a^+(p) := \frac{1}{(2\pi)^{m/2}}\int_{\mathbb{R}^m} dx \exp(i\langle x|p\rangle)a^+(x) \tag{213}$$

denotes the momentum creation operator for any $p \in \mathbb{R}^m$.

Moreover, any functional $\varphi_n^{(\sigma)}(\rho) \in \Phi_n^{(\sigma)}, n \in \mathbb{N}$, can be uniquely represented as

$$\varphi_n^{(\sigma)}(\rho) := \int_{\mathbb{R}^{m\times n}}\prod_{j=\overline{1,n}} dp_j\tilde{\varphi}_n^{(\sigma)}(p_1,p_2,...,p_n)\varphi_{p_1,p_2,...,p_n}^{(\sigma)}(\rho) \tag{214}$$

for $\tilde{\varphi}_n^{(\sigma)} \in L_2^{(s)}(\mathbb{R}^{m\times n};\mathbb{C})$, since the following condition

$$\left(B_{p_{n+1}}(\rho)\prod_{j=\overline{1,n}} B_{p_j}(\rho)\cdot 1\right)\Bigg|_{\rho=a^+(x)a(x)} |\varphi_n^{(\sigma)}) = 0 \tag{215}$$

holds identically for all $p_j \in \mathbb{R}^m, j = \overline{1,n+1}$, and arbitrary state $|\varphi_n^{(\sigma)}) \in \Phi_F, n \in \mathbb{N}$.

Remark 7. *The condition (215) jointly with the constraint $\int_{\mathbb{R}^m}\rho(x)dx = n$ in $\Phi_n^{(\sigma)}, n \in \mathbb{N}$, should be, in general, naturally satisfied for any current algebra representation space Φ_ρ, if and only if $\rho(x) = \sum_{j=\overline{1,n}}\delta(x - c_j) \in \mathcal{M}_n$ for arbitrary $n \in \mathbb{N}$.*

As a result of the construction above, we can state that the Hilbert spaces $\Phi_n^{(\sigma)}, n \in \mathbb{N}$, embed, respectively, isomorphically into the related Fock subspaces $\Phi_n^{(\sigma)}, n \in \mathbb{N}$. As a consequence, we derive that the Hilbert space Φ_ρ allows an isomorphic embedding into the related Fock space Φ_F.

Consider now, following [5,72,139], the action of the current operator (209) on the basic vectors $\varphi_n^{(\sigma)}(\rho) \in \Phi_n^{(\sigma)}, n \in \mathbb{N}$:

$$J^{(\sigma)}(x)\varphi_n^{(\sigma)}(\rho) = \bar{\varphi}_0^{(\sigma)}(\rho)[-i\rho(x)\nabla\frac{\delta}{\delta\rho(x)} + i\sigma\nabla\rho(x)]\varphi_n^{(\sigma)}(\rho), \tag{216}$$

from which one ensues easily at $\sigma = 1/2$ its n-particle representation on the functional manifold $\mathcal{M}_n$:

$$\begin{aligned} &J^{(1/2)}(x)\varphi_n^{(1/2)}(\rho)\Big|_{\rho(y)=\sum_{j=\overline{1,n}}\delta(y-c_j)} = \\ &= \textstyle\sum_{j=\overline{1,n}}\frac{1}{2}[-i\delta(x-c_j)\nabla_{c_j} + i\nabla_{c_j}\circ\delta(x-c_j)]\tilde{f}_n^{(1/2)}(c_1,c_2,...,c_n), \end{aligned} \tag{217}$$

where we took into account that $\bar{\varphi}_0^{(1/2)}(\rho) = 1$ for all densities $\rho : \Phi \to \Phi$ and have put, by definition, the Fourier transform

$$\tilde{f}_n^{(1/2)}(c_1, c_2, ..., c_n) := \int_{\mathbb{R}^{m\times n}} \prod_{j=\overline{1,n}} dp_j f_n^{(1/2)}(p_1, p_2, ..., p_n) \exp(i \sum_{j=\overline{1,n}} \langle p_j | c_j \rangle) \tag{218}$$

for any fixed particle position vectors $c_j \in \mathbb{R}^n, j = \overline{1,n}$, and for arbitrary $n \in \mathbb{N}$. The expression (217), in particular, means that the current operator $J^{(1/2)}(x) : \Phi_\rho \to \Phi_\rho$ is symmetric with respect to the measure $d\mu_n^{(1/2)} := \beta_n \prod_{j=\overline{1,n}} dc_j$ on each functional submanifold $\mathcal{M}^n$ for all $n \in \mathbb{N}$, where the constants $\beta_n \in \mathbb{R}_+, n \in \mathbb{N}$, can be determined from the normalization condition $||\varphi_n^{(1/2)}(\rho)||_{\Phi_n^{(1/2)}} = (\varphi_n^{(1/2)} | \varphi_n^{(1/2)})^{1/2}_{\Phi_n^{(1/2)}}, n \in \mathbb{N}$. The latter gives rise [1,4,57,68,71,72,139] to the following symbolic measure expression

$$d\mu_n^{(1/2)} := \prod_{x\in\mathbb{R}^m} \delta\left(\rho(x) - \sum_{j=\overline{1,n}} \delta(x - c_j)\right) \prod_{j=\overline{1,n}} \frac{dc_j}{(2\pi)^m} \tag{219}$$

for all $c_j \in \mathbb{R}^n, j = \overline{1,n}$, and arbitrary $n \in \mathbb{N}$.

Remark 8. *As was aptly observed in [72], the choice $\sigma = 1/2$ makes it possible to realize the current algebra representation on the space $\mathcal{M}$ of analytic functions, which will be a priori assumed for further, that is the corresponding measure can be symbolically expressed as*

$$d\mu_n := \prod_{x\in\mathbb{R}^m} \delta\left(\rho(x) - \sum_{j=\overline{1,n}} \delta(x - c_j)\right) \prod_{j=\overline{1,n}} \frac{dc_j}{(2\pi)^m} \tag{220}$$

on the subspace $\mathcal{M}_n$ for any $n \in \mathbb{N}$.

5.2. The Current Algebra Representation and Hamiltonian Reconstruction: A Many-Dimensional Quantum Oscillator Model

As a classical application of the construction above, one can consider a density current algebra representation of the quantum Hamiltonian operator

$$\mathrm{H}^{(\omega)} = \frac{1}{2}\int_{\mathbb{R}^m} \langle K(x)^+ | \rho(x)^{-1} K(x) \rangle dx + \frac{1}{2}\int_{\mathbb{R}^m} \langle \omega x | \omega x \rangle \rho(x) dx \tag{221}$$

on the corresponding representation Hilbert space Φ_ρ of the generalized quantum N-particle oscillatory Hamiltonian

$$H_N^{(\omega)} = \frac{1}{2} \sum_{j=\overline{1,N}} \left(\langle \nabla_{x_j} | \nabla_{x_j} \rangle + \langle \omega x_j | \omega x_j \rangle \right) \tag{222}$$

for $N \in \mathbb{Z}_+$ bose-particles in the m-dimensional space $\mathbb{R}^m$ under the external oscillatory potential, parametrized by the positive definite frequency matrix $\omega \in End\, \mathbb{R}^m$.

Having shifted the representation Hilbert space Φ_ρ by the functional $\bar{\varphi}_0^{(1/2)}(\rho) := \exp[-\frac{1}{2}\int_{\mathbb{R}^m} \langle x | \omega x \rangle \rho(x) dx] \in \Phi_\rho$, the corresponding current operator (209) becomes

$$J^{(\omega)}(x) \;=\; -i\rho(x)\nabla\frac{\delta}{\delta\rho(x)} + \frac{i}{2}\nabla\rho(x) - i\omega x\rho(x), \tag{223}$$

simultaneously entailing the related K-operator changing

$$K(x) = \rho(x)\nabla\frac{\delta}{\delta\rho(x)} \to K^{(\omega)}(x) = \rho(x)\nabla\frac{\delta}{\delta\rho(x)} + \omega x\rho(x) \tag{224}$$

for any $x \in \mathbb{R}^m$. The latter gives rise, respectively, to the following equivalent current algebra functional representation of the oscillatory Hamiltonian (221):

$$\mathrm{H}^{(\omega)} = \frac{1}{2}\int_{\mathbb{R}^m} \langle K^{(\omega)}(x)^{+} | \rho(x)^{-1} K^{(\omega)}(x) \rangle dx + \frac{1}{2}\mathrm{tr}\omega \int_{\mathbb{R}^m} \rho(x) dx \tag{225}$$

for any positive defined matrix $\omega \in End\,\mathbb{R}^m$. The shifted current operator (223) makes it possible to construct the suitably deformed free particle measure

$$d\mu_1^{(\omega)}(\rho) := \exp\Big(-\int_{\mathbb{R}^m} dx \rho(x)\langle x|\omega x\rangle\Big) d\mu_1^{(1/2)}(\rho) \tag{226}$$

on the one-particle functional manifold $\mathcal{M}_1$, for which the following expression

$$(\Omega|\mathrm{H}^{(\omega)}|U(\mathrm{f})|\Omega) = \int_{\mathcal{M}} \exp[i\rho(\mathrm{f})] d\mu_1^{(\omega)}(\rho) \tag{227}$$

holds for any test function $\mathrm{f} \in F$. The latter, jointly with the related ground state condition $|\Omega) = 1 \in \Phi_\rho$, makes it possible to easily calculate the scalar product elements

$$(U(\mathrm{f}_1)\Omega|\mathrm{H}^{(\omega)}|U(\mathrm{f}_2)|\Omega) = \int_{\mathbb{R}^m} \exp[i\mathrm{f}_1(c) + i\mathrm{f}_2(c)] \exp(-\langle c|\omega c\rangle) \frac{dc}{(2\pi)^m} \tag{228}$$

for any test functions $\mathrm{f}_1, \mathrm{f}_2 \in F$. The expression (228) makes it possible to successfully calculate the matrix elements $(\rho(\mathrm{f}_{p_1})\Omega|\mathrm{H}^{(\omega)}|\rho(\mathrm{f}_{p_2})|\Omega)$ of the Hamiltonian $\mathrm{H}^{(\omega)} : \Phi_\rho \to \Phi_\rho$ on the corresponding eigenvectors $\rho(\mathrm{f}_p)|\Omega) \in \Phi_\rho$ for arbitrary $p = p_1, p_2 \in \mathbb{N}$ and, therefore, to find its spectrum.

Consider now the operator (55), taking into account the analytical current representation (216) at $\sigma = 1/2$:

$$\begin{aligned} K(x)\varphi_n^{(1/2)}(\rho) = {} & [\rho(x)\nabla\tfrac{\delta}{\delta\rho(x)} - 1/2\nabla\rho(x)]\varphi_n^{(1/2)}(\rho) + \\ & + 1/2\nabla\rho(x)\varphi_n^{(1/2)}(\rho) = \rho(x)\nabla\tfrac{\delta}{\delta\rho(x)}\varphi_n^{(1/2)}(\rho) \end{aligned} \tag{229}$$

for any $n \in \mathbb{N}$. Having substituted instead of $\varphi_n^{(1/2)}(\rho) \in \Phi_\rho^{(n)}, n \in \mathbb{N}$, the ground state eigenfunction $\Omega(\rho) = 1 \in \Phi_\rho$, we can easily retrieve the before derived expression (61). Moreover, based on the representation (224) and the definition (54), one can calculate that

$$\begin{aligned} K^{(\omega)}(x)\bar{\varphi}_0^{(1/2)}(\rho) = {} & \Big[\rho(x)\nabla\frac{\delta}{\delta\rho(x)} + \omega x \rho(x)\Big]\bar{\varphi}_0^{(1/2)}(\rho) = 0 = \\ = {} & \mathrm{A}^{(\omega)}(x;\rho)\bar{\varphi}^{(1/2)}(\rho), \end{aligned} \tag{230}$$

where $\bar{\varphi}_0^{(1/2)}(\rho) = \exp[-\frac{1}{2}\int_{\mathbb{R}^m}\langle x|\omega x\rangle\rho(x)dx] \in \Phi_\rho^{(1/2)} \simeq \Phi_\rho$. The latter means, in particular, that the corresponding multiplication operator $\mathrm{A}^{(\omega)}(x;\rho) = 0$, or, respectively,

$$K(x)\bar{\varphi}_0^{(1/2)}(\rho) := \mathrm{A}(x;\rho)\bar{\varphi}_0^{(1/2)}(\rho) = -\omega x \rho(x)\bar{\varphi}_0^{(1/2)}(\rho), \tag{231}$$

where $\bar{\varphi}_0^{(1/2)}(\rho) := |\Omega(\rho)) \in \Phi_\rho$ is the corresponding ground state vector in Φ_ρ for the oscillatory Hamiltonian operator (222). Making use of the operator (226), based on expression (64), one can present a special solution to the functional Equation (63) in the form

$$\mathcal{L}(\mathrm{f}) = \exp\Big(-\int_{\mathbb{R}^m} dx \langle \omega x|x\rangle \frac{1}{2i}\frac{\delta}{\delta \mathrm{f}(x)}\Big) \exp\Big(\bar{\rho}\int_{\mathbb{R}^m}\{\exp[i\mathrm{f}(x)] - 1\}dx\Big), \tag{232}$$

confirming similar statements from [5,6,77].

5.3. Conclusions

In this Section, we have reviewed the development and applications of an effective algebraic scheme of constructing density operator and density functional representations for the local quantum current algebra and its application to quantum Hamiltonian and symmetry operators reconstruction. We analyzed the corresponding factorization structure for quantum Hamiltonian operators, spatially governing many- and one-dimensional integrable dynamical systems. The quantum generalized oscillatory and Calogero–Moser–Sutherland models of spin-less bose-particles were analyzed in detail. The central vector of the density operator current algebra representation proved to be the ground vector state of the corresponding completely integrable factorized quantum Hamiltonian system in the classical Bethe anzatz form. The latter makes it possible to quantum classify completely integrable Hamiltonian systems a priori, allowing the factorized form and those whose groundstate is of the Bethe anzatz from. These and related aspects of the factorized and completely integrable quantum Hamiltonians systems are planned to be studied in other places.

6. The Quantum Current Algebra Quasi-Classical Representations and the Collective Variable Approach in Equilibrium Statistical Physics

Introductory Notes

We consider a large system of $N \in \mathbb{N}$ (one-atomic and spinless) bose-particles with a fixed density $\bar{\rho} := N/\Lambda$ in a volume $\Lambda \subset \mathbb{R}^3$, which is specified by a quantum-mechanical Hamiltonian operator $\hat{H} : L_2^{(sym)}(\mathbb{R}^{3N};\mathbb{C}) \to L_2^{(sym)}(\mathbb{R}^{3N};\mathbb{C})$ of the form:

$$\hat{H} := -\frac{\hbar^2}{2m}\sum_{j=1}^{N}\nabla_j^2 + \sum_{j<k}^{N} V(x_j - x_k), \tag{233}$$

where $\nabla_j := \partial/\partial x_j$, $j = \overline{1,N}$, $\hbar$—the Planck constant, $m \in \mathbb{R}_+$—a particle mass and $V(x-y) := V(|x-y|)$, $x,y \in \Lambda$,—a two-particle potential energy, allowing a partition $V = V^{(l)} + V^{(s)}$, where $V^{(s)}$—a short range potential of the Lennard–Johns type and $V^{(l)}$—a long range potential of the Coulomb type. Making use of the second quantization representation [13,22,56,59,63,140,141], the Hamiltonian (233) as $\Lambda \to \mathbb{R}^3$ and $N \to \infty$ can be written as a sum $\mathrm{H} = \mathrm{H}_0 + \mathrm{V}$, where

$$\begin{aligned} \mathrm{H}_0 &:= -\frac{\hbar^2}{2m}\int_{\mathbb{R}^3} d^3x \psi^+ \nabla_x^2 \psi, \\ \mathrm{V} &:= \frac{1}{2}\int_{\mathbb{R}^3} d^3x \int_{\mathbb{R}^3} d^3y V(x-y)\psi^+(x)\psi^+(y)\psi(y)\psi(x), \end{aligned} \tag{234}$$

and the operator $\mathrm{H} : \Phi_F \to \Phi_F$ acts on the corresponding Fock space Φ_F and $\psi^+(x), \psi(y)$: $\Phi_F \to \Phi_F$ are the creation and annihilation operators at points $x \in \mathbb{R}^3$ and $y \in \mathbb{R}^3$. Assume now that our particle system is in a thermodynamically equilibrium state at an *"inverse"* temperature $\mathbb{R}_+ \ni \beta \to \infty$. Assume also that this equilibrium state is compatible with the respectively constructed quantum current algebra $\mathcal{G}$ representation in a separable Hilbert space Φ_μ [4,6,56,68,141], whose generating cyclic vector $\Omega \in \Phi_\mu$ realizes the ground state of the Hamiltonian operator $\mathrm{H} : \Phi_\mu \to \Phi_\mu$. Then, the corresponding n-particles distribution functions can be written down [56,86,142] as

$$f_n(x_1, x_2, ..., x_n) := (\Omega| : \rho(x_1)\rho(x_1)...\rho(x_n) : \Omega), \tag{235}$$

where $n \in \mathbb{N}$, $\rho(x) : \Phi_\mu \to \Phi_\mu, x \in \mathbb{R}^3$—the density operator acting on the Hilbert space Φ_μ and $: \cdot :$—the related Wick normal ordering, naturally ensued from that defined over the creation and annihilation operators, and $\Omega \in \Phi_\mu$ is the ground state of the Hamiltonian

(234) at the temperature $\beta \to \infty$, normed by the stability condition $(\Omega|\Omega) = 1$. Having introduced the corresponding Bogolubov generating functional

$$\mathcal{L}(\mathrm{f}) := (\Omega|\exp[i\rho(\mathrm{f})]\Omega) \tag{236}$$

for any "test" Schwartz function $\mathrm{f} \in F \simeq \mathcal{S}(\mathbb{R}^3;\mathbb{R})$, where $\rho(\mathrm{f}) := \int_{\mathbb{R}^3} d^3x\mathrm{f}(x)\rho(x)$, then for the n-particle distribution functions (235) one can get the expression

$$f_n(x_1,x_2,...,x_n) =: \frac{1}{i}\frac{\delta}{\delta\mathrm{f}(x_1)}\frac{1}{i}\frac{\delta}{\delta\mathrm{f}(x_2)}...\frac{1}{i}\frac{\delta}{\delta\mathrm{f}(x_n)} : \mathcal{L}(\mathrm{f})|_{\mathrm{f}=0}. \tag{237}$$

Here $x_j \in \mathbb{R}^3$, $j = \overline{1,n}$, $n \in \mathbb{N}$, and the symbol ": $\frac{1}{i}\frac{\delta}{\delta\mathrm{f}(x_1)}\frac{1}{i}\frac{\delta}{\delta\mathrm{f}(x_2)}...\frac{1}{i}\frac{\delta}{\delta\mathrm{f}(x_n)}$:" imitates the normal ordering symbol ": :" action on operator expressions $\rho(x_1)\rho(x_1)...\rho(x_n)$, that is

$$\begin{aligned} &: \frac{1}{i}\frac{\delta}{\delta\mathrm{f}(x_1)} := \frac{1}{i}\frac{\delta}{\delta\mathrm{f}(x_1)}, \\ &: \frac{1}{i}\frac{\delta}{\delta\mathrm{f}(x_1)}\frac{1}{i}\frac{\delta}{\delta\mathrm{f}(x_2)} := \frac{1}{i}\frac{\delta}{\delta\mathrm{f}(x_1)}[\frac{1}{i}\frac{\delta}{\delta\mathrm{f}(x_2)} - \delta(x_1-x_2)], \end{aligned} \tag{238}$$

and so on. Consider now the expression (236) at some $\beta \in \mathbb{R}_+$, making use of the statistical operator $P: \Phi_\mu \to \Phi_\mu$ and the *"shifted"* 'Hamiltonian $\mathrm{H}^{(\lambda)} := \mathrm{H} - \lambda\int_{\mathbb{R}^3} d^3x\rho(x)$ with $\lambda \in \mathbb{R}$ being a suitable *"chemical"* potential:

$$\mathcal{L}(\mathrm{f}) := \mathrm{Tr}(P\exp[i\rho(\mathrm{f})]), \quad P := \frac{\exp(-\beta\mathrm{H}^{(\mu)})}{\mathrm{Tr}\exp(-\beta\mathrm{H}^{(\mu)})}, \tag{239}$$

where "Tr" means the operator trace-operation on the Hilbert space Φ_μ. Keeping in mind within the task of studying distribution functions (235) in the classical statistical mechanics case, we need to calculate the trace in (239) as $\hbar \to 0$. The latter gives rise to the following expressions:

$$\begin{aligned} &\mathcal{L}(\mathrm{f}) = \mathrm{Z}(\mathrm{f})/\mathrm{Z}(0), \quad \mathrm{Z}(\mathrm{f}) := \exp[-\beta\mathrm{V}(\delta)]\mathcal{L}_0(\mathrm{f}), \\ &\mathcal{L}_0(\mathrm{f}) = \exp(\varsigma\int_{\mathbb{R}^3} d^3x\{\exp[i\mathrm{f}(x)] - 1\}), \end{aligned} \tag{240}$$

where $\varsigma := \exp(\beta\lambda)(2\pi\hbar^2\beta m)^{-3/2}$ is the system *"activity"* [56,142], and

$$\mathrm{V}(\delta) := \frac{1}{2}\int_{\mathbb{R}^3} d^3x\int_{\mathbb{R}^3} d^3yV(x-y) : \frac{1}{i}\frac{\delta}{\delta\mathrm{f}(x)}\frac{1}{i}\frac{\delta}{\delta\mathrm{f}(y)} : . \tag{241}$$

Based on expressions (240) and (241) we can formulate the following proposition.

Proposition 8. *The functional (236) satisfies [20,56,86] the following functional Bogolubov type equation:*

$$\begin{aligned} &[\nabla_x - i\nabla_x\mathrm{f}(x)]\frac{1}{i}\frac{\delta\mathcal{L}(\mathrm{f})}{\delta\mathrm{f}(x)} \\ &= -\beta\int_{\mathbb{R}^3} d^3y\nabla_x V(x-y) : \frac{1}{i}\frac{\delta}{\delta\mathrm{f}(x)}\frac{1}{i}\frac{\delta}{\delta\mathrm{f}(y)} : \mathcal{L}(\mathrm{f}), \end{aligned} \tag{242}$$

with the expression (240) being its exact functional-analytic solution.

Below, we will proceed to constructing effective analytic tools allowing the exact functional-analytic solutions to the Bogolubov functional Equation (242) to be found, describing equilibrium many-particle dynamical systems, as well as generalizing the obtained results for the case of non-equilibrium dynamical many particle systems.

7. The Bogolubov-Zubarev "Collective" Variables Transform

Taking into account the two-particle potential energy partition $\mathrm{V} = \mathrm{V}^{(s)} + \mathrm{V}^{(l)}$, owing to the representation (240) one can easily write down the following expression for generating functional $Z(\mathrm{f}), \mathrm{f} \in F$:

$$Z(\mathrm{f}) = \exp[-\beta \mathrm{V}^{(s)}(\delta)]\mathcal{L}^{(l)}(\mathrm{f}), \quad \mathcal{L}^{(l)}(\mathrm{f}) := \exp[-\beta \mathrm{V}^{(l)}(\delta)]\mathcal{L}_0(\mathrm{f}), \tag{243}$$

where we put

$$\begin{aligned} \mathrm{V}^{(l)}(\delta) &:= \frac{1}{2}\int_{\mathbb{R}^3} d^3x \int_{\mathbb{R}^3} d^3y V^{(l)}(x-y) : \frac{1}{i}\frac{\delta}{\delta \mathrm{f}(x)}\frac{1}{i}\frac{\delta}{\delta \mathrm{f}(y)} : , \\ \mathrm{V}^{(s)}(\delta) &:= \frac{1}{2}\int_{\mathbb{R}^3} d^3x \int_{\mathbb{R}^3} d^3y V^{(s)}(x-y) : \frac{1}{i}\frac{\delta}{\delta \mathrm{f}(x)}\frac{1}{i}\frac{\delta}{\delta \mathrm{f}(y)} : \quad . \end{aligned} \tag{244}$$

Needing to calculate the functional $\mathcal{L}^{(l)}(\mathrm{f}), \mathrm{f} \in F$, corresponding to the long range part $\mathrm{V}^{(l)}$ of the full potential energy $\mathrm{V} : \Phi_\mu \to \Phi_\mu$, we will apply the analogue of Bogolubov–Zubarev [143,144] "collective" variables transform within the grand canonical ensemble, suggested before in [20,68,145,146]. Namely, denote by $\mathcal{L}^{(l)}_{(n)}(\mathrm{f}), n \in \mathbb{N}$,—a partial solution to the functional Equation (242), possessing exactly $n \in \mathbb{N}$ particles. Then, owing to the results of [86], for $\mathcal{L}^{(l)}_{(n)}(\mathrm{f}), n \in \mathbb{N}$, there holds the following exact expression:

$$\mathcal{L}^{(l)}_{(n)}(\mathrm{f}) = \int_{\mathbb{R}^3} d^3x_1 \int_{\mathbb{R}^3} d^3x_2 ... \int_{\mathbb{R}^3} d^3x_n \prod_{j=1}^{n} \exp[i\mathrm{f}(x_j)] \exp(-\beta V_n^{(l)}), \tag{245}$$

where $V_n^{(l)}$—the long term part potential energy of an $n-$particle group of the system. Then we get that

$$\mathcal{L}^{(l)}(\mathrm{f}) := \sum_{n \in \mathbb{Z}_+} \frac{z^n}{n!}\mathcal{L}^{(l)}_{(n)}(\mathrm{f})Q_{0,}^{-1} \quad Q_0 := (\sum_{n \in \mathbb{Z}_+} \frac{z^n}{n!}\mathcal{L}^{(l)}_{(n)}(0))^{-1}. \tag{246}$$

The sum in (246) can be calculated exactly, taking into account the expression

$$\mathcal{L}^{(l)}_{(n)}(\mathrm{f}) = \int \mathcal{D}(\omega)\{z \int_{\mathbb{R}^3} d^3x \exp[i\mathrm{f}(x)]g(x;\omega)\}^n J(\omega), \tag{247}$$

where $\mathcal{D}(\omega) := \prod_{k \in \mathbb{R}^3} \frac{i}{2}(d\omega_k^* \wedge d\omega_k),\ \omega_k^* := \omega_{-k} \in \mathbb{C}, k \in \mathbb{R}^3$,

$$\begin{aligned} g(x;\omega) &:= \exp\left[-2\pi i (\int_{\mathbb{R}^3} d^3k \omega_k \exp(ikx) + \frac{\beta}{2}\int_{\mathbb{R}^3} d^3k v(k)\right], \\ J(\omega) &:= \exp\left[-\int_{\mathbb{R}^3} d^3k \frac{2\pi^2}{\beta v(k)}\omega_k \omega_{-k} + \int_{\mathbb{R}^3} d^3k \ln\frac{\pi}{\beta v(k)}\right] \end{aligned} \tag{248}$$

and $v(k) := (2\pi)^{-3}\int_{\mathbb{R}^3} d^3x V^{(l)}(x)\exp(-ikx), k \in \mathbb{R}^3$. Now, from (246)–(248) one easily finds that

$$\mathcal{L}^{(l)}(\mathrm{f}) = \int \mathcal{D}(\omega)\exp(\bar{z}\int_{\mathbb{R}^3} d^3x\{\exp[i\mathrm{f}(x)] - 1\}g(x;\omega))J^{(l)}(\omega)Q^{-1}, \tag{249}$$

where $\bar{\varsigma} := \varsigma \exp(\frac{\beta}{2}\int_{\mathbb{R}^3} d^3k\nu(k)) = \varsigma \exp[\frac{\beta}{2}V^{(l)}(0)]$ and the function $J^{(l)}(\omega), \omega \in \mathbb{R}^3$, allows the following series expansion:

$$J^{(l)}(\omega) := J(\omega)\exp\left[\int_{\mathbb{R}^3} d^3x g(x;\omega)\right] = J(\omega)\exp\left[-\frac{(2\pi)^2}{2!}(2\pi)^3\int_{\mathbb{R}^3} d^3k\omega_k\omega_{-k}\right. \tag{250}$$
$$\left. + \sum_{n\neq 2}\frac{(-2\pi i)^n}{n!}(2\pi)^3\int_{\mathbb{R}^3} d^3k_1\int_{\mathbb{R}^3} d^3k_2 ... \int_{\mathbb{R}^3} d^3k_n\prod_{j=1}^{n}\omega_{k_j}\delta\left(\sum_{J=1}^{N}k_j\right)\right].$$

The expression (249) can now be represented [22,115,117,118] in the following cluster Ursell form:

$$\mathcal{L}^{(l)}(\mathrm{f}) = \exp\left(\sum_{n=1}^{\infty}\frac{\bar{z}^n}{n!}\int_{\mathbb{R}^3} d^3x_1\int_{\mathbb{R}^3} d^3x_2 ... \int_{\mathbb{R}^3} d^3x_n\prod_{j=1}^{n}\{\exp[i\mathrm{f}(x)] - 1\}g_n(x_1, x_2, ..., x_n)\right). \tag{251}$$

Here for any $n \in \mathbb{Z}_+$

$$g_n(x_1, x_2, ..., x_n) := \sum_{\sigma[n]}(-1)^{m+1}(m-1)!\prod_{j=1}^{m}R_{\sigma[j]}(x_k \in \sigma[j]),$$
$$R_n(x_1, x_2, ..., x_n) := \sum_{\sigma[n]}\prod_{j=1}^{m}g_{\sigma[j]}(x_k \in \sigma[j]), \tag{252}$$

where $g_n(x_1, x_2, ..., x_n)$, $n \in \mathbb{N}$, are called the $n-$particle Ursell cluster functions, $R_n(x_1, x_2, ..., x_n)$, $n \in \mathbb{N}$, are suitable "correlation" functions [20,56,68] and $\sigma[n]$ denotes a partition of the set $\{1, 2, ..., n\}$ into non-intersecting subsets $\{\sigma[j] : j = \overline{1, m}\}$, that is $\sigma[j] \cap \sigma[k] = \varnothing$ for $j \neq k = \overline{1, m}$, and $\sigma[n] = \cup_{j=1}^{m}\sigma[j]$. Having separated from the function $J^{(l)}(\omega), \omega \in \mathbb{C}^3$, the natural "Gaussian" part $J_0^{(l)}(\omega), \omega \in \mathbb{C}^3$, one can write down that

$$g_1(x_1) = G(\xi_k^{(1)})/G(0), \ g_2(x_1, x_2) = G(\xi_k^{(2)})/G(0) - g_1(x_1)g_1(x_2), ..., \tag{253}$$

where $\xi_k^{(n)} := -2\pi i\sum_{s=1}^{n}\exp(ikx_s), \ k \in \mathbb{R}^3, n \in \mathbb{N}$,

$$G(\xi_k^{(n)}) := \exp[\mathrm{M}(\xi_k^{(n)})]\int D(\omega)g^{(l)}(\xi_k^{(n)};\omega)J_0(\omega),$$
$$\mathrm{M}(\xi_k^{(n)}) := \sum_{m\neq 2}\frac{(-2\pi i)^m}{m!}(2\pi)^3\int_{\mathbb{R}^3} d^3k_1\int_{\mathbb{R}^3} d^3k_2 ... \int_{\mathbb{R}^3} d^3k_m\delta\left(\sum_{s=1}^{m}k_s\right)\prod_{s=1}^{m}\frac{\delta}{\delta\xi_{k_s}^{(n)}},$$
$$g^{(l)}(\xi_k^{(n)};\omega) := \prod_{j=1}^{n}g(x_j;\omega). \tag{254}$$

Since the integrals $\int\mathcal{D}(\omega)g^{(l)}(\xi_k^{(n)};\omega)J^{(l)}(\omega)$, $n \in \mathbb{N}$, one can calculate exactly, the formulae (251) and (253) are sources of the so called "virial" variables for Ursell–Mayer "cluster" correlation functions $g_n(x_1, x_2, ..., x_n)$, $n \in \mathbb{N}$, having important applications. In particular, from the function $J^{(l)}(\omega), \omega \in \mathbb{C}^3$, one gets right away that the cluster expansion for the functions $g_n(x_1, x_2, ...x_n)$, $n \in \mathbb{N}$, are fulfilled by means of the "screened" potential function $\bar{V}^{(l)}(x - y)$, $x, y \in \mathbb{R}^3$, where

$$\bar{V}^{(l)}(x - y) := \int_{\mathbb{R}^3} d^3k\frac{\nu(k)\exp[ik(x - y)]}{1 + \nu(k)\beta\bar{z}(2\pi)^3}. \tag{255}$$

In particular, from (237) and (251) one easily finds that

$$
\begin{aligned}
f_1(x_1) &= z\int \mathcal{D}(\omega) g(x;\omega) J^{(l)}(\omega)\left[\int \mathcal{D}(\omega) J^{(l)}(\omega)\right]^{-1} = \\
&= \bar{\rho} \simeq \bar{z}\exp\left[\frac{\beta}{2}\int d^3k \frac{\beta v^2(k)(2\pi)^3 \bar{z}}{1+v(k)\beta\bar{z}(2\pi)^3}\right], \\
f_2(x_1,x_2) &= z^2\int \mathcal{D}(\omega) g(x_1;\omega) g(x_2;\omega) J^{(l)}(\omega)\left[\int \mathcal{D}(\omega) J^{(l)}(\omega)\right]^{-1} \simeq \\
&\simeq \bar{\rho}^2\exp[-\beta\bar{V}^{(l)}(x_2-x_1)]\left\{1+\bar{\rho}\int_{\mathbb{R}^3} d^3x_3[\exp\left(-\beta\bar{V}^{(l)}(x_1-x_3)\right)-1\right. \\
&+\beta\bar{V}^{(l)}(x_1-x_3)][\exp\left(-\beta\bar{V}^{(l)}(x_2-x_3)\right)-1+\beta\bar{V}^{(l)}(x_2-x_3] \\
&+\bar{\rho}\int_{\mathbb{R}^3} d^3x_3[-\beta\bar{V}^{(l)}(x_1-x_3)][\exp(-\beta\bar{V}^{(l)}(x_2-x_3))-1+\beta\bar{V}^{(l)}(x_2-x_3)] \\
&\left.+\bar{\rho}\int_{\mathbb{R}^3} d^3x_3[-\beta\bar{V}^{(l)}(x_2-x_3)][\exp(-\beta\bar{V}^{(l)}(x_1-x_3))-1+\beta V^{(l)}(x_1-x_3)]+\right\}...
\end{aligned}
\tag{256}
$$

and so on. The result, presented above, can be obtained by means of slightly formal calculations, based on generalized functions and operator theories [22,115,118,147]. Really, as $\hbar \to 0$ one has that

$$
\mathcal{L}^{(l)}(\mathrm{f}) = \exp[-\beta \mathrm{V}^{(l)}(\delta)]\mathcal{L}_0(\mathrm{f})Q^{-1} = \tag{257}
$$

$$
\begin{aligned}
&= tr\left\{\exp(-\beta \mathrm{H}_0^{(\mu)})\exp\left[-\frac{\beta}{2}\int_{\mathbb{R}^3} d^3k v(k) : \rho_k\rho_{-k} :\right]\exp[i\rho(\mathrm{f})]\right\} \\
&= tr\left\{\exp(-\beta \mathrm{H}_0^{(\mu)})\exp\left[\frac{\beta}{2}\int_{\mathbb{R}^3} d^3k v(k)\int_{\mathbb{R}^3} d^3x\rho(x)\right]\right. \\
&\times\left.\int \mathcal{D}(\omega)\exp\left[-\int_{\mathbb{R}^3} d^3k\frac{2\pi^2}{\beta v(k)}\omega_k\omega_{-k}-\int_{\mathbb{R}^3} d^3k 2\pi i\omega_k\rho_k\right]\exp[i\rho(\mathrm{f})]\right\}Q^{-1} \\
&= \int \mathcal{D}(\omega) J(\omega) tr\left\{\exp(-\beta \mathrm{H}_0^{(\mu)})\exp\left[i\left(\rho, \mathrm{f}-2\pi\int_{\mathbb{R}^3} d^3k\omega_k\exp(ikx)-\frac{i\beta}{2}\int_{\mathbb{R}^3} d^3k v(k)\right)\right]\right\}Q^{-1} \\
&= \int \mathcal{D}(\omega) J(\omega)\mathcal{L}_0(\mathrm{f}-2\pi\int_{\mathbb{R}^3} d^3k\omega_k\exp(ikx)-\frac{i\beta}{2}\int_{\mathbb{R}^3} d^3k v(k))Q^{-1} \\
&= \int \mathcal{D}(\omega) J^{(l)}(\omega)\exp\left(\int_{\mathbb{R}^3} d^3k\{\exp[i\mathrm{f}(x)]-1\}g(x;\omega)\right),
\end{aligned}
$$

where $\mathrm{H}_0^{(\mu)} := \mathrm{H}_0 - \lambda\int_{\mathbb{R}^3} d^3x\rho(x)$, $\rho_k := \int_{\mathbb{R}^3} d^3x\rho(x)\exp(ikx)$, $k \in \mathbb{R}^3$. The expression (257) coincides exactly with that of (251), thereby proving the validity of our expressions (240) and (243) for the N.N. Bogolubov type generating functional $\mathcal{L}(\mathrm{f})$, $\mathrm{f} \in F$, satisfying the functional Equation (242) of Proposition (8).

8. The Functional-Analytic Solution and Its Ursell–Mayer Type Diagram Expansion

Having considered (243) and (249) as starting expressions with just known functions $g_n(x_1, x_2, ...x_n)$, $n \in \mathbb{N}$, for the functional $\mathcal{L}(\mathrm{f})$, $\mathrm{f} \in F$, one can obtain the following expansion:

$$\begin{aligned}\mathcal{L}(\mathrm{f}) &= Z(\mathrm{f})/Z(0), \;\; Z(\mathrm{f}) = \exp[-\beta V^{(s)}(\delta)]\mathcal{L}^{(l)}(\mathrm{f}) \\ &= \exp[-\beta V^{(s)}(\delta)] \exp\left[\sum_{n=1}^{\infty} \frac{z^n}{n!} \int_{\mathbb{R}^3} d^3x_1 \int_{\mathbb{R}^3} d^3x_2 ... \int_{\mathbb{R}^3} d^3x_n \right. \\ &\left. \times \prod_{j=1}^{n} \{\exp[i\mathrm{f}(x_j)] - 1\} g_n(x_1, x_2, ..., x_n)\right] \\ &= \exp\left[\sum_{N=1}^{\infty} \frac{1}{N!} W(G_N^{(c)})\right],\end{aligned} \tag{258}$$

where functionals $W(G_N^{(c)})$, $N \in \mathbb{N}$, are calculated via the following rule. Denote by $G_N^{(c)}$, $N \in \mathbb{N}$, such a connected graph that: it consists of exactly N generalized vertices of $[\gamma(n_j)]$ type, $j = \overline{1, N}$, and $\sum_{j=1}^{N} n_j$ ordinary vertices of $[\alpha]$ type. Moreover, each vertex $[y(n)]$ is necessarily connected with n vertices of type $[\alpha]$ by means of dashed lines each to other, and $[\alpha]$ vertices can be connected arbitrarily by means of uniform lines. If now, to attribute to each generalized $[\gamma(n)]-$vertex—the factor $g_n(x_1, x_2, ... x_n)$, to each simple $[\alpha]-$vertex—the factor $\varsigma \int_{\mathbb{R}^3} d^3x \exp[i\mathrm{f}(x)]$, and to the line connecting them—the factor $\{exp[-\beta V^{(s)}(x_{l_1} - x_{l_2})] - 1\}$, then the obtained resulting expression will be exactly equal to the functional $W(G_N^{(c)})$. The final summing up over all such connected graphs gives rise to the expression (257), where the factor $1/N!$ counts the symmetry order of the graph $G_N^{(c)}$ under the generalized vertices permutations. It is evident that, by representing the factor $\exp[i\mathrm{f}(x)]$, entering the vertex $[\alpha]$, as $\{\exp[i\mathrm{f}(x)] - 1\} + 1$, the expression (257) can easily be resumed into Ursell–Mayer type expressions but already with other suitable g_n—functions, replacing the former ones, giving rise to expansions similar to (256), based already on the "screened" potential (255).

Thereby, we can formulate, taking into account the results of [20,68], the next proposition, characterizing the Bogolubov type generating functional $\mathcal{L}(\mathrm{f})$, $\mathrm{f} \in F$, satisfying the functional Equation (242).

Proposition 9. *Let the Bogolubov type generating functional $\mathcal{L}(\mathrm{f})$, $\mathrm{f} \in \mathcal{S}(\mathbb{R}^3; \mathbb{R})$, represented analytically as a series (258) of graph-generated functionals, satisfy the following conditions:*

(i) continuity with respect to the natural topology on F, $|\mathcal{L}(\mathrm{f})| \leq 1$, $\mathrm{f} \in F$;

(ii) positivity: $\sum_{j,k=1}^{n} c_j c_k^* \mathcal{L}(\mathrm{f}_j - \mathrm{f}_k) \geq 0$ for any $\mathrm{f}_j \in F$ and all $c_j \in \mathbb{C}$, $j = \overline{1, n}$, $n \in \mathbb{N}$;

iii) symmetry and normalization conditions: $\mathcal{L}^*(\mathrm{f}) = \mathcal{L}(-\mathrm{f})$ for all $\mathrm{f} \in F$ and $\mathcal{L}(0) = 1$;

(iv) translational-invariance: $\mathcal{L}(\mathrm{f}) = \mathcal{L}(\mathrm{f}_a)$, where $\mathrm{f}_a(x) := \mathrm{f}(x - a)$, $x, a \in \mathbb{R}^3$, for any $\mathrm{f} \in \mathcal{S}(\mathbb{R}^3; \mathbb{R})$;

(v) cluster condition or, equivalently, the Bogolubov correlation decay: $\lim_{\lambda \to \infty} [\mathcal{L}(\mathrm{f} + \mathrm{g}_{\lambda a}) - \mathcal{L}(\mathrm{f}_a)\mathcal{L}(\mathrm{g}_{\lambda a})] = 0$, $a \in \mathbb{R}^3$, for any $\mathrm{f}, \mathrm{g} \in F$;

(vi) density condition: $\frac{1}{i} \frac{\delta \mathcal{L}(\mathrm{f})}{\delta \mathrm{f}(x)}|_{\mathrm{f}=0} = \bar{\rho} \in \mathbb{R}_+$.

Then the functional (258) solves the Bogolubov type functional equation (242), allowing the positive measure $d\bar{\mu}$, whose Fourier representation on the adjoint tempered generalized functions space F' is exactly

$$\mathcal{L}(\mathrm{f}) = \int_{F'} d\bar{\mu}(\xi) \exp[i\xi(\mathrm{f}), \tag{259}$$

where convolution $\xi(\mathrm{f}) := \int_{\mathbb{R}^3} d^3x \xi(x) \mathrm{f}(x)$ for $\xi \in F'$ and $\mathrm{f} \in F$.

The obtained result makes it possible to find the many-particle distribution functions (237) and apply them to constructing different thermodynamic functions important [56,65] for applications.

Below, following the Bogolubov method [86], we obtain, based on the expression (245), the important Kirkwood–Saltzbourg–Simansic functional equation for the Bogolubov generating functional $\mathcal{L}(\mathrm{f}), \mathrm{f} \in F$. Namely, making use of the expression (245) we can write down the following relationship:

$$\frac{1}{i}\frac{\delta \mathcal{L}_{(N+1)}(\mathrm{f})}{\delta \mathrm{f}(x)} = \exp[i\mathrm{f}(x)]\frac{(N+1)Z_N}{Z_{N+1}}\mathcal{L}_{(N)}(\mathrm{f}(\cdot) + i\beta V(\cdot - x)) \tag{260}$$

for any $x \in \mathbb{R}^3$, where $Z_N := \int_{\mathbb{R}^{3N}} dx_1 dx_2 ... dx_N \exp(-\beta V_N)$, $N \in \mathbb{N}$.

Since, by definition, $\lim_{N\to\infty} \mathcal{L}_{(N)}(\mathrm{f}) = \mathcal{L}(\mathrm{f}), \mathrm{f} \in F$, $\lim_{N\to\infty} \frac{(N+1)Z_N}{Z_{N+1}} := \varsigma \in \mathbb{R}_+$, from (260) one gets right away that

$$\exp[-i\mathrm{f}(x)]\frac{1}{i}\frac{\delta \mathcal{L}(\mathrm{f})}{\delta \mathrm{f}(x)} = \varsigma \mathcal{L}(\mathrm{f}(\cdot) + i\beta V(\cdot - x)), \tag{261}$$

which is called the Kirkwood–Saltzburg–Symanzik functional equation, being very important for proving the Proposition (9) by means of the classical Leray–Schauder fixed point theorem [56,141,148] in some suitably defined Banach space. In particular, at $\mathrm{f} = 0$ from (261) one finds the following important relationship:

$$\bar{\rho} = \varsigma \mathcal{L}(i\beta V(\cdot - x)) \tag{262}$$

for any $x \in \mathbb{R}^3$.

Conclusions

In the article, we have showed that the N.N. Bogolubov generating functional method is a very effective tool for studying distribution functions of both equilibrium and non equilibrium states of classical many-particle dynamical systems. In some cases, the N.N. Bogolubov generating functionals can be represented by means of infinite Ursell–Mayer diagram expansions, whose convergence holds under some additional constraints on a statistical system. We also have shown that the Bogolubov idea [56] to use the Wigner density operator transformation to study the non equilibrium distribution functions proved to be very effective, having proposed a new analytic form of non-stationary solutions to the classical N.N. Bogolubov evolution functional equation.

9. The Wigner Type Current Algebra Representation and Its Application to Non-Equilibrium Classical Statistical Mechanics

9.1. Many-Particle Distribution Functions Space and Its Poissonian Structure

In the case of non-stationary (non-equilibrium) states of the many-particle dynamical systems, the Bogolubov's generating functional (236) does not possess all needed information. To specify this case, we introduce the generating representation functional:

$$\mathcal{L}(\mathrm{f}, \mathrm{g}) = (\Omega| \exp[i\rho(\mathrm{f})] \exp[iJ(\mathrm{g})]\Omega) = \mathrm{Tr}(P \exp[i\rho(\mathrm{f})] \exp[i\, J(\mathrm{g})]), \tag{263}$$

where $\Omega \in \Phi_\mu$ is a cyclic vector of the representation of the current group G, satisfying the following additional conditions:

$$\mathrm{T}\rho(\mathrm{f})\mathrm{T}^{-1} = \rho(\mathrm{f}), \quad \mathrm{T}\Omega = \Omega^*, \quad \mathrm{T}J(\mathrm{g})\mathrm{T}^{-1} = -J(\mathrm{g}), \quad \mathrm{T}\mathrm{H}\mathrm{T}^{-1} = \mathrm{H},$$

with the mapping $T : \mathbb{R} \ni t \to -t \in \mathbb{R}$ being the operator of time inversion, and $f \in \mathcal{J}(\mathbb{R}^3; \mathbb{R})$, $\mathrm{g} \in \mathcal{J}(\mathbb{R}^3; \mathbb{R}^3)$ taken arbitrary. In the N-particle representation of the current Lie algebra $\mathcal{G}$ (29) for any finite $N \in \mathbb{N}$ the functional $\mathcal{L}(f, \mathrm{g})$ (263) allows the following [5–7] standard finite-particle form:

$$\begin{aligned}\mathcal{L}(\mathrm{f}, \mathrm{g}) = \int_{\mathbb{R}^3} dx_1 ... \int_{\mathbb{R}^3} dx_N \Omega^*(x_1, ..., x_N) \prod_{j=1}^{N} \exp[i\mathrm{f}(x_j)] \times \\ \times \exp[i\xi(x_j, \mathrm{g})]\Omega(x_1, ..., x_N),\end{aligned} \tag{264}$$

where $\xi(x,g) = \frac{1}{2i}[g(x)\nabla_x + \nabla_x g(x)]$, $x \in \mathbb{R}^3$, and $\Omega \in L_2(\mathbb{R}^{3N};\mathbb{C})$ is a cyclic state. The operator $\exp[i\xi(x,g)]$ acts on any function $\omega_N \in L_2(\mathbb{R}^{3N};\mathbb{C})$ by the rule:

$$\exp[i\xi(x,g)]\omega_N(x_1,...,x_N) = (\phi^*\omega_N)(x_1,...,x_N)\left(\det\left|\left|\frac{\partial\phi(x)}{\partial x}\right|\right|\right)^{1/2}$$

where $\phi \in Diff(\mathbb{R}^3)$ is a diffeomorphism of $\mathbb{R}^3$, corresponding to the vector field $g \in \mathcal{J}(\mathbb{R}^3;\mathbb{R}^3)$, that is $\phi(x) = \phi_t^g$, where $\frac{d}{dt}\phi_t^g = g(\phi_t^g(x))$, $x \in \mathbb{R}^3$. For $N \to \infty$ the expression (264) becomes

$$\begin{aligned}\mathcal{L}(f,g) = \sum_{n\in\mathbb{Z}_+}\frac{1}{n!}\int_{\mathbb{R}^3} dx_1...\int_{\mathbb{R}^3} dx_n\int_{\mathbb{R}^3} dy_1...\int_{\mathbb{R}^3} dy_n\prod_{j=1}^n[\delta(x_j-y_j)\times \\ \times\{\exp[if(x_j)]\exp[i\xi(x_j,g)]-1\}f_n(y_1,...,y_n;x_1,...,x_n)]\end{aligned} \tag{265}$$

where for all $n \in \mathbb{N}$ Bogolubov's quantum distribution functions [56] are

$$f_n(y_1,...,y_n;x_1,...,x_n) = (\Omega|\psi^+(y_n)...\psi^+(y_1)\psi(x_1)...\psi(x_n)\Omega) \tag{266}$$

and satisfy the compatibility conditions

$$f_n(x_1,...,x_n) := f_n(x_1,...,x_n;x_1,...,x_n), \tag{267}$$

where $x_j = \mathbb{R}^3$, $j = \overline{1,n}$, $n \in \mathbb{N}$.

To proceed with the further study of the classical distribution functions of the many-particle dynamical system, when the inverse temperature $\beta \to 0$, and the Planck constant $\hbar \to 0$. Let us introduce [21,22,149,150] the following quantized selfadjoint Wigner operator $w(x,p) : \Phi_W \to \Phi_W$, $(x,p) \in T^*(\mathbb{R}^3)$

$$w(x,p) = \frac{1}{(2\pi)^n}\int_{\mathbb{R}^3} d\alpha \exp(i\langle\alpha|p\rangle)\psi^+\left(x+\frac{\hbar\alpha}{2}\right)\psi\left(x-\frac{\hbar\alpha}{2}\right), \tag{268}$$

where, by definition, $\Phi_W := \lim_{\beta\to\infty}\Phi_\mu$ is the corresponding Hilbert space for the constructed Wigner type current algebra representation with the generating cyclic vector $\Omega \in \Phi_W$. Performing transformation (268) in the expression (263), we can find that

$$\begin{aligned}\mathcal{L}(f,g) \to \mathcal{L}(\tilde{f}) = \sum_{n\in\mathbb{Z}_+}\frac{1}{n!}\int_{\mathbb{R}^3\times\mathbb{R}^3} dx_1dp_1...\times \\ \times\int_{\mathbb{R}^3\times\mathbb{R}^3} dx_ndp_n\prod_{j=1}^n\{\exp(i\tilde{f}(x_j,p_j))-1\}f_n(x_1,p_1;...;x_n,p_n),\end{aligned} \tag{269}$$

for some functions $\tilde{f} \in \mathcal{J}(\mathbb{R}^3\times\mathbb{R}^3;\mathbb{R}^3)$. From the expression (269) it also follows that

$$\mathcal{L}(f) = (\Omega|\exp[iw(f)]\Omega) = \mathrm{Tr}(P\exp[iw(f)]) \tag{270}$$

where $w(f) = \int_{T^*(\mathbb{R}^3)} dxdp w(x,p)f(x,p)$, $\tilde{f} \in \mathcal{J}(\mathbb{R}^3\times\mathbb{R}^3;\mathbb{R}^3)$, $P : \Phi_W \to \Phi_W$ is the Gibbs statistical operator and Tr: $\mathrm{End}(\Phi_W) \to \mathbb{C}$ is the corresponding trace-operator, defined on the space $\mathcal{B}(\Phi_W)$ of the nuclear operators on the corresponding Hilbert space representation Φ_W. The corresponding quantum current Lie algebra $\mathcal{G}$ suitably transforms [55,56] into the Abelian Lie algebra $\mathcal{G}_W$ of the operator functionals $\{w(f) \in \mathcal{G} : f \in \mathcal{J}(\mathbb{R}^3\times\mathbb{R};\mathbb{R})\}$.

Consider now a quantum dynamical system of many identical particles with the average nonvanishing density $\bar{\rho} = \lim_{\Lambda\nearrow\mathbb{R}^3}(N/A) \in \mathbb{R}^1_+\backslash\{0\}$ as $N \to \infty$ and $\Lambda \nearrow \mathbb{R}^3$ in the

Van Hove's sense [150,151]. Then, according to [1,4,59], the Hamiltonian operator (234) in the Wigner representation (268) looks as

$$\mathrm{H} = \int\limits_{T^*(\mathbb{R}^3)} dz \frac{p^2}{2m} w(z) + \int\limits_{T^*(\mathbb{R}^3)} dz \int\limits_{T^*(\mathbb{R}^3)} dz' V(x-y') : w(z)w(z') :, \tag{271}$$

where $z = (x,p) \in T^*(\mathbb{R}^3), z' = (y,q) \in T^*(\mathbb{R}^3)$ and $dz = dxdp$, $dz' = dydq$ are the standard phase space measures in $T^*(\mathbb{R}^3)$ and the ordering $: :$ operation is naturally inherited from (238). According to the Heisenberg's principle [56], the evolution equation with respect to temporal variable $t \in \mathbb{R}_+$ for an arbitrary observable operator quantity $\mathrm{A} : \Phi_W \to \Phi_W$ in the Wigner type representation space Φ_W is

$$d\mathrm{A}/dt = \frac{i}{\hbar}[\mathrm{H}, \mathrm{A}], \tag{272}$$

where $[\cdot,\cdot]$ is a usual operator commutator, naturally ensued from that on the Hilbert space Φ_μ. Following [20,56,152–154], one can state, that for $\hbar \to 0$ in the weak sense the following theorem is true.

Theorem 5. *Let us denote $\mathcal{M}$ as an algebra of the self-adjoint operators with $A(\mathcal{G})$ in the Wigner representation. Then, the operator bracket $[\cdot,\cdot]_0 = \lim\limits_{\hbar\to 0}[\cdot,\cdot]$ on the algebra $\mathcal{M}$ in the weak sense is equivalent to*

$$\begin{aligned} \left[a_j, a_n\right]_0 = \sum_{k=1}^{\min\{j,n\}} \int\limits_{T^*(\mathbb{R}^3)} dz_1 ... \int\limits_{T^*(\mathbb{R}^3)} dz_k : w(z_1)...w(z_k) \times \\ \times \left\{ \frac{\delta^k a_j}{\delta w(z_1)...\delta w(z_k)}, \frac{\delta^k a_n}{\delta w(z_1)...\delta w(z_k)} \right\}^{(k)} :, \end{aligned} \tag{273}$$

where $\{\cdot,\cdot\}^{(k)}$ is a standard canonical Poisson bracket on the phase space of $k \in \mathbb{N}$ particles.

The statement (273) could be proved by means of the next general Bohr–Dirac correspondence principle in the quasi-classical approach:

$$\lim_{\hbar\to 0} \frac{i}{\hbar}[a,b] = \{a,b\}^{(N)}, \tag{274}$$

where $N \in \mathbb{N}$ is a maximal number of the particle in the system and $a, b \in A(\mathcal{G})$ are operators in N-particle Hilbert space representation $\Phi_N = L_2(\mathbb{R}^{3N};\mathbb{C})$, $F = \sum\limits_{j=1}^{N} \delta(x - x_j)$. Here, it is worth making the following corollary.

Corollary. Algebra of the operators of the observable quantities $A(\mathcal{G})$ for $\hbar \to 0$ allows *"hierarchical"* representation

$$A(\mathcal{G}) = \sum_{j\in\mathbb{Z}_+} A_j(\mathcal{G}) \Rightarrow \mathcal{M} = \bigoplus_{j=\mathbb{Z}_+} A_j(\mathcal{G}) \tag{275}$$

along with Lie bracket $[\![\cdot,\cdot]\!]$, which is inducted by the bracket $[\cdot,\cdot]_0$ (273):

$$[\![a,b]\!] = \bigoplus_{l\in\mathbb{Z}_+} \sum_{j,k\in\mathbb{Z}_+} \left[a_j, b_k\right]_0^{(l)}, \tag{276}$$

where $a, b \in \mathcal{M}$ in the Wigner representation and the following expansions hold

$$a = \sum_{j\in\mathbb{Z}_+} a_j, \quad b = \sum_{j\in\mathbb{Z}_+} b_j, \quad [a_j, b_k]_0 = \sum_{l\in\mathbb{Z}_+} [a_j, b_k]_0^{(l)}. \tag{277}$$

Consider now the following linear mapping $\alpha : \mathcal{M} \to A(\mathcal{G})$, where

$$\alpha(\bigoplus_{j\in\mathbb{Z}_+} a_j) = \sum_{j=\mathbb{Z}_+} a_j \in A(\mathcal{G}), \tag{278}$$

and the Lie bracket $[\![\cdot,\cdot]\!]$ is defined in $\mathcal{M}$, and the corresponding Lie bracket $[\cdot,\cdot]_\alpha$ (273) in $A(\mathcal{G})$. Let us consider the dual to (278) mapping $\alpha^* : A(\mathcal{G})^* \to \mathcal{M}^*$, where

$$\mathcal{M}^* = \bigoplus_{l\in\mathbb{Z}_+} \mathcal{M}_j^*, \quad \mathcal{M} = \bigoplus_{l\in\mathbb{Z}_+} \mathcal{M}_j, \tag{279}$$
$$\mathcal{M}^* = \sum_{j\in\mathbb{Z}_+} \{P \in A_j(\mathcal{G})^* : \; F(a) = \mathrm{Tr}(Pa), \; a \in A(\mathcal{G})\}.$$

Here $P : \Phi_\mu \to \Phi_\mu$ is statistic operator of the initial dynamical system (271), which satisfy the Heisenberg–Liouville equation

$$dP/dt = \frac{i}{\hbar}[P, \mathrm{H}] \tag{280}$$

for all $t \in \mathbb{R}_+$. The expression (280), according to (274), transforms into the quasi-classical Liouville equation in the Wigner representation.

It is easy to check that for element $F \in A(\mathcal{M})^*$ the expression

$$\alpha^* F = (f_1, ..., f_j, ...) = \mathcal{F} \in \mathcal{M}^* \tag{281}$$

defines the representation on the space $\mathcal{M}^*$ of the distribution functions

$$f_j(z_1, ..., z_j) = \mathrm{Tr}(P : w(z_1)...w(z_j) :), \tag{282}$$

where $z_j \in T^*(\mathbb{R}^3)$, $j \in \mathbb{Z}_+$, and for any $a \in \mathcal{M}$

$$a(\mathcal{F}) = \sum_{j\in\mathbb{Z}_+} \int_{T^*(\mathbb{R}^3)} dz_1 ... \int_{T^*(\mathbb{R}^3)} dz_j f_j(z_1, ..., z_j) a_j(z_1, ..., z_j). \tag{283}$$

Let $b(F), c(F) \in D(A(\mathcal{G})^*)$ be linear functionals on $A(\mathcal{G})^*$, then on $D(A(\mathcal{G})^*)$ the standard [59] Lie-Poisson bracket $\{\cdot,\cdot\}_0$ is defined via the rule

$$\{b(F), c(F)\}_0 = F([b, c]_0), \tag{284}$$

where $b, c \in A(\mathcal{G})$ are such that $F(b) = b(F)$, $F(c) = c(F)$, $F \in A^*(\mathcal{G})$. In the same way, the dual Lie–Poisson bracket $\{\{\cdot,\cdot\}\}$ is defined on the set of functionals $D^*(\mathcal{M})$ over the adjoint space $\mathcal{M}$ (279)

$$\{\{b(\mathcal{F}), c(\mathcal{F})\}\} = F([\![b, c]\!]), \tag{285}$$

where $F(b) = b(\mathcal{F})$, $F(c) = c(\mathcal{F})$, $F \in \mathcal{M}^*$.

Definition 16. *It is said that mapping of the Lie algebras $\alpha : \mathcal{M} \to A(\mathcal{G})$ is canonical (or Poissonian [59]), if for all $b(\mathcal{F})$ and $c(\mathcal{F})$ the following equality holds*

$$\alpha^* \{b(\mathcal{F}), c(\mathcal{F})\}_0 = \{\{\alpha^* b(\mathcal{F}), \alpha^* c(\mathcal{F})\}\}, \tag{286}$$

where $\mathcal{F} = \alpha^* F \in \mathcal{M}^*$.

From reasonings presented above we can formulated the following proposition.

Proposition 10. *Let A and $\mathcal{M}$ be two arbitrary Lie algebras and $\alpha : \mathcal{M} \to A$ be a linear mapping. Then dual mapping $\alpha^* : D(A(\mathcal{G})^*) \to D(\mathcal{M}^*)$ is canonical if $\alpha : \mathcal{M} \to A$ is Lie algebras homomorphism.*

As a consequence of the statement above, one derives the next theorem.

Theorem 6. *Dual mapping $\alpha^* : D(A(\mathcal{G})^*) \to D(\mathcal{M}^*)$, which was built by means of the hierarchical Lie algebra of the operators $\mathcal{M}$, is canonical.*

Let us consider the generating functional $\mathcal{L}(\mathrm{f})$, $\mathrm{f} \in \mathcal{J}(T^*(\mathbb{R}^3);\mathbb{R})$, defined by expression (270) in Wigner representation, and apply the developed above algebraic technique to the calculation of the following quantity:

$$\frac{d}{dt}\mathcal{L}(\mathrm{f}) = \lim_{\hbar \to 0} \frac{i}{\hbar} \mathrm{Tr}(P[\mathrm{H}, : \exp[iw(e^{i\mathrm{f}} - 1)]) :]) \tag{287}$$

for the evolution with respect to the temporal parameter $t \in \mathbb{R}$. From (270) one can easily obtain that

$$\frac{d}{dt}\mathcal{L}(\mathrm{f})(\mathcal{F}) = \mathrm{Tr}(P[\mathrm{H}, : \exp(iw(e^{i\mathrm{f}} - 1)) :]) = \alpha^* \{\mathcal{H}(\mathcal{F}), \mathcal{L}(\mathrm{f})(\mathcal{F})\}_0, \tag{288}$$

where for all $\mathcal{F} \in A(\mathcal{G})^*$ the Hamiltonian functional $\mathcal{H}(\mathcal{F}) \in D(A(\mathcal{G})^*)$ is given as

$$\begin{aligned} \mathcal{H}(\mathcal{F}) = \mathrm{Tr}(P\mathrm{H}) = & \int\limits_{T^*(\mathbb{R}^3)} dz T(p) f_1(z) + \\ & + \frac{1}{2} \int\limits_{T^*(\mathbb{R}^3)} dz_1 \int\limits_{T^*(\mathbb{R}^3)} dz_2 V(x_1 - x_2) f_2(z_1, z_2). \end{aligned} \tag{289}$$

Based on (288) and Theorem 6, we immediately obtain the Hamiltonian evolution equation

$$\frac{d}{dt}\mathcal{L}(\mathrm{f})(\mathcal{F}) = \{\{\mathcal{L}(\mathrm{f})(\mathcal{F}), \mathcal{H}(\mathcal{F})\}\}, \tag{290}$$

where $t \in \mathbb{R}$, $\mathcal{L}(\mathrm{f})(\mathcal{F}) = \alpha^* \mathcal{L}(\mathrm{f})(F)$, $\mathcal{H}(\mathcal{F}) = \alpha^* H(F)$ and $\mathcal{F} \in \mathcal{M}^*$ is arbitrary. Thus, the following theorem is stated.

Theorem 7. *The generating Wigner type representation functional $\mathcal{L}(\mathrm{f})(\mathcal{F})$ (270) on the phase space $D(\mathcal{M})$ satisfies the Hamiltonian dynamical system (290) with respect to the Lie–Poisson bracket (285) and Hamiltonian function (289), taken as a smooth functional on $\mathcal{M}^*$.*

Using Equation (290) and formulae (273), (276), we finally get the following non-equilibrium functional Bogolubov's equation [143]

$$\begin{aligned} \frac{d}{dt}\mathcal{L}(\mathrm{f}) = & \int\limits_{T^*(\mathbb{R}^3)} dz \{\frac{1}{i}\frac{\delta\mathcal{L}(\mathrm{f})}{\delta \mathrm{f}(z)}, T(p)\}^{(1)} + \\ & + \frac{1}{2} \int\limits_{T^*(\mathbb{R}^3)} dz_1 \int\limits_{T^*(\mathbb{R}^3)} dz_2 \{: \frac{1}{i}\frac{\delta}{\delta \mathrm{f}(z_1)} \frac{1}{i}\frac{\delta}{\delta \mathrm{f}(z_2)} :, V(x_1 - x_2)\}^{(2)} \mathcal{L}(\mathrm{f}), \end{aligned} \tag{291}$$

where for any $n \in \mathbb{N}$

$$: \frac{1}{i}\frac{\delta}{\delta \mathrm{f}(z_1)} \cdots \frac{1}{i}\frac{\delta}{\delta \mathrm{f}(z_n)} := \prod_{j=1}^{n}\left[\frac{1}{i}\frac{\delta}{\delta \mathrm{f}(z_j)} - \sum_{k=1}^{j}\delta(z_j - z_k)\right] \tag{292}$$

and, by definition, $\{\cdot,\cdot\}^{(j)}$ denotes the standard canonical Poisson bracket on the phase space $T^*(\mathbb{R}^3)^j$ for all $j \in \mathbb{Z}_+$.

Taking into account that for functional $\mathcal{L}(\mathrm{f})$, $\mathrm{f} \in \mathcal{J}(T^*(\mathbb{R}^3);\mathbb{R})$, there exists the unlimited expansion (269):

$$\mathcal{L}(\mathrm{f}) = \sum_{n\in\mathbb{Z}}\frac{1}{n!}\int_{T^*(\mathbb{R}^3)} dz_1 \ldots \int_{T^*(\mathbb{R}^3)} dz_n \prod_{i=1}^{n}\{\exp[i\mathrm{f}(z_j) - 1]\} f_n(z_1,\ldots,z_n), \tag{293}$$

from (291), we obtain the kinetic equations for the hierarchy of the Bogolubov distribution functions [143]:

$$\begin{aligned} \frac{\partial}{\partial t} f_n(z_1,\ldots,z_n) &= \{f_n(z_1,\ldots,z_n), H_n(z_1,\ldots,z_n)\}^{(n)} + \\ &+ \int_{T^*(\mathbb{R}^3)} dz_1 \ldots \int_{T^*(\mathbb{R}^3)} dz_{n+1}\{f_n(z_1,\ldots,z_{n+1}), H_n(z_1,\ldots,z_{n+1}), \sum_{j=1}^{n} V(x_j - x_{n+1})\}^{(n+1)}, \end{aligned} \tag{294}$$

where $z_j \in \mathbb{R}^3$, $j = 1,\ldots,n$, are the coefficients of the n-particle cluster in $\mathbb{R}^3$, $H_n(z_1,\ldots,z_n)$ denotes its corresponding energy:

$$H_n(z_1,\ldots,z_n) = \sum_{j=1}^{n}\frac{p_j^2}{2m} + \frac{1}{2}\sum_{j\neq k=1}^{n} V(x_j - x_k) \tag{295}$$

Thus, the problem of the construction of the kinetic theory by Bogolubov is reduced to finding the special solutions of the unlimited hierarchy of the Equations (294), where the selection criterion is based on Bogolubov's fundamental weakening correlation principle:

$$\lim_{\|\langle n\rangle - \langle m\rangle\| \to \infty} |f_{n+m}(z_1,\ldots,z_{n+m}) - f_n(z_1,\ldots,z_n) f_m(z_{n+1},\ldots,z_{n+m})| \to 0 \tag{296}$$

where $\| \langle n\rangle - \langle m\rangle \| = dist(\{z_i \in T^*(\mathbb{R}^3) : i = 1,\ldots,n\}, \{z_{i+n} \in T^*(\mathbb{R}^3) : i = 1,\ldots,m\})$ is a distance between two clusters with $n \in \mathbb{Z}_+$ and $m \in \mathbb{Z}_+$ numbers of the particles. If a special solution of the hierarchy (289) exists in the functional form

$$f_n(z,\ldots,z_n;t) = f_n(z_1,\ldots,z_n; f_1(z;t)) \tag{297}$$

for all $t \in \mathbb{R}_+$ and $n \in \mathbb{Z}_+$, then the corresponding equation for one-particle distribution function of the system in the external field $V_0 : \mathbb{R}^3 \to$ is the following

$$\frac{\partial}{\partial t} f_1(z;t) + \langle p/m | \nabla_x f_1(z;t)\rangle + \langle \nabla_x V_0(x) | \nabla_p f_1(z;t)\rangle = J(f_1(z;t)), \tag{298}$$

where $J(f_1(z;t))$ is the so called "collision integral" [56,143,149,150,155], and is called the kinetic Boltzmann equation [56,149,156]. Below, we will focus on the such special solutions of the Bogolubov's hierarchy of the Equations (294), using the above developed algebraic method of Bogolubov's generating functional.

9.2. Generating Representation Functional and Its Solution Space Structure

Let us consider Bogolubov's functional Equation (291)

$$\frac{d}{dt}\mathcal{L}(\mathrm{f}) = \int\limits_{T^*(\mathbb{R}^3)} dz \left\{ \frac{1}{i}\frac{\delta\mathcal{L}(\mathrm{f})}{\delta\mathrm{f}(z)}, T(p) \right\}^{(1)} + \\ + \frac{1}{2} \int\limits_{T^*(\mathbb{R}^3)} dz_1 \int\limits_{T^*(\mathbb{R}^3)} dz_2 \left\{ : \frac{1}{i}\frac{\delta}{\delta\mathrm{f}(z_1)} \frac{1}{i}\frac{\delta}{\delta\mathrm{f}(z_2)} : \mathcal{L}(\mathrm{f}), \mathrm{V}(x_1 - x_2) \right\}^{(2)}, \tag{299}$$

generated by the statistical operator evolution

$$P(t, t_0) = \exp\left[\frac{i}{\hbar}(t_0 - t)\mathbb{H}\right] \bar{P} \exp\left[\frac{i}{\hbar}(t - t_0)\mathbb{H}\right] \tag{300}$$

for $t, t_0 \in \mathbb{R}$, solving the Heisenberg evolution equation

$$\frac{dP}{dt} = \frac{i}{\hbar}[P, H], \quad P\Big|_{t=t_0} = \bar{P} \tag{301}$$

for the statistical Gibbs operator $P : \Phi_W \to \Phi_W$ with $\mathrm{Tr}\bar{P} = 1$.

When $\hbar \to 0$ in the Wigner representation, the expression (300), as an explicit solution of the (301), allows the following expansion

$$\mathcal{L}(\mathrm{f}) = \mathrm{Tr}\left(\exp\left[\frac{i}{\hbar}(t_0 - t)\mathrm{H}\right] \bar{P} \exp\left[\frac{i}{\hbar}(t - t_0)\mathrm{H}\right] \exp[iw(\mathrm{f})] \right) = \\ = \mathrm{Tr}\left(\exp\left[\frac{i}{\hbar}(t_0 - t)(\mathrm{H}_0 + \mathrm{V})\right] \bar{P} \exp\left[\frac{i}{\hbar}(t - t_0)(\mathrm{H}_0 + \mathrm{V})\right] \exp[iw(\mathrm{f})] \right)\Big|_{\hbar \to 0} = \\ = \mathrm{Tr}\left(\exp\left[\frac{i}{\hbar}(t_0 - t)\mathrm{H}_0\right] \bar{P} \exp\left[\frac{i}{\hbar}(t - t_0)\mathrm{H}_0\right] \exp[\pi(t, t_0)] \exp[iw(\mathrm{f})] \right), \tag{302}$$

where we denoted $\mathrm{H} = \mathrm{H}_0 + \mathrm{V}$,

$$\mathrm{H}_0 = \int\limits_{T^*(\mathbb{R}^3)} dz \frac{p^2}{2m} w(z), \mathrm{V} = \int\limits_{T^*(\mathbb{R}^3)} dz_1 \int\limits_{T^*(\mathbb{R}^3)} dz_2 V(x_1 - x_2) : w(z_1) w(z_2) :, \\ \exp[\pi(t, t_0)] = P_0(t_0, t) P(t, t_0), \quad P_0(t_0, t) = \exp\left[\frac{i}{\hbar}(t_0 - t)\mathrm{H}_0\right] \bar{P} \exp\left[\frac{i}{\hbar}(t - t_0)\mathrm{H}_0\right]. \tag{303}$$

The operator $\pi(t, t_0)$, $t, t_0 \in \mathbb{R}$, in (303) is called a "cluster operator" and allows the next expansion into the unlimited series:

$$\pi(t_0, t) = \sum_{n \in \mathbb{Z}_+} \frac{1}{n!} \int\limits_{T^*(\mathbb{R}^3)} dz_1 ... \int\limits_{T^*(\mathbb{R}^3)} dz_n \, \pi_n(z_1, ..., z_n; t, t_0) \times \\ \times : w(z_1) ... w(z_n) : \stackrel{def}{=} \pi(t, t_0; w), \tag{304}$$

where the functions $\pi_n(z_1, ..., z_n; t, t_0)$, $n \in \mathbb{N}$, can be defined uniquely form the representation (303) under the condition that the Gibbs operator $\bar{P} : \Phi_W \to \Phi_W$ is defined explicitly in the Wigner representation. Thus, from (302)–(304) we obtain the following expressions for the Bogolubov's generating functional

$$\mathcal{L}(f) = \mathrm{Tr}(P_0 \exp[\pi(t,t_0;w)]\exp(iw))\Big|_{\hbar\to 0} = \\ = \exp\left[\pi\left(t,t_0;\frac{1}{i}\frac{\delta}{\delta \mathrm{f}}\right)\right]\mathrm{Tr}(P_0\exp[iw(\mathrm{f})]) = \exp\left[\pi\left(t,t_0;\frac{1}{i}\frac{\delta}{\delta \mathrm{f}}\right)\right]\mathcal{L}_0(\mathrm{f}), \tag{305}$$

where $\mathcal{L}_0(\mathrm{f})$, $\mathrm{f}\in\mathcal{J}(T^*(\mathbb{R}^3);\mathbb{R})$, is a generating functional if the initial dynamical system of the particles under absent of interaction, that is

$$\mathcal{L}_0(\mathrm{f})(t,t_0) = \sum_{n\in\mathbb{Z}_+}\frac{1}{n!}\int_{T^*(\mathbb{R}^3)}dz_1...\int_{T^*(\mathbb{R}^3)}dz_n\times \\ \times f_n(x_1+\frac{p_1}{m}(t_0-t),p_1;...;x_n+\frac{p_n}{m}(t_0-t),p_1)\prod_{j=1}^{n}\{\exp[i\mathrm{f}(z_j)]-1\}. \tag{306}$$

Applying to (306) when $t_0\to-\infty$ Bogolubov's correlation weakening (296), we obtain that for all $t\in\mathbb{R}_+$

$$\mathcal{L}_0(\mathrm{f})(t) = \exp\left[\int_{T^*(\mathbb{R}^3)}dz f_1(x-\frac{p}{m}t;p)\{\exp[i\mathrm{f}(z)]-1\}\right], \tag{307}$$

where $\mathcal{L}_0(\mathrm{f})(t) = \lim_{t_0\to-\infty}\mathcal{L}_0(\mathrm{f})(t,t_0)$. Now, according to (305) and (307), we find that

$$\mathcal{L}(\mathrm{f})(t) = \exp\left[\pi\left(t,t_0;\frac{1}{i}\frac{\delta}{\delta\mathrm{f}}\right)\mathcal{L}_0(\mathrm{f})(t)\right] \tag{308}$$

is a solution of Bogolubov's functional Equation (299), where

$$\pi\left(t;\frac{1}{i}\frac{\delta}{\delta\mathrm{f}}\right) = \lim_{t_0\to-\infty}\pi\left(t,t_0;\frac{1}{i}\frac{\delta}{\delta t}\right) \tag{309}$$

for all $t\in\mathbb{R}_+$. To specify the form of the operators (309), we note that operator $\xi(t,t_0;w) = \exp[\pi(t,t_0;w)]$ for all $t,t_0\in\mathbb{R}_+$ satisfies under $\hbar\to 0$ the following differential evolution relationship:

$$\frac{d\xi}{dt} = \frac{1}{i\hbar}[\xi,\mathrm{H}]_0 + \lim_{\hbar\to 0}\frac{1}{\hbar}\left(\mathrm{V}-P_0^{-1}\mathrm{V}P_0\right)\xi, \tag{310}$$

where all operators are assumed to be given in the Wigner representation. Expanding the operator $\xi(t,t_0,w)$ into the sum of n-particles components, $n\in\mathbb{N}$, we find

$$\xi(t,t_0;w) = \sum_{n\in\mathbb{Z}_+}\frac{1}{n!}\int_{T^*(\mathbb{R}^3)}dz_1...\int_{T^*(\mathbb{R}^3)}dz_n\,\xi_n(z_1,...,z_n;t,t_0) : w(z_1)...w(z_n):, \tag{311}$$

and there is mutually unambiguous correspondence [56,59] between coefficient functions $\xi_n(z_1,...,z_n;t,t_0)$ in (311) and coefficient functions in the expansion (304)

$$\pi_n(z_1,...,z_n) = \sum_{\sigma\in\Sigma_n}(-1)^{n+i}(\sigma-1)\prod_{j=1}^{\infty}\xi_\sigma(z_{\langle k\rangle}\in\sigma_j), \\ \xi_n(z_1,...,z_n) = \sum_{\sigma\in\Sigma_n}\prod_{j=1}^{\infty}\pi_{\sigma_j}(z_{\langle k\rangle}\in\sigma_j). \tag{312}$$

Here, $\sigma\in\Sigma_n$ is an arbitrary partition of the symmetry group Σ_n of all permutations of the set of numbers $\{1,2,...,n\}$ on the subsets $\{\sigma_j : j=1,...,s\}$, which are not intersect, that is $\bigcup_{j=1}^{n} \sigma_j = \{1, ..., n\}$ and ξ_{σ_j} and $\pi_{\sigma_j}, j = 1, ..., s$, are the corresponding to this partition coefficient functions. In particular,

$$\xi_1(z_1) = \pi_1(z_1), \quad \pi_2(z_1, z_2) = \xi_2(z_1, z_2) - \xi_1(z_1)\xi_1(z_2)$$

and so on. Thus, on the base of the defined operator series (304) or (311), the problem of the explicit calculations of the distribution functions become very simple. Below we will analyze these series by means of the language of Bogolubov's generating functional $\mathcal{L}(\mathrm{f})$, $\mathrm{f} \in \mathcal{J}(T^*(\mathbb{R}^3); \mathbb{R})$, using Bogolubov's functional hypothesis [56,143,149,150,155].

9.3. Bogolubov–Boltzmann Kinetic Equation in the Frame of Functional Hypothesis

The generating functional, as it was stated above, is given by the expression

$$\mathcal{L}(\mathrm{f})(t) = \exp\left[\pi\left(t_0; \frac{1}{i}\frac{\delta}{\delta \mathrm{f}}\right)\right]\mathcal{L}_0(\mathrm{f})(t). \tag{313}$$

Here, $\mathcal{L}_0(\mathrm{f})(t)$, $t \in \mathbb{R}_+$, is a generating functional of the system of non-interacting particles, which is equal to the expression (307) when $t_0 \to -\infty$. From (313), it follows that for all $t \in \mathbb{R}_+$ for the n-particle distribution function $f_n(z_1, ..., z_n; t)$ the general functional relationship holds

$$f_n(z_1, ..., z_n; t) := f_n(z_1, ..., z_n; f_1(z; t)). \tag{314}$$

Respectively, the generating functional (313) satisfies, according to (290) when $t = 0$, the following dynamic equation:

$$\frac{d}{dt}\mathcal{L}(\mathrm{f}) = \int_{T^*(\mathbb{R}^3)} dz \left\{\frac{1}{i}\frac{\delta \mathcal{L}(\mathrm{f})}{\delta \mathrm{f}(z)}, T(p)\right\}^{(1)} + \tag{315}$$
$$+ \frac{1}{2}\int_{T^*(\mathbb{R}^3)} dz_1 \int_{T^*(\mathbb{R}^3)} dz_2 \left\{: \frac{1}{i}\frac{\delta}{\delta \mathrm{f}(z_1)}\frac{1}{i}\frac{\delta}{\delta \mathrm{f}(z_2)} : \mathcal{L}(\mathrm{f}), \mathrm{V}(x_1 - x_2)\right\}^{(2)},$$

Let us put $f_1(z) \to f_1(z; \tau)$, where $\tau \in \mathbb{R}_-$ and that

$$\frac{\partial f_1(z;\tau)}{\partial \tau} = \{f_1(z;\tau), T(p)\}^{(1)} + \int_{T^*(\mathbb{R}^3)} dz_1 \{f_1(z_1) f_1(z), \mathrm{V}(x_1 - x)\}^{(2)}. \tag{316}$$

Then from (315), we also obtain that

$$\frac{d}{d\tau}\mathcal{L}(\mathrm{f}) = \{\{\mathcal{L}(\mathrm{f}), \mathcal{H}(\mathcal{F})\}\} = \int_{T^*(\mathbb{R}^3)} dz \left\{\frac{1}{i}\frac{\delta \mathcal{L}(\mathrm{f})}{\delta \mathrm{f}(z)}, T(p) + \int_{T^*(\mathbb{R}^3)} dz_1 f_1(z_1), \mathrm{V}(x_1 - x)\right\}^{(1)} \tag{317}$$

for all $\tau \in \mathbb{R}_-$. Then Equation (317) can be rewritten in the following way:

$$\frac{d}{d\tau}\mathcal{L}(\mathrm{f}) = \{\{\mathcal{L}(\mathrm{f}), \tilde{\mathcal{H}}(\mathcal{F})\}\} = \int_{T^*(\mathbb{R}^3)} dz \left\{\frac{1}{i}\frac{\delta \mathcal{L}(\mathrm{f})}{\delta \mathrm{f}(z)}, \tilde{H}(f_1)\right\}^{(1)}, \tag{318}$$

where, by definition, $\tilde{H}(f_1) := \frac{\delta}{\delta f_1}\tilde{\mathcal{H}}(\mathcal{F})$ and

$$\tilde{\mathcal{H}}(\mathcal{F}) = \int_{T^*(\mathbb{R}^3)} dz \frac{p^2}{2m} f_1(z) + \frac{1}{2}\int_{T^*(\mathbb{R}^3)} dz_1 \int_{T^*(\mathbb{R}^3)} dz_2 f_1(z_1) f_1(z_2) \mathrm{V}(x_1 - x_2), \tag{319}$$

is the Vlasov-type Hamiltonian of the self-consistent particles interaction. Let us define the following mapping on the phase space of $n \in \mathbb{Z}_+$ particles:

$$S_n(\tau)x_j = x_j(\tau), \quad S_n(\tau)p_j = p_j(\tau), \tag{320}$$

where for all $\tau \in \mathbb{R}_-, j = \overline{1,n}$,

$$\begin{aligned} \frac{dx(\tau)}{dt} &= \{\tilde{H}, x(\tau)\}^{(1)}, \quad \frac{dp(\tau)}{dt} = \{\tilde{H}, p(\tau)\}^{(1)}, \\ \tilde{H} &= \sum_{j=1}^{n} \frac{p_j^2}{2m} + \frac{1}{2}\sum_{j=k}^{n} \mathrm{V}(x_j - x_k). \end{aligned} \tag{321}$$

It easy to see that the system of Equations (321) gives the exact solution [157] for the dual Equation (316) in the form of the sum of δ-functions of $n \in \mathbb{Z}_+$ particles:

$$f_1(z) = \sum_{j=1}^{n} \delta(z - z_j), \tag{322}$$

where $z_j \in \mathbb{R}^3, j = 1, ..., n$, are the coordinates of the cluster. Using (320) from (318) we obtain that for all $\tau \in \mathbb{R}$

$$\begin{aligned} \frac{d}{d\tau}\mathcal{L}(\mathrm{f})(\tau) = \{\{\mathcal{L}(\mathrm{f})(\tau), \tilde{\mathcal{H}}(\mathcal{F})\}\} + \frac{1}{2}\int\limits_{\mathbb{R}^3\times\mathbb{R}^3} dz_1 \int\limits_{T^*(\mathbb{R}^3)} dz_2 \times \\ \times \left\{: \frac{1}{i}\frac{\delta}{\delta \mathrm{f}(z_1)}\frac{1}{i}\frac{\delta}{\delta \mathrm{f}(z_2)} : \mathcal{L}(\mathrm{f}), \mathrm{V}(x_1 - x_2)(\tau)\right\}^{(2)}, \end{aligned} \tag{323}$$

where we denoted

$$\begin{aligned} &\mathcal{L}(\mathrm{f})(\tau) = S(\tau)\mathcal{L}(\mathrm{f}|S(-\tau)f_1), \\ &\mathrm{V}(x_1 - x_2)(\tau) = S(\tau)\mathrm{V}(S(-\tau)(x_1 - x_2)), \\ &f_2(z, z_1)(\tau) = S(\tau)f_2(z_1, z|S(-\tau)f_1) \end{aligned} \tag{324}$$

Integrating the Equation (323) in limits $\tau \in (-\infty, 0)$, we obtain that

$$\begin{aligned} \mathcal{L}(\mathrm{f})|_{\tau=0} = \lim_{\tau\to-\infty} S(\tau)\mathcal{L}(\mathrm{f}|S(-\tau)f_1) + \int\limits_{-\infty}^{0} d\tau\left\{\tfrac{1}{i}\tfrac{\delta}{\delta f(z)}\mathcal{L}(\mathrm{f})(\tau), \tilde{\mathcal{H}}(\mathcal{F}(\tau))\right\}^{(1)} + \\ \left. + \tfrac{1}{2}\int\limits_{-\infty}^{0} d\tau \int\limits_{\mathbb{R}^3\times\mathbb{R}^3} dz_1 \int\limits_{T^*(\mathbb{R}^3)} dz_2 \left\{: \tfrac{1}{i}\tfrac{\delta}{\delta \mathrm{f}(z_1)}\tfrac{1}{i}\tfrac{\delta}{\delta \mathrm{f}(z_2)} : \mathcal{L}(\mathrm{f}), \mathrm{V}(x_1 - x_2)(\tau)\right\}^{(2)}\right] \end{aligned} \tag{325}$$

We should also note here, that due to the Bogolubov's principle of correlations weakening (296) and using (307) the first item in (325) can be represented in the form

$$\begin{aligned} \lim_{\tau\to-\infty} S(\tau)\mathcal{L}(\mathrm{f}|S(-\tau)f_1) = \lim_{\tau\to\infty} \exp\left[\int\limits_{T^*(\mathbb{R}^3)} dz S(\tau)f_1(z)(\tau)\{\exp[i\mathrm{f}(z)] - 1\}\right] = \\ = \exp\left[\int\limits_{T^*(\mathbb{R}^3)} dz f_1(z)\{\exp[i\,\mathrm{f}(z)] - 1\}\right]. \end{aligned} \tag{326}$$

Applying to the expression (325) the different variants of the successive approximations method [56,143,149,150,155], we can get the generating functional $\mathcal{L}(\mathrm{f})$ in explicit form and then, using formula

$$f_n(z_1,...,z_n) =: \frac{1}{i}\frac{\delta}{\delta \mathrm{f}(z_1)}...\frac{1}{i}\frac{\delta}{\delta \mathrm{f}(z_n)} : \mathcal{L}(\mathrm{f})\Big|_{\mathrm{f}=0} \tag{327}$$

for all $n \in \mathbb{Z}_+$ obtain distribution function for any order of perturbation theory. In particular, choosing expansion by the particle density in container $A \in \mathbb{R}^3$ as a small parameter, it is easy to get the modified kinetic Bogolubov–Boltzmann equation for one-particle distribution function $f_1(z;t)$, $z \in T^*(\mathbb{R}^3)$, $t \in \mathbb{R}_+$:

$$\frac{\partial f_1(z_1;t)}{\partial t} + \langle \frac{p}{m} | \nabla_x f_1(z_1;t) \rangle = \int\limits_{T^*(\mathbb{R}^3)} dz_2 \{\tilde{f}_2(z_1,z_2;t), \mathrm{V}(x_1 - x_2)\}^{(2)}, \tag{328}$$

where the function $\tilde{f}_2(z_2,z_1;t)$ is defined according to (325) and (326) by the following expression:

$$\tilde{f}_2(z_1,z_2;t) = f_1(\tilde{z}_1;t) f_1(\tilde{z}_2;t), \tag{329}$$

$$\tilde{z}_j = \lim_{\tau\to\infty} S_2(\tau) S_1(-\tau) z_j \Rightarrow \begin{cases} \tilde{x}_j = \lim\limits_{\tau\to\infty} S_2(-\tau) x_j + \tau \frac{p_j}{m}, \\ \tilde{p}_j = \lim\limits_{\tau\to\infty} S_2(-\tau) p_j, \end{cases}$$

for $j = \overline{1,2}$. Taking into account that the Poisson bracket $\{\cdot,\cdot\}^{(n)}$ is invariant with respect to the mappings $S_n(\tau)$, $n \in \mathbb{Z}_+$, from (329) it is easy to find that

$$\begin{aligned} \{\tilde{f}_2(z_1,z_2;t), \mathrm{V}(x_2 - x_1)\}^{(2)} &= \tfrac{|p_2-p_1|}{m} \tfrac{\partial}{\partial \xi}(f_1(\tilde{z}_1;t) f_1(\tilde{z}_2;t)) - \\ -\langle \tfrac{(\tilde{p}_2-p_1)}{m} | \nabla_{x_1} f_1(\tilde{z}_1;t)\rangle f_1(\tilde{z}_2;t) &+ \langle \tfrac{(\tilde{p}_2-p_1)}{m} | \nabla_{x_2} f_1(\tilde{z}_2;t)\rangle f_1(\tilde{z}_1;t), \end{aligned} \tag{330}$$

where $\xi \in \mathbb{R}^1$ is a parameter of the axis in a cylindrical coordination system which is directed along the vector $(p_2 - p_1) \in \mathbb{E}^3$ and beginning at the point $x_1 \in \mathbb{R}^3$. After substituting (330) into (328), we can get the kinetic Bogolubov–Boltzmann equation [56,143,149,150,155] in the form of (298) with the explicitly defined collision integral $J(f_1)$, obtained from (330) via integration by $\xi \in \mathbb{R}$. Choosing in (326) other approximations of the generating functional $\mathcal{L}(\mathrm{f})$, $\mathrm{f} \in \mathcal{J}(T^*(\mathbb{R}^3);\mathbb{R})$, one can find other forms of Bogolubov–Boltzmann kinetic Equations (298).

We can also make a remark concerning the nature of the operator-functional expression (309) or (304). Namely, it is easy to see that generating functional $\mathcal{L}(\mathrm{f})(t,t_0)$, $\mathrm{f} \in \mathcal{J}(T^*(\mathbb{R}^3);\mathbb{R})$, allows the following operator-functional representation for all $t, t_0 \in \mathbb{R}$:

$$\begin{aligned} \mathcal{L}(\mathrm{f})(t,t_0) = \exp\Bigg[\frac{1}{2} \int\limits_{T^*(\mathbb{R}^3)} dz_1 \int\limits_{T^*(\mathbb{R}^3)} dz_2 \Big\{ : \frac{1}{i}\frac{\delta}{\delta \mathrm{f}(z_1)} \times \\ \times \frac{1}{i}\frac{\delta}{\delta \mathrm{f}(z_2)} :, \mathrm{V}(x_1 - x_2) \Big\}^{(2)} (t - t_0) \Bigg] \mathcal{L}_0(\mathrm{f})(t,t_0). \end{aligned} \tag{331}$$

Comparing the expressions (331) and (305), we find that for arbitrary t, t_0

$$\begin{aligned} \pi\Big(t,t_0;\tfrac{1}{i}\tfrac{\delta}{\delta \mathrm{f}}\Big) &= \tfrac{1}{2}(t-t_0) \times \\ \times \int\limits_{T^*(\mathbb{R}^3)} dz_1 \int\limits_{T^*(\mathbb{R}^3)} dz_2 &\Big\{ : \tfrac{1}{i}\tfrac{\delta}{\delta \mathrm{f}(z_1)} \tfrac{1}{i}\tfrac{\delta}{\delta \mathrm{f}(z_2)} :, \mathrm{V}(x_1 - x_2) \Big\}^{(2)} \end{aligned} \tag{332}$$

since the functional $\mathcal{L}_0(\mathrm{f})(t,t_0)$, $\mathrm{f} \in \mathcal{J}(T^*(\mathbb{R}^3);\mathbb{R})$, is arbitrary. It is easy to see from (332), that operator $\pi\left(t,t_0;\frac{1}{i}\frac{\delta}{\delta \mathrm{f}}\right)$ is not poly-local with respect to the functional derivatives $\frac{1}{i}\frac{\delta}{\delta \mathrm{f}}$, which corresponds to the singularity in the operator expansion (304). Thus, using the expression (331), the arbitrariness of the initial state and the classical Bogolubov weakening correlation condition gives a possibility to find many types of the solutions via the method of successive approximations, which follows from (331) and the Bogolubov functional hypothesis subject to the generating representation functional of distribution functions.

Having analyzed the Bogolubov generating functional (331) within the quasi-classical Wigner density operator representation (287), one can obtain an exact functional-operator solution to the evolution Bogolubov functional Equation (323):

$$\mathcal{L}(f) = Z(f)/Z(0), \qquad Z(f) = \exp[\tilde{\mathrm{V}}(\delta)]\mathcal{L}_0(f) \tag{333}$$

for $f \in \mathcal{J}(T^*(\mathbb{R}^3);\mathbb{R})$. Here we denoted

$$\tilde{\mathrm{V}}(\delta) = \sum_{n\in\mathbb{Z}_+} \frac{1}{n!}\int_{T(\mathbb{R}^3)} dz_1 \int_{T(\mathbb{R}^3)} dz_2 ... \tag{334}$$

$$\times \int_{T(\mathbb{R}^3)} dz_n \Phi_n(z_1,z_2,...,z_n|t) : \frac{1}{i}\frac{\delta}{\delta f(z_1)}\frac{1}{i}\frac{\delta}{\delta f(z_2)}...\frac{1}{i}\frac{\delta}{\delta f(z_n)} : ,$$

$$\mathcal{L}_0(f) = \sum_{n\in\mathbb{Z}_+} \frac{1}{n!}\int_{T(\mathbb{R}^3)} z_1 \int_{T(\mathbb{R}^3)} dz_2 ... \int_{T(\mathbb{R}^3)} dz_n$$

$$\times \bar{f}_n(x_1 - \frac{p_1}{m}t, x_2 - \frac{p_2}{m}t, ..., x_n - \frac{p_n}{m}t; p_1,p_2,...,p_n)\prod_{j=1}^{n}\{\exp[if(z_j)] - 1\},$$

where $\bar{f}_n(\, z_1,z_2,...,z_n)$, $n \in \mathbb{N}$, are given $n-$particle distribution functions at $t = 0$, that is, owing to the definition (237),

$$\begin{aligned} \bar{f}_n(z_1,z_2,...,z_n) &:= tr(\bar{P} : w(z_1)wz_2)...w(z_n) :) \\ &=: \frac{1}{i}\frac{\delta}{\delta f(z_1)}\frac{1}{i}\frac{\delta}{\delta f(z_2)}...\frac{1}{i}\frac{\delta}{\delta f(z_n)} : \mathcal{L}(f)|_{t=0,f=0}, \end{aligned} \tag{335}$$

and $\Phi_n(x_1,x_2,...,x_n;p_1,p_2,...,p_n|t)$, $n \in \mathbb{Z}_+$, are so-called cluster potential functions, determined recursively by means of the following functional-operator relationships:

$$\begin{aligned} \log(P_0^{-1}P) &:= \textstyle\sum_{n\in\mathbb{Z}_+}\frac{1}{n!}\int_{T(\mathbb{R}^3)} dz_1 \int_{T(\mathbb{R}^3)} dz_2 ... \int_{T(\mathbb{R}^3)} dz_n \\ &\times \tilde{V}_n(z_1,z_2,...,z_n|t) : w(z_1)w(z_2)...w(z_n) : \end{aligned} \tag{336}$$

with

$$P_0 = \exp(-\frac{it}{\hbar}\mathrm{H}_0)\bar{P}\exp(\frac{it}{\hbar}\mathrm{H}_0) \tag{337}$$

being the statistical operator of the non-interacting particle system.

If the initial distribution at $t = 0$ is "chaotic", that is for all $n \in \mathbb{N}$, the following relationships

$$\bar{f}_n(z_1,z_2,...,z_n) = \prod_{j=1}^{n}\bar{f}_1(z_j) \tag{338}$$

hold, one easily gets from (334) and (338) that

$$\mathcal{L}_0(f) = \exp\left(\int_{T(\mathbb{R}^3)} dz f_1(x - \frac{p}{m}t;p)\{\exp[if(z)] - 1\}\right). \tag{339}$$

If the "chaotic" condition is not fulfilled, we can proceed to the usual cluster Ursell–Mayer type representation [20,22,115,118] for the Bogolubov generating functional (333), where

$$\mathcal{L}_0(f) = \exp\left(\sum_{n\in\mathbb{Z}_+}\frac{1}{n!}\int_{T(\mathbb{R}^3)}dz_1\int_{T(\mathbb{R}^3)}dz_2...\int_{T(\mathbb{R}^3)}dz_n\times\right. \tag{340}$$

$$\left.\times\bar{g}_n\,(x_1-\frac{p_1}{m}t, x_2-\frac{p_2}{m}t,...,x_n-\frac{p_n}{m}t; p_1,p_2,...,p_n)\prod_{j=1}^{n}\{\exp[if(z_j)]-1\}\right),$$

where "*cluster*" distribution functions $\bar{g}_n(z_1,z_2,...,z_n)$, $n\in\mathbb{N}$, have the form

$$\bar{g}_n(z_1,z_2,...,z_n) := \sum_{\sigma[n]}(-1)^{m+1}(m-1)!\prod_{j=1}^{m}\bar{F}_{\sigma[j]}(z_k\in\sigma[j]),$$

$$\bar{f}_n(z_1,z_2,...,z_n) := \sum_{\sigma[n]}\prod_{j=1}^{m}\bar{g}_{\sigma[j]}(z_k\in\sigma[j]), \tag{341}$$

and $\sigma[n]$ denotes a partition of the set $\{1,2,...,n\}$ into non-intersecting subsets $\{\sigma[j] : j=\overline{1,m}\}$, that is $\sigma[j]\cap\sigma[k]=\varnothing$ for $j\neq k=\overline{1,m}$, and $\sigma[n]=\cup_{j=1}^{m}\sigma[j]$. In particular,

$$\begin{aligned}\bar{g}_1(z_1) &= \bar{f}_1(z_1),\\ \bar{g}_2(z_1,z_2) &= \bar{f}_2(z_1,z_2)-\bar{f}_1(z_1)\bar{f}_1(z_2),...,\end{aligned} \tag{342}$$

and so on. The classical Bogolubov generating functional (333), owing to (334) and (340), allows a natural infinite series expansion, whose coefficients can be represented as above, by means of the usual Ursell–Mayer type diagram expressions, which can be effectively used for studying the kinetic properties of our many-particle statistical system.

9.4. The Kinetic Equations for Many-Particle Distribution Functions, Their Lie-Algebraic Structure and Invariant Reductions

It is well known that the classical Bogolubov–Boltzmann kinetic equations under the condition of many-particle correlations [56,86,142,149–151,155,157–161] at weak short range interaction potentials describe long waves in a dense gas medium. In general, based on the Liouville equations of a finite number of particles in a fixed volume, it is easy to get for these distribution functions a finite chain of the corresponding kinetic equations, within which one can formally proceed to the statistical mechanics limit and get a chain of equations for the limiting distribution functions. There will be strong difficulties here if we try to mathematically justify the correctness of this limiting transition in a chain of multi-particle kinetic equations. If we do not pay attention to this complex problem, and consider a fairly weak interaction between particles under appropriate initial conditions, one can obtain the related Boltzmann equation, characterizing the process of establishing statistical equilibrium. Many of the problems related to these limiting distribution functions can be omitted if the infinite particle statistical physics ensemble is worked with from the very beginning, making use of the secondly quantized representation [56,76,151,162] of the particle states in the corresponding Fock type space.

Relating to the Boltzmann kinetic equation, the same equation, called the Vlasov equation, as it was shown by N. Bogolubov [157], also describes exact microscopic solutions of the infinite Bogolubov chain [86] for the many-particle distribution functions, which was widely studied, making use of both classical approaches in [20,21,56,76,77,142,161–179], and making use of the generating Bogolubov functional method and the related quantum current algebra representations.

A.A. Vlasov proposed his kinetic equation [180] for electron-ion plasma, based on general physical reasonings that in contrast to the short range interaction forces between neutral gas atoms, interaction forces between charged particles slowly decrease with distance, and therefore the motion of each such particle is determined not only by its pair-wise interaction with either particle, but also by the interaction with the whole ensemble of charged particles. In this case, the Bogolubov equation for distribution functions in a domain $\Lambda \subset \mathbb{R}^3$

$$\frac{\partial f_1(z;t)}{\partial t} + \langle \frac{p}{m} | \nabla_x f_1(z;t) \rangle = \int_{T^*(\Lambda)} dz' \{ f_2(z,z';t), V(x-x') \}^{(2)}, \tag{343}$$

where $z := (x,p) \in T^*(\Lambda), t \in \mathbb{R}_+$ is the temporal evolution parameter, $\{\cdot,\cdot\}^{(m)}$ denotes the canonical Poisson bracket [56,122,181] on the product $T^*(\Lambda)^m, m \in \mathbb{N}$, and $V(x - x'), x, x' \in \Lambda$, is an interparticle interaction potential, - reduces to the Vlasov equation if to put in (343)

$$f_2(z,z';t) = f_1(z;t) f_1(z';t), \tag{344}$$

that is to assume that the two-particle correlation function [142,158,161,179] vanishes:

$$g_2(z,z';t) = f_2(z,z';t) - f_1(z;t) f_1(z';t) = 0 \tag{345}$$

for all $z, z' \in T^*(\Lambda)$ and $t \in \mathbb{R}_+$. Then one easily obtains from (343) that

$$\frac{\partial f_1(z;t)}{\partial t} + \langle \frac{p}{m} | \nabla_x f_1(z;t) \rangle = \langle \frac{\partial f_1(z;t)}{\partial p} | \nabla_x \int_{T^*(\Lambda)} dz' \, V(x-x') f_1(z';t) \rangle \tag{346}$$

for all $z \in T^*(\Lambda)$ and $t \in \mathbb{R}_+$. We remark here that the Equation (346) is reversible under the time reflection $\mathbb{R}_- \ni -t \rightleftarrows t \in \mathbb{R}_+$, thus it is obvious that it can not describe thermodynamically stable limiting states of the particle system in contrast to the classical Bogolubov–Boltzmann kinetic equations [20,56,86,142,149,151,161,166], being *a priori* time non-reversible owing to the choice of boundary conditions in the correlation weakening form. This means that in spite of the Hamiltonicity of the Bogolubov chain for the distribution functions, the Bogolubov–Boltzmann equation *a priori* is not reversible. It is also evident that the condition (345) does not break the Hamiltonicity—the Equation (346) is Hamiltonian with respect to the following Lie–Poisson–Vlasov bracket:

$$\{\{a(f), b(f)\}\} := \int_{T^*(\Lambda)} dz \, f(z) \{ \mathrm{grad}\, a(f)(z), \mathrm{grad}\, b(f)(z) \}^{(1)}, \tag{347}$$

where $\mathrm{grad}(\cdot) := \delta(\cdot)/\delta f, f \in D(T^*(\Lambda)) := M_{f_1}$, respectively $a, b \in D(M_{f_1})$ are smooth functionals on the functional manifold M_{f_1}, consisting of functions fast decreasing at the boundary $\partial\Lambda$ of the domain $\Lambda \subset \mathbb{R}^3$. The statement above easily ensues from the following proposition.

Proposition 11. *Let $M_{\mathcal{F}}$ denote a set of many-particle distribution functions. Then the classical Bogolubov–Poisson bracket [20,21,86,172] on the functional space $D(M_{\mathcal{F}})$ reduces invariantly on the subspace $D(M_{f_1}) \subset D(M_{\mathcal{F}})$ to the Lie–Poisson–Vlasov bracket (347).*

Concerning the general case when we work with an infinite Bogolubov chain of kinetic equations on the many-particle distribution functions and are forced to break it at some place, numbered by some natural number $N \in \mathbb{N}$, the usual approaches always give rise to the resulting inconsistency [155,158] of the chain and, as a result, to the nonphysical solutions. The most successful approach to obtaining the Boltzmann kinetic equation for the one-particle distribution function was suggested many years ago by N. Bogolubov [86,149], based on the effective application of the so called weak correlation condition. So far, regretfully, this approach, being conjugated with the complex problem of solving functional equations, also gives rise to inconsistency of the higher order kinetic equations. Nonetheless, being inspired by former studies [20,76,162] of these problems, based on the geometrical interpretation of the Bogolubov kinetic equations chain, we devised a new functional analytic approach [23] to constructing its compatible reduction a priori free of any non-physical consequences. We also succeeded in constructing a reduced set of kinetic equations, based on a suitably devised Dirac type invariant reduction scheme of the corresponding many-particle Lie–Poisson phase space. The approach to solving this problem and its different consequences will be analyzed in more detail in sections to follow below.

9.5. The Classical Lie-Poisson-Vlasov Bracket and Kinetic Equation For The One-Particle Distribution Function

The bracket expression (347) allows a slightly different Lie-algebraic interpretation, based on considering the functional space $D(M_{f_1})$ as a Poisson manifold, related with the canonical symplectic structure on the diffeomorphism group $\mathrm{Diff}(\Lambda)$ of the domain $\Lambda \subset \mathbb{R}^3$, first described [31,32] in 1887 by Sophus Lie. Namely, the following classical theorem holds.

Theorem 8. *The Lie-Poisson bracket at point $(\mu;\eta) \in T^*_\eta(\mathrm{Diff}(\Lambda))$ on the coadjoint space $T^*_\eta(\mathrm{Diff}(\Lambda)), \eta \in \mathrm{Diff}(\Lambda)$, is equal to the expression*

$$\{f,g\}(\mu) = (\mu|[\delta g(\mu)/\delta\mu, \delta f(\mu)/\delta\mu])_c \tag{348}$$

for any smooth right-invariant functionals $f, g \in C^\infty(T^*_\eta(\mathrm{Diff}(\Lambda));\mathbb{R})$.

Proof. By classical definition [31,32,122,131,181] of the Poisson bracket of smooth functions $(\mu|a)_c, (\mu|b)_c \in C^\infty(T^*_\eta(\mathrm{Diff}(\Lambda));\mathbb{R}), a, b \in \mathrm{diff}(\Lambda) \simeq T_\eta(\mathrm{Diff}(\Lambda))$ on the symplectic space $T^*_\eta(\mathrm{Diff}(\Lambda))$, it is easy to calculate that

$$\begin{aligned} &\{\mu(a), \mu(b)\} := \delta\alpha(X_a, X_b) = \\ &= X_a(\alpha|X_b)_c - X_b(\alpha|X_a)_c - (\alpha|[X_a, X_b])_c, \end{aligned} \tag{349}$$

where $X_a := \delta(\mu|a)_c/\delta\mu = a \in \mathrm{diff}(\Lambda), X_b := \delta(\mu|b)_c/\delta\mu = b \in \mathrm{diff}(\Lambda)$. Since the expressions $X_a(\alpha|X_b)_c = 0$ and $X_b(\alpha|X_a)_c = 0$ owing the right-invariance of the vector fields $X_a, X_b \in T_\eta(\mathrm{Diff}(\Lambda))$, the Poisson bracket (349) transforms into

$$\begin{aligned} &\{ (\mu|a)_c, (\mu|b)_c\} = -(\alpha|[X_a, X_b])_c = \\ &= (\mu|[b,a])_c = (\mu|[\delta(\mu|b)_c/\delta\mu, \delta(\mu|a)_c/\delta\mu])_c \end{aligned} \tag{350}$$

for all $(\mu;\eta) \in T^*_\eta(\mathrm{Diff}(\Lambda)) \simeq \mathrm{diff}^*(\Lambda)$, and any $a, b \in \mathrm{diff}(\Lambda)$. The Poisson bracket (350) is easily generalized to

$$\{f,g\}(\mu) = (\mu|[\delta g(\mu)/\delta\mu, \delta f(\mu)/\delta\mu])_c \tag{351}$$

for any smooth functionals $f, g \in C^\infty(\mathrm{diff}^*(\Lambda);\mathbb{R})$, finishing the proof. □

Concerning our special problem of describing evolution equations for one-particle distribution functions, we will consider the one particle cotangent space $T^*(\Lambda)$ over a domain $\Lambda \subset \mathbb{R}^3$ and the canonical Poisson bracket $\{\cdot,\cdot\} := \{\cdot,\cdot\}^{(1)}$ on $T^*(\Lambda)$, for which, by definition, for any $f, g \in M_{f_1}$

$$\{f,g\}(z) := \langle\frac{\partial f}{\partial p}|\frac{\partial g}{\partial x}\rangle - \langle\frac{\partial g}{\partial p}|\frac{\partial f}{\partial x}\rangle, \tag{352}$$

where $z = (x,p) \in T^*(\Lambda)$. We denote now by $\mathcal{G} := (M_{f_1};\{\cdot,\cdot\})$ the related functional Lie algebra and $\mathcal{G}^*$ its adjoint space with respect to the standard bilinear symmetric form $(\cdot|\cdot) : M_{f_1} \times M_{f_1} \to \mathbb{R}$ on the product $M_{f_1} \times M_{f_1}$, where

$$(f|g) := \int_{T^*(\Lambda)} f(z)g(z)dz. \tag{353}$$

The constructed Lie algebra $\mathcal{G}$ with respect to the bilinear symmetric form (353) proves to be metrized, that is $\mathcal{G} \simeq \mathcal{G}^*$ and

$$(\{f,g\}|h) = (f|\{g,h\}) \tag{354}$$

for any f, g and $h \in \mathcal{G}$. $If\ \gamma \in D(\mathcal{G}^*)$ is a smooth functional on $\mathcal{G}^*$, its gradient grad $\gamma(f) \in \mathcal{G}$ at point $f \in \mathcal{G}^*$ is naturally defined via the limiting expression

$$(g|\operatorname{grad}\gamma(f)) := \frac{d}{d\varepsilon}\gamma(f+\varepsilon g)\bigg|_{\varepsilon=0} \tag{355}$$

for arbitrary element $g \in \mathcal{G}^*$. Now we define the Poisson structure $\{\{\cdot,\cdot\}\} : \mathcal{G}^* \times \mathcal{G}^* \to \mathbb{R}$ by means of the standard Lie–Poisson [31,32,59,122,159,181–184] expression:

$$\{\{\gamma,\mu\}\} := (f|\{\operatorname{grad}\gamma(f), \operatorname{grad}\gamma(f)\}) \tag{356}$$

for arbitrary functionals $\gamma, \mu \in D(\mathcal{G}^*)$. It is evident that the expression (356) identically coincides with the Poisson bracket (347).

Consider a functional $\gamma \in D(\mathcal{G}^*)$ and the related coadjoint action of the element grad $\gamma(f) \in \mathcal{G}$ at a fixed element $f := f_1 \in \mathcal{G}^*$:

$$\partial f_1/\partial t := ad^*_{\operatorname{grad}\gamma(f_1)} f_1, \tag{357}$$

where $t \in \mathbb{R}$ is the corresponding evolution parameter. It is easy observe that

$$\partial f_1/\partial t = \{\{\gamma, f_1\}\} \tag{358}$$

is a Hamiltonian equation with the functional $\gamma \in D(\mathcal{G}^*)$ taken as its Hamiltonian, being simultaneously equivalent to the following canonical Hamiltonian flow:

$$\partial f_1/\partial t = \{f_1, \operatorname{grad}\gamma(f_1)\}, \tag{359}$$

if to choose as a Hamiltonian the following functional

$$\gamma(f_1) := \int_{T^*(\Lambda)} dz_1 \frac{p_1^2}{2m} f_1(z_1) + \frac{1}{2}\int_{T^*(\Lambda)^2} dz_1 dz_2 V(x_1 - x_2) f_1(z_1) f_1(z_2), \tag{360}$$

where $V(x_1 - x_2)$ is a two-particle interaction potential, $x_1, x_2 \in \Lambda$. It is easy to observe here that the Hamiltonian (360) is obtained from the corresponding classical Hamiltonian expression

$$\mathcal{H}(\mathcal{F}) := \int_{T^*(\Lambda)} dz_1 \frac{p_1^2}{2m} f_1(z_1) + \frac{1}{2}\int_{T^*(\Lambda)^2} dz_1 dz_2 V(x_1 - x_2) f_2(z_1, z_2), \tag{361}$$

where $\mathcal{F} = (f_1, f_2, ...) \in M_{\mathcal{F}}$ denotes an infinite vector from the space $M_{\mathcal{F}} := \prod_{j\in\mathbb{N}} M_{f_j}$ of multiparticle distribution functions, and if to impose on it the constraint (344). Thus, we have stated the following proposition.

Proposition 12. *The Boltzmann–Vlasov kinetic Equation (346) is a Hamiltonian system on the functional manifold $\mathcal{G}^* \simeq \mathcal{G} = (M_f; \{\cdot,\cdot\})$ with respect to the canonical Lie–Poisson structure (356) with Hamiltonian (360). As a consequence, the flow (346) is time reversible.*

9.6. Boltzmann–Vlasov Kinetic Equations and Microscopic Exact Solutions

Proposition 11, stated above, claims that the Boltzmann–Vlasov Equation (346) is a suitable reduction of the whole Bogolubov chain upon the invariant functional subspace $M_{f_1} \subset M_{\mathcal{F}}$. Moreover, this invariance in no way should be compatible *a priori* [20,21,24,157,166,171,173,174] with the other kinetic equations from the Bogolubov chain, and can even be contradictory. Nonetheless, as it was stated [157] by N. Bogolubov, namely owing to this invariance of the subspace $M_{f_1} \subset M_{\mathcal{F}}$ the Boltzmann–Vlasov Equation (346) in the case of the Boltzmann–Enskog hard sphere approximation of the inter-particle potential possesses exact microscopical solutions which are compatible with the whole hierarchy of the Bogolubov kinetic equations. The latter is, obviously, equivalent to its Hamiltonicity on the manifold M_{f_1} with respect to the Lie–Poisson bracket (356). The Boltzmann–Enskog kinetic equation [151,157,158,161,179] equals

$$\begin{gathered}\frac{\partial f_1(z;t)}{\partial t} + \langle \frac{p}{m}|\nabla_x f_1(z;t)\rangle = \\ = a^2 \int_{\mathbb{S}^2} dn \int_{\mathbb{E}^3} dp' \, \langle p'|n\rangle\langle \frac{\tilde{p}'}{m}|n\rangle [f_2(x,\tilde{p};x+an,\tilde{p}';t) - f_2(x,p;x-an,p';t)]\end{gathered}, \qquad (362)$$

where $\tilde{p} := p + n\langle p'-p|n\rangle, \tilde{p}' := p - n\langle p'-p|n\rangle, a > 0-$ a particle diameter, $n \in \mathbb{S}^2$ – a unit vector, $\langle n|n\rangle = 1$, and, by definition, $f_2(z,z';t) = 0$ for all $z,z' \in T^*(\Lambda), t \in \mathbb{R}$, satisfying the condition $||z-z'|| < a$. The Equation (362) easily reduces to the Vlasov–Enskog equation

$$\begin{gathered}\frac{\partial f_1(z;t)}{\partial t} + \langle \frac{p}{m}|\nabla_x f_1(z;t)\rangle = J_{V-E}(f), \\ J_{V-E}(f) = a^2 \int_{\mathbb{S}^2} dn \int_{\mathbb{E}^3} dp' \, \langle p'|n\rangle\langle \frac{\tilde{p}'}{m}|n\rangle \times \\ \times [f_1(x,\tilde{p};t) f_1(x+an,\tilde{p}';t) - f_1(x,p;t) f_1(x-an,p';t)]\end{gathered} \qquad (363)$$

for all $(z;t) \in T^*(\Lambda) \times \mathbb{R}$ owing to its Hamiltonicity on the space $M_{f_1} \subset M_{\mathcal{F}}$. If, in addition, there exists a nontrivial interparticle potential, the equation above is naturally generalized to the kinetic equation

$$\begin{gathered}\frac{\partial f_1(z;t)}{\partial t} + \langle \frac{p}{m}|\nabla_x f_1(z;t)\rangle = J_{V-E}(f) + \\ + \int_{T^*(\Lambda)} dz' \{f_1(z;t) f_1(z';t), V(x-x')\}^{(2)},\end{gathered} \qquad (364)$$

which remains to be Hamiltonian on M_{f_1} and possesses, in particular, the following exact singular solution:

$$f_1(z;t) = \sum_{j=\overline{1,N}} \delta(z - z_j(t)), \qquad (365)$$

where $z_j(t) \in T^*(\Lambda), j = \overline{1,N}$—phase space coordinates in $T^*(\Lambda)^N$ of $N \in \mathbb{N}$ interacting particles in the domain $\Lambda \subset \mathbb{R}^3$. Specified above the Hamiltonicity problem and the existence of exact solutions to the Boltzmann–Vlasov kinetic Equation (364) is deeply related to that of describing correlation functions [142,161,179], suitably breaking the infinite Bogolubov chain [20,76,77,86,142,161] of many-particle distribution functions. Namely, if to introduce many-particle correlation functions [142,161,179] for related Bogolubov distribution functions as

$$\begin{gathered}g_1(z_1) = 0, g_2(z_1,z_2) = f_2(z_1,z_2) - f_1(z_1)f_1(z_2), \\ g_3(z_1,z_2,z_3) = f_3(z_1,z_2,z_3) - f_1(z_1)f_1(z_2)f_1(z_3) - f_1(z_1)g_2(z_2,z_3) - \\ - f_1(z_2)g_2(z_3,z_1) - f_1(z_3)g_2(z_1,z_2), \ldots,\end{gathered} \qquad (366)$$

where $z_j \in T^*(\Lambda), j \in N$, then the Vlasov Equation (364) is obtained from the Bogolubov hierarchy at $n = 1$ and $g_2(z_1,z_2) = 0$ for all $z_1,z_2 \in T^*(\Lambda)$.

As it was mentioned above, the constraint imposed on the infinite Bogolubov hierarchy is compatible with its Hamiltonicity. Yet in many practical cases, this closedness procedure by means of imposing the conditions like

$$g_{m+1}(z_1, z_2, \ldots, z_{m+1}) = 0 \tag{367}$$

for all $z_s \in T^*(\Lambda), s = \overline{1, m+1}$ at some fixed $m \geq 2$ gives rise to some serious dynamical problems related to its mathematical correctness. Namely, if to close the infinite Bogolubov chain of kinetic equations on many-particle distribution functions in this way, one easily checks that the imposed constraint (367) does not persist in time subject to the evolution of the distribution functions $f_j(z_1, z_2, \ldots, z_j), z_j \in T^*(\Lambda), j = \overline{1, m}$. This means that these naively reduced kinetic equations are written down somehow incorrectly, as the reduced functional submanifold $M_{\mathcal{F}}^{(m)} := \{\mathcal{F} \in M_{\mathcal{F}} : g_{m+1} = 0\}$ should remain invariant in time. To dissolve this problem, we are forced to consider the whole Bogolubov hierarchy of kinetic equations on multiparticle distribution functions as a Hamiltonian system on the functional manifold $M_{\mathcal{F}}$ and correctly reduce it on the constructed above functional submanifold $M_{\mathcal{F}}^{(m)} \subset M_{\mathcal{F}}$ via the classical Dirac type [11,59,63,122,181] procedure. The kinetic equations obtained this way by means of the reduced Lie–Poisson–Bogolubov structure will evidently differ from those naively obtained by means of the direct substitution of the imposed constraint (367) into the Bogolubov chain of kinetic equations, and in due course will conserve the functional submanifold $M_{\mathcal{F}}^{(m)} \subset M_{\mathcal{F}}$ invariant.

9.7. *The Invariant Reduction of the Bogolubov Distribution Functions Chain*

Consider the constructed before Hamiltonian functional $\mathcal{H}(\mathcal{F}) \in D(M_{\mathcal{F}})$ (361)

$$\mathcal{H}(\mathcal{F}) = \int_{T^*(\Lambda)} dz_1 \frac{p_1^2}{2m} f_1(z_1) + \frac{1}{2} \int_{T^*(\Lambda)^2} dz_1 dz_2 V(x_1 - x_2) f_2(z_1, z_2) \tag{368}$$

and calculate the evolution of the distribution functions vector $\mathcal{F} \in M_{\mathcal{F}}$ under the simplest constraint (367) at $m = 1$, that is

$$g_2(z_1, z_2) = f_2(z_1, z_2) - f_1(z_1) f_1(z_2) = 0 \tag{369}$$

for all $z_1, z_2 \in T^*(\Lambda)$. To perform this reduction on $M_{\mathcal{F}}^{(1)} \subset M_{\mathcal{F}}$, we need [11,59,63] to constraint the λ-extended Hamiltonian expression

$$\mathcal{H}_\lambda(\mathcal{F}) := \mathcal{H}(\mathcal{F}) + \frac{1}{2} \int_{T^*(\Lambda)^2} dz_1 dz_2 \lambda(z_1, z_2)[f_2(z_1, z_2) - f_1(z_1) f_1(z_2)] \tag{370}$$

for some smooth function $\lambda \in D(T^*(\Lambda)^2)$ and next to determine it from the submanifold $M_{\mathcal{F}}^{(1)}$ invariance condition

$$\begin{aligned} \tfrac{\partial g_2(z_1, z_2)}{\partial t} &= \{\{\mathcal{H}_\lambda(\mathcal{F}), g_2(z_1, z_2)\}\} = \\ &= \tfrac{\partial f_2(z_1, z_2)}{\partial t} - \tfrac{\partial f_1(z_1)}{\partial t} f_1(z_2) - f_1(z_1) \tfrac{\partial f_1(z_2)}{\partial t} = 0 \end{aligned} \tag{371}$$

for all $z_1, z_2 \in T^*(\Lambda)$ and $t \in \mathbb{R}$. To effectively calculate the condition (371), let us first calculate the evolutions for distribution functions f_1 and $f_2 \in M_{\mathcal{F}}$:

$$\begin{aligned} \frac{\partial f_1(z_1)}{\partial t} &= \{\{\mathcal{H}_\lambda(\mathcal{F}), f_1(z_1)\}\} = \left\{ f_1(z_1), \frac{\delta \mathcal{H}_\lambda(\mathcal{F})}{\delta f_1(z_1)} \right\}^{(1)} + \\ &+ \int_{T^*(\Lambda)} dz_2 \left\{ f_2(z_1, z_2), \frac{\delta \mathcal{H}_\lambda(\mathcal{F})}{\delta f_2(z_1, z_2)} \right\}^{(1)}, \end{aligned} \tag{372}$$

and

$$\frac{\partial f_2(z_1,z_2)}{\partial t} = \{\{\mathcal{H}_\lambda(\mathcal{F}), f_2(z_1,z_2)\}\} = \left\{f_2(z_1,z_2), \frac{\delta\mathcal{H}_\lambda(\mathcal{F})}{\delta f_1(z_1)} + \frac{\delta\mathcal{H}_\lambda(\mathcal{F})}{\delta f_1(z_2)}\right\}^{(2)} + \\ + \left\{f_2(z_1,z_2), \frac{\delta\mathcal{H}_\lambda(\mathcal{F})}{\delta f_2(z_1,z_2)}\right\}^{(2)} + \int_{T^*(\Lambda)} dz_3 \left\{f_3(z_1,z_2,z_3), \frac{\delta\mathcal{H}_\lambda(\mathcal{F})}{\delta f_2(z_1,z_3)} + \frac{\delta\mathcal{H}_\lambda(\mathcal{F})}{\delta f_2(z_2,z_3)}\right\}^{(2)}, \tag{373}$$

which can be rewritten equivalently as follows:

$$\begin{aligned} \frac{\partial f_1(z_1)}{\partial t} &= -\langle \frac{\partial f_1(z_1)}{\partial p_1} | \int_{T^*(\Lambda)} dz_2 \frac{\partial \lambda(z_1,z_2)}{\partial x_1} f_1(z_2) - \\ &- \langle \frac{p_1}{m} - \int_{T^*(\Lambda)} dz_2 \frac{\partial \lambda(z_1,z_2)}{\partial p_1} f_1(z_2) | \frac{\partial f_1(z_1)}{\partial x_1} \rangle + \\ &+ \frac{1}{2} \int_{T^*(\Lambda)} dz_2 \langle \frac{\partial}{\partial x_1}[V(x_1-x_2)+\lambda(z_1,z_2)] | \frac{\partial f_2(z_1,z_2)}{\partial p_1} \rangle - \\ &- \frac{1}{2} \int_{T^*(\Lambda)} dz_2 \langle \frac{\partial \lambda(z_1,z_2)}{\partial p_1} | \frac{\partial f_2(z_1,z_2)}{\partial x_1} \rangle \end{aligned} \tag{374}$$

and

$$\begin{aligned} \frac{\partial f_2(z_1,z_2)}{\partial t} &= -\langle \frac{\partial f_2(z_1,z_2)}{\partial p_1} | \int_{T^*(\Lambda)} dz_2 \frac{\partial \lambda(z_1,z_2)}{\partial x_1} f_1(z_2) \rangle - \\ &- \langle \frac{\partial f_2(z_1,z_2)}{\partial p_2} | \int_{T^*(\Lambda)} dz_1 \frac{\partial \lambda(z_1,z_2)}{\partial x_2} f_1(z_1) \rangle - \\ &- \langle \frac{\partial f_2(z_1,z_2)}{\partial x_1} | \frac{p_1}{m} - \int_{T^*(\Lambda)} dz_2 \frac{\partial \lambda(z_1,z_2)}{\partial p_1} f_1(z_2) \rangle - \\ &- \langle \frac{\partial f_2(z_1,z_2)}{\partial x_2} | \frac{p_2}{m} - \int_{T^*(\Lambda)} dz_1 \frac{\partial \lambda(z_1,z_2)}{\partial p_2} f_1(z_1) \rangle + \\ &+ \frac{1}{2} \langle \frac{\partial f_2(z_1,z_2)}{\partial p_1} | \frac{\partial}{\partial x_1}[V(x_1-x_2)+\lambda(z_1,z_2)] \rangle + \\ &+ \frac{1}{2} \langle \frac{\partial f_2(z_1,z_2)}{\partial p_2} | \frac{\partial}{\partial x_2}[V(x_1-x_2)+\lambda(z_1,z_2)] \rangle - \\ &- \frac{1}{2} \langle \frac{\partial f_2(z_1,z_2)}{\partial x_1} | \frac{\partial \lambda(z_1,z_2)}{\partial p_1} \rangle - -\frac{1}{2} \langle \frac{\partial f_2(z_1,z_2)}{\partial x_2} | \frac{\partial \lambda(z_1,z_2)}{\partial p_2} \rangle + \\ &+ \frac{1}{2} \langle \int_{T^*(\Lambda)} dz_3 \frac{\partial f_3(z_1,z_2,z_3)}{\partial p_1} | \frac{\partial}{\partial x_1}[V(x_1-x_3)+\lambda(z_1,z_3)] \rangle + \\ &+ \frac{1}{2} \langle \int_{T^*(\Lambda)} dz_3 \frac{\partial f_3(z_1,z_2,z_3)}{\partial p_2} | \frac{\partial}{\partial x_2}[V(x_2-x_3)+\lambda(z_2,z_3)] \rangle - \\ &- \frac{1}{2} \langle \int_{T^*(\Lambda)} dz_3 \frac{\partial f_3(z_1,z_2,z_3)}{\partial x_1} | \frac{\partial \lambda(z_1,z_2)}{\partial p_1} \rangle - \frac{1}{2} \langle \int_{T^*(\Lambda)} dz_3 \frac{\partial f_3(z_1,z_2,z_3)}{\partial x_2} | \frac{\partial \lambda(z_1,z_2)}{\partial p_2} \rangle \end{aligned} \tag{375}$$

Having now substituted temporal derivatives (374) and (375) into the equality (371) in their explicit form, one obtains the following functional relationship:

$$\begin{aligned} &\tfrac{1}{2} \langle f_1(z_2) \tfrac{\partial f_1(z_1)}{\partial p_1} | \tfrac{\partial}{\partial x_1} (V(x_1-x_2)+\lambda(z_1,z_2) - \\ &- \textstyle\int_{T^*(\Lambda)} dz_3 f_1(z_3) [V(x_1-x_3)+\lambda(z_1,z_3)]) \rangle + \\ &+ \tfrac{1}{2} \langle f_1(z_1) \tfrac{\partial f_1(z_2)}{\partial p_2} | \tfrac{\partial}{\partial x_2} (V(x_2-x_1)+\lambda(z_2,z_1) - \\ &- \textstyle\int_{T^*(\Lambda)} dz_3 f_1(z_3) [V(x_2-x_3)+\lambda(z_2,z_3)]) \rangle = 0, \end{aligned} \tag{376}$$

which is satisfied if

$$\lambda(z_1,z_2) = -V(x_1-x_2) \tag{377}$$

for all $z_1, z_2 \in T^*(\Lambda)$. Taking into account the result (377), one easily obtains from the Equation (374) the invariantly reduced on the submanifold $M_{\mathcal{F}}^{(1)} \subset M_{\mathcal{F}}$ kinetic equation on the one-particle distribution function:

$$\frac{\partial f_1(z_1)}{\partial t} + \langle p_1/m | \frac{\partial f_1(z_1)}{\partial x_1} \rangle = \langle \frac{\partial f_1(z_1)}{\partial p_1} | \frac{\partial}{\partial x_1} \int_{T^*(\Lambda)} dz_2 f_1(z_2) V(x_1 - x_2) \rangle, \tag{378}$$

which can be rewritten in the following compact form:

$$\frac{\partial f_1(z_1)}{\partial t} = \left\{ f_1(z_1), \frac{\delta \tilde{\mathcal{H}}(\mathcal{F})}{\delta f_1(z_1)} \right\}^{(1)}, \tag{379}$$

where we put, by definition,

$$\tilde{\mathcal{H}}(\mathcal{F}) := \int_{T^*(\Lambda)} dz_1 \frac{p_1^2}{2m} f_1(z_1) + \frac{1}{2} \int_{T^*(\Lambda)^2} dz_1 dz_2 V(x_1 - x_2) f_1(z_1) f_1(z_2). \tag{380}$$

The kinetic Equation (378) naturally coincides exactly with that obtained previously from the naively reduced evolution equation

$$\frac{\partial f_1(z_1)}{\partial t} = \{\{\mathcal{H}(\mathcal{F}), f_1(z_1)\}\}|_{M_{\mathcal{F}}^{(1)}} \tag{381}$$

on the submanifold $M_{\mathcal{F}}^{(1)} \subset M_{\mathcal{F}}$, as it is globally invariant [20,172] with respect to the classical Lie–Poisson–Bogolubov structure on $M_{\mathcal{F}}$.

The obtained result can be formulated as the following proposition.

Proposition 13. *The first correlation function Dirac type reduction on the functional submanifold $M_{\mathcal{F}}^{(1)} \subset M_{\mathcal{F}}$, formed by relationships (369), reduces the corresponding Bogolubov chain of many-particle kinetic equations to the well known classical Vlasov kinetic equation.*

Remark 9. *It is worth mentioning here that the well known classical Bogolubov approximation of the many-particle distribution functions as $f_n(z_1, z_2, \ldots, z_n) := \varphi_n(z_1, z_2, \ldots, z; f_1), z_j \in T(\Lambda), j = \overline{2, n}$, with mapping $\varphi_n : (\ldots) \times M_{f_1} \to \mathbb{R}$, $n \in \mathbb{N} \backslash \{1\}$, presenting smooth nonlinear functionals, independent of the temporal parameter $t \in \mathbb{R}_+$, define a suitably different functional submanifold $\tilde{M}_{\mathcal{F}}^{(1)} \subset M_{\mathcal{F}}$, upon which the reduced evolution flow*

$$\frac{\partial f_1(z_1)}{\partial t} = \{\{\mathcal{H}(\mathcal{F}), f_1(z_1)\}\}|_{\tilde{M}_{\mathcal{F}}^{(1)}} \tag{382}$$

gives rise to a new Boltzmann type kinetic equation, being compatible with evolution equations for higher distribution functions, free of evolution inconsistencies and completely different from that derived previously by Bogolubov [86].

In the same way as above, one can explicitly construct the system of invariantly reduced kinetic equations

$$\frac{\partial f_1(z_1)}{\partial t} = \{\{\mathcal{H}(\mathcal{F}), f_1(z_1)\}\}|_{M_{\mathcal{F}}^{(2)}}, \frac{\partial f_2(z_1, z_2)}{\partial t} = \{\{\mathcal{H}(\mathcal{F}), f_2(z_1, z_2)\}\}|_{M_{\mathcal{F}}^{(2)}} \tag{383}$$

on the submanifold $M_{\mathcal{F}}^{(2)} \subset M_{\mathcal{F}}$, which already is not *a priori* globally invariant with respect to the Hamiltonian evolution flows on $M_{\mathcal{F}}$ and whose detail structure and analysis are postponed to another place.

9.8. Conclusions and Perspectives

We presented a review of the Boltzmann type kinetic equations in statistical physics as analytical objects based on the non-relativistic current algebra symmetry approach to constructing the Bogolubov generating functional of many-particle distribution functions. We then applied it to an important classical problem of describing Boltzmann–Bogolubov and Boltzmann–Vlasov type kinetic equations, naturally related with an invariantly reduced canonical Hamiltonian system on the infinite-dimensional space of distribution functions subject to the constraints imposed on suitably chosen many-particle correlation functions. As an interesting introductory example of deriving Boltzman–Vlasov type kinetic equations, we considered a quantum-mechanical model of spinless particles with delta-type interaction, having applications [159,185,186] for describing so called Benney type hydrodynamic particle flows. We also reviewed new results on a special class of dynamical systems of Boltzmann–Bogolubov and Boltzmann–Vlasov type on infinite dimensional functional manifolds modeling kinetic processes in many-particle media. There was demonstrated construction of the classical Bogolubov generating functional method in non-equilibrium statistical mechanics within the classical Wigner quasi-classical representation. We also analyzed and presented the kinetic Boltzmann type equation in non-equilibrium statistical mechanics in the frame of the Bogolubov functional hypothesis. Moreover, the Hamiltonian analysis of the infinite hierarchy of many-particle distribution functions was reviewed, and the algebraic structure of the Boltzmann–Bogolubov kinetic equations and their invariant Poissonian reductions were analyzed in detail, including the derivation of the related Boltzmann–Vlasov kinetic equations. Based on the methods and devised techniques, an approach was proposed to invariant reduction of the chain of Bogolubov distribution functions on suitably chosen correlation function constraints, which allowed the derivation of the related modified Boltzmann–Bogolubov kinetic equations for a finite set of multi-particle distribution functions.

We also elaborated in detail effective enough invariant analytical tools reducing the infinite Boltzmann–Bogolubov hierarchy of kinetic equations upon the two-particle correlation function constraint. Within this aspect of invariant reduction of the infinite Boltzmann–Bogolubov hierarchy of kinetic equations that has very important applications, there stays an interesting problem of analytically presenting this reduction upon the three-particle correlation function constraint and deriving a closed system of the Boltzmann type kinetic equations on the corresponding one- and two-particle distribution functions. The similar, yet much more complicated, analytical problem for the future analysis consists of deriving invariantly reduced kinetic equations under the Bogolubov functional hypothesis and its modified versions.

10. The Current Algebra Functional Representations and Geometric Structure of Quasi-Stationary Hydrodynamic Flows

10.1. Introductory Notes

This section is devoted to compressible liquid or gas motions in which entropy remains locally constant throughout the flow field, i.e., the flow for which the entropy of a moving element along a streamline remains constant, is called isentropic. This means that along different streamlines, the entropy changes normal to the streamlines. As a typical example, one can mention the flow field behind a curved shock wave, where streamlines, passing through different locations along the curved shock wave, experience different increases in entropy. Hence, downstream from this shock, the entropy can be constant along a given streamline but differs from one streamline to another. This type of flow, with entropy constant along streamlines, is defined as isentropic. Flow with entropy constant everywhere is then called homentropic. Here we need to remark that owing to the second law of thermodynamics, an isentropic flow does not strictly exist. We know from thermodynamics that an isentropic flow is defined to be along streamlines both adiabatic and reversible. Yet, all real flows always experience to some extent the irreversible phenomena of friction, thermal conduction, and diffusion. For instance, any non-equilibrium, chemically reacting flow is always irreversible, when considered to be a closed system. Nonetheless, there are a large number of liquid and gas dynamic problems with entropy increase negligibly slight, which for the purpose of analysis are assumed to be isentropic. Examples are flows through subsonic and supersonic nozzles, as in wind tunnels and rocket engines, or shock-free flows over a wing, fuselage, or other aerodynamic shapes. For all of them, except for a flow near the thin boundary-layer region, adjacent to the surface where friction and thermal conduction effects can be strong, the outer inviscid flow can be considered isentropic. In contrast, if shock waves exist in the flow, the entropy increase across these shocks destroys the assumption of isentropic flow, although the flow along streamlines between shocks may persist to be isentropic.

As an isentropic flow is governed by thermodynamically reversible processes, being adiabatic along a streamline, it needs to be specified with locally defined [187] thermodynamical parameters, such as the medium density ρ, the specific entropy σ, the local medium absolute temperature T, the pressure p and the specific energy e. All these quantities are related to each other in some way, which can be retrieved following the classical Gibbs reasonings. We assume from the very beginning that the reversible thermodynamical state of the medium under regard is completely locally described by means of the following first pair: (p-local *pressure*, ρ-*specific density*) of thermodynamical parameters. Assume now that the same thermodynamical state of this medium can also be simultaneously described by means of the following second pair: (T-local absolute temperature, σ-specific entropy). The latter, in particular, means that a suitable functional transformation from one parameter to another, if smooth, is diffeomorphic, which is the Jacobian $J_{(\sigma,T)}(p,\rho)$ of this transformation $\mathbb{R}^2_+ \ni (\sigma,T) \to (p,\rho) \in \mathbb{R}^2_+$ is not degenerate everywhere, i.e.,

$$J_{(\sigma,T)}(p,\rho) = \frac{\partial(p,\rho)}{\partial(\sigma,T)} := \det\begin{pmatrix} \frac{\partial p}{\partial \sigma} & \frac{\partial p}{\partial T} \\ \frac{\partial \rho}{\partial \sigma} & \frac{\partial \rho}{\partial \sigma} \end{pmatrix} \neq 0 \tag{384}$$

at all points $(\sigma,T) \in \mathbb{R}^2_+$. Taking into account that the local absolute temperature T and the adiabatic σ parameters are, in general, defined with some scaling ambiguity, we can always put, by definition, that $J_{(\sigma,T)}(p,\rho) = \rho^2 \neq 0$ everywhere. As a simple consequence of multiplying this expression by the unity Jacobian $J_{(\sigma,\rho)}(\sigma,\rho) = 1$ one easily derives that

$$\begin{aligned} &J_{(\sigma,T)}(p,\rho) \times J_{(\sigma,\rho)}(\sigma,\rho) = \\ &= \tfrac{\partial(p,\rho)}{\partial(\sigma,T)}\tfrac{\partial(\sigma,\rho)}{\partial(\sigma,\rho)} = \tfrac{\partial(\rho,p)}{\partial(\sigma,\rho)}\tfrac{\partial(\sigma,\rho)}{\partial(\sigma,T)} = \rho^2, \end{aligned} \tag{385}$$

or, equivalently,

$$\frac{\partial(p,\rho)}{\partial(\sigma,\rho)} = \rho^2\frac{\partial(\sigma,T)}{\partial(\sigma,\rho)} \Longleftrightarrow \left.\frac{\partial(p/\rho^2)}{\partial\sigma}\right|_\rho = \left.\frac{\partial T}{\partial\rho}\right|_\sigma \tag{386}$$

at all points $(\sigma,\rho) \in \mathbb{R}^2_+$. The equality of partial derivatives above simply means, owing to the well known Montel–Menchoff–Young theorem [188–190], the existence of such a differentiable thermodynamic state function $\mathbb{R}^2_+ \ni (\rho,\sigma) \to e \in \mathbb{R}$, that its differential satisfies the following equality:

$$\delta e(\rho,\sigma) = T\delta\sigma + p\delta\rho/\rho^2. \tag{387}$$

The latter expression presents exactly the written down second thermodynamic law with respect to the locally defined variables, if the smooth function $\mathbb{R}^2_+ \ni (\rho,\sigma) \to e \in \mathbb{R}$ is interpreted as the specific medium energy of the system at the internal absolute temperature $T = T(\rho,\sigma)$ and pressure $p(\rho,\sigma)$ at suitably fixed state parameters $(\rho,\sigma) \in \mathbb{R}^2_+$. Taking into account that our medium is embedded into some domain $M \subset \mathbb{R}^3$, moving in space-time, our next task is to adequately describe the related motion spatial phase space variables, compatible with the corresponding Euler evolution equations.

10.2. Diffeomorphism Group Structure and Functional Phase Space Description

It is well known that the same physical system is often described using different sets of variables, related with their different physical interpretation. Simultaneously, this same system is endowed with different mathematical structures deeply depending on the geometric scenario used for its description. In general, these structures prove to be not equivalent but some special way connected to each other. In particular, such double descriptions commonly occur in systems with distributed parameters such as hydrodynamics, magnetohydrodynamics and diverse gauge systems, which are effectively described by means of both symplectic and Poisson structures on suitable phase spaces. In particular, it was observed [25–33] that these structures are canonically related to each other. Mathematical properties, lying in a background of their analytical description, make it possible to study additional important parameters [34–50] of different hydrodynamic and magnetohydrodynamic systems, amongst which we will mention integral invariants, describing such internal fluid motion peculiarities as vortices, topological singularities [51] and other different instability states, strongly depending [52,53] on imposed isentropic fluid motion constraints. Being interested in their general properties and mathematical structures, responsible for their existence and behavior, we present a detailed enough differential geometrical approach to investigating thermodynamically quasi-stationary isentropic fluid motions, paying more attention to analytical argumentation of tricks and techniques used during the presentation.

In particular, we consider a compressible liquid filling a compact linearly-connected domain $M \subset \mathbb{R}^3$ with smooth boundary ∂M, and moving free of external forces. A configuration of this fluid is called the reference or Lagrangian configuration, its points are called material or Lagrangian points and are denoted by $X \in M$ and are referred to as material, or Lagrangian coordinates. We shall not for now be specific about the correct choices of the related functional spaces to be used and refer to works [191,192], where this is discussed in great detail. The manifold $M \subset \mathbb{R}^3$, thought of as the target space of a configuration $\eta \in \mathrm{Diff}(M)$ of the fluid at a different time, is called the spatial or Eulerian configuration, whose points, called spatial or Eulerian points, will be denoted by small letters $x \in M$. Then a motion of the fluid is a time dependent family [26,29,41,48,122,192–195] of diffeomorphisms written as

$$M \ni x_t = \eta(X,t) := \eta_t(X) \in M \tag{388}$$

for any initial configuration $X \in M$ and some mapping $\eta_t \in \mathrm{Diff}(M), t \in \mathbb{R}$. We also are given the mass density $\rho_0 \in \mathcal{R}(M) \subset C^\infty(M;\mathbb{R}_+)$ and the specific entropy $\sigma_0 \in \Sigma(M) \subset C^\infty(M;\mathbb{R}_+)$ of the fluid in the reference configuration, changing in time in such a way that

$$\rho_0(X) = \rho_t(x_t)J_{\eta_t}(x_t), \sigma_0(X) = \sigma_t(x_t), \tag{389}$$

where $J_{\eta_t}(x_t)$ denotes the standard Jacobian determinant of the motion $\eta_t \in \mathrm{Diff}(M)$ at $x_t \in M$ and $\sigma_t(x_t)$ denotes the specific entropy for any $x_t = \eta_t(X) \in M$ and $t \in \mathbb{R}$. For a motion $x_t = \eta_t(X) \in M$ and arbitrary $X \in M, t \in \mathbb{R}$, one usually defines three velocities:

the material or Lagrangian velocity

$$V(X,t) = V_t(X) := \partial\eta_t(X)/\partial t, \tag{390}$$

the spatial or Eulerian velocity

$$v(x_t,t) = v_t(x_t) := v_t \circ \eta_t(X) \tag{391}$$

and convective or body velocity

$$\mathcal{V}(X,t) = \mathcal{V}_t(X) := -\partial X(x_t,t)/\partial t = -\partial\eta_t^{-1}(x_t)/\partial t, \tag{392}$$

being equivalent to the expression $\mathcal{V}_t = \eta_{t,*}^{-1} v_t$ for all $t \in \mathbb{R}$. Since the velocity $v_t : M \in T(M)$ is tangent to M for all $t \in \mathbb{R}$ at $x_t = \eta_t(X) \in M$, it determines a time dependent vector field on M. On the other hand, tangency of $V_t(X)$ and $\eta_t(X), X \in M$, means that the velocity V_t is a vector field over a configuration $\eta_t \in \mathrm{Diff}(M)$ on M, that is $V_t : M \to T(M)$ is such a map that $V_t(X)$ is tangent to M not at $X \in M$, but at point $x_t = \eta_t(X) \in M$. Simultaneously, the velocity $\mathcal{V}_t(X)$ is a tangent vector to M at $X \in M$, that is $\mathcal{V}_t$ is also a time dependent vector field on M. In what will follow, we will think of the fluid as moving smoothly in the domain $M \subset \mathbb{R}^3$, at any time filling it and producing no shocks and cavitation.

We present in the introductory section a modern differential geometric description of the isentropic fluid motion phase space and featuring diffeomorphism group structure, modelling the related dynamics, as well as its compatibility with the quasi-stationary thermodynamical constraints. The next section is devoted to the Hamiltonian analysis of the adiabatic liquid dynamics, within which, following the general approach of [28,41,194], we explain the nature of the related Poissonian structure on the fluid motion phase space, as a semidirect Banach groups product, and a natural reduction of the canonical symplectic structure on its cotangent space to the classical Lie–Poisson bracket on the adjoint space to the corresponding semidirect Lie algebras product. A modification of the Hamiltonian analysis in case of the isothermal liquid dynamics is presented in the next section. We proceed further to studying the differential-geometric structure of the adiabatic magneto-hydrodynamic superfluid phase space and its related motion within the Hamiltonian analysis and invariant theory. We construct there an infinite hierarchy of different kinds of integral magneto-hydrodynamic invariants, generalizing those previously constructed in [194,196], and analyzing their differential-geometric origins. The last section presents charged fluid dynamics on the phase space invariant with respect to an Abelian gauge group transformation.

10.3. A Modified Current Algebra, Its Functional Representation And Geometric Description of the Ideal Liquid Dynamics

It is well known that the motion of an ideal compressible and isentropic fluid is governed by the Euler equations

$$\begin{aligned} \partial v/\partial t + \langle v|\nabla\rangle v + \rho^{-1}\nabla p^{(0)} = 0, \\ \partial\rho/\partial t + \langle\nabla|\rho v\rangle = 0, \partial\sigma/\partial t + \langle v|\nabla\rangle\sigma = 0, \end{aligned} \tag{393}$$

where $p_0 : M \to \mathbb{R}$ is the internal fluid pressure, $\sigma = \sigma(x_t, t) = \sigma_t(x_t)$ is the specific entropy at a spatial point $x_t = \eta_t(X) \in M$ for any $t \in \mathbb{R}$, which is fixed owing to the Euler Equation (393), $\nabla := \partial/\partial x$ is the usual gradient on the space of smooth functions $C^\infty(M;\mathbb{R})$ and $\langle\cdot|\cdot\rangle$ denotes the usual convolution on $T(M) \times T(M)$ subject to the usual metric in $\mathbb{R}^3$, reduced on the submanifold M. The evolution (393) is considered to be *a priori* thermodynamic equilibrium and quasi-stationary, meaning that the following *infinitesimal heat convective* and strictly mathematical relationship (387), derived above in the Introduction,

$$\delta e_t(\rho_t(x_t), \sigma_t(x_t)) = T_t(x_t)\delta\sigma_t(x_t) + p_t^{(0)}(x_t)\rho_t^{-2}(x_t)\delta\rho_t(x_t) \tag{394}$$

holds for all $x_t \in M$ and $t \in \mathbb{R}$, where $e_t : \mathcal{R}(M) \times \Sigma(M) \to C^\infty(M \times \mathbb{R};\mathbb{R})$ denotes the internal specific fluid energy, $T_t : M \to \mathbb{R}_+$ denotes the internal fluid absolute temperature, $p_t^{(0)} : M \to \mathbb{R}$ is the internal liquid pressure and the variation sign "δ" means the change subject to both the temporal variable $t \in \mathbb{R}$ and the spatial variable $x_t \in M$.

Let us now analyze the internal mathematical structure of quantities $(\rho_t, \sigma_t) \in \mathcal{R}(M) \times \Sigma(M)$ as the *physical observables* subject to their evolution (393) with respect to the group diffeomorphisms $\eta_t \in \text{Diff}(M), t \in \mathbb{R}$, generated by the liquid motion vector field $dx_t/dt = v_t(x_t), x_t := \eta_t(X), t \in \mathbb{R}, X \in M$:

$$\begin{gathered}\mathcal{L}_{d/dt}(\rho_t d^3x_t \langle v_t|dx_t\rangle) = \rho_t d^3x_t(-\rho_t^{-1}dp_t^{(0)} + d|v_t|^2/2), \\ \mathcal{L}_{d/dt}(\rho_t d^3x_t) = 0, \ \ \mathcal{L}_{d/dt}\sigma_t = 0,\end{gathered} \tag{395}$$

where $\mathcal{L}_{d/dt} : \Lambda(M) \to \Lambda(M)$ denotes the corresponding Lie derivative with respect to the vector field $d/dt := \partial/\partial t + \langle v_t|\nabla\rangle \in \Gamma(M \times \mathbb{R}; T(M)), t \in \mathbb{R}$. The relationships (395) here simply mean that at every fixed $t \in \mathbb{R}$ the space of physical observables, being by definition, the adjoint space $\mathcal{G}^* := (\Lambda^1(M) \otimes \Lambda^3(M)) \oplus (\Lambda^3(M) \oplus \Lambda^0(M))$ to the vector space $\mathcal{G} := \Gamma(M; T(M)) \times (\Lambda^0(M) \oplus \Lambda^3(M)) \simeq T_{Id}(G)$, the tangent space at the identity Id to the extended differential-functional current group manifold $G := \text{Diff}(M) \times (\Lambda^0(M) \times \Lambda^3(M)) \simeq \text{Diff}(M) \times (\mathcal{R}(M) \times \Sigma(M))$, where we have naturally identified the Abelian group product $\Lambda^0(M) \times \Lambda^3(M)$ with its direct tangent space sum $T(\Lambda^0(M)) \oplus T(\Lambda^3(M))$.

Consider now the natural action $\text{Diff}(M) \times G \to G$ of the $\text{Diff}(M)$-group on the constructed differential-functional manifold G:

$$\begin{gathered}(\eta \circ \varphi)(X) := \varphi(\eta(X)), (\eta \circ r)(X) := r(\eta(X)), \\ \eta \circ (s(X)d^3X) := \eta^*(s(X)d^3X)\end{gathered} \tag{396}$$

for $\eta \in \text{Diff}(M), X \in M$ and any $(\varphi; r, s) \in \text{Diff}(M) \times (\mathcal{R}(M) \times \Sigma(M))$. Then, taking into account the suitably extended action (396) on the differential-functional manifold G, one can formulate the following easily checkable and crucial for what will follow further proposition.

Proposition 14. *The functional manifold $G := \text{Diff}(M) \times (\mathcal{R}(M) \times \Sigma(M))$ in Eulerian coordinates is a smooth symmetry Banach group $G := \text{Diff}(M) \ltimes (\mathcal{R}(M) \times \Sigma(M))$, equal to the semidirect product of the diffeomorphism group $\text{Diff}(M)$ and the direct product $\mathcal{R}(M) \times \Sigma(M)$ of the Abelian functional $\mathcal{R}(M) \simeq \Lambda^0(M)$, and density $\Sigma(M) \simeq \Lambda^3(M)$ group, endowed in Eulerian variables with the following right group multiplication law:*

$$\begin{gathered}(\varphi_1; r_1, s_1 d^3x) \circ (\varphi_2; r_2, s_2 d^3x) = \\ = (\varphi_2 \cdot \varphi_1; r_1 + r_2 \cdot \varphi_1, s_1 d^3x + (s_2 d^3x) \cdot \varphi_1)\end{gathered} \tag{397}$$

for arbitrary elements $\varphi_1, \varphi_2 \in \text{Diff}(M), r_1, r_2 \in \Lambda^0(M)$ and $s_1 d^3x, s_2 d^3x \in \Lambda^3(M)$.

This proposition allows a simple enough interpretation, namely, it means that the adiabatic mixing of the $G \ni (\varphi_2; r_2, s_2 d^3x)$-liquid configuration with the $G \ni (\varphi_1; r_1, s_1 d^3x)$-liquid configuration amounts to summation of their densities and entropies, simultaneously changing the common specific density owing to the fact that some space of the domain M is already occupied by the first liquid configuration and the second one should be diffeomorphically shifted from this configuration to another free part of the spatial domain M, whose volume is assumed to be fixed and bounded.

The second important observation concerns the variational one-form (394), which can be naturally interpreted as some constraint on the group manifold G for any fixed initial extended Lagrangian configuration $(\eta; \rho_0, \sigma_0 d^3X) \in G$, as it follows from the conditions (389):

$$J_{\eta_t}(X)\rho_t \circ \eta_t(X) := \rho_0(X), \ \sigma_t \circ \eta_t(X) := \sigma_0(X) \tag{398}$$

for all $X \in M$, $\eta_t \in \text{Diff}(M)$ and $t \in \mathbb{R}$. In addition, if to determine, owing to (394) and the streamline adiabatic constraint $\delta\sigma_t(x_t) = 0$ for all $t \in \mathbb{R}$, the specific energy density

$$e_t(\rho_t, \sigma_t) := w_t^{(0)}(\rho_t, \sigma_t) + c_t(\sigma_t) \tag{399}$$

for some still unknown mapping $c_t: \Sigma(M) \to C^\infty(M \times \mathbb{R}; \mathbb{R})$ and the internal potential energy function $w_t^{(0)}: \mathcal{R}(M) \times \Sigma(M) \to C^\infty(M; \mathbb{R})$ of the liquid under regard, the local energy conservation property

$$\frac{d}{dt} \int_{D_t} e_t(\rho_t, \sigma_t) \rho_t(x_t) d^3 x_t = -\int_{D_t} \langle \nabla | p_t^{(0)}(x_t) v_t(x_t) \rangle d^3 x_t \tag{400}$$

holds for all $t \in \mathbb{R}$ and the domain $D_t := \eta_t(D) \subset M$, where a smooth submanifold $D \subset M$ is chosen arbitrarily and $\eta_t: M \to M$ denotes the corresponding evolution subgroup of the diffeomorphism group $\text{Diff}_0(M)$, generated by the Euler evolution Equation (393), becomes compatible with constraint (394) if there holds the following equality:

$$p_t^{(0)}(x_t) = \rho_t(x_t)^2 \partial w_t^{(0)}(\rho_t, \sigma_t) / \partial \rho_t \tag{401}$$

for all $x_t \in M$ and $t \in \mathbb{R}$. In particular, from (400) and (401) the following global internal energy functional

$$H := \int_M [w_t^{(0)}(\rho_t, \sigma_t) + c_t(\sigma_t)] \, \rho_t(x_t) d^3 x_t \tag{402}$$

is conserved that is $dH/dt = 0$ for all $t \in \mathbb{R}$.

As the extended Lagrangian configuration $(\eta; \rho_0, \sigma_0 d^3 X) \in G$ is fixed for all whiles of time $t \in \mathbb{R}$ and the dynamical variables $\rho_t \in \mathcal{R}(M)$ and $\sigma_t \in \Sigma(M)$ depend only on the evolution diffeomorphisms $\eta_t \in \text{Diff}(M)$, $t \in \mathbb{R}$, it is reasonable to consider the constraint (394) as an element of the cotangent space $T^*_{\eta_t}(\text{Diff}(M))$ to the diffeomorphism group $\text{Diff}(M)$ at the point $\eta_t \in \text{Diff}(M)$ for any $t \in \mathbb{R}$.

Determine first the tangent space $T_\eta(G)$ to the group manifold G at point $(\eta; \rho_0, \sigma_0 d^3 X) \in G$, which will be the direct product of the tangent spaces $T_\eta(\text{Diff}(M)), T_{\rho_0}(\Lambda^0(M))$ and $T_{\sigma_0 d^3 X}(\Lambda^3(M))$. The last two tangent spaces are isomorphic, respectively, to themselves, that is $T_{\rho_0}(\Lambda^0(M)) \simeq \Lambda^0(M)$ and $T_{\sigma_0 d^3 X}(\Lambda^3(M)) \simeq \Lambda^3(M)$ at any $X \in M$. Their adjoint spaces are naturally determined as suitably constructed density and functional spaces on the manifold $M: T^*_{\rho_0}(\Lambda^0(M)) \simeq \Lambda^3(M)$ and $T^*_{\sigma_0 d^3 X}(\Lambda^3(M)) \simeq \Lambda^0(M)$. Concerning the tangent space $T_\eta(\text{Diff}(M))$ at a configuration $\eta \in \text{Diff}(M)$ we will make use of the construction, devised before in [122,181,194]. Namely, let $\eta \in \text{Diff}(M)$ be a Lagrangian configuration and determine the tangent space $T_\eta(\text{Diff}(M))$ at $\eta \in \text{Diff}(M)$ as the collection of left invariant vectors $\xi_\eta := L_{\eta,*}\xi$ at $\eta \in \text{Diff}(M)$, where $L_\eta: \text{Diff}(M) \to \text{Diff}(M)$ is, by definition, the left shift on the diffeomorphism group $\text{Diff}(M)$ and $\xi \in T_{Id}(\text{Diff}(M))$ is a tangent vector at the unity $Id \in \text{Diff}(M)$. It is obvious that for all reference points $X \in M$ and any smooth curve $\mathbb{R} \ni \tau \to \eta_\tau \in \text{Diff}(M)$ of diffeomorphisms of M, the set of right invariant vectors $\xi(X) = (\eta^{-1} \circ d\eta_t / d\tau)(X))|_{\tau=0} \in T_X(M)$ at point $X \in M$ defines a smooth vector field $\xi: M \to T(M)$ on the manifold M. Since, by definition, the tangent space $T_{Id}(\text{Diff}(M))$ coincides with the Lie algebra $\text{Diff}(M)$ of the diffeomorphism group $\text{Diff}(M)$, strictly isomorphic to the Lie algebra $\Gamma(T(M))$ of right invariant vector fields on M, the dual space $T^*_{Id}(\text{Diff}(M))$ can be naturally determined from the geometric point of view as the space $diff^*(M)$, consisting of analytic functions on $diff(M)$ and coinciding with the set of one-form densities on M:

$$diff^*(M) \simeq \Lambda^1(M) \otimes |\Lambda^3(M)|. \tag{403}$$

Similarly, the cotangent space $T^*_\eta(\text{Diff}(M))$ consists of all one-form densities on M over $\eta \in \text{Diff}(M)$:

$$T^*_\eta(\text{Diff}(M)) = \{\alpha_\eta : M \to T^*(M) \otimes |\Lambda^3(M)| : \alpha_\eta(X) \in T^*_{\eta(X)}(M) \otimes |\Lambda^3(M)|\} \tag{404}$$

subject to the canonical nondegenerate convolution $(\cdot|\cdot)_c$ on $T^*_\eta(\mathrm{Diff}(M)) \times T_\eta(\mathrm{Diff}(M))$: if $\alpha_\eta \in T^*_\eta(\mathrm{Diff}(M))$, $\xi_\eta \in T_\eta(\mathrm{Diff}(M))$, where $\alpha_\eta|_X = \langle \alpha_\eta(X)|dx\rangle \otimes d^3X$, $\xi_\eta|_X = \langle \xi_\eta(X)|\partial/\partial x\rangle$, then

$$(\alpha_\eta|\xi_\eta)_c := \int_M \langle \alpha_\eta(X)|\xi_\eta(X)\rangle d^3X. \tag{405}$$

The construction above makes it possible to identify the cotangent bundle $T^*_\eta(\mathrm{Diff}(M))$ at the fixed Lagrangian configuration $\eta \in \mathrm{Diff}(M)$ to the tangent space $T_\eta(\mathrm{Diff}(M))$, insomuch as the tangent space $T(M)$ is endowed with the natural internal tangent bundle metric $\langle\cdot|\cdot\rangle_g$ at any point $\eta(X) \in M$, identifying $T(M)$ with $T^*(M)$ via the metric isomorphism $\sharp : T^*(M) \to T(M)$. The latter can also be naturally lifted to $T^*_\eta(\mathrm{Diff}(M))$ at $\eta \in \mathrm{Diff}(M)$, namely: for any elements $\alpha_\eta, \beta_\eta \in T^*_\eta(\mathrm{Diff}(M)), \alpha_\eta|_X = \langle \alpha_\eta(X)|dx\rangle \otimes d^3X$ and $\beta_\eta|_X = \langle \beta_\eta(X)|dx\rangle \otimes d^3X \in T^*_\eta(\mathrm{Diff}(M))$ we can define the metric

$$(\alpha_\eta|\beta_\eta)_g := \int_M \rho_0(X)\langle \alpha^\sharp_\eta(X)|\beta^\sharp_\eta(X)\rangle_g d^3X, \tag{406}$$

where, by definition, $\alpha^\sharp_\eta(X) := \sharp(\rho_0(X)^{-1}\langle \alpha_\eta(X)|dx\rangle)$, $\beta^\sharp_\eta(X) := \sharp(\rho_0(X)^{-1}\langle \beta_\eta(X)|dx\rangle)$ $\in T_{\eta(X)}(M)$ for any $X \in M$.

The diffeomorphism group $\mathrm{Diff}(M)$ can be naturally restricted to the factor-group $\mathrm{Diff}_0(M) := \mathrm{Diff}(M)/Diff_{\rho_0,\sigma_0}(M)$ subject to the stationary normal symmetry subgroup $\mathrm{Diff}_{\rho_0,\sigma_0}(M) \subset \mathrm{Diff}(M)$, where

$$Diff_{\rho_0,\sigma_0}(M) := \{\varphi \in \mathrm{Diff}(M) : \rho_0(X) = J_{\varphi(X)}\rho_0(\varphi(X)), \sigma_0(X) = \sigma_0(\varphi(X))\} \tag{407}$$

for any $X \in M$. Based on the construction above, one can proceed to constructing smooth flows and functionals on the specially extended group manifold $G_0 := \mathrm{Diff}_0(M) \ltimes (\Lambda^0(M) \times \Lambda^3(M))$ and consider their coadjoint action on the cotangent bundle $T^*_{g_\eta}(G_0), g_\eta := (\eta; \rho_0, \sigma_0) \in G_0$, and relate them in some way to the evolution with respect to the Euler Equation (393). Moreover, as the cotangent bundle $T^*_{g_\eta}(G_0), g_\eta \in G_0$, is a priori endowed with the canonical Poisson structure, one can study both the Hamiltonian flows on it, related with the Euler Equation (393), and a hidden geometrical meaning of the differential constraints like (394).

10.4. The Hamiltonian Analysis and Related Adiabatic Liquid Dynamics

We observed above that the liquid motion is adequately described by means of the symmetry diffeomorphism group $\mathrm{Diff}_0(M)$, acting on the target manifold $M \subset \mathbb{R}^3$, and in this way the modeling liquid motion, generated by suitable vector fields on $\mathrm{Diff}_0(M)$. This also means that the fluid motion strongly depends on the constraint (394) on the cotangent bundle $T^*_{g_\eta}(G_0), g_\eta \in G_0$, and *a priori* possesses the canonical Poisson structure on it. Taking into account that the diffeomorphism group $\mathrm{Diff}_0(M)$ acts on the extended group density manifold $G_0 := \mathrm{Diff}_0(M) \ltimes (\Lambda^0(M) \times \Lambda^3(M))$, fixed by the element $(\eta; \rho_0, \sigma_0 d^3X) \in G$, one can suitably construct the canonical Poisson bracket on the cotangent bundle $T^*_{g_\eta}(G_0), g_\eta \in G_0$, using the canonical coordinate variables on it. Namely, let $(\mu_\eta; \rho_0 d^3X, \sigma_0)$ $\in T^*_{g_\eta}(G_0), g_\eta \in G_0$, be coordinates on $T^*_{g_\eta}(G_0)$, where

$$\begin{aligned} \mu_\eta(X) &= \rho_0(X)[V^\flat_\eta(X)]d^3X|_{x=\eta(X)} = \\ &= \rho_0(X)v^\flat(\eta(X))J_{\eta^{-1}}(x)d^3x := \rho(x)v(x)d^3x, \\ r_\eta(X) &= \rho_0(X)d^3X = \rho_0(X)d^3X|_{x=\eta(X)} := \rho(x)d^3x, \\ s_\eta(X) &= \sigma_0(X) = \sigma(\eta(X))|_{x=\eta(X)} := \sigma(x) \end{aligned} \tag{408}$$

and $\flat := \sharp^{-1}$, being suitably represented into the Eulerian spatial variables on $T^*_{g_\eta}(G_0)$ at point $(\eta; \rho, \sigma d^3x) \in G_0$. In particular, the quantities $\mu(x) := \rho(x)v(x)d^3x = (\eta^*\mu_\eta)(X)$,

$r(x) := \rho(x)d^3x = (\eta^* r_\eta)(X)$ and $s(x) := \sigma(x) = (\eta^* s_\eta)(X)$ are called, respectively, the Eulerian momentum density, the Eulerian fluid density and entropy variables at point $x = \eta(X) \in M$. The corresponding metric on $T^*_{g_\eta}(G_0)$ is given by the expression

$$((\alpha_{\eta,1}; r_{\eta,1}s_{\eta,1})|(\alpha_{\eta,2}; r_{\eta,2}s_{\eta,2})) := (\alpha_{\eta,1}|\alpha_{\eta,2})+ \\ +(r_{\eta,1}|r_{\eta,2}) + (s_{\eta,1}|s_{\eta,2}), \tag{409}$$

where $(\alpha_{\eta,1}|\alpha_{\eta,2})$ for $\alpha_{\eta,1}, \alpha_{\eta,2} \in T^*_\eta(\mathrm{Diff}_0(M))$ is determined by (406) and for any $r_{\eta,1}, r_{\eta,2} \in T^*_\eta(\Lambda^3(M))$ and $s_{\eta,1}, s_{\eta,2} \in T^*_\eta(\Lambda^0(M))$ one determines, respectively, as

$$(r_{\eta,1}|r_{\eta,2}) := \int_M (\rho_1(x)\rho_2(x))d^3x, \; (s_{\eta,1}|s_{\eta,2}) := \int_M (\sigma_1(x)\sigma_2(x))d^3x. \tag{410}$$

Consider now the cotangent bundle $T^*_{g_\eta}(G_0)$ at point $g_\eta = (\eta; \rho, \sigma d^3x) \in G_0$ as a smooth manifold endowed with the canonical symplectic structure on it, equivalent to the corresponding canonical Poisson bracket on $T^*_{g_\eta}(G_0)$. Taking into account that the manifold $T^*_{g_\eta}(G_0)$, shifted by the right $R_{\eta^{-1}}$-action to the manifold $T^*_{Id}(G_0)$, $Id \in G_0$, becomes diffeomorphic to the adjoint space $\mathcal{G}^*$ to the Lie algebra $\mathcal{G}$ of the group G_0, as was stated [30–33,41] by S. Lie in 1887, this canonical Poisson bracket on $T^*_\eta(G_0)$ transforms [26,31,32,41,181,195] into the classical Lie–Poisson bracket on the adjoint space $\mathcal{G}^*$. Moreover, the orbits of the group G_0 on $T^*_{g_\eta}(G_0), g_\eta = (\eta; \rho, \sigma d^3x) \in G_0$, transform into the corresponding coadjoint orbits on the adjoint space $\mathcal{G}^*$, generated by elements of the Lie algebra $\mathcal{G}$. To construct this Lie–Poisson bracket, we formulate the following preliminary proposition.

Proposition 15. *The Lie algebra $\mathcal{G} \simeq \Gamma(M; T(M)) \ltimes (\Lambda^0(M)) \oplus \Lambda^3(M))$ is determined by the following Lie commutator relationships:*

$$[(a_1; r_1, s_1), (a_2; r_2, s_2)] = ([a_1, a_2]; \\ \langle a_1|\nabla r_2\rangle - \langle a_2|\nabla r_1\rangle, \langle \nabla|a_1 s_2\rangle - \langle \nabla|a_2 s_1\rangle) \tag{411}$$

for any vector fields $a_1, a_2 \in diff_0(M) \simeq \Gamma(M; T(M))$ and scalar quantities $r_1, r_2 \in \Lambda^0(M)$ and $s_1, s_2 \in \Lambda^3(M)$ on the manifold M.

Proof. Proof of the commutation relationships (411) easily follows from the group multiplication (397), if to take into account that tangent spaces $T(\Lambda^0(M)) \simeq \Lambda^0(M)$ and $T(\Lambda^3(M)) \simeq (\Lambda^3(M))$. □

As an example, we calculate, for brevity, the Poisson bracket on the cotangent space $T^*_\eta(\mathrm{Diff}(\mathbb{T}^n))$ at any $\eta \in \mathrm{Diff}(\mathbb{T}^n)$. Consider the cotangent space $T^*_\eta(\mathrm{Diff}(\mathbb{T}^n)) \simeq diff^*(\mathbb{T}^n)$, the adjoint space to the tangent space $T_\eta(\mathrm{Diff}(\mathbb{T}^n))$ of left invariant vector fields on $\mathrm{Diff}(\mathbb{T}^n)$ at any $\eta \in \mathrm{Diff}(\mathbb{T}^n)$, and take the canonical symplectic structure on $T^*_\eta(\mathrm{Diff}(\mathbb{T}^n))$ in the form $\omega^{(2)}(\mu, \eta) := \delta\alpha(\mu, \eta)$, where the canonical Liouville form $\alpha(\mu, \eta) := (\mu|\delta\eta)_c \in \Lambda^1_{(\mu,\eta)}(T^*_\eta(\mathrm{Diff}(\mathbb{T}^n)))$ at a point $(\mu, \eta) \in T^*_\eta(\mathrm{Diff}(\mathbb{T}^n))$ is defined *a priori* on the tangent space $T_\eta(\mathrm{Diff}(\mathbb{T}^n)) \simeq \Gamma(T(M))$ of right-invariant vector fields on the torus manifold $\mathbb{T}^n$. Having calculated the corresponding Poisson bracket of smooth functions $(\mu|a)_c, (\mu|b)_c \in C^\infty(T^*_\eta(\mathrm{Diff}(\mathbb{T}^n)); \mathbb{R})$ on $T^*_\eta(\mathrm{Diff}(\mathbb{T}^n)) \simeq diff^*(\mathbb{T}^n)$, $\eta \in \mathrm{Diff}(\mathbb{T}^n)$, one can formulate the following proposition.

Proposition 16. *The Lie–Poisson bracket on the coadjoint space $T^*_\eta(\mathrm{Diff}(\mathbb{T}^n)) \simeq diff^*(\mathbb{T}^n)$ is equal to the expression*

$$\{f, g\}(\mu) = (\mu|[\delta f(\mu)/\delta\mu, \delta g(\mu)/\delta\mu])_c \tag{412}$$

for any smooth functionals $f, g \in C^\infty(\mathcal{G}^*; \mathbb{R})$.

Proof. By definition [26,122] of the Poisson bracket of smooth functions $(\mu|a)_c, (\mu|b)_c \in C^\infty(T^*_\eta(\mathrm{Diff}(\mathbb{T}^n)); \mathbb{R})$ on the symplectic space $T^*_\eta(\mathrm{Diff}(\mathbb{T}^n))$, it is easy to calculate that

$$\begin{gathered}\{\mu(a), \mu(b)\} := -\delta\alpha(X_a, X_b) = \\ = -X_a(\alpha|X_b)_c + X_b(\alpha|X_a)_c + (\alpha|[X_a, X_b])_c,\end{gathered} \tag{413}$$

where $X_a := \delta(\mu|a)_c/\delta\mu = a \in diff(\mathbb{T}^n), X_b := \delta(\mu|b)_c/\delta\mu = b \in diff(\mathbb{T}^n)$. Since the expressions $X_a(\alpha|X_b)_c = 0$ and $X_b(\alpha|X_a)_c = 0$ owing the right-invariance of the vector fields $X_a, X_b \in T_\eta(\mathrm{Diff}(\mathbb{T}^n))$, the Poisson bracket (412) transforms into

$$\begin{gathered}\{(\mu|a)_c, (\mu|b)_c\} = (\alpha|[X_a, X_b])_c = \\ = (\mu|[a, b])_c = (\mu|[\delta(\mu|a)_c/\delta\mu, \delta(\mu|b)_c/\delta\mu])_c\end{gathered} \tag{414}$$

for all $(\mu, \eta) \in T^*_\eta(\mathrm{Diff}(\mathbb{T}^n)) \simeq diff^*(\mathbb{T}^n), \eta \in \mathrm{Diff}(\mathbb{T}^n)$ and any $a, b \in diff(\mathbb{T}^n)$. The Poisson bracket (412) is easily generalized to

$$\{f, g\}(\mu) = (\mu|[\delta f(\mu)/\delta\mu, \delta g\mu)/\delta\mu])_c \tag{415}$$

for any smooth functionals $f, g \in C^\infty(\mathcal{G}^*; \mathbb{R})$, finishing the proof. □

Proceed now to the Grassmann algebra $\Lambda(M)$ endowed with Hodge [197] star-isomorphism $* : \Lambda(M) \to \Lambda(M)$ subject to the usual metric on the tangent space $T(M)$ and determine the adjoint space to the Abelian subalgebra $\mathcal{R}(M) \oplus \Sigma(M) \simeq \Lambda^0(M) \oplus \Lambda^3(M)$ as the space $*\Lambda^3(M) \oplus *\Lambda^0(M)$ with respect to the following scalar product on $\Lambda(M)$:

$$(\alpha^{(n)}|\beta^{(m)}) := \delta_{mn} \int_M (\alpha^{(n)} \wedge *\beta^{(m)}) \tag{416}$$

for any $\alpha^{(n)}, \beta^{(m)} \in \Lambda(M), m, n = \overline{0, 3}$. Then the adjoint space $\mathcal{G}^*$, owing to the expressions (409) and (389), is described by means of the Eulerian variables $(\mu; \rho d^3x, \sigma) \in \mathcal{G}^* \simeq (\Lambda^1(M) \otimes |\Lambda^3(M)|) \ltimes (\Lambda^3(M) \oplus \Lambda^0(M))$. The latter makes it possible to calculate the corresponding Lie–Poisson bracket on the adjoint space $\mathcal{G}^*$ at a point $l := (\mu; \rho d^3x, \sigma) \in \mathcal{G}^*$, generalizing the Poisson bracket (414):

$$\begin{gathered}\{f, g\}(l) = (l|[\delta f/\delta l, \delta g/\delta l])_c = \\ = \int_M d^3x \Big\langle m \Big| \Big[\Big\langle \tfrac{\delta f}{\delta m} |\nabla \Big\rangle \tfrac{\delta g}{\delta m} - \Big\langle \tfrac{\delta g}{\delta m} |\nabla \Big\rangle \tfrac{\delta f}{\delta m} \Big] \Big\rangle + \\ + \int_M \rho d^3x \Big[\Big\langle \tfrac{\delta f}{\delta m} \Big| \nabla \tfrac{\delta g}{\delta \rho} \Big\rangle - \Big\langle \tfrac{\delta g}{\delta m} \Big| \nabla \tfrac{\delta f}{\delta \rho} \Big\rangle \Big] + \\ + \int_M \sigma \Big[\Big\langle \nabla \Big| \tfrac{\delta f}{\delta m} \tfrac{\delta g}{\delta \sigma} \Big\rangle - \Big\langle \nabla \Big| \tfrac{\delta g}{\delta m} \tfrac{\delta f}{\delta \sigma} \Big\rangle \Big] d^3x\end{gathered} \tag{417}$$

for any smooth functionals $f, g \in C^\infty(\mathcal{G}^*; \mathbb{R})$, where we put, by definition, $\mu(x) := \langle m(x)|dx\rangle \otimes d^3x$, $m(x) = \rho(x)v(x) \in T^*(M)$ for all $x \in M$ and any $t \in \mathbb{R}$.

Return now to the constraint (394) in the following variational form:

$$\delta e_t(\rho_t, \sigma_t)/\delta t = T_t(x_t)\delta\sigma_t(x_t)/\delta t + p_t^{(0)}(x_t)\rho_t^{-2}(x_t)\delta\rho_t(x_t)/\delta t, \tag{418}$$

which should hold at any $x_t \in M$ for all $t \in \mathbb{R}$. Insomuch as, owing to the Euler Equation (393), the full (convective) derivative $\delta\sigma_t(x_t)/\delta t = 0$ at any $x_t \in M$ for all $t \in \mathbb{R}$, one checks once more that the expression (399) holds at any $x_t \in M$ for all $t \in \mathbb{R}$. To determine the energy density function (399), we consider the Euler Equation (393) and check their Hamiltonian structure subject to the Poisson bracket (417), i.e., the existence of a Hamiltonian functional $H : \mathcal{G}^* \rightarrow \mathbb{R}$, for which

$$\frac{\partial}{\partial t}(m;\rho,\sigma)^\intercal = \{H,(m,\rho,\sigma)^\intercal\} \tag{419}$$

at any element $l = (m := \rho v;\rho,\sigma)^\intercal \in \mathcal{G}^*$. By means of easy calculations, one obtains from the system (419) the variational gradient vector

$$\delta H(l)/\delta l = (m\rho^{-1}; -|m|^2/(2\rho^2) + w^{(0)}\ (\rho,\sigma) + \rho\partial w^{(0)}(\rho,\sigma)/\partial\rho, \rho\partial w^{(0)}(\rho,\sigma)/\partial\sigma), \tag{420}$$

from which one derives [11,59,120] via the Volterra homotopy mapping

$$H = \int_0^1 (\delta H(\lambda l)/\delta l|l)_c d\lambda \tag{421}$$

the exact Hamiltonian expression

$$H = \int_M (|m|^2/(2\rho) + \rho w^{(0)}(\rho,\sigma)]d^3x, \tag{422}$$

coinciding with the expression (402) at $c(\sigma) := |m|^2/(2\rho^2) = |v|^2/2$, as $m := \rho v$ for $v \in T(M)$. Thus, we obtain the internal energy density functional (399) as

$$e_t(\rho_t,\sigma_t) = |v_t|^2/2 + w_t^{(0)}(\rho_t,\sigma_t), \tag{423}$$

for all $\rho := \rho_t \in \mathcal{R}(M), \sigma := \sigma_t \in \Sigma(M)$ and $v_t \in T(M)$, satisfying simultaneously both the constraint (394) and the Euler evolution Equation (393) for all $t \in \mathbb{R}$. Moreover, from the condition (400) one easily finds [194] the following important local differential relationship:

$$\begin{gathered}\partial[\rho_t(x_t)e_t(\rho_t,\sigma_t)]/\partial t + \langle\nabla|\rho_t(x_t)v_t(x_t)(e_t(\rho_t,\sigma_t)+ \\ +\rho_t(x_t)\partial w_t^{(0)}(\rho_t,\sigma_t)/\partial\rho_t\ )\rangle = 0,\end{gathered} \tag{424}$$

satisfied for all $x_t \in M$ and $t \in \mathbb{R}$, also confirming the energy conservation (422).

10.5. The Hamiltonian Analysis and Related Isothermal Liquid Dynamics

Consider a liquid motion governed by the Euler equations

$$\begin{gathered}\partial v/\partial t + \langle v|\nabla\rangle v + \rho^{-1}\nabla p^{(0)} = 0, \\ \partial\rho/\partial t + \langle\nabla|\rho v\rangle = 0, \partial T/\partial t + \langle v|\nabla T\rangle = 0,\end{gathered} \tag{425}$$

and describing the ideal compressible and isothermal motion of an ideal compressible and adiabatic fluid in a spatial domain $M \subset \mathbb{R}^3$, as the temperature $T_t(x_t) = T_0(x_t)$ at any evolution point $x_t := \eta_t(X) \in M$ for all $X \in M$ and $t \in \mathbb{R}$. The evolution (425) is considered to be *a priori* thermodynamically quasi-stationary, i.e., the following, *infinitesimal convective* energy relationship

$$\delta\tilde{h}_t(\rho_t,T_t) = -\sigma_t(x_t)\ \delta T_t + p_t^{(0)}(x_t)\ \rho_t^{-2}\delta\rho_t \tag{426}$$

holds for all densities $\rho_t \in \mathcal{R}(M)$, temperature $T_t \in \mathcal{T}(M)$ and specific entropy $\sigma_t \in \Sigma(M)$, where $\tilde{h} : \mathcal{R}(M) \times \mathcal{T}(M) \rightarrow \mathbb{R}$ denotes the corresponding internal specific fluid "energy"and the variation sign "δ" means the change subject to both the temporal variable $t \in \mathbb{R}$ and the spatial variable $x_t \in M$. Under the imposed *isothermal condition* $\delta T_t = 0$ the expression (426) transforms into

$$\tilde{h}_t(\rho_t,T_t) = |v_t|^2/2 + \tilde{w}_t^{(0)}(\rho_t,T_t), \tag{427}$$

where $\tilde{w}_t^{(0)}(\rho_t, T_t) := w_t^{(0)}(\rho_t, \sigma_t)|_{\sigma_t := \tilde{\sigma}(\rho_t, T_t)} - T_t\sigma_t(\rho_t, T_t)$, is the specific potential liquid energy for *the isothermal flow*, determined at $\sigma_t := \sigma_t(\rho_t, T_t)$, solving the functional relation $T_t = \partial w_t^{(0)}(\rho_t, \sigma_t)/\partial\sigma_t \in \mathcal{T}(M)$ subject to the entropy argument $\sigma_t \in \Sigma(M)$, if the condition $\partial^2 w_t^{(0)}(\rho_t, \sigma_t)/\partial\sigma_t^2 \neq 0$ holds for all densities $\rho_t \in \mathcal{R}(M)$ and $t \in \mathbb{R}$.

Observe now that the third equation of (425) is exactly equivalent to the internal average fluid kinetic energy conservation integral relationship

$$\frac{d}{dt}\int_{D_t} \rho_t(x_t)T_t(x_t)d^3x_t = 0 \tag{428}$$

over the domain $D_t := \eta_t(D) \subset M$, where a smooth submanifold $D = D_t|_{t=0} \subset M$ is chosen arbitrarily and $\eta_t : M \to M, t \in \mathbb{R}$, denotes the corresponding evolution subgroup of the diffeomorphism group $\mathrm{Diff}_0(M)$, generated by the Euler evolution Equation (425). The relationship (428) simply means that if the density function $\rho_t \in \mathcal{R}(M)$ transforms under diffeomorphism group $\mathrm{Diff}_0(M)$ action as the Abelian functional group $\mathcal{R}(M) \simeq \Lambda^0(M)$, the corresponding transformation of the temperature $T_t \in \mathcal{T}(M)$ is induced by the diffeomorphism group $\mathrm{Diff}_0(M)$ action on the related Abelian group $\mathcal{T}(M) \simeq \Lambda^3(M)$. Concerning the energy density (427) one easily obtains the following differential relationship:

$$\partial[\rho_t(x_t)\tilde{h}_t(\rho_t, T_t)]/\partial t + \langle\nabla|\rho_t(x_t)v_t(x_t)\left[\tilde{h}_t(\rho_t, T_t)\rangle + \rho_t\partial\tilde{w}_t^{(0)}(\rho_t, T_t)/\partial\rho_t\right]\rangle = 0, \tag{429}$$

satisfied for all $t \in \mathbb{R}$. As a simple consequence of the relationship (429), one obtains that the following functional

$$\tilde{H} = \int_{D_t} \rho_t(x_t)\tilde{h}_t(\rho_t, T_t)d^3x_t \tag{430}$$

is conserved over the domain $D_t := \eta_t(D) \subset M$, $t \in \mathbb{R}$, where a smooth submanifold $D = D_t|_{t=0} \subset M$ is chosen arbitrarily.

Similarly to the reasoning above, one can now construct the differential-functional group space $\mathrm{Diff}(M) \times (\mathcal{R}(M) \times \mathcal{T}(M))$ and formulate the following easily checkable proposition. The differential-functional group functional manifold $\mathrm{Diff}(M) \times (\mathcal{R}(M) \times \mathcal{T}(M))$ in Eulerian coordinates is a smooth Banach group $G := \mathrm{Diff}(M) \ltimes (\mathcal{R}(M) \times \mathcal{T}(M))$, equal to the semidirect product of the diffeomorphism group $\mathrm{Diff}(M)$ and the direct product $\mathcal{R}(M) \times \mathcal{T}(M)$ of Abelian functional $\mathcal{R}(M) \simeq \Lambda^0(M)$ and density $\mathcal{T}(M) \simeq \Lambda^3(M)$ groups, endowed with the following group multiplication law:

$$\begin{gathered}(\varphi_1; r_1, \tau_1 d^3x\ ) \circ (\varphi_2; r_2, \tau_2 d^3x) = \\ +(\varphi_2 \cdot \varphi_1; r_1 + r_2 \cdot \varphi_1, \tau_1 d^3x + (\tau_2 d^3x) \cdot \varphi_1)\end{gathered} \tag{431}$$

for arbitrary elements $\varphi_1, \varphi_2 \in \mathrm{Diff}(M), r_1, r_2 \in \Lambda^0(M)$ and $\tau_1 d^3x, \tau_2 d^3x \in \Lambda^3(M)$.

This proposition allows a simple enough interpretation, namely, it means that the adiabatic mixing of the $G \ni (\varphi_2; r_2, \tau_2 d^3x)$ liquid configuration with the $G \ni (\varphi_1; r_1, \tau_1 d^3x)$ liquid configuration amounts to summation of their spatially shifted densities, simultaneously changing the common specific kinetic energy, proportional [55,198,199] to the liquid temperature, owing to the fact that some space in the domain M is already occupied by the first liquid configuration and the second one should be diffeomorphically shifted from this configuration to another free part of the spatial domain M with fixed and bounded volume. The diffeomorphism group $\mathrm{Diff}(M)$ can be naturally restricted to the factor-group $\mathrm{Diff}_0(M) := \mathrm{Diff}(M)/\mathrm{Diff}_{\rho_0, T_0}(M)$ subject to the stationary normal symmetry subgroup $\mathrm{Diff}_0(M) := \mathrm{Diff}_{\rho_0, T_0}(M) \subset \mathrm{Diff}(M)$, where

$$\mathrm{Diff}_{\rho_0, T_0}(M) := \{\varphi \in \mathrm{Diff}(M) : \rho_0(X) = J_{\varphi(X)}\rho_0(\varphi(X)), T_0(X) = T_0(\varphi(X))\} \tag{432}$$

for any $X \in M$. Based on the construction above one can proceed to studying the extended Banach group $G := \mathrm{Diff}_0(M) \ltimes (\Lambda^0(M) \times \Lambda^3(M))$ action on the cotangent bundle $T^*_{g_\eta}(G)$

at $g_\eta := (\eta;\rho_0,T_0) \in G_0$, generated by the fluid evolution with respect to the Euler Equation (425). The related fluid motion is naturally modelled by means of the coadjoint action of the corresponding Lie algebra $\mathcal{G} \simeq T_{g_\eta}(G_0) \simeq \Gamma(M;T(M)) \ltimes (\Lambda^0(M) \oplus \Lambda^3(M))$ of the group G_0, $g_\eta = Id \in G_0$, on its adjoint space $\mathcal{G}^* \simeq (\Lambda^1(M) \otimes \Lambda^3(M)) \ltimes (*\Lambda^0(M) \oplus *\Lambda^3(M)) = (\Lambda^1(M) \otimes \Lambda^3(M)) \ltimes (\Lambda^3(M) \oplus \Lambda^0(M))$.

The related Lie structure on $\mathcal{G}$ easily ensues from the action (431):

$$\begin{gathered}[(a_1;r_1,\tau_1),(a_2;r_2,\tau_2)] = ([a_1,a_2]; \\ \langle a_1|\nabla r_2\rangle - \langle a_2|\nabla r_1\rangle, \langle\nabla|a_1\tau_2\rangle - \langle\nabla|a_2\tau_1\rangle)\end{gathered} \tag{433}$$

for any representative elements $(a_1;r_1,\tau_1)$ and $(a_2;r_2,\tau_2) \in \mathcal{G}$. Moreover, as the cotangent bundle $T^*_{g_\eta}(G_0)$ at $g_\eta = Id \in G_0$ is diffeomorphic to the adjoint space $\mathcal{G}^*$ to the Lie algebra $\mathcal{G}$ of the Banach group G_0, it is *a priori* endowed with the canonical Lie–Poisson structure

$$\begin{gathered}\{f,g\}(l) = (l|[\delta g/\delta l, \delta f/\delta l])_c = \\ = \int_M d^3x\Big\langle m\Big|\Big[\Big\langle\tfrac{\delta f}{\delta m}|\nabla\Big\rangle\tfrac{\delta g}{\delta m} - \Big\langle\tfrac{\delta g}{\delta m}|\nabla\Big\rangle\tfrac{\delta f}{\delta m}\Big]\Big\rangle + \\ + \int_M \rho d^3x\Big[\Big\langle\tfrac{\delta f}{\delta m}\Big|\nabla\tfrac{\delta g}{\delta \rho}\Big\rangle - \Big\langle\tfrac{\delta g}{\delta m}\Big|\nabla\tfrac{\delta f}{\delta \rho}\Big\rangle\Big] + \\ + \int_M T\Big[\Big\langle\nabla\Big|\tfrac{\delta f}{\delta m}\tfrac{\delta g}{\delta T}\Big\rangle - \Big\langle\nabla\Big|\tfrac{\delta g}{\delta m}\tfrac{\delta f}{\delta T}\Big\rangle\Big]d^3x\end{gathered} \tag{434}$$

for any smooth functional $f,g \in C^\infty(\mathcal{G}^*;\mathbb{R})$, where we put, by definition, an element $l := (m;\rho,T) \simeq (\mu;\rho d^3x,T) \in \mathcal{G}^*$, $\mu(x) := \langle m(x)|dx\rangle \otimes d^3x$, $m(x) = \rho(x)v(x) \in T^*(M)$ for all $x \in M$ and $t \in \mathbb{R}$, one can easily check that the flow (425) is Hamiltonian:

$$dl/dt = \{\tilde{H}, l\} \tag{435}$$

subject to the adjusted Hamiltonian functional (430):

$$\tilde{H} := \int_M \rho_t h_t(\rho_t,T_t)d^3x_t = \int_M \rho_t(|m_t|^2/2\rho_t^2 + \tilde{w}_t^{(0)}(\rho_t,T_t))d^3x. \tag{436}$$

satisfying the conservative condition $d\tilde{H}/dt = 0$ for all $t \in \mathbb{R}$, following simultaneously both from (435) and from the differential relationship (429).

10.6. The Hamiltonian Analysis and Adiabatic Magneto-Hydrodynamic Superfluid Motion

We start with considering a quasi-neutral superfluid contained in a domain $M \subset \mathbb{R}^3$ and interacting with a "frozen" sourceless magnetic field $B \in \mathcal{B}(M) \subset C^\infty(M;\mathbb{E}^3)$, satisfying the superconductivity conditions

$$\tilde{E} := E + v \times B = 0,\ \partial E/\partial t = \nabla \times B, \tag{437}$$

where $\tilde{E} : M \to \mathbb{E}^3$ is the internal net superfluid electric field, $E = -\partial A/\partial t : M \to \mathbb{E}^3$ and $B = \nabla \times A : M \to \mathbb{E}^3$ are the internal electric and magnetic fields, respectively, generated by the corresponding magnetic vector field potential $A : M \to \mathbb{E}^3$, $v : M \longrightarrow T(M)$ is the superfluid velocity and "$\times$" denotes the usual vector product in the Euclidean space $\mathbb{E}^3$. The following natural boundary conditions $\langle n|v\rangle|_{\partial M} = 0$ and $\langle n|B\rangle|_{\partial M} = 0$ are imposed on the superfluid flow, where $n \in T^*(M)$ is the vector normal to the boundary ∂M, which is considered to be smooth almost everywhere.

Then, in adiabatic magneto-hydrodynamics (MHD), quasi-neutral superconductive superfluid motion is described by the following system of evolution equations:

$$\begin{gathered}\partial v/\partial t + \langle v|\nabla\rangle v + \rho^{-1}\nabla p - \rho^{-1}(\nabla \times B) \times B = 0, \\ \partial\rho/\partial t + \langle\nabla|\rho v\rangle = 0, \partial\sigma/\partial t + \langle u|\nabla\sigma\rangle = 0, \partial B/\partial t = \nabla \times (v \times B),\end{gathered} \tag{438}$$

where, as before, $\rho := \rho_t \in \mathcal{R}(M)$ is the superfluid density, $B := B_t : M \longrightarrow \mathbb{E}^3$ is the "frozen" into the superfluid magnetic field, $p := p_t : M \longrightarrow \mathbb{R}$ is the internal liquid pressure and $\sigma := \sigma_t : M \longrightarrow \mathbb{R}$ is the specific superfluid entropy at time $t \in \mathbb{R}$. The latter is related to the internal MHD superfluid specific energy function $e = e_t(\rho_t, \sigma_t)$ owing to the first thermodynamic law:

$$T_t(\rho_t, \sigma_t)\, \delta\sigma_t = \delta e_t(\rho_t, \sigma_t) - p_t \rho_t^{-2} \delta\rho_t, \tag{439}$$

satisfied for any admissible variations of the phase space parameters $\rho_t \in \mathcal{R}(M)$, $\sigma_t \in \Sigma(M)$, where $T_t = T_t(\rho_t, \sigma_t)$ is the internal absolute temperature in the superfluid for $t \in \mathbb{R}$. The isentropic condition $\delta\sigma_t(x_t) = 0$, where $x_t := \eta_t(X) \in M$ for all $X \in M$ and that related to (438) evolution diffeomorphism $\eta_t \in \mathrm{Diff}(M), t \in \mathbb{R}$, entails the following expression for the specific internal energy

$$e_t(\rho_t, \sigma_t) = w_t^{(0)}(\rho_t, \sigma_t) + c_t(\rho_t, B_t), \tag{440}$$

where $w_t^{(0)} : \mathcal{R}(M) \times \Sigma(M) \to C^\infty(M; \mathbb{R})$ is the corresponding internal potential specific energy density and $c_t : \mathcal{R}(M) \times \mathcal{B}(M) \to C^\infty(M; \mathbb{R})$ is some still unknown function, depending in general on the imposed magnetic field $B_t : M \longrightarrow \mathbb{E}^3, t \in \mathbb{R}$.

Let us now analyze, as before, the mathematical structure of quantities $(\rho_t, \sigma_t, B_t) \in \mathcal{R}(M) \times \Sigma(M) \times \mathcal{B}(M)$ as the *physical observables* subject to their evolution (438) with respect to the group diffeomorphisms $\eta_t \in \mathrm{Diff}(M), t \in \mathbb{R}$, generated by the liquid motion vector field $dx_t/dt = v_t(x_t), x_t := \eta_t(X), t \in \mathbb{R}, X \in M$:

$$\begin{gathered} \mathcal{L}_{d/dt}(\langle \rho_t v_t | dx_t \rangle d^3x_t) = [-dp_t^{(0)} + \rho_t^{-1} d|v_t|^2/2 + \langle B_t|\nabla\rangle\langle B_t|dx_t\rangle)]\rho_t d^3x_t, \\ \mathcal{L}_{d/dt}(\rho_t d^3x_t) = 0, \quad \mathcal{L}_{d/dt}\sigma_t = 0, \quad \mathcal{L}_{d/dt}(*\langle B_t|dx_t\rangle) = 0, \end{gathered} \tag{441}$$

where $\mathcal{L}_{d/dt} : \Lambda(M) \to \Lambda(M)$ denotes the corresponding Lie derivative with respect to the vector field $d/dt := \partial/\partial t + \langle v_t|\nabla\rangle \in \Gamma(M \times \mathbb{R}; T(M)), t \in \mathbb{R}$. The relationships (441) mean that the space of physical observables, being by definition, the adjoint space $\mathcal{G}_{em}^* := \Lambda^1(M)d^3x \times (\Lambda^3(M) \oplus \Lambda^0(M) \oplus \Lambda^2(M))$ to the extended configuration space is equal to $\mathcal{G}_{em} := diff(M) \times (\Lambda^0(M) \oplus \Lambda^3(M) \oplus \Lambda^1(M)) \simeq T_{Id}(G_{em})$, the tangent space at the identity Id to the extended differential-functional group manifold $G_{em} := \mathrm{Diff}(M) \times \Lambda^0(M) \times \Lambda^3(M) \times \Lambda^1(M) \simeq \mathrm{Diff}(M) \times \mathcal{R}(M) \times \Sigma(M) \times \mathcal{B}(M)$, where we have naturally identified the Abelian group product $\Lambda^0(M) \times \Lambda^3(M) \times \Lambda^1(M)$ with its direct tangent space sum $T(\Lambda^0(M)) \oplus T(\Lambda^3(M)) \oplus T(\Lambda^1(M))$.

Consider now the constructed differential-functional current group manifold G_{em} in Eulerian variables, on which one naturally acts the $\mathrm{Diff}(M)$-group $\mathrm{Diff}(M) \times G_{em} \to G_{em}$ the standard way:

$$\begin{gathered} (\eta \circ \varphi)(X) := \varphi(\eta(X)), (\eta \circ r)(X) := r(\eta(X)), \\ \eta \circ\ (s(X)d^3X\ ) := \eta^*(s(X)d^3X\ ), \\ \eta \circ\ \langle b(X)|dX\rangle\ := \eta^*\langle b(X)|d^3X\rangle \end{gathered} \tag{442}$$

for $\eta \in \mathrm{Diff}(M), X \in M$ and any $(\varphi; r, s, b) \in \mathrm{Diff}(M) \times \mathcal{R}(M) \times \Sigma(M) \times \mathcal{B}(M)$. Then, taking into account the suitably extended action (442) on the differential-functional manifold G_{em},, one can formulate the following easily checkable further proposition that is crucial for what will follow.

Proposition 17. *The differential-functional current group manifold $G_{em} := \mathrm{Diff}(M) \times \mathcal{R}(M) \times \Sigma(M) \times \mathcal{B}(M)$ in Eulerian coordinates is a smooth symmetry Banach group $G_{em} := \mathrm{Diff}(M) \ltimes (\mathcal{R}(M) \times \Sigma(M) \times \mathcal{B}(M))$, equal to the semidirect product of the diffeomorphism group $\mathrm{Diff}(M)$ and the direct product $\mathcal{R}(M) \times \Sigma(M) \times \mathcal{B}(M)$ of abelian functional $\mathcal{R}(M) \simeq \Lambda^0(M)$, density $\Sigma(M) \simeq \Lambda^3(M)$ and one-form $\mathcal{B}(M) \simeq \Lambda^1(M)$ groups, endowed with the following group multiplication law in Eulerian variables:*

$$\begin{gathered} (\varphi_1; r_1, s_1 d^3x, \langle b_1|dx\rangle) \circ (\varphi_2; r_2, s_2 d^3x, \langle b_2|dx\rangle) = \\ = (\varphi_2 \cdot \varphi_1; r_1 + r_2 \cdot \varphi_1, s_1 d^3x + (s_2 d^3x) \cdot \varphi_1, \langle b_1|dx\rangle + \langle b_2|dx\rangle \circ \varphi_1) \end{gathered} \tag{443}$$

for arbitrary elements $\varphi_1, \varphi_2 \in \mathrm{Diff}(M), r_1, r_2 \in \Lambda^0(M)$, $s_1 d^3x, s_2 d^3x \in \Lambda^3(M)$ *and* $\langle b_1|dx\rangle$, $\langle b_2|dx\rangle \in \Lambda^1(M)$.

Thus, one can proceed to studying the corresponding coadjoint action of the Lie algebra $\mathcal{G}_{em} \simeq T_{Id}(G_{em})$, $Id \in G_{em}$, on the adjoint space $\mathcal{G}^*_{em}$. As the Lagrangian configuration $\eta_0 \in \mathrm{Diff}(M)$ and the entropy $\sigma_0 \in \Sigma(M)$ are assumed to be invariant under the Banach diffeomorphism group action $\mathrm{Diff}(M)$, the resulting group action can be reduced to the factor-group $Diff_0(M) := \mathrm{Diff}(M)/Diff_{\eta_0,\sigma_0}(M)$ action on the semidirect group product $G_{em,0} := Diff_0(M) \ltimes (\mathcal{R}(M) \times \Sigma(M) \times \mathcal{B}(M)$. Based on the multiplication law (443), one easily calculates the following Lie algebra commutation relationships:

$$
\begin{aligned}
&[(a_1; r_1, s_1, b_1), (a_2; r_2, s_2, b_2)] = ([a_1, a_2]; \langle a_1|\nabla r_2\rangle - \\
&\quad \langle a_2|\nabla r\rangle, \langle \nabla|a_1 b_2\rangle - \langle \nabla|a_2 s_1\rangle, \langle a_1|\nabla\rangle b_2 - \\
&\quad -\langle a_2|\nabla\rangle b_1 + \langle b_2|\circ \nabla a_1\rangle - \langle b_1|\circ \nabla a_2\rangle)
\end{aligned}
\tag{444}
$$

for any elements $a_1, a_2 \in diff(M) \simeq T(M), r_1, r_2 \in \mathcal{R}(M) \simeq \Lambda^0(M)$, $s_1, s_2 \in \Sigma(M) \simeq \Lambda^3(M)$ and $b_1, b_2 \in \mathcal{B}(M) \simeq \Lambda^1(M)$.

The adjoint space to the semidirect product Lie algebra $\mathcal{G}_{em,0} = diff(M) \ltimes (\mathcal{R}(M) \oplus \Sigma(M) \oplus \mathcal{B}(M))$ can be, naturally, written symbolically as the space $\mathcal{G}^*_{em,0} = (\Lambda^1(M) \otimes \Lambda^3(M)) \times (*\Lambda^0(M) \oplus *\Lambda^3(M) \oplus * \Lambda^1(M)) = diff^*(M) \times (\Lambda^3(M) \oplus \Lambda^0(M) \oplus \Lambda^2(M))$, whereas before, the mapping $* : \Lambda(M) \to \Lambda(M)$ denotes the Hodge isomorphism. Then, taking into account the adjoint space $\mathcal{G}^*_{em,0}$ to the current Lie algebra $\mathcal{G}_{em,0}$ is endowed with the following [27,28,41,177,194,200] canonical Lie–Poisson bracket

$$
\begin{aligned}
&\{f, g\} := \int_M \langle m|\langle \tfrac{\delta f}{\delta m}|\nabla\rangle \tfrac{\delta g}{\delta m} - \langle \tfrac{\delta g}{\delta m}|\nabla\rangle \tfrac{\delta f}{\delta m}\rangle d^3x + \\
&+ \int_M \rho\Big(\langle \tfrac{\delta f}{\delta m}|\nabla \tfrac{\delta g}{\delta \rho}\rangle - \langle \tfrac{\delta g}{\delta m}|\nabla \tfrac{\delta f}{\delta \rho}\rangle\Big) d^3x + \int_M \sigma\langle \nabla|(\tfrac{\delta f}{\delta m}\tfrac{\delta g}{\delta \sigma} - \tfrac{\delta g}{\delta m}\tfrac{\delta f}{\delta \sigma})\rangle d^3x + \\
&+ \int_M \Big(\langle B|\langle \tfrac{\delta f}{\delta m}|\nabla\rangle \tfrac{\delta g}{\delta B} - \Big\langle \tfrac{\delta g}{\delta m}|\nabla\Big\rangle \tfrac{\delta f}{\delta B}\rangle + \langle \tfrac{\delta f}{\delta B}|\langle B|\nabla\rangle \tfrac{\delta g}{\delta m}\rangle - \langle \tfrac{\delta g}{\delta B}|\langle B|\nabla\rangle \tfrac{\delta f}{\delta m}\rangle\Big) d^3x
\end{aligned}
\tag{445}
$$

for any smooth functionals $f, g \in \mathcal{D}(\mathcal{G}^*_{em,0})$ on the adjoint space $\mathcal{G}^*$, where, as before, we denoted by $m := \rho v \in T^*(M)$ the specific momentum of the superfluid. The bracket (445) naturally ensues from the canonical symplectic structure on the cotangent phase space $T^*(G_{em,0})$, as it was previously demonstrated in the section above.

We now write down the first two equations of the Euler MHD system (438) as the local fluid mass and momentum conservation laws in the integral Ampere–Newton form

$$
\begin{aligned}
&\tfrac{d}{dt}\int_{D_t} \rho_t d^3x_t = 0, \quad \tfrac{d}{dt}\int_{D_t} \rho_t v_t \, d^3x_t + \\
&+ \int_{\partial D_t} p_t^{(0)}(x_t) d^2S_t - \int_{D_t} \langle B_t(x_t)|\nabla\rangle B_t(x_t) d^3x_t = 0,
\end{aligned}
\tag{446}
$$

which is completely equivalent to the relationships (441) and where $p_t^{(0)} : M \to \mathbb{R}_+$ is the net internal superfluid pressure, $(\nabla \times B_t(x_t)) \times B_t(x_t) : M \to C^\infty(M; \mathbb{E}^3)$ is the spatially distributed Lorentz force on the superfluid, d^2S_t is the respectively oriented surface measure on the boundary ∂D_t for the domain $D_t := \eta_t(D) \subset M, t \in \mathbb{R}$, and a smooth submanifold $D \subset M$ is chosen arbitrarily. Taking into account that $(\nabla \times B_t(x_t)) \times B_t(x_t) = \langle B_t|\nabla\rangle B_t - \nabla\langle B_t|B_t\rangle/2$ for any $B_t \in \mathcal{B}(M)$, the second integral relationship (446) becomes equivalent to the following:

$$
\partial v_t/\partial t + \langle v_t|\nabla\rangle v_t + \rho_t^{-1}\nabla p_t^{(0)}(\rho_t, \sigma_t) - \rho_t^{-1}\langle B_t|\nabla\rangle B_t = 0,
\tag{447}
$$

where we have represented the internal superfluid pressure quantity as

$$
p_t(x_t) := p_t^{(0)}(\rho_t, \sigma_t) - \langle B_t|B_t\rangle/2
\tag{448}
$$

for some mapping $p_t^{(0)} : \mathcal{R}(M) \times \Sigma(M) \to C^\infty(M;\mathbb{R})$, strictly depending only on the internal liquid configuration $\eta_t \in \mathrm{Diff}(M)$ for all $t \in \mathbb{R}$.

Based on the Poisson bracket expression (445), one can now easily determine the Hamiltonian function $H : M \to \mathbb{R}$, corresponding to the Euler evolution equation (438) on the adjoint space $\mathcal{G}^*$:

$$H = \int_M \rho_t(|m_t|^2/(2\rho_t^2) + w_t^{(0)}(\rho_t,\sigma_t) + \\ +|B_t|^2/(2\rho_t))dx_t^3 := \int_M \rho(x_t)e_t(\rho_t,\sigma_t)d^3x_t, \tag{449}$$

where the quantity

$$e_t(\rho_t,\sigma_t) = |m_t|^2/(2\rho_t^2) + w_t^{(0)}(\rho_t,\sigma_t) + \\ +|B_t|^2/(2\rho_t) := |m_t|^2/(2\rho_t^2) + w_t(\rho_t,\sigma_t) \tag{450}$$

denotes the specific internal superfluid energy, modified by means of the "frozen" magnetic field $B_t \in \mathcal{B}(M), t \in \mathbb{R}$, replacing the previously defined internal specified potential energy $w_t^{(0)}(\rho_t,\sigma_t)$ by the shifted specified potential energy quantity $w_t(\rho_t,\sigma_t) := w_t^{(0)}(\rho_t,\sigma_t) + |B_t|^2/(2\rho_t)$. In particular, the Equation (447) reduces to the equivalent Hamilton expression

$$\partial m_t/\partial t = \{H, m_t\} \tag{451}$$

for $m_t \in T^*(M) \simeq diff^*(M)$ and all $t \in \mathbb{R}$. It is also seen that if $B_t \to 0$ uniformly with respect to time $t \in \mathbb{R}$, the internal energy expression (450) brings about that (423). Recall now that the following quasi-stationary second thermodynamic energy conservation law

$$\delta e_t(\rho_t,\sigma_t) = \rho_t^{-2}p_t(x_t)\delta\rho_t + T_t(x_t)\delta\sigma_t \tag{452}$$

holds for all admitted superfluid variations $\delta\rho_t \in \mathcal{R}(M)$ and $\delta\sigma_t \in \Sigma(M), t \in \mathbb{R}$. As, by isentropic assumption, $\delta\sigma_t = 0$ for all $t \in \mathbb{R}$ along fluid streamlines, for the internal pressure one easily obtains the expression $p_t(x_t) = \rho_t^2 \partial w_t^{(0)}(\rho_t,\sigma_t)/\partial\rho_t - \langle B_t|B_t\rangle/2$, exactly coinciding with that of (448).

The Hamiltonian function (449) evidently satisfies the conservation condition $dH/dt = 0$ for all $t \in \mathbb{R}$. To check this directly, it is enough to observe [194] that the following differential relationship

$$\partial e_t(\rho_t,\sigma_t)/\partial t + \langle \nabla|\rho_t v_t \Big[e_t(\rho_t,\sigma_t) + \rho_t \partial w_0(\rho_t,\sigma_t)/\partial\rho_t - |B_t|^2/2\Big]\rangle = 0 \tag{453}$$

holds for all $t \in \mathbb{R}$ and whose integration over the domain $M \subset \mathbb{R}^3$ easily gives rise to the conservation of the Hamiltonian function (449).

10.7. A Modified Current Lie Algebra, Magneto-Hydrodynamic Invariants and Their Geometry

The importance of spatial invariants describing the stability [194] of MHD superfluid motion was previously stated long ago [181,193,194,201]. Based on the modern symplectic theory of differential–geometric structures on manifolds, we devise a unified approach to study MHD invariants of compressible superfluid flow, related with specially constructed symmetry structures and commuting to each other vector fields on the phase space.

We start from a useful differential-geometric observation that the magneto-hydrodynamic Euler equations $\Gamma(M; T(M))$ action on the adjoint space to the Lie algebra $\mathcal{G}$ of the modified Banach current group $G = \mathrm{Diff}(M) \ltimes (\Lambda^0(M) \oplus \Lambda^3(M) \oplus *^1(M))$, generated by the following vector field differential relationship:

$$dx_t/dt = v_t(x_t), \tag{454}$$

where $x_t = \eta_t(X) \in M$, $X \in M$, and $v_t : M \to T(M)$, $t \in \mathbb{R}$, is an acceptable time-dependent vector field on the domain M, describing the adiabatic superfluid and superconductive motion via the diffeomorphism subgroup mappings $\eta_t \in \mathrm{Diff}(M), \eta_t|_{t=0} = \eta_0 \in \mathrm{Diff}_0(M)$. Taking into account that the initial superfluid configuration $\eta_0 \in \mathrm{Diff}(M)$ is fixed, one can define, following reasonings from [81], a new differential relationship

$$dx_\tau/dt = u_t(x_\tau) \tag{455}$$

on the domain M with respect to the evolution variable $\tau \in \mathbb{R}$, parameterized by the time parameter $t \in \mathbb{R}$, where $u_t : M \to T(M)$, is a τ-independent vector field on M, generating the diffeomorphism subgroup $\psi_t \in \mathrm{Diff}(M)$, $x_\tau := \psi_\iota(\eta_0(X)), X \in M$, commuting to that generated by the vector field (454), i.e., $\eta_t \circ \psi_\iota = \psi_t \circ \eta_\iota$ for all $t, \tau \in \mathbb{R}$. The action of the diffeomorphism subgroup $\psi_t \in \mathrm{Diff}(M)$ at any fixed time $t \in \mathbb{R}$ can be naturally interpreted as rearranging the particle configurations in the superfluid, not changing its other dynamic characteristics. If to denote the corresponding Lie derivatives with respect to the vector fields (454) and (455) by differential expressions $\mathcal{L}_{d/dt} := \partial/\partial t + \langle v_t|\nabla\rangle\circ : C^\infty(M;\mathbb{R}) \to C^\infty(M;\mathbb{R})$ and $\mathcal{L}_{u_t} := \langle u_t|\nabla\rangle\circ : C^\infty(M;\mathbb{R}) \to C^\infty(M;\mathbb{R})$, the commutation condition $\eta_t \circ \psi_\iota = \psi_t \circ \eta_\iota$ for all $t, \tau \in \mathbb{R}$ is equivalently rewritten as the operator commutator

$$[\mathcal{L}_{d/dt}, \mathcal{L}_{u_t}] = 0. \tag{456}$$

Consider now an arbitrary integral invariant of the MHD superfluid, governed by the Euler system (438):

$$I = \int_{D_t} \rho_t(x_t)\gamma_t(m_t;\rho_t,\sigma_t,B_t)d^3x_t, \tag{457}$$

generated by some specific density functional $\gamma_t : \mathcal{G}^* \to C^\infty(M\times\mathbb{R};\mathbb{R})$ and held over the domain $D_t = \eta_t(D)$ for any domain $D \subset M$, corresponding to the diffeomorphism subgroup $\eta_t \in \mathrm{Diff}(M), t \in \mathbb{R}$, generated by flow (454). Taking into account that there holds the following density relationship

$$\mathcal{L}_{d/dt}(\rho_t(x_t)d^3x_t) = 0 \tag{458}$$

for any $t \in \mathbb{R}$, one easily derives from (457) and (458) that also

$$\mathcal{L}_{d/dt}\gamma_t(m_t;\rho_t,\sigma_t,B_t) = 0 \tag{459}$$

for any $t \in \mathbb{R}$. Thus, based on the commutation relationship (456) one can formulate the following important lemma.

Lemma 1. *Let vector fields (454) and (455) commute to each other and a density functional $\gamma_0 : \mathcal{G}^* \times \mathbb{R} \to C^\infty(M\times\mathbb{R};\mathbb{R})$ satisfies for all $t \in \mathbb{R}$ the condition*

$$\mathcal{L}_{d/dt}\gamma_0(m_t;\rho_t,\sigma_t,B_t) = 0, \tag{460}$$

then the following expressions

$$I_{n,k} = \int_{D_t} \rho_t(\mathcal{L}^n_{u_t}\gamma_0(m_t;\rho_t,\sigma_t,B_t)^k d^3x_t \tag{461}$$

over the domain $D_t = \eta_t(D)$, generated by corresponding to the flow (454) diffeomorphism subgroup $\eta_t \in \mathrm{Diff}(M), t \in \mathbb{R}$, and arbitrary domain $D \subset M$, are for all integers $n \in \mathbb{Z}_+, k \in \mathbb{Z}$, the MHD invariants of the superfluid flow (438).

Proof. A proof easily follows from the commutation condition (456) and the superfluid density relationship (458). □

As examples, let us take following [81,194], the vector field $u_t := \rho_t^{-1}B_t \in \Gamma(T(M))$, commuting to the vector field $v_t \in \Gamma(T(M))$, and $\gamma_0 = i_{u_t}\langle A_t\,|dx_t\rangle = \langle A_t\,|\rho_t^{-1}B_t\rangle \in C^\infty(M\times\mathbb{R};\mathbb{R})$, where the magnetic vector potential $A_t \in C^\infty(M;\mathbb{R}), t \in \mathbb{R}$, satisfies the classical Maxwell relationships: the magnetic field $B_t = \nabla \times A_t$ and the electric field $E_t = -\partial A_t/\partial t = -v_t \times B_t$, owing to the net electric field superconductivity (437) condition $\tilde{E}_t = E_t + v_t \times B_t = 0$. Really, the commutativity condition (456) means that

$$\mathcal{L}_{d/dt}(\rho_t^{-1} B_t) - \langle \rho_t^{-1} B_t | \nabla > v_t = 0, \tag{462}$$

which is satisfied, owing to the second and fourth equations of the Euler MHD system (438), as well as to the invariance

$$\begin{gathered} \mathcal{L}_{d/dt}\gamma_0 = \mathcal{L}_{d/dt} i_{u_t} \langle A_t \, | dx_t \rangle = [\mathcal{L}_{d/dt}, i_{u_t}] \langle A_t \, | dx_t \rangle + \\ + i_{u_t} \mathcal{L}_{d/dt} \langle A_t \, | dx_t \rangle = i_{[d/dt, u_t]} \langle A_t \, | dx_t \rangle + i_{u_t} \mathcal{L}_{d/dt} \langle A_t \, | dx_t \rangle = 0, \end{gathered} \tag{463}$$

which holds owing to the algebraic relationship

$$[\mathcal{L}_{d/dt}, i_{u_t}] = i_{[\partial/\partial t + v_t v_t, u_t]}, \tag{464}$$

commutativity of vector fields u_t and $v_t \in \Gamma(M)$ and the integral relationship

$$\begin{aligned} &\tfrac{d}{dt} \textstyle\int_{\partial S_t} \langle A_t \, | dx_t \rangle = \int_{\partial S_t} \mathcal{L}_{d/dt} \langle A_t \, | dx_t \rangle = \\ &= \textstyle\int_{\partial S_t} [\langle \mathcal{L}_{d/dt} A_t \, | dx_t \rangle + \langle A_t \, | \mathcal{L}_{d/dt} dx_t \rangle] = \\ &= \textstyle\int_{\partial S_t} [\langle \mathcal{L}_{d/dt} A_t \, | dx_t \rangle + \langle A_t \, | dv_t \rangle] = \\ &= \textstyle\int_{\partial S_t} [\langle v_t \times B + \langle v_t | \nabla \rangle A_t \, | dx_t \rangle + \langle A_t \, | dv_t \rangle] = \\ &= \textstyle\int_{\partial S_t} [\langle v_t \times (\nabla \times A) + \langle v_t | \nabla \rangle A_t \, | dx_t \rangle + \langle A_t \, | dv_t \rangle] = \\ &= \textstyle\int_{\partial S_t} [\langle dA_t | v_t \rangle + \langle A_t \, | dv_t \rangle] = \int_{\partial S_t} [d \langle A_t \, | v_t \rangle] = 0, \end{aligned} \tag{465}$$

equivalent to the condition $\mathcal{L}_{d/dt} \langle A_t \, | dx_t \rangle = 0$ for all $t \in \mathbb{R}$. The same statement we obtain from the slightly simpler reasoning:

$$\begin{aligned} &\tfrac{d}{dt} \textstyle\int_{\partial S_t} \langle A_t \, | dx_t \rangle = \tfrac{d}{dt} \int_{S_t} \langle \nabla \times A_t | dS_t^2 \rangle = \\ &= \tfrac{d}{dt} \textstyle\int_{S_t} \langle B_t | dS_t^2 \rangle := - \int_{\partial S_t} \langle \tilde{E}_t | dx_t \rangle = 0, \end{aligned} \tag{466}$$

following from the net electric field $\tilde{E}_t = 0$ superconductivity condition (437) along the boundary ∂S_t, where $S_t := \eta_t(S_0) \subset M$ is the surface, generated by the diffeomorphism subgroup $\eta_t \in \mathrm{Diff}(M), t \in \mathbb{R}$, and an arbitrarily chosen surface $S_0 = S_t|_{t=0} \subset M$. The latter is, evidently, equivalent to the equality $\mathcal{L}_{d/dt} \langle A_t \, | dx_t \rangle = 0$ modulo the gauge transformation $A_t \to A_t + \nabla \xi_t$, where $\mathcal{L}_{d/dt} \xi_t + \langle A_t | v_t \rangle = 0$ for some function $\xi_t \in C^\infty(M; \mathbb{R})$ and all $t \in \mathbb{R}$. Thus, one can formulate [81,194] the following proposition.

Proposition 18. *The functionals*

$$I_{n,k}^{(B)} = \int_{D_t} \rho_t \Big(\mathcal{L}^n_{\rho_t^{-1} B_t} \langle A | \rho_t^{-1} B_t \rangle \Big)^k d^3 x_t \tag{467}$$

over the domain $D_t = \eta_t(D)$, generated by corresponding to the flow (454) diffeomorphism subgroup $\eta_t \in \mathrm{Diff}(M), t \in \mathbb{R}$, and arbitrary domain $D \subset M$, are for all integers $n \in \mathbb{Z}_+, k \in \mathbb{Z}$, the MHD invariants of the superfluid flow (438). In particular, the following relationships $\{H, I_{n,k}^{(B)}\} = 0$ hold for all $n \in \mathbb{Z}_+$.

It is natural here to mention [194,196] that the specific entropy functional $\gamma_0 = \sigma_t : M \to C^\infty(M \times \mathbb{R}; \mathbb{R})$ satisfies the sufficient condition $\mathcal{L}_{d/dt} \sigma_t = 0, t \in \mathbb{R}$, a priori generates for the superfluid flow (438) the infinite hierarchy

$$I_{n,k}^{(\sigma)} = \int_{D_t} \rho_t \Big(\mathcal{L}^n_{\rho_t^{-1} B_t} \sigma_t(x_t) \Big)^k d^3 x_t, \tag{468}$$

$n \in \mathbb{Z}_+, k \in \mathbb{Z}$, of the MHD invariants over the domain $D_t = \eta_t(D)$, generated by the corresponding to the flow (454) diffeomorphism subgroup $\eta_t \in \mathrm{Diff}(M), t \in \mathbb{R}$, and arbitrary domain $D \subset M$.

To construct other MHD invariants, depending on the superfluid velocity $v_t \in \Gamma(T(M)),\ t \in \mathbb{R}$, let us consider, following [81], two differential one-forms $\langle\alpha_t|dx_t\rangle, \langle\beta_t|dx_t\rangle \in \Lambda^1(M),\ x_t := \eta_t(X),\ X \in M$, satisfying for all $t \in \mathbb{R}$ the following identity:

$$\mathcal{L}_{d/dt}\langle\alpha_t|dx_t\rangle = dh_t \ + \mathcal{L}_{u_t}\langle\beta_t|dx_t\rangle, \tag{469}$$

for some function $h_t \in \Lambda^0(M)$, where the vector field

$$dx_t/d\tau = u_t(x_t) \tag{470}$$

is uniform with respect to the evolution parameter $\tau \in \mathbb{R}$ and satisfies the following constraints:

$$[\mathcal{L}_{d/dt}, \mathcal{L}_{u_t}] = 0, \quad \langle\nabla|\rho_t u_t\ \rangle = 0 \tag{471}$$

and $u_t \parallel \partial M$ at almost all points $x_t \in \partial M$ for all evolution parameters $t, \tau \in \mathbb{R}$. Then one can formulate the following general proposition.

Proposition 19. *The following integral expressions*

$$\begin{aligned} I_0^{(\alpha,\beta)} &= \int_M \rho_t\langle\alpha_t|u_t\rangle d^3x_t,\ I_1^{(\alpha,\beta)} = \int_M \rho_t[\langle\alpha_t|v_t\rangle + h_t]d^3x_t, \\ I_2^{(\alpha,\beta)} &= \int_M \rho_t\langle\mathcal{L}_{d/dt}\alpha_t|u_t\rangle d^3x_t \end{aligned} \tag{472}$$

over the whole domain $M \subset \mathbb{R}^3$ are for all integers $n \in \mathbb{Z}_+, k \in \mathbb{Z}$, the global MHD invariants.

Proof. Consider, for example, a proof that $I_0^{(\alpha,\beta)} : \mathcal{G} \to \mathbb{R}$ is an invariant: taking into account that $\mathcal{L}_{d/dt}(\rho_t d^3x_t) = 0$, one obtains the expression:

$$\begin{gathered} dI_0^{(\alpha,\beta)}/dt = \int_M \rho_t\mathcal{L}_{d/dt}\langle\alpha_t|u_t\rangle d^3x_t \ = \\ = \int_M \rho_t i_{u_t}(dh_t + \mathcal{L}_{u_t}\langle\beta_t|dx_t\rangle)d^3x_t = \\ = \int_M \rho_t(i_{u_t}dh_t + i_{u_t}(i_{u_t}d + di_{u_t})\langle\beta_t|dx_t\rangle)d^3x_t = \\ = \int_M \rho_t i_{u_t}d(h_t + \langle\beta_t|u_t\rangle)d^3x_t = \\ = \int_M\langle\nabla|\tilde{h}_t\rho_t u_t\rangle d^3x_t = \int_{\partial M}\langle\tilde{h}_t\rho_t u_t|dS_t^2\rangle = 0 \end{gathered} \tag{473}$$

for all $t \in \mathbb{R}$, where we put, by definition, $\tilde{h}_t := \ (h_t + \langle\beta_t|u_t\rangle)$, denoted dS_t^2 the surface measure on the boundary ∂M, used the Cartan representation $\mathcal{L}_{u_t} = i_{u_t}d + di_{u_t}$ and the natural boundary tangency condition $\rho_t u_t \parallel \partial M$, thus proving the proposition. Exactly similar calculations ensue for the next two invariant $I_k^{(\alpha,\beta)} : \mathcal{G} \to \mathbb{R}, k = \overline{1,2}$, on which we will not stop here. □

As a simple example, one can put $\alpha_t^{(0)} := \flat(v_t) \simeq v_t, \beta_t := B_t$, the vector field $u_t = \rho_t^{-1}B_t : M \to T(M), t \in \mathbb{R}$, and show by easy calculations, using the variational equality (439) that

$$\mathcal{L}_{d/dt}\langle v_t|dx_t\rangle = d(|v_t|^2/2\ - h_t - |B_t|^2/\rho_t) + \mathcal{L}_{u_t}\langle B_t|dx_t\rangle + T_t d\sigma_t, \tag{474}$$

where, we have denoted the specific enthalpy [55,198,199] function $h_t := e_t + p_t\rho_t^{-1}$. As a consequence of equality (474), under the spatial temperature constancy $\nabla T_t = 0$ condition for all $t \in \mathbb{R}$, one obtains the following MHD superfluid invariant:

$$I_0^{(v,B)} := \int_M \langle v_t|B_t\rangle d^3x_t = \int_M \langle m_t|\rho_t^{-1}B_t\rangle, \tag{475}$$

where $m_t \simeq \langle m_t(x_t)|dx_t\rangle \otimes d^3x_t \in diff^*(M)$ and $\rho_t^{-1}B_t \simeq \langle \rho_t^{-1}(x)B_t|\partial/\partial x\rangle \in T(M)$, coinciding with the MHD invariant, presented before in [81,194]. If the above temperature condition does not hold, the equality (474) reduces to the differential relationship

$$\partial\langle v_t|B_t\rangle/\partial t + \langle\nabla|[v_t\langle v_t|B_t\rangle + B_t(h_t - |v_t|^2/2]\rangle + \rho_t T_t\langle\rho_t^{-1}B_t|\nabla\sigma_t\rangle, \tag{476}$$

satisfied for all $x_t \in M$ and $t \in \mathbb{R}$.

Remark. It is worth remarking here that the following baroclinic relationship

$$\nabla\rho_t^{-1} \times \nabla p_t = -\nabla T_t \times \nabla\sigma_t \tag{477}$$

holds for all $x_t \in M$ and $t \in \mathbb{R}$.

Similarly, we also easily obtain the following invariant

$$I_1^{(v,B)} = \int_M \rho_t[|m_t|^2/\left(2\rho_t^2\right) + w_t^{(0)}(\rho_t,\sigma_t) + |B_t|^2/(2\rho_t)]d^3x_t = H, \tag{478}$$

coinciding exactly with the Hamiltonian function for the flow (438) on the phase space $\mathcal{G}^*$. The third invariant is, eventually, closely related to the vorticity vector $\xi_t := \nabla \times v_t : M \to \mathbb{E}^3, t \in \mathbb{R}$, and needs a more detailed analysis.

It is instructive now to analyze the existence of integral invariants for the pure hydrodynamic case when the magnetic field $B_t = 0, t \in \mathbb{R}$, following the approach, devised before in [81]. In particular, owing to the relationship (477), there holds the following integral expression for the vorticity $\xi_t := \nabla \times v_t, t \in \mathbb{R}$:

$$\mathcal{L}_{d/dt}\xi_t - \langle\xi_t|\nabla\rangle v_t = \nabla T_t \times \nabla\sigma_t \tag{479}$$

and define the vector field

$$u_t := \rho_t^{-1}\xi_t \exp f_t(x_t) \tag{480}$$

for some scalar smooth mapping $f_t : M \to \mathbb{R}$, which we will choose from the assumed commutation condition

$$[\mathcal{L}_{d/dt}, \mathcal{L}_{u_t}] = 0. \tag{481}$$

The latter gives rise to the equality $\xi_t\mathcal{L}_{d/dt}f_t(x_t) = -\nabla T_t \times \nabla\sigma_t$ at any $x_t := \eta_t(X) \in M$, $X \in M$, or

$$\dot{f}_t\,(\nabla \times v_t) + \nabla T_t \times \nabla\sigma_t = 0, \tag{482}$$

where we took into account that $\mathcal{L}_{d/dt}f_t(x_t) = df_t(x_t)/dt := \dot{f}_t(x_t),\ x_t \in M$, with respect to the temporal parameter $t \in \mathbb{R}$. From (482), one obtains that the mapping $f_t : M \to \mathbb{R}$ should satisfy the following constraints:

$$\nabla\dot{f}_t = k_t v_t, \quad \dot{f}_t v_t = \rho_t^{-1}\nabla p(t) + \nabla\omega_t \tag{483}$$

for some scalar smooth functions k_t and $\omega_t : M \to \mathbb{R}, t \in \mathbb{R}$. It is easy to check that the system (483) is compatible, i.e., the quasi-stationary thermodynamic relationship $p_t^{(0)} = \rho_t^2\partial w_0(\rho_t,\sigma_t)/\partial\rho_t$ jointly with the Euler Equation (393) make it possible to determine these unknown scalar smooth functions k_t and $\omega_t : M \to \mathbb{R}$ for all $t \in \mathbb{R}$.

Consider now, following [81], a slightly modified expression (474) at the magnetic field $B_t = 0$:

$$\mathcal{L}_{d/dt}\langle v_t \exp f_t|dx_t\rangle = \exp f_t d(\omega_t + |v_t|^2/2) \tag{484}$$

and calculate the related integral expression:

$$\begin{aligned} \tfrac{d}{dt}\textstyle\int_M \rho_t(i_{u_t}\langle v_t|dx_t\rangle)d^3x_t &= \textstyle\int_M \rho_t\mathcal{L}_{d/dt}(i_{u_t}\langle v_t|dx_t\rangle)d^3x_t = \\ = \textstyle\int_M \rho_t(i_{u_t}\mathcal{L}_{d/dt}\langle v_t|dx_t\rangle)d^3x_t &= \textstyle\int_M \rho_t(i_{u_t}d\tilde{h})d^3x_t = \\ = \textstyle\int_M (i_{\rho_t u_t}d\tilde{h})d^3x_t = \int_M\langle\nabla\tilde{h}_t|\rho_t u_t\rangle d^3x_t &= \textstyle\int_M\langle\nabla\tilde{h}_t|\xi_t \exp f_t(x_t)\rangle d^3x_t, \end{aligned} \tag{485}$$

where we put, by definition, the function $\tilde{h}_t := \omega_t + |v_t|^2/2$.

If now to put that the mapping $f_t : M \to \mathbb{R}$ satisfies for all $t \in \mathbb{R}$ the constraint $\langle \nabla f_t|\xi_t\rangle = 0$, the integral expression (485) reduces to

$$\begin{aligned}\tfrac{d}{dt}\int_M \rho_t\,(i_{u_t}\langle v_t|dx_t\rangle)d^3x_t = \int_M \langle \nabla|(\exp f_t(x_t)\tilde{h}_t\xi_t)\rangle d^3x_t = \\ = \int_{\partial M}\langle \exp f_t(x_t)\tilde{h}_t\xi_t|d^2S_t\rangle = 0,\end{aligned} \tag{486}$$

where the vorticity vector tangency $\xi_t||\partial M$ constraint is assumed. Thus, under conditions assumed above, the following vortex functional

$$I = \int_M \langle v_t|\nabla \times v_t\rangle d^3x_t \tag{487}$$

persists to be conserved for all $t \in \mathbb{R}$.

If the function $f_t : M \to \mathbb{R}$, being defined by relationships (483), satisfies for all $t \in \mathbb{R}$ the scalar constraint $\langle \nabla f_t|\xi_t\rangle = 0$, one easily derives the following differential relationship:

$$\begin{aligned}\mathcal{L}_{d/dt}\langle \nabla f_t|\xi_t\rangle &= k_t\langle v_t|\xi_t\rangle + \langle \nabla|f_t\nabla T_t \times \nabla\sigma_t\rangle = \\ &=< \nabla \dot{f}_t|\xi_t\rangle + \langle \nabla|f_t\nabla T_t \times \nabla\sigma_t\rangle = 0,\end{aligned} \tag{488}$$

or, equivalently, in the integral form

$$\begin{aligned}\tfrac{d}{dt}\int_{D_t}\langle \nabla f_t|\xi_t\rangle\rho_t d^3x_t = \int_{D_t}\mathcal{L}_{d/dt}\langle \nabla f_t|\xi_t\rangle\rho_t d^3x_t = \\ = \int_{D_t}[\langle \nabla \dot{f}_t|\xi_t\rangle + \langle \nabla|f_t\nabla T_t \times \nabla\sigma_t\rangle]\rho_t d^3x_t = \\ = \int_{D_t}\Big[\langle \nabla \dot{f}_t|\xi_t\rangle - \langle \nabla f_t|\nabla \times \rho_t^{-1}\nabla p_t^{(0)}\rangle\Big]\rho_t d^3x_t \\ = \int_{D_t}\Big[\langle \nabla \dot{f}_t|\xi_t\rangle\rho_t - \rho_t\langle \nabla\rho_t^{-1}|\nabla \times p_t^{(0)}\nabla f_t\rangle\Big]d^3x_t = \\ = \int_{D_t}\Big[\langle \nabla \dot{f}_t|\xi_t\rangle\rho_t + \langle \nabla \ln\rho_t|\nabla \times p_t^{(0)}\nabla f_t\rangle\Big]d^3x_t = \\ = \int_{D_t}\langle \nabla \dot{f}_t|\xi_t\rangle\rho_t d^3x_t,\end{aligned} \tag{489}$$

where we took into account that for the isentropic fluid flow under regard there holds the tangency $\nabla\rho_t||\partial D_t$ condition for all $t \in \mathbb{R}$. If the right hand side of (489) proves to be zero, i.e., $\langle \nabla\ \dot{f}_t|\xi\ _t\rangle\ = 0, t \in \mathbb{R}$, this will mean that the constraint $\langle \nabla f_t|\xi_t\rangle = 0$ for all $t \in \mathbb{R}$, if $\langle \nabla f_t|\xi_t\rangle|_{t=0} = 0$ at $t = 0$, thus producing the vortex conservation functional (487).

11. A Modified Current Lie Algebra Symmetry on Torus, Its Lie-Algebraic Structure and Related Integrable Heavenly Type Dynamical Systems

11.1. Introductory Notes

The main object of our study is integrable multidimensional dispersionless differential equations, which possess modified Lax–Sato type representations, related with their hidden Hamiltonian structures. Equations of this type arise and are widely applied in mechanics, general relativity, differential geometry and the theory of integrable systems. Among the most mentioned are the Boyer–Finley equations, heavenly type Plebański equations, which are descriptive of a class of self-dual 4-manifolds, as well as the dispersionless Kadomtsev–Petviashvili (dKP) equation, also known as the Khokhlov–Zabolotskaya equation, which arises in non-linear acoustics and the theory of Einstein–Weyl structures. Their integrability have been investigated by a whole variety of modern techniques including symmetry analysis, differential-geometric and algebrogeometric methods, dispersionless $\bar{\partial}$-dressing, factorization techniques, Virasoro constraints, hydrodynamic reductions, etc. The first important examples of the related Hamiltonian structures were previously demonstrated in [202–206] and later were developed in [207–214] , where many examples of dispersionless differential equations were analyzed in detail as flows on orbits of the coadjoint action of loop vector field algebras $\widetilde{\mathrm{diff}}(\mathbb{T}^n)$, generated by specially chosen seed elements $\tilde{l} \in \widetilde{\mathrm{diff}}(\mathbb{T}^n)^*$. In these works, it was observed that many integrable multidimensional dispersionless differential equations are generated by seed elements of a very special structure, namely for them there exist such analytical functional elements $\tilde{\eta}, \tilde{\rho} \in \Omega^0(\mathbb{T}^n) \otimes \mathbb{C}$ that $\tilde{l} = \tilde{\eta} d\tilde{\rho}$. As the latter naturally generates the symplectic structure $\tilde{\omega}^{(2)} := \int_{\mathbb{T}^n} d\tilde{\eta} \wedge d\tilde{\rho} \in \Omega^2(\mathbb{T}^n) \otimes \mathbb{C}$ on the moduli space [215,216] of flat connections on $\mathbb{T}^n$, related to coadjoint actions of the corresponding Casimir functionals, the geometric nature of many integrable multidimensional dispersionless differential equations can be also studied using cohomological techniques, devised in [215,217] for the case of Riemannian surfaces. It is also worth mentioning that in [207–209] a deep connection of the related Hamiltonian flows on $\widetilde{\mathrm{diff}}(\mathbb{T}^n)^*$ was revealed with the well known in classical mechanics Lagrange–d'Alembert principle.

In this section, developing the approach, devised in [202,203,218], we will describe a Lie algebraic structure and integrability properties of a generalized hierarchy of the Lax-Sato type compatible systems of Hamiltonian flows and related integrable multidimensional dispersionless differential equations. Such systems are called the heavenly type equations and were first introduced by Plebański in [219]. The heavenly type equations were analyzed in many articles (see, e.g., [203,218,220–227]) using several different approaches. In [131,207,209,228] the heavenly type equations were analyzed by using non-associative and non-commutative current algebras on the torus $\mathbb{T}^m, m \in \mathbb{N}$. We also mention that [229,230] B. Szablikowski and A. Sergyeyev developed some generalizations of the classical AKS-algebraic and related R-structures [11,17–19]. In [203,218] and recently in [207,231], these ideas were applied to a semi-direct Lie algebra $\tilde{\mathcal{G}} := \widetilde{\mathrm{diff}}(\mathbb{T}^n) \ltimes \widetilde{\mathrm{diff}}(\mathbb{T}^n)^*$ of the loop Lie algebra $\widetilde{\mathrm{diff}}(\mathbb{T}^n) := \widetilde{Vect}(\mathbb{T}^n)$ of vector fields on the torus $\mathbb{T}^n, n \in \mathbb{Z}_+$, and its dual space $\widetilde{\mathrm{diff}}(\mathbb{T}^n)^*$. Several interesting and deep results about the orbits of the corresponding coadjoint actions on the space $\tilde{\mathcal{G}}^* \simeq \tilde{\mathcal{G}}$ and the classical Lie–Poisson type structures on them were presented. It is worth especially remarking here that the AKS-algebraic scheme is naturally embedded into the classical R-structure approach via the following construction.

Let a $(\tilde{\mathcal{G}}; [\cdot,\cdot])$ denote a Lie algebra over $\mathbb{C}$ and $\tilde{\mathcal{G}}^*$ be its natural adjoint space. Take some tensor element $r \in \tilde{\mathcal{G}} \otimes \tilde{\mathcal{G}} \simeq \mathrm{Hom}(\tilde{\mathcal{G}}^*; \tilde{\mathcal{G}})$ and consider its splitting into symmetric and antisymmetric parts

$$r = k \oplus \sigma, \tag{490}$$

respectively, and assume that the symmetric tensor $k \in \tilde{\mathcal{G}} \otimes \tilde{\mathcal{G}}$ does not degenerate. That allows the definition on the Lie algebra $\tilde{\mathcal{G}}$ of a symmetric non-degenerate bi-linear product $(\cdot|\cdot) : \tilde{\mathcal{G}} \otimes \tilde{\mathcal{G}} \to \mathbb{C}$ via the expression

$$(a|b) := k^{-1}a(b) \tag{491}$$

for any $a, b \in \tilde{\mathcal{G}}$. The composed mapping $R := \sigma \circ k^{-1} : \tilde{\mathcal{G}} \to \tilde{\mathcal{G}}$, following the scheme

$$\tilde{\mathcal{G}} \stackrel{k^{-1}}{\to} \tilde{\mathcal{G}}^* \stackrel{\sigma}{\to} \tilde{\mathcal{G}}, \tag{492}$$

defines the following R-structure on the Lie algebra $\tilde{\mathcal{G}}$:

$$[a, b]_R := [Ra, b] + [a, Rb] \tag{493}$$

for all elements $a, b \in \tilde{\mathcal{G}}$. The following theorem, defining the related Poissonian structure [19,121,207,217,232,233] on the adjoint space $\tilde{\mathcal{G}}$ holds.

Theorem 9. *Let $\alpha, \beta \in \tilde{\mathcal{G}}^*$ be arbitrary and define the bracket*

$$\{\alpha, \beta\} := ad^*_{r\alpha}\beta - ad^*_{r\beta}\alpha. \tag{494}$$

Then the bracket (494) is Poissonian if the R-structure on the Lie algebra $\tilde{\mathcal{G}}$ defines the Lie structure on $\tilde{\mathcal{G}}$, that is there holds the Yang–Baxter equation

$$[Ra, Rb] - R[a, b]_R = -[a, b] \tag{495}$$

for any $a, b \in \tilde{\mathcal{G}}$.

Remark 10. *The above theorem makes it possible to consider the Hamiltonian flows on the coadjoint space* $\tilde{\mathcal{G}}^*$ *as those determined on the Lie algebra* $\tilde{\mathcal{G}}$. *The latter is exceptionally useful if for the scalar product (491) there exists such a trace-type* $\mathrm{Tr}(\cdot)$ *symmetric and ad-invariant functional (of Killing type) that*

$$\mathrm{Tr}(ab) := (a|b), \quad (a|[b,c]) = (([a,b]|,c) \tag{496}$$

for any a, b *and* $c \in \tilde{\mathcal{G}}$. *Then any Hamiltonian flow of an element* $a \in \tilde{\mathcal{G}}$ *is representable in the standard Lax type form*

$$da/dt = [\mathrm{grad}(h), a], \tag{497}$$

where $\mathrm{grad}(h) \in \tilde{\mathcal{G}}$ *is generated by the corresponding smooth Hamiltonian function* $h \in \mathrm{D}(\tilde{\mathcal{G}})$.

Concerning the loop Lie algebra $\tilde{\mathcal{G}} := \widetilde{\mathrm{diff}}(\mathbb{T}^n)$ on the torus $\mathbb{T}^n$, it is well known that such a trace-type functional on $\tilde{\mathcal{G}}$ does not exist, thus we need to study the Hamiltonian flows on the adjoint loop space $\tilde{\mathcal{G}}^* \simeq \Omega^1(\mathbb{T}^n)$ of meromorphic differential forms on the torus $\mathbb{T}^n$ and obtain, as a result, integrable dispersionless differential equations as compatibility conditions for the related loop vector fields, generated by Casimir functionals on $\tilde{\mathcal{G}}^*$. This procedure is much more complicated for analysis than the standard one and employs more geometrical tools and considerations about the orbit space structure of the seed elements $\tilde{l} \in \tilde{\mathcal{G}}^*$, generating a hierarchy of integrable Hamiltonian flows. The latter, in part, is deeply related to its reduction properties, guaranteeing the existence of nontrivial Casimir invariants on its coadjoint orbits. By applying and extending these ideas to central extensions of Lie algebras, we construct new classes of commuting Hamiltonian flows on an extended adjoint space $\hat{\mathcal{G}} := \tilde{\mathcal{G}}^* \oplus \mathbb{C}$. These Hamiltonian flows are generated by seed elements $(\tilde{a} \ltimes \tilde{l}; \alpha) \in \hat{\mathcal{G}}^*$ and specially constructed Casimir invariants on the corresponding orbits of $\hat{\mathcal{G}}^*$. In most cases, these seed elements appeared to be represented as specially factorized differential objects, whose real geometric nature is still much hidden and not clear. Moreover, we found that the corresponding compatibility condition of constructed Hamiltonian flows coincides exactly with the compatibility condition for a system of related three Lax–Sato type linear vector field equations. As examples, we found and described new multidimensional generalizations of the Mikhalev–Pavlov and Alonso–Shabat type integrable dispersionless equation, whose seed elements possess a special factorized structure, allowing to extend them to the multidimensional case of arbitrary dimension.

11.2. Differential-Geometric Setting: The Diffeomorphism Group $\mathrm{Diff}(\mathbb{T}^n)$ *and Its Description*

Consider the n-dimensional torus $\mathbb{T}^n$ and call points $X \in \mathbb{T}^n$ as the Lagrangian variables of a configuration $\eta \in \mathrm{Diff}(\mathbb{T}^n)$. The manifold $\mathbb{T}^n$, thought of as the target space of a configuration $\eta \in \mathrm{Diff}(\mathbb{T}^n)$, is called the spatial or Eulerian configuration, whose points, called spatial or Eulerian points, will be denoted by small letters $x \in \mathbb{T}^n$. Then any one-parametric configuration of $\mathrm{Diff}(\mathbb{T}^n)$ is a time $t \in \mathbb{R}$ dependent family [41,122,181,192,193] of diffeomorphisms written as

$$\mathbb{T}^n \ni x = \eta(X, t) := \eta_t(X) \in \mathbb{T}^n \tag{498}$$

for any initial configuration $X \in \mathbb{T}^n$ and some mappings $\eta_t \in \mathrm{Diff}(\mathbb{T}^n), t \in \mathbb{R}$.

Being interested in studying flows on the space of Lagrangian configurations $\eta \in \mathrm{Diff}(\mathbb{T}^n)$ with respect to the temporal variable $t \in \mathbb{R}$, generated by group diffeomorphisms $\eta_t \in \mathrm{Diff}(\mathbb{T}^n)$, $t \in \mathbb{R}$, let us proceed to describing the structure of tangent $T_{\eta_t}(\mathrm{Diff}(\mathbb{T}^n))$ and cotangent $T^*_{\eta_t}(\mathrm{Diff}(\mathbb{T}^n))$ spaces to the diffeomorphism group $\mathrm{Diff}(\mathbb{T}^n)$ at the points $\eta_t \in \mathrm{Diff}(\mathbb{T}^n)$ for any $t \in \mathbb{R}$. Determine first the tangent space $T_{\eta_t}(\mathrm{Diff}(\mathbb{T}^n))$ to the diffeomorphism group manifold $\mathrm{Diff}(\mathbb{T}^n)$ at point $\eta \in \mathrm{Diff}(\mathbb{T}^n)$ for which we will make use of the construction, devised before in [122,181,194]. Namely, let $\eta \in \mathrm{Diff}(\mathbb{T}^n)$ be a Lagrangian configuration and try to determine the tangent space $T_\eta(\mathrm{Diff}(\mathbb{T}^n))$ at $\eta \in \mathrm{Diff}(\mathbb{T}^n)$ as the collection of vectors $\xi_\eta := d\eta_\tau/d\tau|_{\tau=0}$, where $\mathbb{R} \ni \tau \to \eta_\tau \in \mathrm{Diff}(\mathbb{T}^n)$, $\eta_\tau|_{\tau=0} = \eta$, is a smooth curve on $\mathrm{Diff}(\mathbb{T}^n)$, and for arbitrary reference point $X \in \mathbb{T}^n$ there holds $\xi_\eta(X) = d\eta_\tau(X)/d\tau|_{\tau=0}$. The latter equivalently means that the vectors $\xi_\eta(X) \in T_{\eta(X)}(\mathbb{T}^n)$, $X \in \mathbb{T}^n$, represent a vector field $\xi : \mathbb{T}^n \to T(\mathbb{T}^n)$ on the manifold $\mathbb{T}^n$ for any $\eta \in \mathrm{Diff}(\mathbb{T}^n)$. Thus, the tangent space $T_\eta(\mathrm{Diff}(\mathbb{T}^n))$ coincides with the set of vector fields on $\mathbb{T}^n$:

$$T_\eta(\mathrm{Diff}(\mathbb{T}^n)) \simeq \{\xi_\eta \in \Gamma(T(\mathbb{T}^n)) : \xi_\eta(X) \in T_{\xi(X)}(\mathbb{T}^n)\} \tag{499}$$

and similarly, the cotangent space $T^*_\eta(\mathrm{Diff}(\mathbb{T}^n))$ consists of all one-form densities on $\mathbb{T}^n$ over $\eta \in \mathrm{Diff}(\mathbb{T}^n)$:

$$T^*_\eta(\mathrm{Diff}(\mathbb{T}^n)) = \{\alpha_\eta \in \Omega^1(\mathbb{T}^n) \otimes \Omega^3(\mathbb{T}^n) : \alpha_\eta(X) \in T^*_{\eta(X)}(\mathbb{T}^n) \otimes |\Omega^3(\mathbb{T}^n)|\} \tag{500}$$

subject to the canonical non-degenerate pairing $(\cdot|\cdot)_c$ on $T^*_\eta(\mathrm{Diff}(\mathbb{T}^n)) \times T_\eta(\mathrm{Diff}(\mathbb{T}^n))$: if $\alpha_\eta \in T^*_\eta(\mathrm{Diff}(\mathbb{T}^n))$, $\xi_\eta \in T_\eta(\mathrm{Diff}(\mathbb{T}^n))$, where $\alpha_\eta|_X = \langle \alpha_\eta(X)|dx\rangle \otimes d^3X$, $\xi_\eta|_X = \langle \xi_\eta(X)|\partial/\partial x\rangle$, then

$$(\alpha_\eta|\xi_\eta)_c := \int_{\mathbb{T}^n} \langle \alpha_\eta(X)|\xi_\eta(X)\rangle d^3X. \tag{501}$$

The construction above makes it possible to identify the cotangent bundle $T^*_\eta(\mathrm{Diff}(\mathbb{T}^n))$ at the fixed Lagrangian configuration $\eta \in \mathrm{Diff}(\mathbb{T}^n)$ to the tangent space $T_\eta(\mathrm{Diff}(\mathbb{T}^n))$, as the tangent space $T(\mathbb{T}^n)$ is endowed with the natural internal tangent bundle metric $\langle\cdot|\cdot\rangle$ at any point $\eta(X) \in \mathbb{T}^n$, identifying $T(\mathbb{T}^n)$ with $T^*(\mathbb{T}^n)$ via the related metric isomorphism $\sharp : T^*(\mathbb{T}^n) \to T(\mathbb{T}^n)$. The latter can be also naturally lifted to $T^*_\eta(\mathrm{Diff}(\mathbb{T}^n))$ at $\eta \in \mathrm{Diff}(\mathbb{T}^n)$, namely: for any elements $\alpha_\eta, \beta_\eta \in T^*_\eta(\mathrm{Diff}(\mathbb{T}^n))$, $\alpha_\eta|_X = \langle \alpha_\eta(X)|dx\rangle \otimes d^3X$ and $\beta_\eta|_X = \langle \beta_\eta(X)|dx\rangle \otimes d^3X \in T^*_\eta(\mathrm{Diff}(\mathbb{T}^n))$ we can define the metric

$$(\alpha_\eta|\beta_\eta) := \int_{\mathbb{T}^n} \langle \alpha^\sharp_\eta(X)|\beta^\sharp_\eta(X)\rangle d^3X, \tag{502}$$

where, by definition, $\alpha^\sharp_\eta(X) := \sharp\langle \alpha_\eta(X)|dx\rangle)$, $\beta^\sharp_\eta(X) := \sharp\langle \beta_\eta(X)|dx\rangle \in T_{\eta(X)}(\mathbb{T}^n)$ for any $X \in \mathbb{T}^n$. Based on the construction above, one can proceed to constructing smooth invariant functionals on the cotangent bundle $T^*(\mathrm{Diff}(\mathbb{T}^n))$ subject to the corresponding coadjoint actions of the diffeomorphism group $\mathrm{Diff}(\mathbb{T}^n)$. Moreover, as the cotangent bundle $T^*(\mathrm{Diff}(\mathbb{T}^n))$ is *a priori* endowed with the canonical symplectic structure, equivalent [11,18,19,26,28,41,122,181,195] to the corresponding Poisson bracket on the space of smooth functionals on $T^*(\mathrm{Diff}(\mathbb{T}^n))$, one can study both the related Hamiltonian flows on it and their adjoint symmetries and complete integrability.

Consider now the cotangent bundle $T^*(\mathrm{Diff}(\mathbb{T}^n))$ as a smooth manifold endowed with the canonical symplectic structure [26,122] on it, equivalent to the corresponding canonical Poisson bracket on the space of smooth functionals on it. Taking into account that the cotangent space $T^*_\eta(\mathrm{Diff}(\mathbb{T}^n))$ at $\eta \in \mathrm{Diff}(\mathbb{T}^n)$, shifted by the right $R_{\eta^{-1}}$- action to the space $T^*_{Id}(\mathrm{Diff}(\mathbb{T}^n))$, $Id \in \mathrm{Diff}(\mathbb{T}^n)$, becomes diffeomorphic to the adjoint space $\mathrm{diff}^*(\mathbb{T}^n)$ to the Lie algebra $\mathrm{diff}(\mathbb{T}^n) \simeq \Gamma(T(\mathbb{T}^n))$ of vector fields on $\mathbb{T}^n$, as there was stated [31–33,41] still by S. Lie in 1887 this canonical Poisson bracket on $T^*_\eta(\mathrm{Diff}(\mathbb{T}^n))$ transforms [26,31,32,41,181,195] into the classical Lie-Poisson bracket on the adjoint space $\mathcal{G}^*$. Moreover, the orbits of the diffeomorphism group $\mathrm{Diff}(\mathbb{T}^n)$ on $T^*(\mathrm{Diff}(\mathbb{T}^n))$ respectively transform into the coadjoint orbits on the adjoint space $\mathcal{G}^*$, generated by suitable elements of the Lie algebra $\mathcal{G}$. To construct in detail this Lie–Poisson bracket, we formulate the following preliminary simple lemma.

Lemma 2. *The Lie algebra* $\mathrm{diff}(\mathbb{T}^n) \simeq \Gamma(T(\mathbb{T}^n))$ *is determined by the following Lie commutator relationships:*

$$[a_1, a_2] = \langle a_1|\nabla\rangle a_2 - \langle a_2|\nabla\rangle a_1 \tag{503}$$

for any vector fields $a_1, a_2 \in \Gamma(T(\mathbb{T}^n))$ *on the manifold* $\mathbb{T}^n$.

Proof. Proof of the commutation relationships (503) easily follows from the group multiplication

$$(\varphi_{1,t} \circ \varphi_{2,t})(X) = \varphi_{2,t}(\varphi_{1,t}(X)) \tag{504}$$

for any local group diffeomorphisms $\varphi_{1,t}, \varphi_{2,t} \in \text{Diff}(\mathbb{T}^n), t \in \mathbb{R}$, and $X \in \mathbb{T}^n$ under condition that $a_j(X) := d\varphi_{j,t}(X)/dt|_{t=0}$ and $\varphi_{j,t}|_{t=0} = Id \in \text{Diff}(\mathbb{T}^n), j = \overline{1,2}$. □

To calculate the Poisson bracket on the cotangent space $T^*_\eta(\text{Diff}(\mathbb{T}^n))$ at any $\eta \in \text{Diff}(\mathbb{T}^n)$, let us consider the cotangent space $T^*_\eta(\text{Diff}(\mathbb{T}^n)) \simeq \text{diff}^*(\mathbb{T}^n)$, the adjoint space to the tangent space $T_\eta(\text{Diff}(\mathbb{T}^n))$ of left invariant vector fields on $\text{Diff}(\mathbb{T}^n)$ at any $\eta \in \text{Diff}(\mathbb{T}^n)$, and take the canonical symplectic structure on $T^*_\eta(\text{Diff}(\mathbb{T}^n))$ in the form $\omega^{(2)}(\mu,\eta) := \delta\alpha(\mu,\eta)$, where the canonical Liouville form $\alpha(\mu,\eta) := (\mu|\delta\eta)_c \in \Omega^1_{(\mu,\eta)}(T^*_\eta(\text{Diff}(\mathbb{T}^n)))$ at a point $(\mu,\eta) \in T^*_\eta(\text{Diff}(\mathbb{T}^n))$ is defined *a priori* on the tangent space $T_\eta(\text{Diff}(\mathbb{T}^n)) \simeq \Gamma(T(M))$ of right-invariant vector fields on the torus manifold $\mathbb{T}^n$. Having calculated the corresponding Poisson bracket of smooth functions $(\mu|a)_c, (\mu|b)_c \in C^\infty(T^*_\eta(\text{Diff}(\mathbb{T}^n));\mathbb{R})$ on $T^*_\eta(\text{Diff}(\mathbb{T}^n)) \simeq \text{diff}^*(\mathbb{T}^n), \eta \in \text{Diff}(\mathbb{T}^n)$, one can formulate the following proposition.

Proposition 20. *The Lie Poisson bracket on the coadjoint space $T^*_\eta(\text{Diff}(\mathbb{T}^n)) \simeq \text{diff}^*(\mathbb{T}^n), \eta \in M$, is equal to the expression*

$$\{f,g\}(\mu) = (\mu|[\delta g(\mu)/\delta\mu, \delta f(\mu)/\delta\mu])_c \tag{505}$$

for any smooth functionals $f,g \in C^\infty(\mathcal{G}^*;\mathbb{R})$.

Proof. By definition [26,122] of the Poisson bracket of smooth functions $(\mu|a)_c, (\mu|b)_c \in C^\infty(T^*_\eta(\text{Diff}(\mathbb{T}^n));\mathbb{R})$ on the symplectic space $T^*_\eta(\text{Diff}(\mathbb{T}^n))$, it is easy to calculate that

$$\begin{aligned} &\{\mu(a),\mu(b)\} := \delta\alpha(X_a,X_b) = \\ &= X_a(\alpha|X_b)_c - X_b(\alpha|X_a)_c - (\alpha|[X_a,X_b])_c, \end{aligned} \tag{506}$$

where $X_a := \delta(\mu|a)_c/\delta\mu = a \in \text{diff}(\mathbb{T}^n), X_b := \delta(\mu|b)_c/\delta\mu = b \in \text{diff}(\mathbb{T}^n)$. Since the expressions $X_a(\alpha|X_b)_c = 0$ and $X_b(\alpha|X_a)_c = 0$ owing the right-invariance of the vector fields $X_a, X_b \in T_\eta(\text{Diff}(\mathbb{T}^n))$, the Poisson bracket (506) transforms into

$$\begin{aligned} &\{(\mu|a)_c,(\mu|b)_c\} = -(\alpha|[X_a,X_b])_c = \\ &= (\mu|[b,a])_c = (\mu|[\delta(\mu|b)_c/\delta\mu, \delta(\mu|a)_c/\delta\mu])_c \end{aligned} \tag{507}$$

for all $(\mu,\eta) \in T^*_\eta(\text{Diff}(\mathbb{T}^n)) \simeq \text{diff}^*(\mathbb{T}^n), \eta \in \text{Diff}(\mathbb{T}^n)$ and any $a,b \in \text{diff}(\mathbb{T}^n)$. The Poisson bracket (506) is easily generalized to

$$\{f,g\}(\mu) = (\mu|[\delta g(\mu)/\delta\mu, \delta f(\mu)/\delta\mu])_c \tag{508}$$

for any smooth functionals $f,g \in C^\infty(\mathcal{G}^*;\mathbb{R})$, finishing the proof. □

Based on the Lie–Poisson bracket (505), one can naturally construct Hamiltonian flows on the adjoint space $\text{diff}^*(\mathbb{T}^n)$ via the expressions

$$\partial l/\partial t = -ad^*_{\text{grad}\, h(l)} l \tag{509}$$

for any element $l \in \text{diff}^*(\mathbb{T}^n), t \in \mathbb{R}$, where, by definition, $\frac{d}{d\varepsilon}h(l+\varepsilon m)|_{\varepsilon=0} := (m|\,\text{grad}\, h(l))_c$, for some smooth Hamiltonian function $h \in C^\infty(\text{diff}^*(\mathbb{T}^n);\mathbb{R})$. If the system possesses enough additional invariants except the Hamiltonian function, one can expect its simplification often reducing to its complete integrability. Below, we proceed to developing an effective enough analytical scheme, before suggested in [207,209,234] for suitably constructed holomorphic loop diffeomorphism groups on tori, allowing to generate infinite hierarchies of such completely integrable Hamiltonian systems on related functional phase spaces.

11.3. A Modified Current Lie Algebra and Related Symmetry Analysis on Functional Manifolds

Consider a smooth manifold $M \subset \mathbb{R}^n, n \in \mathbb{N}$, endowed with the generalized quantum current group [26,181,216] group G as the semidirect product $\mathrm{Diff}(M) \ltimes (\Lambda^0(M) \times \Lambda^1(M))$ of the diffeomorphism group $\mathrm{Diff}(M)$ with the Abelian groups $\Omega^0(M)$ and $\Omega^1(M)$, defined by the natural $\mathrm{Diff}(M)$—group action $\mathrm{Diff}(M) \times G \to G$:

$$\begin{aligned} (\eta \circ \varphi)(X) := \varphi(\eta(X)), (\eta \circ r)(X) := r(\eta(X)), \\ \eta \circ \langle b(X)|dX\rangle := \eta^* \langle b(X)|dX\rangle \end{aligned} \tag{510}$$

for $\eta \in \mathrm{Diff}(M), X \in M,$ and any $(\varphi; r, b) \in \mathrm{Diff}(M) \times (\Omega^0(M) \times \Omega^0(M)$. The semidirect product group G is endowed with the following internal right group multiplication subject to the Eulerian variable $x := \eta(X) \in M$:

$$\begin{aligned} (\varphi_1; r_1, \langle b_1|dx\rangle) \circ (\varphi_2; r_2, \langle b_2|dx\rangle) = \\ = (\varphi_2 \cdot \varphi_1; r_1 + r_2 \cdot \varphi_1, \langle b_1|dx\rangle + \langle b_2|dx\rangle \circ \varphi_1) \end{aligned} \tag{511}$$

at a fixed point $\eta \in \mathrm{Diff}(M)$ and arbitrary elements $\varphi_1, \varphi_2 \in \mathrm{Diff}(M), r_1, r_2 \in \Omega^0(M)$ and $b_1 \simeq \langle b_1|dx\rangle, b_2 \simeq \langle b_2|dx\rangle \in \Omega^1(M)$.

Let $\mathcal{G} \simeq T_{Id}(G) = diff(M) \ltimes (\Omega^0(M) \times \Omega^1(M)), Id \in G$, denote the Lie algebra of the current group G, where we took into account that $T(\Omega^0(M)) \simeq \Omega^0(M), T(\Omega^1(M)) \simeq \Omega^1(M)$, and proceed to studying its coadjoint action on the adjoint space $\mathcal{G}^*$. Using (511), one can easily write down that

$$[(a_1; r_1, b_1), (a_2; r_2, b_2)] = (\mathcal{L}_{a_1} a_2; \mathcal{L}_{a_2} r_1 - \mathcal{L}_{a_1} r_2, \mathcal{L}_{a_2} \langle b_1|dx\rangle - \mathcal{L}_{a_1} \langle b_2|dx\rangle), \tag{512}$$

where $\mathcal{L}_a$ denotes the standard [122,123,181] Lie derivative with respect to a vector field $a \in diff(M)$. From (512) one easily ensues the following current Lie algebra $\mathcal{G}$ commutation relationships:

$$\begin{aligned} [(a_1; r_1, b_1), (a_2; r_2, b_2)] = (\langle \big(\langle a_1|\tfrac{\partial}{\partial x}\rangle a_2 - \langle a_2|\tfrac{\partial}{\partial x}\rangle a_1\big)|\tfrac{\partial}{\partial x}\rangle; \langle a_2|\tfrac{\partial}{\partial x} r_1\rangle - \\ -\langle a_1|\tfrac{\partial}{\partial x} r_2\rangle, \langle\langle a_2|\tfrac{\partial}{\partial x}\rangle b_1|dx\rangle - \langle\langle a_1|\tfrac{\partial}{\partial x}\rangle b_2|dx\rangle + \langle b_1|\langle dx|\tfrac{\partial}{\partial x}\rangle a_2\rangle - \langle b_2|\langle dx|\tfrac{\partial}{\partial x}\rangle a_1\rangle), \end{aligned} \tag{513}$$

for any elements $a_1, a_2 \in diff(M) \simeq T(M), r_1, r_2 \in \Omega^0(M)$ and $b_1, b_2 \in \Omega^1(M)$, where we have also denoted the gradient vector $\frac{\partial}{\partial x} := \left(\frac{\partial}{\partial x_1}, \frac{\partial}{\partial x_2}, \ldots, \frac{\partial}{\partial x_n}\right)^{\intercal}$ at $x \in M$. The adjoint space $\mathcal{G}^*$ to the semidirect product Lie algebra $\mathcal{G} = diff(M) \ltimes (\Omega^0(M) \oplus \Omega^1(M))$ can be written symbolically as $\mathcal{G}^* = (\Omega^1(M) \otimes \Omega^n(M)) \times (*\Omega^0(M) \oplus *\Omega^1(M)) = diff^*(M) \times (\Omega^n(M) \oplus \Omega^{n-1}(M))$, where $* : \Omega(M) \to \Omega(M)$ denotes the corresponding Hodge isomorphism with respect to the natural scalar product

$$(\alpha^{(k)}|\gamma^{(s)}) := \delta_{sk} \int_M (\alpha^{(k)} \wedge *\gamma^{(s)}) \tag{514}$$

for any forms $\alpha^{(k)} \in \Omega^k(M)$ and $\gamma^{(s)} \in \Omega^s(M)$ $,k,s = \overline{1,n}$. Then, taking into account that the adjoint space $\mathcal{G}^*$ is endowed [27,28,41,177,194,200] with the canonical Lie–Poisson bracket

$$\begin{aligned}\{f,h\}(l) := (l|[\nabla f(l), \nabla h(l)]) = \int_M \Big(\langle \mu | \langle \frac{\delta f}{\delta \mu} | \frac{\partial}{\partial x} \rangle \frac{\delta h}{\delta \mu} - \langle \frac{\delta h}{\delta \mu} | \frac{\partial}{\partial x} \rangle \frac{\delta f}{\delta \mu} \rangle \Big) d^n x + \\ + \int_M \rho \Big(\langle \frac{\delta f}{\delta \mu} | \frac{\partial}{\partial x} \frac{\delta h}{\delta \rho} \rangle - \langle \frac{\delta h}{\delta \mu} | \frac{\partial}{\partial x} \frac{\delta f}{\delta \rho} \rangle \Big) d^n x + \\ + \int_M \Big(\langle \beta | \langle \frac{\delta f}{\delta \mu} | \frac{\partial}{\partial x} \rangle \frac{\delta h}{\delta \beta} - \Big\langle \frac{\delta h}{\delta \mu} | \frac{\partial}{\partial x} \Big\rangle \frac{\delta f}{\delta \beta} \rangle + \\ + \langle \frac{\delta f}{\delta \beta}, \langle \beta | \frac{\partial}{\partial x} \rangle \frac{\delta h}{\delta \mu} \rangle - \langle \frac{\delta h}{\delta \beta}, \langle \beta | \frac{\partial}{\partial x} \rangle \frac{\delta f}{\delta \mu} \rangle \Big) d^n x\end{aligned} \tag{515}$$

for any smooth functionals $f, g \in \mathcal{D}(\mathcal{G}^*)$ on the $\mathcal{G}^*$, where we have denoted by $l := (\langle \mu | dx \rangle \otimes d^n x;\ \rho d^n x, *\langle \beta | dx \rangle \otimes d^n x) \in \mathcal{G}^*$ and by $\nabla(\circ)(l) := \Big(\langle \frac{\delta(\circ)}{\delta \mu} | \frac{\partial}{\partial x} \rangle; \frac{\delta(\circ)}{\delta \rho}, \langle \frac{\delta(\circ)}{\delta \rho} | dx \rangle \Big)$ the corresponding functional gradient.

Remark 11. *We remark here that the bracket (515) naturally derives, as it was demonstrated in [29,31,32,41], from the canonical symplectic structure on the cotangent phase space $T^*(G)$.*

Based on the Lie–Poisson bracket, one can construct the Hamiltonian system

$$\frac{\partial}{\partial t}(\mu, \rho, \beta)^{\intercal} = \{H, (\mu, \rho, \beta)^{\intercal}\}, \tag{516}$$

where $t \in \mathbb{R}$ is the related evolution parameter and $H \in \mathcal{D}(\mathcal{G}^*)$ is some suitably constructed Hamiltonian function. For the evolution flow (516) to be integrable, it should possess [11,122,181,235] enough commuting to each of the other invariant functionals $H_j \in \mathcal{D}(\mathcal{G}^*), j \in \mathbb{N}$, which is in most cases a very complicated problem. Thereby, taking this into account, we will proceed the following way: we will construct a set *a priori* commuting to each of the other invariants $h_j \in \mathcal{D}(\tilde{\mathcal{G}}^*), j \in \mathbb{N}$, defined on the coadjoint space $\tilde{\mathcal{G}}^*$ to a suitably generalized Lie algebra $\tilde{\mathcal{G}}$.

Namely, let us consider a group $\tilde{G} := \tilde{G}_+ \times \tilde{G}_-$,where $\tilde{G}_\pm := \widetilde{\mathrm{Diff}}_\pm(M) \ltimes (\Omega^0_\pm(M) \times \Omega^1_\pm(M))$ are subgroups of the smooth loop mappings $\{\mathbb{C} \supset \mathbb{S}^1 \to G\}$, holomorphically extended, respectively, on the interior $\mathbb{D}^1_+ \subset \mathbb{C}$ and on the exterior $\mathbb{D}^1_- \subset \mathbb{C}$ domains of the unit centrally located disk $\mathbb{D}^1 \subset \mathbb{C}^1$ and such that for any $\tilde{g}(\lambda) \in \tilde{G}_-, \lambda \in \mathbb{D}^1_-, \tilde{g}(\infty) = Id \in G$. The corresponding Lie subalgebras $\tilde{\mathcal{G}}_\pm \simeq \widetilde{\mathrm{diff}}_\pm(M) \ltimes (\Omega^0_\pm(M) \times \Omega^0_\pm(M))$ of the loop current subgroups $\tilde{G}_\pm$ consist, in general, of vector fields on $\mathbb{S}^1 \times \mathbb{T}^n$, holomorphically extended, respectively, on regions $\mathbb{D}^1_\pm \subset \mathbb{C}^1$, where for any $\tilde{p}(\lambda) \in \tilde{\mathcal{G}}_-$ the value $\tilde{p}(\infty) = 0$. The loop current Lie algebra splitting $\tilde{\mathcal{G}} = \tilde{\mathcal{G}}_+ \oplus \tilde{\mathcal{G}}_-$, where

$$\begin{aligned}\tilde{\mathcal{G}}_+ &= \bigcup_{m \in \mathbb{Z}_+} \Big\{ \sum_{j=0}^m \lambda^j \langle a_{-j}(x) | \frac{\partial}{\partial x} \rangle \otimes d^n x; \sum_{j=0}^m \lambda^j \rho_{-j}(x), \sum_{j=0}^m \lambda^j \langle b_{-j}(x) | dx \rangle \Big\}, \\ \tilde{\mathcal{G}}_- &= \Big\{ \sum_{j \in \mathbb{N}} \lambda^{-j} \langle a_j(x) | \frac{\partial}{\partial x} \rangle \otimes d^n x; \sum_{j \in \mathbb{N}} \lambda^{-j} \rho_j(x), \sum_{j \in \mathbb{N}} \lambda^{-j} \langle b_j(x) | dx \rangle \Big\},\end{aligned} \tag{517}$$

can be naturally identified with a dense subspace of the dual space $\tilde{\mathcal{G}}^*$ through the pairing

$$(\tilde{l}|\tilde{a}) := \operatorname*{res}_{\lambda \in \mathbb{C}} (l(x;\lambda)|p(x;\lambda))_{H^0} \tag{518}$$

with respect to the scalar product

$$(l(x;\lambda)|p(x;\lambda))_{H^0} := \int\limits_M d^n x[\langle \mu(x;\lambda)|a(x;\lambda)\rangle + \rho(x;\lambda)r(x;\lambda) + \langle \beta(x;\lambda)|b(x;\lambda)\rangle]. \quad (519)$$

on the usual Hilbert space $H^0 := L_2(M;\mathbb{C}^{n+1} \times \mathbb{C}^1 \times \mathbb{C}^{n+1})$ for any elements $\tilde{l} := (\tilde{\mu};\tilde{\rho},\tilde{\beta}) \in \tilde{\mathcal{G}}^*$ and $\tilde{p} := (\tilde{a};\tilde{r},\tilde{b}) \in \tilde{\mathcal{G}}$, naturally represented in their component wise canonical form as

$$\begin{aligned} \tilde{p} := (\tilde{a};\tilde{r},\tilde{b}) = \left(\left\langle \mathrm{a}(x;\lambda)|\frac{\partial}{\partial \mathrm{x}}\right\rangle; r(x;\lambda), \langle \mathrm{b}(x;\lambda)|d\mathrm{x}\rangle\right), \tilde{l} := (\tilde{\mu};\tilde{\rho},\tilde{\beta}) = \\ = (\langle \mu(x;\lambda)|d\mathrm{x}\rangle \otimes d^n x; \rho(x;\lambda)d^3x, *\langle \beta(x;\lambda)|d\mathrm{x}\rangle \otimes d^n x), \end{aligned} \quad (520)$$

where for any $\mathrm{x} := (x;\lambda) \in \mathbb{C}\times M$ we have denoted, for brevity, the gradient operator $\frac{\partial}{\partial \mathrm{x}} := (\frac{\partial}{\partial \lambda};\frac{\partial}{\partial x}) = \left(\frac{\partial}{\partial \lambda};\frac{\partial}{\partial x_1},\frac{\partial}{\partial x_2},\ldots,\frac{\partial}{\partial x_n}\right)^\intercal$ in the Euclidean space $(\mathbb{E}^n;\langle\cdot,\cdot\rangle)$ and $\tilde{a} := \left\langle \mathrm{a}(x;\lambda)|\frac{\partial}{\partial \mathrm{x}}\right\rangle := a^{(0)}(x;\lambda)\frac{\partial}{\partial \lambda} + \left\langle a(x;\lambda)|\frac{\partial}{\partial x}\right\rangle, \tilde{b} := \langle \mathrm{b}(x;\lambda)|d\mathrm{x}\rangle := b^{(0)}(x;\lambda)d\lambda + \langle b(x;\lambda)|dx\rangle$, $\tilde{\mu} := \langle \mu(x;\lambda)|d\mathrm{x}\rangle := \mu^{(0)}(x;\lambda)d\lambda + \langle \mu(x;\lambda)|dx\rangle$. The corresponding Lie commutator $[\tilde{p}_1,\tilde{p}_2] \in \tilde{\mathcal{G}}$ of any vectors $\tilde{p}_1 = (\tilde{a}_1;\tilde{r}_1,\tilde{b}_1), \tilde{p}_2 = (\tilde{a}_2;\tilde{r}_2,\tilde{b}_2) \in \tilde{\mathcal{G}}$ is calculated the standard way, using (513), and equals

$$\begin{aligned} [(\tilde{a}_1;\tilde{r}_1,\tilde{b}_1),(\tilde{a}_2;\tilde{r}_2,\tilde{b}_2)] = \Big(\Big\langle (\langle \mathrm{a}_1|\tfrac{\partial}{\partial \mathrm{x}}\rangle \mathrm{a}_2 - \langle \mathrm{a}_2|\tfrac{\partial}{\partial \mathrm{x}}\rangle \mathrm{a}_1)|\tfrac{\partial}{\partial \mathrm{x}}\Big\rangle; \\ \langle \mathrm{a}_2|\tfrac{\partial}{\partial x}\mathrm{r}_1\rangle - \langle \mathrm{a}_1|\tfrac{\partial}{\partial x}\mathrm{r}_2\rangle, \langle \mathrm{a}_2|\tfrac{\partial}{\partial \mathrm{x}}\rangle\langle \mathrm{b}_1|d\mathrm{x}\rangle - \\ -\langle \mathrm{a}_1|\tfrac{\partial}{\partial \mathrm{x}}\rangle\langle \mathrm{b}_2|d\mathrm{x}\rangle + \langle \mathrm{b}_1|\langle d\mathrm{x}|\tfrac{\partial}{\partial \mathrm{x}}\rangle \mathrm{a}_2\rangle - \langle \mathrm{b}_2|\langle d\mathrm{x}|\tfrac{\partial}{\partial \mathrm{x}}\rangle \mathrm{a}_1\rangle\Big). \end{aligned} \quad (521)$$

The expression (521) makes it possible to construct the related Lie–Poisson bracket on the adjoint space $\tilde{\mathcal{G}}^*$, modifying that of (515):

$$\begin{aligned} \{f,h\} := res_\lambda \int_M \langle \mu|\langle \frac{\delta f}{\delta \mu}|\frac{\partial}{\partial \mathrm{x}}\rangle \frac{\delta h}{\delta \mu} - \langle \frac{\delta h}{\delta \mu}|\frac{\partial}{\partial \mathrm{x}}\rangle \frac{\delta f}{\delta \mu}\rangle d^n x + \\ + res_\lambda \int_M \rho\left(\langle \frac{\delta f}{\delta \mu}|\frac{\partial}{\partial \mathrm{x}}\frac{\delta h}{\delta \rho}\rangle - \langle \frac{\delta h}{\delta \mu}|\frac{\partial}{\partial \mathrm{x}}\frac{\delta f}{\delta \rho}\rangle\right) d^n x + \\ + res_\lambda \int_M \left(\langle \beta|\langle \frac{\delta f}{\delta \mu}|\frac{\partial}{\partial \mathrm{x}}\rangle \frac{\delta h}{\delta \beta} - \left\langle \frac{\delta h}{\delta \mu}|\frac{\partial}{\partial \mathrm{x}}\right\rangle \frac{\delta f}{\delta \beta}\rangle + \right. \\ \left. + \langle \frac{\delta f}{\delta \beta}|\langle \beta|\frac{\partial}{\partial \mathrm{x}}\rangle \frac{\delta h}{\delta \mu}\rangle - \langle \frac{\delta h}{\delta \beta}|\langle \beta|\frac{\partial}{\partial \mathrm{x}}\rangle \frac{\delta f}{\delta \mu}\rangle\right) d^n x \end{aligned} \quad (522)$$

for any smooth functionals $f,h \in \mathcal{D}(\tilde{\mathcal{G}}^*)$.

The Lie–Poisson bracket (522) is strongly degenerate and possesses a lot of Casimir invariants $h_j \in \mathcal{D}(\tilde{\mathcal{G}}^*), j \in \mathbb{Z}_+$, satisfying the condition

$$\{f,h_j\} = 0 \quad (523)$$

for all smooth functionals $f \in \mathcal{D}(\tilde{\mathcal{G}}^*)$ and $j \in \mathbb{Z}_+$. As the Lie algebra $\tilde{\mathcal{G}}$ acts on its adjoint space $\tilde{\mathcal{G}}^*$ for any $\tilde{p} = (\tilde{a};\tilde{r},\tilde{b}) \in \tilde{\mathcal{G}}$ and $\tilde{l} = (\tilde{\mu};\tilde{\rho},\tilde{\beta}) \in \tilde{\mathcal{G}}^*$ as $ad^* : \tilde{\mathcal{G}} \times \tilde{\mathcal{G}}^* \to \tilde{\mathcal{G}}^*$, where

$$\begin{aligned} ad^*_{\tilde{p}}\tilde{l} = \Big(-\langle \frac{\partial}{\partial \mathrm{x}} \circ |\mathrm{a}\rangle\langle \mu|d\mathrm{x}\rangle \otimes d^n x - \langle \mu|\langle d\mathrm{x}|\frac{\partial}{\partial \mathrm{x}}\rangle \mathrm{a}\rangle \otimes d^n x + \\ + \rho\langle d\mathrm{x}|\frac{\partial \mathrm{r}}{\partial \mathrm{x}}\rangle \otimes d^n x + \langle \beta|\langle d\mathrm{x}|\frac{\partial}{\partial \mathrm{x}}\mathrm{x}\rangle \otimes d^n x - \langle \frac{\partial}{\partial \mathrm{x}} \circ |\beta\rangle\langle \mathrm{x}|d\mathrm{x}\rangle \otimes d^n x; \\ \langle \frac{\partial}{\partial \mathrm{x}}|\rho \mathrm{a}\rangle \otimes d^n x, *\langle\langle \frac{\partial}{\partial \mathrm{x}} \circ |\mathrm{x}\rangle \beta|d\mathrm{x}\rangle \otimes d^n x - *\langle\langle \beta|\frac{\partial}{\partial \mathrm{x}}\rangle \mathrm{a}|d\mathrm{x}\rangle \otimes d^n x\Big), \end{aligned} \quad (524)$$

the latter condition (523) is easily rewritten as

$$ad^*_{\nabla h(\tilde{l})}\,\tilde{l} = 0, \tag{525}$$

where $\nabla h(\tilde{l}) := \left(\langle \frac{\delta h}{\delta \mu}|\frac{\partial}{\partial x}\rangle; \frac{\delta h}{\delta \rho}, \langle \frac{\delta h}{\delta \beta}|dx\rangle\right)^{\intercal} \in \tilde{\mathcal{G}}$, being equivalent, owing to (524), to the following three differential-functional relationships:

$$\begin{aligned}&\langle \tfrac{\partial}{\partial x} \circ |\tfrac{\delta h}{\delta \mu}\rangle \mu + \langle \mu| \circ \tfrac{\partial}{\partial x}\tfrac{\delta h}{\delta \mu}\rangle - \langle \beta| \circ \tfrac{\partial}{\partial x}\tfrac{\delta h}{\delta \beta}\rangle + \langle \tfrac{\partial}{\partial x} \circ |\beta\rangle \tfrac{\delta h}{\delta \beta} - \\ &-\rho \tfrac{\partial}{\partial x}\tfrac{\delta h}{\delta \rho} = 0, \ \langle \tfrac{\partial}{\partial x}|\rho \tfrac{\delta h}{\delta \mu}\rangle = 0, \ \langle \tfrac{\partial}{\partial x} \circ |\tfrac{\delta h}{\delta \mu}\rangle \beta - \langle \beta|\tfrac{\partial}{\partial x}\rangle \tfrac{\delta h}{\delta \mu} = 0\end{aligned} \tag{526}$$

for any $(\tilde{\mu}; \tilde{\rho}, \beta) \in \tilde{\mathcal{G}}^*$. Recall now that the constructed above loop Lie algebra $\tilde{\mathcal{G}} = \tilde{\mathcal{G}}_+ \oplus \tilde{\mathcal{G}}_-$, as the direct sum of its subalgebras, possesses the additional Lie commutator

$$[\tilde{p}_1, \tilde{p}_2]_R := [R\tilde{p}_1, \tilde{p}_2] + [\tilde{p}_1, R\tilde{p}_2] = [\tilde{p}_{1,+}, \tilde{p}_{2,+}] - [\tilde{p}_{1,-}, \tilde{p}_{2,-}] \tag{527}$$

for any $\tilde{p}_1, \tilde{p}_2 \in \tilde{\mathcal{G}}$, where, by definition, the linear homomorphism $R := (P_+ - P_-)/2$, projectors $P_\pm : \tilde{\mathcal{G}} \to \tilde{\mathcal{G}}_+$, and $\tilde{p}_{j,+} := P_\pm \tilde{p}_j \in \tilde{\mathcal{G}}_\pm, j = \overline{1,2}$. Based on the second Lie commutator (527) we can construct, in the same way as above, the second Lie–Poisson bracket on the adjoint space $\tilde{\mathcal{G}}^*$ as

$$\begin{aligned}\{f,h\}_R := {} & res_\lambda \int_M \langle \mu| \langle R\frac{\delta f}{\delta \mu}|\frac{\partial}{\partial x}\rangle \frac{\delta h}{\delta \mu} - \langle R\frac{\delta h}{\delta \mu}|\frac{\partial}{\partial x}\rangle \frac{\delta f}{\delta \mu}\rangle d^n x + \\ & + res_\lambda \int_M \langle \mu| \langle \frac{\delta f}{\delta \mu}|\frac{\partial}{\partial x}\rangle R\frac{\delta h}{\delta \mu} - \langle \frac{\delta h}{\delta \mu}|\frac{\partial}{\partial x}\rangle R\frac{\delta f}{\delta \mu}\rangle d^n x + \\ & + res_\lambda \int_M \rho \left(\langle R\frac{\delta f}{\delta \mu}|\frac{\partial}{\partial x}\frac{\delta h}{\delta \rho}\rangle - \langle R\frac{\delta h}{\delta \mu}|\frac{\partial}{\partial x}\frac{\delta f}{\delta \rho}\rangle \right) d^n x + \\ & + res_\lambda \int_M \rho \left(\langle \frac{\delta f}{\delta \mu}|\frac{\partial}{\partial x}R\frac{\delta h}{\delta \rho}\rangle - \langle \frac{\delta h}{\delta \mu}|\frac{\partial}{\partial x}R\frac{\delta f}{\delta \rho}\rangle \right) d^n x + \\ & + res_\lambda \int_M \left(\langle \beta| \langle R\frac{\delta f}{\delta \mu}|\frac{\partial}{\partial x}\rangle \frac{\delta h}{\delta \beta} - \left\langle R\frac{\delta h}{\delta \mu}|\frac{\partial}{\partial x}\right\rangle \frac{\delta f}{\delta \beta}\rangle + \right. \\ & + \langle \beta| \langle \frac{\delta f}{\delta \mu}|\frac{\partial}{\partial x}\rangle R\frac{\delta h}{\delta \beta} - \left\langle \frac{\delta h}{\delta \mu}|\frac{\partial}{\partial x}\right\rangle R\frac{\delta f}{\delta \beta}\rangle + \\ & + \langle \frac{\delta f}{\delta \beta}| \langle \beta|\frac{\partial}{\partial x}\rangle R\frac{\delta h}{\delta \mu}\rangle - \langle \frac{\delta h}{\delta \beta}| \langle \beta|\frac{\partial}{\partial x}\rangle R\frac{\delta f}{\delta \mu}\rangle + \\ & \left. + \langle R\frac{\delta f}{\delta \beta}| \langle \beta|\frac{\partial}{\partial x}\rangle \frac{\delta h}{\delta \mu}\rangle - \langle R\frac{\delta h}{\delta \beta}| \langle \beta|\frac{\partial}{\partial x}\rangle \frac{\delta f}{\delta \mu}\rangle \right) d^n x\end{aligned} \tag{528}$$

11.4. A New Modified Spatially Four-Dimensional Mikhalev–Pavlov Heavenly Type Integrable System

Let a seed element $\tilde{a} \ltimes \tilde{l} \in \tilde{\mathcal{G}}^*$ be chosen as

$$\tilde{a} \ltimes \tilde{l} = ((u_x + v_x\lambda - \lambda^2)\partial/\partial x \ltimes (w_x + \zeta_x\lambda)dx, \tag{529}$$

where $u, v, w, \zeta \in C^2(\mathbb{R}^2 \times (\mathbb{S}^1 \times \mathbb{T}^1); \mathbb{R})$. The asymptotic splits for the components of the gradient of the corresponding Casimir functional $h \in \mathrm{I}(\tilde{\mathcal{G}}^*)$, as $|\lambda| \to \infty$ have the following forms:

$$\begin{aligned}&\nabla h_{\tilde{l}} \sim 1 - v_x\lambda^{-1} - u_x\lambda^{-2} - v_z\lambda^{-3} - (u_z + v_xv_z - 2(\partial_x^{-1}v_{xx}v_z))\lambda^{-4} + \\ &\quad + v_y\lambda^{-5} - (-u_y - v_xv_y + 2(\partial_x^{-1}v_{xx}v_y))\lambda^{-6} + \ldots, \\ &\nabla h_{\tilde{a}} \sim \zeta_x\lambda^{-1} + w_x\lambda^{-2} + \zeta_z\lambda^{-3} + (w_z - \zeta_xv_z + 2v_x\zeta_z - (\partial_x^{-1}v_x\zeta_x)_z)\lambda^{-4} - \\ &\quad - \zeta_y\lambda^{-5} + (-w_y + \zeta_xv_y - 2v_x\zeta_y + (\partial_x^{-1}v_x\zeta_x)_y)\lambda^{-6} + \ldots.\end{aligned}$$

In the case when

$$\nabla h_{\tilde{l},+}^{(y)} := \lambda^4 - v_x\lambda^3 - u_x\lambda^2 - v_z\lambda - (u_z + v_xv_z - 2(\partial_x^{-1}v_{xx}v_z)),$$

$$\nabla h_{\tilde{a},+}^{(y)} := \zeta_x\lambda^3 + w_x\lambda^2 + \zeta_z\lambda + (w_z - \zeta_xv_z + 2v_x\zeta_z - (\partial_x^{-1}v_x\zeta_x)_z),$$

and

$$\begin{aligned}\nabla h_{\tilde{l},+}^{(t)} &= \lambda^6 - v_x\lambda^5 - u_x\lambda^4 - v_z\lambda^3 - (u_z + v_xv_z - 2(\partial_x^{-1}v_{xx}v_z))\lambda^2 + \\ &+ v_y\lambda - (-u_y - v_xv_y + 2(\partial_x^{-1}v_{xx}v_y)),\end{aligned} \tag{530}$$

$$\begin{aligned}\nabla h_{\tilde{a},+}^{(t)} &= \zeta_x\lambda^5 + w_x\lambda^4 + \zeta_z\lambda^3 + (w_z - \zeta_xv_z + 2v_x\zeta_z - (\partial_x^{-1}v_x\zeta_x)_z)\lambda^2 - \\ &- \zeta_y\lambda + (-w_y + \zeta_xv_y - 2v_x\zeta_y + (\partial_x^{-1}v_x\zeta_x)_y),\end{aligned}$$

the compatibility condition of the Hamiltonian vector flows leads to the system of new integrable evolution equations:

$$\begin{aligned}u_{zt} + u_{yy} &= -u_yu_{xz} + u_zu_{xy} - v_yv_{xy} + v_zv_{xt} - u_zv_yv_{xx} + u_yv_zv_{xx} - \\ &- v_x^2v_zv_{xy} + v_x^2v_yv_{xz} - 2e_xu_{xy} - 2s_xu_{xz} + 2e_{xt} - 2s_{xy} + 2e_xv_yv_{xx} + 2s_xv_zv_{xx}, \\ v_{zt} + v_{yy} &= -u_yv_{xz} + u_zv_{xy} - v_yu_{xz} + v_zu_{xy} - 2e_xv_{xy} - 2s_xv_{xz} - 2v_xv_yv_{xz} + 2v_xv_zv_{xy}, \\ -u_{xy} - u_{zz} &= u_xu_{xz} - u_zu_{xx} - u_{xx}v_xv_z + u_xv_{xz}v_x - u_xv_{xx}v_z + (v_xv_z)_z + 2u_{xx}e_x - 2e_{xz}, \\ -v_{xy} - v_{zz} &= u_{xz}v_x - u_zv_{xx} - u_{xx}v_z + u_xv_{xz} - 2v_{xx}v_xv_z + v_x^2v_{xz} + 2v_{xx}e_x, \\ -u_{xt} + u_{yz} &= -u_xu_{xy} + u_yu_{xx} + u_{xx}v_xv_y - u_xv_{xy}v_x + u_xv_{xx}v_y - (v_xv_y)_z + 2u_{xx}s_x - 2s_{xz}, \\ -v_{xt} + v_{yz} &= -u_{xy}v_x + u_yv_{xx} + u_{xx}v_y - u_xv_{xy} + 2v_{xx}v_xv_y - v_x^2v_{xy} + 2v_{xx}s_x,\end{aligned} \tag{531}$$

where

$$e_{xx} = v_{xx}v_z, \quad s_{xx} = -v_{xx}v_y.$$

Under the constraint $v = 0$, one obtains a new spatially four-dimensional system

$$\begin{aligned}u_{zt} + u_{yy} &= -u_yu_{xz} + u_zu_{xy}, \\ -u_{xy} - u_{zz} &= u_xu_{xz} - u_zu_{xx}, \\ -u_{xt} + u_{yz} &= -u_xu_{xy} + u_yu_{xx},\end{aligned} \tag{532}$$

which reduces to the Mikhalev–Pavlov [204,208,223] integrable heavenly type equation, if to put $z = x \in \mathbb{R}$.

Here, we can observe that the seed element (529) can be presented in the following special compact form:

$$\tilde{a} \ltimes \tilde{l} := \frac{d\tilde{\eta}}{dx}\partial/\partial x \ltimes d\tilde{\rho}, \; \tilde{\eta} = u + v\lambda - \lambda^2x, \; \tilde{\rho} = w + \zeta\lambda, \tag{533}$$

deeply connected with the geometry of the related moduli space of flat connections, related to the coadjoint actions of the corresponding Casimir functionals. Its possible generalization to multidimensional Mikhalev–Pavlov type equations can be done by the seed element

$$\tilde{a} \ltimes \tilde{l} := \langle \nabla\tilde{\eta}|\nabla\rangle \ltimes d\tilde{\rho} \tag{534}$$

for some elements $\tilde{\eta}, \tilde{\rho} \in \Omega^0(\mathbb{T}^n) \otimes \mathbb{C}, n \in \mathbb{N}$. An analysis of the case (534) and corresponding systems of multidimensional Mikhalev–Pavlov type equations is planned to be done in a separate study.

11.5. A Modified Martinez Alonso-Shabat Heavenly Type Integrable System

If the seed element $\tilde{a} \ltimes \tilde{l} \in \tilde{\mathcal{G}}^*$ is chosen as

$$\tilde{a} \ltimes \tilde{l} = (((u_{x_1} + cu_{x_2}) + \lambda)\partial/\partial x_1 + ((v_{x_1} + cv_{x_2}) + c\lambda)\partial/\partial x_2) \ltimes \\ \ltimes ((w_{x_1} + cw_{x_2})dx_1 + (\zeta_{x_1} + c\zeta_{x_2})dx_2), \tag{535}$$

where $u, v, w, \zeta \in C^2(\mathbb{R}^2 \times \mathbb{S}^1 \times \mathbb{T}^2; \mathbb{R})$, $c \in \mathbb{R} \setminus \{0\}$, one has the following asymptotic splits for the components of the gradients of the corresponding Casimir functionals $h^{(1)}, h^{(2)} \in \mathrm{I}(\tilde{\mathcal{G}}^*)$ as $|\lambda| \to \infty$:

$$\nabla h^{(1)}_{\tilde{l}} \sim \begin{pmatrix} 1 + (u_{x_1} + cu_{x_2})\lambda^{-1} - u_z\lambda^{-2} + \dots \\ c + (v_{x_1} + cv_{x_2})\lambda^{-1} - v_z\lambda^{-2} + \dots \end{pmatrix},$$
$$\nabla h^{(1)}_{\tilde{a}} \sim \begin{pmatrix} (w_{x_1} + cw_{x_2})\lambda^{-1} - w_z\lambda^{-2} + \dots \\ (\zeta_{x_1} + c\zeta_{x_2})\lambda^{-1} - \zeta_z\lambda^{-2} + \dots \end{pmatrix},$$

and

$$\nabla h^{(2)}_{\tilde{l}} \simeq \begin{pmatrix} 1 + (u_{x_1} - cu_{x_2})\lambda^{-1} + \chi\lambda^{-2} + \dots \\ -c + (v_{x_1} - cv_{x_2})\lambda^{-1} + \omega\lambda^{-2} + \dots \end{pmatrix},$$
$$\nabla h^{(2)}_{\tilde{a}} \simeq \begin{pmatrix} (w_{x_1} - cw_{x_2})\lambda^{-1} + \varrho\lambda^{-2} + \dots \\ (\zeta_{x_1} - c\zeta_{x_2})\lambda^{-1} + \chi\lambda^{-2} + \dots \end{pmatrix},$$

where

$$\chi_{x_1} + c\chi_{x_2} = -(u_{zx_1} - cu_{zx_2}) + 2c(u_{x_1}u_{x_1x_2} - u_{x_2}u_{x_1x_1} + v_{x_1}u_{x_2x_2} - v_{x_2}u_{x_1x_2}), \tag{536}$$
$$\omega_{x_1} + c\omega_{x_2} = -(v_{zx_1} - cv_{zx_2}) + 2c(u_{x_1}v_{x_1x_2} - u_{x_2}v_{x_1x_1} + v_{x_1}v_{x_2x_2} - v_{x_2}v_{x_1x_2}),$$

and

$$\rho_{x_1} + c\rho_{x_2} = -(w_{zx_1} - cw_{zx_2}) + 2c(u_{x_1}w_{x_1x_2} - u_{x_2}w_{x_1x_1} + 2w_{x_2}u_{x_1x_1} - \\ - 2w_{x_1}u_{x_1x_2} + v_{x_1}w_{x_2x_2} - v_{x_2}w_{x_1x_2} + w_{x_2}v_{x_1x_2} - w_{x_2}v_{x_2x_2} + \zeta_{x_2}v_{x_1x_1} - \zeta_{x_1}v_{x_1x_2}),$$
$$\chi_{x_1} + c\chi_{x_2} = -(\zeta_{zx_1} - c\zeta_{zx_2}) + 2c(v\zeta_{x_2x_2} - v_{x_2}\zeta_{x_1x_2} + 2\zeta_{x_2}v_{x_1x_2} - \\ - 2\zeta_{x_1}v_{x_2x_2} + u_{x_1}\zeta_{x_1x_2} - u_{x_2}\zeta_{x_1x_1} + \zeta_{x_2}u_{x_1x_1} - \zeta_{x_1}u_{x_1x_2} + w_{x_2}u_{x_1x_2} - w_{x_1}u_{x_2x_2}).$$

In the case when the reduced Casimir gradients are equal to the expressions

$$\nabla h^{(y)}_{\tilde{l},+} = \begin{pmatrix} \lambda^2 + (u_{x_1} + cu_{x_2})\lambda - u_z \\ c\lambda^2 + (v_{x_1} + cv_{x_2})\lambda - v_z \end{pmatrix}, \nabla h^{(y)}_{\tilde{a},+} = \begin{pmatrix} (w_{x_1} + cw_{x_2})\lambda - w_z \\ (\zeta_{x_1} + c\zeta_{x_2})\lambda - \zeta_z \end{pmatrix},$$

and

$$\nabla h^{(t)}_{\tilde{l},+} = \begin{pmatrix} \lambda^2 + (u_{x_1} - cu_{x_2})\lambda + \chi \\ -c\lambda^2 + (v_{x_1} - cv_{x_2})\lambda + \omega \end{pmatrix}, \nabla h^{(t)}_{\tilde{a},+} = \begin{pmatrix} (w_{x_1} - cw_{x_2})\lambda + \rho \\ (\zeta_{x_1} - c\zeta_{x_2})\lambda + \chi \end{pmatrix},$$

the Lax–Sato compatibility condition of the Hamiltonian vector flows leads to the system of evolution equations:

$$
\begin{aligned}
& u_{zt} + \chi_y = -u_{zx_1}\chi - u_{zx_2}\omega + u_{zx_1} + v_{zx_2}, \\
& v_{zt} + \omega_y = -v_{zx_1}\chi - v_{zx_2}\omega + u_z\omega_{x_1} + v_z\omega_{x_2}, \\
& u_{yx_1} + cu_{yx_2} = -(u_{x_1} + cu_{x_2})u_{zx_1} - (v_{x_1} + cv_{x_2})u_{zx_2} + (u_{x_1x_1} + cu_{x_1x_2})u_z + \\
& + (u_{x_1x_2} + cu_{x_2x_2})v_z - u_{zz}, \\
& v_{yx_1} + cv_{yx_2} = -(u_{x_1} + cu_{x_2})v_{zx_1} - (v_{x_1} + cv_{x_2})v_{zx_2} + (v_{x_1x_1} + cv_{x_1x_2})u_z + \\
& + (v_{x_1x_2} + cv_{x_2x_2})v_z - v_{zz}, \\
& u_{tx_1} + cu_{tx_2} = (u_{x_1} + cu_{x_2})\chi_{x_1} + (v_{x_1} + cv_{x_2})\chi_{x_2} - (u_{x_1x_1} + cu_{x_1x_2})\chi - \\
& - (u_{x_1x_2} + cu_{x_2x_2})\omega + \chi_z, \\
& v_{tx_1} + cv_{tx_2} = (u_{x_1} + cu_{x_2})\omega_{x_1} + (v_{x_1} + cv_{x_2})\omega_{x_2} - (v_{x_1x_1} + cv_{x_1x_2})\chi - \\
& - (v_{x_1x_2} + cv_{x_2x_2})\omega + \omega_z,
\end{aligned}
\tag{537}
$$

generalizing the Martinez Alonso–Shabat heavenly type integrable system. Thus, the following proposition holds.

Proposition 21. *The constructed system of heavenly type Equations (536) and (537) has the Lax–Sato vector field representation with the "spectral" parameter $\lambda \in \mathbb{C}$, which is related to the element $\tilde{a} \ltimes \tilde{l} \in \tilde{\mathcal{G}}^*$ in the form (535).*

The system of Equations (536) and (537) admits the reduction when $u = v$. In this case, under $c = 1$ one obtains such a system:

$$
\begin{aligned}
u_{zt} + \chi_y &= -(u_{zx_1} + u_{zx_2})\chi + u_z(\chi_{x_1} + \chi_{x_2}), \\
\chi_{x_1} + \chi_{x_2} &= -(u_{zx_1} - u_{zx_2}) - 2(u_{x_1}u_{x_2})_{x_1} - 2(u_{x_1}u_{x_2})_{x_2}.
\end{aligned}
\tag{538}
$$

The additional constraint $u_z = u_{x_1} + u_{x_2}$ transforms the system (538) into the following interesting integro-differential equation:

$$
\begin{aligned}
& (u_{\tilde{t}x_1} + u_{\tilde{t}x_2}) - (u_{\tilde{y}x_1} - u_{\tilde{y}x_2}) = u_{x_1x_2}(u_{x_1} - u_{x_2}) - u_{x_1x_1}u_{x_2} + u_{x_2x_2}u_{x_1} - \\
& - u_{x_1x_2}(u_{x_1}^2 - u_{x_2}^2) - u_{x_1x_1}u_{x_2}(u_{x_1} + u_{x_2}) + u_{x_2x_2}u_{x_1}(u_{x_1} + u_{x_2}) - \\
& - 2(\mathcal{P}(u_{x_1}u_{x_2})_{\tilde{y}}) + (u_{x_1x_1} + 2u_{x_1x_2} + u_{x_2x_2})(\mathcal{P}u_{x_1}u_{x_2}), \\
& \qquad \mathcal{P} = (\partial/\partial x_1 + \partial/\partial x_2)^{-1}(\partial/\partial x_1 - \partial/\partial x_2),
\end{aligned}
$$

where $\tilde{t} = 2t$ and $\tilde{y} = 2y$. Thus, the Equation (538) is integrable and can be considered as some multi-dimensional generalization of the Martinez Alonso–Shabat system [236].

11.6. A Modified Current Loop Algebra and Multidimensional Heavenly Type Integrable Equations: The Generalized Lie-Algebraic Structures

A further generalization of the multi-dimensional case related to the loop group $\widetilde{\mathrm{Diff}}(\mathbb{T}^n)$ on the torus $\mathbb{T}^n$, $n \in \mathbb{Z}_+$ can be developed [207–209] by the following approach. Since the Lie algebra $\widetilde{\mathrm{diff}}(\mathbb{T}^n)$ consists of the loop group elements, analytically continued from the circle $\mathbb{S}^1 := \partial\mathbb{D}^1$, being the boundary of the disk $\mathbb{D}^1 \subset \mathbb{C}$, by means of the complex "spectral"variable $\lambda \in \mathbb{C}$ both into the interior $\mathbb{D}^1_+ \subset \mathbb{C}$ and the exterior $\mathbb{D}^1_- \subset \mathbb{C}$ parts of the disk $\mathbb{D}^1 \subset \mathbb{C}$, one can take into account its analytical invariance to the circle diffeomorphism group $\mathrm{Diff}(\mathbb{S}^1)$. The latter gives rise to the naturally extended holomorphic Lie algebra $\mathrm{diff}(\mathbb{T}^n \times \mathbb{C}) = \widetilde{\mathrm{diff}}(\mathbb{T}^n \times \mathbb{D}^1_+) \oplus \widetilde{\mathrm{diff}}(\mathbb{T}^n \times \mathbb{D}^1_-)$ on the torus $\mathbb{T}^n \times \mathbb{C}$, whose elements are representable as $\tilde{a}(x;\lambda) := \left\langle a(x;\lambda), \frac{\partial}{\partial x} \right\rangle = \sum_{j=1}^{n} a_j(x;\lambda)\frac{\partial}{\partial x_j} + a_0(x;\lambda)\frac{\partial}{\partial \lambda}$ for some holomorphic in $\lambda \in \mathbb{D}^1_\pm$ vectors $a(x;\lambda) \in \mathbb{E} \times \mathbb{E}^n$ for all $x \in \mathbb{T}^n$, and where we denoted by $\frac{\partial}{\partial \mathrm{x}} := (\frac{\partial}{\partial \lambda}, \frac{\partial}{\partial x_1}, \frac{\partial}{\partial x_2}, ..., \frac{\partial}{\partial x_n})^{\intercal}$ the generalized Euclidean vector gradient with respect to the vector variable $\mathrm{x} := (\lambda, x) \in \mathbb{C} \times \mathbb{T}^n$.

Let us construct a modified current loop Lie algebra $\bar{\mathcal{G}}$ as the semi-direct sum $\bar{\mathcal{G}} := \mathrm{diff}(\mathbb{T}^n \times \mathbb{C}) \ltimes \mathrm{diff}(\mathbb{T}^n \times \mathbb{C})^*$ of the Lie algebra $\mathrm{diff}(\mathbb{T}^n \times \mathbb{C})$ and its adjoint space $\mathrm{diff}(\mathbb{T}^n \times \mathbb{C})^*$, taking into account their natural pairing

$$(\bar{l}|\bar{a}) := \underset{\lambda \in \mathbb{C}}{res} (l(\mathrm{x})|a(\mathrm{x}))_{H^0} \tag{539}$$

for any $\bar{l} \in \mathrm{diff}(\mathbb{T}^n \times \mathbb{C})^*$ and $\bar{a} \in \mathrm{diff}(\mathbb{T}^n \times \mathbb{C})$. The corresponding Lie commutator on the loop Lie algebra $\bar{\mathcal{G}}$ is given for any $\bar{a}_1 \ltimes \bar{l}_1, \bar{a}_2 \ltimes \bar{l}_2 \in \bar{\mathcal{G}}$ by

$$[\bar{a}_1 \ltimes \bar{l}_1, \bar{a}_2 \ltimes \bar{l}_2] := [\bar{a}_1, a_2] \ltimes ad^*_{\bar{a}_1} \bar{l}_2 - ad^*_{\bar{a}_2} \bar{l}_1. \tag{540}$$

The Lie algebra $\bar{\mathcal{G}}$ also splits into the direct sum of two subalgebras:

$$\bar{\mathcal{G}} = \bar{\mathcal{G}}_+ \oplus \bar{\mathcal{G}}_-, \tag{541}$$

allowing the introduction of the classical R-structure:

$$[\bar{a}_1 \ltimes \bar{l}_1, \bar{a}_2 \ltimes \bar{l}_2]_R := [R(\bar{a}_1 \ltimes \bar{l}_1),\ \bar{a}_2 \ltimes \bar{l}_2] + [\bar{a}_1 \ltimes \bar{l}_1, R(\bar{a}_2 \ltimes \bar{l}_2)] \tag{542}$$

for any $\bar{a}_1 \ltimes \bar{l}_1, \bar{a}_2 \ltimes \bar{l}_2 \in \bar{\mathcal{G}}$, where, by definition,

$$R := (P_+ - P_-)/2, \tag{543}$$

and

$$P_\pm \bar{\mathcal{G}} := \bar{\mathcal{G}}_\pm \subset \bar{\mathcal{G}}. \tag{544}$$

The space $\bar{\mathcal{G}}^*$ (adjoint to the Lie algebra $\bar{\mathcal{G}}$) can be identified with the space $\bar{\mathcal{G}}$ by using the symmetric and non-degenerate form

$$(\bar{a} \ltimes \bar{l}|\bar{r} \ltimes \bar{m}) := \underset{\lambda \in \mathbb{C}}{res} (\bar{a} \ltimes \bar{l}|\bar{r} \ltimes \bar{m})_{H^0}, \tag{545}$$

where, by definition,

$$(\bar{a} \ltimes \bar{l}|\bar{r} \ltimes \bar{m})_{H^0} = (\bar{m}|\bar{a})_{H^0} + (\bar{l}|\bar{r})_{H^0} \tag{546}$$

for any pair of elements $\bar{a} \ltimes \bar{l},\ \bar{r} \ltimes \bar{m} \in \bar{\mathcal{G}}$.

Remark 12. *The above constructed Lie algebra $\bar{\mathcal{G}}$, being metrized by means of the symmetric, nondegenerate bilinear form (545), is owing to the construction described in the introduction, to uniquely represent the coadjoint orbits on $\bar{\mathcal{G}}^* \simeq \bar{\mathcal{G}}$ in the standard Lax type form on $\bar{\mathcal{G}}$, that will be used further.*

Owing to the convolution (546), the Lie algebra $\bar{\mathcal{G}}$ becomes metrized. For arbitrary smooth functions $f, g \in \mathrm{D}(\bar{\mathcal{G}}^*)$ one can naturally determine two Lie–Poisson brackets

$$\{f, g\} := (\bar{a} \ltimes \bar{l}|[\nabla f(\bar{l}, \bar{a}), \nabla g(\bar{l}, \bar{a})]) \tag{547}$$

and

$$\{f, g\}_R := (\bar{a} \ltimes \bar{l}|[\nabla f(\bar{l}, \bar{a}), \nabla g(\bar{l}, \bar{a})]_R), \tag{548}$$

where at any seed element $\bar{a} \ltimes \bar{l} \in \bar{\mathcal{G}}^* \simeq \bar{\mathcal{G}}$ the gradient element $\nabla f(\bar{l}, \bar{a}) := \nabla f_{\bar{l}} \ltimes \nabla f_{\bar{a}} \simeq \langle \nabla f(l, a)|(\partial/\partial \mathrm{x}, d\mathrm{x})^{\intercal}\rangle \in \bar{\mathcal{G}}$ and $\nabla f_{\bar{l}} = \langle \nabla f_l|\partial/\partial \mathrm{x}\rangle$, $\nabla f_{\bar{a}} = \langle \nabla f_a|d\mathrm{x}\rangle$, and, similarly, the gradient element $\nabla g(\bar{l}, \bar{a}) := \nabla g_{\bar{l}} \ltimes \nabla g_{\bar{a}} \simeq \langle \nabla g(l, a)|(\partial/\partial \mathrm{x}, d\mathrm{x})^{\intercal}\rangle \in \bar{\mathcal{G}}^*$ and $\nabla g_{\bar{l}} = \langle \nabla g_l|\partial/\partial \mathrm{x}\rangle$, $\nabla g_{\bar{a}} = \langle \nabla g_a|d\mathrm{x}\rangle$ are calculated with respect to the metric (546).

Let us now assume that a smooth function $h \in \mathrm{I}(\bar{\mathcal{G}}^*)$ is a Casimir invariant, that is

$$ad^*_{\nabla h(\bar{l},\bar{a})}(\bar{a} \ltimes \bar{l}) = 0 \tag{549}$$

for a chosen seed element $\bar{a} \ltimes \bar{l} \in \bar{\mathcal{G}}^* \simeq \bar{\mathcal{G}}$. Since for an element $\bar{a} \ltimes \bar{l} \in \bar{\mathcal{G}}^* \simeq \bar{\mathcal{G}}$ and an arbitrary $f \in \mathrm{D}(\bar{\mathcal{G}}^*)$ the adjoint mapping is

$$ad^*_{\nabla f(\bar{l},\bar{a})}(\bar{a} \ltimes \bar{l}) = ([\nabla h_{\bar{l}}, \bar{a}] \ltimes (ad^*_{\nabla h_{\bar{l}}}\bar{l} + ad^*_{\bar{a}}\nabla h_{\bar{a}})), \tag{550}$$

the condition (549) can be rewritten as

$$[\nabla h_{\bar{l}}, \bar{a}] = 0, \quad ad^*_{\nabla h_{\bar{l}}}\bar{l} + ad^*_{\bar{a}}\nabla h_{\bar{a}} = 0, \tag{551}$$

and one can easily obtain that the Casimir functional $h \in \mathrm{I}(\bar{\mathcal{G}}^*)$ satisfies the system of determining equations

$$\begin{aligned} &\langle \nabla h_l | \partial/\partial x \rangle a - \langle a | \partial/\partial x \rangle \nabla h_l = 0, \\ &\langle \partial/\partial x | \circ \nabla h_l \rangle l + \langle l | (\partial/\partial x \nabla h_l) \rangle + \\ &+ \langle \partial/\partial x | \circ a \rangle \nabla h_a + \langle a | (\partial/\partial x \nabla h_a) \rangle = 0. \end{aligned} \tag{552}$$

For the Casimir functional $h \in \mathrm{D}(\bar{\mathcal{G}}^*)$ the Equation (552) should be be solved analytically. In the case when an element $\bar{l} \ltimes \bar{a} \in \bar{\mathcal{G}}^*$ is singular as $|\lambda| \to \infty$, one can consider the general asymptotic expansion

$$\nabla h^{(p)}(l,a) \sim \lambda^p \sum_{j \in \mathbb{Z}_+} (\nabla h^{(p)}_{l,j}; \nabla h^{(p)}_{a,j}) \lambda^{-j} \tag{553}$$

for some suitably chosen $p \in \mathbb{Z}_+$, which is substituted into the Equation (552). The latter is then solved recurrently giving rise to a set of gradient expressions for the Casimir functionals $h^{(p)} \in \mathrm{D}(\bar{\mathcal{G}}^*)$ at the specially found integers $p \in \mathbb{Z}_+$.

Assume now that $h^{(y)}, h^{(t)} \in \mathrm{I}(\bar{\mathcal{G}}^*)$ are such Casimir functionals for which the Hamiltonian vector field generators

$$\nabla h^{(y)}(\bar{l},\bar{a})_+ := (\nabla h^{(p_y)}(\bar{l},\bar{a}))_+, \quad \nabla h^{(t)}(\bar{l},\bar{a})_+ := (\nabla h^{(p_t)}(\bar{l},\bar{a}))_+, \tag{554}$$

are, respectively, defined at some specially found integers $p_y, p_t \in \mathbb{Z}_+$. These invariants generate owing to the Lie–Poisson bracket (548) the following commuting to each other Hamiltonian flows:

$$\begin{aligned} \frac{\partial}{\partial y}(\bar{a} \ltimes \bar{l}) &= -ad^*_{\nabla h^{(y)}(\bar{l},\bar{a})_+}(\bar{a} \ltimes \bar{l}), \\ \frac{\partial}{\partial t}(\bar{a} \ltimes \bar{l}) &= -ad^*_{\nabla h^{(t)}(\bar{l},\bar{a})_+}(\bar{a} \ltimes \bar{l}), \end{aligned} \tag{555}$$

on an element $\bar{a} \ltimes \bar{l} \in \bar{\mathcal{G}}^* \simeq \bar{\mathcal{G}}$ with respect to the corresponding evolution parameters $t, y \in \mathbb{R}$. Owing to the construction, the flows (554) can be rewritten equivalently as

$$\begin{aligned} \partial l/\partial t &= -\left\langle \frac{\partial}{\partial x} | \circ \nabla h_l^{(p_t)} \right\rangle l - \left\langle l | (\frac{\partial}{\partial x} \nabla h_l^{(p_t)}) \right\rangle - \left\langle \frac{\partial}{\partial x} | \circ a \right\rangle \nabla h_a^{(p_t)} - \left\langle a | (\frac{\partial}{\partial x} \nabla h_a^{(p_t)}) \right\rangle, \\ \partial l/\partial y &= -\left\langle \frac{\partial}{\partial x} | \circ \nabla h_l^{(p_y)} \right\rangle l - \left\langle l | (\frac{\partial}{\partial x} \nabla h_l^{(p_y)}) \right\rangle - \left\langle \frac{\partial}{\partial x} | \circ a \right\rangle \nabla h_a^{(p_y)} - \left\langle a | (\frac{\partial}{\partial x} \nabla h_a^{(p_y)}) \right\rangle, \\ \partial a/\partial t &= -\left\langle \nabla h_l^{(p_t)} | \frac{\partial}{\partial x} \right\rangle a + \left\langle a | \frac{\partial}{\partial x} \right\rangle \nabla h_l^{(p_t)}, \ \partial a/\partial y = -\left\langle \nabla h_l^{(p_y)} | \frac{\partial}{\partial x} \right\rangle a + \left\langle a | \frac{\partial}{\partial x} \right\rangle \nabla h_l^{(p_y)}, \end{aligned} \tag{556}$$

where $y, t \in \mathbb{R}$ are the corresponding evolution parameters. Since the invariants $h^{(y)}, h^{(t)} \in \mathrm{I}(\bar{\mathcal{G}}^*)$ are commuting to each other with respect to the Lie–Poisson bracket (548), the flows (556) are commuting too, meaning equivalently that the corresponding Hamiltonian vector field generators

$$\nabla h_+^{(t)} := \left\langle \nabla h_l^{(p_t)}(l)_+ | \frac{\partial}{\partial x} \right\rangle, \quad \nabla h_+^{(y)} := \left\langle \nabla h_l^{(p_y)}(l)_+ | \frac{\partial}{\partial x} \right\rangle \tag{557}$$

satisfy the Lax type compatibility condition

$$\frac{\partial}{\partial y} \nabla h_+^{(t)} - \frac{\partial}{\partial t} \nabla h_+^{(y)} = [\nabla h_+^{(t)}, \nabla h_+^{(y)}] \tag{558}$$

for all $y, t \in \mathbb{R}$. On the other hand, the condition (558) is equivalent to the compatibility condition of two linear equations

$$\left(\frac{\partial}{\partial t} + \nabla h_+^{(t)} \right) \psi = 0, \ \langle a | \frac{\partial}{\partial x} \rangle \psi = 0, \ \left(\frac{\partial}{\partial y} + \nabla h_+^{(y)} \right) \psi = 0 \tag{559}$$

for a function $\psi \in C^2(\mathbb{R}^2 \times \mathbb{T}^n \times \mathbb{C}; \mathbb{C})$, all $y, t \in \mathbb{R}$ and any $x \in \mathbb{T}^n \times \mathbb{C}$. The results obtained above can be formulated as the following proposition.

Proposition 22. *Let a seed element $\bar{a} \ltimes \bar{l} \in \bar{\mathcal{G}}^*$ and $h^{(y)}, h^{(t)} \in \mathrm{I}(\bar{\mathcal{G}}^*)$ are some Casimir functionals subject to the metric $(\cdot|\cdot)$ on the holomorphic current loop algebra $\bar{\mathcal{G}}$ and the natural coadjoint action on the co-algebra $\bar{\mathcal{G}}^* \simeq \bar{\mathcal{G}}$. Then the following dynamical systems*

$$\frac{\partial}{\partial y}(\bar{a} \ltimes \bar{l}) = -ad^*_{\nabla h^{(y)}(\bar{l}, \bar{a})_+}(\bar{a} \ltimes \bar{l}), \quad \frac{\partial}{\partial t}(\bar{a} \ltimes \bar{l}) = -ad^*_{\nabla h^{(t)}(\bar{l}, \bar{a})_+}(\bar{a} \ltimes \bar{l}) \tag{560}$$

are commuting to each other Hamiltonian flows for evolution parameters $y, t \in \mathbb{R}$. Moreover, the compatibility condition of these flows is equivalent to the vector field representation

$$(\partial/\partial t + \nabla h_+^{(t)})\psi = 0, \ \langle a | \partial/\partial x \rangle \psi = 0, \ (\partial/\partial y + \nabla h_+^{(y)})\psi = 0, \tag{561}$$

where $\psi \in C^2(\mathbb{R}^2 \times \mathbb{T}^n \times \mathbb{C}; \mathbb{C})$ and the vector fields $\nabla h_+^{(t)}, \nabla h_+^{(y)} \in \mathrm{diff}(\mathbb{T}^n \times \mathbb{C})$ are given by the expressions (557).

Remark 13. *As it was mentioned above, the expansion (553) is effective if a chosen seed element $\bar{a} \ltimes \bar{l} \in \bar{\mathcal{G}}^*$ is singular as $|\lambda| \to \infty$. In the case when it is singular as $|\lambda| \to 0$, the expression (553) should be respectively replaced by the expansion*

$$\nabla h^{(p)}(\bar{l}, \bar{a}) \sim \lambda^{-p} \sum_{j \in \mathbb{Z}_+} \nabla h_j^{(p)}(\bar{l}, \bar{a}) \lambda^j \tag{562}$$

for suitably chosen integers $p \in \mathbb{Z}_+$, and the reduced Casimir function gradients are then given by the Hamiltonian vector field generators

$$\nabla h^{(y)}(\bar{l}, \bar{a})_- = \lambda(\lambda^{-p_y - 1} \nabla h^{(p_y)}(\bar{l}, \bar{a}))_-, \quad \nabla h^{(t)}(\bar{l}, \bar{a})_- = \lambda(\lambda^{-p_t - 1} \nabla h^{(p_t)}(\bar{l}, \bar{a}))_- \tag{563}$$

for suitably chosen positive integers $p_y, p_t \in \mathbb{Z}_+$ and the corresponding Hamiltonian flows are, respectively, written as

$$\frac{\partial}{\partial t}(\bar{a} \ltimes \bar{l}) = ad^*_{\nabla h^{(t)}(\bar{l}, \bar{a})_-}(\bar{a} \ltimes \bar{l}), \quad \frac{\partial}{\partial y}(\bar{a} \ltimes \bar{l}) = ad^*_{\nabla h^{(y)}(\bar{l}, \bar{a})_-}(\bar{a} \ltimes \bar{l}) \tag{564}$$

for evolution parameters $y, t \in \mathbb{R}$.

As it was demonstrated above, the presented construction of Hamiltonian flows on the adjoint space $\bar{\mathcal{G}}^*$ can be generalized proceeding to the point product $\bar{\mathfrak{G}} := \bar{\mathcal{G}}^{\mathbb{S}^1} = \prod_{z\in\mathbb{S}^1} \bar{\mathcal{G}}$ of the holomorphic current Lie algebra $\bar{\mathcal{G}}$, endowed with the central extension, generated by a two-cocycle $\omega_2 : \bar{\mathfrak{G}}\times\bar{\mathfrak{G}} \to \mathbb{C}$, where

$$\omega_2(\bar{a}_1 \ltimes \bar{l}_1, \bar{a}_2 \ltimes \bar{l}_2) := \int_{\mathbb{S}^1} [(\bar{l}_1, \partial\bar{a}_2/\partial z)_1 - (\bar{l}_2, \partial\bar{a}_1/\partial z)_1] dz \tag{565}$$

for any pair of elements $\bar{a}_1 \ltimes \bar{l}_1, \bar{a}_2 \ltimes \bar{l}_2 \in \mathfrak{G}$. The resulting R-deformed Lie–Poisson bracket for any smooth functionals $h, f \in \mathrm{D}(\widehat{\mathfrak{G}}^*)$ on the adjoint space $\widehat{\mathfrak{G}}^*$ to the centrally extended loop Lie algebra $\widehat{\mathfrak{G}} := \bar{\mathfrak{G}} \oplus \mathbb{C}$ becomes equal to

$$\begin{aligned} \{h, f\}_R :=& (\bar{a} \ltimes \bar{l} | [\nabla h(\bar{l}, \bar{a}), \nabla f(\bar{l}, \bar{a})]_R) + \\ &+ \omega_2(R\nabla h(\bar{l}, \bar{a}), \nabla f(\bar{l}, \bar{a})) + \omega_2(\nabla h(\bar{l}, \bar{a}), R\nabla f(\bar{l}, \bar{a})). \end{aligned} \tag{566}$$

The corresponding Casimir functionals $h^{(p)} \in \mathrm{I}(\widehat{\mathfrak{G}}^*)$ for specially chosen $p \in \mathbb{Z}_+$, are defined with respect to the standard Lie–Poisson bracket as

$$\{h^{(p)}, f\} := (\bar{a} \ltimes \bar{l} | [\nabla h^{(p)}(\bar{l}, \bar{a}), \nabla f(\bar{l}, \bar{a})]) + \omega_2(\nabla h^{(p)}(\bar{l}, \bar{a}), \nabla f(\bar{l}, \bar{a})) = 0 \tag{567}$$

for all smooth functionals $f \in \mathrm{D}(\widehat{\mathfrak{G}}^*)$. Based on the equality one easily finds that the gradients $\nabla h^{(p)} \in \widehat{\mathfrak{G}}$ of the Casimir functionals $h^{(p)} \in \mathrm{I}(\widehat{\mathfrak{G}}^*), p \in \mathbb{Z}_+$, satisfy the following equations:

$$[\nabla h_{\bar{l}}, \bar{a}] - \frac{\partial}{\partial z}\nabla h_{\bar{l}} = 0, \;\; ad^*_{\nabla h_{\bar{l}}}\bar{l} + ad^*_{\bar{a}}\nabla h_{\bar{a}} - \frac{\partial}{\partial z}\nabla h_{\bar{a}} = 0 \tag{568}$$

for a chosen element $\bar{a} \ltimes \bar{l} \in \widehat{\mathfrak{G}}^*$. Making use of the suitable Casimir functionals $h^{(y)}, h^{(t)} \in \mathrm{I}(\widehat{\mathfrak{G}}^*)$, one can construct, making use of (566), the following commuting Hamiltonian flows on the adjoint space $\widehat{\mathfrak{G}}^*$:

$$\frac{\partial}{\partial y}(\bar{a} \ltimes \bar{l}) = \{h^{(y)}, \bar{a} \ltimes \bar{l}\}_R, \;\; \frac{\partial}{\partial t}(\bar{a} \ltimes \bar{l}) = \{h^{(t)}, \bar{a} \ltimes \bar{l}\}_R, \tag{569}$$

which are equivalent to the evolution equations

$$\frac{\partial}{\partial y}\bar{a} = -[\nabla h^{(y)}_{\bar{l},+}, \bar{a}] + \frac{\partial}{\partial z}\nabla h^{(y)}_{\bar{l},+}, \;\; \frac{\partial}{\partial t}\bar{a} = -[\nabla h^{(t)}_{\bar{l},+}, \bar{a}] + \frac{\partial}{\partial z}\nabla h^{(t)}_{\bar{l},+} \tag{570}$$

and

$$\begin{aligned} \frac{\partial}{\partial y}\bar{l} &= -ad^*_{\nabla h^{(y)}_{\bar{l},+}}\bar{l} - ad^*_{\bar{a}}(\nabla h^{(y)}_{\bar{a},+}) + \frac{\partial}{\partial z}\nabla h^{(y)}_{\bar{a},+}, \\ \frac{\partial}{\partial t}\bar{l} &= -ad^*_{\nabla h^{(t)}_{\bar{l},+}}\bar{l} - ad^*_{\bar{a}}(\nabla h^{(t)}_{\bar{a},+}) + \frac{\partial}{\partial z}\nabla h^{(t)}_{\bar{a},+}. \end{aligned} \tag{571}$$

The results obtained above are summarized as

Proposition 23. *The Hamiltonian flows (569) on the adjoint space $\widehat{\mathfrak{G}}^*$ generate the separately commuting evolution flows (570) and (571), giving rise to the following unique Lax type compatibility condition:*

$$[\nabla h^{(y)}_{l,+}, \nabla h^{(t)}_{l,+}] - \frac{\partial}{\partial t}\nabla h^{(y)}_{\bar{l},+} + \frac{\partial}{\partial y}\nabla h^{(t)}_{\bar{l},+} = 0, \tag{572}$$

being equivalent to some system of nonlinear heavenly type equations in partial derivatives. Moreover, the system of evolution flows (570) and (571) can be considered as the compatibility condition for the following set of linear vector equations

$$\frac{\partial \psi}{\partial y} + \nabla h^{(y)}_{\tilde{l},+}\psi = 0, \quad \frac{\partial \psi}{\partial z} + \langle a|\partial/\partial \mathrm{x}\rangle\psi = 0, \quad \frac{\partial \psi}{\partial t} + \nabla h^{(t)}_{\tilde{l},+}\psi = 0 \tag{573}$$

for all $(y,t,z;\mathrm{x}) \in (\mathbb{R}^2 \times \mathbb{S}^1) \times \mathbb{T}^n \times \mathbb{C}$ *and a function* $\psi \in C^2((\mathbb{R}^2 \times \mathbb{S}^1) \times \mathbb{T}^n \times \mathbb{C};\mathbb{C})$.

Remark 14. *The Lie-algebraic scheme of constructing heavenly type integrable equations on respectively chosen smooth functional manifolds, applied above for the modified current loop Lie algebra* $\tilde{\mathfrak{G}} := \widetilde{\mathrm{diff}}(\mathbb{T}^n \times \mathbb{C}) \ltimes \widetilde{\mathrm{diff}}(\mathbb{T}^n \times \mathbb{C})^*$ *as the semi-direct sum of the Lie algebra* $\widetilde{\mathrm{diff}}(\mathbb{T}^n \times \mathbb{C})$ *and its dual space* $\widetilde{\mathrm{diff}}(\mathbb{T}^n \times \mathbb{C})^*$, *can be naturally reformulated within a respectively generalized Lagrange–d'Alembert mechanical principle, as was done in the work [214], and which will be analyzed in a separate work under preparation.*

11.7. A New Modified Spatially Four-Dimensional Mikhalev-Pavlov type Heavenly Equation

Let a seed element $\tilde{a} \ltimes \tilde{l} \in \widehat{\mathfrak{G}}^*$ be chosen as

$$\tilde{a} \ltimes \tilde{l} = ((u_x - \lambda)\partial/\partial x + v_x\partial/\partial\lambda) \ltimes (w_x dx + \eta_x d\lambda), \tag{574}$$

where $u, v, w, \eta \in C^2(\mathbb{R}^2 \times (\mathbb{S}^1 \times \mathbb{C});\mathbb{R})$. The asymptotic expressions for the components of the gradients (562) of the corresponding Casimir functionals $h^{(p)} \in \mathrm{I}(\widehat{\mathfrak{G}}^*)$, $p \in \mathbb{Z}_+$, as $|\lambda| \to \infty$ have the following forms:

$$\nabla h_{\tilde{l}} \sim \lambda^p \begin{pmatrix} 1 - u_x\lambda^{-1} + (-u_z + (p-1)v)\lambda^{-2} + (u_y + (p-2)(u_x v + \chi_x))\lambda^{-3} + \dots \\ -v_x\lambda^{-1} - v_z\lambda^{-2} + (v_y - (p-2)v_x v)\lambda^{-3} + \dots \end{pmatrix},$$

$$\nabla h_{\tilde{a}} \sim \lambda^p \begin{pmatrix} w_x\lambda^{-1} + w_z\lambda^{-2} + (-w_y + (p-2)(wv)_x)\lambda^{-3} + \dots \\ \eta_x\lambda^{-1} + (\eta_z + (p-1)w)\lambda^{-2} + (-\eta_y + (p-2)\omega_x)\lambda^{-3} + \dots \end{pmatrix},$$

$p \in \mathbb{Z}_+$, where

$$\chi_{xx} = v_z + u_x v_x, \quad \omega_{xx} = w_z - u_x w_x - v_x\eta_x + v\eta_x.$$

In the case when

$$\nabla h^{(y)}_{\tilde{l},+} := \begin{pmatrix} \lambda^2 - u_x\lambda + (-u_z + v) \\ -v_x\lambda - v_z \end{pmatrix},$$

$$\nabla h^{(y)}_{\tilde{a},+} := \begin{pmatrix} w_x\lambda + w_z \\ \eta_x\lambda + (\eta_z + w) \end{pmatrix},$$

and

$$\nabla h^{(t)}_{\tilde{l},+} = \begin{pmatrix} \lambda^3 - u_x\lambda^2 + (-u_z + 2v)\lambda + (u_y + u_x v + \chi_x) \\ -v_x\lambda^2 - v_z\lambda + (v_y - v_x v) \end{pmatrix},$$

$$\nabla h^{(t)}_{\tilde{a},+} = \begin{pmatrix} w_x\lambda^2 + w_z\lambda + (-w_y + (wv)_x) \\ \eta_x\lambda^2 + (\eta_z + 2w)\lambda + (-\eta_y + \omega_x) \end{pmatrix},$$

the compatibility condition of the Hamiltonian vector flows (569) leads to the system of evolution equations:

$$
\begin{aligned}
&u_{zt} + u_{yy} = -u_y u_{zx} + u_z u_{xy} - u_{xy} v - u_{zz} v - \chi_x u_{xz}, \\
&v_{zt} + v_{yy} = v v_x^2 - v_z^2 - v v_{xy} - v v_{zz} - u_y v_{xz} + u_z v_{xy} - u_z v_x^2 - \chi_x v_{xz}, \\
&- u_{xy} - u_{zz} = u_x u_{xz} - u_z u_{xx} + u_{xx} v, \\
&- v_{xy} - v_{zz} = v_x^2 + v_{xx} v + u_x v_{xz} - u_z v_{xx}, \\
&- u_{xt} + u_{yz} = -u_x u_{xy} + u_y u_{xx} + u_{xz} v + u_{xx} \chi_x, \\
&- v_{xt} + v_{yz} = -u_x v_{xy} + u_y v_{xx} + u_x v_x^2 + v_{xx} v + 2 v_x v_z.
\end{aligned} \tag{575}
$$

Under the constraint $v = 0$ one obtains the modified Michalev–Pavlov type integrable system (532).

Here, we can also observe that the seed element (574) can also be presented in the compact form:

$$
\tilde{a} \ltimes \tilde{l} := \left(\frac{\partial \tilde{\eta}_1}{\partial x}\frac{\partial}{\partial x} + \frac{\partial \tilde{\eta}_0}{\partial \lambda}\frac{\partial}{\partial \lambda}\right) \ltimes d\tilde{\rho}, \tag{576}
$$

$$
\tilde{\eta}_0 = \lambda v_x, \tilde{\eta}_1 = u \ - \lambda x, \ \ \tilde{\rho} = w + \eta_x \lambda,
$$

being closely connected with the geometry of the related moduli space of flat connections, related to the coadjoint actions of the corresponding Casimir functionals. Its suitable generalization to multidimensional Mikhalev–Pavlov type equations can be chosen as

$$
\tilde{a} \ltimes \tilde{l} := (\langle \nabla_x \tilde{\eta} | \nabla_x \rangle + \nabla_\lambda \tilde{\eta}_0 \nabla_\lambda) \ltimes d\tilde{\rho} \tag{577}
$$

for some elements $\tilde{\eta}, \tilde{\eta}_0, \tilde{\rho} \in \Omega^0(\mathbb{T}^n) \otimes \mathbb{C}, n \in \mathbb{N}$. The analysis of corresponding systems of integrable multidimensional Mikhalev–Pavlov type equations is planned to be presented in a separate study.

12. Conclusions

A wide variety of multidimensional completely integrable evolution flows on smooth functional manifolds have been constructed. Our approach was based on a generalized Lie-algebraic Adler–Kostant–Symes scheme, applied to the modified holomorphic current loop algebra $\mathfrak{G} := \widetilde{\text{diff}}(\mathbb{T}^n \times \mathbb{C}) \ltimes \widetilde{\text{diff}}(\mathbb{T}^n \times \mathbb{C})^*$, the semi-direct sum of the loop Lie algebra $\widetilde{\text{diff}}(\mathbb{T}^n \times \mathbb{C}) := \widetilde{Vect}(\mathbb{T}^n \times \mathbb{C})$ of vector fields on the $\mathbb{T}^n \times \mathbb{C}, n \in \mathbb{Z}_+$, and its adjoint space $\widetilde{\text{diff}}(\mathbb{T}^n \times \mathbb{C})^*$. Its relation to the classical R-structure on the loop Lie algebra $\widetilde{Vect}(\mathbb{T}^n \times \mathbb{C})$ is also discussed. The structure of the corresponding seed elements is analyzed, its multidimensional generalizations are presented. We also demonstrated that the obtained Hamiltonian flows are equivalent to the compatibility conditions for the suitably related Lax–Sato type linear vector field equations. We also mentioned a very interesting Lagrange–d'Alembert type mechanical interpretation, naturally related to the devised Lax–Sato vector field equations and their compatibility conditions. As interesting examples, we constructed new modified spatially four-dimensional Mikhalev–Pavlov and Alonso–Shabat type completely integrable equations, appearing in the study of some differential geometric structures on Riemannian spaces with symmetries.

Funding: This research was funded by the Department of Computer Science and Telecommunication of the Cracov University of Technology for a local research grant F-2/370/2018/DS.

Acknowledgments: I would like to convey my warm thanks to Gerald A. Goldin for many discussions of the work and instrumental help in editing a manuscript during the XXVIII International Workshop on "Geometry in Physics", held on 30 June–7 July 2019 in Białowieża, Poland. My special appreciation belongs to Stefan Duplij for friendly encouragement to write this article and to Joel

Lebowitz for the invitation to take part in the 121-st Statistical Mechanics Conference, held 12–14 May 2019 at the Rutgers University, New Brunswick, NJ, USA. I cordially appreciate Joel Lebowitz, Denis Blackmore and Nikolai N. Bogolubov for instructive discussions, useful comments and remarks on the work during the Conference. My warm acknowledgements also belong to my close collaborators Alex A. Balinsky, Radoslaw Kycia, Yarema A. Prykarpatsky, Valeriy H. Samoilenko for the support during my work on manuscript.

Conflicts of Interest: The author declares no conflict of interest.

References

1. Goldin, G.A. Lectures on diffeomorphism groups in quantum physics. In *Contemporary Problems in Mathematical Physics, Proceedings of the Third International Workshop, Helsinki, Finland, 30–31 October 2014*; World Scientific Publishing: Singapore, 2004; pp. 3–93.
2. Goldin, G.A.; Sharp, D.H. Lie algebras of local currents and their representations. In *Group Representations in Mathematics and Physics*; Battelle Seattle 1969 Rencontres, Lecture Nootes in Physics; Springer: Berlin/Heidelberg, Germany, 1970; Volume 6, pp. 300–311.
3. Goldin, G.A.; Sharp, D.H. *Functional Differential Equations Determining Representations of Local Current Algebras in Magic without Magic: John Archibald Wheeler*; Klauder, J.R., Ed.; Freeman: San Francisco, CA, USA, 1972.
4. Goldin, G.A. Nonrelativistic current algebras as unitary representations of groups. *J. Math. Phys.* **1971**, *12*, 462–487. [CrossRef]
5. Goldin, G.A.; Grodnik, J.; Powers, R.T.; Sharp, D. Nonrelativistic current algebra in the N/V-limit. *J. Math. Phys.* **1974**, *15*, 88–100. [CrossRef]
6. Goldin, G.A.; Menikoff, R. Sharp F.H. Diffeomorphism groups, gauge groups, and quantum theory. *Phys. Rev. Lett.* **1983**, *51*, 2246–2249. [CrossRef]
7. Goldin, G.A.; Menikoff, R.; Sharp, F.H. Representations of a local current algebra in nonsimply connected space and the Aharonov-Bohm effect. *J. Math. Phys.* **1981**, *22*, 1664–1668. [CrossRef]
8. Bogolubov N.N., Jr.; Prorok, D.; Prykarpatski, A.K. Integrability Aspects of the Current Algebra Representation and the Factorized Quantum Nonlinear Schrëdinger Type Dynamical Systems. *Phys. Part Nucl.* **2020**, *51*, 434–442. [CrossRef]
9. Prorok, D.; Prykarpatski, A. Quantum Current Algebra Symmetries and Integrable Many-Particle Schrëdinger Type Quantum Hamiltonian Operators. *Symmetry* **2019**, *11*, 975. [CrossRef]
10. Prorok, D.; Prykarpatski, A. The current algebra representations of quantum many-particle Schrëdinger Hamiltonian models, their factorized structure and integrability. *Condens. Matter Phys.* **2019**, *22*, 33101–33130. [CrossRef]
11. Blackmore, D.; Prykarpatsky, A.K.; Samoylenko, V.H. *Nonlinear Dynamical Systems of Mathematical Physics: Spectral and Differential-Geometrical Integrability Analysis*; World Scientific: Singapore, 2011.
12. Bogolyubov, N.N., Jr.; Prikarpatskii, A.K. Quantum current Lie algebra as the universal algebraic structure of the symmetries of completely integrable nonlinear dynamical systems of theoretical and mathematical physics. *Theor. Math. Phys.* **1988**, *75*, 329–339. [CrossRef]
13. Mitropolsky, Y.A.; Bogolubov, N.N.; Prykarpatsky, A.K.; Samoylenko, V.H. Integrable dynamical systems. In *Spectral and Differential Geometric Aspects*; Naukova Dumka: Kyiv, Ukraine, 1987.
14. Kowalski, K. *Methods of Hilbert Spaces in the Theory of Nonlinear Dynamical Systems*; World Scientific: Singapore, 1994.
15. Kowalski, K.; Steeb, W.-H. *Non Linear Dynamical Systems and Carleman Linearization*; World Scientific Singapore, 1991.
16. Prykarpatsky, A.K.; Bogoliubov, N.N., Jr.; Golenia, J.; Taneri, U. Introductive Backgrounds to Modern Quantum Mathematics with Application to Nonlinear Dynamical Systems. *Int. J. Theor. Phys.* **2008**, *47*, 2882–2897. [CrossRef]
17. Faddeev, L.D.; Tahtadjian, L.A. *Hamiltonian Approach in Soliton Theory*; Springer: Berlin/Heidelberg, Germany, 1987.
18. Blaszak, M . *Bi-Hamiltonian Dynamical Systems*; Springer: New York, NY, USA, 1998.
19. Reyman, A.G.; Semenov-Tian-Shansky, M.A. *Integrable Systems*; The Computer Research Institute: Moscow, Russia, 2003. (In Russian)
20. Bogolubov, N.N., Jr.; Prykarpatsky, A.K. Quantum method of Bogolyubov generating functionals in statistical physics: Lie current algebra, its representations and functional equations. *Sov. J. Part. Nucl.* **1986**, *17*, 789–827.
21. Bogolubov, N.N., Jr.; Prykarpatsky, A.K. NN Bogolyubov's quantum method of generating functionals in statistical physics: The current Lie algebra, its representations and functional equations. *Ukr. Mat. Zhurnal* **1986**, *38*, 245–249. [CrossRef]
22. Bogolyubov, N.N., Jr.; Prykarpatsky, A.K. The Wigner quantized operator and N. N. Bogolyubov generating functional method in nonequilibrium statistical physics. *Dokl. Akad. Nauk SSSR* **1985**, *285*, 1365–1370.
23. Ivankiv, L.I.; Prykarpatsky, Y.A.; Samoilenko, V.H.; Prykarpatski, A.K. Quantum Current Algebra Symmetry and Description of Boltzmann Type Kinetic Equations in Statistical Physics. *Symmetry* **2021**, *13*, 1452. [CrossRef]
24. Prykarpatsky, Y.A.; Kycia, R.; Prykarpatski, A.K. On the Bogolubov's chain of kinetic equations, the invariant subspaces and the corresponding Dirac type reduction. *Ann. Math. Phys.* **2021**, *4*, 074–083. [CrossRef]
25. Kupershmidt, B. Hydrodynamical Poisson brackets and local Lie algebras. *Phys. Lett.* **1987**, *21*, 167–174. [CrossRef]
26. Arnold, V.I. Sur la geometrie differerentielle des groupes de Lie de dimension infinie et ses applications a l'hydrodynamique des fluides parfaits. *Ann. Inst. Fourier* **1966**, *16*, 319–361. [CrossRef]

27. Holm, D.; Kupershmidt B. Poisson structures of superfluids. *Phys. Lett.* **1982**, *91*, 425–430. [CrossRef]
28. Kupershmidt, B.A.; Ratiu, T. Canonical Maps between Semidirect Products with Applications to Elasticity and Superfluids. *Commun. Math. Phys.* **1983**, *90*, 235–250. [CrossRef]
29. Marsden, J.; Ratiu, T.; Schmid, R.; Spencer, R.; Weinstein, A. Hamiltonian systems with symmetry, coadjoint orbits, and plasma physics. *Atti Acad. Sci. Torino* **1983**, *117*, 289–340.
30. Marsden, J.; Weinstein, A. The Hamiltonian structure of the Maxwell-Vlasov equations. *Phys. D* **1982**, *4*, 394–406. [CrossRef]
31. Weinstein, A. Sophus Lie and symplectic geometry. *Expos. Math.* **1983**, *1*, 95–96.
32. Weinstein, A. The local structure of Poisson manifolds. *J. Differ. Geom.* **1983**, *18*, 523–557. [CrossRef]
33. Marsden, J.; Weinstein, A. Reduction of symplectic manifolds with symmetry. *Rep. Math. Phys.* **1974**, *5*, 121–130. [CrossRef]
34. Gay-Balmaz, F.; Monastyrsky, M.; Ratiu, T.S. Lagrangian Reductions and Integrable Systems in Condensed Matter. *Commun. Math. Phys.* **2015**, *335*, 609–636. [CrossRef]
35. Gay-Balmaz, F.; Yoshimira, H. Dirac reduction for nonholonomic mechanical systems and semi-direct product. *arXiv* **2014**, arXiv:1410.5394v1.
36. Holm, D.D.; Tronci, C. Euler-Poincare formulation of hybrid plasma models. *arXiv* **2011**, arXiv:1012.0999v2.
37. Khesin, B.; Lenells, J.; Misiolek, G.; Preston, S.C. Geometry of diffeomorphism groups, complete integrability and geometric statistics. *Geom. Funct. Anal.* **2013**, *23*, 334–366. [CrossRef]
38. Kolev, B. Lie groups and mechanics: Introduction. *J. Nonl. Math. Phys.* **2004**, *11*, 480–498. [CrossRef]
39. Kushner, A.; Lychagin, V.; Roop, M. Optimal Thermodynamic Processes for Gases. *Entropy* **2020**, *22*, 448. [CrossRef]
40. Marsden, J.E.; Ratiu, T.S.; Shkoller, S. The geometry and analysis of the averaged Euler equations and a new diffeomorphism group. *Geom. Funct. Anal.* **2000**, *10*, 582–599. [CrossRef]
41. Marsden, J.; Ratiu, T.; Weinstein A. Reduction and Hamiltoninan structures on duals of semidirect product Lie algebras. *Contemp. Math.* **1984**, *28*, 55–100.
42. Mrugala, R. Continuous contact transformations in thermodynamics. *Rep. Math. Phys.* **1993**, 33, 149–154. [CrossRef]
43. Mrugala, R. Lie, Jacobi, Poisson and quasi-Poisson structures in thermodynamics. *Tensor New Ser. 1995*, 56, *37–45*.
44. Preston, S.C. For ideal fluids, Eulerian and Lagrangian instabilities are equivalent. *Geom. Funct. Anal.* **2004**, *14*, 1044–1062. [CrossRef]
45. Schneider, E. Differential invariants. In *Nonlinear PDEs, Their Geometry, and Applications*; Kycia, R.A., Ulan, M., Schneider, E., Eds.; Springer Nature: Cham, Switzerland, 2019.
46. Schneider, E. Differential invariants of measurements, and their connection to central moments. *arXiv* **2020**, arXiv:2005.08895v1.
47. Tronci, C.; Tassi, E.; Camporeale, E.; Morrison, P.J. Hybrid Vlasov-MHD models: Hamiltonian vs. non-Hamiltonian. *arXiv* **2014**, arXiv:1403.2773v2.
48. Vizman, C. Geodesic Equations on Diffeomorphism Groups. *SIGMA* **2008**, *4*, 030. [CrossRef]
49. Blackmore, D.; Balinsky, A.A.; Prykarpatski, A.K. Entropy and Ergodicity of Boole-Type Transformations. *Entropy* **2021**, *23*, 1405. [CrossRef]
50. Nikitin, V.Y.; Tsybenko, S. On Clebsch variables in hydrodynamics of classical fluids and plasmas. *Czechoslov. J. Phys.* **2002**, *52*, 305–309.
51. Jackson, D.M.; Moffatt, I. *An Introduction to Quantum and Vassiliev Knot Invariants*; Springer: Berlin/Heidelberg, Germany, 2019.
52. Esen, O.; Grmela, M.; Gumral, H.; Pavelka, M. Lifts of Symmetric Tensors: Fluids, Plasma, and Grad Hierarchy. *Entropy* **2019**, *21*, 907. [CrossRef]
53. Grmela, M. Contact Geometry of Mesoscopic Thermodynamics and Dynamics. *Entropy* **2014**, *16*, 1652–1686. [CrossRef]
54. Balinsky, A.A.; Blackmore, D.; Kycia, R.; Prykarpatski, A.K. Geometric Aspects of the Isentropic Liquid Dynamics and Vorticity Invariants. *Entropy* **2020**, *22*, 1241. [CrossRef] [PubMed]
55. Berezin F.A. *The Method of Second Quantization (Monographs and Textbooks in Pure and Applied Physics)*; Academic Press: Cambridge, MA, USA, 1966.
56. Bogolubov, N.N.; Bogolubov N.N., Jr. *Introduction to Quantum Statistical Mechanics*; Gordon and Breach: New York, NY, USA; London, UK, 1994.
57. Berezin, F.A.; Shubin, M.A. *Schrëdinger Equation*; Springer Science & Business Media: Berlin/Heidelberg, Germany, 2012; 555p.
58. Faddeev, L.D.; Yakubovskii, O.A. *Lectures on Quantum Mechanics for Mathematics Students*; American Mathematical Society: Providence, RI, USA, 2009.
59. Prykarpatsky, A.; Mykytyuk, I. *Algebraic Integrability of Nonlinear Dynamical Systems on Manifolds: Classical and Quantum Aspects*; Kluwer Academic Publishers: Alphen aan den Rijn, The Newtherlands, 1998.
60. Takhtajan, L.A. *Quantum Mechanics for Mathematicians*; Department of Mathematics, Stony Brook University: Stony Brook, NY, USA, 2008.
61. Berezanskii, Y.M. *Expansions in Eigenfunctions of Selfadjoint Operators*; Translations of Mathematical Monographs; American Mathematical Society: Providence, RI, USA, 1968; 809p.
62. Berezansky, Y.M.; Kondratiev, Y.G. *Spectral Methods in Infinite Dimensional Analysis, v.1 and 2*; Kluwer: Alphen aan den Rijn, The Netherlands, 1995.
63. Dirac, P.A.M. *The Principles of Quantum Mechanics*, 2nd ed.; Clarendon Press: Oxford, UK, 1935.
64. Fock, V.A. Konfigurationsraum und zweite Quantelung. *Zeischrift Phys. Bd.* **1932**, *75*, 622–647. [CrossRef]

65. Prykarpatsky, A.K.; Taneri, U.; Bogolubov, N.N., Jr. *Quantum Field Theory and Application to Quantum Nonlinear Optics*; World Scientific: Singapore, 2002.
66. Gelfand, I.; Vilenkin, N. *Generalized Functions* ; Academic Press: Cambridge, MA, USA, 1964; Volume 4.
67. Balakrishnan, A.V. *Applied Functional Analysis*; Springer: New York, NY, USA, 1981.
68. Bogolubov, N.N., Jr.; Prykarpatsky, A.K. Quantum method of generating Bogolubov functionals in statistical physics: Current Lie algebras, their representations and functional equations. *Phys. Elem. Part. At. Nucl.* **1986**, *17*, 791–827.
69. Reed, M.; Simon, B. *Theory of Operators, v.3*; SpringerBerlin/Heidelberg, Germany, 1987.
70. Albeverio, S.; Kondratiev, Y.G.; Streit, L. *How to Generalize White Noice Analysis to Non-Gaussian Measures*; Bi-Bo-S: Bielefeld, Germany, 1992.
71. Albeverio, S.; Daletsky, A.; Kondratiev, Y.; Lytvynov, E. Laplace operators in de-Rham complexes associated with measures on configuration spaces. *J. Geom. Phys.* **2003**, *47*, 259–302. [CrossRef]
72. Aref'eva, I.Y. Current formalism in nonrelativistic quantum mechanics. *Theoret. Math. Phys.* **1972**, *10*, 146–155. [CrossRef]
73. Parthasarathy, K.R. *Introduction to Probability and Measure*; Hindustan Book Agency: New Delhi, India, 2005.
74. Goldin, G.A.; Sharp, D.H. Rotational generators in two-dimensional space and particles obeying unusual statistics. *Phys. Rev. D* **1983**, *28*, 830–832. [CrossRef]
75. Araki, H. Hamiltonian Formalism and the Canonical Commutation Relations in Quantum Field Theory. *J. Math. Phys.* **1960**, *1*, 492–504. [CrossRef]
76. Menikoff, R. Generating functionals determining representation of a nonrelativistic local current algebra in the N/V-limit. *J. Math. Phys.* **1974**, *15*, 1394–1408. [CrossRef]
77. Menikoff, R.; Sharp, D. Representation of a local current algebra: Their dynamical determination. *J. Math. Phys.* **1975**, *16*, 2341–2352. [CrossRef]
78. Campbell, C.E. Extended Jastrow functions. *Phys. Lett.* **1973**, *44*, 471–477. [CrossRef]
79. Feenberg, E. Ground state of an interacting boson system. *Ann. Phys.* **1974**, *84*, 128–137. [CrossRef]
80. Berezansky, Y.M. A generalization of white noice analysis by means of theory of hypergroups. *Rep. Math. Phys.* **1996**, *38*, 289–300. [CrossRef]
81. Prykarpatsky, A.K.; Bogoliubov, N.N., Jr.; Golenia, J. A symplectic generalization of the Peradzyński helicity theorem and some applications. *Int. J. Theor. Phys.* **2008**, *47*, 1919–1928. [CrossRef]
82. Beckenbach, E.F.; Bellman, R. *Inequalities*; Springer: Berlin/Heidelberg, Germany, 1961.
83. Friedrichs, K. Spektraltheorie halbbeschränkter Operatoren I–III. *Math. Ann.* **1934**, *109*, 465–487. 685–713. [CrossRef]
84. Kato, T. *Perturbations Theory of Linear Operators*; Springer: Berlin/Heidelberg, Germany, 1966.
85. Reed, M.; Simon, B. *Functional Analysis, v.1*; Springer: Berlin/Heidelberg, Germany, 1987.
86. Bogolubov, N.N. *Problems of Dynamical Theory in Statistical Physics*; Geophysics Research Directorate, AF Cambridge Research Laboratories, Air Force Research Division, United States Air Force: Washington, DC, USA, 1960.
87. Onofri, E. A note on coherent state representations of Lie groups. *J. Math. Phys.* **1975**, *16*, 1087. [CrossRef]
88. Twareque-Ali, S.; Antoine, J.-P.; Gazeau, J.-P.; Mueller, U.A. Coherent states and their generalizations: A mathematical overview. *Rev. Math. Phys.* **1995**, *7*, 1013–1104. [CrossRef]
89. Schrëdinger, E. Der stetige Ubergang von der Mikro- zur Makromechanik. *Naturwiss* **1926**, *14*, 664–666. [CrossRef]
90. Glauber, R.J. *Quantum Theory of Optical Coherence*; Selected Papers and Lectures; Wiley-VCH: Weinheim, Germany, 2007.
91. Klauder, J.R. Continuous-representation theory. I. Postulates of continuousrepresentation theory. *J. Math. Phys.* **1963**, *4*, 1055–1058. [CrossRef]
92. Klauder, J.R. Continuous-representation theory. II. Generalized relation between quantum and classical dynamics. *J. Math. Phys.* **1963**, *4*, 1058–1073. [CrossRef]
93. Sudarshan, E.C.G. Equivalence of semiclassical and quantum mechanical descriptions of statistical light beams. *Phys. Rev. Lett.* **1963**, *10*, 277–279. [CrossRef]
94. Klauder, J.R.; Skagerstam, B.S. *Coherent States—Applications in Physics and Mathematical Physics*; World Scientific: Singapore, 1985.
95. Perelomov, A.M. Coherent States for Arbitrary Lie Group. *Commun. Math. Phys.* **1972**, *26*, 222–236. [CrossRef]
96. Gilmore, R. Geometry of symmetrized states. *Ann. Phys.* **1972**, *74*, 391–463. [CrossRef]
97. Gilmore, R. On properties of coherent states. *Rev. Mex. Fis.* **1974**, *23*, 143–187.
98. von Neumann, J. *Mathematische Grundlagen der Quanten Mechanik*; Springer: Berlin/Heidelberg, Germany, 1932.
99. Weyl, H. *The Theory of Groups and Quantum Mechanics*; Dover: Mineola, NY, USA, 1931.
100. Valatin, J.G. Comments on the theory of superconductivity. *Nuovo Cim.* **1958**, *7*, 843–857. [CrossRef]
101. Bogolubov, N.N., Jr.; Blackmore, D.; Prykarpatski, A.K. The Lagrangian and Hamiltonian Aspects of the Electrodynamic Vacuum-Field Theory Models. *Boson J. Mod. Phys.* **2016**, *2*, 105–196.
102. Onofri, E.; Pauri, M. Dynamical Quantization. *J. Math.Phys.* **1972**, *13*, 533. [CrossRef]
103. Bargmann, V. On a Hilbert space of analytic functions and an associated integral transform, Part I, Commun. *Pure Appl. Math.* **1961**, *14*, 187–214. [CrossRef]
104. Sontz, S.B. On the reproducing kernel of the Segal-Bargmann space. *J. Math. Phys.* **1999**, *40*, 1664–1676. [CrossRef]
105. Szafraniec, F.H. *Przestrzenie Hilberta z Jadrem Reprodukcyjnym*; Jagiellonian University Publisher: Kraków, Poland, 2004.

106. Rudin, W. *Functional Analysis*; International Series in Pure and Applied Mathematics, McGraw-Hill Series in Higher Mathematics; Tata McGraw-Hill: New York, NY, USA, 1974.
107. Kondratiev, Y.G.; Streit, L.; Westerkamp, W.; Yan, J.-A. Generalized Functions in Infinite Dimensional Analysis. *Hiroshima Math. J.* **1998**, *28*, 213–260. [CrossRef]
108. Lytvynov, E.W.; Rebenko, A.L.; Shchepaniuk, G.V. Wick calculus on spaces of generalized functions compound Poisson white noise. *Rep. Math. Phys.* **1997**, *39*, 219–247. [CrossRef]
109. Kowalski, K.; Steeb, W.-H. Symmetries and first integrals for nonlinear dynamical systems: Hilbert space approach. I. *Prog. Theor. Phys.* **1991**, *85*, 713–722. [CrossRef]
110. Kostant, B. Quantization and representation. *Lond. Math. Soc. Lect. Notes Ser. A* **1979**, *34*, 287–316.
111. Novikov, S.P. *Theory of Solitons: The Inverse Scattering Method (Monographs in Contemporary Mathematics)*; Springer: Berlin/Heidelberg, Germany, 1984.
112. Bogolyubov, N.N., Jr.; Prikarpatskii, A.K. A bi-local periodic problem for the Sturm-Liouville and Dirac operators and some applications to the theory of nonlinear dynamical systems. *Ukr. Math. J.* **1990**, *42*, 702–707. [CrossRef]
113. Fil, B.N.; Prikarpatskii, A.K.; Pritula, N.N. Quantum Lie algebra of currents—The universal algebraic structure of symmetries of completely integrable dynamical systems. *Ukr. Math. J.* **1988**, *40*, 645–649. [CrossRef]
114. Bogolyubov, N.N., Jr.; Prikarpatskii, A.K. Complete integrability of the nonlinear Ito and Benney-Kaup systems: Gradient algorithm and Lax representation. *Theor. Math. Phys.* **1986**, *67*, 586–596. [CrossRef]
115. Bogolyubov, N.N., Jr.; Prikarpatskii, A.K. Bogolyubov generating functional method in statistical mechanics and the analog of the transformation to collective variables. *Theor. Math.* **1986**, *66*, 305–317. [CrossRef]
116. Bogolyubov, N.N., Jr.; Prikarpatskii, A.K.; Kurbatov, A.M.; Samoilenko, V.G. Nonlinear model of Schrëdinger type: Conservation laws, Hamiltonian structure, and complete integrability. *Theor. Math. Phys.* **1985**, *65*, 1154–1164. [CrossRef]
117. Bogolyubov, N.N., Jr.; Prikarpatskii, A.K. A bilocal periodic problem for Sturm–Liouville and Dirac differential operators, and some applications in the theory of nonlinear dynamical systems. *Dokl. Math.* **1990**, *41*, 21–25.
118. Bogolyubov, N.N., Jr.; Prikarpatskii, A.K. The N.N. Bogolubov generating functional method in statistical mechanics and a collective variables transform analog. *Theor. Math. Phys.* **1986**, *66*, 463–480. (In Russian) [CrossRef]
119. Gelfand, I.M.; Fuchs D.B. Cohomology of the Lie algebra of vector fields on the circle. *Funct. Anal. Appl.* **1968**, *2*, 342–343. [CrossRef]
120. Olver, P. *Applications of Lie Groups to Differential Equations*; Graduate Texts in Mathematics Series 107; Springer: New York, NY, USA, 1986.
121. Fuchssteiner, B.; Fokas, A.S. Symplectic structures, their Bäcklund transformations and hereditary symmetries. *Phys. Nonlinear Phenom.* **1981**, *4*, 47–66. [CrossRef]
122. Abraham, R.; Marsden, J. *Foundations of Mechanics*, 2nd ed.; Benjamin Cummings: San Francisco, CA, USA, 2008.
123. Godbillon, C. *Geometrie Differentielle et Mecanique Analytique*; Hermann: Paris, France, 1969.
124. Sidorenko, Y.N.; Prikarpatskii, A.K. Periodic problem for nonlinear Ablowitz model. *J. Sov. Math.* **1993**, *65*, 1921–1927. [CrossRef]
125. Sidorenko, Y.N. Elliptic bundles and generating operators. *Zap. Nauchn. Semin. LOMI* **1987**, *161*, 76–87. [CrossRef]
126. Lax, P.D. Periodic solutions of the Korteweg-de Vries equation. *Comm. Pure Appl. Math.* **1968**, *21*, 467–490. [CrossRef]
127. Cartan, A. *Differential Forms*; Dover Publisher: Mineola, NY, USA, 1971.
128. Kaup, D.; Newell, A.C. An exact solution for a derivative nonlinear Schrëdinger equation. *J. Math. Phys.* **1987**, *19*, 798–801. [CrossRef]
129. Fokas, A.S.; Santini, P.M. BiHamiltonian formulation of the Kadomtsev–Petviashvili and Benjamin–Ono equations. *J. Math. Phys.* **1988**, *29*, 604–617. [CrossRef]
130. Prykarpatskyj, A.K.; Samoilenko, V.H.; Andrushkiw, R.I. Algebraic structure of the gradient-holonomic algorithm for Lax integrable nonlinear dynamical systems. II. The reduction via Dirac and canonical quantization procedure. *J. Math. Phys.* **1994**, *35*, 4088–4116. [CrossRef]
131. Hentosh, O.E.; Balinsky, A.A.; Prykarpatski, A.K. Poisson structures on (non) associative noncommutative algebras and integrable Kontsevich type Hamiltonian systems. *Ann. Math. Phys.* **2020**, *3*, 001–006. [CrossRef]
132. Lapointe, L.; Vinet, L. Exact operator solution of the Calogero-Sutherland model. *Commun. Math. Phys.* **1996**, *178*, 425–452. [CrossRef]
133. Sergeev, A.N.; Veselov, A.P. Dunkl operators at infinity and Calogero–Moser systems. *arXiv* **2013**, arXiv:1311.0853.
134. An-Chun, J.; Qing, S.; Xie X.C.; Liu W.M. Josephson Effect for Photons in Two Weakly Linked Microcavities. *Phys. Rev. Lett.* **2009**, *102*, 023602.
135. Sklyanin, E.K. Quantum version of the method of inverse scattering problem, Differential geometry, Lie groups and mechanics. Part III. *J. Sov. Math.* **1982**, *19*, 1546–1596. [CrossRef]
136. Sklyanin, E.K.; Takhtadzhyan, L.A.; Faddeev, L.D. Quantum inverse problem method. I. *Theoret. Math. Phys.* **1979**, *40*, 688–706. [CrossRef]
137. Dunkl, C.F. Differential-difference operators associated to reflection groups. *Trans. Am. Math. Soc.* **1989**, *311*, 167–183. [CrossRef]
138. Aniceto, I.; Avan, J.; Jevicki, A. Poisson structures of Calogero–Moser and Ruijsenaars–Schneider models. *J. Phys. A Math. Theor.* **2010**, *43*, 185201. [CrossRef]
139. Pardee, W.J.; Schlessinger, L. Wright. *J. Phys Rev.* **1968**, *165*, 1883.

140. Kac, M. *Some Stochastic Problems in Physics and Mathematics, Colloquium Lectures in Pure and Applied Science;* Magnolia Petroleum Co.: Galveston, TX, USA, 1956; Volume 2.
141. Bogolubov, N.N.; Shirkov, D.V. *Introduction to the Theory of Quantizerd Fields;* Interscience: New York, NY, USA, 1959; 479p.
142. Balescu, R. *Equilibrium and Non-Equilibrium Statistical Mechanics;* Wiley: Hoboken, NJ, USA, 1975.
143. Bogolyubov, N.N. *Problems of Dynamical Theory in Statistical Physics;* Moscow-Leningrad GITTL (State Publishing House for Technical and Theoretical Literature): Moscow, Russia, 1946.
144. Bohm, D. *The General Collective Variables Theory;* Mir: Moscow, Russia, 1964; 187p. (In Russian)
145. Prykarpatsky, A.K.; Kaleniuk, P.I. Gibbs representations of current Lie algebra and quantum functional Bogoliubov equation. *Dokl. Acad. Nauk USSR* **1988**, *301*, 871–876.
146. Prykarpatsky, A.K. The NN Bogolubov generating functional method in statistical mechanics and a collective variables transform analog within the grand canonical ensemble. *Dokl. SSSR* **1985**, *285*, 1096–1101. (In Russian)
147. Vladimirov, V.S. *Generalized Functions in Mathematical Physics;* Nauka: Moscow, Russia 1979; 319p. (In Russian)
148. Granas, A.; Dugunji, J. *Fixed Point Theory;* Springer: New York, NY, USA, 2003; 564p.
149. Akhiezer, A.I.; Peletminsky, S.V. *Methods of Statistical Physics;* Pergamon Press: Oxford, UK, 2013.
150. Bogolubov, N.N., Jr.; Sadovnikov, B.J.; Shumovsky, A.S. *Mathematical Methods of Statistical Mechanical Model Systems,* CRC: Boca Raton, FL, USA, 1984.
151. Petrina, D.Y.; Gerasimenko, V.I.; Malyshev, P.V. *Mathematical Foundations of Classical Statistical Mechanics;* CRC Press Publisher: Boca Raton, FL, USA, 2002.
152. Berezin, F.A.; Sushko, V.N. Relativistic two-dimensional model of a melf-interacting fermion field with non-vanishing rest mass. *Sov. Phys. JETP* **1965**, *21*, 865–873.
153. Tsvetkov, A.A. Integrals of the nonlinear quantum Schrĕdinger equation and the trace of the resolvent of the Dirac operator. *Funkt. Anal. Appl.* **1981**, *15*, 92–93. [CrossRef]
154. Tsvetkov, A.A. On a family of commutative Wick symbols. *Theor. Math. Phys.* **1981**, *47*, 302–309. [CrossRef]
155. Bogolubov, N.N., Jr.; Sadovnikov, B.I. *Some Problems of Statistical Mechanics;* Vyshaya Shkola Publisher: Minsk, Belarus, 1975.
156. Gibbon, J. Collisionless Boltzmann equations and integrable moment equations. *Physica* **1981**, *3*, 502–511. [CrossRef]
157. Bogolubov, N.N. Microscopic solutions of the Boltzmann-Enskog equation in kinetic theory for elastic balls. *Theor. Math. Phys.* **1975**, *24*, 804–807. [CrossRef]
158. Bazarov, I.P.; Gevorkian, E.V.; Nikolaev, P.N. *Nonequilibrium Thermodynamics and Physical Kinetics;* Moscow University Press: Moscow, Russia, 1989.
159. Bogolubov, N.N., Jr.; Prykarpatsky, A.K.; Samoilenko, V.H. *Hamiltonian Structure of Hydrodynamical Benney Type Equations and Associated with Them Boltzmann-Vlasove Equations on Axis;* Preprint of the Institute of Mathematics of NAS of Ukraine: Kiev, Ukraine, 1991; Volume 91, p. 25.
160. Chapman, S.; Cowling, T. *Mathematical Theory of Non-Uniform Gases;* Cambridge University Press: London, UK; New York, NY, USA, 1952.
161. Libov, R. *Introduction to the Theory of Kinetic Equations;* Wiley: Hoboken, NJ, USA, 1969.
162. Mendes, R.V. Current algebra, statistical mechanics and quantum models. *arXiv* **2017**, arXiv:1711.03027v1.
163. Bardos, C.; Besse, N. The Cauchy problem for the Vlasov-Dirac-Benney equation and related issues in fluid mechanics and semi-classical limits. *Kinet. Relat. Model.* **2013**, *6*, 893–917. [CrossRef]
164. Boglolubov, N.N., Jr.; Brankov, J.G.; Zagrebnov, V.A.; Kurbatov, A.M.; Tonchev, N.S. *Approximating Hamiltonian Method in Statistical Physics;* Bulgarian Academy of Sciences Publ.: Sophia, Bulgaria, 1981.
165. Daletsky, Y.L.; Kadobyansky, R.M. The Poisson structures hierarchy and interacting ststems dynamics. *Proceed. Ukr. Sci.* **1994**, *8*, 21–26.
166. Ivankiv, L.I.; Prykarpatski, A.K.; Samulyak, R.V. *Non-Equilibrium Statistical Mechanics of Many-Particle Systems in Bounded Domain with Surface Peculiarities and Adsorption Phenomenon;* Preprint N1-92, Institute for applied Problems of Mechanics and Mathematics of NASU: Lviv, Ukraine, 1992.
167. Kozlov, V.V. *Thermal Equilibrium in the Sense of Gibbs and Poincare;* Inst. Komp'yut. Issled. Publisher: Izhevsk, Russia, 2002.
168. Kozlov, V.V. *Gibbs Ensembles and Nonequilibrium Statistical Mechanics;* Regulyarnaya i Khaoticheskaya Dinamika Publisher: Izhevsk, Russia, 2008.
169. Kozlov, V.V. The Vlasov kinetic equation, dynamics of continuum and turbulence. *Regul. Chaotic Dyn.* **2011**, *16*, 602–622. [CrossRef]
170. Lions, P.L.; Perthame, B. Propagation of Moments and Regularity for the 3-Dimensional Vlasov-Poisson System. *Invent. Math.* **1991**, *105*, 415–430. [CrossRef]
171. Mandjavidze, J.; Sissakian, A. Generating functional method of N.N. Bogolubov and multiple production physics. *arXiv* **2000**, arXiv:hep-ph/0003039v1.
172. Marsden, J.E.; Morrison, P.J.; Weinstein, A. The Hamiltonian structure of the BBBGKY hierarchy equations. *Contemp. Math.* **1984**, *28*, 115–124.
173. Mikhaylov, A.I. The functional mechanics: Evolution of the moments of distribution function and the Poincare recurrence theorem. *Vestn. Samar. Gos. Tekh. Univ. Fiz.-Mat. Nauk.* **2011**, *1*, 124–133. [CrossRef]

174. Mikhaylov, A.I. The functional mechanics: Evolution of the moments of distribution function and the Poincare recurrence theorem. *P-Adic Numbers Ultrametr. Anal. Appl.* **2011**, *3*, 205–211. [CrossRef]
175. Trushechkin, A.S. Microscopic solutions of kinetic equations and the irreversibility problem. *Proc. SteklovInst. Math.* **2014**, *285*, 251–274. [CrossRef]
176. Villani, C. A review of mathematical topics in collisional kinetic theory. In *Handbook of Mathematical Fluid Dynamics*; Friedlander, S., Serre, D., Eds.; Elsevier Science Publisher: Amsterdam, The Netherlands, 2002; Volume 1.
177. Volovik, G.E. Poisson bracket scheme for vortex dynamics in superfluids and superconductors and the effect of the band structure of the crystal. *J. Exp. Theor. Phys. Lett.* **1996**, *64*, 845–852. [CrossRef]
178. Kruglikov, B.; Morozov, O. Integrable dispersionless PDE in 4D, their symmetry pseudogroups and deformations. *arXiv* **2015**, arXiv:1410.7104v2.
179. Zubarev, D.N. *Nonequilibrium Statistical Thermodynamics*; Consultants Bureau: New York , NY, USA, 1974.
180. Vlasov, A.A. *Statistical Distribution Functions*; Nauka Publisher: Moscow, Russia, 1966.
181. Arnold, V.I. *Mathematical Methods of Classical Mechanics*; Springer: New York, NY, USA, 1978.
182. Bogolyubov, N.N., Jr.; Prikarpatskii, A.K.; Samoilenko, V.G. Bogolyubov's functional equation and the Lie-poisson-Vlasov symplectic structure associated with it. *Ukr. Math. J.* **1986**, *38*, 654–657. [CrossRef]
183. Mokhov, O.I. *Symplectic and Poisson Geometry on Loop Spaces of Smooth Manifolds and Integrable Equations*; Cambridge Scientific Publishers: Cottenham, UK, 2008.
184. Prykarpatsky, Y.A. Canonical reduction on cotangent symplectic manifolds with group action and on associated principal bundles with connections. *J. Nonlinear Oscil.* **2006**, *9*, 96–106. [CrossRef]
185. Bogoliubov, N.N.; Blackmore, D.; Prykarpatsky, A.K. On Benney Type Hydrodynamical Systems and Their Boltzmann-Vlasov Equations Kinetic Models. Preprint IC/2006/006. The Abdus Salam International Center for Theoretical Physics, United Nations Educational Scientific and Cultural Organization and International Atomic Energy Agency. Available online: http://www.ictp.it/pub_off (accessed on 1 December 2021).
186. Lebedev, D.R.; Manin, Y.I. Benney's long wave equations: Lax representation and conservation laws. In *Zapiski Nauchnykh Seminarov LOMI.-1980-96*; Boundary Value Problems of Mathematical Physics and Adjacent Function Theory Questions; Elsevier: Amsterdam, The Netherlands, 2019; pp. 169–178. (In Russian)
187. Marsden, J.; Chorin, A. *Mathematical Foundations of the Mechanics of Liquid*; Springer: New York, NY, USA, 1993.
188. Chernoff, P.R.; Royden, H.F. The equation $\partial f/\partial dx = \partial f/\partial y$. *Am. Math. Mon.* **1975**, *82*, 530–531. [CrossRef]
189. Montel, P. Sur differentielles totales et les fontions monogenes. *CR Acad. Sc. Paris* **1913**, *156*, 1820–1822.
190. Tolstoff G. Sur la differentielle totale. *Recl. Math.* **1941**, *9*, 461–468.
191. Ebin, D.; Marsden, J. Groups of diffeomorphisms and the motion of an incompressible fluid. *Ann. Math.* **1970**, *125*, 102–163. [CrossRef]
192. Kambe, T. Geometric theory of fluid flows and dynamical systems. *Fluid Dyn. Res.* **2002**, *30*, 331–378. [CrossRef]
193. Arnold, V.I.; Khesin, B.A. *Topological Methods in Hydrodynamics*; Springer: New York, NY, USA, 1998.
194. Holm, D.; Marsden, J.; Ratiu, T.; Weinstein, A. Nonlinear stability of fluid and plasma equilibria. *Phys. Rep.* **1985**, *123*, 1–116. [CrossRef]
195. Kuznetsov, E.A.; Mikhailov, A.V. On the topological meaning of canonical Clebsch variables. *Phys. Lett. A* **1980**, *77*, 37–38. [CrossRef]
196. Henyea, F. Gauge groups and Noether's theorem for continuum mechanics. *AIF Conf. Proc.* **1982**, *88*, 85–90.
197. Warner, F.W. *Foundations of Diffderentiable Manifolds and Lie Groups*; Springer: New York, NY, USA, 1983.
198. Huang, K. *Statistical Mechanics*; John Wiley and Sons Inc.: New York, NY, USA; London, UK, 1963.
199. Minlos, R.A. *Introduction to Mathematical Statistical Physics*; University Lecture Series 19; American Mathematical Society: Providence, RI, USA, 1999.
200. Holm, D.; Kupershmidt, B. Superfluid plasmas: Multivelocity nonlinear hydrodynamics of superfluid solutions with charged condensates coupled electromagnetically. *Phys. Rev.* **1987**, *36*, 3947–3956. [CrossRef]
201. Moffat, H.K. The degree of knottedness of tangled vortex lines. *J. Fluid Mech.* **1969**, *35*, 117–129. [CrossRef]
202. Kulish, P.P. An Analogue of the Korteweg-de Vries Quation for the Superconformal Algebra, Differential Geometry, Lie Groups and Mechanics. *Zap. Nauchnykh Semin. POMI* **1986**, *155*, 142–149.
203. Ovsienko, V. Bi-Hamilton nature of the equation $u_{tx} = u_{xy}u_y - u_{yy}u_x$. *arXiv* **2008**, arXiv:0802.1818v1.
204. Mikhalev, V.G. On the Hamiltonian formalism for Korteweg-de Vries type hierarchies. *Funct. Anal. Its Appl.* **1992**, *26*, 140–142. [CrossRef]
205. Misiolek, G. A shallow water equation as a geodesic flow on the Bott-Virasoro group. *J. Geom. Phys.* **1998**, *24*, 203–208. [CrossRef]
206. Sheftel, M.B.; Malykh, A.A.; Yazıcı, D. Recursion operators and bi-Hamiltonian structure of the general heavenly equation. *arXiv* **2016**, arXiv:1510.03666v3.
207. Hentosh, O.E.; Prykarpatsky, Y.A.; Blackmore, D.; Prykarpatski, A.K. Pfeiffer-Sato solutions of Buhl's problem and a Lagrange-D'Alembert principle for Heavenly equations. In *Nonlinear Systems and Their Remarkable Mathematical Structures*; Euler, N., Ed.; CRC Press: Boca Raton, FL, USA, 2018; pp. 187–232. ISBN 9780429470462.
208. Hentosh, O.Y.; Prykarpatsky, Y.A.; Blackmore, D.; Prykarpatski, A. Generalized Lie-algebraic structures related to integrable dispersionless dynamical systems and their application. *J. Math. Sci. Model.* **2018**, *1*, 105–130. [CrossRef]

209. Hentosh, O.Y.; Prykarpatsky, Y.A.; Blackmore, D.; Prykarpatski, A. Lie-algebraic structure of Lax–Sato integrable heavenly equations and the Lagrange–d'Alembert principle. *J. Geom. Phys.* **2017**, *120*, 208–227. [CrossRef]
210. Prykarpatski, A.K. On the Linearization Covering Technique and its Application to Integrable Nonlinear Differential Systems. *Symmetry Integr. Geom. Methods Appl.* **2018**, *14*, 023. [CrossRef]
211. Prykarpatskyy, Y. On the Integrable Chaplygin Type Hydrodynamic Systems and Their Geometric Structure. *Symmetry* **2020**, *12*, 697. [CrossRef]
212. Doubrov, B.; Ferapontov, E.V. On the integrability of symplectic Monge-Ampère equations. *J. Geom. Phys.* **2010**, *60*, 1604–1616. [CrossRef]
213. Ferapontov, E.V.; Moss, J. Linearly degenerate PDEs and quadratic line complexes. *arXiv* **2012**, arXiv:1204.2777v1.
214. Prykarpatski, A.K.; Hentosh, O.E.; Prykarpatsky, A.K. Geometric Structure of the Classical Lagrange-d'Alembert Principle and its Application to Integrable Nonlinear Dynamical Systems. *Mathematics* **2017**, *5*, 75. [CrossRef]
215. Hertling, C. *Frobenius Manifolds and Moduli Spaces for Singularities*; Cambridge University Press: Cambridge, UK, 2002.
216. Pressley, A.; Segal, G. *Loop Groups*; Clarendon Press: London, UK, 1986.
217. Audin, M. Lectures on gauge theory and integrable systems. In *Gauge Theory and Symplectic Geometry*; Hurtubise, J., Lalonde, F., Eds.; Kluwer: Alphen aan den Rijn, The Netherlands, 1997; pp. 1–48.
218. Ovsienko, V.; Roger, C. Looped Cotangent Virasoro Algebra and Non-Linear Integrable Systems in Dimension 2 + 1. *Commun. Math. Phys.* **2007**, *273*, 357–378. [CrossRef]
219. Plebański, J.F. Some solutions of complex Einstein equations. *J. Math. Phys.* **1975**, *16*, 2395–2402. [CrossRef]
220. Dunajski, M. Anti-self-dual four-manifolds with a parallel real spinor. *Proc. Roy Soc. A* **2002**, *458*, 1205. [CrossRef]
221. Dunajski, M.; Mason, L.J.; Tod, P. Einstein-Weyl geometry, the dKP equation and twistor theory. *J. Geom. Phys.* **2001**, *37*, 63–93. [CrossRef]
222. Manakov, S.V.; Santini, P.M. On the solutions of the second heavenly and Pavlov equations. *J. Phys. A Mat. Theor.* **2009**, *42*, 404013. [CrossRef]
223. Pavlov, M.V. Integrable hydrodynamic chains. *J. Math. Phys.* **2003**, *44*, 4134–4156. [CrossRef]
224. Schief, W.K. Self-dual Einstein spaces via a permutability theorem for the Tzitzeica equation. *Phys. Lett. A* **1996**, *223*, 55–62. [CrossRef]
225. Schief, W.K. Self-dual Einstein spaces and a discrete Tzitzeica equation. A permutability theorem link. In *Symmetries and Integrability of Difference Equations*; Clarkson, P., Nijhoff, F., Eds.; London Mathematical Society, Lecture Note Series 255; Cambridge University Press: Cambridge, UK, 1999; pp. 137–148.
226. Takasaki, K.; Takebe, T. SDiff(2) Toda equation— Hierarchy, Tau function, and symmetries. *Lett. Math. Phys.* **1991**, *23*, 205–214. [CrossRef]
227. Takasaki, K.; Takebe, T. Integrable Hierarchies and Dispersionless Limit. *Rev. Math. Phys.* **1995**, *7*, 743–808. [CrossRef]
228. Strachan, I.A.B.; Szablikowski, B.M. Novikov algebras and a classification of multicomponent Camassa-Holm equations. *Stud. Appl. Math.* **2014**, *133*, 84–117. [CrossRef]
229. Sergyeyev, A.; Szablikowski, B.M. Central extensions of cotangent universal hierarrchy: (2+1)-dimensional bi-Hamiltonian systems. *Phys. Lett. A* **2008**, *372*, 7016–7023. [CrossRef]
230. Szablikowski, B. Hierarchies of Manakov-Santini Type by Means of Rota-Baxter and Other Identities. *SIGMA* **2016**, *12*, 022. [CrossRef]
231. Prykarpatski, A.K.; Balinsky, A.A. On Symmetry Properties of Frobenius Manifolds and Related Lie-Algebraic Structures. *Symmetry* **2021**, *13*, 979. [CrossRef]
232. Adler M. On a trace functional for formal pseudo-differential operators and the symplectic structure of the Korteweg-de Vries equation. *Invent. Math.* **1979**, *50*, 219–248. [CrossRef]
233. Semenov-Tian-Shansky, M. What is a classical R-matrix? *Funct. Anal. Appl.* **1983**, *17*, 259–272. [CrossRef]
234. Hentosh, O.E.; Prykarpatsky, Y.A.; Balinsky, A.A.; Prykarpatski, A.K. The dispersionless completely integrable heavenly type Hamiltonian flows and their differential-geometric structure. *Ann. Math. Phys.* **2019**, *2*, 011–025. [CrossRef]
235. Thirring, W. *Classical Mathematical Physics*, 3rd ed.; Springer: Berlin, Germany, 1992.
236. Alonso, L.M.; Shabat, A.B. Hydrodynamic reductions and solutions of a universal hierarchy. *Theoret. Math. Phys.* **2004**, *104*, 1073–1085. [CrossRef]

MDPI

Article

Einstein Field Equation, Recursion Operators, Noether and Master Symmetries in Conformable Poisson Manifolds

Mahouton Norbert Hounkonnou [1,2,*], Mahougnon Justin Landalidji [1,2] and Melanija Mitrović [1,2,3]

[1] International Chair in Mathematical Physics and Applications (ICMPA-UNESCO Chair), University of Abomey-Calavi, Cotonou 072 BP 50, Benin; landalidjijustin@cipma.net (M.J.L.); melanija.mitrovic@masfak.ni.ac.rs (M.M.)

[2] International Center for Research and Advanced Studies in Mathematical and Computer Sciences and Applications (ICRASMCSA), Cotonou 072 BP 50, Benin

[3] Center of Applied Mathematics, Faculty of Mechanical Engineering (CAM-FMEN), University of Niš, A. Medvedeva 14, 18000 Niš, Serbia

* Correspondence: norbert.hounkonnou@cipma.uac.bj

Abstract: We show that a Minkowski phase space endowed with a bracket relatively to a conformable differential realizes a Poisson algebra, confering a bi-Hamiltonian structure to the resulting manifold. We infer that the related Hamiltonian vector field is an infinitesimal Noether symmetry, and compute the corresponding deformed recursion operator. Besides, using the Hamiltonian–Jacobi separability, we construct recursion operators for Hamiltonian vector fields in conformable Poisson–Schwarzschild and Friedmann–Lemaître–Robertson–Walker (FLRW) manifolds, and derive the related constants of motion, Christoffel symbols, components of Riemann and Ricci tensors, Ricci constant and components of Einstein tensor. We highlight the existence of a hierarchy of bi-Hamiltonian structures in both the manifolds, and compute a family of recursion operators and master symmetries generating the constants of motion.

Keywords: Einstein field equation; recursion operator; Noether symmetry; master symmetry; conformable differential; Poisson manifold

Citation: Hounkonnou, M.N.; Landalidji, M.J.; Mitrović, M. Einstein Field Equation, Recursion Operators, Noether and Master Symmetries in Conformable Poisson Manifolds. *Universe* **2022**, *8*, 247. https://doi.org/10.3390/universe8040247

Academic Editors: Steven Duplij and Michael L. Walker

Received: 8 March 2022
Accepted: 6 April 2022
Published: 17 April 2022

1. Introduction

Conformable fractional calculus has a long and rich history. In 1695, Gottfried Leibniz asked Guillaume l'Hôspital if the (integer) order of derivatives and integrals could be extended [1]. Would it be possible if the order was some irrational, fractional or complex number? This idea motivated many mathematicians, physicists and engineers to develop the concept of fractional calculus in diverse fields of science and engineers (see, e.g., [2–9], and references therein). Over four centuries, many famous mathematicians contributed to this development. It is still nowadays one of the most intensively developing areas of mathematical analysis, including several definitions of fractional operators like Riemann–Liouville, Caputo, Grünwald–Letnikov, Riesz and Weyl definitions [5,10–12]. Two of these definitions, namely Riemann–Liouville and Caputo, are famous. Mathematicians prefer the Riemann–Liouville fractional derivative while physicists and engineers use the Caputo fractional one. Indeed, the Riemann–Liouville fractional derivative of a constant is not zero, and it requires fractional initial conditions that are not generally specified [5]. In contrast, the Caputo derivative of a constant is zero, and a fractional differential equation expressed in terms of a Caputo fractional derivative requires standard boundary conditions. Unfortunately, the Riemann–Liouville derivative and Caputo derivative do not obey the Leibniz rule and chain rule, which sometimes prevents us from applying these derivatives to ordinary physical systems with a standard Newton derivative. In 2014, Khalil et al. [13] introduced the new fractional derivative called the conformable fractional derivative and the integral obeying the Leibniz rule and chain rule. One year later, i.e., in 2015, Chung [5] used this conformable fractional derivative and integral to discuss the fractional version of the Newtonian mechanics. In that work, he constructed the fractional Euler–Lagrange equation from the fractional version of the calculus of variations and used this equation to discuss some mechanical problems such as fractional harmonic oscillator problem, the fractional damped oscillator problem and the forced oscillator problem. In 2017, Chung et al. [14] discussed the dynamics of a particle in a viscoelastic medium using the conformable fractional derivative of order α with respect to time. Further, in 2019, the same authors [15] discussed the fractional classical mechanics and applied it to the anomalous diffusion relation from the α-deformed Langevin equation. During the same year, Kiskinov et al. [16] investigated the Cauchy problem for nonlinear systems with conformable derivatives and variable delays. Furthermore, Khalil et al. gave the geometric meaning of a conformable derivative via fractional cords in 2019 [17]. In 2020, Chung et al. [18] studied the deformed special relativity based on α-deformed binary operations. In that work, they gave the α-translation invariant distance (α-distance) of infinitesimally close space-time based on the definition of α-translation invariant infinitesimal displacement and α-translation invariant infinitesimal time interval.

In addition, in the last few decades, there was a renewed interest in completely integrable Hamiltonian systems (IHS), the concept of which goes back to Liouville in 1897 [19] and Poincaré in 1899 [20]. In short, IHS are defined as nonlinear differential equations admitting a Hamiltonian description and possessing enough constants of motion so that they can be integrated by quadratures [21]. This Liouville formalism does not provide a method for obtaining the integrals of motion; it has therefore been necessary to elaborate different methods for obtaining constants of motion (Hamilton–Jacobi separability, Lax pairs formalism, Noether symmetries, Hidden symmetries, etc). A relevant progress in the analysis of the integrability was the important remark that many of these systems are Hamiltonian dynamics with respect to two compatible symplectic structures [22–24], permitting a geometrical interpretation of the so-called recursion operator [25–27]. A description of integrability working both for systems with finitely many degrees of freedom and for field theory can be given in terms of an invariant, diagonalizable mixed $(1,1)$-tensor field, having bidimensional eigenspaces and vanishing Nijenhuis torsion. One of the powerful methods of describing IHS with involutive Hamiltonian functions or constants of motion uses the recursion operator admitting a vanishing Nijenhuis torsion. In 2015, Takeuchi constructed recursion operators of Hamiltonian vector fields of geodesic flows for some Riemannian and Minkowski metrics [28], and obtained related constants of motion. In his work, Takeuchi used five particular solutions of the Einstein equation in the Schwarzschild, Reissner–Nordström, Kerr, Kerr–Newman, and FLRW metrics, and constructed recursion operators inducing the complete integrability of the Hamiltonian functions. Further, in 2019, we investigated the same problem in a noncommutative Minkowski phase space [29].

In the present work, we investigate Noether symmetry and recursion operators induced by a conformable Poisson algebra in a Minkowski phase space. We construct recursion operators using conformable Schwarzschild and Friedmann–Lemaître–Robertson–Walker (FLRW) metrics and discuss their relevant master symmetries.

The paper is organized as follows. In Section 2, we give the notion of conformable differential and related formulation of the wellknown Takeuchi Lemma [28]. In Section 3, we construct a conformable Poisson algebra and the Lie algebra of deformed vector fields, prove the existence of infinitesimal Noether symmetry and bi-Hamiltonian structure, and compute the corresponding recursion operator in a conformable Minkowski phase space. In Section 4, we construct recursion operators for Hamiltonian vector fields, related constants of motion, Christoffel symbols, components of Riemann and Ricci tensors, Ricci constant, and components of Einstein tensor in the framework of conformable Schwarzschild and FLRW metrics. In Section 5, we derive a hierarchy of master symmetries and compute the conserved quantities. In Section 6, we end with some concluding remarks.

2. Conformable Differential and Formulation of Takeuchi Lemma

A Hamiltonian system is a triple $(\mathcal{Q}, \omega, H)$, where $(\mathcal{Q}, \omega)$ is a symplectic manifold and H is a smooth function on $\mathcal{Q}$, called Hamiltonianor Hamiltonian function [30].

Given a general dynamical system defined on the $2n$-dimensional manifold $\mathcal{Q}$ [31,32], its evolution can be described by the equation

$$\dot{x}(t) = X(x), \quad x \in \mathcal{Q}, \quad X \in \mathcal{TQ}. \tag{1}$$

If the system (1) admits two different Hamiltonian representations:

$$\dot{x}(t) = X_{H_1,H_2} = \mathcal{P}_1 dH_1 = \mathcal{P}_2 dH_2, \tag{2}$$

its integrability as well as many other properties are subject to Magri's approach. The bi-Hamiltonian vector field X_{H_1,H_2} is defined by two pairs of Poisson bivectors $\mathcal{P}_1, \mathcal{P}_2$ and Hamiltonian functions H_1, H_2. Such a manifold $\mathcal{Q}$ equipped with two Poisson bivectors is called a double Poisson manifold, and the quadruple $(\mathcal{Q}, \mathcal{P}_1, \mathcal{P}_2, X_{H_1,H_2})$ is called a bi-Hamiltonian system. $\mathcal{P}_1$ and $\mathcal{P}_2$ are two compatible Poisson bivectors with a vanishing Schouten–Nijenhuis bracket [33]:

$$[\mathcal{P}_1, \mathcal{P}_2]_{NS} = 0. \tag{3}$$

A recursion operator $T : \mathcal{TQ} \longrightarrow \mathcal{TQ}$ is defined by

$$T := \mathcal{P}_2 \circ \mathcal{P}_1^{-1}. \tag{4}$$

A Noether symmetry is a diffeomorphism $\Phi : \mathcal{Q} \longrightarrow \mathcal{Q}$ such that [34]:

$$\Phi^* \omega = \omega, \quad \Phi^* H = H. \tag{5}$$

An infinitesimal Noether symmetry is a vector field $Y \in \mathfrak{X}(\mathcal{Q})$ (the set of all differentiable vector fields on $\mathcal{Q}$) such that:

$$\mathcal{L}_Y \omega = 0, \quad \mathcal{L}_Y H = 0. \tag{6}$$

Definition 1. *Consider the map g and its inverse g^{-1}:*

$$\begin{aligned} g : \mathbb{R}^{2n}_{\alpha} &\longrightarrow \mathbb{R}^{2n} & g^{-1} : \mathbb{R}^{2n} &\longrightarrow \mathbb{R}^{2n}_{\alpha} \\ z &\longmapsto g(z) = |z|^{\alpha-1} z = \mathbf{Z} & \mathbf{Z} &\longmapsto g^{-1}(\mathbf{Z}) = |\mathbf{Z}|^{(1/\alpha)-1} \mathbf{Z} = z, \end{aligned} \tag{7}$$

where $g(0) = 0$, $g(1) = 1$, and $g(\pm\infty) = \pm\infty$. Then, for this map, the α-addition, α-subtraction, α-multiplication, and α-division are given by:

$$\begin{aligned} a \oplus_{\alpha} b &= |a|a|^{\alpha-1} + b|b|^{\alpha-1}|^{(1/\alpha)-1}(a|a|^{\alpha-1} + b|b|^{\alpha-1}), \\ a \ominus_{\alpha} b &= |a|a|^{\alpha-1} - b|b|^{\alpha-1}|^{(1/\alpha)-1}(a|a|^{\alpha-1} - b|b|^{\alpha-1}), \\ a \otimes_{\alpha} b &= ab, \\ a \oslash_{\alpha} b &= \frac{a}{b}, \end{aligned}$$

where $a, b \in \mathbb{R}^{2n}_{\alpha}$.

Definition 2. *Let h be a differentiable coordinates function on $\mathbb{R}^{2n}_{\alpha}$. The conformable differential, also called α-differential in the sequel, with respect to the position q and its associated momentum p is defined by:*

$$\begin{aligned} d_{\alpha} : \mathbb{R}^{2n}_{\alpha} &\longrightarrow \mathbb{R}^{2n} \\ h &\longmapsto d_{\alpha} h := \sum_{\mu=1}^{2n} \alpha |x_{\mu}|^{\alpha-1} \frac{\partial}{\partial x_{\mu}} h, \quad (x_{\nu} = q^{\nu}, x_{\nu+n} = p_{\nu}, \; n = 4, \; \nu = 1,2,3,4) \end{aligned} \tag{8}$$

satisfying the following properties:

(i) $d_{\alpha}(ah + bf) = a d_{\alpha} h + b d_{\alpha} f$ *for all* $a, b \in \mathbb{R}$*;*
(ii) $d_{\alpha}(h^m) = m h^{m-1} d_{\alpha} h$*, for all* $m \in \mathbb{R}$*;*
(iii) $d_{\alpha}(c) = 0$*, for all constant functions* $h(q,p) = c$*;*
(iv) $d_{\alpha}(hf) = h d_{\alpha} f + f d_{\alpha} h$*;*

(iv) $d_\alpha\left(\frac{h}{f}\right) = \frac{f d_\alpha h - h d_\alpha f}{f^2}$, *where f is also a differentiable coordinates function on* $\mathbb{R}^{2n}_\alpha$.

The α-differential produces a new deformed phase space called a *conformable phase space.* The ordinary differential is obtained for $\alpha = 1$. Using the α-addition and α-subtraction, we obtain the following infinitesimal distance between two points of coordinates $(x_i, \ldots, x_n)$ and $(x_i \oplus_\alpha d_\alpha x_i, \ldots, x_n \oplus_\alpha d_\alpha x_n)$

$$d_\alpha s = (d^2_\alpha x_i + \ldots + d^2_\alpha x_n)^{\frac{1}{2}}. \tag{9}$$

In the $\mathbb{R}^{2n}_\alpha$, Takeuchi Lemma [28] takes the following form:

Lemma 1. *Consider the conformable vector fields*

$$X_{\alpha_i} = -|x_i|^{(1-\alpha)}|x_{n+i}|^{(1-\alpha)}\frac{\partial}{\partial x_{n+i}}, i = 1, \ldots, n \tag{10}$$

on $\mathbb{R}^{2n}_\alpha$ *and*

$$T_\alpha = \sum_{i=1}^{n} |x_i|^{(\alpha-1)}|x_i\left(\frac{\partial}{\partial x_i} \otimes dx_i + \frac{\partial}{\partial x_{n+i}} \otimes dx_{n+i}\right), \tag{11}$$

a $(1,1)$*-tensor field on* $\mathbb{R}^{2n}_\alpha$. *Then, we have that the Nijenhuis torsion of* T_α *is vanishing, i.e.,* $\mathcal{N}_{T_\alpha} = 0$ *and* $\mathcal{L}_{X_{\alpha_i}} T_\alpha = 0$, *that is, the* $(1,1)$*-tensor field* T_α *is a conformable recursion operator of* X_{α_i}, $(i = 1, \ldots, n)$.

Proof of Lemma 1. We have:

$$\begin{aligned}
\mathcal{L}_{X_{\alpha_i}} T_\alpha &= \mathcal{L}_{X_{\alpha_i}}\left\{\sum_{i=1}^{n} |x_i|^{(\alpha-1)}|x_i\left(\frac{\partial}{\partial x_i} \otimes dx_i + \frac{\partial}{\partial x_{n+i}} \otimes dx_{n+i}\right)\right\} \\
&= \sum_{i=1}^{n}\left\{\mathcal{L}_{X_{\alpha_i}}(|x_i|^{(\alpha-1)}|x_i)\left(\frac{\partial}{\partial x_i} \otimes dx_i + \frac{\partial}{\partial x_{n+i}} \otimes dx_{n+i}\right)\right. \\
&\quad \left. + |x_i|^{(\alpha-1)}|x_i\left(\mathcal{L}_{X_{\alpha_i}}\left[\frac{\partial}{\partial x_i} \otimes dx_i\right] + \mathcal{L}_{X_{\alpha_i}}\left[\frac{\partial}{\partial x_{n+i}} \otimes dx_{n+i}\right]\right)\right\} \\
\mathcal{L}_{X_{\alpha_i}} T_\alpha &= \sum_{i=1}^{n} |x_i|^{(\alpha-1)}|x_i\left(\mathcal{L}_{X_{\alpha_i}}\left[\frac{\partial}{\partial x_i} \otimes dx_i\right] + \mathcal{L}_{X_{\alpha_i}}\left[\frac{\partial}{\partial x_{n+i}} \otimes dx_{n+i}\right]\right)
\end{aligned}$$

because $\mathcal{L}_{X_{\alpha_i}}(|x_i|^{(\alpha-1)}|x_i) = 0$.

Then,

$$\begin{aligned}
\mathcal{L}_{X_{\alpha_i}} T_\alpha &= \sum_{i=1}^{n} |x_i|^{(\alpha-1)}|x_i\left(\mathcal{L}_{X_{\alpha_i}}\left[\frac{\partial}{\partial x_i}\right] \otimes dx_i + \frac{\partial}{\partial x_i} \otimes \mathcal{L}_{X_{\alpha_i}}(dx_i)\right. \\
&\quad \left. + \mathcal{L}_{X_{\alpha_i}}\left[\frac{\partial}{\partial x_{n+i}}\right] \otimes dx_{n+i} + \frac{\partial}{\partial x_{n+i}} \otimes \mathcal{L}_{X_{\alpha_i}}(dx_{n+i})\right) \\
\mathcal{L}_{X_{\alpha_i}} T_\alpha &= 0.
\end{aligned}$$

The components of the Nijenhuis torsion are as follows [28]:

$$\begin{aligned}
(\mathcal{N}_{T_\alpha})^h_{ij} &= (T_\alpha)^k_i \frac{\partial (T_\alpha)^h_j}{\partial x_k} - (T_\alpha)^k_j \frac{\partial (T_\alpha)^h_i}{\partial x_k} + (T_\alpha)^h_k \frac{\partial (T_\alpha)^k_i}{\partial x_j} - (T_\alpha)^h_k \frac{\partial (T_\alpha)^k_j}{\partial x_i} \\
&= |x_i|^{(\alpha-1)}|x_i \frac{\partial (T_\alpha)^h_j}{\partial x_i} - |x_j|^{(\alpha-1)}|x_j \frac{\partial (T_\alpha)^h_i}{\partial x_j} + (T_\alpha)^h_i \frac{\partial(|x_i|^{(\alpha-1)}|x_i)}{\partial x_j} - (T_\alpha)^h_j \frac{\partial(|x_j|^{(\alpha-1)}|x_j)}{\partial x_i} \\
&= |x_i|^{(\alpha-1)}|x_i \frac{\partial (T_\alpha)^h_j}{\partial x_i} - |x_j|^{(\alpha-1)}|x_j \frac{\partial (T_\alpha)^h_i}{\partial x_j} + \alpha (T_\alpha)^h_i |x_i|^{(\alpha-1)} \delta^i_j - \alpha (T_\alpha)^h_j |x_j|^{(\alpha-1)} \delta^j_i.
\end{aligned}$$

1. If $i = j$, we have $\delta^i_j = \delta^j_i = 1$ and we get

$$(\mathcal{N}_{T_\alpha})^h_{ij} = |x_i|^{(\alpha-1)}|x_i\frac{\partial(T_\alpha)^h_i}{\partial x_i} - |x_i|^{(\alpha-1)}|x_i\frac{\partial(T_\alpha)^h_i}{\partial x_i} + \alpha|x_i|^{(\alpha-1)}(T_\alpha)^h_i - \alpha|x_i|^{(\alpha-1)}(T_\alpha)^h_i = 0; \tag{12}$$

2. If $i \neq j$, we have $\delta^i_j = \delta^j_i = 0$ and $\dfrac{\partial(T_\alpha)^h_j}{\partial x_i} = \dfrac{\partial(T_\alpha)^h_i}{\partial x_j} = 0$. Then,

$$(\mathcal{N}_{T_\alpha})^h_{ij} = 0. \tag{13}$$

From (12) and (13), we get $\mathcal{N}_{T_\alpha} = 0$. □

3. Recursion Operator in Conformable Minkowski Phase Space

In this section, we derive the recursion operator of Hamiltonian vector fields of geodesic flow for a free particle in a conformable Minkowski phase space and obtain the associated constants of motion.

3.1. Symplectic Structure, Poisson Bracket and Lie Algebra

We consider our configuration space as a manifold $\mathcal{Q} = \mathbb{R}^4_\alpha \backslash \{0\}$ that is, a four-dimensional real Euclidean vector space with the origin removed. The cotangent bundle $\mathcal{T}^*\mathcal{Q} = \mathcal{Q} \times \mathbb{R}^4_\alpha$ has a natural symplectic structure $\omega_\alpha : \mathcal{T}\mathcal{Q} \longrightarrow \mathcal{T}^*\mathcal{Q}$ which, in local coordinates (q, p), is given by

$$\omega_\alpha = \sum_{\mu=1}^{4} d_\alpha p_\mu \wedge d_\alpha q^\mu = \sum_{\mu=1}^{4} \alpha^2 |p_\mu|^{\alpha-1}|q^\mu|^{\alpha-1} dp_\mu \wedge dq^\mu. \tag{14}$$

Since ω_α is non-degenerate, it induces an inverse map, called bivector field $\mathcal{P}_\alpha$: $\mathcal{T}^*\mathcal{Q} \longrightarrow \mathcal{T}\mathcal{Q}$ (tangent bundle) defined by

$$\mathcal{P}_\alpha = \sum_{\mu=1}^{4} \alpha^{-2}|p_\mu|^{1-\alpha}|q^\mu|^{1-\alpha}\frac{\partial}{\partial p_\mu} \wedge \frac{\partial}{\partial q^\mu}, \quad \omega_\alpha \circ \mathcal{P}_\alpha = \mathcal{P}_\alpha \circ \omega_\alpha = 1, \tag{15}$$

and is used to construct the Hamiltonian vector field X_{α_f} of a Hamiltonian function f by the relation

$$X_{\alpha_f} = \mathcal{P}_\alpha df.$$

We consider now the next conformable Minkowski metric on the manifold $\mathcal{Q}$:

$$d_\alpha s^2 = -\alpha^2|q^1|^{2(\alpha-1)}(dq^1)^2 + \alpha^2|q^2|^{2(\alpha-1)}(dq^2)^2 + \alpha^2|q^3|^{2(\alpha-1)}(dq^3)^2 + \alpha^2|q^4|^{2(\alpha-1)}(dq^4)^2, \tag{16}$$

where $c = 1$ for commodity yielding the tensor metric $(g_{\mu\nu})_\alpha$ and its inverse $(g^{\mu\nu})_\alpha$

$$(g_{\mu\nu})_\alpha = \alpha^2 \begin{pmatrix} -(q^1)^{2(\alpha-1)} & 0 & 0 & 0 \\ 0 & (q^2)^{2(\alpha-1)} & 0 & 0 \\ 0 & 0 & (q^3)^{2(\alpha-1)} & 0 \\ 0 & 0 & 0 & (q^4)^{2(\alpha-1)} \end{pmatrix}, \tag{17}$$

$$(g^{\mu\nu})_\alpha = \frac{1}{\alpha^2} \begin{pmatrix} -(q^1)^{2(1-\alpha)} & 0 & 0 & 0 \\ 0 & (q^2)^{2(1-\alpha)} & 0 & 0 \\ 0 & 0 & (q^3)^{2(1-\alpha)} & 0 \\ 0 & 0 & 0 & (q^4)^{2(1-\alpha)} \end{pmatrix}. \tag{18}$$

In our framework, the equation of the geodesic on the manifold $\mathcal{Q}$ is given by

$$\frac{d^2q^\mu}{dt^2} + (\Gamma^\mu_{\nu\lambda})_\alpha \frac{dq^\nu}{dt}\frac{dq^\lambda}{dt} = 0, \quad (\nu, \mu, \lambda = 1, 2, 3, 4), \tag{19}$$

where

$$(\Gamma^\mu_{\nu\lambda})_\alpha = \frac{1}{2}(g^{\mu\epsilon})_\alpha \left(\frac{\partial(g_{\epsilon\nu})_\alpha}{\partial q^\lambda} + \frac{\partial(g_{\epsilon\lambda})_\alpha}{\partial q^\nu} - \frac{\partial(g_{\nu\lambda})_\alpha}{\partial q^\epsilon} \right) \tag{20}$$

are Christoffel symbols. From (20), we have

$$(\Gamma^1_{11})_\alpha = \frac{\alpha-1}{q^1};\ (\Gamma^2_{22})_\alpha = \frac{\alpha-1}{q^2};\ (\Gamma^3_{33})_\alpha = \frac{\alpha-1}{q^3};\ (\Gamma^4_{44})_\alpha = \frac{\alpha-1}{q^4};\ (\Gamma^\mu_{\nu\lambda})_\alpha = 0, \text{ otherwise,} \quad (21)$$

and obtain that the Riemann tensor components are vanished, i.e., $R_{ijkl} = 0,\ (i,j,k,l = 1,2,3,4)$. Then, the Minkowski phase space endowed with the metric $d_\alpha s^2$ is a flat space. Thus, we notice that this result does not change the geometric structure of the ordinary Minkowski phase space. Further, the presence of the Christoffel symbols $(\Gamma^i_{ii})_\alpha,\ (i = 1,2,3,4)$ means that the parallel displacement of any basic vector of our considered manifold with respect to itself always remains parallel with this same basic vector. The ordinary Minkowski phase space is obtained for $\alpha = 1$.

Since the quantities $(\tilde{\Gamma}^\mu_{\nu\lambda})_\alpha = \frac{1}{\alpha+1}(\Gamma^\mu_{\nu\lambda})_\alpha$ do not change the geometric structure of the Minkowski phase space, we replace $(\Gamma^\mu_{\nu\lambda})_\alpha$ by $(\tilde{\Gamma}^\mu_{\nu\lambda})_\alpha$ in (19). Then, the equation of the geodesic becomes:

$$\frac{d^2q^\mu}{dt^2} + (\tilde{\Gamma}^\mu_{\nu\lambda})_\alpha \frac{dq^\nu}{dt}\frac{dq^\lambda}{dt} = 0,\ (\nu,\mu,\lambda = 1,2,3,4). \quad (22)$$

If we put $v^\mu = \frac{dq^\mu}{dt}$, we have a first order differential equation on the tangent bundle $\mathcal{T}(\mathcal{Q})$ of the manifold $\mathcal{Q}$:

$$\dot{q^\mu} = v^\mu, \quad \dot{v^\mu} = -\frac{1}{\alpha+1}(\Gamma^\mu_{\nu\lambda})_\alpha v^\nu v^\lambda. \quad (23)$$

From the above equations, we get the geodesic spray

$$X_\alpha := v^\mu \frac{\partial}{\partial q^\mu} - \frac{1}{\alpha+1}(\Gamma^\mu_{\nu\lambda})_\alpha v^\nu v^\lambda \frac{\partial}{\partial v^\mu}. \quad (24)$$

By setting $p_\mu = \varepsilon_{\mu\epsilon} v^\epsilon,\ \varepsilon = sgn(-,+,+,+)$, the vector field X_α is equivalently transformed to the vector field X_α on the cotangent bundle $\mathcal{T}^*(\mathcal{Q})$ such that

$$X_\alpha = -p_1\frac{\partial}{\partial q^1} + \sum_{k=2}^{4} p_k \frac{\partial}{\partial q^k} + \left(\frac{\alpha-1}{\alpha+1}\right)\frac{p_1^2}{q^1}\frac{\partial}{\partial p_1} - \sum_{k=2}^{4}\left(\frac{\alpha-1}{\alpha+1}\right)\frac{p_k^2}{q^k}\frac{\partial}{\partial p_k}, \quad (25)$$

The vector field X_α is a Hamiltonian vector field of a certain Hamiltonian function H_α.

Proposition 1. *The set $\mathfrak{F}$ of differentiable functions defined on $\mathcal{T}^*(\mathcal{Q})$ endowed with the bracket*

$$\{f,g\}_\alpha := \sum_{\mu=1}^{4} \alpha^{-2}|p_\mu|^{(1-\alpha)}|q^\mu|^{(1-\alpha)}\left(\frac{\partial f}{\partial p_\mu}\frac{\partial g}{\partial q^\mu} - \frac{\partial f}{\partial q^\mu}\frac{\partial g}{\partial p_\mu}\right) \quad (26)$$

is a conformable Poisson algebra.

Proof of Proposition 1. To prove this Proposition, we just have to prove that the bracket $\{.,.\}_\alpha$ is a conformable Poisson bracket.

Let us consider f, g, and h as the three arbitrary elements of $\mathfrak{F}$.

- **Antisymmetry**

$$\begin{aligned}\{f,g\}_\alpha &= \sum_{\mu=1}^{4} \alpha^{-2}|p_\mu|^{(1-\alpha)}|q^\mu|^{(1-\alpha)}\left(\frac{\partial f}{\partial p_\mu}\frac{\partial g}{\partial q^\mu} - \frac{\partial f}{\partial q^\mu}\frac{\partial g}{\partial p_\mu}\right), \\ &= -\sum_{\mu=1}^{4} \alpha^{-2}|p_\mu|^{(1-\alpha)}|q^\mu|^{(1-\alpha)}\left(\frac{\partial g}{\partial p_\mu}\frac{\partial f}{\partial q^\mu} - \frac{\partial g}{\partial q^\mu}\frac{\partial f}{\partial p_\mu}\right), \\ &= -\{g,f\}_\alpha. \end{aligned} \quad (27)$$

- **Jacobi identity**

$$\{f,\{g,h\}_\alpha\}_\alpha = \left\{f, \sum_{\mu=1}^{4} \alpha^{-2}|p_\mu|^{(1-\alpha)}|q^\mu|^{(1-\alpha)}\left(\frac{\partial g}{\partial p_\mu}\frac{\partial h}{\partial q^\mu} - \frac{\partial g}{\partial q^\mu}\frac{\partial h}{\partial p_\mu}\right)\right\}_\alpha$$
$$= \sum_{\mu,\nu=1}^{4} \alpha^{-4}|p_\nu|^{(1-\alpha)}|q^\nu|^{(1-\alpha)}\left[\frac{\partial f}{\partial p_\nu}\left(\sigma_1|p_\mu|^{(1-\alpha)}(q^\mu)^{-\alpha}\left(\frac{\partial g}{\partial p_\mu}\frac{\partial h}{\partial q^\mu} - \frac{\partial g}{\partial q^\mu}\frac{\partial h}{\partial p_\mu}\right)\right.\right.$$
$$\left. + |p_\mu|^{(1-\alpha)}|q^\mu|^{(1-\alpha)}\left(\frac{\partial^2 g}{\partial q^\nu \partial p_\mu}\frac{\partial h}{\partial q^\mu} + \frac{\partial g}{\partial p_\mu}\frac{\partial^2 h}{\partial q^\nu \partial q^\mu} - \frac{\partial^2 g}{\partial q^\nu \partial q^\mu}\frac{\partial h}{\partial p_\mu} - \frac{\partial g}{\partial q^\mu}\frac{\partial^2 h}{\partial q^\nu \partial p_\mu}\right)\right)$$
$$- \frac{\partial f}{\partial q^\nu}\left(\sigma_2|q^\mu|^{(1-\alpha)}(p_\mu)^{-\alpha}\left(\frac{\partial g}{\partial p_\mu}\frac{\partial h}{\partial q^\mu} - \frac{\partial g}{\partial q^\mu}\frac{\partial h}{\partial p_\mu}\right) + |p_\mu|^{(1-\alpha)}|q^\mu|^{(1-\alpha)}\right.$$
$$\left.\left. \times \left(\frac{\partial^2 g}{\partial p_\nu \partial p_\mu}\frac{\partial h}{\partial q^\mu} + \frac{\partial g}{\partial p_\mu}\frac{\partial^2 h}{\partial p_\nu \partial q^\mu} - \frac{\partial^2 g}{\partial p_\nu \partial q^\mu}\frac{\partial h}{\partial p_\mu} - \frac{\partial g}{\partial q^\mu}\frac{\partial^2 h}{\partial p_\nu \partial p_\mu}\right)\right)\right], \tag{28}$$

$$\{h,\{f,g\}_\alpha\}_\alpha = \left\{h, \sum_{\mu=1}^{4} \alpha^{-2}|p_\mu|^{(1-\alpha)}|q^\mu|^{(1-\alpha)}\left(\frac{\partial f}{\partial p_\mu}\frac{\partial g}{\partial q^\mu} - \frac{\partial f}{\partial q^\mu}\frac{\partial g}{\partial p_\mu}\right)\right\}_\alpha$$
$$= \sum_{\mu,\nu=1}^{4} \alpha^{-4}|p_\nu|^{(1-\alpha)}|q^\nu|^{(1-\alpha)}\left[\frac{\partial h}{\partial p_\nu}\left(\sigma_1|p_\mu|^{(1-\alpha)}(q^\mu)^{-\alpha}\left(\frac{\partial f}{\partial p_\mu}\frac{\partial g}{\partial q^\mu} - \frac{\partial f}{\partial q^\mu}\frac{\partial g}{\partial p_\mu}\right)\right.\right.$$
$$\left. + |p_\mu|^{(1-\alpha)}|q^\mu|^{(1-\alpha)}\left(\frac{\partial^2 f}{\partial q^\nu \partial p_\mu}\frac{\partial g}{\partial q^\mu} + \frac{\partial f}{\partial p_\mu}\frac{\partial^2 g}{\partial q^\nu \partial q^\mu} - \frac{\partial^2 f}{\partial q^\nu \partial q^\mu}\frac{\partial g}{\partial p_\mu} - \frac{\partial f}{\partial q^\mu}\frac{\partial^2 g}{\partial q^\nu \partial p_\mu}\right)\right)$$
$$- \frac{\partial h}{\partial q^\nu}\left(\sigma_2|q^\mu|^{(1-\alpha)}(p_\mu)^{-\alpha}\left(\frac{\partial f}{\partial p_\mu}\frac{\partial g}{\partial q^\mu} - \frac{\partial f}{\partial q^\mu}\frac{\partial g}{\partial p_\mu}\right) + |p_\mu|^{(1-\alpha)}|q^\mu|^{(1-\alpha)}\right.$$
$$\left.\left. \times \left(\frac{\partial^2 f}{\partial p_\nu \partial p_\mu}\frac{\partial g}{\partial q^\mu} + \frac{\partial f}{\partial p_\mu}\frac{\partial^2 g}{\partial p_\nu \partial q^\mu} - \frac{\partial^2 f}{\partial p_\nu \partial q^\mu}\frac{\partial g}{\partial p_\mu} - \frac{\partial f}{\partial q^\mu}\frac{\partial^2 g}{\partial p_\nu \partial p_\mu}\right)\right)\right], \tag{29}$$

$$\{g,\{h,f\}_\alpha\}_\alpha = \left\{g, \sum_{\mu=1}^{4} \alpha^{-2}|p_\mu|^{(1-\alpha)}|q^\mu|^{(1-\alpha)}\left(\frac{\partial h}{\partial p_\mu}\frac{\partial f}{\partial q^\mu} - \frac{\partial h}{\partial q^\mu}\frac{\partial f}{\partial p_\mu}\right)\right\}_\alpha$$
$$= \sum_{\mu,\nu=1}^{4} \alpha^{-4}|p_\nu|^{(1-\alpha)}|q^\nu|^{(1-\alpha)}\left[\frac{\partial g}{\partial p_\nu}\left(\sigma_1|p_\mu|^{(1-\alpha)}(q^\mu)^{-\alpha}\left(\frac{\partial h}{\partial p_\mu}\frac{\partial f}{\partial q^\mu} - \frac{\partial h}{\partial q^\mu}\frac{\partial f}{\partial p_\mu}\right)\right.\right.$$
$$\left. + |p_\mu|^{(1-\alpha)}|q^\mu|^{(1-\alpha)}\left(\frac{\partial^2 h}{\partial q^\nu \partial p_\mu}\frac{\partial f}{\partial q^\mu} + \frac{\partial h}{\partial p_\mu}\frac{\partial^2 f}{\partial q^\nu \partial q^\mu} - \frac{\partial^2 h}{\partial q^\nu \partial q^\mu}\frac{\partial f}{\partial p_\mu} - \frac{\partial h}{\partial q^\mu}\frac{\partial^2 f}{\partial q^\nu \partial p_\mu}\right)\right)$$
$$- \frac{\partial g}{\partial q^\nu}\left(\sigma_2|q^\mu|^{(1-\alpha)}(p_\mu)^{-\alpha}\left(\frac{\partial h}{\partial p_\mu}\frac{\partial f}{\partial q^\mu} - \frac{\partial h}{\partial q^\mu}\frac{\partial f}{\partial p_\mu}\right) + |p_\mu|^{(1-\alpha)}|q^\mu|^{(1-\alpha)}\right.$$
$$\left.\left. \times \left(\frac{\partial^2 h}{\partial p_\nu \partial p_\mu}\frac{\partial f}{\partial q^\mu} + \frac{\partial h}{\partial p_\mu}\frac{\partial^2 f}{\partial p_\nu \partial q^\mu} - \frac{\partial^2 h}{\partial p_\nu \partial q^\mu}\frac{\partial f}{\partial p_\mu} - \frac{\partial h}{\partial q^\mu}\frac{\partial^2 f}{\partial p_\nu \partial p_\mu}\right)\right)\right], \tag{30}$$

where $\sigma_1 = (1-\alpha)(sgn(q^\mu))^{1-\alpha}$ and $\sigma_2 = (1-\alpha)(sgn(p_\mu))^{1-\alpha}$.
Summing (28)–(30), we get

$$\{f,\{g,h\}_\alpha\}_\alpha + \{g,\{h,f\}_\alpha\}_\alpha + \{h,\{f,g\}_\alpha\}_\alpha = 0, \tag{31}$$

which is the Jacobi identity.

- **Derivation**

$$
\begin{aligned}
\{f,gh\}_\alpha &= \sum_{\mu=1}^{4} \alpha^{-2}|p_\mu|^{(1-\alpha)}|q^\mu|^{(1-\alpha)}\left(\frac{\partial f}{\partial p_\mu}\frac{\partial (gh)}{\partial q^\mu} - \frac{\partial f}{\partial q^\mu}\frac{\partial (gh)}{\partial p_\mu}\right) \\
&= \sum_{\mu=1}^{4} \alpha^{-2}|p_\mu|^{(1-\alpha)}|q^\mu|^{(1-\alpha)}\left[\frac{\partial f}{\partial p_\mu}\left(\frac{\partial g}{\partial q^\mu}h + g\frac{\partial h}{\partial q^\mu}\right) - \frac{\partial f}{\partial q^\mu}\left(\frac{\partial g}{\partial p_\mu}h + g\frac{\partial h}{\partial p_\mu}\right)\right] \\
&= \sum_{\mu=1}^{4} \alpha^{-2}|p_\mu|^{(1-\alpha)}|q^\mu|^{(1-\alpha)}\left[g\left(\frac{\partial f}{\partial p_\mu}\frac{\partial h}{\partial q^\mu} - \frac{\partial f}{\partial q^\mu}\frac{\partial h}{\partial p_\mu}\right) + \left(\frac{\partial f}{\partial p_\mu}\frac{\partial g}{\partial q^\mu} - \frac{\partial f}{\partial q^\mu}\frac{\partial g}{\partial p_\mu}\right)h\right],
\end{aligned} \tag{32}
$$

which proves the derivative property: $\{f,gh\}_\alpha = g\{f,h\}_\alpha + \{f,g\}_\alpha h$.

Thus, the bracket $\{.,.\}_\alpha$ is antisymmetric and satisfies the Jacobi identity and the derivation property. Therefore, it is a Poisson bracket and $(\mathfrak{F}, \{.,.\}_\alpha)$ is a conformable Poisson algebra. □

Proposition 2. *The set of Hamiltonian vector fields $\mathfrak{X}_{\alpha_{\mathfrak{F}}}$ endowed with the Lie bracket given by the commutator $[.,.]$ is a conformable Lie algebra.*

Proof of Proposition 2. Using the Jacoby identity, we have:

$$
\{f,\{g,h\}_\alpha\}_\alpha + \{g,\{h,f\}_\alpha\}_\alpha + \{h,\{f,g\}_\alpha\}_\alpha = 0. \tag{33}
$$

The left hand side of this identity can be handled as:

$$
\begin{aligned}
&\{f,\{g,h\}_\alpha\}_\alpha + \{g,\{h,f\}_\alpha\}_\alpha + \{h,\{f,g\}_\alpha\}_\alpha \\
&= \{f,\{g,h\}_\alpha\}_\alpha - \{g,\{f,h\}_\alpha\}_\alpha - \{\{f,g\}_\alpha,h\}_\alpha \\
&= X_{\alpha_f}\{g,h\}_\alpha - \{g,X_{\alpha_f}h\}_\alpha - \{X_{\alpha_f}g,h\}_\alpha \\
&= X_{\alpha_f}X_{\alpha_g}h - X_{\alpha_g}X_{\alpha_f}h - X_{\alpha_{\{f,g\}_\alpha}}h \\
&= [X_{\alpha_f},X_{\alpha_g}]h - X_{\alpha_{\{f,g\}_\alpha}}h
\end{aligned}
$$

leading to

$$
[X_{\alpha_f},X_{\alpha_g}]h = X_{\alpha_{\{f,g\}_\alpha}}h. \tag{34}
$$

Then, the map $f \mapsto X_{\alpha_f} = \{f,.\}_\alpha,\quad \{f,.g\}_\alpha \mapsto X_{\alpha_{\{f,g\}_\alpha}}$ is a conformable Lie algebra morphism $(\mathfrak{F},\{.,.\}_\alpha) \to (\mathfrak{X}_{\alpha_{\mathfrak{F}}},[.,.])$. Therefore, $(\mathfrak{X}_{\alpha_{\mathfrak{F}}},[.,.])$ is a conformable Lie algebra. □

3.2. Noether Symmetry and Recursion Operator

By definition, we have

$$
X_\alpha := \{H_\alpha,.\}_\alpha = \sum_{\mu=1}^{4} \alpha^{-2}|p_\mu|^{(1-\alpha)}|q^\mu|^{(1-\alpha)}\left(\frac{\partial H_\alpha}{\partial p_\mu}\frac{\partial}{\partial q^\mu} - \frac{\partial H_\alpha}{\partial q^\mu}\frac{\partial}{\partial p_\mu}\right). \tag{35}
$$

Using (25) and (35), we obtain the following set of equations:

$$
\begin{cases}
\alpha^{-2}|p_1|^{(1-\alpha)}|q^1|^{(1-\alpha)}\dfrac{\partial H_\alpha}{\partial p_1} &= -p_1 \\
\alpha^{-2}|p_1|^{(1-\alpha)}|q^1|^{(1-\alpha)}\dfrac{\partial H_\alpha}{\partial q^1} &= -\left(\dfrac{\alpha-1}{\alpha+1}\right)\dfrac{p_1^2}{q^1} \\
\alpha^{-2}|p_k|^{(1-\alpha)}|q^k|^{(1-\alpha)}\dfrac{\partial H_\alpha}{\partial p_k} &= p_k, \qquad k = 2,3,4 \\
\alpha^{-2}|p_k|^{(1-\alpha)}|q^k|^{(1-\alpha)}\dfrac{\partial H_\alpha}{\partial q^k} &= \left(\dfrac{\alpha-1}{\alpha+1}\right)\dfrac{p_k^2}{q^k}, \qquad k = 2,3,4
\end{cases} \tag{36}
$$

leading to

$$
H_\alpha = -\frac{\alpha^2}{\alpha+1}|q^1|^{\alpha-1}|p_1|^{\alpha+1} + \sum_{k=2}^{4}\frac{\alpha^2}{\alpha+1}|q^k|^{\alpha-1}|p_k|^{\alpha+1}. \tag{37}
$$

This function is called the Hamiltonian function. For $\alpha = 1$, we naturally obtain the Hamiltonian function of a free particle on the ordinary Minkowski phase space.

The vector field

$$Y_\alpha = -\frac{1}{2\alpha^2}|p_1|^{1-\alpha}p_1^{-2}|q^1|^{1-\alpha}|q^1|^{\frac{1-\alpha}{1+\alpha}}\frac{\partial}{\partial p_1} + \frac{1}{2\alpha^2}\sum_{k=2}^{4}|p_k|^{1-\alpha}p_k^{-2}|q^k|^{1-\alpha}|q^k|^{\frac{1-\alpha}{1+\alpha}}\frac{\partial}{\partial p_k} \tag{38}$$

is a master symmetry, i.e.,

$$[[Y_\alpha, X_\alpha], X_\alpha] = 0, \tag{39}$$

and the following relations hold:

$$L_\alpha := \mathcal{L}_{Y_\alpha} H_\alpha = \frac{1}{2}\left(p_1^{-1}(q^1)^{\frac{1-\alpha}{1+\alpha}} + \sum_{k=2}^{4} p_k^{-1}(q^k)^{\frac{1-\alpha}{1+\alpha}}\right), \tag{40}$$

$$\begin{aligned} \omega_{\alpha_1} &:= \mathcal{L}_{Y_\alpha}\omega_\alpha = d\iota_{Y_\alpha}\omega_\alpha + \iota_{Y_\alpha}d\omega_\alpha \\ &= p_1^{-3}|q^1|^{\frac{1-\alpha}{1+\alpha}}dp_1 \wedge dq^1 - \sum_{k=2}^{4} p_k^{-3}|q^k|^{\frac{1-\alpha}{1+\alpha}}dp_k \wedge dq^k, \end{aligned} \tag{41}$$

$$\begin{aligned} X_{\alpha_1} &:= [X_\alpha, Y_\alpha] \\ &= -\frac{1}{2\alpha^2}\left[\frac{1-\alpha}{(1+\alpha)}G_1|p_1|^{-\alpha}|q^1|^{\frac{1-2\alpha-\alpha^2}{1+\alpha}}\frac{\partial}{\partial p_1} + |p_1|^{1-\alpha}p_1^{-2}|q^1|^{\frac{2-\alpha-\alpha^2}{1+\alpha}}\frac{\partial}{\partial q^1}\right] \\ &- \frac{1}{2\alpha^2}\sum_{k=2}^{4}\left[\frac{1-\alpha}{(1+\alpha)}G_k p_k^{-\alpha}|q^k|^{\frac{1-2\alpha-\alpha^2}{1+\alpha}}\frac{\partial}{\partial p_k} + |p_k|^{1-\alpha}p_k^{-2}|q^k|^{\frac{2-\alpha-\alpha^2}{1+\alpha}}\frac{\partial}{\partial q^k}\right], \end{aligned} \tag{42}$$

where $G_i = sgn(p_i)sgn(q^i), i = 1,2,3,4$.

We notice that X_{α_1} satisfies the relation

$$\iota_{X_{\alpha_1}}\omega_\alpha = -dL_\alpha,$$

where $\iota_{X_{\alpha_1}}\omega_\alpha$ is the interior product of ω_α with respect to the vector field X_{α_1}. Since X_{α_1} is a dynamical symmetry, i.e., $[X_\alpha, X_{\alpha_1}] = 0$, L_α is a first integral, also called a constant of motion. Thus, we arrive at the following property:

Proposition 3. *The vector field X_{α_1} is an infinitesimal Noether symmetry.*

Proof of Proposition 3. We have:

$$\mathcal{L}_{X_{\alpha_1}}\omega_\alpha = d\iota_{X_{\alpha_1}}\omega_\alpha + \iota_{X_{\alpha_1}}d\omega_\alpha = d\iota_{X_{\alpha_1}}\omega_\alpha = -d^2L_\alpha = 0. \tag{43}$$

Since X_{α_1} is a dynamical symmetry, then

$$\mathcal{L}_{X_{\alpha_1}}H_\alpha = X_{\alpha_1}(H_\alpha) = 0. \tag{44}$$

Equations (43) and (44) show that X_{α_1} is both an infinitesimal geometric symmetry, i.e., leaving invariant the geometric structure (the symplectic form ω_α), and an infinitesimal Hamiltonian symmetry leaving invariant the dynamics (the Hamiltonian function H_α). Hence, X_{α_1} is an infinitesimal Noether symmetry. □

In the sequel, we consider the following Poisson bivector

$$\mathcal{P}_{\alpha_1} = p_1^3|q^1|^{\frac{\alpha-1}{1+\alpha}}\frac{\partial}{\partial p_1}\wedge\frac{\partial}{\partial q^1} - \sum_{k=2}^{4}p_k^3|q^k|^{\frac{\alpha-1}{1+\alpha}}\frac{\partial}{\partial p_k}\wedge\frac{\partial}{\partial q^k} \tag{45}$$

and define the conformable Poisson bracket

$$\{f,g\}_{\alpha_1} := p_1^3|q^1|^{\frac{\alpha-1}{1+\alpha}}\left(\frac{\partial f}{\partial p_1}\frac{\partial g}{\partial q^1} - \frac{\partial f}{\partial q^1}\frac{\partial g}{\partial p_1}\right) - \sum_{k=2}^{4}p_k^3|q^k|^{\frac{\alpha-1}{1+\alpha}}\left(\frac{\partial f}{\partial p_k}\frac{\partial g}{\partial q^k} - \frac{\partial f}{\partial q^k}\frac{\partial g}{\partial p_k}\right), \tag{46}$$

with respect to the symplectic form ω_{α_1}.

Thus, the vector field X_α is a bi-Hamiltonian vector field with respect to $(\omega_\alpha, \omega_{\alpha_1})$, i.e.,

$$\iota_{X_\alpha}\omega_\alpha = -dH_\alpha \quad \text{and} \quad \iota_{X_\alpha}\omega_{\alpha_1} = -d\tilde{L}_\alpha, \quad X_\alpha = \{H_\alpha, .\}_\alpha = \{\tilde{L}_\alpha, .\}_{\alpha_1}, \tag{47}$$

where

$$\tilde{L}_\alpha = \sum_{\mu=1}^{4} |q^\mu|^{\frac{1-\alpha}{1+\alpha}} p_\mu^{-1} \tag{48}$$

are first integrals for X_{H_α}.

Therefore, the associated recursion operator T_α is given by:

$$\begin{aligned}
T_\alpha &:= \mathcal{P}_{\alpha_1} \circ \mathcal{P}_\alpha^{-1} \\
&= \left(p_1^3 |q^1|^{\frac{\alpha-1}{1+\alpha}} \frac{\partial}{\partial p_1} \wedge \frac{\partial}{\partial q^1} - \sum_{k=2}^{4} p_k^3 |q^k|^{\frac{\alpha-1}{1+\alpha}} \frac{\partial}{\partial p_k} \wedge \frac{\partial}{\partial q^k} \right) \circ \left(\sum_{\mu=1}^{4} \alpha^2 |p_\mu|^{(\alpha-1)} |q^\mu|^{(\alpha-1)} dp_\mu \wedge dq^\mu \right) \\
&= \alpha^2 p_1^3 |p_1|^{(\alpha-1)} |q^1|^{\frac{-2+\alpha^2+\alpha}{1+\alpha}} \frac{\partial}{\partial p_1} \otimes dp_1 - \alpha^2 \sum_{k=2}^{4} p_k^3 |p_k|^{(\alpha-1)} |q^k|^{\frac{-2+\alpha^2+\alpha}{1+\alpha}} \frac{\partial}{\partial p_k} \otimes dp_k \\
&+ \alpha^2 p_1^3 |p_1|^{(\alpha-1)} |q^1|^{\frac{-2+\alpha^2+\alpha}{1+\alpha}} \frac{\partial}{\partial q^1} \otimes dq^1 - \alpha^2 \sum_{k=2}^{4} p_k^3 |p_k|^{(\alpha-1)} |q^k|^{\frac{-2+\alpha^2+\alpha}{1+\alpha}} \frac{\partial}{\partial q^k} \otimes dq^k,
\end{aligned} \tag{49}$$

providing the constants of motion

$$Tr(T_\alpha^h) = 2^h \alpha^{2h} \left\{ \left(p_1^3 |p_1|^{(\alpha-1)} |q^1|^{\frac{-2+\alpha^2+\alpha}{1+\alpha}} \right)^h + (-1)^h \left(\sum_{k=2}^{4} p_k^3 |p_k|^{(\alpha-1)} |q^k|^{\frac{-2+\alpha^2+\alpha}{1+\alpha}} \right)^h \right\}, \ h \in \mathbb{N}. \tag{50}$$

This work can be considered as a conformable case of previous investigations [28,29]. The only difference resides in the fact that we here use the method of Noether symmetry to obtain the integrals of motion instead of the method of Hamilton–Jacobi separability, developed in [27–29].

4. Conformable Einstein Field Equation

In this section, we investigate the solutions of the Einstein field equation in the conformable Schwarzschild and Friedmann–Lemaître–Robertson–Walker (FLRW) metrics. We consider the Einstein field equation shortly written in the tensor form as:

$$G_\alpha + \Lambda g_\alpha = \kappa T_\alpha, \tag{51}$$

where the tensor

$$G_\alpha = R_\alpha - \frac{1}{2} g_\alpha \mathbf{R}_\alpha \tag{52}$$

is the Einstein tensor, the constant Λ is the cosmological constant, κ is a constant; T_α and R_α are the tress-energy tensor and Ricci tensor measuring the geodesic deviation, respectively. g_α is the metric tensor, and $\mathbf{R}_\alpha$, is the scalar curvature. The energy-momentum tensor T_α, determines how the geometry is.

4.1. Recursion Operator in Conformable Schwarzschild Metric

The Schwarzschild metric is the simplest one among the particular solutions of the Einstein field equation.

Here, we consider the following conformable Schwarzschild metric

$$\begin{aligned}
d_\alpha s^2 = &-\left(1 - \frac{2M}{q^2}\right)(q^1)^{2(\alpha-1)}(dq^1)^2 + \left(1 - \frac{2M}{q^2}\right)^{-1}(q^2)^{2(\alpha-1)}(dq^2)^2 \\
&+ (q^2)^2 (q^3)^{2(\alpha-1)}(dq^3)^2 + (q^2)^2 (q^4)^{2(\alpha-1)} \sin^2 q^3 (dq^4)^2,
\end{aligned} \tag{53}$$

where $t = q^1$, $r = q^2$, $\theta = q^3$, $\phi = q^4$, M is a positive constant representing the mass of the black hole, $t \in (-\infty, \infty)$, $r \in (2M, \infty)$, $\theta \in (0, \pi)$, and $\phi \in (0, 2\pi)$.

The metric is defined on a manifold

$$\begin{aligned}
\mathcal{Q} \ &= \{(q^1, q^2, q^3, q^4) \mid 0 \neq q^1 \in (-\infty, \infty), q^2 \in (2M, \infty), \\
&0 \neq q^3 \in (0, \pi), \text{ and } 0 \neq q^4 \in (0, 2\pi)\}.
\end{aligned} \tag{54}$$

For $\alpha = 1$, we recover the Karl Schwarzschild metric [35].

For our purpose, let us consider the phase space $\mathcal{T}^*\mathcal{Q} \ni (q,p), q \in \mathcal{Q}$, and the Hamiltonian function

$$H_{S\alpha} = -\frac{1}{2}\left(1-\frac{2M}{q^2}\right)^{-1}(q^1)^{2(1-\alpha)}p_1^2 + \frac{1}{2}\left(1-\frac{2M}{q^2}\right)(q^2)^{2(1-\alpha)}p_2^2 + \frac{1}{2(q^2)^2}(q^3)^{2(1-\alpha)}p_3^2 + \frac{1}{2(q^2)^2\sin^2 q^3}(q^4)^{2(1-\alpha)}p_4^2. \tag{55}$$

The Hamiltonian vector field of $H_{S\alpha}$ in a conformable Schwarzschild metric with respect to the canonical symplectic structure $\omega_\alpha = \sum_{\mu=1}^{4} \alpha^2 |p_\mu|^{(\alpha-1)}|q^\mu|^{(\alpha-1)} dp_\mu \wedge dq^\mu$ is given by

$$X_{S\alpha} := \{H_{S\alpha}, \cdot\}_\alpha = \sum_{\mu=1}^{4} \alpha^{-2}|p_\mu|^{(1-\alpha)}|q^\mu|^{(1-\alpha)}\left(\frac{\partial H_\alpha}{\partial p_\mu}\frac{\partial}{\partial q^\mu} - \frac{\partial H_\alpha}{\partial q^\mu}\frac{\partial}{\partial p_\mu}\right) = \sum_{\mu=1}^{4} \alpha^{-2}|p_\mu|^{(1-\alpha)}|q^\mu|^{(1-\alpha)}\left(V_\mu\frac{\partial}{\partial q^\mu} + U_\mu\frac{\partial}{\partial p_\mu}\right), \tag{56}$$

where

$$V_1 = -\left(1-\frac{2M}{q^2}\right)^{-1}\eta_1, \quad V_2 = \left(1-\frac{2M}{q^2}\right)\eta_2, \; V_3 = \frac{1}{(q^2)^2}\eta_3, \; V_4 = \frac{1}{(q^2)^2\sin^2 q^3}\eta_4,$$

$$U_1 = (1-\alpha)\left(1-\frac{2M}{q^2}\right)^{-1}\zeta_1, \; U_3 = -\left(\frac{1-\alpha}{(q^2)^2}\zeta_3 - \frac{\cos q^3}{(q^2)^2\sin^3 q^3}\eta_4 p_4\right), \; U_4 = -\frac{1-\alpha}{(q^2)^2\sin^2 q^3}\zeta_4,$$

$$U_2 = -\left\{\frac{M}{(q^2)^2}\left(1-\frac{2M}{q^2}\right)^{-2}\eta_1 p_1 + (1-\alpha)\left(1-\frac{2M}{q^2}\right)\zeta_2 + \frac{M}{(q^2)^2}\eta_2 p_2 - \frac{1}{(q^2)^3}\eta_3 p_3 - \frac{1}{(q^2)^3\sin^2 q^3}\eta_4 p_4\right\},$$

with $\eta_\nu = (q^\nu)^{2(1-\alpha)}p_\nu$, and $\zeta_\nu = (q^\nu)^{(1-2\alpha)}p_\nu^2$, $\nu = 1,2,3,4$.

Then, we get in conformable Schwarzschild metric, the Christoffel symbols $(\Gamma_{ij}^k)_\alpha$, the components of the Riemann and Ricci tensors $(R_{ii})_\alpha$, the Ricci scalar $\mathbf{R}$, and the components of the Einstein tensor $(G_{ij})_\alpha$, $i,j,k,l = 1,2,3,4$, see Appendix A.

Note that the components of defined geometric objects are obtained in the usual undeformed Schwarzschild metric by setting $\alpha = 1$.

Now, we consider the Hamilton–Jacobi equation for the Hamiltonian function $H_{S\alpha}$

$$E_S = H_{S\alpha}\left(q, \frac{\partial W}{\partial q}\right) = -\frac{1}{2}\left(1-\frac{2M}{q^2}\right)^{-1}(q^1)^{2(1-\alpha)}\left(\frac{\partial W}{\partial q^1}\right)^2 + \frac{1}{2}\left(1-\frac{2M}{q^2}\right)(q^2)^{2(1-\alpha)}\left(\frac{\partial W}{\partial q^2}\right)^2 + \frac{1}{2(q^2)^2}(q^3)^{2(1-\alpha)}\left(\frac{\partial W}{\partial q^3}\right)^2 + \frac{1}{2(q^2)^2\sin^2 q^3}(q^4)^{2(1-\alpha)}\left(\frac{\partial W}{\partial q^4}\right)^2, \tag{57}$$

where E_S is a constant and $W = \sum_{\mu=1}^{4} W_\mu(q_\mu)$ is the generating function. In particular, we put $W_1 = \frac{a}{\alpha}|q^1|^\alpha$, where a is a constant. This equation is a type of separation of variables; then, the above Hamilton–Jacobi equation becomes

$$2E_S(q^2)^2 + \left(1-\frac{2M}{q^2}\right)^{-1}(q^2)^2 a^2 - \left(1-\frac{2M}{q^2}\right)(q^2)^{2(2-\alpha)}\left(\frac{dW_2}{dq^2}\right)^2 = (q^3)^{2(1-\alpha)}\left(\frac{dW_3}{dq^3}\right)^2 + \frac{1}{\sin^2 q^3}(q^4)^{2(1-\alpha)}\left(\frac{dW_4}{dq^4}\right)^2, \tag{58}$$

which can be rewritten through a constant K as:

$$K = 2E_S(q^2)^2 + \left(1 - \frac{2M}{q^2}\right)^{-1}(q^2)^2a^2 - \left(1 - \frac{2M}{q^2}\right)(q^2)^{2(2-\alpha)}\left(\frac{dW_2}{dq^2}\right)^2 \tag{59}$$

$$K = (q^3)^{2(1-\alpha)}\left(\frac{dW_3}{dq^3}\right)^2 + \frac{1}{\sin^2 q^3}(q^4)^{2(1-\alpha)}\left(\frac{dW_4}{dq^4}\right)^2. \tag{60}$$

From the above, we set:

$$\left(K - (q^3)^{2(1-\alpha)}\left(\frac{dW_3}{dq^3}\right)^2\right)\sin^2 q^3 = G \tag{61}$$

$$(q^4)^{2(1-\alpha)}\left(\frac{dW_4}{dq^4}\right)^2 = G. \tag{62}$$

and obtain

$$W_4 = \frac{\sqrt{G}}{\alpha}|q^4|^{\alpha-1}q^4 + A, \tag{63}$$

where A is a constant.

We put the solutions of Equations (59) and (61) in the form:

$$W_2 = W_2(q^2, E_S, a, K), \quad W_3 = W_3(q^3, K, G). \tag{64}$$

Then, a generating function W takes the form:

$$W = \frac{a}{\alpha}|q^1|^\alpha + W_2(q^2, E_S, a, K) + W_3(q^3, K, G) + \frac{\sqrt{G}}{\alpha}|q^4|^{\alpha-1}q^4 + A. \tag{65}$$

Now, we consider the canonical system (Q, P), where

$$Q^1 = E_S,\ Q^2 = a,\ Q^3 = K,\ Q^4 = \sqrt{G}, \tag{66}$$

$$P_1 := -\frac{\partial W}{\partial Q^1} = -\frac{\partial W_2}{\partial Q^1}, \quad P_2 := -\frac{\partial W}{\partial Q^2} = -\frac{a}{\alpha}(q^1)^\alpha - \frac{\partial W_2}{\partial Q^2}, \tag{67}$$

$$P_3 := -\frac{\partial W}{\partial Q^3} = -\frac{\partial W_2}{\partial Q^3} - \frac{\partial W_3}{\partial Q^3}, \text{ and } P_4 := -\frac{\partial W}{\partial Q^4} = -\frac{\partial W_4}{\partial Q^4} - \frac{\partial W_3}{\partial Q^4} = -\frac{1}{\alpha}|q^4|^{\alpha-1}q^4 - \frac{\partial W_3}{\partial Q^4}. \tag{68}$$

In this new canonical system, we define the following Poisson bracket

$$\{f, g\}_\alpha = \sum_{\mu=1}^{4} \alpha^{-2}|P_\mu|^{(1-\alpha)}|Q^\mu|^{(1-\alpha)}\left(\frac{\partial f}{\partial P_\mu}\frac{\partial g}{\partial Q^\mu} - \frac{\partial f}{\partial Q^\mu}\frac{\partial g}{\partial P_\mu}\right), \tag{69}$$

with respect to the symplectic form

$$\omega_\alpha = \sum_{\mu=1}^{4} \alpha^2|P_\mu|^{(\alpha-1)}|Q^\mu|^{(\alpha-1)}dP_\mu \wedge dQ^\mu. \tag{70}$$

Then, the Hamiltonian vector field takes the form:

$$X_{S\alpha} := \{H_{S\alpha}, .\}_\alpha = -\alpha^{-2}|P_1|^{(1-\alpha)}|Q^1|^{(1-\alpha)}\frac{\partial}{\partial P_1}. \tag{71}$$

Now, we consider a $(1,1)$-tensor field $T_{S\alpha}$ as

$$T_{S\alpha} = \sum_{\mu=1}^{4} |Q^\mu|^{\alpha-1}Q^\mu\left(\frac{\partial}{\partial P_\mu} \otimes dP_\mu + \frac{\partial}{\partial Q^\mu} \otimes dQ^\mu\right). \tag{72}$$

We can put $Q^\mu = x_\mu$ and $P_\mu = x_{\mu+n}$, where $n = 4$ in this case and $\mu = 1, 2, 3, 4$. Then, by Lemma 1, $T_{S\alpha}$ satisfies $\mathcal{L}_{X_{S\alpha}}T_{S\alpha} = 0$, $\mathcal{N}_{T_{S\alpha}} = 0$, and $degQ^\mu = 2$. Hence, $T_{S\alpha}$ is a recursion operator of $X_{S\alpha}$. The constants of motion $Tr(T^l_\alpha)$ $(l \in \mathbb{N})$ of the Hamiltonian vector field $X_{S\alpha}$ for the conformable Schwarzschild metric are finally obtained as:

$$Tr(T^l_{F\alpha}) = 2((Q^1)^l + (Q^2)^l + (Q^3)^l + (Q^4)^l), \quad l \in \mathbb{N}. \tag{73}$$

4.2. Recursion Operator in Conformable FLRW Metric

Now, we consider the following conformable Friedmann–Lemaître–Robertson–Walker (FLRW) metric:

$$d_\alpha s^2 = -|q^1|^{2(\alpha-1)}(dq^1)^2 + R^2(q^1)\left\{ \frac{|q^2|^{2(\alpha-1)}}{1-k(q^2)^2}(dq^2)^2 + (q^2)^2\left(|q^3|^{2(\alpha-1)}(dq^3)^2 \right.\right.$$
$$\left.\left. + |q^4|^{2(\alpha-1)} \sin^2 q^3 (dq^4)^2 \right) \right\} \tag{74}$$

defined on the same manifold $\mathcal{Q}$ (54), where $R(q^1)$ is a scale factor and k is a constant representing the curvature of the space. Considering the Hamiltonian function

$$H_{F\alpha} = -\frac{1}{2}(q^1)^{2(1-\alpha)}p_1^2 + \frac{1-k(q^2)^2}{2R^2(q^1)}(q^2)^{2(1-\alpha)}p_2^2 + \frac{(q^3)^{2(1-\alpha)}}{2(q^2)^2R^2(q^1)}p_3^2 + \frac{(q^4)^{2(1-\alpha)}}{2(q^2)^2R^2(q^1)\sin^2(q^3)}p_4^2, \tag{75}$$

we obtain the following Hamiltonian vector field

$$X_{F\alpha} = \sum_{\mu=1}^{4} \alpha^{-2}|p_\mu|^{(1-\alpha)}|q^\mu|^{(1-\alpha)}\left(\tilde{V}_\mu \frac{\partial}{\partial q^\mu} + \tilde{U}_\mu \frac{\partial}{\partial p_\mu} \right), \tag{76}$$

with respect to the symplectic structure

$$\omega_\alpha = \sum_{\mu=1}^{4} \alpha^2 |p_\mu|^{(\alpha-1)}|q^\mu|^{(\alpha-1)} dp_\mu \wedge dq^\mu, \tag{77}$$

where

$$\tilde{V}_1 = \eta_1, \quad \tilde{V}_2 = \frac{1-k(q^2)^2}{2R^2(q^1)}\eta_2, \quad \tilde{V}_3 = \frac{1}{(q^2)^2R^2(q^1)}\eta_3, \quad \tilde{V}_4 = \frac{1}{(q^2)^2R^2(q^1)\sin^2(q^3)}\eta_4,$$
$$\tilde{U}_1 = (1-\alpha)\zeta_1 + \frac{1}{R^3(q^1)}\left((1-k(q^2)^2)\eta_2 p_2 + \frac{1}{(q^2)^2}\eta_3 p_3 + \frac{1}{\sin^2 q^3}\eta_4 p_4 \right)\frac{dR(q^1)}{dq^1},$$
$$\tilde{U}_2 = -\frac{p_2^2}{R^2(q^1)}\left(-kq^2\eta_2 p_2 + (1-\alpha)(1-k(q^2)^2)\zeta_2 \right) + \frac{1}{(q^2)^3R^2(q^1)}\eta_3 p_3 + \frac{1}{(q^2)^3R^2(q^1)\sin^2 q^3}\eta_4 p_4,$$
$$\tilde{U}_3 = -\frac{(1-\alpha)}{(q^2)^2R^2(q^1)}\zeta_3 + \frac{\cos q^3}{(q^2)^2R^2(q^1)sin^3q^3}\eta_4 p_4, \quad \tilde{U}_4 = -\frac{(1-\alpha)}{(q^2)^2R^2(q^1)sin^2q^3}\zeta_4,$$

with $\eta_\nu = (q^\nu)^{2(1-\alpha)}p_\nu$, $\zeta_\nu = (q^\nu)^{(1-2\alpha)}p_\nu^2$, and $\nu = 1,2,3,4$.

Here, we perform in a conformable FLRW metric, the computation of the Christoffel symbols, the components of the Riemann and Ricci tensors, the Ricci scalar and the components of the Einstein tensor, see Appendix A.

Remark that for $\alpha = 1$, we recover the components of these geometric objects in the usual FLRW metric, as expected.

The Hamiltonian–Jacobi equation here takes the form:

$$2E_F = -(q^1)^{2(1-\alpha)}\left(\frac{\partial W}{\partial q^1}\right)^2 + \frac{1-k(q^2)^2}{R^2(q^1)}(q^2)^{2(1-\alpha)}\left(\frac{\partial W}{\partial q^2}\right)^2$$
$$+ \frac{(q^3)^{2(1-\alpha)}}{(q^2)^2R^2(q^1)}\left(\frac{\partial W}{\partial q^3}\right)^2 + \frac{(q^4)^{2(1-\alpha)}}{(q^2)^2R^2(q^1)\sin^2(q^3)}\left(\frac{\partial W}{\partial q^4}\right)^2, \tag{78}$$

where E_F is a constant and $W = \sum_{\mu=1}^{4} W_\mu(q_\mu)$ is the generating function. The above equation can be rewritten as

$$2E_F R^2(q^1) + (q^1)^{2(1-\alpha)}R^2(q^1)\left(\frac{dW_1}{dq^1}\right)^2 = (1-k(q^2)^2)(q^2)^{2(1-\alpha)}\left(\frac{dW_2}{dq^2}\right)^2 + \frac{(q^3)^{2(1-\alpha)}}{(q^2)^2}\left(\frac{dW_3}{dq^3}\right)^2$$
$$+ \frac{(q^4)^{2(1-\alpha)}}{(q^2)^2\sin^2(q^3)}\left(\frac{dW_4}{dq^4}\right)^2,$$

which is of a type of separation of variables. Thus, we can also express them via a constant K as:

$$K = 2E_F R^2(q^1) + (q^1)^{2(1-\alpha)} R^2(q^1) \left(\frac{dW_1}{dq^1} \right)^2, \tag{79}$$

$$K = (1 - k(q^2)^2)(q^2)^{2(1-\alpha)} \left(\frac{dW_2}{dq^2} \right)^2 + \frac{(q^3)^{2(1-\alpha)}}{(q^2)^2} \left(\frac{dW_3}{dq^3} \right)^2 + \frac{(q^4)^{2(1-\alpha)}}{(q^2)^2 \sin^2(q^3)} \left(\frac{dW_4}{dq^4} \right)^2. \tag{80}$$

Moreover, from Equation (80), we get

$$(1 - k(q^2)^2)(q^2)^{2(2-\alpha)} \left(\frac{dW_2}{dq^2} \right)^2 - (q^2)^2 K = -(q^3)^{2(1-\alpha)} \left(\frac{dW_3}{dq^3} \right)^2 - \frac{(q^4)^{2(1-\alpha)}}{\sin^2(q^3)} \left(\frac{dW_4}{dq^4} \right)^2. \tag{81}$$

Since Equation (81) is of a type of separation of variables, we can introduce a constant L, such that

$$L = (q^2)^2 K - (1 - k(q^2)^2)(q^2)^{2(2-\alpha)} \left(\frac{dW_2}{dq^2} \right)^2, \tag{82}$$

$$L = (q^3)^{2(1-\alpha)} \left(\frac{dW_3}{dq^3} \right)^2 + \frac{(q^4)^{2(1-\alpha)}}{\sin^2(q^3)} \left(\frac{dW_4}{dq^4} \right)^2, \tag{83}$$

and the Equation (83) can be expressed as

$$L \sin^2(q^3) - (q^3)^{2(1-\alpha)} \sin^2(q^3) \left(\frac{dW_3}{dq^3} \right)^2 = (q^4)^{2(1-\alpha)} \left(\frac{dW_4}{dq^4} \right)^2. \tag{84}$$

Setting

$$G = L \sin^2(q^3) - (q^3)^{2(1-\alpha)} \sin^2(q^3) \left(\frac{dW_3}{dq^3} \right)^2, \tag{85}$$

$$G = (q^4)^{2(1-\alpha)} \left(\frac{dW_4}{dq^4} \right)^2, \tag{86}$$

we can formulate the solutions of the Equations (79), (82), and (85) as:

$$W_1 = W_1(q^1; E_F, K),\ W_2 = W_2(q^2; K, L),\ W_3 = W_3(q^3; L, G). \tag{87}$$

From (86), we obtain

$$W_4 = \frac{\sqrt{G}}{\alpha} |q^4|^{\alpha-1} q^4 + C, \tag{88}$$

where C is a constant, and, hence,

$$W = W_1(q^1; E_F, K) + W_2(q^2; K, L) + W_3(q^3; L, G) + \frac{\sqrt{G}}{\alpha} |q^4|^{\alpha-1} q^4 + C. \tag{89}$$

Considering now the canonical system (Q, P), where

$$Q^1 = E_F,\ Q^2 = K,\ Q^3 = \sqrt{L},\ Q^4 = \sqrt{G}, \tag{90}$$

$$P_1 := -\frac{\partial W}{\partial Q^1} = -\frac{\partial W_1}{\partial Q^1},\quad P_2 := -\frac{\partial W}{\partial Q^2} = -\frac{\partial W_1}{\partial Q^2} - \frac{\partial W_2}{\partial Q^2}, \tag{91}$$

$$P_3 := -\frac{\partial W}{\partial Q^3} = -\frac{\partial W_2}{\partial Q^3} - \frac{\partial W_3}{\partial Q^3}, \text{ and } P_4 := -\frac{\partial W}{\partial Q^4} = -\frac{\partial W_3}{\partial Q^4} - \frac{\partial W_4}{\partial Q^4} = -\frac{1}{\alpha} |q^4|^{\alpha-1} q^4 - \frac{\partial W_3}{\partial Q^4}, \tag{92}$$

the Hamiltonian vector field $X_{F\alpha}$ and the $(1,1)$-tensor field $T_{F\alpha}$ are given by

$$X_{F\alpha} := \{H_{F\alpha}, .\}_\alpha = -\alpha^{-2} |P_1|^{(1-\alpha)} |Q^1|^{(1-\alpha)} \frac{\partial}{\partial P_1},\ T_{F\alpha} = \sum_{\mu=1}^{4} |Q^\mu|^{\alpha-1} Q^\mu \left(\frac{\partial}{\partial P_\mu} \otimes dP_\mu + \frac{\partial}{\partial Q^\mu} \otimes dQ^\mu \right), \tag{93}$$

respectively.

Similarly, by Lemma 1, $T_{F\alpha}$ satisfies $\mathcal{L}_{X_{F\alpha}}T_{F\alpha} = 0$, $\mathcal{N}_{T_{F\alpha}} = 0$, and $degQ^{\mu} = 2$. Thus, $T_{F\alpha}$ is a recursion operator of $X_{F\alpha}$, and the constants of motion $Tr(T^{l}_{F\alpha})$ $(l \in \mathbb{N})$ of the vector field $X_{F\alpha}$ for the conformable FLRW metric are provided in the form

$$Tr(T^{l}_{F\alpha}) = 2((Q^1)^l + (Q^2)^l + (Q^3)^l + (Q^4)^l), \quad l \in \mathbb{N}. \tag{94}$$

5. Family of Conserved Quantities

In this section, we consider the Hamiltonian system $(\mathcal{T}^*\mathcal{Q}, \omega, Q^1)$, for which the Hamiltonian function H_α, the vector field X_α, the symplectic form ω_α, the bivector field $\mathcal{P}_\alpha$, and the recursion operator T_α are given in both the conformable Schwarzschild and FLRW metrics by: $H_\alpha = Q^1 > 0$,

$$X_\alpha = \{H_\alpha, .\}_\alpha = -\alpha^{-2}|P_1|^{(1-\alpha)}|Q^1|^{(1-\alpha)}\frac{\partial}{\partial P_1}, \; \omega_\alpha = \sum_{\mu=1}^{4}\alpha^2|P_\mu|^{(\alpha-1)}|Q^\mu|^{(\alpha-1)}dP_\mu \wedge dQ^\mu,$$

$$\mathcal{P}_\alpha = \sum_{\mu=1}^{4}\alpha^{-2}|P_\mu|^{(1-\alpha)}|Q^\mu|^{(1-\alpha)}\frac{\partial}{\partial P_\mu} \wedge \frac{\partial}{\partial Q^\mu}, \text{ and } T_\alpha = \sum_{\mu=1}^{4}|Q^\mu|^{\alpha-1}Q^\mu\left(\frac{\partial}{\partial P_\mu} \otimes dP_\mu + \frac{\partial}{\partial Q^\mu} \otimes dQ^\mu\right).$$

In the sequel, we introduce the functions

$$\tilde{H}_{\alpha_j} = -\sum_{\mu=1}^{4}\alpha|Q^\mu|^{\alpha(1-j)-1}Q^\mu|P_\mu|^{\alpha-1}P_\mu \tag{95}$$

and obtain the vector fields $Z_{\alpha_j} \in \mathcal{T}^*\mathcal{Q}$,

$$Z_{\alpha_j} := \{\tilde{H}_{\alpha_j}, .\}_\alpha = \sum_{\mu=1}^{4}|Q^\mu|^{-\alpha j}\left((1-j)P_\mu\frac{\partial}{\partial P_\mu} - Q^\mu\frac{\partial}{\partial Q^\mu}\right). \tag{96}$$

satisfying the relation

$$\iota_{Z_{\alpha_j}}\omega_\alpha = -d\tilde{H}_{\alpha_j}. \tag{97}$$

Then, it is straightforward to notice that the symplectic structure ω_α generates a set of Hamiltonian systems on the same manifold $\mathcal{T}^*\mathcal{Q}$. The Lie bracket between the vector fields X_{α_i} and Z_{α_j} obeys the relations

$$[X_{\alpha_i}, Z_{\alpha_j}] = X_{\alpha_{i+j}}, \; [X_{\alpha_i}, X_{\alpha_{i+j}}] = 0, \; i,j \in \mathbb{N}, \; X_{\alpha_0} = X_\alpha, \tag{98}$$

with

$$X_{\alpha_{i+j}} = -\alpha^{-2}(1-\alpha i)[1-(i+j)\alpha]|Q^1|^{1-\alpha(i+j+1)}|P_1|^{(1-\alpha)}\frac{\partial}{\partial P_1}. \tag{99}$$

These relations are diagrammatically well represented in Figure 1. In terms of differential geometry, Z_{α_j} and $\tilde{H}_{\alpha_j}$ are called *master symmetries* for X_{α_i} and *master integrals*, respectively. For more details on these symmetries, see [36–40].

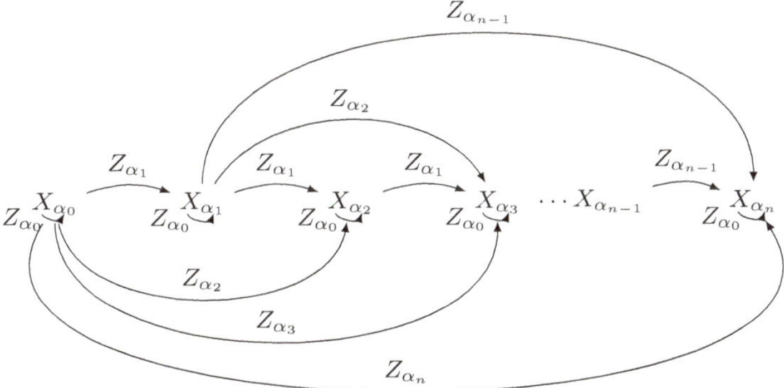

Figure 1. Diagrammatical illustration of Equation (98).

Thus, we can generate a family of Hamiltonian functions:

$$H_{\alpha_{i+j}} := \{H_{\alpha_i}, \tilde{H}_{\alpha_j}\} = (1-\alpha i)(Q^1)^{1-\alpha(i+j)}, \text{ with } H_{\alpha_0} = H_\alpha,\ i,j \in \mathbb{N}. \quad (100)$$

The recursion operator T_α can be written as:

$$T_\alpha = \mathcal{P}_{\alpha_1} \circ \mathcal{P}_\alpha^{-1}, \quad (101)$$

where

$$\mathcal{P}_{\alpha_1} = \sum_{\mu=1}^{4} \alpha^{-2} Q^\mu |P_\mu|^{(1-\alpha)} \frac{\partial}{\partial P_\mu} \wedge \frac{\partial}{\partial Q^\mu} \quad (102)$$

and $\mathcal{P}_\alpha$ are two compatible Poisson bivectors with the vanishing Schouten–Nijenhuis bracket $[\mathcal{P}_\alpha, \mathcal{P}_{\alpha_1}]_{NS} = 0$.

Furthermore, we put $\mathcal{P}_{\alpha_1}^{k+1} = S_{k+1}\mathcal{P}_{\alpha_1} = S_{k+1} \sum_{\mu=1}^{4} \alpha^{-2} Q^\mu |P_\mu|^{(1-\alpha)} \frac{\partial}{\partial P_\mu} \wedge \frac{\partial}{\partial Q^\mu}$, with $S_{k+1} = \frac{1-k\alpha}{1-(k+1)\alpha}$, $k = i+j \in \mathbb{N}$, $(1-(k+1)\alpha) \neq 0$, and introduce the following α-Poisson bracket $\{.,.\}_{\alpha_1}^{k_1}$

$$\{f,g\}_{\alpha_1}^{k+1} := \sum_{\mu=1}^{4} \alpha^{-2} S_{k+1} Q^\mu |P_\mu|^{(1-\alpha)} \left(\frac{\partial f}{\partial P_\mu}\frac{\partial g}{\partial Q^\mu} - \frac{\partial f}{\partial Q^\mu}\frac{\partial g}{\partial P_\mu} \right) \quad (103)$$

with respect to the symplectic form

$$\omega_{\alpha_1}^{k+1} = \sum_{\mu=1}^{4} \alpha^2 S_{k+1}^{-1} (Q^\mu)^{-1} |P_\mu|^{(\alpha-1)} dP_\mu \wedge dQ^\mu \quad (104)$$

and get

$$X_{\alpha_k} = \{H_{\alpha_k}, .\}_\alpha = \{H_{\alpha_{k+1}}, .\}_{\alpha_1}^{k+1}, \quad (105)$$

proving that X_{α_k} are bi-Hamiltonian vector fields defined by the two Poisson bivectors $\mathcal{P}_\alpha$ and $\mathcal{P}_{\alpha_1}^{k+1}$. Then, the quadruple $(\mathcal{Q}, \mathcal{P}_\alpha, \mathcal{P}_{\alpha_1}^{k+1}, X_{\alpha_k})$ is a bi-Hamiltonian system for each k.

The associated recursion operators are given by

$$T_{(k+1)\alpha} := \mathcal{P}_{\alpha_1}^{k+1} \circ \mathcal{P}_\alpha^{-1} = \sum_{\mu=1}^{4} S_k |Q^\mu|^{\alpha-1} Q^\mu \left(\frac{\partial}{\partial P_\mu} \otimes dP_\mu + \frac{\partial}{\partial Q^\mu} \otimes dQ^\mu \right). \quad (106)$$

In addition, we have

$$\mathcal{L}_{Z_{\alpha_0}}(\mathcal{P}_\alpha) = 0,\ (\tilde{\alpha} = 0), \quad \mathcal{L}_{Z_{\alpha_0}}(\mathcal{P}_{\alpha_1}^{k+1}) = -\alpha \sum_{\mu=1}^{4} \alpha^{-2} S_{k+1} Q^\mu |P_\mu|^{(1-\alpha)} \frac{\partial}{\partial P_\mu} \wedge \frac{\partial}{\partial Q^\mu} = -\alpha \mathcal{P}_{\alpha_1}^{k+1},$$
$$(\tilde{\beta} = -\alpha),\ \mathcal{L}_{Z_{\alpha_0}}(H_\alpha) = -Q^1 = -H_\alpha,\ (\tilde{\gamma} = -1)$$

permitting to conclude that the vector field

$$Z_{\alpha_0} = \sum_{\mu=1}^{4} \left(P_\mu \frac{\partial}{\partial P_\mu} - Q^\mu \frac{\partial}{\partial Q^\mu} \right) \quad (107)$$

is a conformal symmetry for $\mathcal{P}_\alpha, \mathcal{P}_{\alpha_1}^{k+1}$ and H_α [39].

Defining now the families of quantities $X_{\alpha_l}^{k+1}$, $Z_{\alpha_l}^{k+1}$, $\mathcal{P}_{\alpha_l}^{k+1}$, $\omega_{\alpha_l}^{k+1}$ and $dH_{\alpha_l}^{k+1}$ by $X_{\alpha_l}^{k+1} := T_{(k+1)\alpha}^l X_\alpha$, $Z_{\alpha_l}^{k+1} := T_{(k+1)\alpha}^l Z_{\alpha_0}$, $\mathcal{P}_{\alpha_l}^{k+1} := T_{(k+1)\alpha}^l \mathcal{P}_\alpha$, $\omega_{\alpha_l}^{k+1} := ((T_{(k+1)\alpha}^l)^*)\omega_\alpha$, $dH_{\alpha_l}^{k+1} := (T_{(k+1)\alpha}^l)^* dH_\alpha$, where $l \in \mathbb{N}$, and $T_{(k+1)\alpha}^* := \mathcal{P}_\alpha^{-1} \circ \mathcal{P}_{\alpha_1}^{k+1}$ denoting the adjoint of $T_{(k+1)\alpha} := \mathcal{P}_{\alpha_1}^{k+1} \circ \mathcal{P}_\alpha^{-1}$, we obtain

$$X_{\alpha_l}^{k+1} = -\alpha^{-2}(S_{k+1})^l (Q^1)^{1+\alpha(l-1)} |P_1|^{(1-\alpha)} \frac{\partial}{\partial P_1}; \quad (108)$$

$$Z_{\alpha_l}^{k+1} = \sum_{\mu=1}^{4} (S_{k+1})^l |Q^\mu|^{l(\alpha-1)} (Q^\mu)^l \left(P_\mu \frac{\partial}{\partial P_\mu} - Q^\mu \frac{\partial}{\partial Q^\mu} \right); \quad (109)$$

$$\mathcal{P}_{\alpha_l}^{k+1} = \sum_{\mu=1}^{4} \alpha^{-2} (S_{k+1})^l (Q^\mu)^l |P_\mu|^{(1-\alpha)} |Q^\mu|^{(1-\alpha)(1-l)} \frac{\partial}{\partial P_\mu} \wedge \frac{\partial}{\partial Q^\mu}; \quad (110)$$

$$\omega_{\alpha_l}^{k+1} = \sum_{\mu=1}^{4} \alpha^2 (S_{k+1})^l (Q^\mu)^l |P_\mu|^{(\alpha-1)} |Q^\mu|^{(\alpha-1)(l+1)} dP_\mu \wedge dQ^\mu; \tag{111}$$

$$dH_{\alpha_l}^{k+1} = (S_{k+1})^l (Q^1)^{\alpha l} dQ^1; \text{ and } H_{\alpha_l}^{k+1} = \frac{1}{l\alpha+1}(S_{k+1})^l (Q^1)^{\alpha l+1} \tag{112}$$

and for each $l \in \mathbb{N}$, we derive the following plethora of conserved quantities:

$$\begin{aligned}\mathcal{L}_{Z_{\alpha_l}^{k+1}}(Z_{\alpha_h}^{k+1}) &= \alpha(l-h)(S_{k+1})^{l+h} \sum_{\mu=1}^{4} |Q^\mu|^{(l+h)(\alpha-1)} (Q^\mu)^{l+h} \left(P_\mu \frac{\partial}{\partial P_\mu} - Q^\mu \frac{\partial}{\partial Q^\mu} \right) \\ &= \alpha(l-h) Z_{\alpha_{l+h}}^{k+1};\end{aligned} \tag{113}$$

$$\begin{aligned}\mathcal{L}_{Z_{\alpha_l}^{k+1}}(X_{\alpha_h}^{k+1}) &= \alpha^{-2}(S_{k+1})^{l+h}(h\alpha+1)(Q^1)^{1+\alpha((l+h)-1)} |P_1|^{(1-\alpha)} \frac{\partial}{\partial P_1} \\ &= -(h\alpha+1) X_{\alpha_{l+h}}^{k+1};\end{aligned} \tag{114}$$

$$\begin{aligned}\mathcal{L}_{Z_{\alpha_l}^{k+1}}(\mathcal{P}_{\alpha_h}^{k+1}) &= \alpha^{-1}(S_{k+1})^{l+h}(l-h)(Q^\mu)^{l+h} |P_\mu|^{(1-\alpha)} |Q^\mu|^{(1-\alpha)(1-(l+h))} \frac{\partial}{\partial P_\mu} \wedge \frac{\partial}{\partial Q^\mu} \\ &= \alpha(l-h) \mathcal{P}_{\alpha_{l+h}}^{k+1};\end{aligned} \tag{115}$$

$$\begin{aligned}\mathcal{L}_{Z_{\alpha_l}^{k+1}}(\omega_{\alpha_h}^{k+1}) &= -\alpha^3 (S_{k+1})^{l+h}(l+h) \sum_{\mu=1}^{4} (Q^\mu)^{l+h} |P_\mu|^{(\alpha-1)} |Q^\mu|^{(\alpha-1)((l+h)+1)} dP_\mu \wedge dQ^\mu \\ &= -\alpha(l+h) \omega_{\alpha_{l+h}}^{k+1};\end{aligned} \tag{116}$$

$$\begin{aligned}< dH_{\alpha_l}^{k+1}, Z_{\alpha_h}^{k+1} > &= -(S_{k+1})^{l+h} \frac{\alpha(l+h)+1}{\alpha(l+h)+1} (Q^1)^{1+\alpha(l+h)} \\ &= -(\alpha(l+h)+1) H_{\alpha_{l+h}}^{k+1};\end{aligned} \tag{117}$$

$$\begin{aligned}\mathcal{L}_{Z_{\alpha_l}^{k+1}}(T_{(k+1)\alpha}) &= -\alpha \sum_{\mu=1}^{4} (S_{k+1})^{l+1} |Q^\mu|^{(\alpha-1)(l+1)} (Q^\mu)^{l+1} \left(\frac{\partial}{\partial P_\mu} \otimes dP_\mu + \frac{\partial}{\partial Q^\mu} \otimes dQ^\mu \right) \\ &= -\alpha T_{(k+1)\alpha}^{l+1},\end{aligned} \tag{118}$$

satisfying the following relations linking the master symmetries Z_{α_j} to the conformal symmetry Z_{α_0} for $\mathcal{P}_\alpha, \mathcal{P}_{\alpha_1}^{k+1}$ and H_α, and to a set of conformal symmetries generated by successive applications of the recursion operator $T_{(k+1)\alpha}$ on Z_{α_0} :

$$\begin{aligned}&\mathcal{L}_{Z_{\alpha_l}^{k+1}}(Z_{\alpha_h}^{k+1}) = (\tilde{\beta}-\tilde{\alpha})(h-l) Z_{\alpha_{l+h}}^{k+1},\ \mathcal{L}_{Z_{\alpha_l}^{k+1}}(X_{\alpha_h}^{k+1}) = (\tilde{\beta}+\tilde{\gamma}+(h-1)(\tilde{\beta}-\tilde{\alpha})) X_{\alpha_{l+h}}^{k+1}, \\ &\mathcal{L}_{Z_{\alpha_l}^{k+1}}(\mathcal{P}_{\alpha_h}^{k+1}) = (\tilde{\beta}+(h-l-1)(\tilde{\beta}-\tilde{\alpha})) \mathcal{P}_{\alpha_{l+h}}^{k+1},\ \mathcal{L}_{Z_{\alpha_l}^{k+1}}(\omega_{\alpha_h}^{k+1}) = (\tilde{\beta}+(l+h-1)(\tilde{\beta}-\tilde{\alpha})) \omega_{\alpha_{l+h}}^{k+1}, \\ &\mathcal{L}_{Z_{\alpha_l}^{k+1}}(T_{(k+1)\alpha}) = (\tilde{\beta}-\tilde{\alpha}) T_{(k+1)\alpha}^{1+l},\ \langle dH_{\alpha_h}^{k+1}, Z_{\alpha_l}^{k+1} \rangle = (\tilde{\gamma}+(l+h)(\tilde{\beta}-\tilde{\alpha})) H_{\alpha_{l+h}}^{k+1}.\end{aligned}$$

This is reminiscent to the well-known Oevel formulas (see [26,31,32,39,41,42]).

Finally, it is worth mentioning a generalization of the conformable Poisson brackets (103), as follows:

$$\{f,g\}_{\alpha_t}^{k+t} := \sum_{\mu=1}^{4} \alpha^{-2} S_{k+t} |Q^\mu|^{1+\alpha(t-1)} |P_\mu|^{(1-\alpha)} \left(\frac{\partial f}{\partial P_\mu} \frac{\partial g}{\partial Q^\mu} - \frac{\partial f}{\partial Q^\mu} \frac{\partial g}{\partial P_\mu} \right), \tag{119}$$

where $S_{k+t} = \dfrac{1-k\alpha}{1-(k+t)\alpha}, k,t \in \mathbb{N},\ (1-(k+t)\alpha) \neq 0$, and $\{f,g\}_{\alpha_0}^0 = \{f,g\}_\alpha$, with $S_0 = 1$, leading to a set of generalized bi-Hamiltonian vector fields

$$X_{\alpha_k} = \{H_{\alpha_k}, .\}_\alpha = \{H_{\alpha_{k+t}}, .\}_{\alpha_t}^{k+t}, \tag{120}$$

the main ingredients governing the Hamiltonian dynamics and pertaining symmetries.

6. Concluding Remarks

In this work, we have proved that a Minkowski phase space endowed with a bracket relatively to a conformable differential realizes a conformable Poisson algebra, conferring a bi-Hamiltonian structure to the resulting manifold. We have deduced that the related conformable Hamiltonian vector field for a free particle is an infinitesimal Noether symmetry. We have computed the corresponding conformable recursion operator. Using the Hamiltonian–Jacobi separability, we have constructed recursion operators in the framework of conformable Schwarzschild and Friedmann–Lemaître–Robertson–Walker (FLRW) metrics, and obtained related constants of motion. We have highlighted the existence of a hierarchy of bi-Hamiltonian structures in both the metrics, and derived a family of conformable recursion operators and master symmetries generating the constants of motion. This study has also shown that Hamiltonian dynamics hint at a connection between the geometry of our physical system, (conformable symplectic manifolds and related Hamiltonian vector fields), and conservation laws. In this connection, the free particle positions on the conformable manifolds are viewed as states and vector fields as laws governing how those states evolve.

Further, we have calculated the conformable Christoffel symbols, Ricci scalar, components of the Riemann, Ricci, and Einstein tensors. This study has revealed that the Christoffel symbols ($(\Gamma^1_{11})_\alpha$, $(\Gamma^2_{22})_\alpha$, $(\Gamma^3_{33})_\alpha$, and $(\Gamma^4_{44})_\alpha$) in conformable Minkowski metric are no longer null, contrary to the ordinary case corresponding to $\alpha = 1$. Similarly, the Christoffel symbols ($(\Gamma^1_{11})_\alpha$, $(\Gamma^3_{33})_\alpha$, and $(\Gamma^4_{44})_\alpha$) are not equal zero in conformable Schwarzschild and FLRW metrics. The existence of these symbols $(\Gamma^i_{ii})_\alpha, (i = 1, 2, 3, 4)$ informs us about the way in which the parallel displacement of any basic vector on the conformable manifolds with respect to itself always remains parallel to the same basic vector.

Author Contributions: Conceptualization, M.N.H., M.J.L. and M.M. All authors contributed equally to the present work in all steps of its conceptualization, computation, draft writing and finalization. All authors have read and agreed to the published version of the manuscript.

Funding: This research received no external funding.

Institutional Review Board Statement: Not applicable.

Informed Consent Statement: Not applicable.

Data Availability Statement: Not applicable.

Acknowledgments: The ICMPA-UNESCO Chair is in partnership with the Association pour la Promotion Scientifique de l'Afrique (APSA), France, and Daniel Iagolnitzer Foundation (DIF), France, supporting the development of mathematical physics in Africa. M. M. is supported by the Faculty of Mechanical Engineering, University of Niš, Serbia, Grant "Research and development of new generation machine systems in the function of the technological development of Serbia".

Conflicts of Interest: The authors declare that they have no conflict of interest.

Appendix A

Table A1. Christoffel symbols $(\Gamma^k_{ij})_\alpha$ in conformable Schwarzschild metric.

$(\Gamma^1_{11})_\alpha = \frac{\alpha-1}{q^1}$	$(\Gamma^2_{11})_\alpha = -\frac{M(2M-q^2)(q^1)^{2(\alpha-1)}}{(q^2)^{2\alpha+1}}$		
$(\Gamma^1_{12})_\alpha = -\frac{M}{q^2(2M-q^2)}$	$(\Gamma^2_{22})_\alpha = \frac{(1-\alpha)q^2+(2\alpha-1)M}{q^2(2M-q^2)}$	$(\Gamma^3_{23})_\alpha = \frac{1}{q^2}$	$(\Gamma^4_{24})_\alpha = \frac{1}{q^2}$
	$(\Gamma^2_{33})_\alpha = -\frac{(2M-q^2)(q^3)^{2(\alpha-1)}}{(q^2)^{2(\alpha-1)}}$	$(\Gamma^3_{33})_\alpha = \frac{\alpha-1}{q^3}$	$(\Gamma^4_{34})_\alpha = \cot(q^3)$
	$(\Gamma^2_{44})_\alpha = \frac{(2M-q^2)(q^4)^{2(\alpha-1)}\sin^2(q^3)}{(q^2)^{2(\alpha-1)}}$	$(\Gamma^3_{44})_\alpha = -\frac{(q^4)^{2(\alpha-1)}\sin(q^3)\cos(q^3)}{(q^3)^{2(\alpha-1)}}$	$(\Gamma^4_{44})_\alpha = \frac{\alpha-1}{q^4}$

Table A2. Components of the Riemann tensor $(R_{ijkl})_\alpha$ in conformable Schwarzschild metric.

$(R_{1212})_\alpha = -\frac{(1+\alpha)\alpha^2 M(q^1)^{2(\alpha-1)}}{(q^2)^3}$	$(R_{1313})_\alpha = -\frac{\alpha^2 M(q^1)^{2(\alpha-1)}(q^3)^{2(\alpha-1)}(2M-q^2)}{(q^2)^{2\alpha}}$	$(R_{1414})_\alpha = -\frac{\alpha^2 M(q^1)^{2(\alpha-1)}(q^4)^{2(\alpha-1)}\sin^2(q^3)(2M-q^2)}{(q^2)^{2\alpha}}$
$(R_{2323})_\alpha = \frac{\alpha^2(M(2\alpha-1)}{2M-q^2} + \frac{(1-\alpha)q^2)(q^3)^{2(\alpha-1)}}{2M-q^2}$	$(R_{2424})_\alpha = \frac{\alpha^2(M(2\alpha-1)}{2M-q^2} + \frac{(1-\alpha)q^2)(q^4)^{2(\alpha-1)}\sin^2(q^3)}{2M-q^2}$	$(R_{3434})_\alpha = -\frac{\alpha^2 q^2(q^4)^{2(\alpha-1)}}{(q^2)^{2(\alpha-1)}q^3}\Big[\sin^2(q^3)(q^2q^3((q^3)^{2(\alpha-1)} - (q^2)^{2(\alpha-1)}) - 2M(q^3)^{2\alpha-1}) + (1-\alpha)(q^2)^{2(\alpha-1)}\sin(q^3)\cos(q^3)\Big]$

Table A3. Components of the Ricci tensor $(R_{ij})_\alpha$ in conformable Schwarzschild metric.

$(R_{11})_\alpha = -\frac{(\alpha-1)M(2M-q^2)(q^1)^{2(\alpha-1)}}{(q^2)^{2(\alpha-1)}}$	$(R_{22})_\alpha = -\frac{(1-\alpha)(3M-2q^2)}{(q^2)^2(-2M+q^2)}$
$(R_{33})_\alpha = \frac{1}{(q^2)^{2(\alpha-1)}q^3\sin(q^3)}\Big[(1-\alpha)(q^2)^{2(\alpha-1)}\cos(q^3) + q^2q^3[(2-\alpha)(q^3)^{2(\alpha-1)} - (q^2)^{2(\alpha-1)}]\sin(q^3) + 2(\alpha-1)M(q^2)^{2\alpha-1}\sin(q^3)\Big]$	$(R_{44})_\alpha = -\frac{1}{(q^2)^{2\alpha-1}(q^3)^{2\alpha-1}}\Big[(q^4)^{2(\alpha-1)}[(\alpha-2)q^2(q^3)^{2\alpha-1} + (q^2)^{2\alpha-1}q^3 + 2(1-\alpha)M(q^3)^{2\alpha-1}]\sin^2(q^3) + (\alpha-1)(q^2)^{2\alpha-1}\sin(q^3)\cos(q^3)\Big]$

Table A4. Ricci scalar **R** in conformable Schwarzschild metric.

$$\mathbf{R} = \frac{2[(1-\alpha)(q^2)^{2\alpha-1}\cos(q^3) + (3(\alpha-1)M(q^3)^{2\alpha-1} + (3-2\alpha)q^2(q^3)^{2\alpha-1} - q^3(q^2)^{2\alpha-1})\sin(q^3)]}{\alpha^2(q^2)^{2\alpha+1}(q^3)^{2\alpha-1}}$$

Table A5. Components of the Einstein tensor G_{ij} in conformable Schwarzschild metric.

$(G_{11})_\alpha = -\frac{1}{(q^1)^{2(1-\alpha)}(q^2)^{2(\alpha+1)}(q^3)^{2\alpha-1}}\Big[(2M-q^2)[(1-\alpha)(q^2)^{2\alpha-1}\cos(q^3) + (4(\alpha-1)M(q^3)^{2\alpha-1} + (3-2\alpha)q^2(q^3)^{2\alpha-1} - q^3(q^2)^{2\alpha-1})\sin(q^3)]\Big]$	$(G_{22})_\alpha = -\frac{q^3\left((q^2)^{2(\alpha-1)} - (q^3)^{2(\alpha-1)}\right)}{q^2(2M-q^2)(q^3)^{2\alpha-1}} - \frac{(\alpha-1)(q^2)^{2(\alpha-1)}\cot(q^3)}{q^2(2M-q^2)(q^3)^{2\alpha-1}}$
$(G_{33})_\alpha = \frac{(\alpha-1)(q^2-M)(q^3)^{2(\alpha-1)}}{(q^2)^{2\alpha-1}}$	$(G_{44})_\alpha = \frac{(\alpha-1)(q^2-M)(q^4)^{2(\alpha-1)}\sin^2(q^3)}{(q^2)^{2\alpha-1}}$

Table A6. Christoffel symbols $(\Gamma^k_{ij})_\alpha$ in conformable FLRW metric.

$(\Gamma^1_{11})_\alpha = \frac{\alpha-1}{q^1}$	$(\Gamma^2_{12})_\alpha = \frac{1}{2}\frac{\frac{dR(q^1)}{dq^1}}{R(q^1)}$	$(\Gamma^3_{13})_\alpha = \frac{1}{2}\frac{\frac{dR(q^1)}{dq^1}}{R(q^1)}$	$(\Gamma^4_{14})_\alpha = \frac{1}{2}\frac{\frac{dR(q^1)}{dq^1}}{R(q^1)}$
$(\Gamma^1_{22})_\alpha = -\frac{1}{2}\frac{\frac{dR(q^1)}{dq^1}(q^2)^{(2\alpha-2)}}{(-1+k\,(q^2)^2)(q^1)^{(2\alpha-2)}}$	$(\Gamma^2_{22})_\alpha = -\frac{\alpha(-1+k\,(q^2)^2)}{q^2\,(-1+k\,(q^2)^2)} - \frac{1-2k\,(q^2)^2}{q^2\,(-1+k\,(q^2)^2)}$	$(\Gamma^3_{23})_\alpha = \frac{1}{q^2}$	$(\Gamma^4_{24})_\alpha = \frac{1}{q^2}$
$(\Gamma^1_{33})_\alpha = \frac{1}{2(q^1)^{(2\alpha-2)}}\frac{dR(q^1)}{dq^1} \times (q^2)^2\,(q^3)^{(2\alpha-2)}$	$(\Gamma^2_{33})_\alpha = \frac{q^2\,(q^3)^{(2\alpha-2)}(-1+k\,(q^2)^2)}{(q^2)^{(2\alpha-2)}}$	$(\Gamma^3_{33})_\alpha = \frac{\alpha-1}{q^3}$	$(\Gamma^4_{34})_\alpha = \cot(q^3)$
$(\Gamma^1_{44})_\alpha = \frac{1}{2(q^1)^{(2\alpha-2)}}\frac{dR(q^1)}{dq^1} \times (q^2)^2\,(q^4)^{(2\alpha-2)}\sin^2(q^3)$	$(\Gamma^2_{44})_\alpha = \frac{q^2\,(q^4)^{(2\alpha-2)}\sin^2(q^3)}{(q^2)^{(2\alpha-2)}} \times (-1+k\,(q^2)^2)$	$(\Gamma^3_{44})_\alpha = -\frac{(q^4)^{(2\alpha-2)}}{(q^3)^{(2\alpha-2)}} \times \sin(q^3)\,\cos(q^3)$	$(\Gamma^4_{44})_\alpha = \frac{\alpha-1}{q^4}$

Table A7. Components of the Riemann tensor $(R_{ijkl})_\alpha$ in conformable FLRW metric.

$R_{1212} = -\frac{\alpha^2\,(q^2)^{(2\,\alpha-2)}}{4(-1+k\,(q^2)^2)\,q^1\,R(q^1)} \times \left[-2\,\frac{d^2R(q^1)}{(dq^1)^2}\,q^1\,R(q^1) + 2\,\frac{dR(q^1)}{dq^1}\,R(q^1)(\alpha-1) + \left(\frac{d\,R(q^1)}{dq^1}\right)^2 q^1\right]$	$R_{1313} = \frac{\alpha^2\,(q^2)^2\,(q^3)^{(2\,\alpha-2)}}{4q^1\,R(q^1)} \times \left[-2\,\frac{d^2R(q^1)}{(dq^1)^2}\,q^1\,R(q^1) + 2\,\frac{dR(q^1)}{dq^1}\,R(q^1)(\alpha-1) + \left(\frac{d\,R(q^1)}{dq^1}\right)^2 q^1\right]$	$R_{1414} = \frac{\alpha^2\,(q^2)^2\,(q^4)^{(2\,\alpha-2)}\,\sin(q^3)^2}{4q^1R(q^1)} \times \left[-2\,\frac{d^2R(q^1)}{(dq^1)^2}\,q^1\,R(q^1) + 2\,\frac{dR(q^1)}{dq^1}\,R(q^1)(\alpha-1) + \left(\frac{dR(q^1)}{dq^1}\right)^2 q^1\right]$
$R_{2323} = \frac{\alpha^2\,(q^3)^{(2\,\alpha-2)}}{4\left((q^1)^{(2\,\alpha-2)}\,(-1+k\,(q^2)^2)\right)} \times \left[-\left(\frac{dR(q^1)}{dq^1}\right)^2 (q^2)^2\,(q^2)^{(2\,\alpha-2)} + 4\alpha\,R(q^1)\,(q^1)^{(2\,\alpha(-2))}\,(-1+k(q^2)^2) + 4\,R(q^1)\,(q^1)^{(2\,\alpha-2)}(1-2k\,(q^2)^2)\right]$	$R_{2424} = \frac{\alpha^2\,(q^4)^{(2\,\alpha-2)}\,\sin(q^3)^2}{4((q^1)^{(2\,\alpha-2)}\,(-1+k\,(q^2)^2))} \times \left[-\left(\frac{d}{dq^1}\,R(q^1)\right)^2 (q^2)^2\,(q^2)^{(2\,\alpha-2)} + 4\alpha\,R(q^1)\,(q^1)^{(2\,\alpha-2)}\,(-1+k(q^2)^2) + 4\,R(q^1)\,(q^1)^{(2\,\alpha-2)}(1-2k\,(q^2)^2)\right]$	$R_{3434} = \frac{\alpha^2\,(q^2)^2\,(q^4)^{(2\,\alpha-2)}\,\sin^2(q^3)}{4(q^1)^{(2\,\alpha-2)}\,(q^2)^{(2\,\alpha-2)}\,q^3} \times \left[4\,R(q^1)\,(q^1)^{(2\,\alpha-2)}\left((q^2)^{(2\,\alpha-2)} - (q^3)^{(2\,\alpha-2)}\right)q^3 + \left(\frac{dR(q^1)}{dq^1}\right)^2 (q^2)^2\,(q^3)^{(2\,\alpha-2)}\,(q^2)^{(2\,\alpha-2)}\,q^3 + 4\,R(q^1)(q^1)^{(2\,\alpha-2)}(q^3)^{(2\,\alpha-2)}\,q^3\,k\,(q^2)^2 + 4\,(\alpha-1)\,R(q^1)\,(q^1)^{(2\,\alpha-2)}\,(q^2)^{(2\,\alpha-2)}\,\cot(q^3)\right]$

Table A8. Components of the Ricci tensor $(R_{ii})_\alpha$ in conformable FLRW metric.

$R_{11} = -\frac{3}{4R(q^1)^2\, q^1}\Big[-2\,\frac{d^2R(q^1)}{(dq^1)^2}\, q^1\, R(q^1) + 2\,\frac{dR(q^1)}{dq^1}\, R(q^1)\,(\alpha-1) + \left(\frac{dR(q^1)}{dq^1}\right)^2 q^1\Big]$	$R_{22} = -\frac{1}{4(q^1)^{(2\alpha-2)}\,(-1+k\,(q^2)^2)\, q^1\, R(q^1)\,(q^2)^2}\times \Big[-2\,(q^2)^{(2\alpha-2)}\,(q^2)^2\,\frac{d^2R(q^1)}{(dq^1)^2}\, q^1\, R(q^1) + 2\,(q^2)^{(2\alpha-2)}\,(q^2)^2\,\frac{dR(q^1)}{dq^1}\, R(q^1)\,(\alpha-1) - (q^2)^{(2\alpha-2)}\,(q^2)^2\left(\frac{dR(q^1)}{dq^1}\right)^2 q^1 + 8\, q^1\, R(q^1)\,(q^1)^{(2\alpha-2)}\,\alpha(-1+k\,(q^2)^2) + 8\, q^1\, R(q^1)\,(q^1)^{(2\alpha-2)}(1-2k\,(q^2)^2)\Big]$
$R_{33} = \frac{1}{4\, q^1\, R(q^1)\,(q^1)^{(2\alpha-2)}\,(q^2)^{(2\alpha-2)}\, q^3}\times \Big[4\, q^1\, R(q^1)\,(q^1)^{(2\alpha-2)}\,(q^2)^{(2\alpha-2)}\Big(\cot(q^3)(1-\alpha)-q^3\Big) + 4\, q^1\, R(q^1)\,(q^1)^{(2\alpha-2)}\, q^3 (q^3)^{(2\alpha-2)}\times \Big(2-3k\,(q^2)^2+\alpha(-1+k\,(q^2)^2)\Big) - 2\,(q^2)^2\,(q^3)^{(2\alpha-2)}\,(q^2)^{(2\alpha-2)}\, q^3\,\frac{d^2R(q^1)}{(dq^1)^2}\, q^1\, R(q^1) + \frac{dR(q^1)}{dq^1}\,(q^2)^2\,(q^3)^{(2\alpha-2)}\,(q^2)^{(2\alpha-2)}\, q^3\times \Big(-q^1\,\frac{dR(q^1)}{dq^1}+2(\alpha-1)R(q^1)\Big)\Big]$	$R_{44} = -\frac{(q^4)^{(2\alpha-2)}\sin^2(q^3)}{4q^1\, R(q^1)\,(q^1)^{(2\alpha-2)}(q^2)^{(2\alpha-2)}\,(q^3)^{(2\alpha-2)}\, q^3}\times \Big[2\,(q^2)^2\,(q^3)^{(2\alpha-2)}\,(q^2)^{(2\alpha-2)}\, q^3\, R(q^1)\times \left(\frac{d^2(R(q^1))}{d(q^1)^2}\, q^1-(\alpha-1)\frac{d(R(q^1))}{dq^1}\right) + 4\,(q^3)^{(2\alpha-2)}\, q^1\, q^3\, R(q^1)\,(q^1)^{(2\alpha-2)}\Big(\alpha(1-k\,(q^2)^2)-2+3k\,(q^2)^2\Big) + 4\, q^1\, R(q^1)\,(q^1)^{(2\alpha-2)}\,(q^2)^{(2\alpha-2)}\Big(q^3+(\alpha-1)\cot(q^3)\Big) + q^1\left(\frac{dR(q^1)}{dq^1}\right)^2 (q^2)^2\,(q^3)^{(2\alpha-2)}\,(q^2)^{(2\alpha-2)}\, q^3\Big]$

Table A9. Ricci scalar **R** in conformable FLRW metric.

$$\mathbf{R} = \frac{1}{(R(q^1)\,(q^2)^2\alpha^2\,q^1\,(q^1)^{(2\alpha-2)}\,(q^2)^{(2\alpha-2)}\,(q^3)^{(2\alpha-2)}\,q^3)}\Big[2\,(1-\alpha)q^1\,\cot(q^3)\,(q^1)^{(2\alpha-2)}\,(q^2)^{(2\alpha-2)}$$
$$+\,3(\alpha-1)\,(q^2)^2\,(q^3)^{(2\alpha-2)}\,(q^2)^{(2\alpha-2)}\,q^3\,\frac{dR(q^1)}{dq^1}$$
$$+\,4\alpha\,(q^3)^{(2\alpha-2)}\,q^1\,q^3\,(q^1)^{(2\alpha-2)}\,(-1+k\,(q^2)^2)+2\,q^1\,(q^1)^{(2\alpha-2)}\,q^3\Big(3(q^3)^{(2\alpha-2)}-(q^2)^{(2\alpha-2)}\Big)$$
$$-\,3\,(q^2)^2\,(q^3)^{(2\alpha-2)}\,(q^2)^{(2\alpha-2)}\,q^3\,\frac{d^2R(q^1)}{(dq^1)^2}\,q^1-10\,q^1\,(q^3)^{(2\alpha-2)}\,(q^1)^{(2\alpha-2)}\,q^3\,k\,(q^2)^2,\Big]$$

Table A10. Components of the Einstein tensor $(G_{ij})_\alpha$ in conformable FLRW metric.

$G_{11} = -\frac{1}{4(R(q^1)^2\,(q^2)^2\,(q^2)^{(2\alpha-2)}\,(q^3)^{(2\alpha-2)}\,q^3)} \times \Big[3\left(\frac{dR(q^1)}{dq^1}\right)^2 (q^2)^2\,(q^3)^{(2\alpha-2)}\,(q^2)^{(2\alpha-2)}\,q^3 + 4(\alpha-1)\,R(q^1)\,\cot(q^3)\,(q^1)^{(2\alpha-2)}\,(q^2)^{(2\alpha-2)} + 4\,R(q^1)\,(q^1)^{(2\alpha-2)}\,(q^2)^{(2\alpha-2)}\,q^3 + 4\,R(q^1)\,(q^3)^{(2\alpha-2)}\,(q^1)^{(2\alpha-2)}\,q^3\Big((-3+5k\,(q^2)^2) - 2\alpha(-1+k\,(q^2)^2)\Big)\Big]$	$G_{22} = \frac{1}{4((q^1)^{(2\alpha-2)}\,(-1+k\,(q^2)^2)q^1\,R(q^1)\,(q^2)^2\,(q^3)^{(2\alpha-2)}\,q^3)} \Big[-4\,(q^2)^2\,(q^3)^{(2\alpha-2)}\,(q^2)^{(2\alpha-2)}\,q^3\,\frac{d^2R(q^1)}{(dq^1)^2}\,q^1\,R(q^1) + (q^2)^2\,(q^3)^{(2\alpha-2)}\,(q^2)^{(2\alpha-2)}\,q^3\,\frac{dR(q^1)}{dq^1} \times \left(4(\alpha-1)\,R(q^1) + q^1\frac{dR(q^1)}{dq^1}\right) + 4\,q^1\,R(q^1)(q^1)^{(2\alpha-2)}\Big(-(q^3)^{(2\alpha-2)}\,q^3(-1+k\,(q^2)^2) + (q^2)^{(2\alpha-2)}(-q^3+(1-\alpha)\cot(q^3))\Big)\Big]$
$G_{33} = -\frac{(q^3)^{(2\alpha-2)}}{4q^1\,R(q^1)\,(q^1)^{(2\alpha-2)}\,(q^2)^{(2\alpha-2)}} \times \Big[4(1-\alpha)\,q^1\,R(q^1)\,(q^1)^{(2\alpha-2)} + 4(\alpha-1)\,(q^2)^{(2\alpha-2)}\,(q^2)^2\,\frac{dR(q^1)}{dq^1}\,R(q^1) - 4\,(q^2)^{(2\alpha-2)}\,(q^2)^2\,\frac{d^2R(q^1)}{(dq^1)^2}\,q^1\,R(q^1) - 4(2-\alpha)\,q^1\,R(q^1)\,(q^1)^{(2\alpha-2)}\,k\,(q^2)^2 + (q^2)^{(2\alpha-2)}\,(q^2)^2\left(\frac{dR(q^1)}{dq^1}\right)^2 q^1\Big]$	$G_{44} = \frac{(q^4)^{(2\alpha-2)}\,\sin^2(q^3)}{4q^1\,R(q^1)\,(q^1)^{(2\alpha-2)}\,(q^2)^{(2\alpha-2)}} \times \Big[4(\alpha-1)\,q^1\,R(q^1)\,(q^1)^{(2\alpha-2)} + 4\,(q^2)^{(2\alpha-2)}\,(q^2)^2\,\frac{d^2R(q^1)}{(dq^1)^2}q^1\,R(q^1) - 4(\alpha-1)\,(q^2)^{(2\alpha-2)}\,(q^2)^2\,\frac{dR(q^1)}{dq^1}\,R(q^1) + 4\,q^1\,R(q^1)\,(q^1)^{(2\alpha-2)}\,k\,(q^2)^2(2-\alpha) - q^1\left(\frac{dR(q^1)}{dq^1}\right)^2 (q^2)^2\,(q^2)^{(2\alpha-2)}\Big]$

References

1. Valéro, D.; Machado, J.; Kiryakova, V. Some pioneers of the applications of fractional calculus. *Fract. Calc. Appl. Anal.* **2014**, *17*, 552–578. [CrossRef]
2. Agrawal, O. Formulation of Euler-Lagrange equations for fractional variational problems. *J. Math. Anal. Appl.* **2002**, *272*, 368–379. [CrossRef]
3. Almeida, R.; Torres, D. Calculus of variations with fractional derivatives and fractional integrals. *Appl. Math. Lett.* **2009**, *22*, 1816–1820. [CrossRef]
4. Baleanu, D. Fractional variational principles in action. *Phys. Scr.* **2009**, *T136*, 014006. [CrossRef]
5. Chung, W.S. Fractional Newton mechanics with conformable fractional derivative. *J. Comput. Appl. Math.* **2015**, *290*, 150–158. [CrossRef]
6. Efe, M. Battery power loss compensated fractional order sliding mode control of a quadrotor UAV. *Asian J. Control* **2012**, *14*, 413–425. [CrossRef]
7. Herrmann, R. Gauge invariance in fractional field theories. *Phys. Lett. A* **2008**, *372*, 5515. [CrossRef]
8. Iomin, A. Fractional-time quantum dynamics. *Phys. Rev. E* **2009**, *80*, 022103. [CrossRef]
9. Metzler, R.; Klafter, J. The restaurant at the end of the random walk: Recent developments in the description of anomalous transport by fractional dynamics. *J. Phys. A* **2004**, *37*, R161. [CrossRef]
10. Jahanshahi, S.; Babolian, E.; Torres, D.F.M.; Vahidi, A. Solving Abel integral equations of first kind via fractional calculus. *J. King Saud Univ. Sci.* **2015**, *27*, 161–167. [CrossRef]
11. Machado, J.T.; Kiryakova, V.; Mainardi, F. Recent history of fractional calculus. *Commun. Nonlinear Sci. Numer. Simul.* **2011**, *16*, 1140–1153. [CrossRef]
12. Tarasov, V.E. Lattice fractional calculus. *Appl. Math. Comput.* **2015**, *257*, 12–33. [CrossRef]
13. Khalil, R.; Al Horani, M.; Yousef, A.; Sababheh, M. A new definition of fractional derivative. *J. Comput. Appl. Math.* **2014**, *264*, 65–70. [CrossRef]
14. Chung, W.S.; Hassanabadi H. Dynamics of a Particle in a Viscoelastic Medium with Conformable Derivative. *Int. J. Theor. Phys.* **2017**, *56*, 851. [CrossRef]
15. Chung, W.S. ; Hassanabadi, H. Deformed classical mechanics with α-deformed translation symmetry and anomalous diffusion. *Mod. Phys. Lett.* **2019**, *B33*, 1950368. [CrossRef]
16. Kiskinov, H.; Petkova, M.; Zahariev, A. About the Cauchy Problem for Nonlinear System with Conformable Derivatives and Variable Delays. *AIP Conf. Proc.* **2019**, *2172*, 050006.
17. Khalil, R.; Al Horani, M.; Yousef, A.; Hammad, M.A. Geometric meaning of conformable derivative via fractional cords. *J. Math. Comput. Sci.* **2019**, *19*, 241–245. [CrossRef]
18. Chung, W.S.; Hounkonnou, M.N. Deformed special relativity based on α-deformed binary operations. *arXiv* **2020**, arXiv:2005.11155.
19. Liouville, R. Sur le mouvement d'un corps solide pesant suspendu par l'un de ses points. *Acta Math.* **1897**, *20*, 239–284. [CrossRef]
20. Poincaré, H. Sur les quadratures mécaniques. *Acta Math.* **1899**, *13*, 1. [CrossRef]
21. De Filippo, S.; Marmo, G.; Salerno, M.; Vilasi, G. A New Characterization of Completely Integrable Systems. *Nuovo Cimento B* **1984**, *83*, 97–112. [CrossRef]
22. Gelfand, I.M.; Dorfman, I.Y. The Schouten Bracket and Hamiltonian Operators. *Funct. Anal. Appl.* **1980**, *14*, 71–74. [CrossRef]
23. Magri, F. A simple model of the integrable Hamiltonian equation. *J. Math. Phys.* **1978**, *19*, 1156–1162. [CrossRef]
24. Vilasi, G. On the Hamiltonian Structures of the Korteweg-de Vries and Sine- Gordon Theories. *Phys. Lett. B* **1980**, *94*, 195–198. [CrossRef]
25. Lax, P.D. Integrals of nonlinear equations of evolution and solitary ways. *Commun. Pure Appl. Math.* **1968**, *21*, 467–490. [CrossRef]
26. Hounkonnou, M.N.; Landalidji, M.J.; Mitrović, M. Noncommutative Kepler Dynamics: Symmetry groups and bi-Hamiltonian structures. *Theor. Math. Phys.* **2021**, *207*, 751–769. [CrossRef]
27. Hounkonnou, M.N.; Landalidji, M.J. Hamiltonian dynamics for the Kepler problem in a deformed phase space. In *Trends in Mathematics, Proceedings of the XXXVII Workshop on Geometric Methods in Physics, Bialowieża, Poland, 1–7 July 2018*; Springer Nature Switzerland AG: Cham, Switzerland, 2019; pp. 34–48.
28. Takeuchi, T. A Construction of a Recursion Operator for Some Solutions of Einstein Field Equations. *Proc. Fifteenth Int. Conf. Geom. Integr. Quantization* **2014**, *15*, 249–258.
29. Hounkonnou, M.N.; Landalidji, M.J.; Baloïtcha, E. Recursion Operator in a Noncommutative Minkowski Phase Space. In *Trends in Mathematics, Proceedings of the XXXVI Workshop on Geometric Methods in Physics, Bialowieża, Poland, 2–8 July 2017*; Springer Nature Switzerland AG: Cham, Switzerland, 2019; pp. 83–93.
30. Rudolph, G.; Schmidt, M. *Differential Geometry and Mathematical Physics, Part I. Manifolds, Lie Group and Hamiltonian Systems*; Springer: New York, NY, USA, 2013.
31. Smirnov, R. G. Magri-Morosi-Gel'fand-Dorfman's bi-Hamiltonian constructions in the action-angle variables. *J. Math. Phys.* **1997**, *38*, 6444. [CrossRef]
32. Smirnov, R. G. The action-angle coordinates revisited: Bi-Hamiltonian systems. *Rep. Math. Phys.* **1999**, *44*, 199–204. [CrossRef]

33. Dubrovin, B. Bihamiltonian Structures of PDEs and Frobenius Manifolds, Lectures at the ICTP Summer School "Poisson Geometry", Trieste, 2005. Available online: https://indico.ictp.it/event/a04198/session/47/contribution/26/material/0/0.pdf (accessed on 21 April 2017).
34. Román-Roy, N. A summary on symmetries and conserved quantities of autonomous Hamiltonian systems. *J. Geom. Mech.* **2020**, *12*, 3. [CrossRef]
35. Bretón, N. An introduction to general relativity, black holesand gravitational waves, VIII Workshop of the Gravitation and Mathematical Physics Division of the Mexican Physical Society. *AIP Conf. Proc.* **2011**, *1396*, 5–25.
36. Rañada, M.F. A system of $n = 3$ coupled oscillators with magnetic terms: Symmetries and integrals of motion. *SIGMA* **2005**, *1*, 004. [CrossRef]
37. Caseiro, R. Master integrals, superintegrability and quadratic algebras. *Bull. Sci. Math.* **2002**, *126*, 617–630. [CrossRef]
38. Damianou, P.A. Symmetries of Toda equations. *J. Phys. A* **1993**, *26*, 3791–3796. [CrossRef]
39. Fernandes, R.L. On the master symmetries and bi-Hamiltonian structure of the Toda lattice. *J. Phys. A Math. Gen.* **1993**, *26*, 3797–3803. [CrossRef]
40. Rañada, M.F. Superintegrability of the Calogero-Moser system: Constants of motion, master symmetries, and time-dependent symmetries. *J. Math. Phys.* **1999**, *40*, 236–247. [CrossRef]
41. Hounkonnou, M.N.; Landalidji, M.J.; Mitrović, M. Hamiltonian Dynamics of a spaceship in Alcubierre and Gödel metrics: Recursion operators and underlying master symmetries. *Theor. Math. Phys.* **2022**; *in press*.
42. Oevel, W. A Geometrical Approach to Integrable Systems Admitting Time Dependent Invatiants. In Proceedings of the Conference on Nonlinear Evolution Equations, Solitons and the Inverse Scattering Transform, Oberwolfach, Germany, 27 July–2 August 1986; Ablowitz, M., Fuchssteiner, B., Kruskal, M., Eds.; World Scientific: Singapore, 1987 .

MDPI

Article

Superposition Principle and Kirchhoff's Integral Theorem †

Mikhail I. Krivoruchenko [1,2,3]

[1] National Research Centre "Kurchatov Institute"—KCTEP, B. Cheremushkinskaya 25, 117218 Moscow, Russia; mikhail.krivoruchenko@itep.ru
[2] Moscow Institute of Physics and Technology, 141700 Dolgoprudny, Russia
[3] Bogoliubov Laboratory of Theoretical Physics, Joint Institute for Nuclear Research, 141980 Dubna, Russia
† Extended notes of a lecture delivered to students of Moscow Institute of Physics and Technology.

Abstract: The need for modification of the Huygens–Fresnel superposition principle arises even in the description of the free fields of massive particles and, more extensively, in nonlinear field theories. A wide range of formulations and superposition schemes for secondary waves are captured by Kirchhoff's integral theorem. We discuss various versions of this theorem as well as its connection with the superposition principle and the method of Green's functions. A superposition scheme inherent in linear field theories, which is not based on Kirchhoff's integral theorem but instead relies on the completeness condition, is also discussed.

Keywords: superposition principle; asymptotic conditions; Kirchhoff's integral theorem

1. Introduction

An excellent and detailed explanation of Huygens' principle for undergraduate students, together with the optical-mechanical analogy and the Hamilton–Jacobi method, can be found in the monograph by Arnold [1]. Students are introduced to a generalization of Huygens' principle, viz. the Huygens–Fresnel superposition principle, in the study of general physics (see, e.g., [2]), and this principle is presented in greater detail in the study of theoretical physics (see, e.g., [3]). The method of Green's functions (GF), which has found numerous applications in a large variety of different fields, is discussed in the first volume of a two-volume monograph by Bjorken and Drell [4,5], where, in particular, the superposition principle is used in §§ 21 and 22 to derive the equation for the Green's function. Further development of concepts related to the superposition principle has led to the emergence in quantum theory of the path integral formalism, an excellent overview of which can be found in the monograph by Dittrich and Reuter [6]. A detailed presentation of the superposition principle for electromagnetic fields, its rationale and its generalizations, based on Kirchhoff's integral theory [7], is given in the monograph by Born and Wolf [8].

Thus, it is clear that the superposition principle is closely related to the GF method which, in turn, lies at the heart of quantum field theory and the diagram technique. In the literature, this relationship is typically mentioned only in passing, while the mathematical aspects, modifications, and physical meaning of the generalized schemes of superposition are treated as matters beyond dispute.

A rigorous formulation of the superposition principle is based on Kirchhoff's integral theorem. The generalizations to which it leads are used also in the theory of interacting fields. In this paper, we attempt to specify the precise place of the superposition principle in classical and quantum field theory and discuss its relationship with the GF method and Kirchhoff's integral theorem.

Surprisingly, the answers to the main questions can be obtained by analyzing the dynamics of the one-dimensional oscillator. The oscillator problem from the viewpoint of Kirchhoff's integral theorem, as well as its connections with the superposition principle and the GF method, is discussed in the next section. In Section 3, we consider a free massive scalar field. For massive fields, the superposition scheme includes an integral

Citation: Krivoruchenko, M.I. Superposition Principle and Kirchhoff's Integral Theorem. *Universe* **2022**, *8*, 315. https://doi.org/10.3390/universe8060315

Academic Editor: Steven Duplij

Received: 28 April 2022
Accepted: 23 May 2022
Published: 3 June 2022

over three-dimensional space. Both in the limit of zero mass and for monochromatic fields, the canonical superposition scheme, in which the summation of the sources of secondary waves is limited to a two-dimensional surface, arises. The statement of Kirchhoff's theorem depends on the asymptotic conditions imposed on the propagator at $t \to \pm\infty$. In quantum field theory, the Feynman asymptotic conditions are used. Emphasis is therefore placed on the versions of the theorem that satisfy the Feynman asymptotic conditions. In Section 4, we discuss a charged scalar field in an external electromagnetic field, prove the appropriate version of Kirchhoff's integral theorem, and demonstrate that in an external electromagnetic field, the superposition schemes are not fundamentally modified.

In nonlinear theories, the superposition principle holds in relation to the secondary waves. In Section 5, we consider a class of nonlinear scalar field theories. The physical meaning of Kirchhoff's integral theorem is discussed, including its connections with the GF method and the superposition principle. Vectorial generalizations of Kirchhoff's integral theorem for retarded Green's function are discussed in Appendix A. The conclusions section summarizes the discussion.

The material of this work is intended for students studying quantum field theory and researchers specializing in the theory of the propagation of electromagnetic waves and light phenomena.

2. The Huygens–Fresnel Superposition Principle and Kirchhoff's Integral Theorem in the Oscillator Problem

A free scalar field obeys the Klein–Gordon equation:

$$(\Box + m^2)\phi_0(x) = 0. \tag{1}$$

Of interest are the general features of solutions of the wave equation, which extend to its nonlinear modifications. The main consequences of Kirchhoff's theorem and the physical content of the Fresnel–Huygens superposition principle can be explained using the example of the one-dimensional oscillator; thus, we begin by considering the evolution of a one-dimensional harmonic oscillator. This problem can also be regarded as a problem of the evolution of a free scalar field in momentum space.

2.1. Harmonic Oscillator

We write the equation in the form

$$\left(\frac{d^2}{dt^2} + m^2\right)\phi_0(t) = 0. \tag{2}$$

Here, m is the frequency of the oscillator and $\phi_0(t)$ is its coordinate. If $\phi_0(t)$ is a spatially homogeneous field in the Klein–Gordon equation, then m is the mass of the particle.

2.1.1. Complete Orthonormal Basis Functions

A complete set of solutions to Equation (2) is formed by the two functions

$$f^{(+)}(t) = \frac{e^{-imt}}{\sqrt{2m}} \quad \text{and} \quad f^{(-)}(t) = \frac{e^{imt}}{\sqrt{2m}}. \tag{3}$$

The normalization and completeness conditions are expressed in terms of the Wronskian. If φ and χ are two functions, then their Wronskian is equal to

$$W[\varphi,\chi] = \det\begin{Vmatrix} \varphi & \chi \\ \dot{\varphi} & \dot{\chi} \end{Vmatrix} = \varphi\dot{\chi} - \dot{\varphi}\chi. \tag{4}$$

The notation

$$\varphi \overleftrightarrow{\partial_t} \chi = W[\varphi,\chi]$$

is often used. The normalization and orthogonality of the basis functions are represented as follows:

$$iW[f^{(\pm)*}, f^{(\pm)}] = \pm 1 \quad \text{and} \quad W[f^{(\pm)*}, f^{(\mp)}] = 0. \tag{5}$$

If the functions for which we compute the Wronskian are solutions of Equation (2), then the Wronskian is independent of time. Let $\phi_0(t)$ be a solution of Equation (2). We define the following time-independent complex numbers:

$$a = iW[f^{(+)*}, \phi_0] \quad \text{and} \quad a^* = -iW[f^{(-)*}, \phi_0]. \tag{6}$$

After quantization, the values a and a^* become annihilation and creation operators.

The completeness condition takes the form

$$\phi_0(t) = f^{(+)}(t) iW[f^{(+)*}, \phi_0] - f^{(-)}(t) iW[f^{(-)*}, \phi_0]. \tag{7}$$

This equation also allows for the decomposition of the solution into its positive- and negative-frequency components:

$$\phi_0(t) = \phi_0^{(+)}(t) + \phi_0^{(-)}(t), \tag{8}$$

where

$$\phi_0^{(\pm)}(t) = \pm f^{(\pm)}(t) iW[f^{(\pm)*}, \phi_0]. \tag{9}$$

Equation (7) is valid not only in the linear vector space spanned by the basis functions (3), but also for any function evaluated at time t. The right-hand side of Equation (7) for an arbitrary function $\chi(t)$ has the form

$$\text{r.h.s.} = i\Big(f^{(+)}(t)f^{(+)*}(t) - f^{(-)}(t)f^{(-)*}(t)\Big)\dot{\chi}(t) - i\Big(f^{(+)}(t)\dot{f}^{(+)*}(t) - f^{(-)}(t)\dot{f}^{(-)*}(t)\Big)\chi(t).$$

Using the explicit form of $f^{(\pm)}(t)$, one can see that r.h.s. $= \chi(t)$. Although this property appears fortuitous, it is rather fundamental.

Let us consider the Poisson bracket relations

$$\{\phi_0(t), \phi_0(t)\} = 0, \tag{10}$$
$$\{\phi_0(t), \pi_0(t)\} = 1, \tag{11}$$

where $\pi_0(t) = \dot{\phi}_0(t)$ is the canonical momentum. A simple calculation using Equation (7) gives

$$\begin{aligned} \{\phi_0(t'), \phi_0(t)\} &= f^{(+)}(t) i\{\phi_0(t'), W[f^{(+)*}, \phi_0]\} - f^{(-)}(t) i\{\phi_0(t'), W[f^{(-)*}, \phi_0]\} \\ &= f^{(+)}(t) i\{\phi_0(t'), f^{(+)*}(t')\pi_0(t') - \dot{f}^{(+)*}(t')\phi_0(t')\} \\ &- f^{(-)}(t) i\{\phi_0(t'), f^{(-)*}(t')\pi_0(t') - \dot{f}^{(-)*}(t')\phi_0(t')\} \\ &= i\Big(f^{(+)}(t)f^{(+)*}(t') - f^{(-)}(t)f^{(-)*}(t')\Big), \end{aligned} \tag{12}$$

$$\{\phi_0(t'), \pi_0(t)\} = i\Big(\dot{f}^{(+)}(t)f^{(+)*}(t') - \dot{f}^{(-)}(t)f^{(-)*}(t')\Big). \tag{13}$$

By virtue of Equations (10) and (11),

$$f^{(+)}(t)f^{(+)*}(t) - f^{(-)}(t)f^{(-)*}(t) = 0, \tag{14}$$
$$f^{(+)}(t)\dot{f}^{(+)*}(t) - f^{(-)}(t)\dot{f}^{(-)*}(t) = i, \tag{15}$$
$$\dot{f}^{(+)}(t)f^{(+)*}(t) - \dot{f}^{(-)}(t)f^{(-)*}(t) = -i. \tag{16}$$

Identity r.h.s. $= \chi(t)$ is, therefore, a consequence of the completeness condition (7) for functions $\phi_0(t)$, which are solutions of Equation (2), and the Poisson bracket relations for the canonical variables.

2.1.2. The Green's Functions

A Green's function is defined by the equation

$$\left(\frac{d^2}{dt^2} + m^2\right)\Delta_X(t) = -\delta(t). \tag{17}$$

By performing the Fourier transform in time, we obtain the Green's function in frequency space: $\Delta_X(\omega) = (\omega^2 - m^2)^{-1}$. For the inverse Fourier transformation,

$$\Delta_X(t) = \int_{-\infty}^{+\infty} \frac{d\omega}{2\pi} e^{-i\omega t} \frac{1}{\omega^2 - m^2}, \tag{18}$$

it is necessary to bypass the poles on the real axis that arise for $\omega = \pm m$. There are four possibilities, which correspond to four Green's functions:

$$\begin{aligned}
\Delta_F(t'-t) &= \int_{-\infty}^{+\infty} \frac{d\omega}{2\pi} e^{-i\omega(t'-t)} \frac{1}{\omega^2 - m^2 + i0} \\
&= -i\Big(f^{(+)}(t')f^{(+)*}(t)\theta(t'-t) + f^{(-)}(t')f^{(-)*}(t)\theta(-t'+t)\Big), \qquad (19) \\
\Delta_F^c(t'-t) &= \int_{-\infty}^{+\infty} \frac{d\omega}{2\pi} e^{-i\omega(t'-t)} \frac{1}{\omega^2 - m^2 - i0} \\
&= i\Big(f^{(-)}(t')f^{(-)*}(t)\theta(t'-t) + f^{(+)}(t')f^{(+)*}(t)\theta(-t'+t)\Big), \qquad (20) \\
\Delta_{\rm ret}(t'-t) &= \int_{-\infty}^{+\infty} \frac{d\omega}{2\pi} e^{-i\omega(t'-t)} \frac{1}{\omega^2 - m^2 + i0\,{\rm sgn}(\omega)} \\
&= -i\Big(f^{(+)}(t')f^{(+)*}(t) - f^{(-)}(t')f^{(-)*}(t)\Big)\theta(t'-t), \qquad (21) \\
\Delta_{\rm adv}(t'-t) &= \int_{-\infty}^{+\infty} \frac{d\omega}{2\pi} e^{-i\omega(t'-t)} \frac{1}{\omega^2 - m^2 - i0\,{\rm sgn}(\omega)} \\
&= i\Big(f^{(+)}(t')f^{(+)*}(t) - f^{(-)}(t')f^{(-)*}(t)\Big)\theta(-t'+t). \qquad (22)
\end{aligned}$$

Each of these functions satisfies Equation (17). The difference between any two Green's functions is a solution of the free Equation (2).

It is instructive to verify by the direct calculation that the representation (19) satisfies Equation (17). With the help of equation

$$f(x)\delta'(x) = f(0)\delta'(x) - f'(0)\delta(x),$$

one finds

$$\begin{aligned}
\left(\frac{d^2}{dt'^2} + m^2\right) i\Delta_F(t'-t) &= 2\Big(\dot f^{(+)}(t')f^{(+)*}(t) - \dot f^{(-)}(t')f^{(-)*}(t)\Big)\delta(t'-t) \\
&+ \Big(f^{(+)}(t')f^{(+)*}(t) - f^{(-)}(t')f^{(-)*}(t)\Big)\delta'(t'-t) \\
&= \Big(\dot f^{(+)}(t')f^{(+)*}(t) - \dot f^{(-)}(t')f^{(-)*}(t)\Big)\delta(t'-t) \\
&+ \Big(f^{(+)}(t)f^{(+)*}(t) - f^{(-)}(t)f^{(-)*}(t)\Big)\delta'(t'-t). \qquad (23)
\end{aligned}$$

Using Equations (14) and (16), we arrive at Equation (17).

In terms of quantized variables, the Feynman propagator is defined by

$$i\Delta_F(t'-t) = \langle 0|T\hat\phi_0(t')\hat\phi_0(t)|0\rangle. \tag{24}$$

The T product entering this expression occurs naturally in solutions of the evolution equation $i\partial_t\Psi(t) = \hat H(t)\Psi(t)$ of systems with a time-dependent Hamiltonian. If, at various times, $\hat H$ does not commute with itself, namely, $[\hat H(t'), \hat H(t)] \neq 0$, then the solution $\Psi(t) = U(t,0)\Psi(0)$

is expressed in terms of the time-ordered exponential $U(t,0) = T\exp(-i\int_0^t \hat{H}(t')dt')$. In perturbation theory, $\Delta_F(t'-t)$ then arises by Wick's theorem, which explains why $\Delta_F(t'-t)$ plays a special role in quantum theory. The definition (24) is consistent with the definition (19).

2.1.3. Superposition Principle from Kirchhoff's Integral Theorem

Let us compute the Wronskian of the Feynman propagator $\Delta_F(t'-t)$ and a solution $\phi_0(t)$ of Equation (2). By taking the derivative with respect to t of $W[\Delta_F(t'-t),\phi_0(t)]$ and integrating the result over the interval (t_1,t_2), the following equation is obtained for $t_1 < t' < t_2$:

$$\phi_0(t') = W[\Delta_F(t'-t_2),\phi_0(t_2)] - W[\Delta_F(t'-t_1),\phi_0(t_1)]. \tag{25}$$

This relation is the harmonic oscillator analog of Kirchhoff's integral theorem. Despite the drastic simplification, the fundamental meaning is maintained and is amenable to interpretation. According to Equation (25), the coordinate $\phi_0(t)$ is determined by both the past and the future. From the past, the Wronskian selects the positive-frequency component of $\phi_0(t_1)$ and propagates it into the future up to the moment $t = t' > t_1$. From the future, the Wronskian selects the negative-frequency component of $\phi_0(t_2)$ and propagates it into the past up to the moment $t = t' < t_2$. The result is a superposition of the two *waves*. Equation (2) is commonly regarded as the equation of motion of a particle (oscillator) in the one-dimensional space. A less obvious interpretation of this equation as an evolution equation of a wave in the zero-dimensional space is also possible. Equation (25) underlines the second interpretation.

The analogy with quantum field theory is apparent: particles are identified with positive-frequency solutions of wave equations, and antiparticles are identified with negative-frequency solutions. Particles move *forward in time*, whereas antiparticles move *backward in time*. In accordance with the Huygens–Fresnel superposition principle adapted here for the Feynman asymptotic conditions, the wave $\phi_0(t')$ is equal to the sum of the negative-frequency component of $\phi_0(t_2)$, propagating backward in time, and the positive-frequency component of $\phi_0(t_1)$, propagating forward in time. Equation (25) can thus be interpreted both in the spirit of the Huygens–Fresnel superposition principle and in the spirit of the GF method, thereby establishing the close relationship between them.

According to Equation (25), the coordinate $\phi_0(t')$ is determined by its value and its first derivative at the other two time points. Arguing reversely, this suggests that the evolution equation contains time derivatives of no higher than second order.

If $t' \notin (t_1,t_2)$, then there is a zero on the left-hand side of Equation (25):

$$0 = W[\Delta_F(t'-t_2),\phi_0(t_2)] - W[\Delta_F(t'-t_1),\phi_0(t_1)]. \tag{26}$$

Equations (25) and (26) remain valid after the replacement Δ_F with any other propagator. For the retarded Green's function, the analog of Equations (25) and (26) for $t_2 \to +\infty$ reads

$$\phi_0(t')\theta(t'-t_1) = -W[\Delta_{\mathrm{ret}}(t'-t_1),\phi_0(t_1)]. \tag{27}$$

Here, the positive- and negative-frequency components propagate forward in time, corresponding to the usual formulation of the Huygens–Fresnel superposition principle, so that $\phi_0(t)$ is determined by the past only.

2.1.4. Superposition Principle from the Completeness Condition

Here, we present a different formulation of the superposition principle. To begin, let us find the Wronskian W of $\Delta_F(t'-t)$ and $\phi_0(t)$. The expression (19), when substituted into W, yields

$$\begin{aligned}
W[\Delta_F(t'-t),\phi_0(t)] &= -if^{(+)}(t')W[f^{(+)*}(t)\theta(t'-t),\phi_0(t)] \\
&\quad -if^{(-)}(t')W[f^{(-)*}(t)\theta(t-t'),\phi_0(t)] \\
&= -if^{(+)}(t')\theta(t'-t)W[f^{(+)*}(t),\phi_0(t)] \\
&\quad -if^{(-)}(t')\theta(t-t')W[f^{(-)*}(t),\phi_0(t)] \\
&\quad +\Delta(t'-t)\phi_0(t)\delta(t'-t),
\end{aligned} \tag{28}$$

where

$$i\Delta(t'-t) = f^{(+)}(t')f^{(+)*}(t) - f^{(-)}(t')f^{(-)*}(t). \tag{29}$$

By virtue of Equation (12),

$$\Delta(t'-t) = \{\phi_0(t'),\phi_0(t)\}.$$

In the transition to the last lines of Equation (28), the properties of the Wronskian and the definitions of the basis functions (3) are used. According to Equation (14), the term $\sim \Delta(t'-t)\delta(t'-t)$ vanishes, yielding

$$\phi_0^{(+)}(t')\theta(t'-t) - \phi_0^{(-)}(t')\theta(t-t') = -W[\Delta_F(t'-t),\phi_0(t)]. \tag{30}$$

Equation (30) can be regarded as an equation for $\Delta_F(t'-t)$. By taking the time (t) derivative of both sides, we obtain Equation (17). The superposition principle, formalized as in (30), thus determines the Green's function up to a solution of the free equation. To obtain a unique Green's function, the asymptotic behavior must be fixed. By taking the differences between both sides of Equation (30) for $t = t_2$ and $t = t_1 < t_2$, we obtain Equation (25), provided that $t' \in (t_1,t_2)$. If the inverse condition, $t' \notin (t_1,t_2)$, holds, then we obtain Equation (26). Finally, by taking the time (t') derivative, we obtain the superposition principle for the canonical momentum $\pi_0(t) = \dot{\phi}_0(t)$:

$$\pi_0^{(+)}(t')\theta(t'-t) - \pi_0^{(-)}(t')\theta(t-t') = -W[\Delta_F(t'-t),\pi_0(t)]. \tag{31}$$

The proof of Equation (30) is not based on Kirchhoff's theorem, nor its obvious modification. For the retarded Green's function, the completeness condition does not lead to a new equation (compared with (27)). In quantum field theory, the diagram technique is based on the Feynman propagator; thus, what is of interest to us here is the superposition principle formalized as in (25), (26) and (30).

2.1.5. Path Integral

Kirchhoff's integral theorem can also be used as a starting point for developing path integral method.

To show this, we note a useful relation

$$\begin{aligned}
iW[\Delta_F(t_3-t_2),\Delta_F(t_2-t_1)] &= -\theta(t_3-t_2)\theta(t_2-t_1)f^{(+)}(t_3)f^{(+)*}(t_1) \\
&\quad +\theta(t_1-t_2)\theta(t_2-t_3)f^{(-)}(t_3)f^{(-)*}(t_1).
\end{aligned} \tag{32}$$

This relation indicates that a wave propagating toward the future continues to propagate forward in time. A similar property holds for waves propagating backward in time. We choose a sequence of the intervals $(t_1,t_2) \subset (t_3,t_4) \subset \ldots \subset (t_{2n-1},t_{2n})$ and consider $t' \in (t_1,t_2)$. Equation (25) being iterated n times gives

$$\phi_0(t') = W[\Delta_F(t'-t_2), W[\Delta_F(t_2-t_4), W[\ldots, W[\Delta_F(t_{2n}-t_{2n+2}), \phi_0(t_{2n+2})]\ldots]]]$$
$$+ \quad (-)^{n+1} W[\Delta_F(t'-t_1), W[\Delta_F(t_1-t_3), W[\ldots, W[\Delta_F(t_{2n-1}-t_{2n+1}), \phi_0(t_{2n+1})]\ldots]]]. \tag{33}$$

According to this equation, $\phi_0(t_{2n+2})$ generates a secondary wave that propagates into the past. In the neighboring instant of time $t = t_{2n} < t_{2n+2}$, it generates new secondary wave, and so on. The same interpretation is valid for the wave propagating forward in time. Equation (25) is reproduced with $n = 0$ for $t_{-1} = t_0 = t'$. The mixed terms containing forward and backward propagation do not arise as a consequence of (32). In the limit of $n \to \infty$, $t_2 - t_1 \to 0$ and $(t_{l+3} - t_{l+2}) \to (t_{l+1} - t_l)$, we arrive at the continuous product *over history*. Equation (33) can be regarded as a path-integral representation in the space $\mathbb{R}^{1,0}$.

Path integral in the space $\mathbb{R}^{1,3}$ is discussed in Section 3.5.

2.2. Harmonic Oscillator with a Time-Dependent Frequency

A field theoretical version of the evolution problem with a time-dependent oscillator frequency, in light of the superposition principle, is discussed in Section 4, where proofs are presented. Here, we restrict ourselves to statements of the main assertions.

We consider the equation

$$\left(\frac{d^2}{dt^2} + m^2 + \Delta m^2(t)\right)\phi(t) = 0, \tag{34}$$

where $\Delta m^2(\pm\infty) = 0$. The perturbation $\Delta m^2(t)$ is switched on and off adiabatically. Let $\Delta_F(t', t)$ be the Feynman propagator for Equation (34). The following superposition schemes hold: As a consequence of Kirchhoff's integral theorem,

$$\begin{aligned} \phi(t') &= W[\Delta_F(t', t_2), \phi(t_2)] - W[\Delta_F(t', t_1), \phi(t_1)] \quad \text{for} \quad t' \in (t_1, t_2), \\ 0 &= W[\Delta_F(t', t_2), \phi(t_2)] - W[\Delta_F(t', t_1), \phi(t_1)] \quad \text{for} \quad t' \notin (t_1, t_2), \end{aligned}$$

and, as a consequence of the completeness condition,

$$\phi^{(+)}(t')\theta(t'-t) - \phi^{(-)}(t')\theta(t-t') = -W[\Delta_F(t', t), \phi(t)],$$

where $\phi^{(\pm)}(t) \sim f^{(\pm)}(t)$ at $t \to \pm\infty$. The expansion of $\phi(t)$ into positive- and negative-frequency components $\phi^{(\pm)}(t)$ has an objective meaning because the evolution equation is linear.

2.3. Anharmonic Oscillator

In nonlinear theories, the superposition principle requires reformulation. Its generalization, based on Kirchhoff's integral theorem, preserves the idea in relation to secondary waves. The main technical points can be illustrated by the example of anharmonic oscillator.

We add to the oscillator potential an arbitrary potential $V(\phi)$. The equation of motion takes the form

$$\left(\frac{d^2}{dt^2} + m^2\right)\phi(t) = -V'(\phi(t)). \tag{35}$$

2.3.1. Secondary Waves beyond Fresnel's Superposition Scheme

Equation (25) is modified as follows:

$$\phi(t') = W[\Delta_F(t'-t_2), \phi(t_2)] - W[\Delta_F(t'-t_1), \phi(t_1)] + \int_{t_1}^{t_2} dt \Delta_F(t'-t) V'(\phi(t)). \tag{36}$$

The propagator $\Delta_F(t)$ is determined from Equation (17). On the interval (t_1, t_2), the sum of the first two terms satisfies the evolution equation of the harmonic oscillator. We denote this sum as

$$\phi_0(t') \equiv W[\Delta_F(t'-t_2), \phi(t_2)] - W[\Delta_F(t'-t_1), \phi(t_1)]. \quad (37)$$

The solution takes the form

$$\phi(t') = \phi_0(t') + \int_{t_1}^{t_2} dt \Delta_F(t'-t) V'(\phi(t)). \quad (38)$$

Given that the Green's function properties of the harmonic oscillator are known, the solution can be written immediately. If $t' \notin (t_1, t_2)$, then we obtain

$$0 = \phi_0(t') + \int_{t_1}^{t_2} dt \Delta_F(t'-t) V'(\phi(t)). \quad (39)$$

The last two equations constitute a version of Kirchhoff's integral theorem for the one-dimensional anharmonic oscillator.

Equation (38) cannot be interpreted canonically. Although the first term has the standard meaning under the Fresnel superposition scheme, the second term indicates that a component arises among the secondary waves that is generated continuously in time.

According to the Huygens–Fresnel superposition principle, to describe the propagation of a wave, it is sufficient to know its phase and amplitude at a fixed time. However, this is true only in linear theories. In nonlinear theories, the propagation of a wave is determined by its entire history (for retarded solutions, its prehistory), even if the original wave equation is local. The dependence of the wave observables on the entire history of the wave indicates, in general, the nonlocal nature of its evolution. Only a narrow family of representations that contain an integral over time correspond to local but nonlinear theories.

The derivative of the potential is an additional source of secondary waves (corrections to the coordinate), and the potential depends on the exact coordinate. This means that Equation (38) is self-consistent and that its solution is obvious only in the context of perturbation theory.

In quantum field theory, an equation similar to Equation (38) serves as the starting point for the development of the diagram technique (see, e.g., [4]). The equations obtained by replacing the Feynman propagator in Equation (38) with the retarded and advanced propagators are used to develop the axiomatic scattering theory (see, e.g., [5]).

2.3.2. Positive- and Negative-Frequency Solutions

In the theory of interacting fields, the decomposition of solutions into positive- and negative-frequency components makes sense only asymptotically for outgoing and incoming states. We assume that the nonlinear interaction is adiabatically switched on at $t \to -\infty$ and adiabatically switched off at $t \to +\infty$. If positive- and negative-frequency components $\phi^{(\pm)}(t)$ are somehow defined, then the subsequent modification of Equation (30) is obvious:

$$\phi^{(+)}(t')\theta(t'-t) - \phi^{(-)}(t')\theta(t-t') = -W[\Delta_F(t'-t), \phi(t)] + \int_t^{t'} d\tau \Delta_F(t'-\tau) V'(\phi(\tau)). \quad (40)$$

By taking the time (t) derivative, after some simple transformations, we obtain $\phi(t) = \phi^{(+)}(t) + \phi^{(-)}(t)$ and Equation (17). The difference in this equation at two unequal time points leads to Equations (36)–(39). It might seem, therefore, that Equation (40) is no less general than Equation (36)–(39). However, we do not have an independent definition of the decomposition into positive- and negative-frequency components. We are forced, therefore, to regard Equation (40) as a definition of $\phi^{(\pm)}(t)$. According to this equation, $\phi^{(\pm)}(t) \sim f^{(\pm)}(t)$ at $t \to \pm\infty$.

The calculation of the first derivative of Equation (40) in t' leads to the superposition principle for the canonical momentum

$$\pi^{(+)}(t')\theta(t'-t) - \pi^{(-)}(t')\theta(t-t') = -W[\Delta_F(t'-t), \pi(t)] + \int_t^{t'} d\tau \Delta_F(t'-\tau) V''(\phi(\tau))\pi(\tau). \quad (41)$$

This equation is consistent with the evolution equation for $\pi^{(\pm)}(t) = \dot{\phi}^{(\pm)}(t)$.

Obviously, in nonlinear theories, a full generalization of (30) does not exist.

A field theoretical version of the anharmonic oscillator problem is discussed in Section 5.

2.3.3. Numerical Example

We use a numerical example to demonstrate the application of the superposition scheme (38) for the description of radial motion in the Keplerian problem. After separation of the angular variables, the evolution problem reduces to solving a problem of one-dimensional motion in an effective potential

$$U = -\frac{\alpha}{r} + \frac{L^2}{2\mu r^2},$$

where $\alpha = GM_{\odot}\mu$, $M_{\odot}$ is the solar mass, μ is the mass of a celestial body, and L is the angular momentum. We add and subtract from the potential U an oscillator potential

$$U_{osc} = \frac{1}{2}\mu m^2 (r-a)^2$$

and treat U_{osc} as the undisturbed potential. The perturbation potential is thus $V = U - U_{osc}$. In order to improve convergence and eliminate the need to determine optimized U_{osc}, the frequency parameter m is chosen in agreement with the exact solution (see, e.g., [1]): $m = 2\pi/T$, where $T = 2\pi\mu ab/L$ is the orbital period, $a = (r_{\min} + r_{\max})/2$ and $b = \sqrt{pa}$ are the major and minor semi-axes of the ellipse and $L = \sqrt{p\alpha\mu}$; the variable r lies in the interval $(r_{\min}, r_{\max})$, where $r_{\min} = p/(1+e)$, $r_{\max} = p/(1-e)$, p is the semi-latus rectum, and e is the eccentricity.

As a zeroth-order approximation for $\phi(t) \equiv r(t) - a$, we choose a free solution

$$\phi^{[0]}(t) = C_+^{[0]} f^{(+)}(t) + C_-^{[0]} f^{(-)}(t) \quad (42)$$

with unknown coefficients $C_\pm^{[0]}$ and $f^{(\pm)}(t)$ defined by Equation (3). The motion begins at perihelion $\phi^{[0]}(0) = r_{\min} - a$, with the vanishing velocity $\dot{\phi}^{[0]}(0) = 0$. These conditions allow $C_\pm^{[0]}$ to be fixed.

Given the lth-order approximation, $r^{[l]}(t) = a + \phi^{[l]}(t)$ can be substituted in place of the argument of V' in Equation (38) to produce the next-order iteration

$$\phi^{[l+1]}(t) = C_+^{[l+1]} f^{(+)}(t) + C_-^{[l+1]} f^{(-)}(t) + \int_{t_1}^{t_2} d\tau \Delta_F(t-\tau) V'(a + \phi^{[l]}(\tau)), \quad (43)$$

where $\Delta_F(t)$ is defined by (19). The interval (t_1, t_2) covers an interval within which we seek the solution. $C_\pm^{[l+1]}$ are fixed by the conditions $\phi^{[l+1]}(0) = r_{\min} - a$ and $\dot{\phi}^{[l+1]}(0) = 0$.

Table 1. Expansion coefficients of free solutions in the unperturbed potential for the first two iterations and for the exact solution ($l = \infty$).

l	$C_+^{[l]}$	$C_-^{[l]}$
0	−0.142872	−0.142872
1	−0.155969 − i0.040544	−0.155322 − i0.068246
∞	−0.151619 − i0.033743	−0.151875 − i0.033990

The numerical convergence of the recursion is a subtle issue that should be studied separately. Assuming the convergence of the approximate sequence, we should obtain an identity when using $r(t)$ to evaluate the integral in Equation (38):

$$\phi^{[\infty]}(t) = C_{+}^{[\infty]} f^{(+)}(t) + C_{-}^{[\infty]} f^{(-)}(t) + \int_{t_1}^{t_2} d\tau \Delta_F(t-\tau) V'(a+\phi(\tau)). \tag{44}$$

The exact solutions are parameterized in terms of the eccentric anomaly E: $r = a(1 - e\cos E)$ and $t = \sqrt{ma^3/\alpha}(E - e\sin E)$, where t is time. For our numerical estimates, we choose $\alpha = \mu = p = 1$ and $e = 0.2$. The values t_1 and t_2 are taken arbitrarily; they correspond to $E_1 = -1$ and $E_2 = 7.2$. The coefficients $C_{\pm}^{[l]}$ for $l = 0, 1, \infty$ found as described above are presented in Table 1. Table 2 shows $r^{[0]}$, $r^{[1]}$ and $r^{[\infty]}$ for seven values of $E \in [0, 2\pi]$. The inclusion of the secondary waves generated by the nonlinear source V' reduces the standard deviation $\chi^2 = \sum(r^{[l]} - r)^2$ from 0.0038 to 0.0015, whereas $r^{[\infty]}$ coincides with r.

Equation (38) can also be derived directly, under the assumption of $t \in (t_1, t_2)$, by using the GF method, whereas Equations (37) and (39) are specific consequences of Kirchhoff's integral theorem. We verified that the free term in Equation (44) fulfills, numerically, Equation (37) and checked Equation (39) for a sample set of time points $t \notin (t_1, t_2)$, as well.

In summary, the idea of Kirchhoff's integral theorem was explained in this section with a one-dimensional toy model (a harmonic oscillator). Such a pedagogical approach illustrates formalism, while the attempt to draw a physical analogy with well-known phenomena leads to the seemingly paradoxical observation: no waves in the $\mathbb{R}^{1,0}$ space, but the superposition principle is there, and even the problem of celestial mechanics was solved using Kirchhoff's integral theorem in a technically consistent manner. A parallelism between classical mechanics and geometrical optics was regarded as purely formal until the advent of quantum mechanics. The possibility of solving the problems of classical mechanics using the methods of wave optics seems to be a surprising circumstance.

Table 2. First two iterations $r^{[l]}$ for the approximate solution of the radial equation of motion as compared to the exact solution $r = r^{[\infty]}$ for seven values of $E \in [0, 2\pi]$.

E	$r^{[0]}$	$r^{[1]}$	$r^{[\infty]}$
0	0.8333	0.8333	0.8333
$\pi/3$	0.9079	0.9410	0.9375
$2\pi/3$	1.1131	1.1619	1.1458
π	1.2500	1.2500	1.2500
$4\pi/3$	1.1132	1.1232	1.1458
$5\pi/3$	0.9079	0.9094	0.9375
2π	0.8333	0.8333	0.8333

3. Kirchhoff's Integral Theorem for a Free Scalar Field

3.1. Complete Orthonormal Basis Functions

A complete set of solutions to the Klein–Gordon equation is formed by the functions

$$f_{\mathbf{k}}^{(+)}(x) = \frac{e^{-ikx}}{\sqrt{2\omega_{\mathbf{k}}}} \quad \text{and} \quad f_{\mathbf{k}}^{(-)}(x) = \frac{e^{ikx}}{\sqrt{2\omega_{\mathbf{k}}}},$$

where $k = (\omega_{\mathbf{k}}, \mathbf{k})$, $\omega_{\mathbf{k}} = \sqrt{\mathbf{k}^2 + m^2}$, $x = (t, \mathbf{x}) \in \mathbb{R}^{1,3}$, and $kx = \omega_{\mathbf{k}} t - \mathbf{kx}$. These functions correspond to the positive- and negative-frequency solutions in the oscillator problem. The orthonormality conditions are

$$\begin{aligned} i\int d\mathbf{x} W[f_{\mathbf{k}'}^{(\pm)*}(x), f_{\mathbf{k}}^{(\pm)}(x)] &= \pm(2\pi)^3 \delta(\mathbf{k}' - \mathbf{k}), \\ \int d\mathbf{x} W[f_{\mathbf{k}'}^{(\mp)*}(x), f_{\mathbf{k}}^{(\pm)}(x)] &= 0. \end{aligned} \tag{45}$$

For any function $\phi_0(x)$ that is a solution of the Klein–Gordon equation,

$$\phi_0(x) = \int \frac{d\mathbf{k}}{(2\pi)^3} \Big(f_{\mathbf{k}}^{(+)}(x) i \int d\mathbf{y} W[f_{\mathbf{k}}^{(+)*}(y), \phi_0(y)] - f_{\mathbf{k}}^{(-)}(x) i \int d\mathbf{y} W[f_{\mathbf{k}}^{(-)*}(y), \phi_0(y)] \Big). \tag{46}$$

After the second quantization, the time-independent quantities

$$a(\mathbf{k}) = i \int d\mathbf{y} W[f_{\mathbf{k}}^{(+)*}(y), \phi_0(y)] \quad \text{and} \quad a^*(\mathbf{k}) = -i \int d\mathbf{y} W[f_{\mathbf{k}}^{(-)*}(y), \phi_0(y)]$$

become annihilation and creation operators.

The first and the second terms in Equation (46) are identified with the positive- and negative-frequency components of $\phi_0(x)$. According to the completeness condition (46), the solutions of the free equation thereby split into the sum

$$\phi_0(x) = \phi_0^{(+)}(x) + \phi_0^{(-)}(x).$$

This decomposition is analogous to the decomposition of Equation (8). The orthonormality conditions (45) and the completeness condition (46) are the generalized equivalents to Equations (5) and (7), respectively, for the oscillator problem.

Using the analogy with Equations (10)–(16) and the Poisson bracket relations

$$\{\phi_0(x), \phi_0(y)\}|_{x^0=y^0} = 0, \tag{47}$$

$$\{\phi_0(x), \pi_0(y)\}|_{x^0=y^0} = \delta(\mathbf{x}-\mathbf{y}), \tag{48}$$

one can prove that

$$\int \frac{d\mathbf{k}}{(2\pi)^3} \Big(f_{\mathbf{k}}^{(+)}(x) f_{\mathbf{k}}^{(+)*}(y) - f_{\mathbf{k}}^{(-)}(x) f_{\mathbf{k}}^{(-)*}(y) \Big)|_{x^0=y^0} = 0, \tag{49}$$

$$\int \frac{d\mathbf{k}}{(2\pi)^3} \Big(f_{\mathbf{k}}^{(+)}(x) \dot{f}_{\mathbf{k}}^{(+)*}(y) - f_{\mathbf{k}}^{(-)}(x) \dot{f}_{\mathbf{k}}^{(-)*}(y) \Big)|_{x^0=y^0} = i\delta(\mathbf{x}-\mathbf{y}), \tag{50}$$

$$\int \frac{d\mathbf{k}}{(2\pi)^3} \Big(\dot{f}_{\mathbf{k}}^{(+)}(x) f_{\mathbf{k}}^{(+)*}(y) - \dot{f}_{\mathbf{k}}^{(-)}(x) f_{\mathbf{k}}^{(-)*}(y) \Big)|_{x^0=y^0} = -i\delta(\mathbf{x}-\mathbf{y}). \tag{51}$$

Equations (49) and (50) can be used to show that the completeness condition (46) holds for arbitrary functions at $x^0 = y^0$.

3.2. Feynman Propagator

The equation for the Feynman propagator is

$$(\Box + m^2)\Delta_F(x) = -\delta^4(x). \tag{52}$$

It is easiest to find the solution in four-momentum space and then apply the Fourier transform to convert it into coordinate space. Here, as in the oscillator problem, we must shift the contour of the integral over k^0 from the real axis in the vicinity of $k^0 = \pm\omega_{\mathbf{k}}$. The four possible ways to do so correspond to four Green's functions.

The Feynman propagator can be written as follows:

$$\begin{aligned} \Delta_F(x-y) &= \int \frac{d^4k}{(2\pi)^4} \frac{e^{-ik(x-y)}}{k^2 - m^2 + i0} \qquad (53) \\ &= -i \int \frac{d\mathbf{k}}{(2\pi)^3} \Big(f_{\mathbf{k}}^{(+)}(x) f_{\mathbf{k}}^{(+)*}(y)\theta(x^0 - y^0) + f_{\mathbf{k}}^{(-)}(x) f_{\mathbf{k}}^{(-)*}(y)\theta(-x^0 + y^0) \Big). \end{aligned}$$

In comparison with Equation (19), the phase space integral is added here. After the replacement $f^{(\pm)}(t) \to f_{\mathbf{k}}^{(\pm)}(x)$ and the integration over the phase space in Equations (20)–(22), the form of the other propagators is restored. Using the analogy with Equation (23) and Equations (49) and (51), one can verify that the propagator (53) satisfies Equation (52).

3.3. *Superposition Principle from Kirchhoff's Integral Theorem*

3.3.1. General form of the Superposition Principle

We start from the identity

$$\phi_0(\xi)\delta^4(\xi - x) = \Delta_F(x-\xi)\Big((\Box_\xi + m^2)\phi_0(\xi)\Big) - \Big((\Box_\xi + m^2)\Delta_F(x-\xi)\Big)\phi_0(\xi). \tag{54}$$

The right-hand side can be written in divergence form as follows:

$$\phi_0(\xi)\delta^4(\xi - x) = \frac{\partial}{\partial \xi_\mu}\left(\Delta_F(x-\xi)\frac{\overleftrightarrow{\partial}}{\partial \xi^\mu}\phi_0(\xi)\right). \tag{55}$$

By taking the integral over a four-dimensional region Ω and transforming the right-hand side into a surface integral, the equation

$$\phi_0(x)\theta(x \in \Omega) = \int_{\partial\Omega} dS^\mu_\xi \left(\Delta_F(x-\xi)\frac{\overleftrightarrow{\partial}}{\partial \xi^\mu}\phi_0(\xi)\right), \tag{56}$$

is obtained, where $\theta(x \in \Omega)$ is the indicator function of Ω:

$$\theta(x \in \Omega) = \begin{cases} 1, & x \in \Omega, \\ 0, & x \notin \Omega. \end{cases}$$

By choosing for the surface $\partial\Omega$ a hyperplane $\xi^0 = y^0$ in the past, i.e., three-dimensional space at a time $\xi^0 = y^0 < x^0$, and a three-dimensional space $\xi^0 = z^0$ at a time $\xi^0 = z^0 > x^0$ in the future, and then combining these spaces at infinity, where the integral vanishes, we arrive at

$$\phi_0(x) = \int d\mathbf{z} W[\Delta_F(x-z), \phi_0(z)] - \int d\mathbf{y} W[\Delta_F(x-y), \phi_0(y)]. \tag{57}$$

If $x \notin \Omega$, we obtain

$$0 = \int d\mathbf{z} W[\Delta_F(x-z), \phi_0(z)] - \int d\mathbf{y} W[\Delta_F(x-y), \phi_0(y)]. \tag{58}$$

Equation (57) states that $\phi_0(x)$ is determined by its past and future. Equation (58) suggests that the interference of secondary waves outside the interval (y^0, z^0) is strictly destructive.

Equation (56) and its consequences (57) and (58) constitute a version of Kirchhoff's theorem in the most general form; these equations hold for any choice of propagator.

3.3.2. Monochromatic Field

The Fourier transform simplifies the superposition scheme of secondary waves. We restrict ourselves to the case of monochromatic, spatially inhomogeneous waves. Consider the following Fourier transforms in time of the scalar field and the Green's function:

$$\phi_0(\omega, \mathbf{x}) = \int_{-\infty}^{+\infty} dt e^{i\omega t}\phi_0(t, \mathbf{x}), \quad \Delta_F(\omega, \mathbf{x}) = \int_{-\infty}^{+\infty} dt e^{i\omega t}\Delta_F(t, \mathbf{x}). \tag{59}$$

They satisfy the equations

$$(\Delta + \mathbf{k}^2)\phi_0(\omega, \mathbf{x}) = 0, \quad (\Delta + \mathbf{k}^2)\Delta_F(\omega, \mathbf{x}) = \delta(\mathbf{x}),$$

where $\mathbf{k}^2 = \omega^2 - m^2$. The right-hand side of the identity

$$\begin{aligned}\phi_0(\omega, \boldsymbol{\xi})\delta(\boldsymbol{\xi} - \mathbf{x}) &= -\Delta_F(\omega, \mathbf{x} - \boldsymbol{\xi})\Big((\Delta_\xi + \mathbf{k}^2)\phi_0(\omega, \boldsymbol{\xi})\Big) \\ &+ \Big((\Delta_\xi + \mathbf{k}^2)\Delta_F(\omega, \mathbf{x} - \boldsymbol{\xi})\Big)\phi_0(\omega, \boldsymbol{\xi}).\end{aligned} \tag{60}$$

can be represented as the divergence

$$\phi_0(\omega,\boldsymbol{\xi})\delta(\boldsymbol{\xi}-\mathbf{x}) = -\frac{\partial}{\partial\xi^\alpha}\left(\Delta_F(\omega,\mathbf{x}-\boldsymbol{\xi})\frac{\overleftrightarrow{\partial}}{\partial\xi^\alpha}\phi_0(\omega,\boldsymbol{\xi})\right).$$

Integrating over the region Ω_3, we obtain von Helmholtz's theorem for the monochromatic field [9]:

$$\phi_0(\omega,\mathbf{x})\theta(\mathbf{x}\in\Omega_3) = -\int_{\partial\Omega_3} dS^\alpha_\xi \Delta_F(\omega,\mathbf{x}-\boldsymbol{\xi})\frac{\overleftrightarrow{\partial}}{\partial\xi^\alpha}\phi_0(\omega,\boldsymbol{\xi}), \tag{61}$$

which is a particular case of the third Green's identity [10] and a precursor of Kirchhoff's integral theorem. The integration is performed over the surface $\partial\Omega_3$, which is the boundary of Ω_3. The equation shows that the field at the point $\mathbf{x}$ is determined by its values on any surrounding surface. This surface is not required to be the wave surface. If the point $\mathbf{x}$ lies outside the closed surface, then the integral vanishes. Regardless of the specific form of $\Delta_F(\omega,\mathbf{x})$, we can conclude from the form of the equation alone that if the field $\phi_0(\omega,\mathbf{x})$ satisfies a differential equation, then this equation contains derivatives over the spatial coordinates that are no higher than second order.

Equation (61) is used to describe the diffraction phenomena of light [3,8].

In the monochromatic, spatially inhomogeneous case, the integration is over the surface rather than over the volume, as in Equation (57). However, because we are discussing the calculation of the Fourier transform in time, an implicit time integration enters the problem.

3.3.3. Massless Field

For massless particles, the interference scheme for secondary waves is simplified. Let us apply the inverse Fourier transform in Equation (61):

$$\phi_0(t,\mathbf{x})\theta(\mathbf{x}\in\Omega_3) = -\int_{\partial\Omega_3} dS^\alpha_\xi \int_{-\infty}^{+\infty} dt'\Delta_F(t-t',\mathbf{x}-\boldsymbol{\xi})\frac{\overleftrightarrow{\partial}}{\partial\xi^\alpha}\phi_0(t',\boldsymbol{\xi}). \tag{62}$$

This equation follows from Equation (56) if we select for Ω an infinite cylinder whose spatial section Ω_3 is covered by the surface of integration $\partial\Omega_3$ and the axis is parallel to the time axis.

As is well known, the propagator $\Delta_F(t,\mathbf{x})$ does not vanish outside of the light cone $t^2-\mathbf{x}^2<0$. This property does not generally violate causality, as $\Delta_F(t,\mathbf{x})$ also describes the propagation of the wave surfaces at which the phase remains constant. In the relativistic theory, the phase velocity $v_p \equiv \omega_\mathbf{k}/|\mathbf{k}| \geq 1$ is greater than the speed of light; however, it is the group velocity $v_g \equiv \partial\omega_\mathbf{k}/\partial|\mathbf{k}| = |\mathbf{k}|/\omega_\mathbf{k} \leq 1$ with which the propagation of signals is associated.

In the limit of zero mass, the propagator $\Delta_F(t,\mathbf{x})$ takes the following form (see [5], Appendix B or Equation (29) in [11] in the massless limit):

$$\Delta_F(t,\mathbf{x}) = \frac{i}{4\pi^2}\frac{1}{t^2-|\mathbf{x}|^2-i0}. \tag{63}$$

Substituting (63) into (62) and taking into account that

$$\int_{-\infty}^{+\infty} dt'\Delta_F(t-t',\mathbf{x})e^{\mp i\omega_\mathbf{k}t'} = \frac{1}{4\pi|\mathbf{x}|}e^{\mp i\omega_\mathbf{k}(t\mp|\mathbf{x}|)},$$

we obtain

$$\phi_0(t,\mathbf{x})\theta(\mathbf{x}\in\Omega_3) = \frac{1}{4\pi}\int_{\partial\Omega_3} dS^{\alpha}_{\xi}\Bigg[-\frac{1}{\rho}\frac{\partial}{\partial\xi^{\alpha}}\left(\phi_0^{(+)}(t-\rho/c,\boldsymbol{\xi})+\phi_0^{(-)}(t+\rho/c,\boldsymbol{\xi})\right) + \left(\frac{\partial}{\partial\xi^{\alpha}}\frac{1}{\rho}\right)\left(\phi_0^{(+)}(t-\rho/c,\boldsymbol{\xi})+\phi_0^{(-)}(t+\rho/c,\boldsymbol{\xi})\right) - \frac{1}{\rho}\frac{\partial\rho}{\partial\xi^{\alpha}}\frac{\partial}{c\partial t}\left(\phi_0^{(+)}(t-\rho/c,\boldsymbol{\xi})-\phi_0^{(-)}(t+\rho/c,\boldsymbol{\xi})\right)\Bigg], \quad (64)$$

where $\rho = |\boldsymbol{\xi} - \mathbf{x}|$ and in the first term, the differentiation with respect to ξ^{α} does not apply to ρ. The dependence on the speed of light c is here made explicit.

Equation (64) represents a general form of Kirchhoff's integral theorem for the Feynman asymptotic conditions. The function is determined by its values on the selected arbitrary closed surface, taking into account the delay of the positive-frequency component and the advancement of the negative-frequency component. This representation is possible because massless particles travel at the speed of light, regardless of their momentum. [1] By contrast, the speed of a massive particle depends on its momentum; therefore, the more general representation (62) includes the integral over time delay and advance. Kirchhoff's theorem is a precise mathematical formulation of the Huygens–Fresnel superposition principle. A special feature of the Feynman asymptotic conditions is that the negative-frequency components are determined by the future. An analogue of Equation (64) for the retarded solutions is the original version of Kirchhoff's integral theorem. It is briefly outlined in Appendix A and discussed in detail in Reference [8].

3.4. Superposition Principle from the Completeness Condition

As a formalization of the superposition principle for the Feynman asymptotic conditions, by analogy with Equation (30), we can consider

$$\phi_0^{(+)}(x)\theta(x^0-y^0) - \phi_0^{(-)}(x)\theta(-x^0+y^0) = -\int d\mathbf{y}W[\Delta_F(x-y),\phi_0(y)]. \quad (65)$$

The physical content of this equation is quite traditional: At the moment y^0, the wave is a source of secondary waves, and the propagation from point y to point x is described by $\Delta_F(x-y)$. To construct the positive-frequency waves, the past $y^0 < x^0$ must be known, and to construct the negative-frequency waves, the future $x^0 < y^0$ must be known. This property is reflected in the presence of the theta functions on the left-hand side of the equation.

The proof of Equation (65) is similar to the proof of Equation (30). It is not based on Kirchhoff's theorem, but instead relies on the completeness condition (46) and the expansion of the Feynman propagator into plane waves. Given that Equation (65) is postulated, the Green's function is uniquely determined. Indeed, let us take the derivative over y^0 on both sides of the equation. After the transformation of the integrand, we obtain Equation (52); it must then be supplemented by asymptotic conditions.

We take the difference (65) for two instants of time, z^0 and y^0, such that $y^0 < x^0 < z^0$. The result is Equation (57). If $x^0 \notin (y^0, z^0)$, then we obtain (58).

According to Equations (65) and (57), the field is determined by its values and first derivatives at two time points. This property indicates the local nature of the evolution equation. Arguing backward, since the initial conditions required to determine the field are the field values and the first derivatives, the evolution equation may contain time derivatives of no higher than second order. Additionally, in Equation (56), a hypersurface $\partial\Omega$ in the form of an infinite cylinder with its axis parallel to the time axis, can be chosen. In such a case, Kirchhoff's theorem would assert that the wave is determined by its values and gradients on a two-dimensional surface at all times. This version of the theorem indicates the local nature of the evolution equation in the spatial coordinates. The corresponding differential equation may contain derivatives of the spatial coordinates of no higher than second order.

3.5. Path Integral

The path integral representation is a consequence of Equation (56). We choose a set of four-dimensional regions $\Omega_1 \subset \Omega_2 \subset \ldots \subset \Omega_n \subset \mathbb{R}^{1,3}$. By iterating Equation (56), we obtain

$$\begin{aligned} \phi_0(x)\theta(x \in \Omega_1) &= \int_{\partial\Omega_1} dS^{\mu_1}_{\xi_1} \int_{\partial\Omega_1} dS^{\mu_2}_{\xi_2} \ldots \int_{\partial\Omega_n} dS^{\mu_n}_{\xi_n} \\ &\times \; \Delta_F(x-\xi_1)\frac{\overleftrightarrow{\partial}}{\partial\xi_1^{\mu_1}}\Delta_F(\xi_1-\xi_2)\frac{\overleftrightarrow{\partial}}{\partial\xi_2^{\mu_2}}\ldots\Delta_F(\xi_{n-1}-\xi_n)\frac{\overleftrightarrow{\partial}}{\partial\xi_n^{\mu_n}}\phi_0(\xi_n). \end{aligned} \tag{66}$$

There exists considerable freedom in choosing Ω_i. A similar freedom exists in the factorization of unitary evolution operator $U(t_2,t_1)$ in quantum mechanics, where the equation $U(t_2,t_1) = U(t_2,t)U(t,t_1)$ holds for any instant of time $t \in (t_1,t_2)$. While the evolution operator is factorized in time, the integration in the path integral goes over the coordinates in three-dimensional space. Such a representation easily follows from Equation (66). Indeed, choosing Ω_i to be cylinders with infinite radii and axes parallel to the time axis, we arrive at a representation of this kind. The broken lines connecting the points x and $\xi_n \in \partial\Omega_n$ through $\xi_i \in \partial\Omega_i$ $(i = 1,\ldots,n-1)$ form in the continuum limit the class of paths over which the continual integral is defined. The comparison of Equations (56) and (66) also yields, in the limit of $n \to \infty$, an integral representation for the Green's function in the form of a continual integral.

4. Charged Scalar Field in an External Electromagnetic Field

Equations (56) and (65) and their particular cases were obtained for a free field. The following question arises: which relations can be generalized in the presence of an external field? We restrict ourselves to scalar electrodynamics.

4.1. Complete Orthonormal Basis Functions

Substituting the normal derivatives with respect to the space-time coordinates in the Klein–Gordon equation with gauge covariant derivatives,

$$\partial_\mu \to D_\mu = \partial_\mu + ieA_\mu \tag{67}$$

yields the evolution equation for a complex scalar field in an external electromagnetic field,

$$(D_\mu D^\mu + m^2)\phi(x) = 0. \tag{68}$$

The external field is adiabatically switched on at $t \to -\infty$ and off at $t \to +\infty$. The set of positive- and negative-frequency asymptotic solutions $f^{(\pm)}_{\mathbf{k}}(x)$ is complete and orthonormal. The second-order Equation (69) has a set of independent solutions $F^{(\pm)}_{\mathbf{k}}(x)$. The asymptotic conditions can be taken as

$$F^{(\pm)}_{\mathbf{k}}(x) \to f^{(\pm)}_{\mathbf{k}}(x) \equiv \frac{e^{\mp ikx}}{\sqrt{2\omega_{\mathbf{k}}}} \;\; \text{for} \;\; t \to -\infty.$$

All other solutions of Equation (68) are expressed as linear superpositions of the basis functions $F^{(\pm)}_{\mathbf{k}}(x)$.

It would be natural to use the prescription (67) for extending the Huygens–Fresnel superposition principle. It can be assumed that in an external electromagnetic field, the suitable generalization of the Wronskian is given by

$$W_A[\varphi^*,\chi] \equiv \varphi^*(\overleftrightarrow{\partial}_t + 2ieA_0)\chi = \varphi^*(D_t\chi) - (D_t\varphi)^*\chi.$$

We note a useful property:

$$\begin{aligned} \partial_t W_A[\varphi^*,\chi] &= \partial_t(\varphi^*(D_t\chi) - (D_t\varphi)^*\chi) \\ &= \varphi^*(D_tD_t\chi) - (D_tD_t\varphi)^*\chi. \end{aligned} \tag{69}$$

It is not difficult to show that if φ and χ are two solutions of Equation (68), then the following condition holds:

$$\partial_t \int d\mathbf{x} W_A[\varphi^*,\chi] = 0.$$

This condition allows us to calculate the normalization integral by sending the time variable to negative infinity, where solutions are represented as plane waves. The orthonormality conditions thus take the form

$$\begin{aligned} i\int d\mathbf{x} W_A[F^{(\pm)*}_{\mathbf{k}'}(x), F^{(\pm)}_{\mathbf{k}}(x)] &= \pm(2\pi)^3\delta(\mathbf{k}'-\mathbf{k}), \\ \int d\mathbf{x} W_A[F^{(\pm)*}_{\mathbf{k}'}(x), F^{(\mp)}_{\mathbf{k}}(x)] &= 0. \end{aligned}$$

The completeness condition is obvious:

$$\phi(x) = \int \frac{d\mathbf{k}}{(2\pi)^3}\Big(F^{(+)}_{\mathbf{k}}(x) i\int d\mathbf{y} W_A[F^{(+)*}_{\mathbf{k}}(y),\phi(y)] - F^{(-)}_{\mathbf{k}}(x) i\int d\mathbf{y} W_A[F^{(-)*}_{\mathbf{k}}(y),\phi(y)]\Big). \tag{70}$$

In the theory of a charged scalar field, the canonical momenta are defined by the equations $\pi^*(x) = D_t\phi(x)$ and $\pi(x) = (D_t\phi(x))^*$. The canonically conjugate variables satisfy

$$\{\phi(x),\pi(y)\}|_{x^0=y^0} = \{\phi(x)^*,\pi^*(y)\}|_{x^0=y^0} = \delta(\mathbf{x}-\mathbf{y}), \tag{71}$$

while other pairs have the vanishing Poisson bracket. The generalization of the corresponding relations of a free scalar field can be written as follows

$$\int \frac{d\mathbf{k}}{(2\pi)^3}\Big(F^{(+)}_{\mathbf{k}}(x)F^{(+)*}_{\mathbf{k}}(y) - F^{(-)}_{\mathbf{k}}(x)F^{(-)*}_{\mathbf{k}}(y)\Big)|_{x^0=y^0} = 0, \tag{72}$$

$$\int \frac{d\mathbf{k}}{(2\pi)^3}\Big(F^{(+)}_{\mathbf{k}}(x)D^*_tF^{(+)*}_{\mathbf{k}}(y) - F^{(-)}_{\mathbf{k}}(x)D^*_tF^{(-)*}_{\mathbf{k}}(y)\Big)|_{x^0=y^0} = i\delta(\mathbf{x}-\mathbf{y}), \tag{73}$$

$$\int \frac{d\mathbf{k}}{(2\pi)^3}\Big(D_tF^{(+)}_{\mathbf{k}}(x)F^{(+)*}_{\mathbf{k}}(y) - D_tF^{(-)}_{\mathbf{k}}(x)F^{(-)*}_{\mathbf{k}}(y)\Big)|_{x^0=y^0} = -i\delta(\mathbf{x}-\mathbf{y}). \tag{74}$$

Equations (72) and (73) show that the completeness condition (70) holds for arbitrary functions evaluated at $x^0 = y^0$.

In conclusion, we note that the zeroth component of vector potential can be removed by a gauge transformation, in which case, $W_A = W$ and other relations and their proofs take the form more similar to the free case.

4.2. Feynman Propagator

The decomposition of the Feynman propagator over the basis functions has the form

$$\Delta_F(x,y) = -i\int \frac{d\mathbf{k}}{(2\pi)^3}\Big(F^{(+)}_{\mathbf{k}}(x)F^{(+)*}_{\mathbf{k}}(y)\theta(x^0-y^0) + F^{(-)}_{\mathbf{k}}(x)F^{(-)*}_{\mathbf{k}}(y)\theta(-x^0+y^0)\Big). \tag{75}$$

The use of Equations (72) and (74) allows to verify by the direct calculation that

$$(D_\mu D^\mu + m^2)\Delta_F(x,\xi) = -\delta^4(x-\xi). \tag{76}$$

4.3. Superposition Principle from Kirchhoff's Integral Theorem

To derive Equation (55), the identity (54) was used. After recapitulating the arguments used in the proof of Equation (69), we rewrite the divergence of

$$\varphi \overset{\leftrightarrow}{D}_\mu \chi \equiv \varphi(D_\mu \chi) - (D^*_\mu \varphi)\chi,$$

where φ and χ are arbitrary functions, in the form

$$\partial_\mu(\varphi \overset{\leftrightarrow}{D^\mu} \chi) = \varphi(D_\mu D^\mu \chi) - (D^*_\mu D^{*\mu}\varphi)\chi.$$

Substituting $\Delta_F(x,\xi)$ and $\phi(\xi)$ in place of φ and χ, respectively, we obtain

$$\phi(\xi)\delta^4(x-\xi) = \frac{\partial}{\partial \xi_\mu}\Big(\Delta_F(x,\xi)(D_\mu \phi(\xi)) - (D^*_\mu \Delta_F(x,\xi))\phi(\xi)\Big). \tag{77}$$

By choosing as the integration region a four-dimensional space with the variable ξ^0 running in the interval (y^0, z^0), we find for $x^0 \in (y^0, z^0)$

$$\phi(x) = \int d\mathbf{z} W_A[\Delta_F(x,z), \phi(z)] - \int d\mathbf{y} W_A[\Delta_F(x,y), \phi(y)]. \tag{78}$$

In the opposite case of $x^0 \notin (y^0, z^0)$ the left-hand side vanishes.

4.4. Superposition Principle from the Completeness Condition

The linearity of the evolution equation allows for the generalization of the superposition principle (65) in the presence of an external electromagnetic field. The completeness condition leads to the following scheme:

$$\phi^{(+)}(x)\theta(x^0-y^0) - \phi^{(-)}(x)\theta(-x^0+y^0) = -\int d\mathbf{y} W_A[\Delta_F(x,y), \phi(y)]. \tag{79}$$

Under the integral sign, the derivative entering W_A also generates the term

$$\Delta(x,y) = -i\int \frac{d\mathbf{k}}{(2\pi)^3}\Big(F^{(+)}_{\mathbf{k}}(x)F^{(+)*}_{\mathbf{k}}(y) - F^{(-)}_{\mathbf{k}}(x)F^{(-)*}_{\mathbf{k}}(y)\Big)$$

multiplied by $\phi(y)\delta(x^0-y^0)$. In view of the relationship $x^0 = y^0$ and Equation (72), this term vanishes. By calculating the derivative of Equation (79) with respect to y^0, one can prove that the propagator obeys equation

$$(D^*_\mu D^{\mu *} + m^2)\Delta_F(x,\xi) = -\delta(x-\xi), \tag{80}$$

where the differentiation is over ξ. This equation is equivalent to Equation (76), where D^μ acts on x.

The superposition scheme for the retarded propagator is as follows

$$\phi(x)\theta(x^0-y^0) = -\int d\mathbf{y} W_A[\Delta_{\rm ret}(x,y), \phi(y)]. \tag{81}$$

This equation is the analog of Equation (27). It can also be derived from Equation (77).

To conclude, the superposition schemes for a free scalar field are fundamentally valid for a scalar complex field in an external electromagnetic field.

5. Nonlinear Field Theory

The superposition principle for secondary waves, which is the consequence of the GF method, should be distinguished from the superposition principle as a manifestation of the linearity of the problem. In linear theory, the wave is a source of secondary waves. In nonlinear theory, two sources of secondary waves exist: the wave itself plus a function $V'(\phi)$.

In both cases, secondary waves satisfy free linear wave equations, so the superposition principle applies to secondary waves universally.

5.1. Secondary Waves beyond Fresnel's Superposition Scheme

For a Lagrangian $\mathcal{L} = \mathcal{L}_{\text{free}} - V$, that contains a term $V = V(\phi)$ of a general form, the identity (54) is modified as follows:

$$\begin{aligned} \phi(\xi)\delta^4(\xi - x) &= \Delta_F(x-\xi)((\Box_\xi + m^2)\phi(\xi) + V'(\phi(\xi))) - ((\Box_\xi + m^2)\Delta_F(x-\xi))\phi(\xi) \\ &= \frac{\partial}{\partial \xi_\mu}\left(\Delta_F(x-\xi)\frac{\overleftrightarrow{\partial}}{\partial \xi^\mu}\phi(\xi)\right) + \Delta_F(x-\xi)V'(\phi(\xi)). \end{aligned} \tag{82}$$

For $x^0 \in (y^0, z^0)$, this equation gives

$$\phi(x) = \phi_0(x) - \int d^4\xi \Delta_F(x-\xi)V'(\phi(\xi)), \tag{83}$$

where the integration over ξ^0 runs over $\xi^0 \in (y^0, z^0)$ and the integral in $\boldsymbol{\xi}$ extends over all space. The field $\phi_0(x)$ is defined by the relation

$$\phi_0(x) = \int d\mathbf{z} W[\Delta_F(x-z), \phi(z)] - \int d\mathbf{y} W[\Delta_F(x-y), \phi(y)]. \tag{84}$$

For $x^0 \in (y^0, z^0)$, $\phi_0(x)$ satisfies the free Klein–Gordon equation. If $x^0 \notin (y^0, z^0)$, then

$$0 = \phi_0(x) - \int d^4\xi \Delta_F(x-\xi)V'(\phi(\xi)). \tag{85}$$

In quantum field theory, Equation (83) in the infinite limits $(y^0, z^0) = (-\infty, +\infty)$ is used in the development of perturbation theory. Unlike in the canonical formulation of the Fresnel superposition scheme, the integrand contains the nonlinear term $V'(\phi(\xi))$ as an additional source of secondary waves and the integration spans the entire four-dimensional space.

Equations (83)–(85) in nonlinear scalar field theory are analogous to Equations (36)–(38) in the anharmonic oscillator problem.

The mass term of $\mathcal{L}$ can be attributed either to $\mathcal{L}_{\text{free}}$ or to the potential V. In the last case, $\mathcal{L}_{\text{free}}$ describes massless particles. This might seem disadvantageous, because asymptotic states of $\mathcal{L}$ are massive in general. The positive feature is that the retarded Green's function of massless particles, being localized on the light cone (see Equation (A1)), ensures reduction of four-dimensional integrals in Equations (83) and (85) to three-dimensional integrals and transformation of integrals in Equation (84) to surface integrals.

5.2. Positive- and Negative-Frequency Solutions

Interacting fields can be decomposed into a sum of positive- and negative-frequency solutions only asymptotically. In Section 2.3.2, we demonstrated that the straightforward generalization of the Fresnel superposition scheme to nonlinear dynamical systems is possible and consistent; however, its value is limited to only providing the definitions of positive- and negative-frequency solutions for arbitrary t. For the sake of completeness, we present here a field theoretical version of the nonlinear superposition scheme (40):

$$\begin{aligned} \phi^{(+)}(x)\theta(x^0 - y^0) - \phi^{(-)}(x)\theta(-x^0 + y^0) = &- \int d\mathbf{y} W[\Delta_F(x-y), \phi(y)] \\ &+ \int d^4\xi \Delta_F(x-\xi)V'(\phi(\xi)), \end{aligned} \tag{86}$$

where the integral over ξ^0 runs from y^0 to x^0.

The derivative over y^0 leads to the relation $\phi(t) = \phi^{(+)}(t) + \phi^{(-)}(t)$ and Equation (52). The difference in Equation (86) at two different time points leads to Equations (83)–(85).

Equation (86) ensures that $\phi^{(\pm)}(x)$ is a linear superposition in $\mathbf{k}$ of the basis functions $f_{\mathbf{k}}^{(\pm)}(x)$ at $t \to \pm\infty$.

The calculation of the first derivative of Equation (86) in x^0 yields a superposition scheme for the canonical momentum:

$$\pi^{(+)}(x)\theta(x^0 - y^0) - \pi^{(-)}(x)\theta(-x^0 + y^0) = - \int d\mathbf{y} W[\Delta_F(x-y), \pi(y)] + \int d^4\xi \Delta_F(x-\xi) V''(\phi(\xi))\pi(\xi), \quad (87)$$

where the integral over ξ^0 runs from y^0 to x^0.

6. Conclusions

The evolution of the ideas underlying the Huygens–Fresnel superposition principle from geometrical and wave optics to the theory of interacting fields is highly instructive.

In geometrical optics, a *wave front* refers to the two-dimensional surface that defines the farthest extent to which the wave has arrived after a certain period of time. Huygens' principle (1678), based on the Fermat principle, allows for the determination of how the wave front is propagating.

In wave optics, the term wave front has no strict definition. Instead, the term *wave surface* is used. The wave surface is the two-dimensional surface on which the phase of the wave is constant. A.-J. Fresnel proposed the principle of superposition (1816), which details the wave process. A wave is a result of interference of secondary waves emitted at an earlier time. At any fixed point, it is determined by the phase and amplitude at a wave surface corresponding to a preceding instant of time. The wave surface in the past can be chosen arbitrarily. The superposition principle anticipates informal content of the GF method (1828).

Kirchhoff's integral theorem (1883) is a dynamic, four-dimensional extension of Green's third identity of the static potential theory. More than half a century separates this theorem from Green's major work [10], which introduced the basic concepts of the GF method.[2] Kirchhoff's integral theorem provides a mathematical proof of the superposition principle, clarifying and quantifying it.

First, the theorem demonstrates that the amplitude of the secondary waves is determined by the Wronskian of the Green's function and the field at a previous time.

Second, the wave surfaces are not highlighted; this is perfectly consistent with the fact that they are not necessarily observable (in the massive theory, e.g., the speed of a wave surface of a plane wave is always greater than the speed of light). The surface must be closed and contain the point at which the wave is calculated; otherwise, it can be arbitrarily chosen. Outside the closed surface, the interference of the secondary waves is strictly destructive: for any exterior point, the calculation of the surface integral yields zero.

The reasoning used in the proof can be regarded as a standard piece of the GF method; it is of high generality, goes beyond the problem of propagation of electromagnetic waves and allows for an understanding of how the superposition principle should be modified in the theory of interacting fields. Note the most significant modifications:

(i) According to the Huygens–Fresnel superposition principle, a wave at a given point is expressed as a superposition of secondary waves emitted from centers located on a two-dimensional surface. This property arises only in massless theories, including the theory of electromagnetic fields, where the group and phase velocities coincide with the speed of light, which is the necessary condition for the integral over time delay and advance to not be available in Equation (64). Kirchhoff's integral theorem for massive particles, Equation (62), states that a wave is determined by its values on a closed surface at all times. The physical interpretation of this fact is quite transparent. The Fourier expansion of a massive field contains components of various momenta corresponding to various group and phase velocities, which leads to a spread in time lags. As a result, the two-dimensional integral over the sources of secondary waves is transformed into a three-dimensional integral;

(ii) In the nonlinear theory, there is a need for a more extensive modification of the superposition scheme. In addition to the wave itself, a nonlinear function of the field $V'(\phi(\xi))$ becomes the source of secondary waves. The summation runs over distributed sources: from a two-dimensional surface in theories with massless particles to a two-dimensional surface and the time axis in theories with massive particles and the entirety of four-dimensional space. This type of representation holds for both local nonlinear and nonlocal theories.

We see that after each modification, the effectiveness of the superposition principle weakens. In the most general nonlinear case, the modified principle certainly does not promise fast results. To determine the field, it is necessary to calculate a four-dimensional integral in a self-consistent manner. In linear theories, the superposition principle solves the evolution problem, but in nonlinear theories, it only offers a different formulation of the problem. Nevertheless, relations of this type are still useful when searching for solutions within the framework of perturbation theory, when the non-linearity is small. In other cases, the solutions found using other techniques can be checked. The four-dimensional representation given by Equation (83) is a consequence of Kirchhoff's integral theorem, but in quantum field theory, it is typically derived directly from the properties of Green's functions.

In the context of a field theory, the original form of the superposition principle only has heuristic value. The superposition schemes for the secondary waves that are used to solve specific problems are unified by Kirchhoff's integral theorem, which exploits the properties of the Wronskian of the Green's functions and solutions of the wave equation under consideration. The spectrum of such problems is quite comprehensive: from the harmonic oscillator to scalar electrodynamics and nonlinear field theories.

In addition to the use of the GF method, which has found a variety of applications in quantum field theory, Kirchhoff's theorem has a wider range of corollaries. Equation (62), which represents one version of Kirchhoff's theorem, does not arise in quantum field theory because of the boundary conditions, which are atypical of a scattering problem. However, the superposition scheme represented by Equation (65), which is not based on Kirchhoff's theorem, is not sufficiently general because it does not extend to theories with interaction.

The statement of Kirchhoff's integral theorem depends on the asymptotic conditions imposed on Green's function. In the main part of the paper, as we were interested in the place of this remarkable principle and well-known theorem in quantum field theory, we applied the Feynman asymptotic conditions almost universally.

Funding: The author was supported in part by the Russian Foundation for Basic Research (RFBR) Grants Nos. 16-02-01104 and 18-02-00733 and Grant No. HLP-2015-18 of the Heisenberg-Landau Program.

Data Availability Statement: All data generated or analyzed during this study are included in this published article.

Conflicts of Interest: The author declares no conflict of interest.

Appendix A. Kirchhoff's Integral Theorem and Its Vector Extensions with the Retarded Green's Function

In the main sections of the paper, emphasis is placed on the Feynman asymptotic conditions, which play a special role in quantum field theory. Here, we formulate Kirchhoff's integral theorem and its vectorial generalizations for the retarded Green function.

Appendix A.1. Free Massless Scalar Field

Equation (56) is essentially the third Green identity for time-dependent solutions of the wave equation; its proof is outlined in Section 3.1. As noted earlier, Equation (56) holds for any Green function.

The retarded Green function in the coordinate representation has the following form (see, e.g., [5])

$$
\begin{aligned}
\Delta_{\rm ret}(t,\mathbf{x}) &= \int \frac{d^4q}{(2\pi)^4} e^{-iqx} \frac{1}{q^2 + i0\,{\rm sgn}(q_0)} \\
&= -\frac{1}{4\pi|\mathbf{x}|}(\delta(|\mathbf{x}|-t) - \delta(|\mathbf{x}|+t))\theta(t).
\end{aligned} \tag{A1}
$$

The product of generalized functions of a single variable is not defined. The propagator depends on four space-time coordinates. Generalized functions of four variables allow for products of up to four generalized functions of one variable, provided their arguments are independent. $\Delta_{\rm ret}(t,\mathbf{x})$ is thus a well-defined generalized function.

$\Delta_{\rm ret}(t,\mathbf{x})$ is localized on the upper half of the light cone $t^2 - \mathbf{x}^2 = 0$. Substituting (A1) in place of $\Delta_F(t,\mathbf{x})$ in Equation (62), one arrives at the original Kirchhoff representation [7,8]

$$
\begin{aligned}
\phi_0(t,\mathbf{x})\theta(\mathbf{x}\in\Omega_3) = \frac{1}{4\pi}\int_{\partial\Omega_3} dS^{\alpha}_{\xi}\Big[-\frac{1}{\rho}\frac{\partial}{\partial\xi^{\alpha}}\phi_0(t-\rho/c,\boldsymbol{\xi}) + \left(\frac{\partial}{\partial\xi^{\alpha}}\frac{1}{\rho}\right)\phi_0(t-\rho/c,\boldsymbol{\xi}) \\
- \frac{1}{\rho}\frac{\partial\rho}{\partial\xi^{\alpha}}\frac{\partial}{c\partial t}\phi_0(t-\rho/c,\boldsymbol{\xi})\Big],
\end{aligned} \tag{A2}
$$

where $\rho = |\boldsymbol{\xi} - \mathbf{x}|$ and where the differentiation in ξ^{α} does not affect ρ in the first term. The dependence on the speed of light c is here made explicit. The wave $\phi_0(t,\mathbf{x})$ at $\mathbf{x}\in\Omega_3$ is determined by its values on the closed surface $\partial\Omega_3$ considering the delay ρ/c. Linear second-order hyperbolic partial differential equations possessing this property are well-studied from a mathematical point of view [13–15].

Appendix A.2. Monochromatic Electromagnetic Fields with Sources

A generalization of Kirchhoff's integral theorem, which takes into account vectorial character of the electromagnetic field and the electromagnetic currents, was obtained by von Ignatowsky [16]. First, however, we consider a generalization of von Helmholtz's theorem, following Stratton and Chu [17].

Most methods used in Section 3.3.2 for a free monochromatic scalar field apply to a monochromatic electromagnetic field with sources after some slight modifications. We replace scalar field ϕ_0 by the electromagnetic field tensor $F^{\mu\nu} = \partial^{\nu}A^{\mu} - \partial^{\mu}A^{\nu}$. In the Lorentz gauge $\partial_{\mu}A^{\mu} = 0$, the evolution equations $\partial_{\nu}F^{\mu\nu} = j^{\mu}$ become $\Box A^{\mu} = j^{\mu}$, where j^{μ} is the electromagnetic current. It is assumed that the fields are harmonic and that all quantities contain a factor $\exp(-i\omega t)$, so that $\partial^0 j^{\mu} = -i\omega j^{\mu}$, $(\omega^2 + \triangle)A^{\mu} = -j^{\mu}$, and

$$
(\omega^2 + \triangle)F^{\mu\nu} = -\partial^{\nu}j^{\mu} + \partial^{\mu}j^{\nu}.
$$

Since the right-hand side of this equation is different from zero, the analogue of Equation (60) takes a more complicated form:

$$
\begin{aligned}
F^{\mu\nu}(\omega,\boldsymbol{\xi})\delta(\boldsymbol{\xi}-\mathbf{x}) &= \Delta_{\rm ret}(\omega,\mathbf{x}-\boldsymbol{\xi})(-\partial^{\nu}j^{\mu}(\omega,\boldsymbol{\xi}) + \partial^{\mu}j^{\nu}(\omega,\boldsymbol{\xi})) \\
&\quad - \frac{\partial}{\partial\xi^{\alpha}}\left(\Delta_{\rm ret}(\omega,\mathbf{x}-\boldsymbol{\xi})\frac{\overleftrightarrow{\partial}}{\partial\xi^{\alpha}}F^{\mu\nu}(\omega,\boldsymbol{\xi})\right).
\end{aligned} \tag{A3}
$$

The sum in α runs from 1 to 3, while $\mu,\nu = 0,1,2,3$. By integrating over a three-dimensional region Ω_3, one obtains

$$
\begin{aligned}
F^{\mu\nu}(\omega,\mathbf{x})\theta(\mathbf{x}\in\Omega_3) &= \int_{\Omega_3} d\boldsymbol{\xi}\Delta_{\rm ret}(\omega,\mathbf{x}-\boldsymbol{\xi})(-\partial^{\nu}j^{\mu}(\omega,\boldsymbol{\xi}) + \partial^{\mu}j^{\nu}(\omega,\boldsymbol{\xi})) \\
&\quad - \int_{\partial\Omega_3} dS^{\beta}_{\xi}\Delta_{\rm ret}(\omega,\mathbf{x}-\boldsymbol{\xi})\frac{\overleftrightarrow{\partial}}{\partial\xi^{\beta}}F^{\mu\nu}(\omega,\boldsymbol{\xi}).
\end{aligned} \tag{A4}
$$

The dependence on the derivatives of $F^{\mu\nu}$ can be eliminated [17].

Appendix A.3. Non-Monochromatic Electromagnetic Fields with Sources

In the presence of external currents, electromagnetic fields satisfy the identity

$$\begin{aligned} F^{\mu\nu}(\xi)\delta^4(\xi - x) &= \Delta_{\text{ret}}(x-\xi)(-\partial^\nu j^\mu(\xi) + \partial^\mu j^\nu(\xi)) \\ &\quad - \frac{\partial}{\partial \xi_\sigma}\left(\Delta_{\text{ret}}(x-\xi) \frac{\overleftrightarrow{\partial}}{\partial \xi^\sigma} F^{\mu\nu}(\xi) \right). \end{aligned} \tag{A5}$$

The sum in σ runs from 0 to 3. By taking the integral over a four-dimensional region Ω, we obtain

$$\begin{aligned} F^{\mu\nu}(x)\theta(x \in \Omega) &= \int_\Omega d^4\xi \Delta_{\text{ret}}(x-\xi)(-\partial^\nu j^\mu(\xi) + \partial^\mu j^\nu(\xi)) \\ &\quad - \int_{\partial\Omega} dS^\sigma_\xi \left(\Delta_{\text{ret}}(x-\xi) \frac{\overleftrightarrow{\partial}}{\partial \xi^\sigma} F^{\mu\nu}(\xi) \right). \end{aligned} \tag{A6}$$

The representation becomes linear in $F^{\mu\nu}$ after replacing j^μ with $\partial_\nu F^{\mu\nu}$. The field derivatives are assumed to be smooth.

Equation (A6) can be simplified by choosing Ω to be an infinite cylinder, $\Omega = \mathbb{R}^1 \otimes \Omega_3$, whose cross section is a three-dimensional space-like region Ω_3. With the use Equation (A1), the integration over the time coordinate gives [16]

$$\begin{aligned} F^{\mu\nu}(t,\mathbf{x})\theta(\mathbf{x} \in \Omega_3) &= -\frac{1}{4\pi}\int_{\Omega_3} d\boldsymbol{\xi}\frac{1}{\rho}(-\frac{\partial}{\partial \xi_\nu} j^\mu(t-\rho,\boldsymbol{\xi}) + \frac{\partial}{\partial \xi_\mu} j^\nu(t-\rho,\boldsymbol{\xi})) \\ &\quad + \frac{1}{4\pi}\int_{\partial\Omega_3} dS^\alpha_{\boldsymbol{\xi}} \left[-\frac{1}{\rho}\frac{\partial}{\partial \xi^\alpha} F^{\mu\nu}(t-\rho,\boldsymbol{\xi}) \right. \\ &\quad \left. + \left(\frac{\partial}{\partial \xi^\alpha}\frac{1}{\rho}\right) F^{\mu\nu}(t-\rho,\boldsymbol{\xi}) - \frac{1}{\rho}\frac{\partial \rho}{\partial \xi^\alpha}\frac{\partial}{\partial t} F^{\mu\nu}(t-\rho,\boldsymbol{\xi}) \right], \end{aligned} \tag{A7}$$

where $\rho = |\boldsymbol{\xi} - \mathbf{x}|$. In the first two lines, the differentiation with respect to ξ^α does not apply to ρ.

Kirchhoff's integral theorem (A2) and Equation (A4) extend von Helmholtz's theorem (61) in different directions. Equation (A7) constitutes, on one hand, the generalization of Kirchhoff's integral theorem by taking into account the vectorial character of electromagnetic field and including the effect of electromagnetic currents and, on the other hand, the generalization of Equation (A4) by going beyond the monochromatic field assumption.

Notes

1 In Euclidean space of dimension $n \geq 3$, Green's function has the form $\Delta(x) \sim 1/(x^2)^{(n-2)/2}$. Performing a Wick rotation, we find that the Green's function as an analytic function of the variable $t = x^0$ has two isolated poles in the spaces of even dimension and two root branching points in the spaces of odd dimension. This means that in the massless case, the Green's function is effectively localized on the light cone in the spaces of even dimension only. Here, an analogue of the representation (64) holds. In the spaces of odd dimension, the superposition scheme involves the integration over all spatial coordinates. This property of the Green's function suggests that the requirement of equal phase and group velocities and the speed of light is a necessary but not sufficient condition for the representation of superposition scheme in the form of a surface integral.

2 In 1839, G. Green came closely to the notion of the four-dimensional Green's function. The value of the GF method in quantum field theory is highly appreciated [12].

References

1. Arnold, V.I. *Mathematical Methods of Classical Mechanics*, 2nd ed.; Springer: New York, NY, USA, 1989.
2. Saveliev, I.V. Electricity. In *A Course of General Physics*; Nauka: Moscow, Russia, 1982; Volume 2. (In Russian)
3. Landau, L.D.; Lifshitz, E.M. The Classical Theory of Fields. In *A Course of Theoretical Physics*, 4th ed.; Pergamon Press: Oxford, UK, 1975; Volume 2.

4. Bjorken, J.D.; Drell, S.D. *Relativistic Quantum Mechanics*; McGraw-Hill: New York, NY, USA, 1964.
5. Bjorken, J.D.; Drell, S.D. *Relativistic Quantum Fields*; McGraw-Hill: New York, NY, USA, 1965.
6. Dittrich, W.; Reuter, M. *Classical and Quantum Dynamics: From Classical Paths to Path Integrals*, 3rd ed.; Springer: Berlin/Heidelberg, Germany, 2001.
7. Kirchhoff, G. Zur Theorie der Lichtstrahlen. *Ann. Phys.* **1883**, *254*, 663–695. [CrossRef]
8. Born, M.; Wolf, E. *Principles of Optics. Electromagnetic Theory of Propagation, Interference and Diffraction of Light*, 7th ed.; Cambridge University Press: Cambridge, UK, 1999.
9. von Helmholtz, H. Theorie der Luftschwingungen in Röhren mit offenen Enden. *J. Math.* **1860**, *57*, 1–72.
10. Green, G. *An Essay on the Application of Mathematical Analysis to the Theories of Electricity and Magnetism*; T. Wheelhouse: Nottingham, UK, 1828.
11. Zavialov, O.I. *Renormalized Feynman Diagrams*; Nauka: Moscow, Russia, 1979. (In Russian)
12. Schwinger, J. The greening of quantum field theory: George and I. *arXiv* **1993**, *preprint*. arXiv:hep-ph/9310283.
13. Courant, R.; Hilbert, D. *Methods of Mathematical Physics*; John Wiley & Sons: New York, NY, USA, 1962; Volume II.
14. Günther, P. Huygens' principle and Hadamard's conjecture. *Math. Intell.* **1991**, *13*, 56–63. [CrossRef]
15. Duistermaat, J.J. Huygens' principle for linear partial differential equations. In *Huygens' Principle 1690–1990: Theory and Applications*; Blok, H., Ferwerda, H.A., Kuiken, H.K., Eds.; North Holland: Amsterdam, The Netherlands, 1992; pp. 273–297.
16. von Ignatowsky, W. Diffraktion und Reflexion, abgeleitet aus den Maxwellschen Gleichungen. *Ann. Phys.* **1907**, *328*, 875–904. [CrossRef]
17. Stratton, J.A.; Chu, L.J. Diffraction Theory of Electromagnetic Waves. *Phys. Rev.* **1939**, *56*, 99. [CrossRef]

MDPI

Article

Linear Superposition as a Core Theorem of Quantum Empiricism

Yurii V. Brezhnev

Department of Quantum Field Theory, Tomsk State University, 634050 Tomsk, Russia; brezhnev@phys.tsu.ru

Abstract: Clarifying the nature of the quantum state $|\Psi\rangle$ is at the root of the problems with insight into counter-intuitive quantum postulates. We provide a direct—and math-axiom free—empirical derivation of this object as an element of a vector space. Establishing the linearity of this structure—quantum superposition—is based on a set-theoretic creation of ensemble formations and invokes the following three principia: (**I**) quantum statics, (**II**) doctrine of the number in the physical theory, and (**III**) mathematization of matching the two observations with each other (quantum covariance). All of the constructs rest upon a formalization of the minimal experimental entity—the registered micro-event, detector click. This is sufficient for producing the $\mathbb{C}$ -numbers, axioms of linear vector space (superposition principle), statistical mixtures of states, eigenstates and their spectra, and non-commutativity of observables. No use is required of the spatio-temporal concepts. As a result, the foundations of theory are liberated to a significant extent from the issues associated with physical interpretations, philosophical exegeses, and mathematical reconstruction of the entire quantum edifice.

Keywords: quantum foundations; non-axiomaticity; detector clicks; ensembles; superposition principle; arithmetic; numbers; vector space; abstracting; interpretations; self-referentiality

Citation: Brezhnev, Y.V. Linear Superposition as a Core Theorem of Quantum Empiricism. *Universe* **2022**, *8*, 217. https://doi.org/10.3390/universe8040217

Academic Editors: Steven Duplij and Michael L. Walker

Received: 8 February 2022
Accepted: 21 March 2022
Published: 28 March 2022

1. Introduction and Summary

> ... somewhat curious that, even after nearly a full century, physicists still do not quite agree on what the theory tells us ...—G. 't Hooft ([1], p. 5)

> It is almost a crying shame that we are nowhere close to that with quantum mechanics, given that it is over 70 years old now—C. Fuchs ([2], p. 32)

The contradiction between the fundamental nature of quantum theory (QT) and the phenomenological feature of its mathematics [3] is likely to never cease instigating the attempts to overcome it. As H. Putnam had said, "Human curiosity will not rest until ... questions [of the nature of the QT-formalism] are answered".

The subject-matter and leitmotiv of what follows is that the linear superposition and theory's axioms have an origin—they are derivable, and it is entirely empirical. The theory is thereby demystified, and the interpretative challenges that accompany the exegeses of QT are a *nonexistent* problem coming from "a confusion of categories" ([4], p. 89), i.e., from the "semantic confusion" ([5], p. 10). A direct outgrowth of this ideology is not only a derivation of the superposition principle (page 35) but also the axiom-free production of the "chief" quantum formula—the Born rule $p = |\alpha|^2$ [6].

1.1. On the Foundations of Quantum Theory

The debates concerning the foundations of quantum mechanics (QM) hitherto "show no sign of abating" ([7], p. 222, [8,9]), and despite widespread scepticism [10–15], it is generally acknowledged that the problem is a real one [9,16–19]; it is not something made up or "just a dispute over words" ([20], p. 5) and sometimes "has been regarded as a very serious one" ([15], p. 418, [21,22]). Say, R. Penrose has expressed (2004) an even more radical "conviction that present-day quantum mechanics has no credible ontology, so that it *must* be seriously modified".

In recent decades, the discussions have even worsened [2,23], and current research has intensified due to the tremendously increased and formerly inconceivable technological possibilities of operating with individual micro-objects and the urge to implement the idea of quantum computing [20,24].

The reason for this state of affairs remains the same as it was before. Unlike the classical theories, e.g., thermodynamics or relativity theory, "Ma di assiomatizzazioni della teoria quantistica ce ne sono moltissime" ([17], p. 30) and the QM-axiomatics itself seems wholly divorced from human language [5,8,9,15,17,25–32]. Quantum postulates are not merely formal. They are phrased in terms of linear operators on a complex Hilbert space $\mathbb{H}$ [4,10,13,25,33–37] and, with that, literally not a single word here can be brought into conjunction with reality by means that have *at least some kind* of relationship with the classical description. What is more, it is very well known that the abstract character of these terms is required by the essence of the point (covariance) and, at the same time, that the attempt to link them with physical images is imposed by a decree and results in the famous paradoxes associated with concepts, such as causality, (non)locality, and realism [9,27,28,38–45]. All of that causes a problem with interpretations of QM.

It is well known that the theory has steadfastly resisted any unique ontological reading and, in particular, reconciliation between interpretations. This is reflected not only in the voluminousness of the literature. The differences in viewpoint are often based on points of principle [3,8,14,15,46–51], and even highly qualified publications face criticism [52–55]. Among other things, we encounter appeals [3,12,16,17,43,56–58] (there is even a manifesto(s) [50], (p. 990), [59],), striking titles such as "scandal of quantum mechanics" [60,61], "QUANTUM OUTCOME: ALLAH WILLED IT?" ([62], p. 188; Wheeler), "the Oxford Questions ... to two clouds" ([63], p. 6), "The Canon for Most of the Quantum Churches" ([50], p. 988), "Quantum mechanics for the Soviet naval officers" ([64], p. 161), "the patron saint of heretics in the One True Church of Copenhagen" (about D. Bohm), "A Feminist Approach to Teaching Quantum Physics" ([2], p. 182), "Church of the Larger Hilbert Space" (J. Smolin) [2,12], and also April Fools' [65] and the medical jokes about "the "state of health of the quantum patient'" ([66], p. vii), political parallels with "Marxism ... [and] the Cold War" [67], and many more [3,9,27,68–71].

An interesting fact is that Cambridge University Press has published a 500-page-long book [2] containing an arresting electronic correspondence—D. Mermin called it "samizdat" (self-published)—between C. Fuchs and modern researchers and philosophers in the field of quantum foundations. This correspondence has continued ([23], over 2300 pages) and now covers 1995–2011. It characterizes the state of affairs in the field, and does not merely add to one's impression of the unending discussions about quantum matters (see introductory sections in [50] (!) and in [64]), it also represents, due to the lack of formality, a plentiful source of ideas and of valuable thoughts. Schlosshauer's very informative "quantum interviews" [16] pursue the same goals.

It is worth mentioning that the quantum challenges had led, quite a while ago, to attempts to revise, even formalizing, the logic of our thinking [72,73]—a very nice mathematical theory dating back to von Neuman in the 1930s ([25], Section III.5) termed quantum logic [74]. There are handbooks on that subject [75], and this topic is still under intensive investigation now. See also the last paragraph in Section 6.3.1.

The lack of transparent motivations for mathematics—a pressing requirement of physics—means that QM-formalism is hard to distinguish from a "cook book of procedures and rituals" (J. Nash), a "user-manual" ([32], p. 1690), [76], ([27], p. xiii), or from "a library of ... tricks and intuitions" [21]. Therefore, the "dissatisfaction regarding comprehension" and the "need for interpretation that is alien to an exact science" ([77], pp. 7–8) lead to the fact that "we admit, be it willingly or not, that quantum mechanics is not a physical theory but a mathematical model" ([32], p. 1701) or that "nature imitates a mathematical scheme" ([78], p. 347; Heisenberg). De facto, QT "is in a sense like a traditional herbal medicine used by "witch doctors". We do not REALLY understand what is happening" (J. Nash) and "we have essentially *no* grasp on why the theory takes the precise structure that it does" ([2],

p. 32), which raises the suspicion that "something is wrong with the theory" (H. Putnam) and that "this quantum skyscraper is built on very shaky ground" ([64], p. 8). (Throughout the text, the *italic* and *slanted* type in "quotations" is original, unless otherwise indicated.)

At the same time, well-founded opinions have long been known to the effect that "quantum theory needs no "interpretation'" [43], in refs. [3,12,60] or that "only consequences of the basic tenets of quantum mechanics can be verified by experiment, and not its basic laws" ([11], p. 16). In other words, the discrepancies between opinions are significant and often radical: from epithets such as "schizoid, ... situation is desperate" ([15], p. 420), ([79], p. 424) to supporting the rationale for quantum computations [24] and whole books written on the subject [8]. Concerning the "schizoid", the case in point is the many-world conception by Everett–DeWitt. See also pages 158, 161, 176–179 in [80] regarding the "state of schizophrenia" and "explanations" as to why "schizophrenia cannot be blamed on quantum mechanics" ([80], p. 182).

In any case, the controversy between "the warring factions, ..., many [quantum] faiths, ... and instrumentalist camps" ([16], pp. 60–61), ([81], Section 5.5), [30,33]—"[t]hey all declare to see the light, the ultimate light" ([50], p. 987)—cannot be considered as an acceptable state of affairs (see also Section 11.1) for the simple reason that the quantum philosophy issues turn into an "industry"of interpretations—an unhealthy state of affairs—while, at the same time, the very same philosophers call for its denial: "interpretation of QM emerged as a growth industry" ([82], p. 92).

1.2. Formula of Superposition

Conversely, the "dominant role of mathematics in constructing quantum mechanics" has led to the conclusion that mathematical "assumptions are usually considered to be physical" ([32], p. 1691). That is to say, "there has been a substitution of concepts" ([76], p. 295) and "one of the consequences of quantum revolution was the replacement of explanations of physical phenomena by their mathematical description" ([76], p. 296). These characteristics convey, in the best possible way, the dissatisfaction with the fact that quantum physics "was actually reduced to a physical interpretation of the Hilbert space theory" ([32], p. 1690). The $\mathbb{H}$-space in itself is a fairly cumbersome mathematical structure and even determines a crucial principle: the superposition of states [26]. It is thus not surprising that this principle becomes "one of the vague points ... the [Dirac] argument is difficult to consider as rational ... the physical principle simply fits underneath it" (excerpt from the preface to the Russian edition of [83]).

The mathematics of the $\mathbb{H}$-space contains three constituents: a vector space, the inner-product add-on over it, and topology. The two latter ones invoke the first one, which is completely independent (algebra) and begins with the formula

$$|\psi\rangle = \mathfrak{a} \cdot |\varphi\rangle + \mathfrak{b} \cdot |\chi\rangle \,. \tag{1}$$

This is the pivotal expression of quantum theory. Comprehending its genesis is tantamount to comprehending the nature of the *linearity* of QM.

In Formula (1), there occur the complex numbers $\mathfrak{a}, \mathfrak{b} \in \mathbb{C}$, symbols of operations $\cdot$ and $+$, and also vectors $|\psi\rangle$, $|\varphi\rangle$, $|\chi\rangle \in \mathbb{H}$. It is clear that until an empirical basis for all these devices is found, the interpretation of Abstraction (1) and questions of the kind "Quantum States: What the Hell Are They?" (55 times in [23]) will remain a problem. To all appearances, the problem is considered so difficult—"quantum states ... cannot be "found out'" ([8], p. 428)—that the non-axiomatic meaning of these symbols was not even discussed in the literature. In the meantime, not only is the situation far from hopeless, but it also admits a solution. The present work is devoted to gradual progress towards an understanding of Formula (1). Stated differently, Equality (1) becomes an empirical "theorem" (p. 56).

- The main part of the challenge is not only to ascertain what is being added/multiplied in Formula (1), but also to realize primarily *what "to add/multiply" is*, and "Where Mathematics Comes from" [84] at all.

"What does it mean, physically, to "add" things?" [2], (p. 178; D. Darling). More than that, aside from the symbols $\{\mathfrak{a}, \mathfrak{b}, |\psi\rangle, |\varphi\rangle, |\chi\rangle, \cdot, +\}$, Expression (1) contains the sign of equality $=$ (see also [85] (pp. 29, 30 (!), [86]), and, surprising as it may seem, it conceals one of the key points—the third principium of quantum theory (**III**, p. 29).

The guiding observation is based on the fact that the only thing that we have access to are the microscopic events, and therefore, "we have little to begin with other than what an experimental physicist would call experiments with a single microsystem" ([87], p. 5).

> "[W]e must recognize that the focusing on individual elements whatever these may be is absolutely indispensable for all our thinking. ... What may be regarded as an individual event?"
>
> R. Haag ([88], p. 302)

Consequently, we must begin from individual events and from collecting them into ensemble formations. It is precisely in this context that we will use the word empiricism—quantum empiricism of micro-acts of perception—and it is in this respect that QT has a statistical nature. As Einstein put it, "It may be a correct theory of statistical laws, but an inadequate conception of individual elementary processes" ([30], p. 156; Einstein); see also ([25], ([30], Chs. 7–8), [70], [89], pp. 38–40), ([89], p. 40). Such a viewpoint has been long championed by L. Ballentine [90] and H. Groenewold ([91], p. 468) and justified in detail by G. Ludwig [87,92–94]. A. Leggett proposes accordingly "extreme statistical interpretation"[16] (p. 79) , [95], in the sense that "to seek any further "meaning" in the formalism is pointless and can only generate pseudoquestions". With that, he overtly applies such characteristics as "complete gibberish" ([95], p. 70) and "verbal window dressing" ([16], p. 79).

The difficulty is, of course, in creating the object $|\psi\rangle$ itself. A step-by-step characterization of this procedure (Sections 3–8) and key words to what follows have been reflected in the (sub)section titles listed in the Contents.

1.3. *Physics $\rightleftarrows$ Mathematics; Doctrine of Numbers*

Thus, the situation appears to be one whereby the physics itself faces inconsistencies in its foundations and the mathematical superstructures are difficult to reconcile with its motivations (physical principles) [96]. However, on the other hand, attempts to axiomatize an interface between them [97] only conceal a deeper insight [22]. M. Born had called attention to the fact that "probable refinements of mathematical methods will not suffice to produce a satisfactory theory, but that somewhere in our doctrine is hidden a concept" and T. Maudlin was more definite: "physicists have been misled by the mathematical language they use to represent the physical world".

In other words, we observe an overemphasis on the role of the ready-made math-structures—algebras, spaces, and the like—and an under-evaluation of "seemingly naïve" empirical aspects voiced in the ordinary language [98]. The situation is no different from that which H. Weil had characterized in the introductory section to ([99], p. 10) as follows.

> "All beginnings are obscure. Inasmuch as the mathematician operates with their conceptions along strict and formal lines, he, above all, must be reminded from time to time that the origins of things lie in greater depths than those to which their methods enable them to descend".

The "origins" are expressible of course only in the natural language; Section 2 is devoted to this.

What we propose below is an implementation of the idea that the postulational view must be abandoned and replaced by a negation of the prior existence of both the physical "preconceived notions" [92] (p. 328) and the mathematical structures. Physics and mathematics should be created "from scratch". Paul Benioff calls this idea "a coherent theory of physics and mathematics" [2] (p. 33; P. Benioff), [96] (p. 639), Then, due to the initial absence of mathematics, introducing mathematical structures is almost ruled out, proofs must be replaced by an empirical inference, and semantics of physics—the language of physical reasoning—is initially under a linguistic ban. It cannot exist a priori. That is to say, even the natural-language conjunction of mathematical terms with physical adjectives (and verbs [100] (p. 3102; "to happen, to be, to exist")) becomes far from being free, as with the classical description's language (Sections 2.1, 5.4, 6.4 and 6.5). R. Haag, on the first page of the work [101], emphasizes:

- "we should not consider ["vocabulary of Quantum Theory"] as sacrosanct. … every word in the vocabulary is subject to criticism".

Returning to the ensemble formations, it is only they that have to come to the fore, and argumentation should be subordinated only to the low-level microscopic empiricism. The predominance of the empirical over the theoretical will then immediately touch on the closest creature of the latter—the notion of a number—since numbers do not come "from the sky", and the theory will have to be a quantitative one.

Despite the overflow of abstracta in QT, the *doctrine of number*—$\lceil$number × unit$\rceil$, to be precise (Sections 7 and 9)—has, it seems, not yet entered foundational discussions [102]. Consequently, the numbers turn into a kind of "problem of numbers" (principium **II**), and we are thus led to the necessity of revising the take on the foundations themselves:

$$\lceil\text{quantum fundamentals}\rceil \quad \dashrightarrow \quad \lceil\text{the problem/doctrine of the number}\rceil\,.$$

This paradigm shift is a unique trait of the quantal (not the classical) view of things and a substantial part of the following is devoted specifically to that.

In the outline of the present work, the workflow will constitute re-creating the structure of a linear vector space. More precisely, producing an *a priori unknown* mathematics, which *will be* an algebra of such a space with a complex conjugation. As a matter of fact, we provide an answer to Haag's question "How do we translate the description of an experimental arrangement into mathematical symbols?" in the context of their own "idea of basing the interpretation of quantum theory on the concept of "events" which may be considered as facts independent of the consciousness of an observer" ([88], p. 295).

The main point to be immediately emphasized is that the mathematization of the discrete micro-acts of observations is quite a nontrivial procedure (105), and further, the strategy, along with the structure of this article, can be schematized as follows.

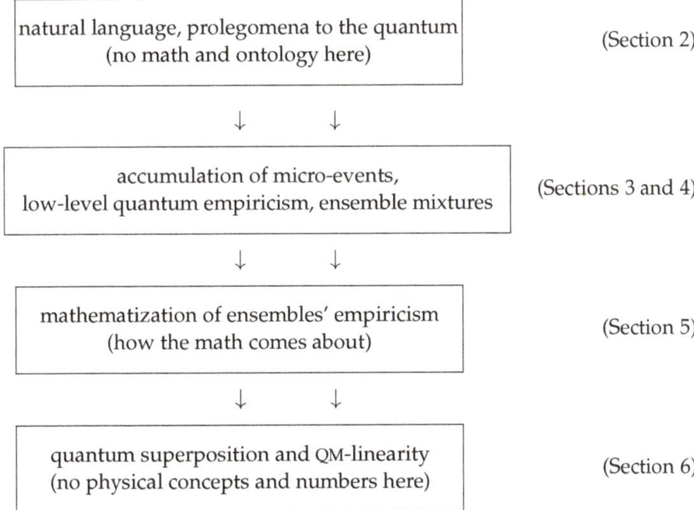

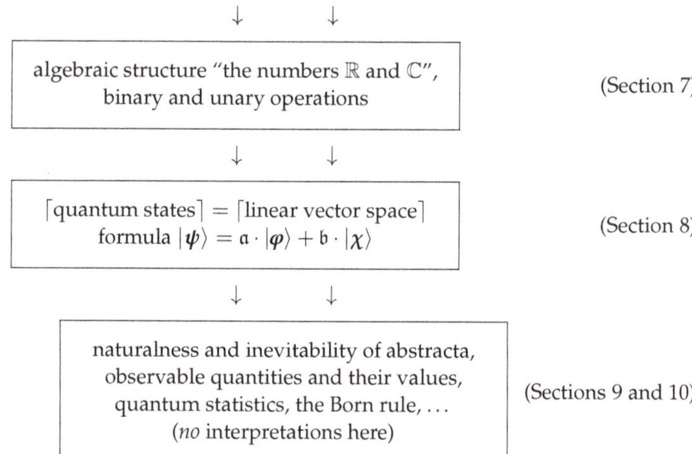

This box diagram cannot be reduced or restructured. For example,

- *Superposition foregoes numbers*, and measurement and physical properties follow *strictly after* the |`ket`⟩-vectors have been created.

By and large, the aforesaid ideology is supported by the common belief—often certainty even [12]—that QM is *not* perturbative, its linearity is *not* associated with linear approximation of something else, and, in general, it is *not* extensible (ultimate [103] and non-deformable) and must be free of interpretations [12,43]. All of these concerns, in one way or another, are directly related to the derivation of Formula (1).

2. Points of Departure

> In the Beginning was the Word—A. Zeilinger ([39], 01:05′47″)

> Most of the time the apparatus is empty and sometimes you have a photon coming through—A. Zeilinger ([39], 12′39″)

Since empiricism is in essence supra-mathematical [104], i.e., it is concerned with *meta*mathematics [105,106], its mathematization, i.e., theory construction, should begin not with postulates and definitions, but rather with the semantic formation of an object language (of "the Quanta") of vocabulary that "may only be described by "words" and not by a theory" [58], [87] (p. 106), [93]. As A. Peterson and K. Popper had observed, "Math can never be used in phys until have words" ([107], p. 209), i.e., "we cannot construct theories without using words" ([108], p. 12). Therefore, relying on the established understanding of the underlying causes for the quantum eye on physics [13,25,27,33], up until the end of this section, we will adopt the natural-language meaning of the words observation, system, state, numbers (!), plus and to divide, physical influence, transition, large/small, micro/macro, etc. Their contents will later be defined more precisely or entirely changed. For instance, the sense of the word "state" will be drastically transformed, to which we are drawing attention in advance. Accordingly, a degree of informality—it has been clarified in Remark 10—is inevitable here, but "the lack of precision ... is a necessity" ([109], p. 48) at the moment.

2.1. *Variations as Micro-Level Transitions*

We will (and "must" [105] (Ch. 3)) first view the concept of a system at an intuitive level ([110], Section 1.1)—there is what is referred to as an isolated system $\mathcal{S}$.

S Let us tentatively (a priori) relate the concept of a *state* to the associated context describable by the words "the system $\mathcal{S}$ can *vary*, *be different*, or *in different states*". That is to say, system $\mathcal{S}$ is always in a certain state $\underline{\Psi}$ belonging to the set $\mathfrak{T} = \{\underline{\Psi}, \underline{\Phi}, \ldots\}$, each element of which is admissible for $\mathcal{S}$, and all of them are different from each other: $\underline{\Psi} \neq \underline{\Phi}$.

In other words, the concept of a quantum system may not have a precise/axiomatical definition at the moment. Otherwise, if it comes to that, the system is what is being constantly varied when observed, and "varied" is the key word here.

The statement "states are different" does not require a consideration when $\underline{\Psi}$ and $\underline{\Phi}$, referred to as state, are the abstract elements of an abstract set $\{\underline{\Psi}, \underline{\Phi}, \ldots\}$. However, in order to tie its elements to reality, we have to introduce the criteria of coincidence/distinguishability of one from the other. Criteria may not come from observational procedures, without which it is impossible to either detect states or claim that they differ, coincide, or that they are, if any.

On the other hand, the nature of micro-phenomena shows that observations are always associated with an irreducible intervention in the system, manifesting in what is known as *transition* $\underline{\Psi} \rightsquigarrow \underline{\Psi}'$ (or destruction). As an example, observations at accelerators are literally the destructions, and bulk at that. Due to a lack of criteria, there is no sense in attributing to this concept the adjectives small/large, (in)significant/partial, or collocations such as "comparison of destructions at instants t_1, t_2". Let us proceed from the idea that initially there is nothing but the transition. Transitions may actually occur without destruction $\underline{\Psi} \rightsquigarrow \underline{\Psi}$, however.

Two different $\underline{\Psi}$, $\underline{\Phi}$ may be destroyed into new $\underline{\Psi}'$, $\underline{\Phi}'$, as well as into the combinations of the old/new. Thus, strictly speaking, the sense of words "different, new, ..." eludes us in this case, which is why even the identification of $\underline{\Psi}$-elements and the $\mathfrak{T}$ itself, as a set, becomes questionable. Therefore, besides the formal writings $\underline{\Psi} = \underline{\Phi}$ and $\underline{\Psi} \neq \underline{\Phi}$ for $\underline{\Psi}, \underline{\Phi} \in \mathfrak{T}$, the *physical* distinguishability/equivalence (recognizability $\not\approx/\approx$) needs to be established. As to the identification (and to the identity) in this regard, see von Neumann's reasoning: "`One might object against` II ..." on page 302 of their book [25]. The sole thing that distinguishability may rely on is the transition acts. In turn, variation is a key element in transitions, which is why we will begin constructing with distinguishability.

Let us take the still virtually unlimited way $\mathscr{A}$ of intervening $\stackrel{\mathscr{A}}{\rightsquigarrow}$ in $\mathcal{S}$ and attempt to introduce distinguishability $\underline{\Psi} \not\approx \underline{\Phi}$ as $\mathscr{A}$-distinguishability. Due to the fact that micro-transition $\underline{\Psi} \stackrel{\mathscr{A}}{\rightsquigarrow} \underline{\Psi}'$ is not pre-determined, initial states $\underline{\Psi}$ undergo arbitrarily free changes. Next time, the results will be different and absolutely arbitrary (the term "different" is understood to mean $\neq$), and each act is indiscernible from a case in which it contains ones similar to itself within itself. It would be natural to associate such a case to the absurd, which is unrelated to the meaning of the words "physical observation", and to discard the given $\mathscr{A}$.

Non-meaninglessness arises only if we impose the negation of random combinations of $\neq$ and $=$ in transitions, at least for a part of $\mathfrak{T}$, i.e., introduce the preservation acts $\underline{\Psi} \stackrel{\mathscr{A}}{\rightsquigarrow} \underline{\Psi}$. The "preservation" should be read here as indestructibility of state, i.e., as a $(=)$-coincidence under the secondary act $\underline{\Psi} \rightsquigarrow \underline{\Psi} \rightsquigarrow \underline{\Psi}$. Otherwise, the vanishing difference between "preservation" and "variation" leads to linguistic chaos ([111], p. 232). This means that the destruction $\underline{\Psi} \rightsquigarrow \underline{\Psi}'$ may not be considered as a one-fold one. State $\underline{\Psi}'$ on the right should be examined for changeability and transform into the left part of the subsequent transition: $\underline{\Psi} \rightsquigarrow \underline{\Psi}' \rightsquigarrow \underline{\Psi}''$. Thereby, the structure $\underline{\Psi}' \rightsquigarrow \underline{\Psi}''$ with the *binate* entity "before/after" or "on the left/right" becomes the key one, and we consider it an initial object in subsequent constructs. The preserved states are, by definition, those that pass the reproducibility test.

Thus, logic requires beginning with the transition compositions

$$\underline{\Psi} \stackrel{\mathscr{A}}{\rightsquigarrow} \underline{\Psi}' \stackrel{\mathscr{A}}{\rightsquigarrow} \underline{\Psi}'' \stackrel{\mathscr{A}}{\rightsquigarrow} \cdots ,$$

wherein the cases such as

$$\cdots \quad \underline{\Psi}' \stackrel{\mathscr{A}}{\rightsquigarrow} \underline{\Psi}' \stackrel{\mathscr{A}}{\rightsquigarrow} \underline{\Psi}'' \quad \cdots \tag{2}$$

are ruled out (a ban on changing of what has been unchanged), and the never-ending sequence

$$\underline{\Psi} \overset{\mathscr{A}}{\dashrightarrow} \underline{\Psi}' \overset{\mathscr{A}}{\dashrightarrow} \cdots \overset{\mathscr{A}}{\dashrightarrow} \underline{\Psi}'' \overset{\mathscr{A}}{\dashrightarrow} \cdots \tag{3}$$

(non-recognisability of states) must be terminated

$$\underline{\Psi} \overset{\mathscr{A}}{\dashrightarrow} \underline{\Psi}' \overset{\mathscr{A}}{\dashrightarrow} \cdots \overset{\mathscr{A}}{\dashrightarrow} \underline{\Psi}'' \overset{\mathscr{A}}{\dashrightarrow} \underline{\alpha} \overset{\mathscr{A}}{\dashrightarrow} \underline{\alpha}\,, \tag{4}$$

yielding a "finiteness" (= realisticness) and the concept of conserved/distinctive $\underline{\alpha}$-states. The terminology $\underline{\alpha}$-event [12] could be used instead.

Freedom of elements in Sequence (4), including the choice of $\underline{\alpha}$-states, is not limited by anything besides the ban on Equation (2). Therefore, this arbitrariness, which is physically never recognizable, curtails the generic chain from Sequence (4) into the shortened one

$$\underline{\Psi} \overset{\mathscr{A}}{\dashrightarrow} \boxed{\;\cdots\cdots\;} \overset{\mathscr{A}}{\dashrightarrow} \underline{\alpha} \overset{\mathscr{A}}{\dashrightarrow} \underline{\alpha}\,, \tag{5}$$

which is identical to the scheme

$$\boxed{\cdots\underline{\Psi}\cdots} \overset{\mathscr{A}}{\dashrightarrow} \underline{\alpha} \tag{6}$$

with certain $\underline{\alpha} \in \mathfrak{T}$.

Discussions on "what happens ... [and] "how" ([25], p. 217) at the very microscopic level are extremely widespread in the literature [18,41,44,45,53,56,77] (see [19,33,40,112] for the exhaustive references), although it is not difficult to predict the fact that the attempts to understand the inner structure of Box (6) will only lead back to an identical box; so, the "turtles all the way down" (ascribed to W. James), followed by the great Wheeler's slogan "No tower of turtles" (1989).

Indeed, the uncontrollability of micro-changes is universally known, yet describing them as a process in time $t \mapsto t+\varepsilon$ will start employing language terminology—functions, arithmetic operations, the physical words, etc.—that has not yet been created even for the fixed instants t_1, t_2. However, what may be associated with fixed time are only non-temporal entities, for which we have nothing but transitions (Equation (5)). The attempt to manage them, i.e., to control intervention in $\mathcal{S}$, results in looping or "measuring the measurement", in addition to the ambiguity of this term itself.

> "[I]t is not meaningful to speak of a measurement "at time $\tilde{t}$." ... the real physical meaning of the time parameter ... has nothing to do with the notion "time of measurement"". "[T]he description of the measurement process in quantum mechanics in terms of "pre-theories" is not possible"
>
> G. Ludwig ([87], p. 288), ([92], p. 340)

See also [58] (p. 100), [92] (p. 365), [94] (p. 150), [113] (pp. 644–646), and [114]. Just as before, the physical assessments such as "abrupt", "(ir)reversible", "(non)simultaneous", "immediately following ..." [25] (pp. 231, 410), or the "weak/nondemolishing" (measurements [53]), etc.are unacceptable here. No temporal process may be present in the foundations of the theory (([87], Sections VII.4, 6), ([92], Chs. III, XVII), [93]) since it is immediately not clear: "Furthermore, what exactly are we having at instants t_1 or t_2?". In the reverse direction—⌈time ⇢ measurement⌉—the situation is also rather indefinite since the ""Time" is not an entity to which the operations of measurement, direct or indirect, apply" ([114], p. 5).

Remark 1. *All the information stated above means that attempts to deduce* QM *dynamically ([16], 10 · Reconstructions) are beforehand doomed to vicious circles "round the boxes" and time t, such as attempts to dynamically "vindicate" Lorenz's contraction instead of kinematic postulates of the relativity theory [69]. A consistent theory must rest either on "irreducible" elements (6) or upon "boxes" of a different kind. In the latter case, the theory becomes a particular* model with interpretation; *e.g., the Lindblad equations [115,116], decoherence [112,117–119], stochastic dynamics,* *and other statistic-dynamical models [40,77]. Anyway, an ability to model and understanding are not the same thing, and this point was repeatedly emphasized in the literature ([16], ([17], Section I.2), [32]) with regard to* QT.

That said, if theory is built *as a fundamental one rather than as a model* ([16], p. 144), with a primary entity *changeability* $\overset{\mathscr{A}}{\rightsquigarrow}$, Box (6) may only be involved in it as the initial starting point and as an *indescribable* object with the absolute rather than with a relative sense. Elements of reality, in whatever understanding—say, Bell's "*be*ables" [28]—may not exist before/after/inside/outside of the box. It can only be the *structureless abstractio*. Accordingly, the notions of preparation, of measurement, of "interaction with", and of a physical process are meaningless without the construction of Box (6).

These statements are clearly in agreement with the fact that any reasoning must not contradict the formal logical rules [105]; hence, there must exist [96,106,120] the empirically undefinable logical atoms. A. Peres writes ([121], p. 173): "While quantum theory can in principle describe *anything*, a quantum description cannot include *everything*. In every physical situation *something* must remain unanalyzed". Moreover, as Pauli put it, "Like the ultimate fact without any cause, the individual outcome of a measurement is … not comprehended by laws". Specifically, the set $\mathfrak{T}$ and transitions arrows $\overset{\mathscr{A}}{\rightsquigarrow}$ are also the atoms. It is a "… preexisting concept … We cannot formulate the theory without this concept", concludes B. Englert ([12], p. 2). From the aforesaid, we may formulate the following tenet.

I Quantum statics should forego quantum dynamics.

(***The first principium of quantum theory***)

The rationales do not end here and will be later amplified once we begin to exploit the terminology that is usually taken for granted from the outset, viz., the quantitative descriptions ([2], p. 178). If they arise not as numerical interpretations of something but out of an experiment, then observation should be the beginning, and the "manufacture of numbers'—the end. In other words, the model "theory with boxes" other than Boxes (5) and (6) implicitly implies the logical sequence ⌈model of process⌉ $\rightarrowtail$ ⌈numerical interpretation⌉, in which empiricism holds a role other than primary. It is clear that, regardless of the model, such a situation will always remain unsatisfactory in the physical respect.

2.2. Observation

The sequences addressed above lead to the following outcome.

- Any meaningful micro-act $\overset{\mathscr{A}}{\rightsquigarrow}$ either saves a state ($\underline{\alpha} \overset{\mathscr{A}}{\rightsquigarrow} \underline{\alpha}$) or turns it into a conserved one ($\underline{\Psi} \overset{\mathscr{A}}{\rightsquigarrow} \underline{\alpha}$).

The two extremes do not contradict this fact. The first—maximally rough observations—is when all states are destroyed into a certain one: $\underline{\Psi} \rightsquigarrow \underline{\Psi}_0$ ("whatever and however we watch, all we see is one and the same"). In this, the state $\underline{\Psi}_0$ is not destroyed: $\underline{\Psi}_0 \rightsquigarrow \underline{\Psi}_0$. Another extreme is when none of the states are destroyed: $\underline{\Psi} \rightsquigarrow \underline{\Psi}$. This is the case of ideal (quantum) observation, but, due to the absence of any changes, it is indistinguishable from the case where observations are entirely absent.

Situated in between these extremes lies the simplest case with two distinctive states

$$\underline{\alpha}_1 \overset{\mathscr{A}}{\rightsquigarrow} \underline{\alpha}_1\,, \qquad \underline{\alpha}_2 \overset{\mathscr{A}}{\rightsquigarrow} \underline{\alpha}_2\,. \tag{7}$$

Of course, these are prohibited from transitioning into each other. Because there is still the free admissibility of transitions $\underline{\Psi} \overset{\mathscr{A}}{\rightsquigarrow} \underline{\alpha}_1$, $\underline{\Psi} \overset{\mathscr{A}}{\rightsquigarrow} \underline{\alpha}_2$, we can turn the semantic sequence

$$\lceil \text{arbitrariness} \quad \rightarrowtail \quad \text{preservation} \quad \rightarrowtail \quad \text{distinctive } \underline{\alpha}\text{'s} \rceil$$

into the more rigorous scheme

$$\lceil \mathfrak{T} = \{\underline{\Psi}, \underline{\Phi}, \dots\} \rceil + \lceil \mathscr{A}\text{-observations} \rceil \quad \rightarrowtail \quad \{\underline{\alpha}_1, \underline{\alpha}_2, \dots\} =: \mathfrak{T}_{\mathscr{A}} \subset \mathfrak{T}\,, \tag{8}$$

which gives, even though partially, rise to the concept of a physical distinguishability ("distinguo"). It is formally defined only on the subset $\mathfrak{T}_{\mathscr{A}}$: the statement $\underline{\alpha}_1 \not\approx \underline{\alpha}_2$ is equivalent to (7). To avoid overloading the further notation, we do not use symbols such as $\approx_{\mathscr{A}}$ and $\not\approx_{\mathscr{A}}$; the context is always obvious.

O By a *physical observation* $\mathscr{A}$ or, in short, *observation* we will mean such interventions $\overset{\mathscr{A}}{\dashrightarrow}$, in which the "never-ending" chaos (3) is replaced by chaos with the notion of preservation, i.e., "chaos with rule (6)":

$$\underline{\Psi} \overset{\mathscr{A}}{\dashrightarrow} \underline{\alpha}, \quad \text{where} \quad \underline{\alpha} \overset{\mathscr{A}}{\dashrightarrow} \underline{\alpha}\,. \tag{9}$$

The set of $\underline{\alpha}$-objects $\mathfrak{T}_{\mathscr{A}}$ with the property

$$\underline{\alpha}_1 \overset{\mathscr{A}}{\dashrightarrow} \underline{\alpha}_1, \qquad \underline{\alpha}_2 \overset{\mathscr{A}}{\dashrightarrow} \underline{\alpha}_2, \qquad \ldots \tag{10}$$

is discrete, and the $\underline{\alpha}_s$ themselves are termed *the eigen* (proper) *for observation* $\mathscr{A}$. They define $\mathscr{A}$ and do not depend on $\mathcal{S}$. No logical connection between $\underline{\Psi}$ (the left of (9)), family $\mathfrak{T}_{\mathscr{A}}$, and system $\mathcal{S}$ exists.

(The comprehensive terminology here is this: a micro-act of observation by instrument $\mathscr{A}$. The zig-zag arrow $\rightsquigarrow$ is replaced with the straight one $\dashrightarrow$.) Expressed another way, the introduction of the concept "the eigen" is equivalent to the following informal, yet minimal, motivation: *at least some* certainty instead of *total* arbitrariness.

Two instruments $\mathscr{A}$ and $\mathscr{B}$ may have arbitrarily different eigen-states $\{\underline{\alpha}_1, \ldots, \underline{\alpha}_n\} \neq \{\underline{\beta}_1, \ldots, \underline{\beta}_m\}$. Accordingly, as regards observation $\mathscr{B}$, the (distinctive) states $\{\underline{\alpha}_s\}$ do not differ, in general, from the "regular" $\underline{\Psi}$'s, i.e., from those chaotically destroyable into the $\mathscr{B}$-eigen states: $\underline{\alpha}_j \overset{\mathscr{B}}{\dashrightarrow} \underline{\beta}_k$. All kinds of instruments $\{\mathscr{A}, \mathscr{B}, \ldots\}$ are thus defined by aggregates $\{\mathfrak{T}_{\mathscr{A}}, \mathfrak{T}_{\mathscr{B}}, \ldots\}$. The number $|\mathfrak{T}_{\mathscr{A}}|$ of corresponding $\underline{\alpha}$-objects therein may be an arbitrary integer. There are also no (logical) grounds for restricting/prescribing the composition of $\mathfrak{T}_{\mathscr{A}}$. Any element of $\mathfrak{T}$ may be the conserved one for a certain instrument. Parenthetically, the notion of an eigen-state—in different forms—is sometimes present in axiomatics of QM [18,72,122].

In a generic case, the chaos present in Rule (9) leaves open the problem of correlating the recognizability $\underline{\Psi} \not\approx \underline{\Phi}$ (or $\underline{\Psi} \approx \underline{\Phi}$) with physics. Clearly, the issue is linked to the ambiguity of the term $\underline{\Psi}$-state itself, which is used in pt. **S**—an important point—because we need to start with something since building the mathematical description without some sort of a set is impossible.

Remark 2. *Informally, metalinguistic semantics—the association of meanings with texts [58]—is in general as follows. Inasmuch as we are receiving* different *$\underline{\alpha}$-responses to each micro-act $\overset{\mathscr{A}}{\dashrightarrow}$, let us say that "on the other side from us there is something that can also be different", and all of that is to be described. This reflects our intuitive perception of reality, which, both at the micro-level and the macro-level, boils down to pt.* **S** *and to an ineradicable pair: ⌈something outside⌉ + ⌈that which can be different for us⌉. If we give up either of these semantic premises—"something outside" ($\underline{\Psi}$) or "can be different" ($\underline{\alpha}_j$)—then, as above, we face a linguistic dead end, as the possibility for reasoning disappears. There must be two sides present. Because of this, the arrow $\overset{\mathscr{A}}{\dashrightarrow}$ must be accompanied by "some things" to the left and right of it. The low-level set $\mathfrak{T} = \{\underline{\Psi}, \underline{\Phi}, \ldots, \underline{\alpha}_1, \underline{\alpha}_2, \ldots\}$ does arise. Then, the arbitrary elements $\underline{\Psi} \in \mathfrak{T}$ (unrestricted chaos) are assigned to the left of this act instead of "some thing" and the micro $\not\approx$-distinguishable $\underline{\alpha}$-objects ($\underline{\alpha}_j \not\approx \underline{\alpha}_s$) to the right. Put another way,*

- *What is being abstracted is not "concrete things" ([13], p. 27) or behavior of things" ([123], p. 414) but a primitive element of perception—a micro-event—the $\underline{\alpha}$-click. Other than "the click", no entities, such as very small objects/particles, fields, or, much less the knowledge, human psychology, "personal judgments", "memory configuration" [52,124], "mysterious interaction … brain of the observer" [108] (p. 11, thesis 3), [113] (p. 645), agents, their belief/consciousness [55,71,125], etc., may exist in empiricism. This is a kind of "*Radical Empiricism*… [by] William James" ([23], pp. 289, …). The "click … and nothing more" ([16],* *p. 42; Č. Brukner) is a kind of experimental zero-principium of* QT*. Therefore, the initial math premises of* QT *should contain nothing but the* $\not\approx_{\mathscr{A}}$*-distinguishability and formalization from* (9) *and* (10).

Ideas of "a click (signal) in a counter" ([126], p. 758) have, time and again, already been expressed in the literature [2] (A. Peres), [123], and we draw attention to answers of Č. Brukner on pp. 41–43 in [16], their work [127] (p. 98), and page 635 in [22]. "Having grown up collecting clicks ... I would start with "clicks" as the only point of contact between observer and observed", wrote J. Summhammer in [23] (p. 261). It may be added that the micro-observation, as such, is terminated at the eigen elements; one and the same $\underline{\alpha}_j$ *has always remained on the right.*

As a result, the minimal entity $\underline{\Psi} \overset{\mathscr{A}}{\dashrightarrow} \underline{\alpha}_j$ constitutes, mathematically, an ordered pair $(\underline{\Psi}, \underline{\alpha}_j)$ of elements of the set $\mathfrak{T}$, which are labeled by the symbol $\mathscr{A}$, which is equivalent to the $\mathfrak{T}_{\mathscr{A}}$-family (8). Accordingly, the customary physical notion of the observation is substituted for a micro-event, an act. "Physics should forget" about processes or time of interaction when observing about the interaction itself and about anything but $\underline{\Psi} \overset{\mathscr{A}}{\dashrightarrow} \underline{\alpha}$. This object represents a completed formalization of the empirical/laboratory notion of a quantum micro-event—a detector click. The click is sometimes considered from an information viewpoint as an information bit [22]. However, it cannot be such a (classical) bit with reified content because it is completely unpredictable. The next (different) click does entirely negate the previous one, and the information bit is in turn a concrete thing—the bit. For the same reason, there cannot be any information behind the single event. It is "too small and too momentary" to possess or to carry information about something inasmuch as even the "something" is composed from elementary clicks—see below.

2.3. Numerical Realizations

Is there a possibility of relying exclusively on the inflexibility of the eigen-type elements (10) or of defining the sought-for ultimate distinguishability $\not\approx$ through the $\mathscr{A}$-(micro)distinguishabilities $\underline{\alpha}_j \not\approx_{\mathscr{A}} \underline{\alpha}_s$? Let us formulate a thesis.

T There is no (linguistic) means of recognizing the system $\mathcal{S}$ to be different (pt. **S**) other than through the results of its destructions into the $\{\underline{\alpha}_1, \underline{\alpha}_2, \ldots\}$-objects of observational instruments $\mathscr{A}$.

Granted, the stringency of this linguistic taboo (**T**) must be accompanied by something constructive, and we will adopt the following program, which reflects the fact that the unequivocal description may only take the form of a quantitative mathematical theory.

R • Out of the primary ("proto")elements $\{\underline{\Psi}, \underline{\alpha}, \ldots\} \in \mathfrak{T}$, one constructs a new set $\mathbb{H}$, of which the elements

$$|\Xi\rangle := \oplus(\mathfrak{a}_1, |\boldsymbol{\alpha}_1\rangle; \mathfrak{a}_2, |\boldsymbol{\alpha}_2\rangle; \ldots) \in \mathbb{H} \tag{11}$$

are said to be (number) *representations* in the "reference frame for instrument $\mathscr{A}$", and $\mathfrak{a}_s$ are the numerical objects. The distinguishability relation $\not\approx_{\mathscr{A}}$ is carried over to $\mathbb{H}$ and admits an $\mathfrak{a}$-coordinate realization there—symbol $\not\approx$.

•• No preferential or preordained observational reference frame $\mathscr{A}\{\underline{\alpha}_1, \underline{\alpha}_2, \ldots\}$—an absolute instrument—exists.

Identification (11) is always tied to a certain family $\mathfrak{T}_{\mathscr{A}}$. Accordingly, images of $\underline{\alpha}_s$—symbols $|\boldsymbol{\alpha}_s\rangle$—are present in Equation (11), and character $\oplus$ is also no more than a symbol here. Even though coordinates $\mathfrak{a}_s$ are declared to be numbers or aggregates of numbers, there is no arithmetic stipulated for them yet. The number is a name for $\mathfrak{a}_s$. The distinguishability $|\boldsymbol{\Psi}\rangle \not\approx |\tilde{\boldsymbol{\Psi}}\rangle$ of two representatives

$$\oplus(\mathfrak{a}_1, |\boldsymbol{\alpha}_1\rangle; \mathfrak{a}_2, |\boldsymbol{\alpha}_2\rangle; \ldots) =: |\boldsymbol{\Psi}\rangle, \qquad \oplus(\tilde{\mathfrak{a}}_1, |\boldsymbol{\alpha}_1\rangle; \tilde{\mathfrak{a}}_2, |\boldsymbol{\alpha}_2\rangle; \ldots) =: |\tilde{\boldsymbol{\Psi}}\rangle$$

by means of numbers $\mathfrak{a}_k \neq \tilde{\mathfrak{a}}_k$ and mathematical implementation of (11) and of the $\mathbb{H}$-space, i.e., a "coordinatization" scheme have yet to be established. This will comprise the meaning of the word "constructs" (Sections 7–8), which may not be even linked to the mathematical term mapping yet, since *no math of* QM *exists at the moment*. It immediately follows that the question about number entities—specifically, about (11)—is nontrivial in physics.

II To speak of an exact correspondence between experiment and mathematics ($\lceil$observation + measurement$\rceil$) makes no sense until there is a detailed *mechanism for the emergence* of what is understood by number.

(***The second principium of quantum theory***)

In other words, we wonder what an empiricist/observer understands (semantics) by the word (syntax) "number". The underlying message here implies that the reliance upon the all-too-familiar arithmetic elucidates nothing. *There is no arithmetic in interferometers/colliders*—there are only clicks there—and the empirical nature arising from this construction (along with the measurement) must be scrutinized.

From pts. **T**, **R**, and **II**, it also follows that the search for a description through hidden variables, over which something is averaged, is indistinguishable from the utopian attempts to find out intrinsic content of boxes (5).

2.4. Macro and Micro

The task becomes more precise at this point. Instead of nonphysical identity/ noncoincidence ($\underline{\Psi} = \underline{\Phi}$ or $\underline{\Psi} \neq \underline{\Phi}$) of two abstract elements $\underline{\Psi}$, $\underline{\Phi}$ of the abstract set $\mathfrak{T}$, we need the concept of a physical $\approx$-equivalence ($\not\approx$-distinguishability) of $\mathbb{H}$-representatives $\{|\boldsymbol{\Psi}\rangle, |\boldsymbol{\Phi}\rangle, \ldots\}$. That is, there must hold either relation $|\boldsymbol{\Psi}\rangle \approx |\boldsymbol{\Phi}\rangle$ or its negation $|\boldsymbol{\Psi}\rangle \not\approx |\boldsymbol{\Phi}\rangle$ for all $|\boldsymbol{\Psi}\rangle, |\boldsymbol{\Phi}\rangle \in \mathbb{H}$. The primitive set $\mathfrak{T}$, initially required by point **S**, must disappear from the ultimate mathematics of symbols $|\boldsymbol{\Psi}\rangle \in \mathbb{H}$. Therefore, elements $\underline{\Psi} \in \mathfrak{T}$ are henceforth named *primitives*.

Let us sum up the fallaciousness of the metaphysical belief in the meaningfulness of the wording "there is a quantum state", i.e., the belief that the existence of a state has some math-numerical form.

- There is no a priori way to endow the term (quantum) state of system $\mathcal{S}$ with any meaning ([15], p. 419). It may not have a definition and any predefined semantics. This term should be created. Meanwhile, one cannot get around the concept of the (micro)observation $\mathscr{A}$ [127] (pp. 98–100), [113] (p. 646), [34,96]. Essentially, no one thing, including $\underline{\Psi}$, $\underline{\alpha}$, or the $\mathfrak{T}$-set itself, can be the primary bearer of data about $\mathcal{S}$. "There is an entirely new idea involved, ... in terms of which one must proceed to build up an exact mathematical theory" (P. Dirac [26] (p. 12)).

There is no escape from quoting K. Popper: "... language for the theory; ... it remains (like every language) to some extent vague and ambiguous. It cannot be made "precise": the meaning of concepts cannot, essentially, be laid down by any definition, whether formal, operational, or ostensive. Any attempt to make the meaning of the conceptual system "precise" by way of definitions must lead to an infinite regress, and to merely *apparent* precision, which is the worst form of imprecision because it is the most deceptive form. (This holds even for pure mathematics.)" ([108], p. 13).

The notions of a physical observable and of its observable values are also ambiguous at this point ([87], p. 5). Their ambiguity is even greater than that of state due to questions such as "what is being measured?" and even 'what is a measurement?'. Nonetheless, up until the end of this section, we will not discard the term state within the context of pt. **S**.

The irreproducibility of outcomes, i.e., the "turnability of $\underline{\Psi}$-primitives into the various", leaves only one option: "to take a look at $\mathcal{S}$ again, once again, ..."—in other words, to seek the source of description in repeatability. It is necessary, then, to move to the subject of macro- rather than micro-observation. This intention fits perfectly with the undefined verb "constructs" in pt. **R**, and the following paradigm should be understood as the macro.

M The only way of handling the uncontrollable micro-level changes is the treatment of the results of repeated destructions, accompanied by what we shall call the common physical macro-setting (experimental context):

$$\begin{array}{ccccccccc} \underline{\Psi} & \cdots & \underline{\Psi} & \underline{\Psi} & \cdots & \underline{\Psi} & & \cdots\cdots & \\ \mathscr{A}\downarrow & & \downarrow & \downarrow & & \downarrow & & \downarrow\mathscr{A} & \\ \underline{\alpha}_1 & \cdots & \underline{\alpha}_1 & \underline{\alpha}_2 & \cdots & \underline{\alpha}_2 & & \cdots\cdots & \end{array} \quad + \quad \ulcorner\text{common macro-environment } \mathbf{M}\urcorner\,. \tag{12}$$

To be precise, we should have to (and we shall do) indicate the different $\{\underline{\Psi}, \underline{\Psi}', \ldots\}$ here because the same ingoing $\underline{\Psi}$'s in (12) is a preassumption, which we eschew throughout the work. This point will be very fully addressed further below (Sections 2.5, 2.6, and 3.1).

The importance of repetitions and distinguishability had long been noted (Bohr, von Neumann et al. [78]), and recently, it was particularly emphasized in the work [128]. The words "copy/repeat... /distin..." occur 90 times therein.

Thus, the empiricism of quantum statics forces us to operate exclusively with such formations of copies $\underline{\alpha}, \ldots, \underline{\Psi}$, and this is the maximum amount of data provided by the supra-mathematical problem setup. ***All*** further mathematical structures may come only from constructions such as (12) and from nothing else. Getting ahead of ourselves, let us once again turn our attention to the fact that the implementation of this idea *is not short-length*—"the mathematization process (cor) is not simple" ([58], p. 24), and Sections 3–9 are devoted specifically to this—see, e.g., the chain (105).

One can once more repeat (Section 1.3) that much of what follows does not and cannot contain the mathematical definienda and proofs as they are usually present in the literature on quantum foundations. Instead, there appears a step-by-step *inference* of objects as they result themselves: numbers, operations, groups, algebras, etc. The only instrument that may be applicable here is the empirical inference.

The common macro-environment **M** in (12) is also viewed as a supra-mathematical notion [106], the mathematical implementation of which is yet to be created. The same considerations regarding qualitative adjectives are applicable as to the physical convention **M** as well as the transition acts in Section 2.1. Representations (11) will be the formalization of the meaning $\ulcorner$observation$\urcorner$ + $\ulcorner$data on system $\mathcal{S}\urcorner$, but now with no references to the elementary acts in (12). The physical distinguishability criteria $|\boldsymbol{\Psi}\rangle \not\approx |\boldsymbol{\Phi}\rangle$ may not be formulated yet because the physical attributes are not yet available, but $|\boldsymbol{\alpha}_s\rangle$-elements have already appeared in (11) as prototypes of explicitly distinguishable $\underline{\alpha}_s$.

Remark 3. *The dual form of the typical quantum statements such as "$\mathcal{S}$ is a* micro-system *and $\mathscr{A}$-instrument is a* macro-object*" (N. Bohr) is identical to the initial premise "observation does always destroy a system". It follows that there is actually little need for that terminology. Indeed,* QM*-micro has no internal structure and, hence, an oft-discussed issue about boundary (and limit (According to A. Zeilinger, "... no limit. The limit is only a question of ... money and of experience" ([39], 13'09").)) between micro and macro [8,30,33,90] is devoid of sense; "The notions of 'microscopic' and 'macroscopic' defy precise definition" ([28], p. 215). Therefore, this may be a matter only of "different macro", either "smaller/bigger", i.e., when they describe certain models.*

*As a (partially philosophical) note, what is understood by observational randomness does, in fact, boil down to distinguishability, and more specifically, to postulating the micro-chaos (9). In considering the denial of (9) as an impossible proposition, we arrive at the **M**-paradigm and conclude that the only way to deal with that which is contemplated for the subject-matter of a physical description must be the treatment of micro-acts as assemblages ([129], Lect. 6). In other words, and in accordance with the outline of the clicks' analysis set out below, the determinism of micro-processes (micro-ontology)—much less the microscopic time-arrow—is meaningless as a concept since they are not processes but rather structureless acts that have not even any relationship to each other. Since there are no physical phenomena as of yet, the claim that "phenomenon-1 appears to be the cause that precedes phenomenon-2 as the effect" is no more than a collection of words. To attribute physical content and mathematical formulation at the micro-level to them is* *impossible in principle—the "problem of boxes" noted above. Accordingly, the cause of (classical) macro-indeterminism is the absurdity of the notion of its twin concept—micro-determinism—and the unavoidable repetition of the arrows* $\dashrightarrow$ *(**M**). N. Bohr puts the point very definitely: "there can be no question of causality in the ordinary sense of the word" ([78], p. 351), and Heisenberg adds that "l'indeterminismo, ... ë necessario, e non solo consistentemente possibile" ([17], Section IX.4). See also ([129], p. 223).*

2.5. Quantum Ensembles and Statistics

Let us call the upper row in Scheme (12), as a collection of the $\underline{\Psi}$-copies, a (quantum) homogeneous *ensemble* (Kollektiv, by von Mises [129]). We will designate it, simplifying when needed, by

$$\{\underbrace{\underline{\Psi}\,\underline{\Psi}\cdots\underline{\Psi}}_{N\text{ times}}\} \equiv \{\underline{\Psi}\cdots\underline{\Psi}\}_N \equiv \{\underline{\Psi}\}_N\,,$$

where N is understood to be an arbitrary large number. Scheme (12) also dictates considering the generic ensembles

$$\big\{\{\underline{\alpha}_1\cdots\underline{\alpha}_1\}_{n_1}\,\{\underline{\alpha}_2\cdots\underline{\alpha}_2\}_{n_2}\,\cdots\cdots\big\},\qquad \{\cdots\underline{\Psi}\cdots\underline{\Psi}\,\underline{\Phi}\cdots\underline{\Phi}\,\underline{\Theta}\cdots\underline{\Theta}\cdots\} \tag{13}$$

as collections of homogeneous sub-ensembles. Ensembles are symbolized in the same manner as sets but, for typographical convenience, without the numerous commas and internal parentheses $\{\}$ in Ensemble (13); for example,

$$\{ab\cdots b\{bca\}\cdots\} = \{a,b,\ldots,b,b,c,a,\ldots\} = \quad\cdots\quad =: \{ab\cdots bbca\cdots\}\,.$$

Scheme (12) is the first point in which numbers emerge in theory, and conversion

$$\lceil\underline{\alpha}\text{-ensemble (13)}\rceil \quad\rightarrowtail\quad (n_1, n_2, \ldots)$$

into the integer collection anticipates a *numerical* $\mathscr{A}$*-measurement of* $\mathcal{S}$. Quantities $n_s \in \mathbb{Z}^+$, however, should not be associated with such, as they are potentially infinite. The minimal way of creating the knowingly finite numbers out of independent and potential infinities n_s (without loss of their independence) is to divide each of them by a greater infinity, which is a "constant" Σ for Ensemble (13). It is clear that one should put

$$\Sigma := n_1 + n_2 + \cdots \quad\text{and}\quad \Big\{\mathtt{f}_1 := \frac{n_1}{\Sigma},\quad \mathtt{f}_2 := \frac{n_2}{\Sigma},\quad \ldots\Big\} \quad (\Sigma \rightsquigarrow \infty)\,, \tag{14}$$

and that Ensemble (13) does not provide any numerical data besides the relative frequencies (14). All the other data are functions of $\mathtt{f}_s$. An independence of the theory from the ensemble's Σ-constant, i.e., the scheme $\Sigma \rightsquigarrow \infty$, is also implied to be a principle, and it can only be the semantic one. Without it—the Σ-postulate of infinity—there can be no question of a rational theory, i.e., empiricism will not turn into a mathematics (Sections 5.1 and 5.2). In turn, the concepts "closely, limit, the limiting frequencies", and the like will arise later when we obtain the state of space as a Hilbert one $\mathbb{H}$ and topology on it [130].

Thus, the **M**-paradigm in Scheme (12) does not only give birth to a concept of numerical data in the theory per se but also converts their $\mathbb{Z}^+$-discreteness into the $\mathbb{R}$-continuum of real measurements. Namely, numbers $\mathtt{f}_s \in \mathbb{R}$ are the statistics $(\mathtt{f}_1, \mathtt{f}_2, \ldots)$ of destructions $\overset{\mathscr{A}}{\dashrightarrow}$ into the ensemble of primitives $\{\{\underline{\alpha}_1\}_{n_1}, \{\underline{\alpha}_2\}_{n_2}, \ldots\}$.

2.6. Distinguishability and Numbers

The distinguishability of the two ensembles now turns out to be the $\mathbb{R}$-numerical, i.e., it is determined by the difference between $\mathtt{f}$-numbers. As a result, and according to pt. **R**, the two elements $|\Psi\rangle \not\approx_{\mathscr{A}} |\tilde{\Psi}\rangle$ of $\mathbb{H}$ will differ in the numbers $\mathfrak{a}_s$ and $\tilde{\mathfrak{a}}_s$ if the latter turn out to be the bearers of different statistics

$$\mathtt{f}_j(\mathfrak{a}_1, \mathfrak{a}_2, \ldots) \neq \tilde{\mathtt{f}}_j(\tilde{\mathfrak{a}}_1, \tilde{\mathfrak{a}}_2, \ldots)\,. \tag{15}$$

As a consequence, distinguishability $\not\approx$ is carried over to $\mathbb{H}$ with an extension to the non-eigen objects, but it is inherently incomplete since it does not take into account the most significant fact—arbitrariness of transitions (6).

The collection $(\mathtt{f}_1,\ldots)$, as a final result of transitions $\{\underline{\Psi} \dashrightarrow \underline{\alpha}_s\}$, actually "knows nothing" about their left-hand side, much less about its uniqueness $\underline{\Psi}$. For instance, if under the equal $\underline{\alpha}$-statistics $\{\mathtt{f}_s\}$ for the two families $\{\underline{\mathbf{?}} \dashrightarrow \alpha_s\}_N$ and $\{\underline{\Psi} \dashrightarrow \alpha_s\}_N$ (collectivity of $\underline{\mathbf{?}}$'s), we would claim $\underline{\mathbf{?}} = \underline{\Psi}$, which would mean a mass control over transitions (9). Instead of a "black box" above, we find that prior to acts $\overset{\mathscr{A}}{\dashrightarrow}$, all the undefined $\underline{\mathbf{?}}$'s were equal to $\underline{\Psi}$. This, however, is the declaration of a property: "prior to observation the system $\mathcal{S}$ was/dwelled in ... ". With any continuation of this sentence, it is pointless and prohibited if one theoretically accepts that, prior to observation, nothing exists, and there are no properties (Section 2.1). The words "initial state of $\mathcal{S}$" thus make no sense. The indeterminacy of the ongoing $\underline{\mathbf{?}}$'s is therefore mandatory, and numbers $(\mathtt{f}_1,\mathtt{f}_2,\ldots)$ required for recognition are manifestly insufficient. Considering that the micro-changeability of single primitives $\underline{\Psi}$ also means nothing [15] (p. 419 (!), left column), [33] (p. 493), [41], only a generic ensemble

$$\{\underline{\mathbf{?}}\cdots\underline{\mathbf{?}}\} \quad \rightarrowtail \quad \{\cdots\underline{\Psi}\cdots\underline{\Psi}\,\underline{\Phi}\cdots\underline{\Phi}\,\underline{\Theta}\cdots\underline{\Theta}\cdots\} =: \mathfrak{A} \tag{16}$$

can be an intermediary in the sought-for translation of $\underline{\Psi}$'s onto representations $|\Xi\rangle \in \mathbb{H}$ under Construction (11).

In the accustomed physical terminology, the above is expressed in the sequence

$$\lceil\text{state}\rceil \overset{\mathscr{A}_{\text{quant}}}{\dashrightarrow} \lceil\text{state}'\rceil \longmapsto \lceil\text{measurement}\rceil\,. \tag{17}$$

The removal of the intermediate component here, i.e., switch to the sequence

$$\lceil\text{state}\rceil \overset{\mathscr{A}_{\text{class}}}{\dashrightarrow} \cdots\cdots \longmapsto \lceil\text{measurement}\rceil \tag{18}$$

amounts to the rejection of micro-destructibility and of unpredictability. Even with the classical framework, this supposition is questionable since the notion of a "change when observed" disappears. The relationships between the dual concepts—(micro/macro)-scopicity, big/middle/small, etc.—do also get lost. That is the reason why, developing Heisenberg's question "... is it ... I can only find in nature situations which can be described by quantum mechanics?"' ([78], p. 325), we conclude that, strictly speaking,

- *All observations*, regardless of (the envisioned physical) macro/meso/micro characteristics, do have the structure (17), i.e., are quantum. *No non-quantal observations exist.*

With their idealized "roughening", the classical description appends numerical $\mathtt{f}$-statistics to (18), which is when the left/right sides of (18) become indistinguishable with respect to the arrow symbols. The arrows may then be replaced with the equivalence

$$\lceil\text{state}\rceil \overset{\mathscr{A}_{\text{class}}}{\leftrightarrows} \lceil\mathtt{f}\text{-statistics numbers}\rceil\,. \tag{19}$$

Supplementing the right-hand side here with the concept of numerical values $\{\alpha_s\}$ for all of the observables $\mathscr{A} = A(q,p)$ (or for phase variables $\{q_1,q_2,\ldots;p_1,p_2,\ldots\}$), this side will turn into an exhaustive numerical realization of the left-hand side. Criterion $\approx$, then, turns into the $\mathbb{R}$-number equality $=$ of all the $\mathscr{A}$-statistics or into an equality of phase distributions $\varrho(q_1,\ldots;p_1,\ldots)$. This is a situation of the classical (statistical) physics (ClassPhys), i.e., when "the physics is initially identified" with quantities being numerical in character: the particle coordinates/numbers, the number values of field functions, etc. The ill-posedness of such a paradigm—the core motive of QT—is discussed further below at greater length in Sections 6.4, 6.5, and 7. Consequently, "classicality" is not and cannot be regarded as a primitive in the logical construct. In both these cases, distinguishability $\not\approx$ depends on the concept of $\underline{\alpha}$-states.

Remark 4. *From this point onward, by state we will strictly mean representations (11). Thus, it makes* no sense to speak of transitions between states, *much less of "transition from possible to actual" ([107], p. 189; Everett), [117–119]. The writing* $|\Xi\rangle \dashrightarrow |\boldsymbol{\alpha}\rangle$ *and its typical wave-function collapse interpretation are not correct. Indeed, in treating transition* $|\Xi\rangle \dashrightarrow |\boldsymbol{\alpha}_1\rangle$ *as a state-to-state destruction, its left-hand side cannot carry any information about* $\mathtt{f}_{(\mathscr{A})}$*-frequencies for other events* $|\Xi\rangle \dashrightarrow |\boldsymbol{\alpha}_s\rangle$, much less *about the amount of destruction from envisioned* $\mathscr{B}$*-observations* $|\Xi\rangle \overset{\mathscr{B}}{\dashrightarrow} |\boldsymbol{\beta}\rangle$*. Such "*$\mathtt{f}_{(\mathscr{B})}$*-amounts" are always present at the experimental interpretation of the* $|\Xi\rangle$*-symbol. For this reason, the concept of a state should not be used as a correct term at all [58]; the terminology, however, has been settled.*

The motivation given above—**S** (system, primitives), **O** (observation), **R** (representations), **T** (taboo), the semantic principia **I** (QM-statics), and **II** (numbers) complemented below with the principium **III**—is sufficient for further creating the basis of the mathematical formalism of QM. These tenets should hardly be regarded as postulates, at least in the common meaning of the phrase "postulates of a physical theory", since they are a natural language and are, as we believe, the points of departure for reasoning whatever the approach to the micro-world. It is clear that they are directly concerned with the familiar dialogs, which reflect, in the words by Bohr, "[Einstein's] feeling of disquietude as regards the apparent lack of firmly laid down principles . . . , in which all could agree" ([131], p. 228).

The underpinning of QT must thus begin, at least to a large extent, with a simplification/reducing the terminology in use and putting the language and the semantics of observations/numbers in order, rather than giving the "improved" postulates or definitions.

> "The task is not to make sense of the quantum axioms by heaping more structure, more definitions, . . . , but to throw them away wholesale"
>
> C. Fuchs ([50], p. 989)

> "Simplicity is implicit in the basic goals of scientific inquiry. . . . only simple theories can attain a rich explanatory depth. . . . the Basic Propert[ies] should indeed be very simple"
>
> N. Chomsky ([132], pp. 4–5)

As was underscored above, these (organizing) principles do not stipulate for predetermined mathematics and physics, with the exception of a linguistic/metamathematical understanding [105,106] of how to look at the mathematical axioms, structures, rational theories, and their interpretations altogether. See also Remarks 7 and 10 and Sections 5 and 10.

3. Ensemble Formations

> Your acquaintance with reality grows literally by buds or drops of perceptions. . . . they come totally or not at all—W. James (1911)

> Are billions upon billions of acts of observer- participancy the foundation of everything?—J. Wheeler ([62], p. 199)

The key corollary of Macro-paradigm (12) is not merely the appearance of numerical data in the theory but also the fact that the further construct cannot rely on isolated primitives but rather on their aggregates being considered as an integrated whole, i.e., as a set. This causes a choice for the ensemble notation.

3.1. Mixtures of Ensembles

Returning to the analysis of transitions $\underline{?} \dashrightarrow \alpha_s$, one obtains that the lower row in (12) actually comes from indeterminacy

$$\begin{matrix} \{\underline{\alpha}_1\}_{n_1}\{\underline{\alpha}_2\}_{n_2}\cdots\cdots \\ \uparrow\ \ \uparrow\ \ \uparrow\ \ \uparrow\ \ \uparrow \\ \underline{?}\cdots\underline{?}\cdots\underline{?}\cdots\cdots \end{matrix}$$

and thus (12), by virtue of (16), should be replaced with the scheme

$$\begin{array}{c} \{\cdots \underline{\Psi} \cdots \underline{\Phi} \cdots \underline{\Theta} \cdots\} \\ \mathscr{A} \downarrow \quad \downarrow \quad \downarrow \quad \downarrow \quad \downarrow \mathscr{A} \\ \{\cdots \underline{\alpha}_1 \cdots \underline{\alpha}_2 \cdots\} \end{array} , \tag{20}$$

wherein the composition of the upper ingoing row may not be predetermined. Fundamentally, according to (17), it may not be withdrawn from (20), yet at the same time, the meaning of the row can in no way be aligned with the adjective "observable" via typical empirical/physical words: properties, readings, quantities/amounts, and other "observable" characteristics. Such non-detectability is the equivalent of a box that may be prepended to Scheme (6):

$$\boxed{\cdots\cdots} \dashrightarrow \underline{\Psi} \overset{\mathscr{A}}{\dashrightarrow} \underline{\alpha} \,. \tag{21}$$

If $\underline{\beta}$'s serve as $\underline{\Psi}$ in (21), then we have the schemes of precedence and of continuation:

$$\boxed{\cdots\cdots} \overset{\mathscr{A}}{\dashrightarrow} \underline{\alpha} \overset{\mathscr{B}}{\dashrightarrow} \underline{\beta} \qquad \text{or} \qquad \boxed{\cdots\cdots} \overset{\mathscr{B}}{\dashrightarrow} \underline{\beta} \overset{\mathscr{A}}{\dashrightarrow} \underline{\alpha} \,.$$

Let an observer capture the fact of any distinguishability in the penultimate $\mathscr{A}$. Section 2.1 tells us that this may only be the distinguishability of objects $\{\underline{\alpha}_1, \underline{\alpha}_2, \ldots\}$; hence, this very $\mathscr{A}$ turns into an observation (pt. **O**). The **M**-paradigm then gives rise to the numbers of $\underline{\alpha}$-events $(n_1, n_2, \ldots)$ and, thereupon, their relative frequencies $(\varrho_1, \varrho_2, \ldots)$ by the rule (14). If subsequently micro-observations $\mathscr{B}$ are to follow, then a composite macro-observation $\mathscr{B} \circ \mathscr{A}$ has been formed, and frequencies $\{\varrho_j\}$ cannot impact statistical characteristics of these later $\mathscr{B}$-observation's micro-events. However, being an ongoing ensemble for $\mathscr{B}$, each homogeneous $\{\underline{\alpha}_s \cdots \underline{\alpha}_s\}_{n_s}$ is indistinguishable from an indefinite ensemble $\{\cdots \underline{\Psi} \cdots \underline{\Phi} \cdots\}_{n_s}$ since the concept of "$\approx_{\mathscr{A}}$-sameness" is unknown for $\mathscr{B}$. Instrument $\mathscr{B}$ is "aware of only its own $\approx_{\mathscr{B}}$ and cannot know *what* it destroys", or that the source-object consists of one and the same $\underline{\alpha}_s$. Rejecting this point brings us once again (p. 9) to attempts at "penetrating the black box" of transitions (5), i.e., to attempts at creating the physics of a more primary level. According to pts. **O** and **M**, an instrument produces nothing more than its own "destruction list"; in this case, $(\{\underline{\beta}_1\}_{m_1}, \{\underline{\beta}_2\}_{m_2}, \ldots)$. This list is completely independent of the preceding one since, according to pt. **R••**, there cannot be restrictions on $\mathfrak{T}_{\mathscr{A}}$ and $\mathfrak{T}_{\mathscr{B}}$. In case the set $\{\underline{\alpha}_s \cdots \underline{\alpha}_s\}_{n_s}$ transits into collection $\{\underline{\beta}_k \cdots \underline{\beta}_k\}_{n_s}$, this means that $\underline{\alpha}_s$ has always transited into one and the same $\underline{\beta}_k$ every time (under the convention $\Sigma \rightsquigarrow \infty$), and merely a coincidence $\underline{\alpha}_s = \underline{\beta}_k$ of eigen-primitives in the lists $\mathfrak{T}_{\mathscr{A}}$ and $\mathfrak{T}_{\mathscr{B}}$ takes place.

If $\mathscr{B} \circ \mathscr{A}$ is proceeded with a third observation $\mathscr{C}$, the preceding analysis is repeated recursively with the same result; only the values $\{\varrho_j\}$ will be changed. As a consequence, only the following two ongoing types for macro-scheme (20) are conceivable:

$$\{\cdots \underline{\Psi} \cdots \underline{\Phi} \cdots \underline{\Theta} \cdots\} \quad \begin{array}{l} \text{indefinite ensemble} \\ \lceil \text{no statistics} \rceil \,, \end{array} \tag{22}$$

$$\{\{\cdots \underline{\Psi} \cdots \underline{\Phi} \cdots \underline{\Theta} \cdots\}^{(\varrho_1)} \{\cdots \underline{\Psi} \cdots \underline{\Phi} \cdots \underline{\Theta} \cdots\}^{(\varrho_2)} \cdots\} \quad \begin{array}{l} \text{ensemble mixture} \\ \lceil \text{with statistics } (\varrho_1, \varrho_2, \ldots) \rceil \,. \end{array} \tag{23}$$

It is reasonable to regard Case (23) as a "non-interfering" mixture of the system's $\mathscr{A}$-preparations

$$\{\mathcal{S}^{(\varrho_1)}, \mathcal{S}^{(\varrho_2)}, \ldots\} \quad \Longleftrightarrow \quad \{\underline{\alpha}_1^{(\varrho_1)}, \underline{\alpha}_2^{(\varrho_2)}, \ldots\} \,,$$

to each of which one assigns the positive number $\varrho_s < 1$ referred to as its statistical weight. These weights—"an element of reality" ([113], p. 649)—are all that is inherited from the preparation $\mathscr{A}$, and subsequent micro-observation acts $\mathscr{B}$ are performed again on indefinite ensembles (22).

It is clear that in the view of transitions $\dashrightarrow$ in scheme (20), this situation is a derivative of (22) and this very type (22) is crucial ([34], p. 53). In other words, if the preparation is regarded as a concept as essential as observation (pt. **O**), we still remain within the framework of the binate essence of the transition:

$$\underline{\Psi} \overset{\mathscr{B}}{\dashrightarrow} \underline{\beta}, \qquad \underline{\alpha} \overset{\mathscr{B}}{\dashrightarrow} \underline{\beta}\,.$$

Its left-hand side should always be seen as an undetermined primitive, even though we treat/call it the preparatory (micro)observation. See also "preparation-measurement reciprocity" in [133].

3.2. Ensemble Brace

According to pts. **R** and **M**, the representations in (11) must reflect all information about the physics of the problem: primitives/incomes, transitions ("arrows" $\dashrightarrow$), and outgoing statistics. All the data are contained in Scheme (20), which is why the maximum that the model of a future mathematical object—it characterizes everything we obtain while watching the $\mathcal{S}$—can rely on is the *ensemble brace*:

$$(\underline{\Xi}) := \begin{pmatrix} \{\cdots\underline{\Psi}\cdots\underline{\Phi}\cdots\underline{\Theta}\cdots\} \\ {\scriptstyle\mathscr{A}}\downarrow \;\downarrow\;\downarrow\;\downarrow\;\downarrow{\scriptstyle\mathscr{A}} \\ \{\cdots\underline{\alpha}_1\cdots\underline{\alpha}_2\cdots\} \end{pmatrix} \tag{24}$$

(or a couple of ensemble bunches).

It is immediately seen that (24) carries the radical difference between situation (17) and its "roughening" (19) because of the upper row. The enormous arbitrariness within the brace and arrows $\overset{\mathscr{A}}{\dashrightarrow}$ is "programmed" to give birth to the different processing rules of statistics and to effects that are typical for QM. Thanks to the maximality of (24), it is only this row that encodes all the sought-for cases of distinguishability $\not\approx$. In particular, by varying the upper row while the lower one remains unchanged, we get into a situation when $\underline{\alpha}$-statistics $(\mathtt{f}_1, \mathtt{f}_2, \ldots)$ are found to be the same for $(\underline{\Xi})$ and $(\tilde{\underline{\Xi}})$, and meanwhile, $(\underline{\Xi}) \not\approx (\tilde{\underline{\Xi}})$.

The problem is thus as follows. With the indefinite $\mathfrak{A}$-ensemble (16) in hand, i.e., with the upper row of (24), is it possible, based on the principles described above, to bring the still incomplete relation $\not\approx$ to the maximal quantum-physical distinguishability of states?

4. Why Does Domain $\mathbb{C}$ Come into Being?

> … quod ideo sint imaginariae, … quod ideo sint … tum certe forent reales ideoque non imaginariae—L. Euler (1736)
>
> (… this is why they are imaginary. Were they …, they would certainly be real and therefore not imaginary.)
>
> … denn die imaginären Größen existierten doch nicht?—D. Hilbert (1926)

The first priority in the $\not\approx$-distinguishability of objects (24) is to separate the closest and unconditional criterion—the outgoing $\underline{\alpha}$-statistics. To do this, let us split the lower row into families $\{\{\underline{\alpha}_1\}_{\infty_1}\{\underline{\alpha}_2\}_{\infty_2}\cdots\}$, where

$$\infty_1 + \infty_2 + \cdots = \infty\,, \tag{25}$$

and, subsequently (rather than the reverse, otherwise (23)), taking into account the "arbitrariness of arrows", we also split the upper row:

$$(\underline{\Xi}) = \begin{pmatrix} \{\{\cdots\underline{\Psi}\cdots\underline{\Phi}\cdots\underline{\Theta}\cdots\}_{\infty_1} & \{\cdots\underline{\Psi}\cdots\underline{\Phi}\cdots\underline{\Theta}\cdots\}_{\infty_2}\cdots\} \\ \downarrow\;\downarrow\;\downarrow\;\downarrow & \downarrow\;\downarrow\;\downarrow\;\downarrow \\ \{\;\{\underline{\alpha}_1\cdots\cdots\underline{\alpha}_1\}_{\infty_1} & \{\underline{\alpha}_2\cdots\cdots\underline{\alpha}_2\}_{\infty_2}\;\cdots\} \end{pmatrix} \tag{26}$$

(the indication of observation $\overset{\mathscr{A}}{\dashrightarrow}$ is omitted further below since it has been mirrored in primitives $\underline{\alpha}$). Hereafter, infinities ∞_j stand for cardinal numbers (a number of elements, possibly finite) of their own ensembles. Therefore, the extension of distinguishability (15) should be produced by comparing the sub-objects such as

$$\begin{array}{c} \{\cdots\underline{\Psi}\cdots\underline{\Phi}\cdots\underline{\Theta}\cdots\}_{\infty_1} \\ \downarrow \;\; \downarrow \;\; \downarrow \;\; \downarrow \\ \{\underline{\alpha}_1\cdots\cdots\underline{\alpha}_1\}_{\infty_1} \end{array}, \tag{27}$$

that differ from each other in the upper-row composition.

4.1. Continuum of Quantum Phases

The cardinality of the $\mathfrak{T}$-set cannot be finite. This would finitely entail many $\underline{\alpha}$-primitives for all kinds of instruments. However, the finiteness of this number $n_{\mathfrak{T}}$ would mean an exclusivity of its value that does not follow from anywhere. At the same time, all the $\mathfrak{A}$-ensembles (16) are subsets of the set $\mathfrak{T}$ (boolean $2^{\mathfrak{T}}$); any finite portion of it is ruled out. Hence, the endless variety of upper rows in (27) is uncountable.

Aside from the number of $\mathfrak{f}$-statistics, program **R** does also require an association of the numerical objects with each row

$$\mathfrak{A} = \{\cdots\underline{\Psi}\cdots\underline{\Psi}\,\underline{\Phi}\cdots\underline{\Phi}\,\underline{\Theta}\cdots\underline{\Theta}\cdots\}_{\infty} \quad \Longleftrightarrow \quad \cdots,$$

because primitive's symbols must disappear in the ultimate description. To avoid introducing the structures ad hoc, we will produce numbers here—the upper row—in the same manner, in which statistics were producing in Section 2.5—the lower row. Indeed, the genesis of the concept of the number must be single in theory. That is, we should again take into account the presence of copies of primitives and write

$$\cdots \quad \Longleftrightarrow \quad \{\underbrace{\{\underline{\Psi}'\}_{\infty'}\{\underline{\Psi}''\}_{\infty''}\cdots}_{K\text{ times}}\}, \tag{28}$$

and numbers per se will come into being by the Σ-convention, such as (14), i.e., through the cardinal ratios

$$\varkappa' := \frac{\infty'}{\infty}, \qquad \varkappa'' := \frac{\infty''}{\infty}, \qquad \ldots. \tag{29}$$

Now, the discreteness of micro-transition acts is embodied in (28) with the sequence $(\underline{\Psi}', \underline{\Psi}'', \ldots)$, and the uncountability of micro-arbitrariness is inherited by attaching the symbolic "quantities"—"countless" characters $(\infty', \infty'', \ldots)$—to elements of this sequence. The global discreteness says that there are no grounds to assume a more than countable infinity $\aleph_o$ for the set $\mathfrak{T}$, i.e., $|\mathfrak{T}| = \aleph_o$. The infinity of the family (28), hence, has the type

$$2^{\aleph_o} = \aleph\,,$$

i.e., it is continual [134]. Parenthetically, the $2^{\aleph_o}$ is the only known way of introducing the continual (more than discrete) mathematical infinity. Which possibilities exist for the form of row (28)?

The trivial case $\mathfrak{A} = \{\{\underline{\Psi}'\}_{\infty'}\}$, i.e., $K = 1$ in (28) drops out at once since element $\underline{\Psi}'$ would always go into the same primitive:

$$\begin{array}{c} \{\underline{\Psi}'\cdots\cdots\underline{\Psi}'\}_{\infty_1} \\ \downarrow \;\; \downarrow \;\; \downarrow \;\; \downarrow \\ \{\underline{\alpha}_1\cdots\cdots\underline{\alpha}_1\}_{\infty_1} \end{array} = \begin{array}{c} \{\underline{\Psi}'\}_{\infty_1} \\ \mathscr{A}\downarrow\cdots\downarrow\mathscr{A} \\ \{\underline{\alpha}_1\}_{\infty_1} \end{array}. \tag{30}$$

However, this is tantamount to the identity $\underline{\Psi}' \equiv \underline{\alpha}_1$, which robs of any meaning the concept of the transition $\underline{\Psi} \overset{\mathscr{A}}{\dashrightarrow} \underline{\alpha}$. We obtain a single number here—the number of $\underline{\alpha}_1$-clicks—and arrive thereby at classical statistics, the physics of which is inadequate with respect to the interference patterns. Hence, the following options are admissible for the formations (28):

$$\underbrace{\{\{\underline{\Psi}'\}_{\infty'}\{\underline{\Psi}''\}_{\infty''}\}}_{K=2}, \quad \ldots, \quad \underbrace{\{\{\underline{\Psi}'\}_{\infty'}\{\underline{\Psi}''\}_{\infty''}\{\underline{\Psi}'''\}_{\infty'''}\cdots\}}_{3\leqslant K<\infty}, \quad \ldots$$
$$\ldots, \quad \underbrace{\{\{\underline{\Psi}'\}_{\infty'}\{\underline{\Psi}''\}_{\infty''}\{\underline{\Psi}'''\}_{\infty'''}\cdots\}}_{K=\infty} \tag{31}$$

with minimal $K = 2$. If some of the infinities $(\infty', \infty'', \ldots)$ are finite here or countable, this does not change the total continuality $\aleph$. The extreme case $K = \infty$—a countable infinity of continuums—also changes this count because of $\aleph + \aleph + \cdots = \aleph$ [135]. All of these infinities may be even countably duplicated without augmenting the continuum since $\aleph \cdot \aleph \cdots = \aleph^{\aleph_0} = \aleph$.

What can one say about relationship of cases (31) between each other? Do we have to deal with their total arbitrariness or with only one of these schemes? The latter case—the sameness/indistinguishability of upper rows in (27)—would correspond to the structural staticity of theory. Otherwise, whether one (unrecognizable upper) row should differ (why?) from another in the number (what?) of defining primitives $\{\underline{\Psi}', \underline{\Psi}'', \ldots\}$ (which ones?)?

Suppose the variability of K. That is, consider the simultaneous existence of, say, the $K = \{2, 3\}$ rows

$$\{\{\underline{\Psi}'\}_{\infty'}\{\underline{\Psi}''\}_{\infty''}\}, \qquad \{\{\underline{\Psi}'\}_{\infty'}\{\underline{\Psi}''\}_{\infty''}\{\underline{\Psi}'''\}_{\infty'''}\}\,.$$

However, each of the 2-row is a particular case of the 3-row with a cardinal number $\infty''' = 0$:

$$\{\{\underline{\Psi}'\}_{\infty'}\{\underline{\Psi}''\}_{\infty''}\} = \{\{\underline{\Psi}'\}_{\infty'}\{\underline{\Psi}''\}_{\infty''}\{\underline{\Psi}'''\}_{\{\infty'''=0\}}\} \subset \{\{\underline{\Psi}'\}_{\infty'}\{\underline{\Psi}''\}_{\infty''}\{\underline{\Psi}'''\}_{\infty'''}\}$$

(the case in point is sets). Therefore, these situations are structurally indistinguishable from each other, and the $K = 2$ theory is a *sub*theory for $K = 3$. So, the cases $K = \{2, 3\}$ are actually not mutually exclusive; rather, they form an embedding. We thus have arrived at the one cumbersome and common construct akin to the Russian dolls $2 \supset 3 \supset 4 \supset \cdots$. Hence, the minimal 2-theory will always be present inside all the higher orders $K > 2$ as an "independent (sub)world". For this reason, the $K = 2$ theory must be created in any way; incidentally, it will enclose the $K = 1$ case.

In the other part, we have no criteria to terminate the sequence $2 \rightarrowtail 3 \rightarrowtail \cdots$ at some intermediate $K < \infty$. Such a cut-off does immediately raise an issue of the questionable empirical exclusivity of a certain "world integer $K \geqslant 3$" that defines the number of "physically inaccessible" $\underline{\Psi}$-objects. Moreover, these options would be related to a certain topological dimension $K \geqslant 3$ that has an unmotivated origin. We thus conclude that the non-minimal options $K = 3$, $K = 4, \ldots$ in (31) should be dismissed.

Remark 5. *A few remarks can be made in connection with the case when $K = \infty$. It is related to a conglomerate of infinities, which has the form of a discretely infinite family of continual infinities $\{\varkappa', \varkappa'', \ldots\}$, and things would have been even "worse" had the staticity of the schemes (31) been changed to variability. Such formations would need to be equipped with topology and with associated concepts of convergence and of limit. However, all this touches on principally unobservable numerical entities, for which it is not clear how to motivate the further reductions to "finite mathematics" as required: dimensions, finite approximations, finite numbers (which ones?), and the like. More to the point, all of that would pertain to the global structural parameters of the theory prior to constructing it per se, not to mention the physical models. To put it plainly, such an assumption would result not in a theory but in a theory of theories, and so on ad infinitum, which should be somewhere terminated in some way. For these reasons, we leave the case $K = \infty$ aside, though it might be worth elaborating on it. However, in Section 7.6, we will give a further justification that the number domain of the theory is what it has already been known in* QM.

As a result, one has a choice: the structural staticity $K = 2$ or entirely non-structured/undetermined set of outgoing primitives $\{\underline{\Psi}_j, \underline{\Psi}_k, \ldots\}$, i.e., extremely complex case $K = \infty$. We do choose $K = 2$. This option might have been adopted even before on the ground that the most minimalistic construction, which set-theoretically gives rise, as a minimum, to the minimal numerical object—a single number—corresponds to the minimal $K = 2$ in (31). The maximal case is problematic, while the mid-ones are ruled out. That is to say, all possible assumptions regarding the upper row structure in (27) are indistinguishable from a case just as if the row contained two primitives only $\{\underline{\Psi}', \underline{\Psi}''\} =: \{\underline{\Psi}, \underline{\Phi}\}$. The functionality of the symbol $\cup$, with regards to the inclusion of the $\{\underline{\Psi}, \underline{\Phi}\}$'s copies, is unchanged (see Section 5.1 further below).

We establish in the following writing of Scheme (27) that

$$\{\underline{\Psi}\cdots\underline{\Psi}\}_{\infty_1'} \cup \{\underline{\Phi}\cdots\underline{\Phi}\}_{\infty_1''} \quad \dashrightarrow \quad \{\underline{\alpha}_1\cdots\underline{\alpha}_1\}_{\infty_1}$$

none of the primitives $\{\underline{\Psi}, \underline{\Phi}\}$ coincide with $\underline{\alpha}_1$. Otherwise, the unrestricted adjunction of identical transitions $\underline{\alpha}_1 \dashrightarrow \underline{\alpha}_1$ to (27) would mean indeterminacy of both the number $\varkappa_1$ and the actual statistics $(\mathtt{f}_1, \mathtt{f}_2, \ldots)$.

Let us take into account that numbers (29) are mathematically generated by the standard scheme: ⌈(ordered) integers⌉ $\rightarrowtail$ ⌈(ordered) rationals⌉ $\rightarrowtail$ ⌈(ordered) continuum ⌉. The natural ordering $<$ is always present here and, as is well known ([136], p. 52), can be isomorphically represented by the set-theoretic inclusion $\subset$ on a certain system of sets. That inclusion (= "to be contained in"), in turn, is directly concerned with the semantics of Section 2. The natural-language term "accumulating"—"the old is being nested into the new"—is formalized to create sets by the cumulative ensembles (see Section 5.1).

We now conclude that all kinds of schemes (27) form an $\aleph$-continuum, for which there is no reasonable rationale for equipping it with a topology other than the standard order topology of the one-dimensional real $\mathbb{R}$-axis or its equivalents. Call the quantity $\varkappa \in \mathbb{R}$ *quantum phase*.

It should be added that in considering some two upper rows in (27) as infinite sets

$$\{\underline{\Psi}\cdots\underline{\Psi}\}_{\infty'} \cup \{\underline{\Phi}\cdots\underline{\Phi}\}_{\infty''} \quad \text{and} \quad \{\underline{\Psi}\cdots\underline{\Psi}\}_{\infty'} \cup \{\underline{\Phi}\cdots\underline{\Phi}\}_{\infty''}\,,$$

one can always establish their formal identity. However, physics requires distinguishing the rows, which is what the numerical part of pt. **R** and comparison of cardinals (∞', ∞'') do "serve".

4.2. Statistics + Phases

Thus, the closest reconciliation of Scheme (26) with the $\mathbf{R}^{\bullet}$-postulate is an ensemble brace of the form

$$(\underline{\Xi}) = \begin{pmatrix} \{\{\underline{\Psi}\}_{\infty_1'}\{\underline{\Phi}\}_{\infty_1''}\} & \{\{\underline{\Psi}\}_{\infty_2'}\{\underline{\Phi}\}_{\infty_2''}\} & \cdots\cdots \\ \downarrow\ \downarrow\ \downarrow\ \downarrow & \downarrow\ \downarrow\ \downarrow\ \downarrow & \\ \{\underline{\alpha}_1\cdots\cdots\underline{\alpha}_1\}_{\infty_1} & \{\underline{\alpha}_2\cdots\cdots\underline{\alpha}_2\}_{\infty_2} & \cdots\cdots \end{pmatrix} \tag{32}$$

followed by the (upper) continual numeration through $\mathbb{R}$-numbers

$$\varkappa_s := \frac{\infty_s'}{\infty_s} \qquad (\infty_s := \infty_s' + \infty_s'')\,. \tag{33}$$

In other words, the quantitative description in the theory is created on the basis of the minimal building bricks

$$\begin{pmatrix} \{\{\underline{\Psi}\}_{\infty'}\{\underline{\Phi}\}_{\infty''}\} \\ \downarrow\ \downarrow\ \downarrow\ \downarrow \\ \{\underline{\alpha}\cdots\cdots\underline{\alpha}\}_{\infty} \end{pmatrix} \qquad (\textit{unitary brace}) \tag{34}$$

with *two* abstract ongoing primitives.

Now, we have had cardinals connected by Relation (25) and Structures (32) and (33). In the above-described context, parentheses { } and symbols $\underline{\Psi}$, $\underline{\Phi}$, $\dashrightarrow$ no longer carry meaning at this point. Therefore, we may omit them as "extraneous" and write (32) as

$$(\underline{\Xi}) \quad \Longleftrightarrow \quad \left(\begin{array}{c|c|c} \varkappa_1 & \varkappa_2 & \cdots \\ \infty_1 & \infty_2 & \cdots \end{array}\right) = \cdots,$$

where $\underline{\alpha}_s$ are well represented by a subscripted numerals; observation $\mathscr{A}$ has been fixed so far. Let us now introduce a statistics from the "embracing infinity" (25):

$$\cdots = \left(\begin{array}{c|c|c} \varkappa_1 & \varkappa_2 & \cdots \\ \mathtt{f}_1 \cdot \infty & \mathtt{f}_2 \cdot \infty & \cdots \end{array}\right) = \left(\begin{array}{c|c|c} \varkappa_1 & \varkappa_2 & \cdots \\ \mathtt{f}_1 & \mathtt{f}_2 & \cdots \end{array}\right) \cdot \infty, \qquad \mathtt{f}_s := \frac{\infty_s}{\infty}.$$

Then, by Σ-postulate, one arrives at a continually numeral labeling of objects (32):

$$(\underline{\Xi}) \quad \Longleftrightarrow \quad \left\{ \binom{\varkappa_1}{\mathtt{f}_1}, \binom{\varkappa_2}{\mathtt{f}_2}, \ldots \right\}.$$

Recall that the arithmetical operations on the emergent pairs $(\mathtt{f}, \varkappa)$ are still out of the question, and Σ-limit does not care the "innards" of $(\underline{\Xi})$. Only one of all the potentially infinite quantities tends to the ∞-infinity—the total cardinality (25) of brace (32). What remains "non-extraneous" in (32) is $\underline{\alpha}$'s, and we return them to their place. Hence, from the viewpoint of observation $\mathscr{A}$, the aggregate of the possible brace (24) is indistinguishable from an order-indifferent two-parametric family of data

$$(\Xi) = \left\{ \binom{\varkappa_1}{\mathtt{f}_1}\underline{\alpha}_1, \binom{\varkappa_2}{\mathtt{f}_2}\underline{\alpha}_2, \ldots \right\}. \tag{35}$$

We drop a lower bar in the symbolic designation $(\underline{\Xi})$, highlighting the fact that the meaning of the (Ξ)-object becomes increasingly divorced from primitives in pt. **S** and gets into the number domain to match program **R**.

As an outcome, despite the freedom of ingoing collection in (26) and quantum micro-arbitrariness, the distinguishability $(\underline{\Xi}) \not\approx (\underline{\tilde{\Xi}})$ is *indeed determinable*, it is determinable not only by statistics, and is the $(\mathbb{R} \times \mathbb{R})$-numerical:

$$(\Xi) \not\approx (\tilde{\Xi}), \quad \text{if} \quad (\mathtt{f}_s, \varkappa_s) \neq (\tilde{\mathtt{f}}_s, \tilde{\varkappa}_s). \tag{36}$$

What is more, the preliminary (classical) $\not\approx$-criterium (15) fits in (35)–(36) as a particular case by omitting the $\varkappa$-numbers and middle link from (17). That is to say, ignoring quantum "$\varkappa$-effects" is only possible via the $(3 \mapsto 2)$ reduction of (17) into (18), with an automatic imposition of the ClassPhys description. A simplified and hypothetical version of QM over $\mathbb{R}^1$ is also ruled out. It would mean a reduction in the two numbers $(\mathtt{f}, \varkappa)$ to a single one. However, they have fundamentally different origins. The construct and reasoning in Section 2.1 also tell us that the attempt at a greater "quantum specification" to (5) and (17) is impossible by virtue of the two-row structure—ingoing/outgoing—of the object $(\underline{\Xi})$, and distinguishability by numerical pairs (36) is the highest possible.

The (Ξ)-objects (35) remain, and they, as a family, exhaustively inherit the problem's physics. The quantities $\mathtt{f}_s$ are the really observable (unitless) numbers—the percentage quantity of events—which are declared by instrument/observer to be the distinguishable $\underline{\alpha}$-objects. The quantities $\varkappa_s$ are the internal and unremovable degrees of freedom. Figuratively speaking, the $\varkappa$'s may be speculatively referred to as phases, but they may not be associated with an actual quantity of something. Not only is any material or the classical treatment of these "amounts" impossible, but it is fundamentally *prohibited* since the converse would have meant endowing the nonexistent boxes (5) and (6) with a notional content or asserting the nature of their origin. Justification is only allowed here for the fact of their existence, which is mirrored by the presence of the left-hand side in the concept of the transition of $\underline{\Psi} \overset{\mathscr{A}}{\dashrightarrow} \underline{\alpha}$ (Remark 2).

In view of numerous ongoing discussions of the meaning to the quantum state [21], note that, for the same reason, any (even merely similar) classical/ontological and causal "visualization mechanisms" ([5], p. 137) as the wave function of a certain real matter, of a hypothetical observable, of an "objective knowledge', or of the classical data (whatever this all means) are—and this we stress with emphasis—pointless. This is why, strictly speaking, without further theoretical conventions,

- It is impossible ([12], p. 13) to make/prepare, observe/read-off, transmit or measure/approximate a state, or to endow it with the property of being known/unknown, or physically recognize/compare/distinguish it from the other.

We will be repeatedly turning back to this matter in Sections 6.3.1, 6.4, 6.5, and 10.2. The present thesis has not undergone a change even with regard to the word "statistics" in the Born rule [6], if only because the rule is a substantial—two-to-one—reduction in the $(\mathtt{f}, \varkappa)$-data. The state will itself, when created as a mathematical object, determine the meaning of all of these words (see Section 5.3) with an appropriate concept of the physical distinguishability (Section 2.4). Cf. the works [53,54] and the "methods to directly measure general quantum states ... by weak measurements" in [137] and, on the other hand, the statements in Section 15.5 of the book [33].

All the $\varkappa_s$ and $\mathtt{f}_s$ are independent of each other, except for relation $\mathtt{f}_1 + \mathtt{f}_2 + \cdots = 1$. Taking into account the admissible renormalization of both $\mathbb{R}$-numbers, the pair $(\mathtt{f}, \varkappa)$ can be topologically identified with a point on the complex plane:

$$\binom{\varkappa}{\mathtt{f}} \rightleftarrows (\lambda, \mu) \in \mathbb{R}^2 =: \mathbb{C}\,.$$

That is, the domain $\mathbb{C}$ is at the moment just a two-dimensional numeric continuum without algebra of complex numbers. Notice that the pairs of $\mathbb{R}$-numbers is a starting point—different from ours—to the QT in ref. [138]. More than that, the impossibility of the real-number QM became a subject of the direct experimental test to distinguish between the complex-number and real-number representations of QT: on photonic systems [139] and the superconducting qubits [140].

The issue of the numerical domain over which the quantum description is being conducted—the real $\mathbb{R}$, the complex $\mathbb{C}$, the quaternions Q, or whatever—is non-trivial and continues to be the subject of study [57,93,138,141,142]. The complexity $\mathbb{C}$ is often motivated by quantum dynamics (Schrödinger's equation) ([36], p. 132; Stueckelberg), [143]; however, such a motivation is inconsistent, and as we have seen, there is no need for it. The rigidity of the $\mathbb{C}$-domain points to the fact that, in particular, the quaternion QM also has no place to originate from ([33], Section 10.1), although it was the object of theoretical constructs in the 1960–1970s [144]. Note that even the most comprehensive works [36] (p. 131), [72] (p. 234), [93] (p. 217), [96,138], and [142] (!) observe a difficulty in the full substantiation of the $\mathbb{C}$-domain in QT. Within the last decade, this theme had also attracted the particular attention of the information-theoretic approach to QT [138,145,146].

The above-outlined emergence of the numerical quantities in theory is a draft at the moment and will be refined further below in Section 7.

5. Empiricism and Mathematics

> Set theory does not seem today to have ... organic interrelationship with physics—P. Cohen and R. Hersh ([147], p. 116)

> ... physics has ... to say about the foundations of mathematics ... "if we believe in ZF there is nothing for physics to say" is not right—P. Benioff ([2], p. 31)

Up to this point, we have dealt, roughly speaking, with a single abstract aggregate $(\underline{\Xi})$ isolated from the others. However, the constructional nature of the ensemble brace (32) entails the following closedness relation between them. Every brace $(\underline{\Xi})$ is composed of some others in infinitely many ways (for remote analogies, see ([40], Section 11.2)), i.e., it is a union

$$(\underline{\Xi}) = (\underline{\Xi}') \cup (\underline{\Xi}''), \tag{37}$$

and, to put it in reverse, any union of two braces is a third object-brace. In assemblages (37), the operation $\cup$, which generates them, is commutative and associative:

$$A \cup B = B \cup A, \qquad (A \cup B) \cup C = A \cup (B \cup C), \tag{38}$$

and these two- and three-term relations not only are not a formal supplement, but should be read as the *structural* properties in general. Let us address the matter more closely.

5.1. Union of Ensembles

Consider the lower $\underline{\alpha}$-rows of brace (26) and experimentally forming the new real $\underline{\alpha}$-ensembles from them. Let the procedure of that forming be denoted by $\mathsf{U}(A, B, \ldots)$, where $(A, B, \ldots)$ are the ensembles per se. Its essence is such that it is comprehensively determined by the following minimum. A rule that involves the *fewest* (i.e., two) number of arguments $\mathsf{U}(A, B) = \mathbf{?}$ and a rule of the *repeated* applying U to itself: $\mathsf{U}(\mathsf{U}(\ldots), \ldots) = \mathbf{?}$. Obviously, we should write

$$\mathsf{U}(A, B) = \mathsf{U}(B, A), \qquad \mathsf{U}\big(\mathsf{U}(A, B), C\big) = \mathsf{U}\big(A, \mathsf{U}(B, C)\big), \tag{39}$$

which is of course merely the empirical rephrasing the standard properties (38) of operation $\cup$. However, the converse is logically preferable: Empiricism (39) is formalized into the abstract properties (38). If we now attach the upper "quantum" primitives to the low $\underline{\alpha}$-rows—a requirement of Section 2.1—then the operationality of actions with the resulting $(\underline{\Xi})$-braces would be just like that of U, i.e., (39). In other words, we carry over (and had already used everywhere) properties (39) to the general operation on $(\underline{\Xi})$-brace, without distinguishing between the essences of symbols $\cup$ and U. "Micro-operationality' of empiricism and its formalization are confined, at most, by the rules (38) and (39).

Let us temporarily discontinue using the numerical terminology as applied to (Ξ)-objects. They differ from each other due to relationships between their "innards", rather than because of our assignment of differed symbols (λ, μ) to them. The brace is comprised of elements that are combined into sets and are added to them. In the language of the abstract logic, we are dealing with the fact that transitions x form the brace $A, B, \ldots$, i.e., they are in the membership relationships $x \in A$, $x \in B, \ldots$ or, when accumulated as micro-acts, "get belonged to them". That is to say, the braces themselves and their formation (accumulation of statistics for the Σ-limit) are equivalent to a huge number of propositional "micro-sentences $x \in A$ `or` $x \in B$ `or` …". However, again, this is nothing but a logically formal equivalent of the union operation $\cup$:

$$A \cup B = \big\{x \,\big|\, (x \in A) \vee (x \in B)\big\}, \tag{40}$$

which is already being constantly exploited above.

Remark 6. *As is well known [134,136], due to properties of logical atoms* $\in$ (`membership`) *and* $\vee$ (`or`), *the properties of sentences such as* (40) *are determined precisely by rules* (38) *for* $\cup$. *Technically, we should also take an idempotence* $A \cup A = A$ *into account, however. At the same time, the need to have a number requires that the duplicates in ensembles have to be taken into consideration. Nevertheless, this situation is easily simulated by the set theory itself. Indeed, look first at the lower row in* (24) *as a strictly abstract set* $\{\underline{\alpha}', \underline{\alpha}'', \ldots\} \subset \mathfrak{T}$. *Then, instrument* $\mathscr{A}$ *"asserts" the distinguishable elements* $\{\underline{\alpha}_1, \underline{\alpha}_2, \ldots\}$ *and those that should be thought of as their equivalents:*

$$\underline{\alpha}_1' \approx \underline{\alpha}_1'' \approx \cdots =: \underline{\alpha}_1, \qquad \underline{\alpha}_2' \approx \underline{\alpha}_2'' \approx \cdots =: \underline{\alpha}_2, \qquad \ldots .$$

This equivalence can be characterized, say, by words "a detector click at one and the same place $\underline{\alpha}_1$". Upon such a formalization, one obtains the formation $\{\underline{\alpha}_1'\underline{\alpha}_1''\cdots\}\{\underline{\alpha}_2'\underline{\alpha}_2''\cdots\}\cdots\approx\{\underline{\alpha}_1\cdots\}\{\underline{\alpha}_2\cdots\}$ $\cdots$, i.e., the very lower row in (26). It is within this context that we think of the union operation without running into inconsistencies. Accordingly, $(\underline{\Xi})\cup(\underline{\Xi})\neq(\underline{\Xi})$, but the standard symbol $\cup$ continues to be used for simplicity.

Therefore if we get back to the numeral labels (35) but ignore the "inner composition" of $(\underline{\Xi})$, i.e., the **M**-paradigm, thus excluding $\cup$ and (38) from the reasoning, then all possible (Ξ)-objects would turn into the semantically "segregated ideograms". Micro-transitions, their mass nature, arbitrariness, $\not\approx$-distinguishability, and the "quantumness" of the task simply disappear. To illustrate, the obvious statement

$$\text{the brace } \{\underline{\Psi} \overset{\mathscr{A}}{\dashrightarrow} \underline{\alpha}\} =: (\underline{\Xi}) \text{ has an empirical "kindred"}$$
$$\text{with its duplication } \{\underline{\Psi} \overset{\mathscr{A}}{\dashrightarrow} \underline{\alpha}, \underline{\Psi} \overset{\mathscr{A}}{\dashrightarrow} \underline{\alpha}\} =: (\underline{\Xi}')$$

becomes pointless because the property $(\underline{\Xi})\cup(\underline{\Xi})=(\underline{\Xi}')$ is missing. Furthermore, this is despite the fact that creating the transition copies in $(\underline{\Xi}')$ is a primary operation for generating the objects and reasoning at all. The construction of the theory would then become possible only with the interpretative introduction of the vanished concepts anew. Therefore, macro-empiricism necessitates that the relationships (38) be operative rules, and with that, the quantumness or classicality of consideration is of no significance.

Remark 7. *Let us take a closer look at the situation on the opposite—mathematical—side. The union of sets $\cup$ is already a fundamental operation at the level of the set-theoretic formalization, e.g., the Zermelo–Fraenkel (ZF) axioms [134]. This is one of the first ways to create sets—the axiom of union. Thus, if we believe in the set-theoretic mode of explaining/creating the quantum rudiments, the quantitative description will inevitably invoke the operationality of the mathematical primitive $\cup$ through rules (38). This would be suffice to declare,*

- *Inasmuch as* we have nothing but $\cup$ and $(\underline{\Xi})$ (taboo **T**), *commutativity/associativity of theory is then postulated from the outset by (38), with the subsequent carrying these structures over to numerical representations, i.e., to $\mathbb{R}$ or $\mathbb{C}$.*

  It is preferable, however, to adhere to the sequence order in ideology more stringently—⌈observation⌉ $\rightarrowtail$ ⌈mathematics⌉, ⌈empiricism⌉ $\rightarrowtail$ ⌈numerical representation⌉—without substituting it for the opposite. At least, if we rely upon the comprehension of the empiricism as a formalization of the zero-principium of QT (Remark 2):

- *Our primordial perceptions are formalized only into sets and set-theoretic $\cup$-abstraction (40).*

See also [2] (p. 178), [58] (Ch. 3), [78] (p. 323), [96,104], [148] (pp. 12, 86, Ch. 4), [149] (Section V.9), and Section 11.1.

Summing up, we detect a kind of junction point: the physical and mathematical fundamentality of operation $\cup$ for describing the elementary acts. That is to say, the mathematics of $(\underline{\Xi})$-brace (32) and of objects (35) may not inherently be exhausted by them as "bare" sets without structures.

Recalling now pr. **II**, we draw a conclusion regarding the very construction of the theory.

- The reconciliation of the **R**-paradigm with empiricism must transform itself into *rewriting* the primary ensemble $\cup$-constructions (26), (32), (34), and relationships between them into the language of numerical symbols.

  More formally, we have the following continuation of pt. **R•**.

R⁺ *Homomorphism of the ensemble-brace properties "onto numbers"*: mutual $\cup$-relationships (38) between the $(\underline{\Xi})$-brace should be carried over to relations between their numerical (Ξ)-representations (35).

Thereby, we once again fix the maximum that is available for the building up of quantum mathematics. One may only handle the $\cup$-aggregates of transitions—constructions (32), (35)—and the minimal modules (34).

5.2. Semigroup

In line with (37), let us split the unitary brace (34) into two or combine two brace into one, then delete the symbols of primitives $\underline{\Psi}$ and $\underline{\Phi}$ from there. As was pointed out above, they are not necessary at this stage. By replacing the notation of upper cardinals (34) with pairs (∞_1', ∞_2') and (∞_1'', ∞_2''), upon the union, one obtains

$$(\infty_1', \infty_2') \cup (\infty_1'', \infty_2'') = (\infty_1' + \infty_1'', \infty_2' + \infty_2'') \,. \tag{41}$$

Here, addition $+$ obviously satisfies the properties (38). If the cardinal "∞-coordinates" are replaced with the "finite percentages" $(\varkappa, \mathfrak{S})$ introduced above, i.e., if one puts

$$\Big\{\varkappa = \frac{\infty_1}{\infty_1 + \infty_2}, \quad \mathfrak{S} = \infty_1 + \infty_2\Big\}, \qquad \Big\{\infty_1 = \varkappa\mathfrak{S}, \quad \infty_2 = (1-\varkappa)\mathfrak{S}\Big\} \tag{42}$$

as in (33), then Rule (41) acquires the form of a number composition:

$$(\varkappa', \mathfrak{S}') \circ (\varkappa'', \mathfrak{S}'') = \left(\frac{\varkappa'\mathfrak{S}' + \varkappa''\mathfrak{S}''}{\mathfrak{S}' + \mathfrak{S}''}, \; \mathfrak{S}' + \mathfrak{S}''\right). \tag{43}$$

The commutativity/associativity properties of operation $\circ$ hold here due to the birationality of (42). Then, the formal application of Σ-postulate $\mathfrak{S}' + \mathfrak{S}'' \to \infty$ breaks, however, the symmetry $(') \leftrightarrow ('')$ and associativity of $\circ$ since

$$(\varkappa', \mathfrak{S}') \circ (\varkappa'', \mathfrak{S}'') \quad \rightarrowtail \quad \varkappa' \circ \varkappa'' = s \cdot \varkappa' + (1-s) \cdot \varkappa'', \qquad s := \frac{\mathfrak{S}'}{\mathfrak{S}' + \mathfrak{S}''} \tag{44}$$

and s is an undefined parameter. The consequence of the same kind holds true for the $\mathtt{f}$-components of pairs (35), for which a convex w-combination of the statistical weights does arise:

$$(\mathtt{f}_1', \mathtt{f}_2', \ldots) \circ (\mathtt{f}_1'', \mathtt{f}_2'', \ldots) =: (\bar{\mathtt{f}}' \circ \bar{\mathtt{f}}'') = w \cdot \bar{\mathtt{f}}' + (1-w) \cdot \bar{\mathtt{f}}'', \qquad w := \frac{\Sigma'}{\Sigma' + \Sigma''} \,. \tag{45}$$

At the same time, the splitting (41) is no more than an "intrinsic reshuffle" of one and the same $\lbrace\underline{\Xi}\rbrace$-brace, which "knows nothing" about the concept of a number (numbers s, w), much less about the concept of observation or its numerical form. Therefore, mathematics of the ensemble structures should be independent of any representation for (37) by such operations as (43). Composition $\lbrace\Xi'\rbrace \circ \lbrace\Xi''\rbrace = \lbrace\Xi\rbrace$ should be determined solely by its constituents $(\mathtt{f}', \varkappa')$ and $(\mathtt{f}'', \varkappa'')$, i.e., such numbers as (s, w) must not appear here.

Remark 8. *In classical statistics, the foregoing has an analog as indifference of data on events to the way of gathering and layout thereof. For example,* $(2,3) + (1,4) \equiv (0,6) + (3,1) \equiv \cdots =: \mathtt{data}$. *Then, the observation proper is being created by the scheme* $\mathtt{data} \rightarrowtail (3,7) \mapsto \big(\frac{3}{3+7}, \frac{7}{3+7}\big) = (0.3, 0.7) = (\mathtt{f}_1, \mathtt{f}_2) =: \mathtt{observ}$. *Parameters such as w can appear in* $\lbrace\Xi\rbrace$ *only if, prior to any of the* $\cup$*-unions (37), a construction similar to (23) has been fixed. That is, the invariantly number-free brace (37)* has been supplemented by an external number w *and ratio* $w : (1-w)$. *The correction* $\lbrace\Xi\rbrace \rightarrowtail \big\{\lbrace\Xi'\rbrace^{(w)}, \lbrace\Xi''\rbrace^{(1-w)}\big\}$ *of the theory, related to this number and to arrays (23), is very well known. This is a w-statistical mixture* $\{(w; \psi'), (1-w; \psi'')\}$ *of wave functions, accompanied by a formalization in terms of the statistical operator* $w \cdot |\psi'\rangle\langle\psi'| + (1-w) \cdot |\psi''\rangle\langle\psi''|$.

Now, to ensure that numerical $(\mathtt{f}, \varkappa)$-realization (35) of ensemble brace (32) inherits quantum empiricism (**O**, **M**) and structural properties (37) and (38) properly, we reassign the quantities $(\mathtt{f}, \varkappa)$ with a "percentage meaning" and replace them with different numbers $[\lambda, \mu]$:

$$(\Xi) = \left\{ \left[\begin{smallmatrix}\mu_1\\ \lambda_1\end{smallmatrix}\right] \underline{\alpha}_1,\ \left[\begin{smallmatrix}\mu_2\\ \lambda_2\end{smallmatrix}\right] \underline{\alpha}_2,\ \ldots \right\} \tag{46}$$

(this important move will be touched upon once again in Section 7.1). In so doing, each pair $\left[\begin{smallmatrix}\mu'\\ \lambda'\end{smallmatrix}\right]$, $\left[\begin{smallmatrix}\mu''\\ \lambda''\end{smallmatrix}\right]$ behaves as a whole, and, under coinciding $\underline{\alpha}_s$, the pairs are endowed with a composition $\left[\begin{smallmatrix}\mu'\\ \lambda'\end{smallmatrix}\right] \oplus \left[\begin{smallmatrix}\mu''\\ \lambda''\end{smallmatrix}\right]$ that is to be commutative. Along with this, if symbol $\uplus$ denotes a composition of objects (46), it should obviously copy properties (38):

$$(\Xi) \uplus (\Psi) = (\Psi) \uplus (\Xi), \qquad \big((\Xi) \uplus (\Psi)\big) \uplus (\Phi) = (\Xi) \uplus \big((\Psi) \uplus (\Phi)\big)\,.$$

The finite ensembles are vanishingly small in their contribution into infinite ones (Σ-postulate), i.e., elements of the $(\underline{\Xi})$-family, as infinite sets, are considered modulo finite ensembles. Once again, the "finitely many" is forbidden in theory. As soon as we put the numbers of $\underline{\alpha}_1$, of $\underline{\alpha}_2$, ... to be finite, we immediately obtain the numerical distinguishability $n_1 \neq n_2, \ldots$, i.e., the act of macro-observation. Let us designate the image of finite ensembles as (0), and, due to property $(\Xi) \uplus (0) = (\Xi)$, it is naturally referred to as zero. The collection (46) itself has also been formed by the $\cup$-combining the ingredients

$$\left\{ \left[\begin{smallmatrix}\mu_1\\ \lambda_1\end{smallmatrix}\right] \underline{\alpha}_1,\ \left[\begin{smallmatrix}\mu_2\\ \lambda_2\end{smallmatrix}\right] \underline{\alpha}_2,\ \ldots \right\} \equiv \left\{ \left[\begin{smallmatrix}\mu_1\\ \lambda_1\end{smallmatrix}\right] \underline{\alpha}_1 \right\} \cup \left\{ \left[\begin{smallmatrix}\mu_2\\ \lambda_2\end{smallmatrix}\right] \underline{\alpha}_2 \right\} \cup \cdots = \cdots\,,$$

which is why the same symbol $\uplus$ may be freely used between objects with different $\underline{\alpha}_s$:

$$\cdots = \left\{ \left[\begin{smallmatrix}\mu_1\\ \lambda_1\end{smallmatrix}\right] \underline{\alpha}_1 \right\} \uplus \left\{ \left[\begin{smallmatrix}\mu_2\\ \lambda_2\end{smallmatrix}\right] \underline{\alpha}_2 \right\} \uplus \cdots\,.$$

For the sake of brevity, we omit the redundant curly brackets further, redefining

$$(\Xi)_{\mathscr{A}} := \left[\begin{smallmatrix}\mu_1\\ \lambda_1\end{smallmatrix}\right] \underline{\alpha}_1 \uplus \left[\begin{smallmatrix}\mu_2\\ \lambda_2\end{smallmatrix}\right] \underline{\alpha}_2 \uplus \cdots\,. \tag{47}$$

As a result, we have had that the set-theoretic prototypes (26) and (27), (32) of states (11) do invariantly exist in the form of every possible $\cup$-decomposition. Thus, in dealing with the only instrument $\mathscr{A}$, one reveals the following property.

- For each observation $\mathscr{A}$, the set of $(\Xi)_{\mathscr{A}}$-objects forms an infinite commutative semigroup $\mathfrak{G}$ with respect to operation $\uplus$.
  An internal (beyond the observation) nature of $(\Xi)_{\mathscr{A}}$-objects (47) is characterized by commutative superpositions $(\Xi')_{\mathscr{A}} \uplus (\Xi'')_{\mathscr{A}}$ thereof, which are independent of the classical composition of observational $\mathtt{f}$-statistics.

5.3. Measurement

The described above numerical (Ξ)-version of the $(\underline{\Xi})$-brace "$\cup$-phenomenology" makes it possible now to preliminarily formalize a concept, the absence of which deprives the theory of its basis. Namely, *measuring statistics by observation $\mathscr{A}$ over $\mathcal{S}$*:

$$\text{QM-measurement:} \qquad \big([\lambda_1, \mu_1], [\lambda_2, \mu_2], \ldots\big) \ \longmapsto\ (\mathtt{f}_1, \mathtt{f}_2, \ldots)\,. \tag{48}$$

That is, the $[\lambda, \mu]$-collection gets mathematically mapped into the $\mathtt{f}$-statistics. This is a maximum of information provided by observation $\mathscr{A}$. Mapping (48) annihilates the pairs $[\lambda, \mu]$. Therefore, the inheritance/homomorphism of operations $\cup$ and $\uplus$ onto anything at all is eliminated. Upon operation (48), both the $(\mathtt{f}, \varkappa)$-sets and $\cup$-unions thereof, $\uplus$-operations, and the semigroup $\mathfrak{G}$ per se disappear. As a well-known result, the distinctive feature subsequently referred to as a superposition will also disappear after measurement. The new numbers $\{\mathtt{f}_s\}$ may be "added up" only as required by the different, i.e., the classical rule: forming the convex combinations (45). We note that the formalization of the measurement does not now depend on how the mathematical map $[\lambda, \mu] \rightarrowtail (\mathtt{f})$ would be further implemented—it is a separate job [6]—or how the t-dynamics would be introduced.

Remark 9. *The incorporation of t-dynamics into the theory is still impossible due to the absence of mathematics to be applied to instants t_1, t_2. Accordingly, no physical t-process, a temporal imitation of the measuring, or its dynamical description may correspond to the mathematical mapping shown in Mapping* (48). *The known "conceptual" problems with the collapse dynamics [1,13,18,45,115] are actually non-existent [15,87,93]. More precisely, they stem from the blurring of meaning that we typically give to the words "states" (what is that?), "ensembles" (what are they comprised of?), and "dynamics/collapse" (of what?). In regard to the latter, the authors of the book [58] speak out in a most definitive manner—the "fairy tales". See also Section 10.3 further below.*

In Section 2.1, the fundamental premise of the $\underline{\alpha}$-symbol-based distinguishability $\not\approx$ was the foundation of the entire subsequent language; "two clicks are never identical" ([126], p. 761). One then observes that the measurement or its outcome will essentially remain a vacuous term "for microsystems nothing can be directly measured" ([92], p. 304) until it invokes the concept of a QM-state, i.e., the (Ξ)- and $\underline{\alpha}$-objects. In the following, we shall see that, as a rough guide, everything that is observable whatsoever is a function of the state and of the state space.

Once again, it is stressed that the concept of the state must precede the notion of measurement, rather than the reverse. "[J.] Bell fulminated against the use of the word "measurement" as a primary term when discussing quantum foundations" ([30], p. 262). See also the entire chapter 23—"Against "measurement'"—in ([28], pp. 213–231).

5.4. Covariance with Respect to Observations ("the same")

Up to this point, we had had no need for the matching of observation $\mathscr{A}$ with observation $\mathscr{B}$, although it is clear that a description based on a certain specified $\mathscr{A}$ will inevitably be non-invariant with respect to the tool $\{\mathscr{A}, \mathscr{B}, \ldots\}$—"observation space"—and unacceptable (pt. **R**) due to the impermissible exclusivity of the set $\{\underline{\alpha}_1, \ldots\}$. At the same time, we do not have anything but $\{\mathscr{A}, \mathscr{B}, \ldots\}$ and micro-acts (12) (pts. **T** and **M**). In the brace, this fact has already been present; transitions $\overset{\mathscr{A}}{\dashrightarrow}$ are combined into integrities (24). Logically, however, the $(\Xi)_{\mathscr{A}}$-, $(\Xi)_{\mathscr{B}}$-objects are incomparable and isolated from each other as carriers of statistics of different origins.

On the other hand, "*the same* is observed by instrument $\mathscr{B}$, as by instrument $\mathscr{A}$". Although this context has not yet been invoked, without it, the application of set-theoretic constructs to physics is devoid of meaning, just like the union of the speeds of an electron and of the Moon into a set $\{v_e, v_M\}$, with the subsequent creating a certain physical characteristic of this "two-body system"—say, the mean velocity $\frac{1}{2}(v_e + v_M)$. Indeed, "The statements of quantum mechanics are meaningful and can be logically combined *only* if one can imagine a *unique experimental context*" ([40], p. 115).

Thus, the global structuredness is required in the set of various (Ξ)-data according to the context "the same, identical" or its negation. Apparently, this addition implies such entities as "the same particle", "in the same preparation/state", "under the same temperature and **M**-environment (12)", "the same closed system $\mathcal{S}$", "in the same external field", "in the same interferometer" with "the same detectors/solenoids", the like [33,40,44]; the short and generalized notation $\langle\!\langle \mathcal{S}, \mathbf{M}, \ldots \rangle\!\rangle$. All the notions here, including the state, are physical conventions, yet their formalization and modeling are called for the creation of a theory (Section 2.4).

The notion "with the same initial data" falls under the same category, if the intention is to use the term time t. Again, the very creation of the (Ξ)-brace as a set "by the piece" is from the outset thought of as (Section 2.4) a creation on the assumption of common $\langle\!\langle \mathcal{S}, \mathbf{M}, \ldots \rangle\!\rangle$. For instance, the $\mathscr{A}$-statistics $(\Xi)_{\mathscr{A}}$ are gathered within "the same" $\langle\!\langle \mathcal{S}, \mathbf{M}, \ldots \rangle\!\rangle$ as the $\mathscr{B}$-statistics $(\Xi)_{\mathscr{B}}$. On its part, any variation is sufficient to obtain "not the same", even if we "*envision it as null*" in the spirit of the widely known "without in any way disturbing a system" ([131], p. 234). To take an illustration, equipment of interferometer (Section 6.5) with additional "which-slit" detectors is already at variance with the notion of "the same

$\mathcal{S}''$. In similar cases, we end up in situations similar to Case (23) since the detectors cause an $\underline{\alpha}$-distinguishability.

Notice that the notions "the same" and "distinguishable" (Remark 2), while antonymous, mutually exclude each other. Semantically, one without the other makes no sense, which closely resembles Bohr's conception of complementarity [78].

It follows from the above that in order to match $\mathscr{A}$ and $\mathscr{B}$, the *metatheoretical* [149] category $\langle\!\langle \mathcal{S}, \mathbf{M}, \dots \rangle\!\rangle$ is required; however, we are only in possession of the ensemble brace $(\Xi)_{\mathscr{A}}$ and $(\Xi)_{\mathscr{B}}$ (pt. **T**). On the other hand, without joint consideration of *the two instruments*, i.e., without introducing a mechanism for the mathematical matching $(\Xi')_{\mathscr{A}} \rightleftarrows (\Xi')_{\mathscr{B}}$, $(\Xi'')_{\mathscr{A}} \rightleftarrows (\Xi'')_{\mathscr{B}}, \dots$, the segregation of the (Ξ)-objects is absolute. (It is clear that the matching of single micro-events $\underline{\Psi} \overset{\mathscr{A}}{\dashrightarrow} \underline{\alpha}_s$ and $\underline{\Psi} \overset{\mathscr{B}}{\dashrightarrow} \underline{\beta}_j$ is also futile.) It is impossible to associate physics with the abstractly segregated $(\Xi)_{\mathscr{A}}$-brace. Otherwise, the solitary object $(\Xi)_{\mathscr{A}}$, generating nothing more than statistics provided by the single instrument $\mathscr{A}$, would yield a description of everything, which is absurd by pt. **R••**. The physical contents (to come) arise precisely through the above-mentioned matching (see Section 6.4 below).

As a result, we adopt a kind of the relativity-principle analogue—a tenet on the quantum observational covariance.

III Theory should introduce a means of equating the macro-observations (pts. **O** + **M**) by differing instruments $\{\underline{\alpha}_1, \dots\}_{\mathscr{A}} \neq \{\underline{\beta}_1, \dots\}_{\mathscr{B}}$ under a common (the same) experimental environment $\langle\!\langle \mathcal{S}, \mathbf{M}, \dots \rangle\!\rangle$.

(***The third principium of quantum theory***)

Cf. [22] (p. 632) and mathematical analogies [85,86].

5.4.1. Semantic Closedness and the Equal Sign =

We are currently returning once again to Section 2.1, falling into a situation when the case in hand does not just entail fundamental theory in the form of ⌈math⌉ + ⌈physical "bla-bla-bla"⌉, while, continuing on an informal note, the mathematics of physics—quantum mathematics—is being created "from scratch". When building up this math, it is impossible to forego the physical conventions $\langle\!\langle \mathcal{S}, \mathbf{M}, \dots \rangle\!\rangle$; meanwhile, any preliminary and the formal characterization for $\langle\!\langle \mathcal{S}, \mathbf{M}, \dots \rangle\!\rangle$ is ruled out.

Indeed, the attempts to mathematically formalize the physical context of observation, rather than observation itself, will not logically manage without another "more fundamental" observation, in this case, of the very experimental environment. The semantic cycling is apparent here, and any of its mathematization will lead to a retrogression of definitions into infinity, which is known as the "von Neumann catastrophe" ([80], pp. 158–…)) or as "trying to swallow itself by the tail" ([28], p. 220). Which is why, once again, the "Box (6) method" prohibitions are required. See also a paragraph containing the capitalized emphasize "CANNOT IN PRINCIPLE" on p. 418 of the work [15]. Sooner or later, it will have to be declared that mathematics will be created for *the* convention $\langle\!\langle \mathcal{S}, \mathbf{M}, \dots \rangle\!\rangle$and that this mathematics will be a mathematical model for this $\langle\!\langle \mathcal{S}, \mathbf{M}, \dots \rangle\!\rangle$. The analogous argument—"mathematics is there to serve physics, and not the other way round" ([16], p. 242; L. Hardy)—has already long been met in the literature [23,33,40]. In connection with the "general contextual models", see the books [64,150] (the Växjö-model, "quantum contextuality") and bibliography therein.

Remark 10 (semantic)**.** *To avoid the just mentioned linguistic closedness—a kind of mathematical "pathology" of the physical and natural languages—a description that lays claim to the role of an unambiguous/rigorous theory requires a careful separation of the object- and meta-languages. For more detail, see [105] (Sections 14–16), [106] (Section V.1), and [136] (Section 3.9). For this reason, the constructs should track the blending of the object* QM*-domain (syntax) and the meta-domain (semantics) and, more generally, the penetration of extra-linguistic elements of thinking [148] into* QM*. The notion of "the same $\langle\!\langle \mathcal{S}, \mathbf{M}, \dots \rangle\!\rangle$", which is intuitive in the natural language, should explicitly be indicated as an external and fundamental category (pr.* ***III****), and its circular re-interpretations/re-translations within the theory should be banned. That is, re-stating "the sameness $\langle\!\langle \mathcal{S}, \mathbf{M}, \dots \rangle\!\rangle$, the* *identical ⟪$\mathcal{S}$, **M**, ...⟫" by way of word or of the equality symbol = between some other entities is forbidden.*

- *"The same" may no have a definition in terms of anything else. It exists prior to theory and has only a meaning (= verbal context), though its natural-language descriptions may be of great variety and be "presented to us in wildly different ways" ([86], p. 2 and the whole of the Section "The awkwardness of equality").*

*One could, e.g., accept the typical verbal vehicle "a complex of conditions, which allows of any number of repetitions" (quotation from the literature). It is clear that the words "complex ... allows ... repetitions" here are just another semantic equivalent to the word "the same ⟪$\mathcal{S}$, **M**, ...⟫". The physics terminology per se (Sections 6.4, 6.5, and 9.1) will become accessible when physical concepts are introduced via the originating—and obligatorily very ascetic—quantum-mechanics language. See also the selected thesis on page 70.*

It is crucial to immediately note that, in the same manner, the classical description contains the cited arguments in their entirety. It is easy to convince that such a description also implies implicitly that which is designated above as ⟪$\mathcal{S}$, **M**, ...⟫; otherwise, the physical reasoning would be entirely impossible. "[W]e often prefer to regard a number of outcomes of distinct physical operations as registering the same property, ... representing the same measurement. ... permitting an unrestricted identification of outcomes would lead to "grammatical chaos"" (Foulis–Randall ([111], p. 232)). More to the point, the physics and mathematics not merely have been closely interwoven with each other. Any recursive procedure of definitions will inevitable result in either a cyclic definition at some level, or a definition that refers outside not only of the physics but even of the math. Hence, the hierarchical arrangement of notions/.../definitions—a property that is frequently uncontrolled and violated in the human thinking—can only be meaningful if at least one knot in the definition network is externally defined. In this work, that basic points are, as a rough guide, the brace $\underline{\Psi} \overset{\mathscr{A}}{\dashrightarrow} \underline{\alpha}$ and the notion of "the same ⟪$\mathcal{S}$, **M**, ...⟫", motivated in Remarks 2 and 10, respectively.

Remark 11. *Here, the situation is similar to the role of the axiom of choice in the* ZF*-system [134,147]. It has been well known for a long time that the axiom is often subconsciously implied ([149], Chs. II, IV); it can also not be either circumvented or ignored. Another counterexample to "infinite retrogression and circularity" in logic comes from the very same system. This is a ban on infinite chain of set memberships* $\in$ *on the left*

$$\| \cdots \in X_n \in \cdots \in X_2 \in X_1 \in X_0$$

(the regularity axiom $[\forall x \in X,\ x \cap X \neq \varnothing] \Rightarrow [X = \varnothing]$*) under the permissibility of the infinite* $(\in)$*-continuing to the right:*

$$X_0 \in X_1 \in X_2 \in \cdots \in X_n \in \cdots \in \cdots$$

(not rigorously, the infinity axiom) [120,134].

The obvious parallels here are the famous Russell paradox [149] or a chaos in the computer file system when the "hard links" from a folder to the parent folder are allowed. Thus, the relations $\in$ *"downwards" to the left and necessarily terminates in something, i.e., in a set* X_0 *that contains nothing:* $\varnothing = X_0 \in \cdots \in X_n \in \cdots$*. Therefore, one needs to give "meaning" to the only set—the empty one* $\varnothing$*. Incidentally, it is these axioms that guarantee the existence of infinitely many ordinal numbers (106) and the uniqueness of this structure. The ordinals and numbers have yet to be dealt with further below in more detail.*

All that remains is to add that no theory in physics is feasible without re-calculations of physical units and of vectors/tensors without transformations in the fiber superstructures over manifolds, etc. Accordingly, the considerations on invariance and on transformations should be present in the quantum case as well, but it, which is its principal difference from the classical case, still lacks the concepts of physical quantities/properties (see Section 6.4). Therefore, such argumentation may only be applied to those objects that we have at our disposal, i.e., to the (Ξ)-brace. The renunciation of pr. **III** would actually be tantamount to the inability to make the physics theories whatsoever.

Now, pr. **III** and the "quantum diversity of the reference frames" $\{\mathscr{A}, \mathscr{B}, \ldots\}$ require a kind of factorization of the entire family $\{(\Xi)_{\mathscr{A}}, (\Xi)_{\mathscr{B}}, \ldots, (\Xi)'_{\mathscr{A}}, (\Xi)'_{\mathscr{B}}, \ldots\}$ with respect to the conception $\langle\!\langle \mathcal{S}, \mathbf{M}, \ldots \rangle\!\rangle$, i.e., the introduction of an operation of equating the results $(\Xi)_{\mathscr{A}}, (\Xi)_{\mathscr{B}}$ that came when observing $\mathcal{S}$. $(\Xi)_{\mathscr{A}} \stackrel{?}{=} (\Xi)_{\mathscr{B}}$ should not be immediately put since these braces are simply different sets. That is why, with isolated semigroups

$$\underbrace{\{\overset{\mathscr{A}}{\uplus}\,;\, (\Xi')_{\mathscr{A}}, (\Xi'')_{\mathscr{A}}, \ldots\}}_{\mathfrak{G}_{\mathscr{A}}}, \quad \underbrace{\{\overset{\mathscr{B}}{\uplus}\,;\, (\Xi')_{\mathscr{B}}, (\Xi'')_{\mathscr{B}}, \ldots\}}_{\mathfrak{G}_{\mathscr{B}}}, \quad \ldots$$

at our disposal, we have to conceive of them as elements of a new set H of objects having a single nature, 1) to carry out the mapping $\{\mathfrak{G}_{\mathscr{A}}, \mathfrak{G}_{\mathscr{B}}, \ldots\} \mapsto H$, assigning new representatives $|\Xi_{\mathscr{A}}\rangle \in H$ to the (Ξ)-brace, and 2) to equip H with an equivalence relation $|\Xi_{\mathscr{A}}\rangle \approx |\Xi_{\mathscr{B}}\rangle$ (the concept "the same" above). Let us implement all of that by the scheme

$$\begin{array}{lll} (\Xi)_{\mathscr{A}} := \left[\begin{smallmatrix}\mu_1\\ \lambda_1\end{smallmatrix}\right] \underline{\alpha}_1 \overset{\mathscr{A}}{\uplus} \left[\begin{smallmatrix}\mu_2\\ \lambda_2\end{smallmatrix}\right] \underline{\alpha}_2 \overset{\mathscr{A}}{\uplus} \cdots & \rightarrowtail \left[\begin{smallmatrix}\mu_1\\ \lambda_1\end{smallmatrix}\right] |\alpha_1\rangle \dotplus \left[\begin{smallmatrix}\mu_2\\ \lambda_2\end{smallmatrix}\right] |\alpha_2\rangle \dotplus \cdots & =: |\Xi_{\mathscr{A}}\rangle \in H\,, \\ (\Xi)_{\mathscr{B}} := \left[\begin{smallmatrix}\mu_1\\ \lambda_1\end{smallmatrix}\right] \underline{\beta}_1 \overset{\mathscr{B}}{\uplus} \left[\begin{smallmatrix}\mu_2\\ \lambda_2\end{smallmatrix}\right] \underline{\beta}_2 \overset{\mathscr{B}}{\uplus} \cdots & \rightarrowtail \left[\begin{smallmatrix}\mu_1\\ \lambda_1\end{smallmatrix}\right] |\beta_1\rangle \dotplus \left[\begin{smallmatrix}\mu_2\\ \lambda_2\end{smallmatrix}\right] |\beta_2\rangle \dotplus \cdots & =: |\Xi_{\mathscr{B}}\rangle \in H\,, \\ \ldots\ldots\ldots & \ldots\ldots & \ldots\ldots\ldots \end{array} \tag{49}$$

In this, the new addition $\dotplus$ must of course homomorphically inherit operations $\overset{\mathscr{A}}{\uplus}$, $\overset{\mathscr{B}}{\uplus}$, …, and the extension of this definition throughout H is then made with the aid of the very equivalence $\approx$:

$$|\Xi'_{\mathscr{A}}\rangle \dotplus |\Xi''_{\mathscr{B}}\rangle = \Big| |\Xi''_{\mathscr{B}}\rangle \approx |\Xi''_{\mathscr{A}}\rangle \Rightarrow \Big| = |\Xi'_{\mathscr{A}}\rangle \dotplus |\Xi''_{\mathscr{A}}\rangle = |\Xi'_{\mathscr{B}}\rangle \dotplus |\Xi''_{\mathscr{B}}\rangle\,.$$

The negation $\not\approx$ of the relation $\approx$, e.g., $|\Xi'_{\mathscr{A}}\rangle \not\approx |\Xi''_{\mathscr{A}}\rangle$, is exactly the very same distinguishability that was discussed in Sections 2 and 3.

For the sake of convenience, we adopt the regular sign $=$ for $\approx$ in order not to introduce yet a further homomorphism, which are already numerous, with more to come. In other words, the physics $\langle\!\langle \mathcal{S}, \mathbf{M}, \ldots \rangle\!\rangle$ is "concentrated" in the sign $=$, turning the empirical structures (49) into the $\mathscr{A}$-, $\mathscr{B}$-implementations of the object $|\Xi\rangle \equiv |\Xi_{\mathscr{A}}\rangle = |\Xi_{\mathscr{B}}\rangle$ under construction. The adequate term for it—the `Info/Data-Source` or "representative of information" (Č. Brukner (2014))—corresponds to the preliminary prototype of the concept of a state, but we will remain within the standard term, disregarding its variance.

6. Quantum Superposition

How come the quantum? … No space, no time—J. Wheeler (1989)

… postulation of something as a Primary Observable is itself a sort of theoretical act and may turn out to be wrong—T. Maudlin ([151], p. 142)

6.1. Representations of States

Let us simplify notation according to the rule $\left[\begin{smallmatrix}\mu\\ \lambda\end{smallmatrix}\right] =: \mathfrak{a}$. The sought-for relationships between $\mathscr{A}, \mathscr{B}, \ldots$ then turn into the key point of further construct—the equalities

$$\left[\begin{array}{c}\text{representations}\\ \text{of } |\Xi\rangle\text{-state}\end{array}\right] : \qquad \mathfrak{a}_1 |\alpha_1\rangle \dotplus \mathfrak{a}_2 |\alpha_2\rangle \dotplus \cdots \stackrel{\text{"the same"}}{=\!=} \mathfrak{b}_1 |\beta_1\rangle \dotplus \mathfrak{b}_2 |\beta_2\rangle \dotplus \cdots = \cdots\,. \tag{50}$$

They furnish *representations* $|\Xi_{\mathscr{A}}\rangle, |\Xi_{\mathscr{B}}\rangle, \ldots$ *of quantum state* $|\Xi\rangle$ *of system* $\mathcal{S}$. By design, this `DataSource` object $|\Xi\rangle$ carries data $(\Xi)_{\mathscr{A}}, (\Xi)_{\mathscr{B}}$ and, more generally, (Ξ)-data (47) from arrays of *any* observations, including the imaginary ones. That is what eliminates the initial need for the $(\Xi)_{\mathscr{A}}$-brace (24) to come from the observation $\mathscr{A}$, which is reflected in the shortening of the term "representation of state" to simply "state" $|\Xi\rangle$. It should be added that the straightforward storing of objects $\{|\Xi_{\mathscr{A}}\rangle, |\Xi_{\mathscr{B}}\rangle, \ldots\}$ in a certain set H, but with the independence of operations $\{\dotplus^{(\mathscr{A})}, \dotplus^{(\mathscr{B})}, \ldots\}$ preserved, would not differ from the tautological substitution of symbols. Accordingly, the semantic autonomy of (Ξ)-brace would also be inherited, whereas covariance **III** requires an elimination of precisely this autonomy. What is more, the set-theoretic original copy for operations $\{\overset{\mathscr{A}}{\uplus}, \overset{\mathscr{B}}{\uplus}, \ldots\}$ and $\dotplus$ is one and the same—the union $\cup$.

The symbols $|\alpha_s)$ and $|\beta_s)$ in (50) are no more than symbols. Hence, the objects' property (50) of being identical must be reflected in terms of their coordinate $\mathfrak{a}, \mathfrak{b}$-components (pt. **R**$^\bullet$). This means that any aggregate $(\mathfrak{a}_1, \mathfrak{a}_2, \ldots)$ is unambiguously calculated by means of a certain transformation $\widehat{U}$ into any other $(\mathfrak{b}_1, \mathfrak{b}_2, \ldots)$ when the two aggregates represent a common $|\Xi\rangle$:

$$(\mathfrak{a}_1, \mathfrak{a}_2, \ldots) = \widehat{U}(\mathfrak{b}_1, \mathfrak{b}_2, \ldots) .$$

The $\widehat{U}$ then becomes an isomorphism between these aggregates (a preimage of the future unitary transformation ([6], p. 14)) and, accordingly, their lengths must coincide. This length—a certain single constant—will be symbolized as D.

6.2. Representations of Devicesand Spectra

Naturally, the instrument is converted to the H-structure language along with (Ξ)-objects. It is a set of symbols $\{|\gamma_1), |\gamma_2), \ldots\}$ in place of the previous $\{\underline{\gamma_1}, \underline{\gamma_2}, \ldots\}$. As has just been shown, their number for any $\mathscr{C}$-instrument should be equal to D. However, generally speaking, $|\mathfrak{T}_{\mathscr{A}}| \neq |\mathfrak{T}_{\mathscr{B}}|$ since $\mathfrak{T}_{\mathscr{A}}$ and $\mathfrak{T}_{\mathscr{B}}$ are assigned in an arbitrary way (pt. **O**). Therefore, if we take an illustration $\mathscr{A}\{\underline{\alpha_1}, \underline{\alpha_2}\}$ and $\mathscr{B}\{\underline{\beta_1}, \underline{\beta_2}, \underline{\beta_3}\}$, then H-representation of instrument $\mathscr{A}$ should appear at least as $\{|\alpha_1), |\alpha_2), |\alpha_3)\}$. Clearly, the already present distinguishability $\underline{\alpha_1} \not\simeq \underline{\alpha_2}$ (Section 2.2) is automatically converted into an abstract distinguishability of new symbols $|\alpha_1) \neq |\alpha_2)$, and empirical $\mathscr{A}$-distinguishability is confined exclusively by these two symbols. In that case, for the purpose of noncontradiction, the added third symbol $|\alpha_3)$, as an adjunction to the abstract relations $|\alpha_3) \neq |\alpha_1)$ and $|\alpha_3) \neq |\alpha_2)$, should be complemented with the notion of its physical *in*discernibility from $|\alpha_1)$ or $|\alpha_2)$. By an extension of this argument, one obtains that every $\mathscr{A}$-instrument should be endowed with the (non)equivalence relation $(\simeq/\not\simeq)$ in terms of the H-structure by its formal $\{|\alpha_1), \ldots\}$-representations. How do we do this?

Let us proceed further from a self-suggested extension of pt. **R**. Let us declare—and it is more than natural—that the number representations α_s are linked not only to observations but to instruments as well. Each α_s is the new object of a numerical type: a number or a collection of numbers. Then, indiscernibility, say $|\alpha_3) \simeq |\alpha_1)$, is recorded by coincidence of the numeral labels $\alpha_3 = \alpha_1$ attached to the symbols $|\alpha_3)$ and $|\alpha_1)$, respectively. The abstract ("old") distinguishability $|\alpha_3) \neq |\alpha_1)$, meanwhile, remains as it is. From here, we have the following formalization of the relationship between $\simeq$ and $=$ by means of dropping/adding the brackets $|\ )$:

$$\left.\begin{array}{lcl} |\alpha_s) \not\simeq |\alpha_k) & \Longleftrightarrow & \alpha_s \neq \alpha_k \\ |\alpha_s) \simeq |\alpha_k) & \Longleftrightarrow & \alpha_s = \alpha_k \end{array}\right\} \quad \text{under} \quad |\alpha_s) \neq |\alpha_k) . \tag{51}$$

Call the quantity α_s (numerical) the *spectral label/marker* of eigen-element $|\alpha_s)$. Then, by the *H-representation* $[\mathscr{A}]$ *of instrument* $\mathscr{A}$, we will mean the set of objects $\{|\alpha_1), \ldots, |\alpha_{\mathrm{D}})\}$ supplemented with the spectral structure (51):

$$[\mathscr{A}] := \left\{|\alpha_1)_{|\alpha_1}, |\alpha_2)_{|\alpha_2}, \ldots\right\} . \tag{52}$$

It is not difficult to see that if $|\alpha_1) \not\simeq |\alpha_2)$, then either $|\alpha_3) \simeq |\alpha_1)$ or $|\alpha_3) \simeq |\alpha_2)$. Otherwise, spectral markers $|\alpha_1 = |\alpha_2$ should coincide, and primary primitives $\underline{\alpha_1} \not\simeq \underline{\alpha_2}$ lose their empirical distinguishability in contrast to (7). The multiple coincidence of $\lfloor\alpha_s$-markers is admissible.

In the presence of relations (51), it is natural to state that instrument $\mathscr{A}$ is coarser (more symmetrical) than $\mathscr{B}$ and, terminologically, to declare that the degeneration of the spectral-label values takes place. In cases of embeddability such as $\mathscr{A}_2\{\underline{\alpha}_1, \underline{\alpha}_2\} \subset \mathscr{A}_3\{\underline{\alpha}_1, \underline{\alpha}_2, \underline{\alpha}_3\}$, instrument $\mathscr{A}_2$ can even be called the same as (coinciding with) $\mathscr{A}_3$, but with a more rough scale. Conversely, $\mathscr{A}_3$ is a more precise extension of $\mathscr{A}_2$. In particular, the natural notion of a device resolution fits here.

All instruments may then be mathematically imagined as having the same resolution, but, perhaps, with degeneration of spectra. The non-coinciding instruments may be interpreted as non-equivalent reference frames $\mathscr{A} \neq \mathscr{B}$ in an observation space. According to pts. **R••** and **III**, they are mandatorily present in the description. The spectral degenerations are also always present since element $\underline{\alpha}_1$ can always be removed from $\mathfrak{T}_{\mathscr{A}}$, and there are no logical foundations to prohibit an observational instrument with family $\mathfrak{T}_{\mathscr{A}} - \{\underline{\alpha}_1\}$. Hence, it follows that introducing the spectra—instrumental readings—is required even formally, without physics. It is of course implied here that spectral (in)discernibility is realized in the same manner as its statistical counterpart in Sections 2.6 and 4, i.e., by numbers. Incidentally, such a property of $\lfloor\alpha_s$—i.e., of being a numerical object—is not at all necessary at the moment. The spectrum $\{\lfloor\alpha_1, \lfloor\alpha_2, \ldots\}$ may be thought of as an abstract set of labels attached to the eigen-elements. As numbers, it is introduced for the subsequent creation of models to classical/macroscopic dynamic, and they are numerical.

Returning to D, we note that, in any case, the toolkit $\{\mathscr{A}, \mathscr{B}, \ldots\} =: \mathcal{O}$ in real use has always been defined, fixed, and is finite. Consequently, the constant

$$\mathsf{D} \geqslant 2 \tag{53}$$

has also been defined and fixed, and it becomes the globally static observable characteristic—an empirically external parameter. Meanwhile, the entire scheme internally contains the natural method of its own extension $\mathsf{D} \mapsto \mathsf{D}+1$, and the potentially all-encompassing choice $\mathsf{D} = \infty$ may be considered the universally preferable one in QT. By freezing the different $\mathsf{D} < \infty$, the theory makes it possible to create models, and they are not only admissible but also well-known. Their efficiency is examined in experiments. Once again:

- The D-constant concept of spectra and their degenerations is created by the $(\mathscr{A}, \mathscr{B})$-covariance requirement, i.e., by principium **III**.

As a result, the structure of H-representations of states and of instruments are liberated from the arbitrariness in assigning the subsets $\mathfrak{T}_{\mathscr{A}}$ in (8). The statistical unitary pre-images (34) and H-elements of the form $\mathfrak{c}|\gamma_s)$ can be associated with any "eigen symbol" $|\gamma_s)$. They are always available because every possible brace (32) is known to contain subfamilies when ongoing $\underline{\Psi}, \underline{\Phi}$-primitives get to a single one, e.g., to $\underline{\gamma}_1$. Therefore, every representation $\mathfrak{a}_1|\alpha_1) \dotplus \cdots$ is always equivalent to a $(\Xi)_{\mathscr{C}}$-brace for some observation $\mathscr{C}$ with a homogeneous outgoing ensemble $\{\underline{\gamma}_1 \cdots \underline{\gamma}_1\}$. That is, one may always write

$$\mathfrak{a}_1|\alpha_1) \dotplus \mathfrak{a}_2|\alpha_2) \dotplus \cdots = \mathfrak{c}_1|\gamma_1) \dotplus 0|\gamma_2) \dotplus \cdots =: \mathfrak{c}_1|\gamma_1)\,, \tag{54}$$

while naturally referring to $\mathfrak{c}_1|\gamma_1)$ as one of the *eigen-states of instrument* $\mathscr{C}$, with an appropriate adjustment of the similar definition in pt. **O**. The construction of the representation-state space is far from being complete since it is still a "bare" semigroup H.

6.3. Superposition of States

Since writings (50) exist for any ensemble (Ξ)-brace, let us consider the following two representations:

$$\begin{aligned} \mathfrak{a}_1|\alpha_1) \dotplus \mathfrak{a}_2|\alpha_2) &= \mathfrak{b}_1|\beta_1) \dotplus \mathfrak{b}_2|\beta_2) \dotplus \cdots\,, \\ \mathfrak{a}_2|\alpha_2) &= \mathfrak{b}'_1|\beta_1) \dotplus \mathfrak{b}'_2|\beta_2) \dotplus \cdots\,. \end{aligned} \tag{55}$$

Comparison of these equalities tells us that the second one is a solution of the first one with respect to $\mathfrak{a}_2|\alpha_2)$. Hence, the semigroup operation $\dotplus$ admits a cancellation of element $\mathfrak{a}_1|\alpha_1)$. This means that there exists an H-element $\tilde{\mathfrak{a}}_1|\alpha_1)$ such that

$$\begin{aligned}
\{\tilde{\mathfrak{a}}_1|\alpha_1) \dotplus \mathfrak{a}_1|\alpha_1)\} \dotplus \mathfrak{a}_2|\alpha_2) &= \tilde{\mathfrak{a}}_1|\alpha_1) \dotplus \{\mathfrak{b}_1|\beta_1) \dotplus \mathfrak{b}_2|\beta_2) \dotplus \cdots\} \\
&\Downarrow \\
0|\alpha_1) \dotplus \mathfrak{a}_2|\alpha_2) &= \tilde{\mathfrak{a}}_1|\alpha_1) \dotplus \mathfrak{b}_1|\beta_1) \dotplus \mathfrak{b}_2|\beta_2) \dotplus \cdots \\
&\Downarrow \quad \text{(due to (54))} \\
\mathfrak{a}_2|\alpha_2) &= \mathfrak{b}_1'|\beta_1) \dotplus \mathfrak{b}_2'|\beta_2) \dotplus \cdots \\
\Downarrow \qquad\qquad & \qquad\qquad \Downarrow \\
|0) := 0|\alpha_1) = \tilde{\mathfrak{a}}_1|\alpha_1) \dotplus \mathfrak{a}_1|\alpha_1), \qquad & \tilde{\mathfrak{a}}_1|\alpha_1) \dotplus \mathfrak{b}_1|\beta_1) \dotplus \cdots = \mathfrak{b}_1'|\beta_1) \dotplus \cdots,
\end{aligned}$$

where $|0)$ stands for a zero in the semigroup H (image (0) of the finite-length brace $(\Xi|)$) and 0 in $0|\alpha_1)$ is a symbol of its $[\lambda,\mu]$-coordinates. By canceling out $\mathfrak{a}_s|\alpha_s)$, one by one, if necessary, one deduces that any element of H does have an inversion. That is, H is actually a group. We re-denote inverse elements $\tilde{\mathfrak{a}}_s|\alpha_s)$ by $(-\mathfrak{a}_s)|\alpha_s)$ and inversions of sums are formed from $(\dotplus)$-sums thereof. Moreover, all the $[\lambda,\mu]$-pairs turn into a set $\{\mathfrak{a},\mathfrak{b},\ldots\}$ equipped with the above-mentioned composition $\oplus$, which follows from an obvious property of unitary brace:

$$\mathfrak{a}|\alpha_1) \dotplus \mathfrak{b}|\alpha_1) = (\mathfrak{a}\oplus\mathfrak{b})|\alpha_1) \tag{56}$$

(inheritance of clossedness under the $\cup$-operation). This composition is also a $\oplus$-operation of a group and of a commutative one:

$$\mathfrak{a}\oplus\mathfrak{b} = \mathfrak{b}\oplus\mathfrak{a}, \qquad (\mathfrak{a}\oplus\mathfrak{b})\oplus\mathfrak{c} = \mathfrak{a}\oplus(\mathfrak{b}\oplus\mathfrak{c}), \qquad \mathfrak{a}\oplus 0 = \mathfrak{a}, \qquad \mathfrak{a}\oplus(-\mathfrak{a}) = 0\,. \tag{57}$$

Therefore, the group nature of semigroup H and the group (57) come from the scheme

$$\begin{array}{|c|}\hline
\begin{bmatrix}\text{single observations}\\ \mathscr{A},\mathscr{B},\ldots\end{bmatrix} \Rightarrow \begin{bmatrix}\text{semigroups}\\ \mathfrak{G}_{\mathscr{A}},\mathfrak{G}_{\mathscr{B}},\ldots\end{bmatrix} \rightarrowtail \\
\rightarrowtail \begin{bmatrix}(\mathscr{A},\mathscr{B})\text{-covariance,}\\ \langle\!\langle \mathcal{S},\mathbf{M},\ldots\rangle\!\rangle \text{ and principium } \mathbf{III}\end{bmatrix} \Rightarrow [\text{group } H] \\ \hline
\end{array}$$

and, technically, from equatings/identifyings (50), i.e., from conception "the same" (Section 5.4). For its part, it is this very structure of algebraic operations—the two- and three-term (and nothing else) axioms of commutativity/associativity, i.e., the group—that comes from properties (39). All of this provides an answer to the key question: where do the (semi)group and the minus sign come from and why?

Thus, handling the $|\Xi\rangle$-objects breaks free from its ties to the notion of observation, and the objects admit the formal writings $\mathfrak{a}|\Psi) \dotplus \mathfrak{b}|\Phi) \dotplus \cdots$. Call them *superpositions*. However, as soon as they or the state are associated in meaning with the word "readings" (this is discussed at greater length in Sections 6.4 and 6.5), this term should be replaced with a non-truncated one, i.e., a representation of the state with respect to a certain observation. Specifically, the statistical weights $\mathtt{f}_j$ are extracted from such expressions only after their conversion into a sum over eigen-states of the form (50); a task of the subsequent mathematical tool.

No superposition $\mathfrak{a}|\Psi) \dotplus \mathfrak{b}|\Phi) \dotplus \cdots$, including (54), has any physical sense in and of itself [5] (p. 137), [12] nor is it preferable to any other one. It merely mirrors the closedness of states with respect to operation $\dotplus$ since any $|\Xi\rangle$ is re-recorded as a sum of various $\{\mathfrak{a}|\Psi)$, $\mathfrak{b}|\Phi), \ldots\}$ in a countless number of ways and is linked to any other such sum. Without a system of $|\alpha_s)$-symbols for instrument $\mathscr{A}$, nothing observable is extractable out of the aggregate of coefficients $\{\pm\mathfrak{a},\pm\mathfrak{b},\ldots\}$ (and, of course, of the $|\Psi)$-letters themselves) in any imaginable way. Accordingly, it is incorrect to speak of—a widespread misconception—

the destruction of the superposition or of the "relative-phase information" ([119], p. 253), associating the word destruction with the physical/observational meanings or processes.

As a result, *even without having a numerical theory yet* and without recourse to the concept of a physical quantity, superposition may not address whatever physical concepts, we arrive at the paramount property, which characterizes the most general type of micro-observation's ensembles (17).

- ***Superposition principle***
  A ($\dotplus$)-composition of quantum states $\mathfrak{a}|\Psi)$ and $\mathfrak{b}|\Phi)$, which are admissible for system $\mathcal{S}$, is an admissible state
$$\mathfrak{a}|\Psi) \dotplus \mathfrak{b}|\Phi) = \mathfrak{c}|\Xi) \tag{58}$$
  and, with that, the set $\{\mathfrak{a}|\Psi), \mathfrak{b}|\Phi), \mathfrak{c}|\Xi), \dots\} =: H$ forms a commutative group with respect to operation $\dotplus$. The family $\{\mathfrak{a}, \mathfrak{b}, \dots\}$ of coordinate $\mathbb{R}^2$-representatives of states (50) is also equipped with the same group structure under the $\oplus$-operation (57) and with the rule of carrying the operation $\dotplus$ over to $\oplus$:
$$\mathfrak{a}|\Psi) \dotplus \mathfrak{b}|\Psi) = (\mathfrak{a} \oplus \mathfrak{b})|\Psi)\,. \tag{59}$$

Let us clarify the transferring of (56) to (59). The union of the state prototypes $\mathfrak{a}|\Psi), \mathfrak{b}|\Psi) \in H$ is known to belong to $\mathfrak{G}$. Thus, the composition $\mathfrak{a}|\Psi) \dotplus \mathfrak{b}|\Psi)$ should be identical to a certain element $\mathfrak{c}|\Psi) \in H$. It is clear that $\mathfrak{c}$ depends on $\mathfrak{a}$, $\mathfrak{b}$ and, hence, $\mathfrak{a}|\Psi) \dotplus \mathfrak{b}|\Psi) = \mathfrak{c}(\mathfrak{a}, \mathfrak{b})|\Psi)$. The exhaustive properties of dependence $\mathfrak{c}(\mathfrak{a}, \mathfrak{b})$ are given by Formulas (57) and (59) under notation $\mathfrak{c}(\mathfrak{a}, \mathfrak{b}) =: (\mathfrak{a} \oplus \mathfrak{b})$.

6.3.1. "Physics" of Superposition

Besides the essentially unphysical nature of the ($\dotplus$)-superpositions, i.e., "we cannot recognize them" ([12], p. 13), the primary and salient property of quantum addition is in the fact that, due to the group subtraction, it is possible to experimentally obtain a "quantum zero" in statistics from "non-zeroes'. With that, these "seem to be" positive, but there are "negative non-zeroes", i.e., negative numbers (Section 9.2). Subtraction manifests by the typical obscurations in interference pictures. S. Aaronson adds to this: "We have got minus signs, and so we have got interference" ([20], p. 220). No classical composition
$$w\varrho_1 + (1-w)\varrho_2 \tag{60}$$
of non-zero statistics ϱ_1, ϱ_2 can provide a zero value since the zero will never be obtained via the $\cup$-unions. The same is true for the pre-superposition in isolated brace $(\Xi)_{\mathscr{A}}$, i.e., when one instrument is in question.

Remark 12. *One cannot help but mention yet another counterexample to the superposition's "physicality": the (in)famous "quantum cat". Any combination of the dead and living animal is meaningless as a statement about new/nonclassical entity such as a "(half-)dead/alive cat" or such statements about particles as "their being neither here nor there but everywhere", especially with the stress on "at the same point in time" (see pr.* **I***). It makes absolutely no sense to add (allegedly in accord with the character* $+$*) to each other the nature's phenomena and notions that have not yet been created and are dynamical ("alive") at that.*

- **What is being added is states, not their denominations** or verbal descriptions of envisioned *("fantasized", "fantastic phantoms" ([12], p. 15))* physical properties *such as spin up/down or dead/alive. Cf. [151] (pp. 134 (!), 135).*

The "cat-box open" is a click, not state, without a notion of "a cat". Accordingly, the word combination "the quantum objects exist in "superpositions" of different possibilities" (a representative excerpt from the literature) is at most an interpretative allegory (Section 10) without physical and mathematical content. That is to say, strictly speaking,

- *No quantum (micro)system has ever been/dwelled in any state, much less in a superposition one, and much less at an instant t. Ludwig, on pp. 16 and 78 of the book [58], insists that it is a "myth" and "a fairy tale, ... the very widespread idea that each microsystem has a real state ... represented by a vector in a Hilbert space", and M. Nielsen remarks in [21] that "Saying* $0.6|0\rangle + 0.8|1\rangle$ *is simultaneously* 0 *and* 1 *makes about as much sense as Lewis Carroll's nonsense poem* Jabberwocky: *...". K. Svozil does also underscore that "'coherent superpositions' just correspond to improper, misleading representations of non-existing aspects of physical reality. They are delusive because they confuse ontology with epistemology" ([152], p. 26).*

The meaning of the word "add" is still being created, including an implementation at objects to be thought of as the "atomic irreducible" entities—the numbers (Section 7).

T. Maudlin notes on p. 133 of the work [151]: "Our job ... is to invent mathematical representations ..., rather than merely linguistic terms such as "z-up." ... we are in some danger of confusing physical items with mathematical items" *(italics supplied). Here is an example of confusion. If we are going to measure the z-spin in one of the* $(\rightleftarrows_x)$*-beams in a Stern–Gerlach device, then why and when does this* "observable" *certainty—say* $|{\rightarrow_x}\rangle$*—get turned into a* $(\uparrow\downarrow_z)$*-uncertainty* $|\uparrow\rangle + |\downarrow\rangle$*? (see [153] (p. 232)). However, what if we are about not to do this? We come up against the question:*

- *What does one mean by an equal-sign = in the orthodox notation* $|\rightarrow\rangle = |\uparrow\rangle + |\downarrow\rangle$*?*

Which state does the system "intend" to fall into: the z-uncertainty or the x-determinacy? Which of the states is it in, after all? Examples to the "physicality of states" may be continued endlessly [154].

A statement about QM-superposition (without $\mathbb{C}$-numbers) as a non-independent axiom can be found in the book ([36], p. 108) but arguments given there are circular: ⌈Hilbert space⌉ $\rightarrowtail$ ⌈quantum logic of propositions⌉ $\rightarrowtail$ ⌈superposition principle⌉. Similarly, in the works [122] and [72] (p. 164), all of that is "derived" from modular lattices [155]. However, the lattices are known to enter QM from the Hilbert space structure and, on the other hand, the purging quantum rudiments of such a space' axiomatics constitutes Birkhoff's 110-th problem ([155], p. 286). Note also that, in connection with the formal logic approaches to the theory construction [9,36,72,111,122,156,157], the issue of vindicating the *matters* that this logic deals *with* (logic of what?) [58,93,158] should not be neglected. What we mean here is the questions on logic: of propositions? [105,106] of relations? of (math-logic) classes/sets? [120] of phenomena/properties? (which ones?) of quantum/classical events? ...? "For example, would one have to develop a quantum set theory?" ([110], p. 17). "If by "logic" we mean something like "correct reasoning," then it would make no sense to think of logic as "just another theory."" ([73], p. 258). The more abstract micro-events and Boolean logic we have used in metamathematical reasoning at the moment ([87], pp. 189, 193) contain nothing that depends on classical physics. That is, quantum foundations do not require [58] a different quantum/non-classical logic. See also [74] (p. 29).

6.3.2. When and What Is Non-Commutativity?

Yet another fact that results from the above constructs is that the availability of a superposition math-structure (58) reflects the presence of at least *two* $\mathscr{A}$, $\mathscr{B}$ with *non-coinciding* families of eigen-primitives $\{\underline{\alpha}_s\}$, $\{\underline{\beta}_k\}$. This consequence of pt. $\mathsf{R}^{\bullet\bullet}$ should be particularly emphasized since it will manifest in the *non-commutativity* of operators $\hat{\mathscr{A}}$ and $\hat{\mathscr{B}}$ in the future. Although the present work does not get to operators as a mathematical structure, it is clear that the emergent eigen-states and spectra have a direct bearing on them. In this context, the "commuting instruments" $\{|\alpha_1\rangle, |\alpha_2\rangle, \ldots\} = \{|\beta_1\rangle, |\beta_2\rangle, \ldots\}$ can be treated, roughly speaking, as coinciding because this fact is independent of the specific spectra $\{\lfloor\alpha_1, \lfloor\alpha_2, \ldots\}$, $\{\lfloor\beta_1, \lfloor\beta_2, \ldots\}$ assigned to them. If they differ, this is merely a different (numerical) graduation of the spectrum scale. It is the same for all instruments, and its length is the parameter D.

Notice that the definition of an $\mathscr{A}$-observation is not different from the formal assignment of the family $\mathfrak{T}_{\mathscr{A}}$ (pt. **O** and (8)), which is why the non-coinciding sets $\mathfrak{T}_{\mathscr{A}}$, $\mathfrak{T}_{\mathscr{B}}$ do always exist. This provides a kind of abstractly deductive existence's proof for the non-commutativity, QM-interference—see Section 6.5 further below—and for the utmost low-level finality of QM altogether [13,33,93]. The whys and wherefores of theory do not require invoking the physical conceptions; cf. [10] (p. 2).

Of no small importance is that this point entails an independence of the (existence/presence of) classical physics or of its formal deformation, which are yet to be created from the quantum one (cf. a selected thesis on page 15). In particular, no use is required of the notion of a certain pretty small—again the classical/physical term—quantity, i.e., the Plank constant $\hbar$ ([123], Section 6.5). (Parenthetically, no numerical value of this constant matters here; it is not dimensionless and its zero limit is not meaningful.) What is more, the quantum paradigm (17)–(19) tells us that the classical description begins,i.e., we do create/introduce, with the notions of a micro-event's average and of time, whereas these conceptions are still absent at the moment and in the present work. Similarly for the notions of locality, causality, the classical event, and the classical object.

6.4. Physical Properties

Now, the "general physics" $\langle\!\langle \mathcal{S}, \mathbf{M}, \dots \rangle\!\rangle$ is mathematized into representations (50) of states $|\Xi\rangle$ of system $\mathcal{S}$. There is, however, an ambiguity, the source of which is the fact that the natural/classical language also lays claims to a similar formulation. This refers to the belief in the existence of mathematics ("bad habit" [3]; see also [38], [58] (p. 122), and [159]) that describes $\mathcal{S}$ as an individual object with properties regardless of observation; an observation that is not a *functioning* attribute of the mathematics itself. In classical description, it is specified by definitions: point $\mathcal{P}$ of a phase space, (q,p)-coordinatization of the point (manifold), and statistical distribution $\varrho(q,p)$.

On the other hand, quantum empiricism provides nothing more to us besides the ensemble brace and $|\Xi\rangle$-states (pt. **T**). Preordained definienda with physical contents are unacceptable, i.e., $\mathcal{S}$ should not be conceived as "something with *physical* properties" or as an "*individual* object"[93,94], [113] (p. 645). However, since the observational data (in the broadest sense of the word) may not originate from anywhere but a certain $|\Xi\rangle$-object, there should subsequently create:

(1) The very concept of physical objects and properties ([160], pp. 211–230);
(2) Their numerical values/characteristics, i.e., the "physical attributes of objects" ([131], p. 238; N. Bohr).

This is habitually referred to as elements/images of reality [27] (p. 194), [40] (Section 10.2), [94] (Section XIII.4.8)—Bell's "beables" [28]—or what we have been calling attributes of a physical system.

- "The very notion of 'phenomenon' or of 'the appearance of things,' ... is a cognitive and perceptual act of abstraction"

M. Wartofsky ([160], p. 220)

That is to say, the physical phenomena per se do not exist [92] (p. 310), [127].

Indeed, the primary ideology of Sections 1.3 and 2.1 tells us that an invasion of physically self-apparent images into the theory should be avoided ([87], p. 69) because "quantum theory not only does not use—it does not even dare to mention—the notion of a "real physical situation"" ([27], p. 198; E. Jaynes). Continuing a quotation from R. Haag on page 5, one requires "the renunciation of the absolute significance of conventional physical attributes of objects" ([131], p. 238; N. Bohr) and of concomitant and accustomed logic in reasoning. In fact, we are led to (re)build the language of the classical description. Therefore, everything, with no exceptions, should be created mathematically: coordinates, momenta, energies, optical spectra, device readings, lengths/distances and time, extension and lifetime of objects, the language of particles, their number/numeration (Fock space), (in)discernibility/individuality (bosons/fermions), the notions of a subsystem of system $\mathcal{S}$ (see (23)), and even a notion of the physical rigor (in reasoning), etc.

Degrees of freedom, the concepts of the field/body/mass/inertia/interaction, the numerical labeling the space-time continuum, Newtonian mechanics with its equations and the concepts of the force, interaction, and the causality of classical events, thermodynamics, the very term "the classical state", the numerical labels of the space-time continuum and numerical forms of what is known as the classical reference frames—coordinates on manifolds—need to also be created. Once more to underscore, the numerical forms of the classical space coordinates and the time (e.g., the metric tensor $g_{\alpha\beta}(x)dx^{\alpha}dx^{\beta}$) have a quantum empirical origin. The latter fact is required for carefully posing the questions of quantum gravity, and it should be noted in passing that the simultaneity is an ill-defined term not only in the (general) relativity theory; in QTit is even worse. In common with the simultaneous measurability, this term appears to have come from the classical framework, which is why it is illegal as a quantum-theoretical primitive (pr. **I** and [87]).

6.4.1. Waves/Particles?

The concept of a (non-elementary) particle, which is conceptually close to the notion of a subsystem/part, is also a physical convention and can only arise from the $|\Xi\rangle$ or its models: Bose-condensates, deformation excitations in crystal lattices, quasi-particles in a superfluid phase, quantum theories of various fields (relativistic or non), and more. Here, by particle we mean the classical kinematic conception. "What do we detect? The presence of a particle? Or the occurrence of a microscopic event?" wondered R. Haag (2013). H. Zeh and G. Ludwig do answer: "There are no particles in reality" [161], "we must abandon the notion of a microscopic "object", one to which we have been accustomed" ([87], p. 69).

Clearly, the QFTs is a subclass of QM rather than its extension; not that we have yet given a definition of QM. In particular, it is common knowledge in QFT that there is no logical way to distinguish a particle from a certain state—normally, a vacuum excitation. One word should therefore be used for both. To this extent, the familiar "dualism of … the *particle picture* and the *wave picture*" [78] (Section 7.2), [91], [108] (p. 28) simply disappears. K. Popper is rather emphatic concerning this "problem" and puts it, in their "thirteen theses" [108], quite rightly in the following terms: *"the great quantum muddle"*, "alleged "duality" or "complementarity", … *this* kind of "understanding" is of little value", "has not the slightest bearing on either physics, … ", "fashionable among quantum theorists, … a vicious doctrine", and the like. As a matter of fact, both the particles and waves are the classical terms [61] and, in quantum language, they turn into the derivatives of the concepts of state and mixture (23).

- Like waves, *the particle is already an appearance*—an observable one (phenomenology, derivative)—rather than a logical primitive or a fundamental substance, which is why it may not exist [161] prior to theory's principles ([126], p. 762 (!)). Paraphrasing Heisenberg, Haag remarks, in the context of their "event theory", that "Particles are the roof of the theory, not its foundation" ([88], p. 300).

Both these notions should be superseded by a mathematics of clicks.

The f-statistics also falls under observable quantities, and constant D, if declared finite, is an example of an already created characteristic: the dimension of a state space to come. A tensorial structure of this space—compound systems—also pertains to the physical properties, but we do not touch upon this point here. As an aside, this compositional structure will provide the means of distinguishing the aforementioned models under $D = \infty$.

In other words, the logic of the above constructs prohibits not only endowing the phraseology "internal state of an individual object $\mathcal{S}$" and "the system is in a (definite) state [4,58,93,94] with a meaning but also indirectly using its numerical forms. That would work in the circumvention of empiricism, assuming the a priori availability of mathematical structures that do not rest on the state space. L. Ballentine remarks in this regard: "the habit of considering an individual particle to have its own wave function is hard to break"

([34], p. 238); cf. "To speak of a single possible initial apparatus state is pure fantasy" ([80], pp. 241–242; N. Graham).

6.5. Interference

Let us go on with comments as to involving the *physics*-related argumentation to explicate the quantal behavior. We have already mentioned above that for this purpose there is simply no language of physics (Sections 2.1 and 6.4) and of mathematics yet (Sections 2.3 and 5). That is why analogies of this sort are not only deceptive but must be prohibited for exactly the same reasons that accompanied boxes (5). The typical examples in this connection are the simultaneous measurability mentioned above and the two-slit interference [5,17].

First and foremost, the two cases—whether one or two slits are open—are utterly "different experimental arrangements" [153] (p. 236), [64] (p. 58):

$$\langle\!\langle \mathcal{S}, \mathbf{M}, \ldots \rangle\!\rangle' \neq \langle\!\langle \mathcal{S}, \mathbf{M}, \ldots \rangle\!\rangle'' \,.$$

There is nowhere to seek a means of their comparison or the transference of one into another ([153], p. 236). Nonetheless, the classical approach, when opening another slit $\langle\!\langle \mathcal{S}, \mathbf{M}, \ldots \rangle\!\rangle'_2$ together with the first one $\langle\!\langle \mathcal{S}, \mathbf{M}, \ldots \rangle\!\rangle'_1$, does literally envision properties for $\langle\!\langle \mathcal{S}, \mathbf{M}, \ldots \rangle\!\rangle''$ (see Section 6.4). In doing so, the transference method itself—"addition of the two 1-slit $\langle\!\langle \mathcal{S}, \mathbf{M}, \ldots \rangle\!\rangle'$-physicae" by the rule of arithmetical addition of statistics (60)—is meanwhile considered self-apparent. Thus, natural questions arise, such as "why/where are the zeroes coming from, they should not be there". In accordance with the aforesaid, everything here is erroneous, including the "natural" questions. There are no rules at the outset whether (non)classical and even quantum, just as there is no addition per se. An a priori assumption that stem from the obvious images for $\langle\!\langle \mathcal{S}, \mathbf{M}, \ldots \rangle\!\rangle'_1$ and $\langle\!\langle \mathcal{S}, \mathbf{M}, \ldots \rangle\!\rangle'_2$ is actually a declaration of the physical properties for $\langle\!\langle \mathcal{S}, \mathbf{M}, \ldots \rangle\!\rangle''$, but they do not follow from anywhere [17], [64] (p. 55), [162]. The (illegal) assumption of the "negligible effect of which-slit detectors" were mentioned on p. 28 is identical with a declaration of a physical property, as well as a solenoid's switch-on/off in the Aharonov–Bohm effect.

Taken alone, the ϱ-distributions—separate for $\langle\!\langle \mathcal{S}, \mathbf{M}, \ldots \rangle\!\rangle'_1$ and $\langle\!\langle \mathcal{S}, \mathbf{M}, \ldots \rangle\!\rangle'_2$—are entirely correct observational pictures, but introducing the rule (60) is indistinguishable from "invention" of physics—a logically prohibited operation. As Slavnov had put it, "to invent the physical exegesis of a ... mathematical scheme" ([76], p. 304). "Our custom of seeing classical mechanics as a no-nonsense description of 'reality as it is' does not seem to be justified. This custom is actually based on a confusion of categories ... " (W. de Muynck ([4], p. 89)). In other words, the mere fact of non-adherence to this rule means that the grammatical conjunction of the verbs "to understand/deduce" with the noun "micro-phenomena" is unacceptable even linguistically. It is the point **T** that prohibits predefined (classical) semantics, and this was faithfully summarized by C. Fuchs: "badly calibrated linguistics is the predominant reason for quantum foundations continuing to exist as a field of research" ([2], p. xxxix). To figure out or deduce (from mathematics) that quantal phenomena are unfeasible ([5], p. 111) and are "*absolutely* impossible, to explain in any classical way" (quotation by Feynman). Just as with the elucidation of the nature of the quantum state on p. 23, any (circum-)classical justification or even motivation are guaranteed to fail here since they are based on significant and implicit assumptions.

The classical theory is a theory of *observational* objects with *observational* properties expressed by *observational* numbers. We possess none of the three items required to create the quantum (= correct) description (Section 2). The adjective "observational" itself is a linguistic notion of the classical vocabulary (Section 2.2). Accordingly, the description can only be changed "to describe in newly created terms". A. Leggett notes [95] that which is understood as common-sense should also be changed (see also [12] (p. 10)). The reason is clear.

- Common-sense operates—and that is perfectly normal—with observational categories rather than with structureless "microscopy" (9) and $\cup$-abstractions of Section 5.1;

cf. Bohr's correspondence principle [78]. In effect, we have dealt with a "fundamental chasm" between the right description—"what is really going on?" ([12], p. 12)—and our ability to give a (naturally speaking) explanation in terms of these categories:

> "All our intuition, all our sense of what constitutes concreteness are based upon our everyday experience, and the terms used to describe a phenomenon concretely are necessarily drawn from that experience. There is no indication that such a language could be used without contradictions for phenomena which are as far removed from it as those of microscopic physics" (A. Messiah. *Quantum mechanics*).

The total dismissal of this has to be at the heart of quantum reconstructing.

6.5.1. Detector Micro-Events

For similar reasons, we may not think or envision that a particle in an interferometer "flies through the slit", "has (not) arrived", "is located somewhere in the region of space" ([26], p. 7)], "here, not there", "now/later", that "the choice of a detector has been delayed" ([27], Wheeler), [62], or that a "photon ... interferes ... with itself" ([26], p. 9), and that, generally, "something is flying along a trajectory", and "something" is a particle at an intuitive understanding. Cf. Dirac's description of "the translational states of a photon" in Section 3 of [26].

- "Photons are just clicks in photon detectors; nothing real is traveling from the source to the detector" (ascribed to A. Zeilinger),

and this point is supported by all the known varieties of interferometers. There has to be an amendment here.

The clicks themselves are not the clicks *of photons/particles*, just "merely clicks". "[T]he click is no ... produced by a particle. ... nothing takes place in the source that could be a cause of the click ..., the genuinely fortuitous click comes without a cause and has no precursor" ([126], pp. 758, 765). Nothing really interferes inside interferometers, nor is anything superposed/reinforced. For example, the fact that the path of "photons" is not represented by trajectories was impressively demonstrated with the nested Mach–Zehnder experimental setup in the work [163]. Asking "where the photons have been" [163] is also the matter of a certain $\underline{\alpha}$-distinguishability. An interferometer—the entire installation—should be perceived as nothing more than a black box $\langle\!\langle \mathcal{S}, \mathbf{M}, \ldots \rangle\!\rangle$—the box (5)—outside of space and time. This is a kind of irreducible element that produces the only entity—distinguishable $\underline{\alpha}$-events, and no other. The box contains no "flying particles". Exempli gratia, none of the words in the typical sentence "photon propagates a definite path" are well-defined. Any assessment of the screen flashes observed within the interferometer, e.g., "is zero statistics possible in any spot?", lacks meaning until the theory's numerical apparatus is presented.

Remark 13. *Thus, Young's interference of the light beams (1803) is* inherently *the quantum not the classical effect: a micro-events' accumulation is usually termed as the light intensity. The classical electromagnetism and optics, in an exact sense, do not explain, only describe, the phenomenon quantitatively with the use of the numerical concepts of the positive and, which is important, the negative values of observable strength-fields $\vec{\mathbf{E}}$ and $\vec{\mathbf{H}}$. (The negative numbers are specifically discussed further in Section 9.2.) Accordingly, operations of their addition/subtraction "rephrase" the effect in words "superposing, suppressing, waves, intensities"', and we call this "the explanation". In a quantum way of looking at it, all of these concepts are not yet available, and the phenomenon per se is no more than statistics of the "positively accumulative" quantal clicks:*

- *There are no particles, waves, or subtractions there.*

The same macroscopic effect, which is visible with the naked eye and "explainable by waves", would take place if we had a "laser" of, say, mono-energetic very slow electrons (a proposal for experimentalists). To put it more precisely: a gun or emitter of something we envision as the "tiny bodily formations" the electrons, molecules, microbes, and the like. It is self-evident that we would have seen the wave-like manifestation even from a single slit.

Criticism of the typical (a common event-space) examination of the two-slit experiment [164] is already abundant in the literature. See, for example, the works [17] (Sections V.1, VI.1–2), [64] (pp. 55–58), [66] (Ch. 2), and [121] (p. 93), [162] (!), [165].

By way of continuing the last sentence in Remark 3, we add the following. To force an electron-click to happen each time at the same (or predictable) place is no different from "completely describing *everything* that we have", i.e., from the precise setting of "the same" and of macro-context $\langle\!\langle \mathcal{S}, \mathbf{M}, \dots \rangle\!\rangle$. It is amply evident that this is a manifest absurdity. Hence, it immediately follows that *the unpredictability of microscopic events must exist in principle* and macro-determinism may be only an idealization through a (math) model: the model description of the $\langle\!\langle \mathcal{S}, \mathbf{M}, \dots \rangle\!\rangle$ itself.

Summing up, it is not the quantum interference that requires interpretative comprehension but its classic "roughening". In other words, a scheme that latently presumes the rule (60) of extrapolation of what is observed in macro and micro ([160] (!), last sentence on p. 101). It is this scheme and not the quantum approach that contradicts the logic and experience. Pauli characterizes this as habits "known as 'ontology' or 'realism'". More than that, the chief component of constructs—⌈observation $\rightarrowtail$ state′⌉—is cast out and replaced with (19) under such a transformation. The `DataSource` object (p. 31) begins to be identified with *observational* and *numerical* characteristics (see a paragraph preceding Remark 4), while the logic of the micro-world requires precisely distancing these two concepts, with no need for the characteristics themselves.

Thus, we should not be deriving the physics of one phenomenon from another ([58], p. 92) and making (super)generalizations, as soon as the incorrectness of the previous derivation method was established.

- Quantum-mathematics is not a physical theory—and that is its distinguishing feature—but rather a single syntactical (meta)*principle of forming the mathematical models being subsequently turned into* (the physical) *theories*. This principle is not subject to any physical validation.

Scott Aaronson was likely the first to advance the line of thought about non-physicality. On page 110 of the book [20], he writes that "it's *not* a physical theory in the same sense as electromagnetism or general relativity ... quantum mechanics sits at a level *between* math and physics ... *is the operating system* ...". Fuchs–Peres provoke: "quantum theory does *not* describe physical reality" ([43], p. 70).

To create the models, we already have a good deal of *latitude*: the toolkit $\mathcal{O} = \{\mathscr{A}, \mathscr{B}, \dots\}$, the parameter D, the families $\{\mathfrak{T}_{\mathscr{A}}, \mathfrak{T}_{\mathscr{B}}, \dots\}$, numbers $\{\varrho_s\}$ of mixtures (23), and—thanks to the notion of covariance **III**—spectra, a structure of a group, and the concept of (different) representations of a mathematical structure. This liberty will be subsequently augmented with the key notions of a mean and of time t and also with the composite systems, the classical Lagrangians/Hamiltonians, their symmetries, gauge fields, and phenomenological constants. This is what is currently termed the quantum phenomenology or a *quantization procedure* of the classical models: the path-integrals, S-matrix, etc.

All that remains is to examine the numerical constituent of quantum mathematics. The further strategy (Sections 7–9) lies in the fact that the numbers need to be created at first as a theoretical concept—arithmetic—and then as the "numerical values for observable quantities"—the observations numbers. Sections 9.1 and 9.2 contain some more explanations along these lines.

7. Numeri

> By number we understand not so much a multitude of unities, as the abstracted ratio of any quantity to another quantity of the same kind, which we take for unity—I. Newton (1707)

- We have no any means of translating the aggregates of micro-acts $\overset{\mathscr{A}}{\dashrightarrow}$ (i.e., macro-observations $\mathbf{M}$) into the numerical language other than through the counting of things [172], i.e., through the natural-language notion of the "quantity of something":

$$
\begin{array}{c}
\downarrow\cdots\downarrow\downarrow\ \mathscr{A}\text{-transitions}\ \downarrow\downarrow\cdots\downarrow\\
\Downarrow \qquad \Downarrow\\
\lceil\text{quantity of}\rceil\ \lceil\text{something}\rceil\quad(\text{replication})\\
\Downarrow \qquad \Downarrow\\
\lceil\text{numbers}\rceil\ \lceil\Psi\text{-primitives, ensembles}\rceil\\
\searrow \ \swarrow\\
\mathsf{n}\{\Psi\}
\end{array}
\quad . \tag{68}
$$

Heisenberg stresses an obligatory relationship with "the natural language because it is only there that we can be certain to touch reality" ([98], pp. 201–202). Otherwise, the quantitative theory would have nowhere to originate even at the level of calculating the natural entities by the $\mathbb{N}$-number tokens. It may be added that arising the numbers is a permanently present (innate) process of creating the thought objects by an abstraction in the human brain: the mental suppressing/neglecting of the inessential and identifying the distinguishable entities—perceptual objects—irrespective of their nature ([84], "forming collections, ... putting objects together"; pp. 99, 251). It is something that humans do all the time without even realizing they are doing it. This process, say,

$$
\begin{aligned}
&\lceil\text{language, words}\rceil \cdots \rightarrowtail \{\mathsf{sheep},\ \Psi,\ \mathsf{verb},\ \Psi,\ \mathsf{theory},\ \ldots\} \rightarrowtail\\
&\quad\{\mathsf{a\ sheep},\ \mathsf{a}\ \Psi,\ \mathsf{a\ verb},\ \ldots\} \cdots \rightarrowtail \lceil\text{something/thing/.../Stücke}\rceil \cdots \rightarrowtail\\
&\quad\{\bullet\mathsf{Stück}, \bullet\mathsf{Stück}, \bullet\mathsf{Stück}, \bullet\mathsf{Stück}, \bullet\mathsf{Stück}, \ldots\} \rightarrowtail \{\bullet,\bullet,\bullet,\bullet,\bullet; \mathsf{Stücke}\} \rightarrowtail\\
&\quad\{1,1,1,1,1; \mathsf{Stücke}\} \rightarrowtail 5\mathsf{Stück} \rightarrowtail 5\ \text{\sout{Stück}} \rightarrowtail 5 \rightarrowtail \lceil\text{abstraction } 5\rceil\,,
\end{aligned}
$$

is akin to Cantor's concept of a Menge ([136], Ch. 1, Section 1.1) and has *no* the mathematical (math-logic) nature. Rather, the math of numbers does originate from it [84] (Ch. 3); see also [173] (Section 2.4.5.1 ARITHMETIC).

Incidentally, the "inessential and identifying" just mentioned have the nature just like the "the same" in Section 5.4. It is with these notions—a key feature of the natural/physical language and of speech—that any abstracting begins: the "abstracting from ...".

On the other hand, the numerical tokens are "affixed" not only to the "atom" $\{\Psi\}$ but also to other objects, *any* at that; for more details, see Remark 16 further below. Therein lies the primary meaning of this still proto-mathematical concept [174] ("Psychologie du nombre"). One might even say, a definition according to which this notion has been conceived ("20 Stück", "half an hour" ...) and is being used universally. Here are a few examples:

$$
\begin{array}{c}
\{\Psi\Psi\Psi\} \equiv 3\{\Psi\}, \qquad \{\Theta\Theta\Phi\Phi\} \equiv 2\big(\{\Theta\}\cup\{\Phi\}\big)\\
\{\Phi\Psi\} \overset{2}{\rightarrowtail} \{\Phi\Psi\Phi\Psi\}, \qquad a \overset{3}{\rightarrowtail} 3a, \qquad c \overset{1}{\rightarrowtail} 1c
\end{array}
\quad . \tag{69}
$$

Accordingly, in between the elements, there arise identities such as $2b \equiv 4a$, $3a \equiv c$, $1c \equiv c$. In other words, as we complete Simplification (67),

- While abstracting the empirical contents of the number entities into math-symbols, they should be defined as unary operations $\{\hat{1}, \hat{2}, \ldots, \widehat{3/4}, \ldots, \hat{\pi}, \ldots\}$ that take action at $\mathfrak{A}$-set (67) as automorphisms: $\{\hat{2}b = \hat{4}a, \hat{1}c = c, \ldots\}$.

  That said, replication is formalized as an operator $\hat{n}$ with its numerical symbol n:

$$
\psi \overset{\hat{n}}{\mapsto} \mathsf{n}\psi, \qquad \psi, \mathsf{n}\psi \in \mathfrak{A}, \qquad \mathsf{n} \in \mathbb{R}\,, \tag{70}
$$

where ψ is understood to be any (sub)ensemble/(sub)set. In the language of the ZF-theory, $\mathsf{n}\psi$ would be formally organized as an ordered pair $(\mathsf{n}, \psi) := \{\{\mathsf{n}\}, \{\mathsf{n}, \psi\}\}$ [134], where n

is a cardinality of a set consisting of copies of the object/set ψ. We will refer to these facts as the implementation of a replication operator by numbers.

Attention is drawn to the fact that the case in point at the moment *is not* a math-logical *definitio/formalization* of the concept of a number, such as (106), but is an introduction of what is understood by number in the empirical/physical theory (**II**). For example, Chomsky says with regard to this point: "When multiplying numbers in our heads, we depend on many factors beyond our intrinsic knowledge of arithmetic" ([132], p. 3).

7.3. QM *and Arithmetica*

We immediately observe the following properties.

The operators are applicable to each other. Being a family $\{\hat{n}, \hat{m}, \hat{p}, \ldots\}$, they are closed with respect to their composition $\hat{n}\,(\hat{m}\psi) = (\hat{n}\circ\hat{m})\psi = \hat{p}\psi$, and among them, there is an identical operator $\hat{\mathbb{1}}\,\psi = \psi$. The empirical meaning of the concept indicates a fractional portion of the ensemble (see (62)) requires that for each $\hat{n}$ there exists its inversion $\hat{n}^{-1}$. Hence, the composition of replications $\hat{n}\circ\hat{n}^{-1}$ must return the former "quantity": $(\hat{n}\circ\hat{n}^{-1})\psi = \hat{\mathbb{1}}\,\psi$. Therein lies "the actual meaning of division. … this [operator] construction really corresponds to division" ([85], p. 37). By virtue of the fact that family $\{\hat{n}, \hat{m}, \hat{p}, \ldots\}$ provides automorphisms of the $\mathfrak{A}$-set, these operators entail the associative identities $((\hat{n}\circ\hat{m})\circ\hat{p})\psi = (\hat{n}\circ(\hat{m}\circ\hat{p}))\psi$. This point is a property, and it has a proof [175] (Section I.1.2). The common nature of the replication and of the $\cup$-union also signifies that there are relations in place that mix the actions of the unary $\hat{n}$'s and the binary union of ensembles. At a minimum, suffice it to define the action of the replicator on a "$\cup$-sum" of replications. Clearly, the case in point is the distributive coordination of $\circ$ and $\cup$:

$$\hat{p}\,(\hat{n}\psi\cup\hat{m}\psi) = (\hat{p}\circ\hat{n})\psi\cup(\hat{p}\circ\hat{m})\psi\,.$$

We now observe that the indication of ψ everywhere in the identities above loses the necessity, and the ψ-label becomes a semblance of a dummy index or the unit symbol (kg), which can be changed. As we omit it, the theory is freed of ψ as a "calculation unit". Then, the last relation, as an example, acquires the form of a property between the operator n-symbols (70), if $\{\cup, \circ\}$ are replaced with the symbols of binary operations $\{+, \times\}$:

$$\mathsf{p}\times(\mathsf{n}+\mathsf{m}) = (\mathsf{p}\times\mathsf{n}) + (\mathsf{p}\times\mathsf{m})\,. \tag{71}$$

Supplementing this relation with other empirically determining properties, one infers that the *unary* operationality of $\hat{n}$-replications (70) is *indistinguishable* from the *binary* operationality on their n-symbols. The latter, in turn, acquires the multiplicative structure of a commutative group

$$\mathsf{n}\times\mathsf{m} = \mathsf{m}\times\mathsf{n}, \qquad (\mathsf{n}\times\mathsf{m})\times\mathsf{p} = \mathsf{n}\times(\mathsf{m}\times\mathsf{p}), \qquad \mathsf{n}\times 1 = \mathsf{n}, \qquad \mathsf{n}\times\mathsf{n}^{-1} = 1\,, \tag{72}$$

and, as for the addition $+$, it is already binary and commutative due to properties of $\cup$ (Section 5.1):

$$\mathsf{n}+\mathsf{m} = \mathsf{m}+\mathsf{n}, \qquad (\mathsf{n}+\mathsf{m})+\mathsf{p} = \mathsf{n}+(\mathsf{m}+\mathsf{p}), \qquad \mathsf{n}+0 = \mathsf{n}\,. \tag{73}$$

Incidentally, the three-term multiplicative associativity relation in (72) has the same operatorial nature and origin as operations $\{\cup, \cup\}$ do in (39). We have already commented on the additive analog in this situation—a determinative structure of the binary operation—after Formula (57).

It is also clear that Rules (71)–(73) must be supplemented with the concept of a negative number

$$\mathsf{n}+(-\mathsf{n}) = 0\,, \tag{74}$$

for such numbers have been fully justified in the superposition principle.

After having acquired Properties (71)–(74)—call them *arithmetica*—symbols $\{\mathsf{n}, \mathsf{m}, \ldots\}$ turn into abstract numbers, although their operator genesis does not go away and is yet to be involved. This is where a full list of requirements for the concept of a real number should be added, and which have to do with ordering $<$, completeness/continuality, *and* their relations with Rules (71)–(74). We will assume that this is conducted axiomatically ([176], pp. 35–38), although the algebraic constituent of this "axiomatics", as we have seen, is not axiomatical but deducible from empiricism. Multiplication $\times$, and also the subsequent $\odot$-multiplication of $\mathbb{C}$-numbers (82), is a most nontrivial part in deriving the structure from "the arithmetic".

As an outcome, we reveal an essential asymmetry in the genesis of the standard binary structures $+$ and $\times$ (cf. [84] (p. 60)), and thereby a greater primacy of QM-consideration even over the (seemingly self-evident) arithmetic. Indeed, binarity may come only from operation $\cup$, which is primordially unique and, thereby, is inherited only to the one natural prototype—addition.

- Multiplication is not featured in the superposition principle, nor does it arise directly as a binary structure. The absence of a multiplication symbol in (58) and (59) is no accident.

The multiplication originates in the closedness of replications $\hat{n} \circ \hat{m}$, and they are required according to the **M**-paradigm (12). In effect, any non-operatorial way of introducing the n-numbers is not self-evidence for empiricism. An operator nature of the number is precisely that which gives rise to the second binary operation. Moreover, without such a comprehension of the number, the "linear nature" of QM (Section 8.1) will remain axiomatic at all times, and, as will be seen below, quantum foundations will be doomed to never-ending interpreting the mathematical symbols. However, the pure axiomatic declaration of Arithmetic (71)–(74) will, in one way or another, require a (reciprocal to (68)) treatment of the number in a context of "the quantity of what?", while its empirical pre-image always appears in the pair ⌈the quantity of⌉ + ⌈something⌉. Another way to put it is:

- In the foundations of theory, there arises a predecessor/analog to the notion of a physical unit,

though the ultimate description is a description in terms of binary structures in Arithmetic (71)–(74). It is carried out by dropping/attaching the symbols such as ψ, which is a quantum generalization to the independence of a physical theory, from the measurement units.

Certainly, when formalized, the $\hat{n}$-replication and its binary n-twin become universally abstract. For example, the $\hat{n}$-operator (70) may be applied to the quantum case in which the object ψ has already an internal structure associated with the presence of $\underline{\Psi}, \underline{\Phi}$-primitives. This changes no the essence of the matter. Another example is when numbers n give birth to really observable quantities. See also Section 5.3, Remark 16, and additional discussion in Section 9. Let us now proceed from the fact that the comprehension/relation of the number and its operator has been formalized as described above. This is "Axioms" (71)–(74).

As concerns the philosophical literature, the issue of numbers was likely discussed [177–179] (see also [172] and non-philosophical book [174]), and it would be appropriate to quote T. Maudlin: "… numbers: they can be added to one another, *perhaps* multiplied by one another, …. However, it is typically obscure what sort of *physical* relation these mathematical operations could possibly represent" ([151], p. 138; first emphasis ours, second in original). Cf. Einstein's remarks regarding the "concepts and propositions" and "the series of integers" on p. 287 in [180].

7.4. Two-Dimensional Numbers

A number in and of itself, as a replication operator, may be applied to any ensemble and to anything at all. However, in the quantum case, the "upper" primitives are attached to every "lower" $\underline{\alpha}$-event. These primitives, as was noted above, have to be discarded. At the same time, the minimal structure associated with the homogeneous array $\{\underline{\alpha}_s \cdots \underline{\alpha}_s\}$

as a whole is a unitary brace $\{\overset{\Psi}{\mathsf{n}},\overset{\Phi}{\mathsf{m}}\}\underline{\alpha}_s$ containing two "upper" primitives $\underline{\Psi}$, $\underline{\Phi}$. Their order, however, is arbitrary there. That is to say, given $(\mathsf{n},\mathsf{m})\underline{\alpha}$, there are two quite equal objects $\{\overset{\Psi}{\mathsf{n}},\overset{\Phi}{\mathsf{m}}\}\underline{\alpha}$ and $\{\overset{\Phi}{\mathsf{n}},\overset{\Psi}{\mathsf{m}}\}\underline{\alpha}$ that are subjected to a replication. Each of them should be in a relationship (see Section 5.1) to any other brace (63), which is already apparent in the example of "one-dimensional" versions $(\mathsf{n},0)\underline{\alpha}$ and $(\mathsf{n}',0)\underline{\alpha}$. We mean that for each pair $\{(\mathsf{n},0)\underline{\alpha},(\mathsf{n}',0)\underline{\alpha}\}$, there always exists the number m such that $\widehat{m}(\mathsf{n},0)\underline{\alpha}=(\mathsf{n}',0)\underline{\alpha}$, i.e., $\mathsf{m}\times\mathsf{n}=\mathsf{n}'$.

As in the classical case (69), the sought-for generalizations of replicators are the transitive automorphisms on *unitary* $\underline{\alpha}$-brace (66), but they *are not* abstract and *not* arbitrary. They are strictly bound to the declared meaning of the number: $\widehat{N}$-operation of creating the copies. Therefore, by virtue of the equal rights of $\underline{\Psi}$ and $\underline{\Phi}$, it is imperative to bring the two one-fold copying acts $\widehat{N}\{\overset{\Psi}{\mathsf{n}},\overset{\Phi}{\mathsf{m}}\}\underline{\alpha}$ and $\widehat{M}\{\overset{\Phi}{\mathsf{n}},\overset{\Psi}{\mathsf{m}}\}\underline{\alpha}$ into play, which differ in the permutation of primitives $\underline{\Psi}\rightleftarrows\underline{\Phi}$. This point will determine a quantum extension of the replication.

As a result, since we have nothing but the copying $\widehat{N}$ and "union" $\widehat{+}$, the most general transformation of the brace $\{\overset{\Psi}{\mathsf{n}},\overset{\Phi}{\mathsf{m}}\}\underline{\alpha}$ into (any) brace $\{\overset{\Psi}{\mathsf{n}'},\overset{\Phi}{\mathsf{m}'}\}\underline{\alpha}$, which has been in a quantum-replication relation with it, is determined by the rule

$$\{\overset{\Psi}{\mathsf{n}},\overset{\Phi}{\mathsf{m}}\}\underline{\alpha}\ \overset{\widehat{(N,M)}}{\rightarrowtail}\ \{\overset{\Psi}{\mathsf{n}'},\overset{\Phi}{\mathsf{m}'}\}\underline{\alpha}\qquad\Rightarrow\qquad\{\overset{\Psi}{\mathsf{n}'},\overset{\Phi}{\mathsf{m}'}\}\underline{\alpha}=\widehat{N}\{\overset{\Psi}{\mathsf{n}},\overset{\Phi}{\mathsf{m}}\}\underline{\alpha}\ \widehat{+}\ \widehat{M}\{\overset{\Phi}{\mathsf{n}},\overset{\Psi}{\mathsf{m}}\}\underline{\alpha}\,. \tag{75}$$

This is the quantum version of Operators (69) and (70), and the foregoing ideology of $\widehat{N}$-operators and of liberation from the $\underline{\Psi}$-symbols remains in force and entails the following. The numeral implementation of replicating the unitary brace (66), along with the (n,m)-representation of itself, is also determined by a certain pair $(\mathsf{N},\mathsf{M})\in\mathbb{R}^2$, i.e., by an operator symbol $\widehat{(N,M)}$.

The aforesaid means that the numerical form $(\mathsf{n},\mathsf{m})\overset{\widehat{(N,M)}}{\rightarrowtail}(\mathsf{n}',\mathsf{m}')$ of Transformation (75) is indistinguishable from a composition of pairs

$$(\mathsf{N},\mathsf{M})\odot(\mathsf{n},\mathsf{m})=(\mathsf{n}',\mathsf{m}')\,,$$

where $\odot$ is a designation for the new binary operation. Its resultant structure is derived from the arithmetical nature (72) of the one-dimensional replication (69) described above, i.e., from the rules

$$\widehat{N}\{\overset{\Psi}{\mathsf{n}},\overset{\Phi}{\mathsf{m}}\}\underline{\alpha}=\{\mathsf{N}\overset{\Psi}{\times}\mathsf{n},\ \mathsf{N}\overset{\Phi}{\times}\mathsf{m}\}\underline{\alpha},\qquad\widehat{M}\{\overset{\Phi}{\mathsf{n}},\overset{\Psi}{\mathsf{m}}\}\underline{\alpha}=\{\mathsf{M}\overset{\Phi}{\times}\mathsf{n},\ \mathsf{M}\overset{\Psi}{\times}\mathsf{m}\}\underline{\alpha}\,. \tag{76}$$

Here, a positivity/negativity of symbols (n,m) in (66) should also be taken into account. Having regard to the foregoing, Rules (75) and (76) generate the Ansatz

$$(\mathsf{N},\mathsf{M})\odot(\mathsf{n},\mathsf{m})=(\pm\mathsf{N}\mathsf{n}\pm\mathsf{M}\mathsf{m},\ \pm\mathsf{N}\mathsf{m}\pm\mathsf{M}\mathsf{n})\,, \tag{77}$$

wherein all four signs $\pm$ are independent of each other, and the $(\times)$-multiplication of one-dimensional numbers in (72) and (76) have been re-denoted by the habitual standard $\mathsf{N}\mathsf{m}:=\mathsf{N}\times\mathsf{m}$. What should the pair-composition rule (77) be?

As was the case previously, the just-emerged binarity for $\odot$ should inherit—due to its operator origin—associativity, the existence of unity $\mathbb{1}$, and inversions. Namely, if the (n,m)-pairs are identified with the notation (57) according to the convention

$$(\mathsf{n},\mathsf{m})=:\mathfrak{a}\,, \tag{78}$$

then the following properties should be declared:

$$(\mathfrak{a}\odot\mathfrak{b})\odot\mathfrak{c}=\mathfrak{a}\odot(\mathfrak{b}\odot\mathfrak{c}),\qquad\mathfrak{a}\odot\mathbb{1}=\mathfrak{a},\qquad\mathfrak{a}\odot\mathfrak{a}^{-1}=\mathbb{1}\,. \tag{79}$$

From (75) and (76), it is not difficult to see that the combining (79) with (57) leads to a distributive coordination of operations $\oplus$ and $\odot$:

$$\mathfrak{c} \odot (\mathfrak{a} \oplus \mathfrak{b}) = (\mathfrak{c} \odot \mathfrak{a}) \oplus (\mathfrak{c} \odot \mathfrak{b}) . \tag{80}$$

However, the direct examination of this property shows that Ansatz (77) satisfies it automatically. More than that, we can even consider Ansatz (77) with parameters $\{\alpha, \beta, \gamma, \delta\}$ instead of $(\pm)$-signs:

$$\mathfrak{a} \odot \mathfrak{b} = (\mathsf{N}, \mathsf{M}) \odot (\mathsf{n}, \mathsf{m}) = (\alpha \mathsf{N}\mathsf{n} + \beta \mathsf{M}\mathsf{m}, \ \gamma \mathsf{N}\mathsf{m} + \delta \mathsf{M}\mathsf{n}) .$$

Then, the straightforward calculation shows that Distributivity (80) holds under the arbitrary $\{\alpha, \beta, \gamma, \delta\}$.

In turn, the examination of associativity—the first equality in (79)—under the same meaning for $\{\alpha, \beta, \gamma, \delta\}$ yields $\alpha = \gamma = \delta$ and free β. Returning to the $(\pm)$-values of these parameters, this associativity particularizes Ansatz (77) into the expression

$$(\mathsf{N}, \mathsf{M}) \odot (\mathsf{n}, \mathsf{m}) = \pm(\mathsf{N}\mathsf{n} \pm \mathsf{M}\mathsf{m}, \mathsf{N}\mathsf{m} + \mathsf{M}\mathsf{n});$$

now, with two independent signs $\pm$. Moreover, in passing, we reveal the commutativity

$$\mathfrak{a} \odot \mathfrak{b} = \mathfrak{b} \odot \mathfrak{a} , \tag{81}$$

though it was not presumed prior to that.

The search for unity $\mathbb{1}$ and subsequent finding of an inversion of the element (n, m) yield:

$$\mathbb{1} = (\pm 1, 0), \qquad (\mathsf{n}, \mathsf{m})^{-1} = \left(\frac{\mathsf{n}}{\Delta}, -\frac{\mathsf{m}}{\Delta}\right), \qquad \Delta := \mathsf{n}^2 \pm \mathsf{m}^2 .$$

Both the $(\pm)$-symbols continue to be independent here. The choice $\Delta = \mathsf{n}^2 - \mathsf{m}^2$ results in the absence of inversions $(\mathsf{n}, \mathsf{n})^{-1}$. This is in conflict with the group property (79) and also causes the unmotivated exclusivity of the unitary brace $\{\overset{\Psi}{\mathsf{n}}, \overset{\Phi}{\mathsf{n}}\}\underline{\mathfrak{a}}$. There remains the case $\Delta = \mathsf{n}^2 + \mathsf{m}^2$, and it reduces the scheme to the form

$$\mathbb{1} = \pm(1, 0), \qquad (\mathsf{N}, \mathsf{M}) \odot (\mathsf{n}, \mathsf{m}) = \pm(\mathsf{N}\mathsf{n} - \mathsf{M}\mathsf{m}, \mathsf{N}\mathsf{m} + \mathsf{M}\mathsf{n})$$

with a single symbol $\pm$. It is a simple matter to see that the choice of sign $+$ or $-$ leads to the models that are isomorphic in regard to which of representatives $(+1, 0)$ or $(-1, 0)$ should be assigned for the identical replication $\hat{\mathbb{1}}$. By virtue of (61), it does not matter, and we declare

$$\boxed{\mathbb{1} := (1, 0), \qquad (\mathsf{N}, \mathsf{M}) \odot (\mathsf{n}, \mathsf{m}) = (\mathsf{N}\mathsf{n} - \mathsf{M}\mathsf{m}, \mathsf{N}\mathsf{m} + \mathsf{M}\mathsf{n})} . \tag{82}$$

This is nothing more nor less than the canonical multiplication of complex numbers $\mathsf{n} + \mathrm{i} \cdot \mathsf{m} = \mathfrak{a} \in \mathbb{C}$, if the following identifications are performed:

$$(1, 0) \rightleftarrows \mathbb{1}, \qquad (0, 1) \rightleftarrows \mathrm{i}, \qquad \{\oplus, \odot\} \rightleftarrows \{+, \cdot\}, \qquad (\mathsf{n}, \mathsf{m}) \rightleftarrows (\mathsf{n} + \mathrm{i} \cdot \mathsf{m}) . \tag{83}$$

Notice that the known fully matrix (over $\mathbb{R}$) equivalent to (82)

$$(\mathsf{n} + \mathrm{i} \cdot \mathsf{m}) \mapsto (\mathsf{n}, \mathsf{m}) \mapsto \begin{pmatrix} \mathsf{n} \\ \mathsf{m} \end{pmatrix} \mapsto \begin{pmatrix} \mathsf{n} & -\mathsf{m} \\ \mathsf{m} & \mathsf{n} \end{pmatrix}, \qquad \begin{pmatrix} \mathsf{n}' & -\mathsf{m}' \\ \mathsf{m}' & \mathsf{n}' \end{pmatrix} = \begin{pmatrix} \mathsf{N} & -\mathsf{M} \\ \mathsf{M} & \mathsf{N} \end{pmatrix} \circ \begin{pmatrix} \mathsf{n} & -\mathsf{m} \\ \mathsf{m} & \mathsf{n} \end{pmatrix}$$

does directly reflect the above ascertained operator essence

$$\widehat{(n', m')} = \widehat{(N, M)} \circ \widehat{(n, m)}$$

of both the number multiplication $\odot$ and the $\mathbb{C}$-number itself.

In view of the paramount importance of the $\mathbb{C}$-number field in QT [96,138,142], let us provide additional substantiations to the rigidity of the emergence of this specific number structure, i.e., of the axiom collection (57), (79)–(82). Among other things, the transpositions $\underline{\Psi} \rightleftarrows \underline{\Phi}$ used above fit more general reasoning.

7.5. Involutions and $\tilde{\mathbb{C}}^*$-Algebra

Apart from a freedom in ordering the primitives $\underline{\Psi} \rightleftarrows \underline{\Phi}$ in brace $\{\overset{\Psi}{\mathsf{n}}, \overset{\Phi}{\mathsf{m}}\}\underline{\alpha}$, there is one more arbitrariness: reappointing them ($\underline{\Psi} \rightarrowtail \underline{\Theta}, \ldots$) as elements of the set $\mathfrak{T}$. However, no physics predetermines any of these degrees of freedom. For, if other ingoing $\mathfrak{T}$-elements $\underline{\Theta}$ $\underline{\Omega}$ were present in (32) instead of $\underline{\Psi}$, $\underline{\Phi}$, then the theory of semigroup $\mathfrak{G}$, strictly, should be declared the segregated theories $\mathfrak{G}_{\Psi\Phi}$, $\mathfrak{G}_{\Theta\Omega}$, etc. It is clear that the labeling the theories, or a family thereof, is a manifest absurdity, and they should be thus factorized with respect to all kinds of ways to label them by $\mathfrak{T}$-primitives. The liberation from the $\underline{\Psi}, \underline{\Phi}$-icons and reconciliation of the result with pt. $\mathbf{R}^{+}$ (p. 25) are then performed by the scheme $\lceil$primitive has changed$\rceil \rightarrowtail \lceil$a number character is changing$\rceil$.

Inasmuch as declaring the $\{\underline{\Psi}, \underline{\Phi}, \underline{\Theta}, \ldots\}$ to be ongoing primitives in (32) is a replacement of one to another, any such an appointment boils down to permutations of no more than *pairs* with two types (inner/outer):

$$\hat{\daleth}_{\Psi\Phi}:\quad (\underline{\Psi}, \underline{\Phi}) \xrightleftharpoons{\Psi\leftrightarrow\Phi} (\underline{\Phi}, \underline{\Psi}), \qquad\qquad \hat{\aleph}_{\Phi\Theta}:\quad (\underline{\Psi}, \underline{\Phi}) \xrightleftharpoons{\Phi\leftrightarrow\Theta} (\underline{\Psi}, \underline{\Theta})\,. \tag{84}$$

However, it is immediately obvious that these reappointments change nothing in the $\cup$-relationships between (32) and are defined by the structural relations $\hat{\daleth}^2_{\Psi\Phi} = \hat{\mathbb{I}}$, $\hat{\aleph}^2_{\Phi\Theta} = \hat{\mathbb{I}}$. Then, the need to indicate the primitives themselves, as required, is eliminated, and their symbols may be thrown away if semigroup $\mathfrak{G}$ is properly furnished with the two abstract involutions $\hat{\daleth}$ and $\hat{\aleph}$. The $\mathfrak{G}$ itself, of course, also possesses involution (61) that turns it into the group H, but this involution has already had a numerical representation (66) by signs $\pm$. To be precise, it suffices to identify here the term "numerica" with the group arithmetic of the $\oplus$-addition (57) coming from the superposition principle realized on pairs (65) and (66). Therefore, the operators' actions (84) should be carried over onto objects defined in precisely this manner; nothing more needs to be assumed.

Operator $\hat{\daleth}_{\Psi\Phi}$ is immediately translated into a numerical form independently of the property that the objects $\{\overset{\Psi}{\mathsf{n}}, \overset{\Phi}{\mathsf{m}}\}\underline{\alpha}$ form a (semi)group. Indeed, since the swap $\underline{\Psi} \rightleftarrows \underline{\Phi}$ in the unordered pair

$$\hat{\daleth}_{\Psi\Phi}:\qquad \{\overset{\Psi}{\mathsf{n}}, \overset{\Phi}{\mathsf{m}}\} \rightarrowtail \{\overset{\Phi}{\mathsf{n}}, \overset{\Psi}{\mathsf{m}}\} = \{\overset{\Psi}{\mathsf{m}}, \overset{\Phi}{\mathsf{n}}\} \qquad \cdots$$

(the $\underline{\alpha}$-label is dropped here as superfluous) is indistinguishable from the permutation of numbers $\mathsf{n} \rightleftarrows \mathsf{m}$, the symbols $\underline{\Psi}$ and $\underline{\Phi}$ may be thrown away, organizing the numbers themselves into ordered pairs

$$\cdots \quad \Rightarrow \quad (\mathsf{n}, \mathsf{m}) \overset{\hat{\daleth}}{\rightarrowtail} (\mathsf{m}, \mathsf{n})\,.$$

When required, the $\underline{\alpha}$-symbol returns hereinafter.

Let us now proceed to the outer involution $\underline{\Phi} \rightleftarrows \underline{\Theta}$ in (84):

$$\hat{\aleph}_{\Phi\Theta}:\qquad \{\overset{\Psi}{\mathsf{n}}, \overset{\Phi}{\mathsf{m}}\} \rightarrowtail \{\overset{\Psi}{\mathsf{n}}, \overset{\Theta}{\mathsf{m}}\}\,.$$

It is indifferent to the (first) $\underline{\Psi}$-element of the pair, and, by extracting it by the rule

$$\{\overset{\Psi}{\mathsf{n}}, \overset{\Phi}{\mathsf{m}}\} = \{\overset{\Psi}{\mathsf{n}}, \overset{\Phi}{\mathsf{0}}\} \,\hat{+}\, \{\overset{\Psi}{\mathsf{0}}, \overset{\Phi}{\mathsf{m}}\}\,,$$

the question boils down to finding a representation of the transformations

$$(\{\overset{\Psi}{\mathsf{n}}, \overset{\Phi}{\mathsf{0}}\} \,\hat{+}\, \{\overset{\Psi}{\mathsf{0}}, \overset{\Phi}{\mathsf{m}}\}) \;\rightarrowtail\; (\{\overset{\Psi}{\mathsf{n}'}, \overset{\Phi}{\mathsf{0}}\} \,\hat{+}\, \{\overset{\Psi}{\mathsf{0}}, \overset{\Theta}{\mathsf{m}'}\}) \qquad (\overset{?}{\mathsf{n}'}, \overset{?}{\mathsf{m}'})\,.$$

The component $\{\overset{\Psi}{\mathsf{n}},\overset{\Phi}{0}\}$ must go into itself since the symbol $\underline{\Psi}$ attached to it has not changed. It means that $\mathsf{n}' = \mathsf{n}$, and one is left with the task

$$\{\overset{\Psi}{0},\overset{\Phi}{\mathsf{m}}\} \overset{?}{\rightleftarrows} \{\overset{\Psi}{0},\overset{\Theta}{\mathsf{m}'}\} \,.$$

However, operation $\widehat{\aleph}_{\Phi\Theta}$ recognizes only the primitive's symbols rather than their numbers. That is, replications $\widehat{m}\{\overset{\Psi}{0},\pm\overset{\Phi}{1}\} = \{\overset{\Psi}{0},\pm\overset{\Phi}{\mathsf{m}}\}$ do formally commute with $\widehat{\aleph}_{\Phi\Theta}$. Hence, by omitting the letters $\{\underline{\Psi},\underline{\Phi},\underline{\Theta}\}$, it will suffice to look for the representation of $\widehat{\aleph}$ by numerical pairs $(0,\pm\mathsf{m})$ factorized with respect to replications $\widehat{m}$, i.e., by the set $\{(0,1),(0,-1)\}$. It, for its part, remains to be transformed into itself, and the replication operators $\widehat{n}$, $\widehat{m}$ will recreate the generic case. The identical transformation $(0,\pm 1) \rightarrowtail (0,\pm 1)$ is ruled out since $\widehat{\aleph}_{\Phi\Theta} \neq \widehat{\mathbb{I}}$; therefore, $(0,\pm 1) \overset{\widehat{\aleph}}{\rightarrowtail} (0,\mp 1)$. Restoring all the symbols that were dropped, the effect of $\widehat{\aleph}$ reduces to the sign change for the second element of the coordinate pair:

$$(\mathsf{n},\mathsf{m}) \overset{\widehat{\aleph}}{\rightarrowtail} (\mathsf{n},-\mathsf{m}) \,. \tag{85}$$

There is no need to change sign for the first element, as this change is the operator $-\widehat{\mathbb{I}} \circ \widehat{\aleph}$. Furthermore, one observes that the already existing group inversion $-\widehat{\mathbb{I}}$ coincides with composition

$$(\widehat{\aleph} \circ \widehat{\beth})^2 = -\widehat{\mathbb{I}} \,, \tag{86}$$

and we may even "forget" about (the "old") subtraction, leaving the equipment

$$\{\oplus,\widehat{\mathbb{I}},\widehat{m},\widehat{\aleph},\widehat{\beth}\} \tag{87}$$

of semigroup $\mathfrak{G}$ as an irreducible set of mathematical structures over it.

In this connection, yet another—more formal—motivation of the passage ⌈semigroup $\rightarrowtail$ group⌉ and thus of the superposition principle does arise. Indeed, the derivation of $\widehat{\aleph}$ above engaged the inversion (61), but the reappointment of primitives $\underline{\Phi} \rightleftarrows \underline{\Theta}$ in (84) is a fully independent act. Therefore, if we forget about "$(-)$-copies of the positive pairs" $(0,\mathsf{m})$, the involutory nature of automorphism $\widehat{\aleph}_{\Phi\Theta}$ would still reproduce the semigroup $\mathfrak{G}$ in numbers by "duplication" $\mathsf{m} \rightarrowtail \pm\mathsf{m}$, i.e., create the negative pairs $(0,-\mathsf{m})$, thus turning $\mathfrak{G}$ into a group H. An analogous reasoning on the symbol "$-$" could be cited even earlier, when the $\mathbb{C}$-field was being derived.

Now, remembering the above-described move to the binarity of $\odot$-multiplication on the (n,m)-pairs, we arrive at the problem of matching it with structures (87). Clearly, one needs only to ascertain the functionality of operators $\widehat{\beth}$ and $\widehat{\aleph}$ that were not available yet.

Relation (86) immediately gives us the correspondence $\widehat{\aleph} \circ \widehat{\beth} \rightleftarrows \mathrm{i}$ since $\mathrm{i}^2 = -1$. Hence, one of these operators, say $\widehat{\beth}$, manifests itself in the imaginary unit i. The origin of this operator—permutation $\widehat{\beth}_{\Psi\Phi}$ in (84)—is the very same permutation $\underline{\Psi} \rightleftarrows \underline{\Phi}$ that generated the i-object in algebra (82) and (83). The second operator, i.e., (85), as is directly seen, is also not related to the binary $\oplus$ and $\odot$ but determines the change $\mathrm{i} \rightarrowtail -\mathrm{i}$. This means that the QM-consideration does not just give birth to the field $\mathbb{C}$ but to a division $\widetilde{\mathbb{C}}^*$-algebra, which is equipped with two *non-binary* operations

$$a \overset{\widehat{\aleph}}{\rightarrowtail} a^*\,, \qquad a \overset{\widehat{\beth}}{\rightarrowtail} \tilde{a} \,.$$

Informally, it defines all the basic actions as "complex quantities" and thereby determines a QM-extension/generalization to the intuitive and habitual arithmetical manipulations (68)–(74) with real things. Consequently, the four binary arithmetical operations—addition/subtraction/multiplication/division—should be supplemented with the two unary ones: conjugation $\widehat{\aleph}$ and swap $\widehat{\beth}$.

Remark 14. *A curious observation for formal complex-number mathematics is appropriate here. None of these operators boil down to involution $-\widehat{\mathbb{I}}$. We mean that each of the pairs $(\widehat{\aleph},-\widehat{\mathbb{I}})$ or*

$(\hat{\beth}, -\hat{\mathbb{I}})$ is expressible through $(\hat{\beth}, \hat{\aleph})$ and not the reverse; see (86). To put it plainly, the self-suggested going from the natural sign-change (i.e., $-\hat{1}$ over $\mathbb{R}$) to the inversion of the two-dimensional $\oplus$-addition (i.e., $-\hat{\mathbb{I}}$ over $\mathbb{C}$) deprives the involution $-\hat{\mathbb{I}}$ of its primary character, as it has taken in the one-dimensional domain $\mathbb{R}$. Furthermore, the second operation $\hat{\beth}$ is, in a sense, more "primitive" than the complex conjugation $\hat{\aleph}$, as this operation has had to conduct it with a formal pair (n, m)—merely transposes it—and does not invoke an arithmetic action, as does $\hat{\aleph}$ when changing the sign $\mathsf{m} \mapsto -\mathsf{m}$ in (85).

The relationship between the operators is by binary multiplication: $\tilde{\mathfrak{a}} = \mathrm{i} \odot \mathfrak{a}^*$. By virtue of this relation, it makes no odds which one of these unary operators is left for $\mathbb{C}$-algebra.

We note—and this is important [6]—that the observational statistics $\mathtt{f}_j$ are unchanged upon both operations $\hat{\beth}$ and $\hat{\aleph}$.

7.6. Naturalness of $\mathbb{C}$-Numbers

Thus, the $\mathfrak{T}$-set primitives have been entirely banished from the theory, with the exception of the eigen-state $\underline{\alpha}_s$-markers, which are needed only for distinguishability (Section 2.1) in $\mathscr{A}$-observations. These markers may be interchanged, but permutability $\underline{\alpha}_j \rightleftarrows \underline{\alpha}_k$ is already reflected by the superposition's commutativity. Taking now into account the fact that reassigning the $\underline{\alpha}$-labels does not touch on the concept of the number, one infers: the covariance attained above is exhaustive. As a result, we draw the following conclusion.

- The coordinate representatives $\{\mathfrak{a}, \mathfrak{b}, \mathfrak{c}, \ldots\}$ of states and superpositions thereof (58) form a complex number field $\tilde{\mathbb{C}}^*$ equipped with the structures of conjugation and swap:

$$(\mathsf{n} + \mathrm{i}\mathsf{m}) \overset{*}{\mapsto} (\mathsf{n} - \mathrm{i}\mathsf{m}), \qquad (\mathsf{n} + \mathrm{i}\mathsf{m}) \overset{\sim}{\mapsto} (\mathsf{m} + \mathrm{i}\mathsf{n}) \,. \tag{88}$$

Statistical weights $\mathtt{f}_j$ in object (35) are invariant with respect to both the involutions $\mathtt{f}_j(\mathfrak{a}^*) = \mathtt{f}_j(\mathfrak{a}) = \mathtt{f}_j(\tilde{\mathfrak{a}})$ for each component $\mathfrak{a}_s$ independently.

What is more, the commentary on the primacy of QM over the abstract arithmetic (see p. 46) has a logical continuation.

- Quantum-theoretic description invokes no $\mathbb{C}$-numbers, nor does it introduce them. It *does create* them together with the $\tilde{\mathbb{C}}^*$-algebra. The $\mathbb{C}$-numbers are in and of themselves the *quantum* numbers.

This fact is remarkable in its own right because the "two-dimensional" numbers arise at the lowest empirical level, not from the need for solving any mathematical problems. Mathematics is still lacking. Therefore, pt. R^+ (p. 25) could have even been weakened by replacing ⌈homomorphism onto numbers⌉, roughly, with the ⌈homomorphism onto continuum⌉. Our minimal points of departure are replications and the ingoing/outgoing structure of brace (32). The imaginary part of the complex number—as a supplement to the real one—comes, as a rough guide, from the left-hand side of the conception $\underline{\Psi} \dashrightarrow \underline{\alpha}$. The theory does not depend on the meanings that will be later attached to the physical concepts—observables, measurement, spectra, means, etc.—their interpretations, or rigorous definitions. At the same time, the interferential "effects of subtraction and of zeroes" are intrinsically present within the construct's foundation itself.

Let us add, in conclusion, two more formal vindications of rigidity of the emerging $\mathbb{C}$-structure. In doing so, one assumes that we have already had the $\mathbb{R}$-numbers.

Unitary brace contain pairs of the form $\{\overset{\Psi}{\mathsf{n}}, \overset{\Phi}{0}\}$. The binary operations $\{\oplus, \odot\}$ on their numerical representatives $(\mathsf{n}, 0)$ are closed and, as easily seen, form a commutative field, which is isomorphic to $\mathbb{R}$. It is a subset of the generic pair set (n, m). From the operator nature of $\odot$, it follows that these pairs form a certain distributive ring with general-group properties (79) and (80). The presence of the field $\mathbb{R}$ contained in it tells us that these pairs can be realized by the elements $\mathsf{n} + \mathsf{m}x$ of, at most, associative algebra A over $\mathbb{R}$. Here,

$\mathsf{n}, \mathsf{m} \in \mathbb{R}$, x is a generator of any ring's element beyond $\mathbb{R}$, and the habitual $+$ replaces the sign $\oplus$. The multiplication of two such elements

$$(\mathsf{n} + \mathsf{m}x) \odot (\mathsf{n}' + \mathsf{m}'x) = \mathsf{n}\mathsf{n}' + (\mathsf{m}\mathsf{n}' + \mathsf{n}\mathsf{m}')x + \mathsf{m}\mathsf{m}'x^2 = \cdots$$

immediately shows that the result does not depend on the order of factors, i.e.,

$$\cdots = (\mathsf{n}' + \mathsf{m}'x) \odot (\mathsf{n} + \mathsf{m}x)\,,$$

due to the permutability of $\{\mathsf{n}, \mathsf{m}, \mathsf{n}', \mathsf{m}'\}$ between each other and of any x with itself. This is a direct consequence of the two-dimensionality of the algebra A; it must be commutative. Invoking now the well-known Frobenius theorem on associative and commutative structures containing the field $\mathbb{R}$ [181], we arrive once again at a multiplication of the form (82). Körner puts this point as follows: "The complex numbers constitute the largest system of objects that most people are content to call numbers" ([172], p. 230).

Topologies on Numbers

Yet another reasoning about exclusivity of $\mathbb{C}$-numbers follows from matching the topological and algebraic properties of the general number systems ([182], Section 27). The case in hand is the uniqueness and non-arbitrariness in the emergence of the topological field $\mathbb{C}$; Pontryagin (1932). In our case, we have two continuums: the numerical symbols n and m, each of which, by the very method of constructing the (Ξ)-objects (34), is equipped only with the natural ordering $<$. Since we do not have any more math-structures yet, the topology, continuity, and limits on each of the continuums can already be introduced with respect to this relation. For example, there is no need to introduce the topology a priori by creating the arithmetical operation of multiplication/divisibility of rationals (and a concept of the prime integer), as is conducted in the p-adic approaches to QM [66,183,184]. The "non-naturalness" of multiplication as compared with addition was already noted above. Moreover, in the p-adic versions for a numerical domain, the topologically and physically required matching between the natural ordering, connection, and continuity ([182], Ch. 4) is destroyed, and the approaches themselves stipulate the existence of the *observations* numbers with a comprehensive arithmetic. At the same time, questions about the "structure" of the physical x-space at Planck's scale and about measurements by rationals (see motivation in [183,184]) have not yet emerged because we are not relying on physical conceptions and are not yet introducing these notions as numerical. The x-space itself is as of yet absent, and D. Mermin [3] overtly claims along these lines that "when I hear that spacetime becomes a foam at the Planck scale, I do not reach for my gun". From the low-level empiricism standpoint, any objects, apart from the $\mathbb{R}^2$-continuality and frequencies $\mathtt{f}$, call for independent axioms. In turn, the primary nature of the $\mathbb{R}$-continuality itself follows directly from the boolean $2^{\mathfrak{T}}$ (p. 19) and Σ-postulate of infinity.

8. State Space

Quantum states ... cannot be "found out"—W. Zurek ([8], p. 428)

... quantum theory refuses to offer any picture of what is actually going on out there—D. Mermin (1988)

8.1. Linear Vector Space

Once the replication $(\widehat{N,M})$ of brace $\{\overset{\Psi}{\mathsf{n}}, \overset{\Phi}{\mathsf{m}}\}\underline{\alpha}$ has acquired a binary character

$$(\widehat{N,M})(\{\overset{\Psi}{\mathsf{n}}, \overset{\Phi}{\mathsf{m}}\}\underline{\alpha}) \quad\Longleftrightarrow\quad ((\mathsf{N},\mathsf{M}) \odot (\mathsf{n},\mathsf{m}))|\alpha) = \mathfrak{a}|\alpha)\,, \tag{89}$$

the difference between "what is replicated" and "how many times" disappears. A symbol $|\alpha)$ of the eigen-state has been attached to the abstract $\mathbb{C}$-number $\mathfrak{a}$. Construing this point as a quantum analog of re-choosing (liberation of) the measurement units (p. 46), we obtain that the two formal states $\mathfrak{a}|\alpha)$ and $\mathfrak{b}|\alpha)$ are always connected by a certain number operator $\hat{\mathfrak{p}}$:

$$\mathfrak{b}|\alpha) = \hat{\mathfrak{p}}\big(\mathfrak{a}|\alpha)\big), \qquad \hat{\mathfrak{p}} \rightleftarrows \mathfrak{b} \odot \mathfrak{a}^{-1}.$$

Manipulating the numbers becomes independent of symbols $|\alpha)$. The way to formalize this is to think of generic states $\mathfrak{a}|\Psi) \in H$ as the "solid characters"

$$\mathfrak{a}|\Psi) \rightarrowtail |\Xi\rangle \in \mathbb{H}, \tag{90}$$

i.e., as the $|\Xi\rangle$-elements of a new set $\mathbb{H}$, which is equipped with the $\hat{\mathfrak{p}}$-replication images represented by the $\mathfrak{p}$-family ($\mathfrak{p} \in \mathbb{C}$) of maps

$$\mathbb{C} \times \mathbb{H} \overset{\bullet}{\mapsto} \mathbb{H}: \qquad \mathfrak{p} \bullet |\Xi\rangle = |\boldsymbol{\Phi}\rangle \in \mathbb{H}, \tag{91}$$

and which is obliged to inherit the structure (89). This inheritance says that the coordination of $\odot$-multiplication in (89) with the replication's $\mathfrak{p}$-realization is performed by a new operation $\bullet$ of the unary kind on $\mathbb{H}$, i.e., (91), which should be subordinated to the rule

$$\mathfrak{p} \bullet \big(\mathfrak{a} \bullet |\boldsymbol{\Psi}\rangle\big) = (\mathfrak{p} \odot \mathfrak{a}) \bullet |\boldsymbol{\Psi}\rangle \qquad (\mathfrak{p}, \mathfrak{a} \in \mathbb{C}, \quad |\boldsymbol{\Psi}\rangle \in \mathbb{H}). \tag{92}$$

Due to this connection between operations $\odot$ and $\bullet$, the latter is usually referred to as "multiplication" as well; however, such an intuition with dropping the word "unary" may have implications ([130], Section 6.2). An analogous rule had already occurred in the relationship (59) between the $\oplus$-number $\mathbb{C}$-structure and the $(\dotplus)$-group superposition, i.e., when the multiplicative structures $\{\odot, \bullet\}$ were not available yet.

Among replication operators $\hat{\mathfrak{p}}$, there exists an identical transformation

$$\hat{\mathfrak{p}} = \hat{\mathbb{I}}: \qquad \mathfrak{a}|\Psi) \overset{\hat{\mathbb{I}}}{\rightarrowtail} \mathfrak{a}|\Psi),$$

to which a symbol of the numerical unity $\mathfrak{p} = \mathbb{1}$ corresponds. From this, in accordance with (90) and (91), there follows the rule

$$\mathbb{1} \bullet |\Xi\rangle = |\Xi\rangle, \qquad \forall |\Xi\rangle \in \mathbb{H}.$$

It is clear that the $(\bullet)$-multiplication needs to be agreed with the $\uplus$-union. Let us make use of the fact that an object of (quantum) replication may be not only the unitary brace $\{\overset{\Psi}{\mathsf{n}}, \overset{\Phi}{\mathsf{m}}\}\underline{\alpha}$, which is equivalent to the eigen-element $\mathfrak{a}|\alpha)$, but a $(\hat{+})$-sum of the like objects and, in general, any constituents of quantum ensembles (see p. 44). Therefore, the $\hat{\mathfrak{p}}$-replication

$$\hat{\mathfrak{p}}\big(\mathfrak{a}|\alpha) \dotplus \mathfrak{b}|\beta)\big) = \cdots \tag{93}$$

is known to have its twin-sum

$$\cdots = \mathfrak{a}'|\alpha) \dotplus \mathfrak{b}'|\beta) = \cdots \tag{94}$$

with certain coefficients $\mathfrak{a}'$, $\mathfrak{b}'$.

Let us, for the moment, give back (93) to the initial language of operators/brace according to the scheme

$$\underbrace{(\widehat{N,M})}_{\mathfrak{p}}, \qquad \underbrace{\{\overset{\Psi}{\mathsf{n}'}, \overset{\Phi}{\mathsf{m}'}\}}_{\mathfrak{a}}\underline{\alpha} \hat{+} \underbrace{\{\overset{\Psi}{\mathsf{n}''}, \overset{\Phi}{\mathsf{m}''}\}}_{\mathfrak{b}}\underline{\beta}. \tag{95}$$

Take into account a pre-image of operation $\hat{+}$ on objects (93), i.e., the $\dotplus$. Then, (95) and the content of Sections 7.4 and 7.5 certainly show that the expression (94) must be of the form

$$\cdots = (\mathfrak{p} \odot \mathfrak{a})|\alpha) \dotplus (\mathfrak{p} \odot \mathfrak{b})|\beta) = \cdots.$$

Reconverting, by (92), expressions such as $(\mathfrak{p}\circ\mathfrak{a})|\alpha)$ into the operatorial $\hat{\mathfrak{p}}(\mathfrak{a}|\alpha))$, we complete the ellipsis

$$\cdots = \hat{\mathfrak{p}}\big(\mathfrak{a}|\alpha)\big) \dotplus \hat{\mathfrak{p}}\big(\mathfrak{b}|\beta)\big)\,.$$

Passing now to the $\mathfrak{p}$-number and to the $|\Xi\rangle$-objects (90), i.e., replacing $\mathfrak{a}|\alpha) \rightarrowtail |\boldsymbol{\Psi}\rangle$ and $\mathfrak{b}|\beta) \rightarrowtail |\boldsymbol{\Phi}\rangle$, one derives an additivity of operation $\cdot$ when acting on a sum:

$$\mathfrak{p}\cdot\big(|\boldsymbol{\Psi}\rangle \hat{+} |\boldsymbol{\Phi}\rangle\big) = \mathfrak{p}\cdot|\boldsymbol{\Psi}\rangle \hat{+} \mathfrak{p}\cdot|\boldsymbol{\Phi}\rangle\,.$$

Here, the *H*-addition $\dotplus$ has been carried over to the group $\mathbb{H}$ as a new symbol $\hat{+}$. This is nothing but a distributive coordination of the $(\cdot)$-multiplication with the group addition $\hat{+}$.

In a similar way, through a number operator, one establishes yet another relation

$$\mathfrak{a}\cdot|\Xi\rangle \hat{+} \mathfrak{b}\cdot|\Xi\rangle = (\mathfrak{a}\oplus\mathfrak{b})\cdot|\Xi\rangle$$

between $\cdot$ and $\hat{+}$. Its origin is equivalent to (59). From the constructs above, it is not difficult to see that we have examined all the possibilities of $\mathbb{C}$-replicating the superpositions (58) or their constituents, which is why we have exhausted all the compatibility rules that stem from the two fundamental operations—replication and union ($\dotplus$).

Thus, having considered the passage (90) and (91) as a final homomorphism of the *H*-group elements $\mathfrak{a}|\Psi)$ onto the objects $|\Xi\rangle \in \mathbb{H}$, i.e., adjusting the previous concept of a state and of `DataSource` (p. 31), we infer the following.

- The minimal and mathematically invariant bearer of the observation's empiricism is an abstract space $\mathbb{H}$ of states $|\boldsymbol{\Psi}\rangle$ of the system $\mathcal{S}$. The structural properties

$$\mathbb{H} := \big\{|\boldsymbol{\Psi}\rangle, |\boldsymbol{\Phi}\rangle, \ldots\big\} \quad \big[\text{commutative group under operation } \hat{+}\big]\,, \tag{96}$$

$$\mathbb{C} := \{\mathfrak{a}, \mathfrak{b}, \ldots\} \quad \big[\text{field of complex numbers (57), (78)–(82)}\big]\,,$$

$$\hat{\mathfrak{a}}|\boldsymbol{\Psi}\rangle =: \mathfrak{a}\cdot|\boldsymbol{\Psi}\rangle \in \mathbb{H} \quad \big[\text{closedness under operation } \cdot \iff \text{operator automorphism } \hat{\mathfrak{a}}|\boldsymbol{\Psi}\rangle\big]\,, \tag{97}$$

$$\begin{aligned} \mathfrak{a}\cdot\big(\mathfrak{b}\cdot|\boldsymbol{\Psi}\rangle\big) &= (\mathfrak{a}\odot\mathfrak{b})\cdot|\boldsymbol{\Psi}\rangle, & \mathfrak{a}\cdot|\boldsymbol{\Psi}\rangle \hat{+} \mathfrak{b}\cdot|\boldsymbol{\Psi}\rangle &= (\mathfrak{a}\oplus\mathfrak{b})\cdot|\boldsymbol{\Psi}\rangle\,,\\ \mathbb{1}\cdot|\boldsymbol{\Psi}\rangle &= |\boldsymbol{\Psi}\rangle, & \mathfrak{a}\cdot\big(|\boldsymbol{\Psi}\rangle \hat{+} |\boldsymbol{\Phi}\rangle\big) &= \mathfrak{a}\cdot|\boldsymbol{\Psi}\rangle \hat{+} \mathfrak{a}\cdot|\boldsymbol{\Phi}\rangle \end{aligned} \tag{98}$$

  of the space coincide with the axioms of a linear vector space (LVS) over the field $\mathbb{C}$.

Attention is drawn to the fact that this is the first place in our construct where the word *"linear"* has appeared, and even the superposition principle, page 35, was formulated without using this term. The "axiom list" (96)–(98) should also be complemented with a declaration of the global D-number value (53) established above.

In a nutshell, the nature of the quantum state space is two-fold: group superposition (58) and operator nature of the "$\mathfrak{a}$-numbering" the elements of the group. It admits the $\mathbb{C}$-field scalars as operators. Relations (98) describe the rules of "interplay" between all the objects. It is known that such formations, while being implemented by a binary algebra of numbers, turn into the vector spaces and modules [175] (Ch. 5), [181] (Sections I(7.1–2), II(13.4)). Concerning the consistency of these rules—say, of numerical distributivity (80)—with relations (98), see the work [185].

Remark 15. *A certain oddity is in place.* QM*-empiricism is such that the standard definition of* LVS *by the all-too-familiar axioms* (96)–(98) *is more "*nonphysical*" by its nature than the "generalistic" abstraction of a group with operator automorphisms of the group H-structure itself [166] (Section I(4.2)), [175]. A point like this might be expected, though. This is because, as noted in Section 1.2, meaning all of the tokens in* (1) *and their origin are entirely unknown, and the linearity of* QM *is radically different from other "linearities" in physics.*

All told, the appearance sequence of the mathematical structures is as follows:

$$\lceil \textit{sets, union } \cup, \dots \rceil \rightarrowtail \lceil \textit{semigroup (Section 5.2)} \rceil$$
$$\rightarrowtail \lceil \textit{group } H \textit{ (Section 6.3)} \rceil \rightarrowtail \lceil \textit{numbers \& arithmetics (Section 7)} \rceil$$
$$\rightarrowtail \lceil \textit{compatibility of the group and numbers (Section 8.1)} \rceil$$
$$\rightarrowtail \lceil \textit{the abstract LVS and its bases} \rceil .$$

This sequence is rigid, such as the box-diagram in Section 1.3, so the structure of LVS *cannot be weakened because we have the two fundamental principia (**II** and **III**) in between the semigroup, the group, and the vector space.*

8.2. Bases, Countability, and Infinities

From a ban on transitions $\underline{\alpha}_s \dashrightarrow \underline{\alpha}_n$ under $s \neq n$, it follows that unitary $\underline{\alpha}_s$-brace (34) corresponds to vectors $\mathfrak{a}_s \cdot |\boldsymbol{\alpha}_s\rangle$ that are linearly inexpressible through each other. Aside from the general ensemble brace (32), no other elements exist, and all of them are in one-to-one correspondence with the vector representations $\mathfrak{a}_1 \cdot |\boldsymbol{\alpha}_1\rangle \hat{+} \mathfrak{a}_2 \cdot |\boldsymbol{\alpha}_2\rangle \hat{+} \cdots$. Each such vector has a statistical pre-image (32), and vice versa; there are no gaps. This means that the system of vectors $\{|\boldsymbol{\alpha}_1\rangle, |\boldsymbol{\alpha}_2\rangle, \dots\}$ forms a basis of $\mathbb{H}$ as the basis of LVS—*basis of an observable $\mathscr{A}$ or $\mathscr{A}$-basis*—and the number of symbols $|\boldsymbol{\alpha}_s\rangle$ is its dimension: $\dim \mathbb{H} = \mathsf{D}$. The $\mathsf{D} = \infty$ case, just like anything associated with infinity, cannot be formalized without topology, and its presence is presumed, but this discussion is dropped. We just remark that even earlier, when arising the two-dimensional continuum, we have silently assumed the $(\mathbb{R} \times \mathbb{R})$-product topology on it. This supposition is natural, inasmuch as it does not involve additional constructions/requirements. Thus, if Properties (96)–(98) are directly accepted as empirical, then the mathematical rigors augment them axiomatically on the outside because one constructs the mathematical theory.

The micro-transition $\overset{\mathscr{A}}{\rightsquigarrow}$ in Section 2.1 is a solitary entity. This means that the number of eigen $\underline{\alpha}_s$-primitives for an actual instrument may be either finite or discretely unbounded. We base this on the fact that continual formations are products of mathematics rather than empiricism (see also [58] (p. 35)). The $\mathfrak{T}$-set, as an example, is also non-continual, but that premise may even be given up because only a discrete portion of this set is present at arguments (transitions $\overset{\mathscr{A}}{\dashrightarrow}$). Notice incidentally that continuum, along with the number, does not feature in the ZF-axioms [134] but is also created, just as "an infinity is actually not given to us at all, but is ... extrapolated through an intellectual process." [105] (p. 55; Hilbert–Bernays); see also the book [84] for the conceptualization of infinity. One obtains a countability of the $\mathscr{A}$-basis. Hence, it follows a completeness of $\mathbb{H}$ and countability of dimension (53), as of the number LVS-invariant:

$$\mathsf{D} = 2, 3, \dots, \aleph_\mathrm{o} \qquad (= \dim \mathbb{H}) . \tag{99}$$

Finally, let us mention the following. The basis is a term that in no way is present in the abstract axiomatics (96)–(98), and LVS on its own account does not contain a motive for introducing that concept. However, empirically, the $\mathbb{H}$-space is arising entirely and ab initio in all possible linear combinations over $|\boldsymbol{\alpha}_j\rangle$, i.e., through $\mathscr{A}$-bases. Because of this, in order for an abstract LVS to become the quantum state space, the LVS should be considered as being accompanied by the concepts bases and changes thereof. Conforming to such a requirement and the formal existence of a basis is given by a nontrivial math theorem invoking the axiom of choice ([166], Section **II**(7.1)).

8.3. The Theorem

The states $|\boldsymbol{\Psi}\rangle$ and sums thereof, at the moment, form a formal family of different elements. Recall that symbols $\{\approx, \not\approx\}$ in pt. **R**, as from the end of Section 5.4, have been replaced with the standard ones $\{=, \neq\}$. The physical aspects of $\langle\!\langle \mathcal{S}, \mathbf{M}, \dots \rangle\!\rangle$ were being left aside so far, and, for example, $|\boldsymbol{\Psi}\rangle$ and $\mathfrak{c} \cdot |\boldsymbol{\Psi}\rangle$ were the different vectors of the $\mathbb{H}$-space. However,

- empiricism (deals with and) yields originally *not states* and superpositions thereof *but* $|\boldsymbol{\alpha}\rangle$*-representations*.

It is these representations (alone) that carry information about statistics $(\mathtt{f}_1, \mathtt{f}_2, \ldots)$ through coefficients $\mathfrak{a}_j$. The replicative character of $\mathfrak{c}$-multipliers and Σ-postulate entail, however, that the two vectors $\mathbb{1}\cdot|\boldsymbol{\Psi}\rangle$ and $\mathfrak{c}\cdot|\boldsymbol{\Psi}\rangle$ should correspond to the one and same statistics $\mathtt{f}_{(\mathscr{D})} = (1, 0, \ldots) = \tilde{\mathtt{f}}_{(\mathscr{D})}$ under an observation $\mathscr{D}$ with the eigen collection $\{\mathbb{1}|\Psi\},\ldots\}$.

Let us write the equalities

$$
\begin{array}{ccccccc}
\mathtt{f}_{(\mathscr{D})} & \twoheadleftarrow & \underbrace{\mathbb{1}\cdot|\boldsymbol{\Psi}\rangle}_{\text{observation } \mathscr{D}} & = & \underbrace{\mathfrak{a}_1\cdot|\boldsymbol{\alpha}_1\rangle \hat{+} \mathfrak{a}_2\cdot|\boldsymbol{\alpha}_2\rangle \hat{+}\cdots}_{\text{observation } \mathscr{A}} & \twoheadrightarrow & \mathtt{f}_{(\mathscr{A})}\,, \\
\tilde{\mathtt{f}}_{(\mathscr{D})} & \twoheadleftarrow & \overbrace{\mathfrak{c}\cdot|\boldsymbol{\Psi}\rangle} & = & \overbrace{\mathfrak{c}\cdot\big(\mathfrak{a}_1\cdot|\boldsymbol{\alpha}_1\rangle \hat{+} \mathfrak{a}_2\cdot|\boldsymbol{\alpha}_2\rangle \hat{+}\cdots\big)} & \twoheadrightarrow & \tilde{\mathtt{f}}_{(\mathscr{A})}
\end{array}
\tag{100}
$$

and look at them in the following order: the first line from right to left and the second in the reverse direction. Their right-hand sides are the carriers of some statistics $\mathtt{f}_{(\mathscr{A})}$ and $\tilde{\mathtt{f}}_{(\mathscr{A})}$. The frequencies $\mathtt{f}_{(\mathscr{A})} = (\mathtt{f}_1, \mathtt{f}_2, \ldots)$ come from the number set $(\mathfrak{a}_1, \mathfrak{a}_2, \ldots)$ under the same environments $\langle\!\langle \mathcal{S}, \mathbf{M}, \ldots \rangle\!\rangle$ that give rise to the statistics $\mathtt{f}_{(\mathscr{D})}$. However, it is also generated by the representative $\mathfrak{c}\cdot|\boldsymbol{\Psi}\rangle$, which is associated with the same $\langle\!\langle \mathcal{S}, \mathbf{M}, \ldots \rangle\!\rangle$; hence, $\tilde{\mathtt{f}}_{(\mathscr{D})} = \mathtt{f}_{(\mathscr{D})}$. By virtue of the second equal sign in (100), the same $\langle\!\langle \mathcal{S}, \mathbf{M}, \ldots \rangle\!\rangle$ are associated with the second $\mathscr{A}$-collection $(\mathfrak{c}\odot\mathfrak{a}_1, \mathfrak{c}\odot\mathfrak{a}_2, \ldots)$. Therefore, the frequencies $\tilde{\mathtt{f}}_{(\mathscr{A})}$ that emanate from it have to be identical to those emanating from the first collection $(\mathfrak{a}_1, \mathfrak{a}_2, \ldots)$. That is to say $\tilde{\mathtt{f}}_{(\mathscr{A})} = \mathtt{f}_{(\mathscr{A})}$, and the scale stretches $|\boldsymbol{\Psi}\rangle \rightarrowtail \mathfrak{c}\cdot|\boldsymbol{\Psi}\rangle$ are not recognized by any $\mathscr{A}$-instruments. A more concise reasoning is that the quantum replication $\mathfrak{c} = \mathsf{n} + \mathsf{i}\mathsf{m}$ may be viewed as one-dimensional replications $\hat{n}\ (\mathfrak{a}|\alpha\}), \hat{\imath}\ (\mathfrak{a}|\alpha\})$ of all the brace $\mathfrak{a}_s|\alpha_s\}$-images and of sums such as $\hat{n}\ (\mathfrak{a}|\alpha\}) \,\hat{+}\, (\hat{\imath}\circ\hat{m})(\mathfrak{a}|\alpha\})$. These replications do not change the superposition statistics as a whole.

The aforesaid gives birth to a universal—stronger than $\approx$ and irrespective of instruments—observational equivalence relation

$$
|\boldsymbol{\Psi}\rangle \approxeq \text{const}\cdot|\boldsymbol{\Psi}\rangle
$$

on the space $\mathbb{H}$, i.e., the "physical" indistinguishability (Section 2.4).

The basis vectors $|\boldsymbol{\alpha}_s\rangle$ and their $(\approxeq)$-equivalents will be referred to as *eigen vectors/states of instrument* $\mathscr{A}$. Clearly, the concepts of instrument and of (macro)-observation (**O**) should now be distinguished. Accordingly, the spectral constructions (51) and (52) should be corrected. Call the data set

$$
\big\{|\boldsymbol{\alpha}_1\rangle_{|\alpha_1}, |\boldsymbol{\alpha}_2\rangle_{|\alpha_2}, \ldots\big\} =: [\mathscr{A}]
\tag{101}
$$

the $[\mathscr{A}]$*-representative* of instrument $\mathscr{A}$ in $\mathbb{H}$. The add-on (101) does not touch on $\mathbb{H}$-space since the spectral labels $|\alpha_j$ are the self-contained objects independent of vectors $|\boldsymbol{\alpha}_k\rangle$. These labels and their degenerations determine internal properties of the formalized notion of an instrument (101). Conversely, any state vector $|\boldsymbol{\Psi}\rangle$ or $\mathfrak{c}\cdot|\boldsymbol{\Psi}\rangle$ may be treated as a $[\mathscr{C}]$-representative for an imagined/actual instrument $\mathscr{C}$ (spectrum is arbitrary) and is a certain $(\hat{+})$-sum of the eigen elements for any other $[\mathscr{A}]$-representative.

Remembering (23), we arrive at the quantum *"kinematic framework"* [69], i.e., at the ultimate conclusion that determines the pre-dynamical theory of macroscopic data on micro-events.

- ***The core first theorem of quantum empiricism:***
  (1) The mathematical representatives of physical observations and of preparations are the quantum states $|\boldsymbol{\Psi}\rangle$ and statistical mixtures of eigen $|\boldsymbol{\alpha}\rangle$-states

$$
\big\{|\boldsymbol{\alpha}_1\rangle^{(\varrho_1)},\ |\boldsymbol{\alpha}_2\rangle^{(\varrho_2)},\ \ldots\big\}, \qquad \varrho_1 + \varrho_2 + \cdots = 1\,.
\tag{102}
$$

(2) Properties (96)–(99) define objects $|\boldsymbol{\Psi}\rangle$ as elements of a (complete separable) linear vector space $\mathbb{H}$ over the algebra of complex numbers $\mathbb{C}^*$.
(3) Dimension $\dim \mathbb{H} = \mathsf{D} \geqslant 2$, representing an observable quantity ($\mathsf{D} < \infty$), is set to the value $\max\{|\mathfrak{T}_{\mathscr{A}}|, |\mathfrak{T}_{\mathscr{B}}|, \ldots\} = \mathsf{D}$ as required by the accuracy of the toolkit $\mathcal{O} = \{\mathscr{A}, \mathscr{B}, \ldots\}$. The eigen $|\boldsymbol{\alpha}\rangle$-vectors for each $[\mathscr{A}]$-representative provide a basis of $\mathbb{H}$ ($\mathsf{D} < \infty$) independently of spectra (101).
(4) The $\mathscr{A}$-bases stand out because the *observational* number-notion has been associated to them—statistics of the micro-events. The frequencies $\mathfrak{f}_k(\mathfrak{a})$ are invariant under involutions (88) and states $|\boldsymbol{\Psi}\rangle$ and $\mathfrak{c} \cdot |\boldsymbol{\Psi}\rangle$ are statistically indistinguishable.
(5) Rules (96)–(99), for a fixed $\mathsf{D} \neq \infty$, are categorical as an axiomatic system; they admit no non-isomorphic models.

The words "complete separable" have been supplemented here for mathematical reasons. This point is partly commented on in [6] and more fully in [130]. Indeed, the algebra constructed above calls for some amendments of a topological nature because the construction contains three infinities: continuum $\mathbb{C}$, continuum $\mathbb{H}$, and dimension D. In this connection, see the book [182]. The term "categorical" may require some explanation, and it is fully given in [130]. Here, one suffices to mention the point that one mathematical axiomatical system can in general have different inequivalent realizations/models [105,106,136]. In turn, the only thing that distinguishes two vector-space models between themselves is their dimension D.

Now, having considered the micro-destruction arrays with empirical rather than a formal take on arithmetic, the ideology of creating the quantitative theory leads to the key feature of quantum states—addition thereof—and the quantities under addition "do amount to" the complex numbers.

Remark 16. *As in Section 7.3, we draw attention to a hidden and (logically) unremovable extension of the physical units' concept.*

- "... units. *Despite the rudimentary nature of units, they are probably the most inconsistently understood concept in all of physics ... where do units come from?".*

  S. Gryb and F. Mercati ([102], p. 91)

Surprisingly, the naïve and straightforward conjunction of this concept with an abstract number seems to contravene the multiplication arithmetic but not the addition one. The typical example illustrates the point:

$$(2\,\mathsf{kg}) \times (3\,\mathsf{kg}) \neq 6\,\mathsf{kg} \qquad (3\,\mathsf{sheep} + 5\,\psi \neq 8\,\mathsf{Stück})$$

(see also [172] (items (4) and (5) on p. 16)). On the other hand, $2 \times (3\,\mathsf{kg}) = 6\,\mathsf{kg}$ and $(2\,\mathsf{kg}) + (3\,\mathsf{kg}) = 5\,\mathsf{kg}$, and the kg may be replaced here with any other entity: the classical meters, the abstract "Quanten Stücke ψ", and the like. They have no any operational significance, but one cannot get by without them.

The numeral characters acquire their usual abstract-numerical meaning—mathematization [58] —only when we throw (Section 7.3) the "units" $\lceil$Stück, °C, sheep, ψ, ...$\rceil$ out of data like $\lceil$5Stück, 5 °C, 5sheep, 5ψ, ...$\rceil$. The symbol "5" in 5 °C is the very same "5" as in $5 \cdot |\boldsymbol{\alpha}\rangle$. It is pointless without such a matching/abstracting. In the Newtonian spirit (epigraph to Section 7), the symbol could be defined as follows:

$$\lceil \textit{the abstraction } 5 \rceil := \frac{\hat{5}\,^{\circ}\mathsf{C}}{\hat{1}\,^{\circ}\mathsf{C}} = \frac{\hat{5}\,|\boldsymbol{\Psi}\rangle}{\hat{1}\,|\boldsymbol{\Psi}\rangle} = \cdots = \frac{\hat{5}\,\mathsf{unit}}{\hat{1}\,\mathsf{unit}}\,.$$

It may be even said that creating the number n (from its operator $\hat{n}$) as an abstracted entity reflects a kind of covariance with respect to attaching the various language tags regardless of whether they are real Stück, sheep, or the abstract ones such as $|\boldsymbol{\Psi}\rangle$. At the same time, the inversion of this abstraction—attaching the $\lceil$Stück, °C, sheep, ψ, ...$\rceil$ to the character 5—is always an interpretation of abstraction: *interpreting "the Stück", "°C-interpretation the Celsius", etc. It is not improbable that this is the only point when the completed* QT—QM/QFT/*quantum-gravity yet to be constructed will resort to word interpretation. See also comments by D. Darling on "sheep, fingers, tokens, numbers, things, to "add" things, abstraction—the process of addition" and the like on p. 178 in the book [2] and in [23] (pp. 263–264).*

Incidentally, within this physical and quantum context:

- The LVS itself should be regarded as no less a primary math-structure than the numbers themselves. Empiricism gives birth to both these structures together. Neither of them is more/less abstract/necessary than the other. Behind them is certainly a commutative group with operator automorphisms over it, and "numbers" is just a shortened term for that operators. Therein lies their nature (Section 7.2).

The habitual physics' construct ⌈number⌉ × ⌈physical unit⌉ exemplifies in effect the simplest (one-dimensional) LVS. However, the structure "the LVS", in contrast to the "bare" arithmetic, simply "does not forget and keeps" an operator nature (unary multiplication $\cdot$) of the structure "the number" and its empirical inseparability from the notion of the unit:

$$\underbrace{\hat{2}\,(\hat{3}\,\text{unit}) \iff 2\cdot(3\cdot\text{unit})}_{\text{vector space}} \;\rightarrowtail \cdots \text{abstracting} \cdots \rightarrowtail\; \underbrace{(2\times 3)\,\cancel{\text{unit}}}_{\text{arithmetic}}\,.$$

A direct corollary of this point is the fact that principium **II** can in no way be given up or disregarded. This would be tantamount to impossibility to introduce the further empirical (and classical) notion of a physical unit. The "forgetfulness" of arithmetic about measuring units even leads to a new way of looking at the classical Pythagoras theorem ([130], Section 6).

At the moment, it is worthwhile to summarize where we stand. As we have seen, nothing above and beyond what was used in constructing the mathematics (96)–(99) is required to explain the nature and meaning of the quantum state. Moreover, we have obtained not merely a completion of construction (11):

$$\oplus(\mathfrak{a}_1, |\alpha_1\rangle; \mathfrak{a}_2, |\alpha_2\rangle; \ldots) = \mathfrak{a}_1 \cdot |\alpha_1\rangle \,\hat{+}\, \mathfrak{a}_2 \cdot |\alpha_2\rangle \,\hat{+} \cdots .$$

In the first place, one establishes *a genesis of the quantal discreteness. Discriminating is an isolated act in the very nature of the perception process*: "one thing is distinct from another", "the controlling the minimal begins with a distincting of something the two", and the like (Section 2.2). Accordingly, "indivisibility, or "individuality", characterizing the elementary processes" ([131], p. 203; N. Bohr) must be formalized into the *"elemental" click*.

- The classical continuality of the perceptual reality—the (3+1)-space, fields $\{u(x,t), \psi(x,t), \ldots\}$, and the $\mathbb{R}$-numbers—is a theorization act, whereas the nature of the perception fundamentally "contains an element of discontinuity" ([4], p. 179). The continuality of the classical-physics mathematics we are used to is a "quantum effect".

The theorization also bears on preparation $\langle\!\langle S, \mathbf{M}, \ldots\rangle\!\rangle$. For example, smoothly reducing the interferometer intensity is not an empiricism but an *imagination* of abstracta the continuity/infinity. Clearly, such an (incorrect) substitution of the perception process should somewhere be replaced with a "correct understanding", such as the introduction of the categories: ⌈isolated micro-events⌉ + ⌈(myriads) assemblages thereof⌉. Granted, the natural language is able to describe the discontinuity only in the classical (the energy) terms—Plank's quantum of action $\hbar$, although the *quantum discreteness is not a discretization of* something classical but a discreteness on its own account.

We also clarify the formalization of measurement/preparation and of genesis of the $\mathbb{C}$-numbers. The well-known $(*)$-conjugation operation also finds its origin. Moreover, it is supplemented with a transposition $\Re(\mathfrak{a}) \rightleftarrows \Im(\mathfrak{a})$ of the real/imaginary part of the $\mathbb{C}$-number, and this transposition should be regarded just as natural operation as the conjugation. The emergent concepts of spectra and of their degenerations and eigen-states provide a nearly comprehensive mathematical image of physical observables. The state becomes devoid of its mysteriousness [21,31,186] since it is explicitly built in terms of the unique model of the "statistical" $|\boldsymbol{\alpha}\rangle$-representatives supplemented with macroscopic mixtures (102).

9. Numbers, Minus, and Equality; Revisited

> ... quas decet numeris negativis exprimantur, additio et subtractio consueto more peracta nullis premitur difficultatibus—L. Euler (1735)
>
> (... if we represent the notions, which are necessary, by negative numbers, then addition and subtraction ... are executed without any difficulty.)

9.1. Separation of the Number Matters

The empirical adequacy of QT can be based only on empirical ensembles (Sections 2.5 and 4). The creation of their mathematics tells us, then, that the "quantity of something" (68)–(70) turns into a formal operational algebra through labeling the operator replications (Sections 7.1 and 7.2) and properly yields the numbers. At first, they appear merely as

$$\overbrace{\underbrace{\text{n-symbols of abstractions (68) and (69) with ordering} <}}\qquad \begin{matrix}\downarrow \quad \downarrow \\ \cdots\cdots\cdots\end{matrix}$$

and then as internal objects of theory:

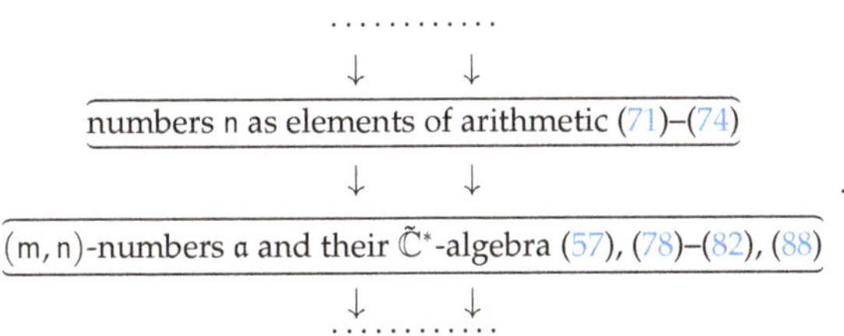

These steps are necessary and mean that not only are the complex numbers far from self-evident, but even the negative ones are; a key place (50), (55) wherein a group arises. All the other structural points, first and foremost the observational quantities, may be further produced (even as concepts) only by way of certain mathematical mappings:

$$\begin{matrix}\cdots\cdots\cdots \\ \downarrow \quad \downarrow \\ \text{the observations numbers } \mathtt{f}_j \text{ and } \alpha_j\text{:} \\ \mathfrak{a} \mapsto \lceil \text{statistics } \mathtt{f}_j \rceil, \quad |\boldsymbol{\alpha}_j\rangle \mapsto \lceil \text{spectra } \alpha_j \rceil \\ \text{tensorial structure of the } \mathbb{H}\text{-space}\end{matrix} \quad . \tag{103}$$

In other words, if a concept is a numerical one already in empiricism—frequencies, spectra, etc.—then its meaningful formalization by means of a mathematical definitio can only resort to mathematics that we have at our disposal: LVS and algebra of numbers.

Thus, numerical quantities in the entire theory are initially divided up by their emergence mechanism (**II**): the intrinsic abstracta and reifications (103). Without such a division, the *circular logic is inevitable*, and the above-mentioned "unit" treatment of numbers would still be supplemented with the task of their observational interpretation complicated by two-dimensionality. This task would be present in formalism not merely as a problem but as an inherently intractable challenge. Actually, *any entity can be identified with numbers,* and this is why, the quantum empiricism and principium **II**—paradigm of the very number in the physical theory—insist on the need to pay the closest possible attention to all these things.

Remark 17. *In this regard, the situation is parallel with the familiar history of electrodynamics of moving bodies, as was pointed out just before principium **III**. Lorentz's contraction theory is inconsistent, if the space-time tags to events are not linked up to the empirically precise and operationally defined concepts in different reference frames: clocks, simultaneity, rigid rods, distances, and the like.*

In the quantum case, the chief subject of empirical definition is a concept of the number and of the "numerical value of ... ". Otherwise, the meaning given to the conception of a quantitative theory itself has been blurred. The "quantum numbers" $\mathbb{C}$ are built up from the reals, and the latter have an operator nature (Section 7.2). However, the complexes $\mathbb{C}$, being also operators and unlike the reals, never act (operationally) on the reified quantities. They do act on the abstract $|\Psi\rangle$-elements of the abstract commutative group $\mathbb{H}$. Recall that this group and superposition principle were arising before the numbers.

We have seen now that "it is quite wrong to try founding a theory on observable magnitudes[/categories] alone" [8] (p. 504; Einstein, in a talk (1926) to Heisenberg), and resorting to the physical notions—the camouflaged **M**-observations—is prohibited (see also Remark 2). The attempts to use statistics at the very beginning of the theory are known [29,34,87,90,93,187,188], and rightly so; they were initiated by H. Margenau (1936) [33] (Ch. 15). However, the scheme just given is rigid. To obviate the premature appearance of the very need for an interpretation, the scheme must not be varied. Being a sequence of steps, it provides in essence an answer to principium **II**.

9.2. Operations on Numbers

The last step in this scheme contains, in particular, the map $\mathtt{a} \mapsto \mathtt{f}$, i.e., measurement (48). Its form should be established in its own right—Born's rule [6]. To illustrate, the naïve transformation of negative numbers $\mathtt{p}$ into the actually perceived quantities by a "seemingly natural" rule such as $|\pm\mathtt{p}|$ is not correct and does not follow from anywhere. For the built algebra (96)–(98), the operation $\pm\mathtt{p} \mapsto |\pm\mathtt{p}|$ is extrinsic and illegal. According to ideology of Section 1.3, not only objects—numbers, vectors, quantities, characteristics, etc.—but also all the math operations should be created because one without the other is meaningless. The numerical object of the theory—the complex pair $(\pm\mathtt{p}, \pm\mathtt{q})$—is as yet single, and it contains a principally "non-materializable" ingredient (Section 7.4) and behaves as a whole. With regards to empiricism, the negative and $\mathbb{C}$-numbers are equally "nonexistent, fictitious entities" since the state's mathematics (96)–(98) has not been supplemented with the doctrine of "empirically perceived" quantities (103). As a matter of fact, the step-by-step transformation of the binary operation $\cup$ to symbols $\uplus$, $\dotplus$, $\hat{+}$ and, finally, to operations $\{\hat{+}, \cdot, \oplus, \odot\}$ does not terminate at states. Algebra (96)–(98) will be further required to create the mathematically correct calculation rules of the proper observational quantities.

The foregoing is amplified by the fact that pr. **II** has been involved in the classical description and in vindication/refutation of, say, the hidden variable theories. Here, numbers are identified with the reified quantities, and subtraction is taken for granted from the outset. However, the negative quantities are also being created here, and they are constructed in the same manner as the "quantum zero" for the *H*-group in Section 6.3.

Indeed,

- the instrument indications and physical quantities are not numbers, nor the ("pointer") states;

"detector ... does not measure a field or an *S*-matrix-element" (R. Haag (2010)). They are no more than notches, and "negative notches" are introduced prior to mathematics of symbols according to the following subconsciously intended scheme. The self-apparent physical conventionality, which has been called "an addition" of two such notches, must produce, in accord with the supra-mathematical requirements of physics, what is named "nought, zero". Two waves at a point, for instance, compensate each other. The result is asserted to be identical with a mathematical zero, and that is the subtraction.

The classical "explanations" are the ones that use compensations/subtractions (see Remark 13), whereas *the minus* we have become accustomed to *is a fairly abstract construction in its own right*. J. Baez and J. Dolan best reflected the situation, observing on page 37 of [85] that "half an apple is easier to understand than a negative apple!"; on the same page, a good discussion of division is given. In this respect, one might state that the very classical physics needs an interpretation in terms of strictly positive "the number of Stücke". The mathematization of empiricism into numbers is not a distinctive feature of a quantum description. However, comprehending "abstracting the minus sign" is not confined to this. A word of explanation is necessary with regard to the situation.

Mathematics formalizes [134] the positive/negative $\pm\mathsf{p}$ into the pairs' classes (m,n) being equivalent with respect to an "adding" of the class (ℓ,ℓ) (the "zero"):

$$\begin{aligned} (+\mathsf{m}) &:= (m,0) \approx (m+\ell, 0+\ell), \qquad (-\mathsf{n}) := (0,n) \approx (0+\ell, n+\ell)\,, \\ \pm\mathsf{p} \quad &\Longleftrightarrow \quad (\mathsf{m}-\mathsf{n}) := (m,n) \approx (m+\ell, n+\ell)\,, \end{aligned} \tag{104}$$

where m,n,ℓ are to be seen as "something strictly positive". This "adding" is yet another tacitly assumed and much more abstract action: the addition of objects of some other kind—"positive couples" (m,n). Technically, at an appropriate place in Section 7, we had to introduce such classes and to assign their own algebraic operations for them. The result might be called the "genuine" arithmetic (of "the positives") and could be enlarged to the "complete arithmetic" with multiplication and division.

As a consequence, the single-token object $(+\mathsf{m})$ or $(-\mathsf{n})$, which we perceive as self-evident (cf. pr. **II**), is a highly unobvious construction—the generic equivalence-class of two-token (m,n)-abstractions (104). The essence of the symbol of a negative number $(-\mathsf{n})$ is revealed only when contrasting the two positive ones. Exactly the same situation has occurred when deriving the superposition principle in (49), (50), and (55).

It is clear that once all the $\pm\mathsf{p}$-numbers and the "normal" positive $+\mathsf{p}$'s among them have been formalized into the equivalences (104), the fact that they possess any "natural meanings", such as the "operation of the quantity $\mathsf{p} \mapsto |\mathsf{p}|$" invented above, becomes more than unnatural; the abstract class operations appear out of nowhere. Similarly with $\mathbb{Q}$-numbers and their $\mathbb{R}$-extension: classes of equivalent pairs $(\mathsf{n}/\mathsf{m}) := (n,m) \approx (n\ell, m\ell)$.

9.3. Naturalness of Abstracta

We thus infer that the rejection or disregard of the similar "naturally abstract" set-theoretic models would be tantamount to the rejection of the minus sign even in elementary physics. This is absurd, but its root is a need for abstracting. On the other hand, the motivated deduction of these models cannot be replaced with (hidden) axiomatic assumptions or with ready-made math-structures. Such an ambivalence is, in our view, one reason why the problem with "decrypting" quantum postulates is so difficult; it touches on the metamathematical and metaphysical aspects of the very thinking [84,105,148,169,179]. The stream of subconsciously abstractive homomorphisms

$$\begin{aligned} &\lceil \text{pt. } \mathbf{S}, \mathfrak{T} \rceil \rightarrowtail \{\underline{\alpha}, \underline{\Psi}\} \rightarrowtail \{\underline{\Psi}, \cup\} \rightarrowtail \cdots \rightarrowtail \{(\Xi)_{\mathscr{A}}, \overset{\mathscr{A}}{\uplus}; \dots\} \rightarrowtail \cdots \rightarrowtail \\ &\qquad \{[{}^{\mu}_{\lambda}]\underline{\alpha}, \uplus\} \rightarrowtail \cdots \rightarrowtail \{\mathsf{a}|\Psi), \pm\} \rightarrowtail \{\{\overset{\Psi}{\mathsf{m}}, \overset{\Phi}{\mathsf{n}}\}\underline{\alpha}, \hat{+}\} \rightarrowtail \{\mathsf{a}\cdot|\alpha\rangle, \hat{+}\} \rightarrowtail \\ &\qquad\qquad \rightarrowtail \{\mathsf{a}, |\boldsymbol{\Psi}\rangle; \hat{+}, \cdot\} \;\Rightarrow\; \mathbb{H} \end{aligned} \tag{105}$$

is considerable and is always larger than it seems. In Sections 3–9, we have described not all of them. Each such homomorphism is a mapping into a representation by a model, and for a philosophical discussion of these representations and the origin of these models, see pages 1–230 in [160]. For another comment concerning abstracting/realism, we refer the reader to the first half of a letter from A. Einstein to H. Samuel in [189] (pp. 157–160); see also [180].

Thus, "difficulties" with complex numbers, stricto sensu, should already be attributed to the level of the usual negative ones. Bearing in mind that the minus comes from the equal sign = [86] and the equality comes from the scheme (49) $\rightarrowtail$ (50), both principia **II** and **III** are very important (and functioning) in the classical case. In quantum case, they are just fundamentally unavoidable for the very creation of the theory. The nature of QM theory, of arithmetic, of complex numbers, and of their algebras is one and the same.

Transferring the reasoning above to the natural numbers $\mathbb{N}$, the degrees of classical and quantum abstractions become indistinguishable. Empirical motivation leads, in one way or another, to the standard von Neumann's representation for ordinals

$$0 := \varnothing, \qquad 1 := \{\varnothing\}, \qquad 2 := \{\varnothing, \{\varnothing\}\}, \qquad 3 := \{\varnothing, \{\varnothing\}, \{\varnothing, \{\varnothing\}\}\}, \qquad \ldots, \tag{106}$$

i.e., to using the ZF-axiom of union: $n+1 := \{n\} \cup n$ [134]. Therefore, the $\mathbb{N}$-numbers are less obvious themselves, followed by the ordering $<$, topologies, extensions, generalizations, etc. The formal characterization of all the experimental values reduces to the successive creation of the set-theoretic atoms—unions of sets—some direct products thereof and mappings into other constructions of the same kind. Hence, both the physical images "being under a ban above" and auxiliary structures—dimensionalities/orders, etc.—should equally become homomorphisms onto certain formal constructions regardless of the description's classicality/quantumness. The presence of, say, non-binary operations (88) does not stand out because their nature does not differ from the one of habitual subtraction and of division. All of these are involutory structures that have been mathematically inherited from the empirical meta-requirements: repetitions (**M**-paradigm), experimental context $\langle\!\langle \mathcal{S}, \mathbf{M}, \ldots \rangle\!\rangle$, and covariance **III** (Section 5.4).

To close the section, we add that the distancing of concepts of state/`DataSource` and of a physical property is the continuation of a more primary idea—detaching the proper macro-perceptions from what is being *represented* theoretically [87,93,160] and from conceptualizing the notions [84]. As B. Mazur has noted in [86] (p. 2), "This issue has been with us … forever: the general question of *abstraction*, as separating what we want from what we are presented with", i.e., the separating the "bare" empiricism from mathematics with Σ-limit and the number.

The atomic constituent of perception—sensory experience—is an elementary quantum event [93,97,123,126], and it begins and terminates in the $(\not\approx_{\mathscr{A}})$-distinguishability of $\underline{\alpha}$-clicks (Section 2.1). Any continual is a "speculative theory" (act of abstracting), not the underlying empiricism. Therefore, all the further matters—numbers, arithmetic, cause/effect, (non)inertial reference frames, the notions of an observer, of a classical event in the Minkowski space, the spacetime concept itself and coordinates in the relativity theory (a quantum view of the equivalence principle), device read-out, tensors, composite systems, symbols $\otimes$, and the like—self-evident as they may seem, are the math add-ons, which could originate only in the "U-theory" of Section 5. Following von Weizsäcker, it might be coined the name "Ur-theory". There are no contradictions in the observations themselves, regardless of whether we call them macro- or micro-scopic. Contradictions do arise in the "mathematicae being constructed".

10. About Interpretations

> It is … not … a question of a re-interpretation … quantum mechanics would have to be objectively false, in order that another description … than the statistical one be possible—J. von Neumann ([25], p. 325)

> … one begins to suspect that all the deep questions about the meaning of measurement are really empty—S. Weinberg

"At this point in time it appears that a stalemate has been reached with regard to the interpretation of quantum mechanics" (E. Tammaro [190] (p. 1)). A "stalemate in which each side refuses to cede territory but is unable to produce a defining argument that would change the hearts and minds of the opponents" (M. Schlosshauer [16] (p. 227)).

10.1. Click, Again

The source of the "foundational skirmishes" [16] (p. 227) and the numerous treatments of QM [7,8], [9] (Ch. 10), [30] (Ch. 10), [33,112]—"the Copenhagen" among them—is the fact that the $\underline{\alpha}$-event and intuitive sense of the $\underline{\Psi}$-primitive (pt. **S**) are a priori endowed with physical properties, observational/determinative characteristics of the `DataSource`, and operationality of the canonical QM-concept of the ket-vector $|\boldsymbol{\Psi}\rangle$. A representative example in this regard is one of the first sentences from Everett's PhD: "The state function ψ is thought of as objectively characterizing the physical system ... at all times ... independently of our state of knowledge of it" [107] (p. 73), [124] (p. 3), and also, on p. 48, "The general validity of pure wave mechanics, ***without any statistical assertions***, is assumed for ***all*** physical systems, including observers and measuring apparata". Furthermore, again the Everett's: "The physical 'reality' is assumed to be the wave function of the whole universe itself" [67] (p. 100), [107] (p. 70). However, none of these initially exist. The primitive $\underline{\alpha}$-events' abstractions $\underline{\Psi} \overset{\mathscr{A}}{\dashrightarrow} \underline{\alpha}$ are all there is.

An important point is that the eigen $\underline{\alpha}$-click (of a photon/electron in the EPR-experiment, say) should not be identified with an $|\boldsymbol{\alpha}\rangle$-state. The latter is re-developable with respect to eigen-states associated with other click-sets of any other instruments:

$$|\boldsymbol{\alpha}_1\rangle = \mathfrak{b}_1 \cdot |\boldsymbol{\beta}_1\rangle \hat{+} \mathfrak{b}_2 \cdot |\boldsymbol{\beta}_2\rangle \hat{+} \cdots = \mathfrak{c}_1 \cdot |\boldsymbol{\gamma}_1\rangle \hat{+} \mathfrak{c}_2 \cdot |\boldsymbol{\gamma}_2\rangle \hat{+} \cdots = \cdots ,$$

which is why it is logically meaningless to attribute one $\underline{\alpha}$-click to that which carries the statistics of other clicks $\underline{\beta}_k$, $\underline{\gamma}_k$, ... and has nothing in common with $\underline{\alpha}$. All the more so the click may not be related with the physical texts or physically descriptive collocations, such as "the measuring act on Bob's electron reveals the spin-up state". As in the "cat case" (Section 6.3.1), the spin-up here is a click-up rather than a state $|\uparrow\rangle$. Similarly, a click (allegedly of a photon) with Alice/Bob has nothing to do with distance between photons (the locality "problem"), with speed of light, nor with a kinematic "understanding" of the photon.

The quantum-detection micro-event is not a classical one as we have been understanding it, say, in the special relativity. The "click does not establish the presence of something" [126] (p. 761), it "is an elementary act of "fact creation."" [23] (Wheeler). The facts and phenomena are made up of clicks. That is to say, the distinguished $\underline{\alpha}$-clicks are not the events spaced at some distance from each other or at different points in time. These are just clicks without accounting to them such descriptive notions, such as distance, coordinates, point of time t, or the picturesque words, such as "dead/alive/... /cat"; of course, the click itself has no size/duration. Exempli gratia, the particles at accelerators and their physical properties are observed not as "material bodies in the proper sense of the word" [109] (p. 62)—this is impossible—but rather through the abstract detector-snaps. Neither the electron in interferometer nor the Higgs boson at a collider are observed in a detector as objects that are finite in extent; they are not observable entities. "A Higgs" is just a frequency 5σ-histogram at LHC.

Likewise, the math-properties of the eigen $|\boldsymbol{\alpha}\rangle$ and of the abstract $|\boldsymbol{\Psi}\rangle$ can in no way be "syncretized" with $\underline{\alpha}$-events when they are still being accumulated. The screen scintillation is not a photon and a photon is not a scintillation. Similarly, "the arrival of an electron" [164] (p. 3) at the screen does not mean "what is here at this point in time with a given coordinate is the materialized particle-electron". $|\boldsymbol{\alpha}\rangle$-states and $\underline{\alpha}$-clicks are accompanied by the phenomenological and dramatic words "up/down/... /alive/dead", and this has nothing to do with physics, which is yet to be created. The click should not be an element of the language in which $|\boldsymbol{\Psi}\rangle$-terminology, numbers, and physical properties have been employed whatsoever.

10.2. Abstraction the State

Then, something subsequently referred to as a state (the abstract) and a measurement (the concrete) is created. However, as already stressed in Section 2.4, the process of abstracting is a rather multistage one (Sections 3–9), and a reduction in the long sequence (105) "for physical reasons" does always contain phenomenological axioms a priori. Clearly, in the reverse direction, we confront hard-to-disentangle assumptions and the well-known axiomatic cycle. The physical considerations and phenomenology should not be present in fundamentals of quantum mathematics.

To avoid paradoxes with "quantum cats", "state vector does not describe ... a single cat" [68] (p. 37)), and "One cannot think about it as in a superposition" [16] (p. 134; D. Greenberger), with "the presence of a particle here and there", or with "quantum bomb-testing" (Elitzur–Vaidman) [27,33]:

- It is imperative to keep a severe conceptual differentiation [3] (first column) between the term "the state" and "physically sounding" adjectives/verbs and the spatiotemporal or cause-effect images.

Similarly T. Maudlin: "... we need to keep the distinction between mathematical and physical entities sharp. Unfortunately, the usual terminology makes this difficult" [151] (p. 129). Even indirect usage of the terminology borrowed from the classical description can be a source of confusion. For example, a so-called exchange interaction as a "cause of correlation" between identical parts of system.

It seems preferable to radicalize the non-connectivity of these categories, i.e., to proclaim it a postulate. For instance, boldface italics in Remark 12 or the selected thesis on page 41. At least, the differentiation between them should not be neglected in reasoning, inasmuch as it seems unrealistic to change the deeply ingrained [59] (p. 7) and ill-defined terminological locutions, such as a "photon is in a certain *state of polarization* ... *one* photon being in a particular place" [26] (p. 5, 9), "an observable has/acquires a (numerical) value when being measured" [92] (p. 310; criticism), "outcome of a measurement" [9,30,33], "quantum parallelism" [30] (p. 282), or "simultaneous measurability"; see Section 2.1 and pr. **I**. (As if we have had some micro-physics prior to math; there is nothing a priori.) With this mixing, the circular logic pointed out in Remark 10 will be present at all times. See, for instance, pages 29–30 in the work [74] and notably an emphasized warning by D. Foulis about *"a mistake, and a serious one!"*, including criticism addressed to von Neumann on p. 29. This

- trap of the "*braket*ting the ClassPhys'—$|\texttt{physical words}\rangle$ or $|\texttt{in}\rangle/|\texttt{out}\rangle$—is the very "somewhere ... hidden a concept" that M. Born spoke of (Section 1.3, p. 4), i.e., the mistaken *"physicality of* $|\psi\rangle$ *and of* $+$" in (1).

Again (see p. 23 and Section 6.4), even the indirect attempts to physically characterize the state function or "reconcile" its non-classicality with any *observational prototypes* are hopeless. "The wave function is in the head and not in nature" ascribed to A. Zeilinger (2014) by A. Khrennikov. The function is the very information `DataSource` around which all sorts of words on physics—readings, frequencies, objects, phenomena, particles, events, and other entities—are only slated to create.

- "We cannot ... manage to make do with such old, familiar, and seemingly indispensible terms" (Schrödinger (1933)) as the "" physikalische Realität " .... " Realität der Aussenwelt ", " Real-Zustand eines Systems "" [89] (p. 34) in the way we are doing this in classical physics, even philosophically. To put it both informally and more precisely, the automatic speech—stereotype—"the system in a state" (pt. **S**) [93,94] (criticism) should be dismissed from QT-fundamentals because the microscopy of quantum $\underline{\alpha}$-clicks shows that this colloquial habit is an unmeaning collocation.

This term may only be a theoretical conventionality in the follow-up *physical* theory. See also the first sentence in [61] and the selected theses on page 23 and at the beginning of Section 2.4.

The principled abstractness of the $|\mathbf{\Psi}\rangle$-object [13] (pp. 27–28) is a core attribute of quantum theory as contrasted to the classical one. This abstraction cannot be "struggled"; it is not an idealization of something phenomenological. It is absolute. An interpretative comprehension such as $|\mathtt{dead}\rangle \hat{+} |\mathtt{alive}\rangle$, even if it is permissible, may issue only from the $|\alpha\rangle$-representations $\mathfrak{a}_1 \cdot |\alpha_1\rangle \hat{+} \mathfrak{a}_2 \cdot |\alpha_2\rangle \hat{+} \cdots$, i.e., from a treatment of the $(\hat{+})$-addition (of quantum amplitudes) as an accumulation of $\underline{\alpha}$-microevents—many "cat boxes".

- In other words, *the* interpretation of the quantum state *is its very definiendum* (96)–(98). Even with the physical terminology created, there may be only one paraphrase for the meaning to the state: an abstract element of the abstract, linear (not Hilbert [130]) vector space over $\mathbb{C}^*$. (Point (4) in ***Theorem*** determines a supplement—the number add-on over the utterly abstract LVS.)

The "not Hilbert" here is because the norm and inner-product are the extra, nonessential math add-ons over $\mathbb{H}$ [130], which come from the follow-up introducing the Born statistics [6]. In and of itself, the state needs none and knows nothing of them. These concepts, similarly to the descriptive physical notions and a measurement, will be required further but not now for the calculation of observable quantities: math-calculus of statistics $\mathtt{f}_k$ and of means.

We may not blend the fundamentally abstract part of quantum mathematics—pre-physics and the structural properties of $\mathbb{H}$—with those in charge of its observational/physical constituent, i.e., we may not ascribe the ontological status [95,186] to everything. In the strict sense, the ontology of/and physics, the classical one included, cannot arise before the statistical processing of quantum micro-events. (Parenthetically, the sixth Hilbert problem on "Mathematical Treatment of the Axioms of Physics" [191] becomes an ill-posed problem ([130], Section 8).) The processing itself begins with the Born rule [6].

Continuing Scheme (68), a certain parallel takes place between the following couples:

$$\begin{array}{rl} \text{observations' language:} & \lceil\mathbb{R}\text{-numbers}\rceil + \lceil\text{physical quantities}/\ldots\rceil \\ & \quad\quad\quad\downarrow\uparrow \quad\quad\quad \downarrow\uparrow \\ \text{quantum language:} & \lceil\mathbb{C}\text{-}, |\mathbf{\Psi}\rangle\text{-objects}\rceil + \lceil\text{physical properties/data}/\ldots\rceil \end{array}.$$

Just as we are not raising the question about the abstractness/treatment of the $\lceil\mathbb{R}$-numbers$\rceil$ in isolation from the $\lceil$physical quantities/...$\rceil$ (Remark 16), so also we should not question a treatment or the physical meaning of the $\lceil\mathbb{C}$-, $|\mathbf{\Psi}\rangle$-objects$\rceil$. By analogy, being torn away from the \$-symbol in \$5, the number 5 in and of itself may carry neither the financial nor any other ("bank/(non)commuting/...") treatment, nor does it contain some hidden "microeconomic" content. The number has no a "retrograde memory".

The first "summands" in the aforementioned (+)-conjunctions are the abstracta of principle. They may exist as the "math-things-in-themselves", and we know that they really do just as we are comprehending the existence of the $\mathbb{N}$-arithmetic that has been constructed in Section 7. The second "summands" are the interpretative supplementations in their own rights. If a theory does not spell out a nature of accounting the second to the first (pr. **II**), then it is impossible to find-out/guess the "true" interpretation or nonexistent physical "protosource" of the abstracta n, $\mathfrak{a}$, and $|\mathbf{\Psi}\rangle$ "ab intra" their algebras (57), (71)–(74), (79)–(81), (88), and (96)–(98) or from the Hilbert-space mathematics. See again Remarks 2 and 16, and warnings by Ludwig of "a mistake. ... false notion that "mathematical objects" must be pictures of physical objects" [94] (p. 228) and of a "reality [of the] word "state," a reality in which one must not believe!" [58] (p. 78). A. Peres also makes special note of the analogous: "... physicists have been tempted to elevate the state vector ψ to the status of substitute of reality" [113] (p. 645); D. Mermin puts this as "a regretable atavistic tendency to reify the quantum state" [23] (p. 144).

10.3. Measurement "Problem"

The most representative example is the (in)famous problem of measurement [25] (Section V.4 and Ch. VI)—"tyranny of thinking of von Neumann measurements" [23] (p. 534) with the collapse postulate. This is the subject of an "endless stream of publications suggesting new theories ... unending discussions ... symposia" [91] (p. 519) and of "the mountains of literature" [94] (p. 118) containing opposing opinions [91] (Ch. 11), [9,18,19,30,45,56,159]. It is, indeed, the source of questions around locality in QM. As we have seen, this problem "is simply not a problem at all!" [50] (p. 1013). It is a nonexistent—"the alleged ... does not exist as a problem of quantum theory" [12] (p. 15)—as well as a pseudo problem and a non-issue [93] (p. 79 (!)), [94] (p. 118 (!)) [110], because

- in measurements, nothing propagates (much less at superluminal speeds) or interacts; nothing collapsed [92] (Section XVII.4.3), [9] (p. 328), [58,68], nullified, or localized; there are no such things as quantum jumps [161]; no "pieces" of the wave function are "cut out" [192] (p. 57, 158).

It is no exaggeration to say that the need to projective postulate—"a fruit of realist thinking" [4] (p. 172)—is much the same as the necessity for the world ether supporting the electromagnetic waves. All the more so because such a view of the theme has been present in the literature for quite a while [4,13,15,32,37,93,94,124,188] even as appeals.

> "There is nothing ... problematic about measurement"
>
> L. Ballentine (1996)

> "... there is no collapse of wave packets in reality. Do not believe in fairy tales!"
>
> G. Ludwig [58] (p. 104)

> "A state vector ... does not evolve continuously between measurements, nor suddenly "collapse" into a new state vector whenever a measurement is performed"
>
> A. Peres [113] (p. 644)

> "This "reduction" ... is not a new fundamental process, and, ... has nothing ... to do with measurement"
>
> L. Ballentine [34] (p. 244)

> "The mystifying notions arise from attributing physical reality to the "jump" at a given time t"
>
> G. Ludwig [92] (p. 327)

> "Really bad books ... claim that the state of the physical system ... collapses into the corresponding u_n. This is sheer nonsense. (Finding appropriate references is left as an exercise for the reader.)"
>
> A. Peres (2003)

Englert [12] (p. 8) does particularly object to the "folklore that "a measurement leaves the system in the relevant eigenstate" ... It is puzzling that some textbook authors consider it good pedagogy to elevate this folklore to an "axiom" of quantum theory". See also the second epigraph to Section 2.

The point here, put very briefly, is that the measuring the "problem" is one of principle not of practice. Expressed by Bell's words, "the word [measurement] has had such a damaging effect on the discussion, that ... it should now be banned altogether in quantum mechanics" [28] (p. 216). J. Bub and I. Pitowsky do insist in the book [8] (p. 453) that presumptions "about the ontological significance of the quantum state and about the dynamical account of how measurement outcomes come about, should be rejected as unwarranted dogmas about quantum mechanics".

Another example of circular logic is the critiqued [30,71,193] meaning of the phrase "an ensemble of similarly prepared systems" [8,30,90]. The revision of this (by and large correct) idea, as was set forth above, does actually demonstrate that, like in the ensemble approaches, "quantum mechanics is a *statistical* theory" [4] (p. 2), [58] (p. 123), [129]

(p. 223), [5,34,40,58,64,90,93,187], with a frequency content of randomness and the classical logic [58] but with a different math-calculus of the statistical weights. The "different" is due to the fact that the theory is not tied, as in the classical description, to the notion of an observable quantity, and the f's are calculated from the "other/abstract" numbers [6]. However, for the same reason, emphasizing a close resemblance with the statistical mechanics [11,194], ref. [124] (p. 72) and "explanations" with playing cards/dice, coins/balls/.../urns/"socks" [28] (Ch. 16) or with the classical phenomena—unusual as it may sound—are in error. The case in point is not a drastic dismissal of the classical ideas, but rather a "quantum audit" of the classical-physics language [130] (Section 8). The correct "audit" of the classical is a *re*creation of the classical:

$$
\begin{aligned}
&\lceil \underset{\sim\sim\sim}{\text{classical phenomena}} \rceil \rightarrowtail \lceil \text{classical events/objects} \rceil \rightarrowtail \lceil \text{micro world} \rceil \rightarrowtail \\
&\quad \lceil \text{micro-event} \rceil \rightarrowtail \lceil \text{quantum micro-event} \rceil \overset{(!)}{\rightarrowtail} \lceil \textit{abstract}\ \text{click} \rceil \rightarrowtail \\
&\quad \lceil \text{abstraction}\ \underline{\Psi} \overset{\mathscr{A}}{\dashrightarrow} \underline{\alpha} \rceil \rightarrowtail \lceil (\underline{\Xi})\text{-brace (24)}, \ldots \rceil \rightarrowtail \lceil \text{state}\ |\boldsymbol{\Psi}\rangle \rceil \rightarrowtail \\
&\quad \lceil \text{observable concepts} \rceil \rightarrowtail \lceil \text{observable numbers} \rceil \rightarrowtail \cdots \rightarrowtail \\
&\quad \lceil \text{statistics, the concept of a mean} \rceil \rightarrowtail \lceil \text{state, objects}, \ldots, \text{physics} \rceil \rightarrowtail \\
&\lceil \underset{\sim\sim\sim}{\text{classical phenomena}} \rceil
\end{aligned}
\tag{107}
$$

and, consequently, the creation of the classical concept of a measuring process. Thus, this scheme along with quanta's statistics and LVS-mathematics all add up to a positive answer to Wheeler's question: "Is the entirety of existence, rather than being built on particles or fields of force or multidimensional geometry, built upon billions upon billions of elementary quantum ..., ... acts of "observer-participancy," ...?" [16] (p. 286), [23].

10.4. Interpretations and Self-Referentiality

Although we have not yet touched on other significant attributes—the means over statistics, operators, and products of $\mathbb{H}$-spaces will be considered in their own rights—it is clear that the need to quest for a description in terms of hidden variables is also eliminated. Even from a formalistic perspective, the proof of the presence/absence [25,27,195] of these "physical" quantities should be attributed to the semantic conclusions of meta-theory (=physics) [106], i.e., to theorems *about* formal theory rather than to theorems of its *inner* calculus. In our case, and more generally, the formal theory is the syntactical axioms of QM. The corollaries of such axioms are inherently unable to lead to statements about interpretations [106] since theorems *about* object-theory itself is not provable by means *of* its object-language [106,120]. In a word, the nature and interpretation of axioms are not recovered from the very axioms or from the replacement thereof by the other ones.

A similar line of reasoning has accompanied QT for quite a while: "claim that the formalism by itself can generate an interpretation is unfounded and misleading" [68] (p. 38). It is known that even the mathematics itself cannot be grounded in a self-contained way [98] (p. 201), [105,149], [173] (!). All of this stands in stark contrast with the known statement of DeWitt to the effect that "mathematical formalism of the quantum theory is capable of yielding its own interpretation" [80] (pp. 160, 165, 168) or that "conventional statistical interpretation of quantum mechanics thus emerges from the formalism itself" [80] (p. 185). In particular, if we take account of the fact that it is not the theory itself but only its formal interpretation that determines the very semantic terms truth/falsity/provability of sentences (K. Gödel). In turn, "interpretation ... allows a certain freedom of choice" [78] (p. 310). See also [96] and specifically Ch. III in [105]. In other words, the subconscious striving for "to interpret" and transporting the macro into the micro is the very thing that prevents us from truly gaining an understanding of quantum mathematics.

In any case, the fact that we were initially constructed the set-theoretic model (cf. [96]) rather than an interpretation simply eliminates the problem or, at most, transfers it into the domain of questions about micro-transitions $\overset{\mathscr{A}}{\dashrightarrow}$ and $\mathfrak{T}$-family as entities being employed (see Remark 2). This is the domain of questions that invoke the set theory and touch on the ontological status of sets at all [149] (Sections V.8 and 9 (!)). Be this as it may, logic—formalized or not—does not allow us to make statements about statements, much less a statement that refers to itself. The self-referentiality ("von Neumann catastrophe") is almost the chief trouble [4,18] encountered in quantum foundations.

All of this, of course, does not depend on whether interpretation is built in a strictly formalized form [120] or in a physically natural one. In effect, the issue of interpretations—in the rigorous definition sense [106] (Ch. 2), [120], ref. [136]—is simply nonexistent. Accordingly, the demystification of the known and the quest for ontological interpretations to $\mathfrak{a}$-coordinates of the $|\mathbf{\Psi}\rangle$-vector [33,165,188]—the wave function—is no longer a problem, and with it, disappears the Feynman question of "the *only* consistent interpretation of this quantity" [164] (p. 22). See also M. Leifer's review [186] and extensive list of references therein.

11. Closing Remarks

> ... quantum mechanics has been a rich source for the invention of fairy tales—G. Ludwig and G. Thurler ([58], p. 122)

> I simply do not know how to change quantum mechanics by a small amount without wrecking it altogether ... any small change ... would lead to logical absurdities—S. Weinberg (1994)

11.1. Language and "Philosophy of Quanta"

Remembering and continuing Section 1.3, it is generally tempting to infer that when creating the theory, we may not rest on any meanings that are tacitly associated with the typical terminology—no matter physical or mathematical—and on the tacit assumption that customary concepts are substantially correct [98].

One should also be very cautious about the wording of statements concerning the phenomena outside the everyday experience. One means that even the very natural utterances—"here/there, electron *with* Alice/Bob" (locality), "big(ger)/small(er)", "let there be a two-particle $\mathcal{S}$" (quantitative statements), "subsystem $\mathcal{S}_1$ in such-and-such system $\mathcal{S}$, consisting of ..." (statements about structure)—are de facto "(apparently) "plausible" conclusions from the observed phenomena" [92] (p. 334). These have comprised an equivalent of a measurement/preparation ([4] (pp. 195–196) and Section 3.1) and of physical (pre)imagery and thereby imitate the way of thinking and schemes of classical mechanics; see also the second epigraph to Section 6.

- The "particle, here/there, big/small, this/that/another one, before/after", and the like are *already* "illegal" observations, numbers of sorts, and a premature arithmetization, i.e., this is *already* the subconscious quantifying the micro-events or the arrays thereof by a theory and classical (18) and (19) at that.

Reality's attributes are only slated to create. Say, when we decrease the particles in experiments and reach the atomic level, we still stay in the atomistic paradigm of the particle and of numbers: the objects having mass, their coordinates, degrees of freedom, etc. This is a mistaken intuition. Very informally, we should "religionize themselves" to the quantum micro-events, while the return to the words "particle/.../macro" must be performed by a new reasoning mechanism. It comprises, apart from the quantum-LVS apparatus (Section 8.1), the re-creation of the very classical concept of the particle, as schematized in (107).

At the other extreme is an attempt to "hurry up" and bring the reasoning to a Hilbert-space theory or to the quantum mixtures (102). As in Section 6.5, all this may well be incorrect [151]. A source of antinomies is implicit, implying, i.e., in the eclectic—this we stress once again—confounding the observations, clicks, numbers, physics, time, math, and imagination, followed by the *uncontrollable* lexical-"branching", such as *replacing the symbol* $\hat{+}$ *with a meaning* taken from reality. For instance, the emerging the word "simultaneously" in the sequence $\lceil$the ($\hat{+}$)-superposition of multiple states$\rceil \rightarrowtail \lceil$simultaneously $\rceil \rightarrowtail \lceil$quantum parallelism$\rceil$ [152] (p. 26). W. James has underscored that the *"viciously privative employment of abstract characters and class names* is, ..., one of the great original sins of the rationalistic mind" [23] (p. 547). This results in the sense messes, well-known no-go (meta)theorems [28,44], the locality "problem" in QM, and paradoxes such as the EPR [92] (Section XVII.4.4) or the jocular Bell question: "Was the world wave function waiting to jump for thousands of millions of years ... for some more highly qualified measurer—with a Ph.D.?" [28] (p. 117), [9] (p. 18), [15,37]. As to the no-go theorems, Ballentine remarks that "the growing number [thereof], combined with some peculiar terminology, has led to confusion ... A woefully common feature, ... each protagonist had some interpretation of the quantum state in mind, but never stated clearly what it was" [133] (pp. 2 and 6). Ludwig, echoing Ballentine, asks: "However, what do we mean by the notion of a state?" [87] (p. 5).

Clearly, the quantum-clicks do not depend on whether personified homo sapiens interpret the arrays thereof or a biological observer such as a "Heisenberg–Zeilinger dog" [196] (pp. 171–174) [12,65] does simply perceive. The observer—without their "subjective features" [98] (pp. 55, 137) or "the anthropomorphic notions "specifying" and "knowing" "[113] (p. 645)—is just a formally logical element **O** in theory. Without numbers, solely a quantitative theory is not possible (Section 6.5) because the entire terminology becomes indefinite.

Thus, once a mathematics and *unambiguous* language—spectra, means, and macroscopic dynamical models—have been created, not only is there no longer a need to call on the "otherworldly", eccentric, or anthropic explanations, but the very presence of a certain share of mysticism, of subjectivity, and of (circum-)philosophy [192]—"a philosophical *Überbau*" [12] (p. 12)—in quantum foundations becomes extremely questionable. Ludwig is much more thoroughgoing in his assessment of the language games, which he refers to as the "philosophical gymnastics" [93] (p. 79).

Eventually, we no longer have any freedom to invent exegeses of the quantum-postulate as "a Bible" or "a sacred text" [23] (p. 1038). Moreover, the liberty to ask questions is no longer there since the created object-language of states, of spectra, and of frequencies narrows down the entire admissible lexicon. It is able to generate questions that are not only ill-posed but must, as in Section 6.5, be qualified as "meaningless" [14] (p. 422). For example, those that are based on (human-beings') *intuition* taking the term observation or questions about "the underlying nature of reality". As we mentioned earlier, the notion of "a physical level of rigor" (in reasoning) and the physical justification will not help us with regard to the grounds of QT. Another example is the attempts at (or "to refrain from") "tying description to a clear hypothesis about the real nature of the world" (Schrödinger (1933)) and, in general, the question of "how it should function" at the micro-level. See also [58] (p. 100) on "reality".

In the classical framework, the language sentences are always interrelated since *all* of them, one way or another, handle the observational notions. In the famous Como address, N. Bohr had remarked that "every word in the language refers to our ordinary perception". These notions, in medias res, form our natural speech when describing experiments but are inadequate in the quantum [98] (!). That is, these concepts do not make clear the fact that behind the QT are some structureless abstracta, rather than an "improved" physicomathematical axiomatics or sophisticated math vehicles, e.g., non-commutative calculus; we believe that these are $\{\underline{\Psi} \overset{\mathscr{A}}{\dashrightarrow} \underline{\alpha}, \cup\}$ and procedures (105)—rather than an 'improved' physicomathematical axiomatics or sophisticated math vehicles; e.g., non-commutative calculus.

Language intuition usually makes it easy for us to do away with paradoxes the semantic closedness causes. However, the quantum situation is just one of a misuse of the vocabulary, i.e., when contradictions are inevitable, and this unlimited source of confusion demands *control over the language itself*. One does create the other ("relative") languages within itself [132]: at first, the language of quantum mathematics and thereafter the language of math-physical description and of classical physics, followed by the language of the semantic interpretations. This is just what we call the metamathematics and math-logic [105], discriminating between metamathematics and philosophy [106]. If this is not the case—the "quantum conclusions" from thinking (even if partly/implicitly) in terms of physical influences between the classical objects (Deutsch's "bad philosophy")—then we obtain an everlasting source of paradoxes since human intuition has roots in the classical world and is a rather problematic and personal category. A. Stairs calls upon "Do not trust intuition" [73] (p. 256) because it is not meant for QM.

Inasmuch as the conceptual autonomy in quantum fundamentals is minimal (Section 2), the quantum scheme of things must commence with an extremely "ascetic" language (Remark 10), and it should be independent of our intuitive knowledge, which "tend to declare war on our deductions" (van Fraassen). To avoid collisions between theory and meta-language, the subconscious striving of the natural language to include one in the other has to be limited. Einstein adds also the situations when "er führt dazu, überhaupt alle sprachlich ausdrückbaren Sätze als sinnleer zu erklären" [89] (p. 33). A. Leggett's comments on "pseudoquestions" and "gibberish" at the end of Section 1.2 may then be strengthened so that the meaninglessness by itself should become a constitutive element of language, including the language of "philosophy of quanta".

- The rudimentary quantum (meta)mathematics creates the notion of a *prohibited* statement/phrase/question, one that is devoid of meaning. These are sentences that involve the classical analogies in the circumvention of 1) the $|\alpha\rangle$-representatives to the non-interpretable abstraction $|\Psi\rangle$ and of 2) the numerical quantities' nature (Section 9.1).

It is appropriate at this point to quote the 't Hooft remark: "I go along with everything [Copenhagen] says, except for one thing, and the one thing is you're not allowed to ask any questions" and the Einstein reasoning on page 669 in the collected articles [131]: "One may not merely ask ... not even ask what this ... *means*". See also Heisenberg's discussion of the problem $\lceil$language $\rightleftarrows$ concepts$\rceil$ on pages 48–54 in [109], their work [197], the pages 234–235 in [131] with Bohr's appeals regarding the "necessity of a radical revision of basic principles for physical explanation ... revision of the foundation for the unambiguous use of elementary concepts", and their comments on words "phenomena", "observations", "attributes", and "measurements" on p. 237.

The literature on this subject, even taking only the qualified sources into account, is vast [1–3,8,9,16,24,27,29–31,33,41,44,57,64,71,77,119,151,154,159,165] and abounds with terminology—"words, *ostensibly* English" (A. Leggett [9] (p. 300; emphasis ours))—that defies translation into the language of events or of concretization: observer's consciousness, parallel/branching universes/worlds, free will, catalogue of knowledge, world branch, and also such collocations as rational agents, information ("*Whose*" and "about *what*?" [28], by "Bell's sardonic comments" [30] (p. 262)) has been recorded/transmitted/(not)reached an observer (Wigner's friend), teleporting a state, many-minds/worlds/words, quantum psychology, psycho-physical parallelism (in this connection, see [148] (p. 86 (!))), and many other "bad words" by Bell. He italicizes them on p. 215 of [28].

Of course, "without philosophy, science would lose its critical spirit and would eventually become a technical device" [33] (p. 800), but, on the other hand, "the concept of the free will cannot be defined by indications on devices" [94] (p. 151), and "one must not confuse physics with philosophy" [12] (p. 12). Furthermore, yet, we should like to remember a Heisenberg attitude [197] on "a misconception ... [and 'possibility'] to avoid philosophical arguments ... and the way of thinking of ... physicists who insisted on not dealing with philosophy". Namely, "[w]e can not avoid using a language bound up with the traditional philosophy". One cannot but mention the Rovelli article [198] that is entirely devoted to this topic. Therefore, "[i]t must be our task to adapt our thinking and speaking—indeed our scientific philosophy—to the new situation" with regard to the *abstract* meaning of the linear quantum addition $\hat{+}$ and quantum math altogether; all of the quotations are from pages 32 and 37–38 of the work [197].

As concerns the attitudes towards QM—at the suggestion of M. Tegmark in the 1990s, polls and statistical analysis of their correlations were even carried out [7]. There are also known attempts to involve here the biology of consciousness/brain [71], [119] (Ch. 9), [125],

[199] (Section 6). Regarding them, however, there have been not merely skeptical but quite the opposite opinions [94] (Section XII.5 (!)), [200] (Sections 17.5–6). Of special note are Ballentine's remark "to stop talking about "consciousness" or "free will"" on the last page of the preprint [133] and Popper's criticism of "the alleged … *intrusion of the observer, or the subject,* [or of consciousness] *into quantum theory* … based on bad philosophy and on a few very simple mistakes" [108] (pp. 11, 17, 42; everything as in the original) with an appeal "to exorcize the ghost called "consciousness" or "the observer" from quantum mechanics" [108] (p. 7). "[Q]uantum mechanics is a physical theory, not psychology" [4] (p. 83).

11.2. *Math-"Assembler" of Quantum Theory*

As a result, we gain "a contribution to philosophy, but not to physics" [82] (p. 86). At the same time, the proposed math "$\cup$-assembler"

$$\underline{\alpha}_j \not\approx \underline{\alpha}_k, \qquad \underline{\Psi} \overset{\mathscr{A}}{\dashrightarrow} \underline{\alpha}, \qquad (\underline{\Xi})\text{-brace (32)}, \qquad (\vee, \in, \cup)\text{-logic (37)–(40)}$$

is quite sufficient for creating the object-language. Giving a natural form to it would be acceptable; however, it is clear that the set-theoretic $\cup$-base of the language cannot be avoided [96,149]. Nevertheless, the syntactically more formal description of the sequence $\lceil$transitions $\rightarrowtail$ brace $\rightarrowtail$ numbers$\rceil$ is surely of interest until the way of looking at quanta's mathematics is harmonized with the math-logic. This would turn, however, all the above material into a pure-logic text, which we eschew in the present work. It is probably for this reason that the very important and extremely thorough works (Pre-theories, 76 axioms [93] (p. 241), ordered sets, morphisms, absence of the word superposition in monographs [87,93], the (valid) criticism of "theories of … so-called states" [58] (p. 78), etc.) by Günther Ludwig [87,92–94] and by their school are often left out of the literature on quantum foundations. Among other things, in spite of explicitly pointing out a "solution in principle of the measuring problem" in [93] (p. V) and "Derivation of Hilbert Space Structure" of [93], this author has not been mentioned in the detailed reviews [112,118,186] or even in the books [2,5,8,18,31,119].

11.3. *Well, Where is Probability?*

An answer to this question in *quantum* elements is brief enough—nowhere. "There is no probability meter" [8] (p. 185; S. Saunders), and the relationship of this concept with empiricism is unique [34] (p. 46)—the statistical proportions $\mathtt{f}_k$. Cf. the famous de Finetti's (1970) claim that "probability does not exist" [2] and A. Khrennikov's remarks to the effect that "*the only bridge between "reality" and our subjective description is given by relative frequencies*" [23] (p. 139) and that "Experimenters are only interested in … frequencies" [150] (p. 36). Moreover, more carefully stated by von Mises' words,

> "If we base the concept of probability, *not on* the notion of relative frequency, … at the end of the calculations, the meaning of the word 'probability' ***is silently changed*** from that adopted at the start to a definition based on the concept of frequency" ([129], p. 134; all the emphasis ours).

Indeed, suppose that the word "frequencies" has been banned [19] (p. 44) in substantiating the QT-elements and so have the usage of the words "over/repetition/…/statistics". Then, the questions do immediately arise: why the Kolmogorovian axiomatic, and why does it have this very quantification? In other words, why zero/one/…/positive? Why not the $(-1\ldots1)$-interval? Whence the single-case probability postulates? … subjectivity? Well, what is the quantification thereof, and what does subjectivity do in the natural-scientific theory?

> "… it is very doubtful that quantum probabilities can be introduced as a measure of our personal belief. Well, it may be belief, but belief based on frequency information"
>
> A. Khrennikov; Växjö Conference (2001)

One way or the other, quantum foundations *would demand an interpretation* of Kolmogorov's axioms (besides, these are not categorical in contrast to LVS), and the latter, in turn, demand interpreting the concept of the number—an axiomatic add-on *over* the ZF theory [134].

Bearing in mind the primary nature of numbers and nontriviality of their emergence in physical theory (Section 7.2), it is not just impossible to avoid the statistical weights $\mathtt{f}_k$ [121] (p. 25). Logic also forbids them from being subsidiary with reference to probability in any definition: "probability is the picture for reproducible frequencies; and it is the [only] *prescription* for a *correct* experiment" [94] (p. 144). Pauli, among the few, had been "convinced that

- the concept of 'probability' should not occur in the fundamental laws of a satisfying physical theory".

(an excerpt from their 1925 letter to Bohr)

Ensemble empiricism, for its part, is self-sufficient, and the only conventionality within it is an infinite number of repetitions. In this connection, we cannot agree with a statement of theorem III in van Kampen's work [56] (p. 99) and with further comment as to "a single system" and "calculation of spectra". At the same time, for formalizing the infinite, there is an appropriate axiom *in* the ZF-theory [134,135].

To say all this still informally, any non-statistical/non-ensemble framework for what we have been calling QM-probability does explicitly or implicitly—if the expression may be tolerated—"parasitize" on statistics by addressing the words "repetitions, multiplied, ..." and, at the same time, does "attract the empirically vague justifications" in terms of anthropomorphic surrogates: potentiality, tendency, propensity, the amount of ignorance, subjective uncertainty [8] or likelihood, degree of belief, and the like [165]. However, even from a philosophical point of view "probability is a deeply troublesome notion" [16] (p. 78; L. Hardy), which is supported by the vast literature on this subject [17,24,33], [66] (!), [81] (pp. 41–43), [170] (Chs. 3–4), [196], [201] (!). According to Deutsch, D. Papineau calls "this state of affairs ... a scandal" [8] (p. 550).

An Einsteinian "scientific instinct" [30] (p. 174) against the probability is very well known [78], and Pauli, again, had been recollecting their (Einstein's) frequent remarks in this regard: "One cannot make a theory out of a lot of "maybe's" [= probably] ... deep down it is wrong, even if it is empirically and logically right". More to the point, the question of what exactly is meant by a probability *event*, i.e., "Probability of *what* exactly?" [28] (p. 228), is also a matter of principle. The answer to it, as seen above, is this: "not of the classical events", i.e., "[n]ot of the ... *being*" [28] (p. 228) such as "QM-cats", "particle is here/there", "roll of the dice", and the like. An excellent text about probability and the aspects of the probability-physical constructs is the work [202]. Its "verdict" concerning the treatment of this concept [202] (Sections 4.5 and 8) is clearly Misesian [129], i.e. the "*ensemble* and *frequency*" [66] (p. xiii).

Thus, to sum up, the philosophy/axioms of probability or its "quantum deformations" should not be present in *quantum* foundations. There cannot be hidden details underlying the quantum probability because the "details" imply some terminology with a classical content. Quantum probability is the statistical regularity. It comes from Kollektivs [129] of abstracta (32) and may only be a shortened term for the relative "frequencies in long runs" (von Neumann) or "the Einstein hypothesis" by M. Jammer [91] (p. 441). The realistic/physical/.../pictorial adjectives and descriptive supplementations to the term "long runs" are prohibited. This is why the conventional tractability of the quantum-postulates' mathematics, i.e., the calculation of *probabilities for the classical* events to occur in the reality—"alive/cat/.../imploded/bomb"—"is not adequate" neither as a doctrinal point of departure nor as a post-math interpretation. It presents us with a circulus vitiosus of re-exegeses. Fuchs, referring to de Finetti's words in an interview with Quanta Magazine (4 June 2015), prognosticates that this conception "will go the way of phlogiston". The "not adequate" is a R. Haag quotation, and he expresses this "conviction", applying it even to "the conceptual structure of standard Quantum Theory" [101] (p. 743).

The ultimate conclusion completes Remarks 2 and 7. If we accept the set-theoretic eye on things, Section 5.1, by all appearances, provides a positive answer to the question about the rigidity of QT [103]—"change any one aspect, and the whole structure collapses" [57] (p. 1); see also the second epigraph to this section. At least, it is hard to imagine *what any other* axiom-free way of turning empiricism into quantum mathematics would look like, as soon as we abandon the primitive minimality of the scheme

$$\lceil \text{distinguishable micro } \underline{\alpha}\text{-events} \rceil + \lceil \text{ensembles of abstracta } \underline{\Psi} \overset{\mathscr{A}}{\dashrightarrow} \underline{\alpha} \rceil \,.$$

Funding: This research received no external funding.

Institutional Review Board Statement: Not applicable.

Informed Consent Statement: Not applicable.

Acknowledgments: The author wishes to express their gratitude to the QFT-department staff of TSU for stimulating conversations. A special word of thanks is due to Ivan Gorbunov and to Professor V. Bagrov, who initiated considering the matters on quantum logic [72,74,111]. The work was supported by the Tomsk State University Development Programme (Priority–2030).

Conflicts of Interest: The authors declare no conflict of interest.

References

1. Hooft, G. *The Cellular Automaton Interpretation of Quantum Mechanics*; Findamental Theories of Physics 185; Springer: Berlin/Heidelberg, Germany, 2016.
2. Fuchs, C.A. *Coming of Age with Quantum Information*; Cambridge University Press: New York, NY, USA, 2011.
3. Mermin, N.D. What's bad about this habit. *Phys. Today* **2009**, *62*, 8–9. [CrossRef]
4. De Muynck, W.M. *Foundations of Quantum Mechanics, an Empiricist Approach*; Kluwer Academic Publishers: Dordrecht, The Netherlands, 2002.
5. Silverman, M.P. *Quantum Superposition. Counterintuitive Consequences of Coherence, Entanglement, and Interference*; Springer: Berlin/Heidelberg, Germany, 2008.
6. Brezhnev, Y.V. The Born rule as a statistics of quantum micro-events. *Proc. R. Soc. A* **2020**, *476*, 20200282. [CrossRef]
7. Schlosshauer, M.; Kofler, J.; Zeilinger, A. A snapshot of foundational attitudes toward quantum mechanics. *Stud. Hist. Phil. Mod. Phys.* **2013**, *44*, 222–230. [CrossRef]
8. Saunders, S.; Barrett, J.; Kent, A.; Wallace, D. (Eds.) *Many worlds? Everett, Quantum Theory, and Reality*; Oxford University Press: New York, NY, USA, 2010.
9. Laloë, F. *Do We Really Understand Quantum Mechanics?*; Cambridge Unversity Press: Cambridge, UK, 2012.
10. Landau, L.D.; Lifshitz, E.M. *Quantum Mechanics*; Pergamon Press: Oxford, UK, 1965.
11. Faddeev, L.D.; Yakubovskiĭ, O.A. *Lectures on Quantum Mechanics for Mathematics Students*; AMS Student Mathematical Library: Providence, RI, USA, 2009; Volume 47.
12. Englert, B.-G. On Quantum Theory. *Eur. Phys. J. D* **2013**, *67*, 238. [CrossRef]
13. Gottfried, K.; Yan, T.-M. *Quantum Mechanics: Fundamentals*; Springer: Berlin/Heidelberg, Germany, 2003.
14. Nakhmanson, R.S. Physical interpretation of quantum mechanics. *Phys. Uspekhi* **2001**, *44*, 421–424. [CrossRef]
15. Lipkin, A.I. Does the phenomenon of 'reduction of the wave function' exist in measurements in quantum mechanics? *Phys. Uspekhi* **2001**, *44*, 417–421. [CrossRef]
16. Schlosshauer, M. (Ed.) *Elegance and Enigma. The Quantum Interviews*; Springer: Berlin/Heidelberg, Germany, 2011.
17. Accardi, L. *Urne e camaleonti: dialogo sulla realtà, le leggi del caso e l'interpretazione della teoria quantistica*, Il Saggiatore: Milano, Italy, 1997.
18. Mittelstaedt, P. *The Interpretation of Quantum Mechanics and the Measurement Process*; Springer: Berlin/Heidelberg, Germany, 2004.
19. Busch, P.; Lahti, P.J.; Mittelstaedt, P. *The Quantum Theory of Measurement*; Springer: Berlin/Heidelberg, Germany, 1996.
20. Aaronson, S. *Quantum Computing since Democritus*; Cambridge University Press: New York, NY, USA, 2013.
21. Nielsen, M.A. What Does the Quantum State Mean? 2018. Available online: http://cognitivemedium.com/qm-interpretation (accessed on 6 February 2022).
22. Zeilinger, A. A Foundational Principle for Quantum Mechanics. *Found. Phys.* **1999**, *29*, 631–643. [CrossRef]
23. Fuchs, C.A. My Struggles with the Block Universe. *arXiv* **2014**, arXiv:1405.2390.
24. Deutsch, D. *The Fabric of Reality*; Penguin Books: London, UK, 1997.
25. Von Neumann, J. *Mathematical Foundations of Quantum Mechanics*; Princeton University Press: Princeton, NJ, USA, 1955.
26. Dirac, P.A.M. *The Principles of Quantum Mechanics*; Clarendon Press: Oxford, UK, 1958.

27. Greenstein, G.; Zajonc, A.G. *The Quantum Challenge. Modern Research on the Foundations of Quantum Mechanics*; Jones and Bartlett Publishers: Burlington, MA, USA, 1997.
28. Bell, J.S. *Speakable and Unspeakable in Quantum Mechanics*; Cambridge University Press: Cambridge, UK, 2004.
29. d'Espagnat, B. *Conceptual Foundations of Quantum Mechanics*; Perseus Books: New York, NY, USA, 1999.
30. Home, D.; Whitaker, A. *Einstein's Struggles with Quantum Theory*; Springer: New York, NY, USA, 2007.
31. Ney, A.; Albert, D.Z. (Eds.) *The Wave Function. Essays on the Metaphysics of Quantum Mechanics*; Oxford University Press: Oxford, UK, 2013.
32. Slavnov, D.A. The possibility of reconciling quantum mechanics with classical probability theory. *Theor. Math. Phys.* **2006**, *149*, 1690–1701. [CrossRef]
33. Auletta, G. *Foundations and Interpretation of Quantum Mechanics*; World Scientific: Singapore, 2001.
34. Ballentine, L.E. *Quantum Mechanics. A Modern Development*; World Scientific: Singapore, 2000.
35. David, F. *The Formalism of Quantum Mechanics*; Lecture Notes in Physics 893; Springer: Cham, Switzerland, 2015.
36. Jauch, J.M. *Foundations of Quantum Mechanics*; Addison-Wesley: Boston, MA, USA, 1968.
37. Klyshko, D.N. Basic quantum mechanical concepts from the operational viewpoint. *Phys. Uspekhi* **1998**, *41*, 885–922. [CrossRef]
38. Gröblacher, S.; Paterek, T.; Kaltenbaek, R.; Brukner, Č.; Żukowski, M.; Aspelmeyer, M.; Zeilinger, A. An experimental test of non-local realism. *Nature* **2007**, *446*, 871–875. [CrossRef] [PubMed]
39. Zeilinger, A. Quantum Information and the Foundations of Quantum Mechanics. Newton Lecture in Institue of Physics, London, 17 June 2008. Available online: https://www.youtube.com/watch?v=3Dzd4J7kE-8 (accessed on 20 March 2022).
40. Allahverdyan, A.E.; Balian, R.; Nieuwenhuizen, T.M. Understanding quantum measurement from the solution of dynamical models. *Phys. Rep.* **2013**, *525*, 1–166. [CrossRef]
41. Alter, O.; Yamamoto, Y. Can we measure the wave function of a single wave packet of light? In *Fundamental Problems in Quantum Theory*; Greenberg, D.M., Zeilinger, A., Eds; New York Academy of Science: New York, NY, USA, 1995; pp. 103–109.
42. Ansmann, M.; Wang, H.; Bialczak, R.C.; Hofheinz, M.; Lucero, E.; Neeley, M.; O'Connell, A.D.; Sank, D.; Weides, M.; Wenner, J.; et al. Violation of Bell's inequality in Josephson phase qubits. *Nature* **2009**, *461*, 504–506. [CrossRef] [PubMed]
43. Fuchs, C.A.; Peres, A. Quantum Theory Needs No 'Interpretation'. *Phys. Today* **2000**, *53*, 70–71. [CrossRef]
44. Greenberg, D.M.; Zeilinger, A. (Eds.) *Fundamental Problems in Quantum Theory*; New York Academy of Science: New York, NY, USA, 1995.
45. London, F.; Bauer, E. The theory of observation in quantum mechanics. In *Quantum Theory of Measurement*; Wheeler, J.A., Zurek, W.H., Eds.; Princeton University Press: Princeton, NJ, USA, 1983; pp. 217–259.
46. Ballentine, L.E. Classicality without Decoherence: A Reply to Schlosshauer. *Found. Phys.* **2008**, *38*, 916–922. [CrossRef]
47. Schlosshauer, M. Classicality, the ensemble interpretation, and decoherence: Resolving the Hyperion dispute. *Found. Phys.* **2008**, *38*, 796–803. [CrossRef]
48. Schlosshauer, M. Implications of the Pusey–Barrett–Rudolph Quantum No-Go Theorem. *Phys. Rev. Lett.* **2012**, *108*, 260404. [CrossRef]
49. Halataei, S.M.H. Testing the reality of the quantum state. *Nat. Phys.* **2014**, *10*, 174. [CrossRef]
50. Fuchs, C.A. Quantum mechanics as quantum information, mostly. *J. Mod. Opt.* **2003**, *50*, 987–1023. [CrossRef]
51. Spekkens, R.W. Why I Am Not a Psi-Ontologist. Talk at Perimeter Institute. PIRSA# 12050021. 2012. Available online: https://pirsa.org/12050021 (accessed on 6 February 2022).
52. Everett, H., III. "Relative State" Formulation of Quantum Mechanics. *Rev. Mod. Phys.* **1957**, *29*, 454–462. [CrossRef]
53. Lundeen, J.S.; Sutherland, B.; Patel, A.; Stewart, C.; Bamber, C. Direct measurement of the quantum wavefunction. *Nature* **2011**, *474*, 188–191. [CrossRef] [PubMed]
54. Pusey, M.F.; Barrett, J.; Rudolph, T. On the reality of the quantum state. *Nat. Phys.* **2012**, *8*, 475–478. [CrossRef]
55. Menskiĭ, M.B. Quantum mechanics: New experiments, new applications, and new formulations of old questions. *Phys. Uspekhi* **2000**, *43*, 585–600. [CrossRef]
56. Van Kampen, N.G. Ten theorems about quantum mechanical measurements. *Physica A* **1988**, *153*, 97–113. [CrossRef]
57. Aaronson, S. Is Quantum Mechanics An Island In Theoryspace? *arXiv* **2004**, arXiv:quant-ph/0401062.
58. Ludwig, G.; Thurler, G. *A New Foundation of Physical Theories*; Springer: Berlin/Heidelberg, Germany, 2005.
59. De Touzalin, A.; Marcus, C.; Heijman, F.; Cirac, I.; Murray, R.; Calarco, T. Quantum Manifesto. A New Era of Technology. 2016. Available online: http://qurope.eu/system/files/u7/93056_Quantum%20Manifesto_WEB.pdf (accessed on 6 February 2022).
60. Van Kampen, N.G. The scandal of quantum mechanics. *Am. J. Phys.* **2008**, *76*, 989–990. [CrossRef]
61. Henry, R.C. The real scandal of quantum mechanics. *Am. J. Phys.* **2009**, *77*, 869–870. [CrossRef]
62. Wheeler, J.A.; Zurek, W.H. (Eds.) *Quantum Theory of Measurement*; Princeton University Press: Princeton, NJ, USA, 1983.
63. Briggs, G.A.D.; Butterfield, J.N.; Zeilinger, A. The Oxford Questions on the foundations of quantum physics. *Proc. R. Soc. A* **2013**, *469*, 0299. [CrossRef]
64. Khrennikov, A.Y. *Introduction to Quantum Theory of Information*; Fizmatlit: Moscow, Russia, 2008. (In Russian)
65. Bergou, J.A.; Englert, B.-G. Heisenberg's dog and quantum computing. *J. Mod. Opt.* **1998**, *45*, 701–711. [CrossRef]
66. Khrennikov, A. *Interpretations of Probability*; Walter de Gruyter: Berlin/Heidelberg, Germany, 2009.
67. Freire, O., Jr. *The Quantum Dissidents. Rebuilding the Foundations of Quantum Mechanics (1950–1990)*; Springer: Berlin/Heidelberg, Germany, 2015.

68. Ballentine, L.E. The formalism is not the interpretation. *Phys. Today* **1971**, *24*, 36–38. [CrossRef]
69. Bub, J.; Pitowsky, I. Two Dogmas About Quantum Mechanics. In *Many worlds? Everett, Quantum Theory, and Reality*; Saunders, S., Barrett, J., Kent, A., Wallace, D., Eds.; Oxford University Press: New York, NY, USA, 2010; pp. 433–459.
70. Stacey, B.C. Von Neumann was not a Quantum Bayesian. *Philos. Trans. R. Soc. A* **2016**, *374*, 20150235. [CrossRef] [PubMed]
71. Stapp, H.P. A theory of mind and matter. In *Fundamental Problems in Quantum Theory*; Greenberg, D.M., Zeilinger, A., Eds.; New York Academy of Science: New York, NY, USA, 1995; pp. 822–833.
72. Beltrametti, E.G.; Cassinelli, G. *The Logic of Quantum Mechanics*; Encyclopedia of Mathematics and Its Applications 15; Addison-Wesley: Boston, MA, USA, 1981.
73. Stairs, A. Kriske, Tupman and Quantum Logic: The Quantum Logician's Conundrum. In *Physical Theory and its Interpretation; Essays in Honor of Jeffrey Bub*; Demopoulus, W., Pitowsky, I., Eds.; Springer: Dordrecht, The Netherlands, 2006; pp. 253–272.
74. Foulis, D.J. A Half-Century of Quantum Logic. What Have We Learned? In *Quantum Structures and the Nature of Reality*; Aerts, D., Pykacz, J., Eds.; Kluwer Academic Publishers: Dordrecht, The Netherlands, 1999; pp. 1–36.
75. Engesser, K.; Gabbay, D.M.; Lehmann, D. *Handbook of Quantum Logic and Quantum Structures*; Elsevier: Amsterdam, The Netherlands, 2007.
76. Slavnov, D.A. Measurements and the mathematical apparatus of quantum physics. *Phys. Element. Part. Atom. Nuclei* **2007**, *38*, 295–359. (In Russian)
77. Kadomtsev, B.B. *Dynamics and Information*; Physics Uspekhi Press: Moscow, Russia, 1997. (In Russian)
78. Jammer, M. *The Conceptual Development of Quantum Mechanics*; McGraw-Hill Book Company: New York, NY, USA, 1966.
79. Pilan, A.M. Reality and the main question of quantum information. *Phys. Uspekhi* **2001**, *44*, 424–427. [CrossRef]
80. DeWitt, B.S.; Graham, N. (Eds.) *The Many-Worlds Interpretation of Quantum Mechanics*; Princeton University Press: Princeton, NJ, USA, 1973.
81. Sudbery, A. *Quantum Mechanics and the Particles of Nature: An Outline for Mathematicians*; Cambridge University Press: Cambridge, UK, 1986.
82. MacKinnon, E. Why Interpret Quantum Physics? *Open J. Philos.* **2016**, *6*, 86–102. [CrossRef]
83. Schwinger, J. *Quantum Kinematics and Dynamics*; W. A. Benjamin, Inc.: San Francisco, CA, USA, 1970.
84. Lakoff, G.; Núñez, R.E. *Where Mathematics Comes from. How the Embodied Mind Brings Mathematics into Being*; Basic Books: New York, NY, USA, 2000.
85. Baez, J.C.; Dolan, J. From Finite Sets to Feynman Diagrams. In *Mathematics Unlimited—2001 and Beyond I*; Engquist, B., Schmid, W., Eds.; Springer: Berlin/Heidelberg, Germany, 2001; pp. 29–50.
86. Mazur, B. When is One Thing Equal to Some Other Thing? 2007, pp. 1–24. Available online: http://www.math.harvard.edu/~mazur/preprints/when_is_one.pdf (accessed on 6 February 2022).
87. Ludwig, G. *Foundations of Quantum Mechanics I*; Springer: Berlin/Heidelberg, Germany, 1983.
88. Haag, R. Some people and some problems met in half a century of commitment to mathematical physics. *Eur. Phys. J. H* **2010**, *35*, 263–307. [CrossRef]
89. Einstein, A. Elementare Überlegungen zur Interpretation der Grundlagen der Quanten-Mechanik. In *Scientific Papers, Presented to Max Born*; Oliver and Boyd: Edinburgh, UK, 1953; pp. 33–40.
90. Ballentine, L.E. The Statistical Interpretation of Quantum Mechanics. *Rev. Mod. Phys.* **1970**, *42*, 358–381. [CrossRef]
91. Jammer, M. *The Philosophy of Quantum Mechanics*; John Wiley & Sons, Inc.: Hoboken, NJ, USA, 1974.
92. Ludwig, G. *Foundations of Quantum Mechanics II*; Springer: Berlin/Heidelberg, Germany, 1985.
93. Ludwig, G. *An Axiomatic Basis for Quantum Mechanics 1*; Springer: Berlin/Heidelberg, Germany, 1985.
94. Ludwig, G. *An Axiomatic Basis for Quantum Mechanics 2*; Springer: Berlin/Heidelberg, Germany, 1987.
95. Leggett, A.J. Probing quantum mechanics towards the everyday world: Where do we stand? *Phys. Scr.* **2002**, *T102*, 69–73. [CrossRef]
96. Benioff, P.A. Models of Zermelo Frankel set theory as carriers for the mathematics of physics. I, II. *J. Math. Phys.* **1976**, *17*, 618–628. 629–640. [CrossRef]
97. Hartkämper, A.; Neumann, H. (Eds.) *Foundations of Quantum Mechanics and Ordered Linear Spaces*; Lecture Notes in Physics 29; Springer: Berlin/Heidelberg, Germany, 1974.
98. Heisenberg, W. Language and Reality in Modern Physics. In *Physics and Philosophy*; Anshen, R.N., Ed.; Harper & Brothers Publishers: New York, NY, USA, 1958; pp. 167–186.
99. Weyl, H. *Space-Time-Matter*; Dover Publications, Inc.: New York, NY, USA, 1950.
100. Hartle, J.B. Quantum physics and human language. *J. Phys. A Math. Theor.* **2007**, *40*, 3101–3121. [CrossRef]
101. Haag, R. An Evolutionary Picture for Quantum Physics. *Comm. Math. Phys.* **1996**, *180*, 733–743. [CrossRef]
102. Gryb, S.; Mercati, F. Right About Time? In *Questioning the Foundations of Physics*; Aguirre, A., Foster, B., Merali, Z., Eds.; Springer: Cham, Switzerland, 2015; pp. 87–102.
103. Colbeck, R.; Renner, R. No extension of quantum theory can have improved predictive power. *Nat. Commun.* **2011**, *2*, 411. [CrossRef] [PubMed]
104. Bueno, O.; French, S.; Ladyman, J. On Representing the Relationship between the Mathematical and the Empirical. *Philos. Sci.* **2002**, *69*, 497–518. [CrossRef]
105. Kleene, S.C. *Introduction to Metamathematics*; Wolters-Noordhoff Publishing: Groningen, The Netherlands, 1971.

106. Rasiowa, H.; Sikorski, R. *The Mathematics of Metamathematics*; Panstwowe Wydawnictwo Naukowe: Warsaw, Poland, 1963.
107. Barrett, J.A.; Byrne, P. *The Everett Interpretation of Quantum Mechanics. Collected Works 1955–1980 with Commentary*; Princeton University Press: Princeton, NJ, USA, 2012.
108. Popper, K.R. Quantum Mechanics without "The Observer". In *Quantum Theory and Reality*; Bunge, M., Ed.; Studies in the Foundations Methodology and Philosophy of Science 2; Springer: Berlin/Heidelberg, Germany, 1967; pp. 7–44.
109. Heisenberg, W. *Philosophic Problems of Nuclear Science Eight Lectures*; Cambridge University Press: Cambridge, UK, 1952.
110. Edwards, D.A. The mathematical foundations of quantum mechanics. *Synthese* **1979**, *42*, 1–70. [CrossRef]
111. Foulis, D.J.; Randall, C.H. The empirical logic approach to the physical sciences. In *Foundations of Quantum Mechanics and Ordered Linear Spaces*; Hartkämper, A., Neumann, H., Eds.; Lecture Notes in Physics 29; Springer: Berlin/Heidelberg, Germany, 1974; pp. 230–249.
112. Schlosshauer, M. Decoherence, the measurement problem, and interpretations of quantum mechanics. *Rev. Mod. Phys.* **2004**, *76*, 1267–1305. [CrossRef]
113. Peres, A. What is a state vector? *Am. J. Phys.* **1984**, *52*, 644–650. [CrossRef]
114. Reynolds, T. Is direct measurement of time possible? *J. Phys. Conf. Ser.* **2017**, *880*, 012066. [CrossRef]
115. Weinberg, S. Collapse of the State Vector. *Phys. Rev. A* **2012**, *85*, 062116. [CrossRef]
116. Weinberg, S. What Happens in a Measurement? *Phys. Rev. A* **2016**, *93*, 032124. [CrossRef]
117. Joos, E.; Zeh, H.D.; Kiefer, C.; Giulini, D.; Kupsch, J.; Stamatescu, I.-O. *Decoherence and the Appearance of a Classical World in Quantum Theory*; Springer: Berlin/Heidelberg, Germany, 2003.
118. Zurek, W.H. Decoherence, einselection, and the quantum origins of the classical. *Rev. Mod. Phys.* **2003**, *75*, 715–775. [CrossRef]
119. Schlosshauer, M. *Decoherence and the Quantum-to-Classical Transition*; Springer: Berlin/Heidelberg, Germany, 2007.
120. Shoenfield, J. *Mathematical Logic*; Addison-Wesley Publishing Company: Boston, MA, USA, 1967.
121. Peres, A. *Quantum Theory: Concepts and Methods*; Kluwer Academic Publishers: Dordrecht, The Netherlands, 2002.
122. Piron, C. *Foundations of Quantum Physics*; W. A. Benjamin, Inc.: San Francisco, CA, USA, 1976.
123. Bohr, A.; Mottelson, B.R.; Ulfbeck, O. The Principle Underlying Quantum Mechanics. *Found. Phys.* **2004**, *34*, 405–417. [CrossRef]
124. Everett, H., III. The Theory of the Universal Wave Function. Ph.D. Thesis, Princeton University, Princeton, NJ, USA, 1956.
125. Tegmark, M. The importance of quantum decoherence in brain processes. *Phys. Rev. E* **2000**, *61*, 4194–4206. [CrossRef] [PubMed]
126. Ulfbeck, O.; Bohr, A. Genuine Fortuitousness. Where Did That Click Come From? *Found. Phys.* **2001**, *31*, 757–774. [CrossRef]
127. Brukner, Č. On the Quantum Measurement Problem. In *Quantum [Un]Speakables II*; Berltmann, R., Zeilinger, A., Eds.; Springer: Cham, Switzerland, 2017; pp. 95–117.
128. Zurek, W.H. Wave-packet collapse and the core quantum postulates: Discreteness of quantum jumps from unitarity, repeatability, and actionable information. *Phys. Rev. A* **2013**, *87*, 052111. [CrossRef]
129. Von Mises, R. *Probability, Statistics and Truth*; Dover: New York, NY, USA, 1981.
130. Brezhnev, Y.V. Why and whence the Hilbert space in quantum theory? *arXiv* **2021**, arXiv:2110.05932.
131. Schilpp, P.A. (Ed.) *Albert Einstein: Philosopher-Scientist*; The Library of Living Philosophers VII; MJF Books: New York, NY, USA, 1970.
132. Chomsky, N. The Galilean Challenge. Linguistics. Critical Essay. *Inference Int. Rev. Sci.* **2017**, *3*, 1–7. [CrossRef]
133. Ballentine, L.E. Ontological Models in Quantum Mechanics: What do they tell us? *arXiv* **2014**, arXiv:1402.5689.
134. Kuratowski, K.; Mostowski, A. *Set Theory*; North-Holland Publishing Company: Amsterdam, The Netherlands, 1967.
135. Hausdorff, F. *Set Theory*; AMS Chelsea Publishing: Providence, RI, USA, 1957.
136. Stoll, R.R. *Sets, Logic, and Axiomatic Theories*; Dover Publications, Inc.: New York, NY, USA, 1979.
137. Lundeen, J.S.; Bamber, C. Procedure for Direct Measurement of General Quantum States Using Weak Measurement. *Phys. Rev. Lett.* **2012**, *108*, 070402. [CrossRef]
138. Goyal, P.; Knuth, K.H.; Skilling, J. Origin of Complex Quantum Amplitudes and Feynman's Rules. *Phys. Rev. A* **2010**, *81*, 022109. [CrossRef]
139. Li, Z.-D.; Mao, Y.-L.; Weilenmann, M.; Tavakoli, A.; Chen, H.; Feng, L.; Yang, S.-J.; Renou, M.-O.; Trillo, D.; Le, T.P.; et al. Testing Real Quantum Theory in an Optical Quantum Network. *Phys. Rev. Lett.* **2022**, *128*, 040402. [CrossRef]
140. Chen, M.-C.; Wang, C.; Liu, F.-M.; Wang, J.-W.; Ying, C.; Shang, Z.-X.; Wu, Y.; Gong, M.; Deng, H.; Liang, F.-T.; et al. Ruling Out Real-Valued Standard Formalism of Quantum Theory. *Phys. Rev. Lett.* **2022**, *128*, 040403. [CrossRef]
141. Hardy, L. Quantum Theory From Five Reasonable Axioms. In *Quantum Theory: Reconsideration of Foundations*; Khrennikov, A., Ed.; Vaxjo University Press: Smolan, Sweden, 2002; pp. 117–130.
142. Baez, J.C. Division Algebras and Quantum Theory. *Found. Phys.* **2012**, *42*, 819–855. [CrossRef]
143. Stueckelberg, E.C.G. Quantum Theory in Real Hilbert Space. *Helv. Phys. Acta* **1960**, *33*, 727–752.
144. Finkelstein, D.; Jauch, J.M.; Schminovich, S.; Speiser, D. Foundations of Quaternion Quantum Mechanics. *J. Math. Phys.* **1962**, *3*, 207–220. [CrossRef]
145. Müller, M.P. Probabilistic theories and reconstructions of quantum theory. *SciPost Phys. Lect. Notes* **2021**, *28*, 1–41. [CrossRef]
146. D'Ariano, G.M.; Chiribella, G.; Perinotti, P. *Quantum Theory from First Principles. An Informational Approach*; Cambridge University Press: New York, NY, USA, 2017.
147. Cohen, P.J.; Hersh, R. Non-Cantorian Set Theory. *Sci. Am.* **1967**, *217*, 104–116. [CrossRef]
148. Chomsky, N. *Language and Mind*; Cambridge University Press: New York, NY, USA, 2008.

149. Fraenkel, A.A.; Bar-Hillel, Y. *Foundations of Set Theory*; North-Holland Publishing Company: Amsterdam, The Netherlands, 1958.
150. Khrennikov, A. *Contextual Approach to Quantum Formalism*; Fundamental Theories of Physics 160; Springer: Berlin/Heidelberg, Germany, 2009.
151. Maudlin, T. The Nature of the Quantum State. In *The Wave Function. Essays on the Metaphysics of Quantum Mechanics*; Ney, A., Albert, D.Z., Eds.; Oxford University Press: New York, NY, USA, 2013; pp. 126–153.
152. Svozil, K. Quantum hocus-pocus. *Ethics Sci. Environ. Polit.* **2016**, *16*, 25–30. [CrossRef]
153. Ballentine, L.E. Can the Statistical Postulate of Quantum Theory be Derived?—A Critique of the Many-Universes Interpretation. *Found. Phys.* **1973**, *3*, 229–240. [CrossRef]
154. Barrett, J.A. *The Quantum Mechanics of Minds and Worlds*; Oxford University Press: Oxford, UK,1999.
155. Birkhoff, G. *Lattice Theory*; American Mathematical Society: Providence, RI, USA, 1967.
156. Varadarajan, V.S. *Geometry of Quantum Theory*; Springer: Berlin/Heidelberg, Germany, 2007.
157. Zierler, N. Axioms for non-relativistic quantum mechanics. *Pac. J. Math.* **1961**, *11*, 1151–1169. [CrossRef]
158. Spekkens, R.W. Evidence for the epistemic view of quantum states: A toy theory. *Phys. Rev. A* **2007**, *75*, 032110. [CrossRef]
159. Mermin, N.D. In praise of measurement. *Quant. Inform. Process.* **2006**, *5*, 239–260. [CrossRef]
160. Wartofsky, M.W. *Models. Representation and the Scientific Understanding*; D. Reidel Publishing Co.: Dordrecht, The Netherlands, 1979.
161. Zeh, H.D. There are no quantum jumps, nor are there particles! *Phys. Lett. A* **1993**, *172*, 189–195. [CrossRef]
162. Szabó, L. Quantum Structures Do Not Exist in Reality. *Int. J. Theor. Phys.* **1998**, *37*, 449–456. [CrossRef]
163. Danan, A.; Farfurnik, D.; Bar-Ad, S.; Vaidman, L. Asking Photons Where They Have Been. *Phys. Rev. Lett.* **2013**, *111*, 240402. [CrossRef] [PubMed]
164. Feynman, R.P.; Hibbs, A.R. *Quantum Mechanics and Path Integrals*; McGraw-Hill : New York, NY, USA, 1965.
165. Fine, A. Probability and the Interpretation of Quantum Mechanics. *Br. J. Philos. Sci.* **1973**, *24*, 1–37. [CrossRef]
166. Bourbaki, N. *Elements of Mathematics. Algebra I: Chapters 1–3*; Springer: Berlin, Germany, 1974.
167. Clifford, A.H.; Preston, G.B. The algebraic theory of semigroups I. *AMS Math. Surv.* **1961**, *7*, 1967.
168. Mal'cev, A.I. *Algebraic Systems*; Springer: Berlin/Heidelberg, Germany, 1973.
169. Kharin, N.N. *Mathematical Logic and the Set Theory (On a Relationship Between the Abstract and the Concrete)*; RosVuzIzdat: Moscow, Russia, 1963. Available online: https://biblioclub.ru/index.php?page=book_red&id=428668 (accessed on 6 February 2022). (In Russian)
170. Sklar, L. *Physics and Chance*; Cambridge University Press: Cambridge, UK, 1993.
171. Wallace, D. *The Emergent Multiverse. Quantum Theory According to the Everett Interpretation*; Oxford University Press: Oxford, UK, 2012.
172. Köerner, T.W. *Where Do Numbers Come From?*; Cambridge Unversity Press: Cambridge, UK, 2020.
173. Gray, J. *Plato's Ghost: The Modernist Transformation of Mathematics*; Princeton University Press: Princeton, NJ, USA, 2008.
174. Santerre, S. *Psychologie du Nombre et des Opérations Élémentaires de l'Arithmétique*; Octave Doin: Paris, France, 1907.
175. Kurosh, A.G. *Lectures on General Algebra*; Chelsea Publishing Company: New York, NY, USA, 1963.
176. Zorich, V.A. *Mathematical Analysis I*; Springer: Berlin/Heidelberg, Germany, 2004.
177. Russell, B. *Introduction to Mathematical Philosophy*; George Allen & Unwin, Ltd.: Crows Nest, Australia, 1920.
178. Shapiro, S. *Philosophy of Mathematics: Structure and Ontology*; Oxford University Press: Oxford, UK, 1997.
179. Knott, A. The Process of Mathematisation in Mathematical Modelling of Number Patterns in Secondary School Mathematics. Ph.D. Thesis, Stellenbosch University, Stellenbosch, South Africa, 2014.
180. Einstein, A. Remarks on Bertrand Russel's theory of knowledge. In *The Philosophy of Bertrand Russel*; Schilpp, P.A., Ed.; The Library of Living Philosophers V; Northwestern University: Carbondale, IL, USA; Southern Illinois University: Carbondale, IL, USA, 1971; pp. 277–291.
181. Van der Waerden, B.L. *Algebra I, II*; Springer: Berlin/Heidelberg, Germany, 1970.
182. Pontryagin, L. *Topological Groups*; Gordon and Breach: London, UK, 1986.
183. Khrennikov, A.Y. *Non-Archimedean Analysis and Its Applications*; Fizmatlit: Moscow, Russia, 2003. (In Russian)
184. Vladimirov, V.S.; Volovich, I.V.; Zelenov, E.I. *p-adic Analysis and Mathematical Physics*; Fizmatlit: Moscow, Russia, 1994. (In Russian)
185. Rigby, J.F.; Wiegold, J. Independent axioms for vector spaces. *Math. Gazette* **1973**, *57*, 56–62. [CrossRef]
186. Leifer, M.S. Is the Quantum State Real? An Extended Review of ψ-ontology Theorems. *Quanta* **2014**, *3*, 67–155. [CrossRef]
187. Moyal, E. Quantum mechanics as a statistical theory. *Proc. Camb. Philos. Soc.* **1949**, *45*, 99–124. [CrossRef]
188. Hartle, J.B. Quantum mechanics of individual systems. *Am. J. Phys.* **1968**, *36*, 704–712. [CrossRef]
189. Samuel, H.L. *Essays in Physics*; Hartcourt, Brace and Company: San Diego, CA, USA, 1952.
190. Tammaro, E. Why Current Interpretations of Quantum Mechanics are Deficient. *arXiv* **2014**, arXiv:1408.2093.
191. Accardi, L.; Degond, P.; Gorban, A. Hilbert's sixth problem. *Phil. Trans. Royal Soc. A* **2018**, *376*, 20170238.
192. Ivanov, M.G. *How to Comprehend Quantum Mechanics*; R&C Dynamics: Moscow–Izhevsk, Russia, 2015. Available online: https://mipt.ru/students/organization/mezhpr/biblio/q-ivanov.php (accessed on 6 February 2022). (In Russian)
193. Stapp, H.P. *S*-matrix interpretation of quantum theory. *Phys. Rev. D* **1971**, *3*, 1303–1320. [CrossRef]
194. Mackey, G.W. *The Mathematical Foundations of Quantum Mechanics*; W. A. Benjamin, Inc.: San Francisco, CA, USA, 1963.

195. Dmitriev, N.A. Von Neumann's Theorem on the Impossibility of Introducing Hidden Parameters in Quantum Mechanics. *Theor. Math. Phys.* **2005**, *143*, 848–853. [CrossRef]
196. Khrennikov, A. *Probability and Randomness. Quantum versus Classical*; Imperial College Press: London, UK, 2016.
197. Heisenberg, W. The nature of elementary particle. *Phys. Today* **1976**, *29*, 32–39. [CrossRef]
198. Rovelli, C. Physics Needs Philosophy. Philosophy Needs Physics. *Found. Phys.* **2018**, *48*, 481–491. [CrossRef]
199. Schlosshauer, M. Experimental motivation and empirical consistency in minimal no-collapse quantum mechanics. *Ann. Phys.* **2006**, *321*, 112–149. [CrossRef]
200. Wiseman, H.M.; Eisert, J. Nontrivial quantum effects in biology: A skeptical physicist' view. In *Quantum Aspects of Life*; Abbott, D., Davies, P.C.W., Pati, A.K., Eds.; Imperial College Press: London, UK, 2008; pp. 381–401.
201. Szabó, L. Objective probability-like things with and without objective indeterminism. *Stud. Hist. Phil. Mod. Phys.* **2007**, *38*, 628–634. [CrossRef]
202. Alimov Yu, I.; Kravtsov Yu, A. Is probability a 'normal' physical quantity? *Sov. Phys. Uspekhi* **1992**, *35*, 606–622. [CrossRef]

MDPI

Article

Abelianized Structures in Spherically Symmetric Hypersurface Deformations

Martin Bojowald

Institute for Gravitation and the Cosmos, The Pennsylvania State University, 104 Davey Lab, University Park, State College, PA 16802, USA; bojowald@gravity.psu.edu

Abstract: In canonical gravity, general covariance is implemented by hypersurface-deformation symmetries on thephase space. The different versions of hypersurface deformations required for full covariance have complicated interplays with one another, governed by non-Abelian brackets with structure functions. For spherically symmetric space-times, it is possible to identify a certain Abelian substructure within general hypersurface deformations, which suggests a simplified realization as a Lie algebra. The generators of this substructure can be quantized more easily than full hypersurface deformations, but the symmetries they generate do not directly correspond to hypersurface deformations. The availability of consistent quantizations therefore does not guarantee general covariance or a meaningful quantum notion thereof. In addition to placing the Abelian substructure within the full context of spherically symmetric hypersurface deformation, this paper points out several subtleties relevant for attempted applications in quantized space-time structures. In particular, it follows that recent constructions by Gambini, Olmedo, and Pullin in an Abelianized setting fail to address the covariance crisis of loop quantum gravity.

Keywords: canonical gravity; covariance; black holes

Citation: Bojowald, M. Abelianized Structures in Spherically Symmetric Hypersurface Deformations. *Universe* **2022**, *8*, 184. https://doi.org/10.3390/universe8030184

Academic Editor: Steven Duplij

Received: 17 February 2022
Accepted: 11 March 2022
Published: 15 March 2022

1. Introduction

Canonical gravity describes the 4-dimensional, generally covariant structure of space-time by canonical fields defined on the slices of a spatial foliation. The evolution of these fields in time as well as transformations between different foliations are described by the geometrical structure of hypersurface deformations. In a canonical theory, these transformations are generated by certain phase-space functions, the diffeomorphism and Hamiltonian constraints. In spherically symmetric models, which will be considered here, the full set of constraints can be written as $D[M]$ and $H[N]$ with arbitrary spatial functions M (of density weight -1) and N. The constraint equations $D[M]=0$ and $H[N]=0$, valid for any M and N, restrict the phase-space degrees of freedom, given by the spatial metric and its momentum related to extrinsic curvature.

At the same time, the constraints generate (i) time evolution,

$$\mathcal{L}_{t(N,M)}f=\{f,H[N]+D[M]\} \tag{1}$$

for a phase-space function f along a time-evolution vector field $t^a=Nn^a+Ms^a$ in space-time with the unit normal n^a to a spatial slice and the tangent vector field $s^a=(\partial/\partial x)^a$ within the radial manifold (with coordinate x) of a spatial slice, and (ii) gauge transformations

$$\delta_{\xi(\eta,\epsilon)}f=\{f,H[\eta]+D[\epsilon]\} \tag{2}$$

along a space-time vector field

$$\xi^a=\eta n^a+\epsilon s^a \tag{3}$$

where ϵ, like M, has density weight -1.

The reference to normal and tangential directions relative to a foliation implies crucial differences between the mathematical formulation of hypersurface deformations in canonical gravity and the more common formulation of general covariance in terms of space-time tensors. In space-time, vector components ξ^a transform, by definition, in such a way that $\xi^a\partial/\partial x^a$ determines a unique direction independent of coordinate choices. Similarly, the spatial vector $\epsilon s^a = \epsilon\partial/\partial x$ defines a coordinate-independent direction because a scalar of density weight -1 in one dimension transforms like a 1-form dual to $\partial/\partial x$. The normal deformation, however, cannot be introduced in this way because the canonical setting does not provide a time coordinate or the corresponding $\partial/\partial t$. Moreover, even if such a coordinate could be introduced by hand, for instance by using t merely as a parameter as it also appears in Hamilton's equations, it would be impossible to endow η with a density weight -1 in the time direction because, canonically, there is no time manifold. The only alternative is given by the procedure that has been used since [1,2] and formalized in [3]: The normalization of n^a as a unit vector (with respect to the space-time metric, which is available in the canonical setting through the spatial metric on a slice as well as lapse N and shift M) associates a unique normal displacement to any given function η (without density weight).

The normal can be made unit only by reference to the metric, which provides some of the canonical degrees of freedom. The geometrical meaning of normal hypersurface deformations and their commutators depend on the spatial metric, resulting in structure functions in the canonical bracket relations. As a consequence, the canonical symmetries do not form a Lie algebra. This property is responsible for several complications well-known in attempts of canonical quantizations of the theory, starting with [4]. It also makes it harder to develop suitable mathematical structures for transformations generated by the constraints, in particular in an off-shell manner when one does not insist on solving the constraint equations. In [3], for instance, it was shown that a direct composition of transformations generated by the constraints is meaningful in the sense of path independence (a notion introduced in there) only on-shell.

The full structure of transformations is nevertheless required for general covariance to be implemented properly in the solutions of a canonical theory of gravity, in particular one that has been quantized, modified or deformed by new physical effects. While the restricted on-shell behavior may be easier to handle, the off-shell structure is important to make sure that the theory has a well-defined space-time structure, independently of the dynamics. Only in this case can the theory be considered a geometrical effective theory of some deeper and as yet unknown quantum space-time, just as different dynamical versions of gravity given by higher-curvature effective actions make use of the same Riemannian form of space-time. Because of its importance for covariance and the classification of meaningful effective theories, we will review the structure of hypersurface deformations in the beginning of our first section below, combining classic results from gravitational physics with more recent mathematical developments [5,6].

We will focus on aspects of hypersurface deformations of importance for a suggested simplification of the hypersurface-deformation brackets in spherically symmetric models, given by a partial Abelianization [7], but our statements will apply also to a variety of other reformulations that rely on phase-space dependent lapse and shift. Analyzing a partial Abelianization in the context of hypersurface deformations, we will show that this construction captures only a certain subset of these transformations and, upon modification or quantization, does not guarantee that invariance under hypersurface deformations or general covariance are still realized. This conclusion may be surprising because, at first sight, a partial Abelianization appears to implement the same number of symmetry generators as standard hypersurface deformations and uses only a linear redefinition of the generators. However, the coefficients of these linear redefinitions are phase-space dependent, complicating their mathematical description [5,6]. (Heuristically, phase-space dependent linear redefinitions of the generators introduce new structure functions or modify existing ones.) It is then a non-trivial question whether the redefinitions can be inverted. If they cannot be inverted, the redefined theory is not invariant under full hypersurface deformations and its solutions violate general covariance. An additional construction is therefore needed in a partially Abelianized model (or other reformulations of standard hypersurface deformations) in order to recover all space-time transformations. As shown by explicit examples, this is not always possible if the generators have been modified by quantum corrections.

A recent paper [8] claims that it may be possible to realize general covariance in partial Abelianizations of spherically symmetric models with different types of quantum modifications, such as a spatial discretization. The claim is not accompanied by a successful reconstruction of hypersurface deformations and instead relies on a technical and so far incomplete case-by-case study of quantities that should be invariant in a covariant theory. Using our results about general hypersurface deformation structures, we will explain why the covariance claims of [8] cannot hold.

2. Hypersurface Deformations

Space-time vector fields with their standard Lie bracket generate the Lie algebra of diffeomorphisms. Similarly, the transformations generated by the canonical constraints form an algebraic structure. They are labeled by the components η and ϵ of a vector field ξ used in (3) in a basis (n^a, s^a) adapted to a spatial foliation, rather than a coordinate basis. Their commutators

$$\begin{aligned} & \delta_{\xi_2}(\delta_{\xi_1} f) - \delta_{\xi_1}(\delta_{\xi_2} f) \\ = \; & \{\{f, H[\eta_1] + D[\epsilon_1]\}, H[\eta_2] + D[\epsilon_2]\} - \{\{f, H[\eta_2] + D[\epsilon_2]\}, H[\eta_1] + D[\epsilon_1]\} \\ = \; & \{f, \{H[\eta_1] + D[\epsilon_1], H[\eta_2] + D[\epsilon_2]\}\} \end{aligned} \tag{4}$$

are determined by Poisson brackets $\{H[\eta_1] + D[\epsilon_1], H[\eta_2] + D[\epsilon_2]\}$ of the constraints (using the Jacobi identity). Because the unit normal n^a is normalized by using the space-time metric, including the spatial components q_{ab} on a slice, the brackets of two canonical gauge transformations [1,2,9] turn out to depend on the metric. In spherically symmetric models, in which the radial part of the metric is determined by a single function, q (of density weight 2), we have

$$\{H[\eta_1] + D[\epsilon_1], H[\eta_2] + D[\epsilon_2]\} = H[\epsilon_1\eta_2' - \epsilon_2\eta_1'] + D[\epsilon_1\epsilon_2' - \epsilon_2\epsilon_1' + q^{-1}(\eta_1\eta_2' - \eta_2\eta_1')] \,. \tag{5}$$

In general, the metric components are spatial functions independent of the components η and ϵ that label different gauge transformations. Unlike the Lie bracket of two space-time vector fields, the bracket of two pairs δ_{ξ_i}, $i = 1, 2$, implied by the Poisson bracket (5) does not form a Lie algebra because coefficients determined by spatial fields q_{ab} or q cannot be considered structure constants.

2.1. Algebroids

Instead, the brackets have structure functions or, in a suitable mathematical formulation, form the higher algebraic structure of an L_∞-algebroid rather than a Lie algebra [10–12]. An L_∞-algebroid is defined as a vector bundle over a base manifold M with fiber F and bracket relations on bundle sections together with suitable anchor maps that map bundle sections to objects in the tangent bundle of M. A Lie algebroid [13], for instance, has a Lie bracket $[\cdot, \cdot]$ on its sections and an anchor ρ that maps (as a homomorphism) bundle sections to vector fields on the base manifold, such that the Lie bracket of vector fields is compatible with the algebroid bracket. The anchor map also appears in the Leibniz rule

$$[s_1, f s_2] = f[s_1, s_2] + s_2 \mathcal{L}_{\rho(s_1)} f \tag{6}$$

where s_1 and s_2 are sections and f is a function on the base manifold. The anchor brings abstract algebraic relations on bundle sections in correspondence with geometrical transformations as vector fields on the base manifold. While an anchor that maps any section to the zero vector field is always consistent with the Lie-algebroid axioms (in which case the Lie algebroid is a bundle of Lie algebras given by the fibers), non-trivial transformations on the base require a larger image of the anchor. A Lie algebroid with a non-trivial anchor generalizes bundles of Lie algebras. Yet more generally, and in particular in the case of structure functions, the brackets of bundle sections obey the axioms of an L_∞-algebra, a generalized form of a Lie algebra in which the Jacobi identity is not required to hold strictly.

The introduction of the base manifold makes it possible to formalize brackets with structure functions in terms of an L_∞-algebroid. In particular for gravity, the base manifold is (a suitable extension [6]) of the canonical phase space, given by the spatial metrics and momenta related to extrinsic curvature. The fibers are parameterized by the components η and ϵ of a gauge transformation. A section is then an assignment of spatial functions η and ϵ to any metric (or a pair of a metric and its momentum). In this way, the q-dependent structure function in (5) finds a natural home as a bracket of sections over the space of metrics (and momenta).

Constant sections, given by pairs of η and ϵ that are functions on space but do not depend on the phase-space degrees of freedom, have a bracket, implied by (4), that can be realized as a special case of sections of a Lie algebroid [5]. General, non-constant sections of this Lie algebroid have a bracket that may differ from what hypersurface deformations would suggest. Non-constant sections over phase space, discussed in more detail in [6], either violate some of the Lie-algebra relations on sections (in the controlled way of a specific L_∞-structure, as it follows from a BV-BFV extension of general relativity [14,15]) or require a base manifold that extends the phase space of canonical gravity in a way that is not smooth. (The latter can be formulated by using the notion of a Lie-Rinehart algebra [16] in which functions on the base manifold are replaced with a suitable commutative algebra.

Phase-space dependent functions η and ϵ are also important for physics. They are often considered in specific gravitational applications, as in the simple case of cosmological evolution written in conformal time where the lapse function equals the scale factor, a metric component. More importantly for our purposes, the partial Abelianization of [7] relies on an application of phase-space dependent ϵ and η. Hypersurface deformations with such non-constant sections form a Lie algebroid only on-shell [6] when the constraints are solved. The partial Abelianization is therefore able to describe the solution space to all constraints and its covariance transformations, but it is not guaranteed that it correctly captures off-shell transformations which are relevant for general covariance.

Since the standard derivation of the brackets (5) assumes that η and ϵ are not phase-space dependent, the general brackets must be extended by additional terms that, heuristically, result from Poisson brackets of constraints with phase-space dependent η and ϵ. (A complete derivation is based on the BV-BFV analysis of [14,15]). The Poisson bracket of two diffeomorphism constraints, for instance, can still be written in the compact form

$$\{D[\epsilon_1], D[\epsilon_2]\} = D[\epsilon_2\epsilon_1' - \epsilon_1\epsilon_2'] \tag{7}$$

but with an application of the chain rule in the derivatives. Similarly, the mixed Poisson bracket of a Hamiltonian and a diffeomorphism constraint in general form reads

$$\{H[\eta], D[\epsilon]\} = H[-\epsilon\eta'] + D[\eta\mathcal{L}_n\epsilon] \tag{8}$$

where the normal derivative $\mathcal{L}_n$ of a spatial function is defined by the Poisson bracket with the Hamiltonian constraint, $\eta_1\mathcal{L}_n\eta_2 = \{H[\eta_1], \eta_2\}$. For two Hamiltonian constraints, we have the Poisson bracket

$$\{H[\eta_1], H[\eta_2]\} = D[q^{-1}(\eta_1\eta_2' - \eta_2\eta_1')] + H[\eta_1\mathcal{L}_n\eta_2 - \eta_2\mathcal{L}_n\eta_1]. \tag{9}$$

In general, the extra terms implied by phase-space dependent η and ϵ, such as those in $\epsilon' = \partial_x\epsilon + (\partial_x q_i)(\partial_{q_i}\epsilon) + (\partial_x k_i)(\partial_{k_i}\epsilon)$ summing over the two independent components q_i, $i = 1, 2$, of a spherically symmetric spatial metric as well as two components k_i of extrinsic curvature, introduce further structure functions, such as $\partial_x q_i$ and $\partial_x k_i$, that depend on the metric as well as its momenta.

While these Poisson brackets illustrate the additional complications encountered with phase-space dependent ϵ and η, they do not immediately show the algebraic nature of general non-constant sections of hypersurface deformations. In particular, Poisson brackets do not directly mirror relevant L_∞-structures. In our following discussion, we will not need the full algebraic structure and instead perform a comparison of different versions of constant and non-constant sections in gravitational applications.

2.2. Partial Abelianization

As noticed in [7], certain linear combinations of $H[\eta]$ and $D[\epsilon]$ have vanishing Poisson brackets in spherically symmetric models. In order to specify these combinations, we have to refer to explicit variables that determine the spatial metric and its momenta. Following Refs. [17–19], this is conveniently done in triad variables (E^x, E^φ) such that the spatial metric is given by the line element

$$ds^2 = \frac{(E^\varphi)^2}{E^x}dx^2 + E^x(d\vartheta^2 + \sin^2\vartheta d\varphi^2) \tag{10}$$

in standard spherical coordinates. (For our purposes, it is sufficient to assume $E^x > 0$, fixing the orientation of the triad.) The triad components are canonically conjugate (up to constant factors) to components of extrinsic curvature, (K_x, K_φ), such that

$$\{K_x(x), E^x(y)\} = 2G\delta(x,y) \quad , \quad \{K_\varphi(x), E^\varphi(y)\} = G\delta(x,y) \tag{11}$$

with Newton's constant G. (We keep a factor of two in the first relation. As implicitly done in [7,8], this factor can easily be eliminated by a rescaling of K_x. Since this procedure would not affect the main equations and conclusions shown below, we do not make use of this rescaling and instead keep the original components of extrinsic curvature).

The delta functions disappear in Poisson brackets of integrated (smeared) expressions, resulting in well-defined brackets. In particular, the diffeomorphism constraint

$$D[M] = \frac{1}{G}\int dx M(x)\left(-\frac{1}{2}(E^x)'K_x + K_\varphi' E^\varphi\right), \tag{12}$$

and Hamiltonian constraint

$$H[N] = \frac{-1}{2G}\int dx N(x)\left(|E^x|^{-1/2}E^\varphi K_\varphi^2 + 2|E^x|^{1/2}K_\varphi K_x + |E^x|^{-1/2}(1-\Gamma_\varphi^2)E^\varphi + 2\Gamma_\varphi'|E^x|^{1/2}\right) \tag{13}$$

where $\Gamma_\varphi = -(E^x)'/(2E^\varphi)$ have Poisson brackets

$$\{D[M_1], D[M_2]\} = D[M_1M_2'] \tag{14}$$
$$\{H[N], D[M]\} = -H[MN'] \tag{15}$$
$$\{H[N_1], H[N_2]\} = D[E^x(E^\varphi)^{-2}(N_1N_2' - N_2N_1')] \tag{16}$$

(for spatial functions M_i and N_i, $i = 1, 2$, that do not depend on the phase-space variables) of the correct form for hypersurface deformations in spherically symmetric space-times.

Simple algebra and integration by parts shows that the linear combinations

$$C[L] = H[(E^x)'(E^\varphi)^{-1}\textstyle\int E^\varphi L dx] - 2D[K_\varphi\sqrt{E^x}(E^\varphi)^{-1}\textstyle\int E^\varphi L dx], \tag{17}$$

where $\int E^\varphi L dx$ is understood as a function of x obtained by integrating $E^\varphi L$ from a fixed starting point up to x, have zero Poisson brackets with one another for different L:

$$\{C[L_1], C[L_2]\} = 0 \tag{18}$$

for all functions L_1 and L_2 on a spatial slice. To see this, it is sufficient to notice that the combination eliminates any dependence on K_x and on spatial derivatives of E^φ. The antisymmetric nature of the Poisson bracket then implies that it must vanish. Explicitly, the new combination of constraints takes the form

$$C[L] = -\frac{1}{G}\int \mathrm{d}x L(x) E^{\varphi}\left(\sqrt{|E^x|}\left(1+K_{\varphi}^2-\Gamma_{\varphi}^2\right)+\text{const.}\right). \tag{19}$$

A free constant appears because a constant $\int E^{\varphi} L \mathrm{d}x$ implies a non-vanishing lapse function in (17), and therefore a non-trivial constraint, but corresponds to a vanishing $E^{\varphi}L$ in (19). The new constraint $C[L]$ therefore constrains one degree of freedom less than the original $H[N]$. The free constant in (19) can be determined through boundary conditions, which would also restrict the lapse functions allowed in gauge transformations.

At first sight, it seems that the partial Abelianization eliminates structure functions from the brackets and may simplify quantization and the preservation of symmetries and therefore covariance. However, the importance of metric-dependent structure functions in the standard brackets, which make sure that deformations are defined with respect to a unit normal that is in fact normalized, raises the question of whether an elimination of these structure functions and their metric dependence by redefined generators can still capture the full picture of general covariance. To answer this question, it is instructive to place the partial Abelianization of the brackets in the context of the hypersurface-deformation structure. Several features of the full mathematical construction are then relevant.

First, the integration of $E^{\varphi}L$ required to define $C[L]$ as a combination of $H[N]$ and $D[M]$ may seem unusual, but while this means that the relevant N and M are non-local in space, they are local within both the fiber (spatial functions N and M) and the base (the gravitational phase space with independent functions E^x, E^{φ}, K_x and K_{φ} or a suitable extension) that may be used to construct a corresponding L_{∞}-algebroid. The combination (17) therefore defines an admissible set of sections.

Secondly, while the section defined by (17) makes use of phase-space dependent N and M in the Hamiltonian and diffeomorphism constraints, which are therefore not constant over the base manifold, an Abelian bracket (18) is obtained only for functions L_1 and L_2 that do not have the full phase-space dependence allowed for general sections. In particular, if L_1 or L_2 are allowed to depend on $(E^{\varphi})'$ or K_x, the bracket $\{C[L_1], C[L_2]\}$ no longer vanishes, and it can then have structure functions. Partial Abelianization is therefore obtained for a restricted class of sections, defined such that L does not depend on $(E^{\varphi})'$ and K_x (while it may still have an unrestricted spatial dependence). If L does not depend on $(E^{\varphi})'$ and K_x but on the other independent phase-space variables, K_{φ} as well as E^x or on E^{φ} but not its derivatives, the bracket $\{C[L_1], C[L_2]\}$ remains zero, but there are then structure functions in the bracket of $C[L]$ with the diffeomorphism constraint, analogously to (8). Therefore, structure functions are eliminated from the brackets only for a restricted class of sections. This observation raises the question whether full covariance can still be realized.

A restriction to constant sections over the base manifold is not unusual, for certain purposes. A similar assumption is made in the standard form (14)–(16) of hypersurface-deformation brackets, in which case the original N and M are often assumed to be constant over the base (while their spatial dependence remains unrestricted). There is, however, a crucial difference between assuming constant N and M over the base and assuming constant L over the base: In the former case, allowing for non-constant sections produces additional terms in the brackets, shown in (7)–(9) , that follow directly from an application of the product rule of Poisson brackets. The partial Abelianization, however, relies on cancellations between different structure functions in the original brackets that are no longer realized once non-constant sections with phase-space dependent L are allowed.

In particular, allowing for phase-space dependent L and M in the $(D[M], C[L])$ system makes the transformation from (N, M) to (M, L) invertible. It is then possible to write the original $H[N]$ as a combination of $D[M]$ and $C[L]$ in the partial Abelianization, regaining the full non-Abelian brackets with metric-dependent structure functions. Restricting the system to phase-space independent L, by contrast, implies that the transformation from the original hypersurface-deformation structure to the brackets of $D[M]$ and $C[L]$ is not invertible. It is then unclear whether hypersurface deformations and general covariance can be recovered from a partial Abelianization, in particular if the latter has been modified by quantum corrections.

2.3. Modified Deformations

It has been known for some time [20–22] that spherically symmetric hypersurface deformations can be modified consistently, maintaining closed brackets while modifying the structure functions. The dependence on K_φ in (13) can be generalized to

$$H[N] = \frac{-1}{2G}\int \mathrm{d}x N(x)\left(|E^x|^{-1/2}E^\varphi f_1(K_\varphi) + 2|E^x|^{1/2}f_2(K_\varphi)K_x + |E^x|^{-1/2}(1-\Gamma_\varphi^2)E^\varphi + 2\Gamma'_\varphi|E^x|^{1/2}\right) \quad (20)$$

where f_1 and f_2 are functions of K_φ related by

$$f_2(K_\varphi) = \frac{1}{2}\frac{\mathrm{d}f_1(K_\varphi)}{\mathrm{d}K_\varphi}. \quad (21)$$

If this equation is satisfied, the bracket of two Hamiltonian constraints is still closed,

$$\{H[N_1], H[N_2]\} = D[\beta(K_\varphi)E^x(E^\varphi)^{-2}(N_1N_2' - N_2N_1')] \quad (22)$$

for phase-space independent N_1 and N_2. In this bracket, $D[M]$ is the unmodified diffeomorphism constraint, but the structure function is multiplied by a new factor of

$$\beta(K_\varphi) = \frac{\mathrm{d}f_2(K_\varphi)}{\mathrm{d}K_\varphi} = \frac{1}{2}\frac{\mathrm{d}^2 f_1(K_\varphi)}{\mathrm{d}K_\varphi^2}. \quad (23)$$

Additional terms in the bracket for non-constant sections follow immediately from the product rule for Poisson brackets.

Similarly, the Abelianized constraint $C[L]$ can be generalized in its dependence on K_φ, using the same function f_1 as before:

$$C[L] = -\frac{1}{G}\int \mathrm{d}x L(x)E^\varphi\left(\sqrt{|E^x|}\left(1 + f_1(K_\varphi) - \Gamma_\varphi^2\right) + \text{const.}\right). \quad (24)$$

Its brackets remain Abelian for phase-space independent L. There is no obvious term in $C[L]$ where the second function f_2 might appear or the important consistency condition (21). It therefore seems easier to modify (or quantize) the constraint $C[L]$ compared with $H[N]$. However, for full hypersurface deformations and covariance to be realized in the modified setting, we still have to make sure that the transformation from (N, M) to (L, M) can be inverted. As shown in [23], this is possible only if we also modify the transformation (17) to

$$C[L] = H[(E^x)'(E^\varphi)^{-1}\textstyle\int E^\varphi L\mathrm{d}x] - 2D[f_2(K_\varphi)\sqrt{E^x}(E^\varphi)^{-1}\textstyle\int E^\varphi L\mathrm{d}x] \quad (25)$$

where f_2 obeys the same consistency condition with f_1, (21), as derived from the modified Hamiltonian constraint. The partial Abelianization and the original form of hypersurface deformations therefore imply equivalent results, provided one makes sure that the transformation of sections can be inverted. Only then can access to full hypersurface deformations and covariance be realized.

3. Non-Covariant Modifications of Abelianized Brackets

A recent paper [8] by Gambini, Olmedo and Pullin (GOP) argues that general covariance can be realized in modified versions of spherically symmetric models, for which a partial Abelianization of the brackets plays a crucial role: As the abstract claims, "We show explicitly that the resulting space-times, obtained from Dirac observables of the quantum theory, are covariant in the usual sense of the way—they preserve the quantum line element—for any gauge that is stationary (in the exterior, if there is a horizon). The construction depends crucially on the details of the Abelianized quantization considered, the satisfaction of the quantum constraints and the recovery of standard general relativity in the classical limit and suggests that more informal polymerization constructions of possible semi-classical approximations to the theory can indeed have covariance problems."

These claims raise several questions. For instance, how can the construction depend "crucially on the details of the Abelianized quantization considered" if a partial Abelianization is either completely equivalent to the non-Abelian orignal version of hypersurface deformations (if the transformation is made sure to be invertible) or gives access to only a subset of hypersurface deformations (if the transformation is not invertible owing to a restriction to a subset of sections)?

A closer inspection of technical calculations performed by GOP shows that spherically symmetric hypersurface deformations are, in fact, violated in the construction. GOP use two different kinds of modifications, a generalized dependence of $C[L]$ on K_φ of the form (24), and a spatial discretization of phase-space functions and their derivatives. Because the authors use a certain combination of solutions to the constraints and gauge-fixing conditions, it turns out that only the latter modification survives in the final expressions for line elements that are supposed to be invariant.

However, also the former (a generalized dependence on K_φ) is relevant because, as we have seen, the correct form of a modification must appear in two different places, in the constraint $C[L]$ and in the transformation back to unrestricted hypersurface deformations. These two appearances are clear but somewhat implicit in [8]: The modified $C[L]$ is implied by the modified solutions in Equation (14) in [8] (or, equivalently, (21) there, referring to the preprint version) where $f_1(K_\varphi) = \sin^2(\rho K_\varphi)/\rho^2$ with a spatial function ρ. The modified transformation back to unrestricted hypersurface deformations is implied by Equation (20) in [8] which in our notation amounts to replacing K_φ in (17) with $\sqrt{f_1(K_\varphi)}$. Using the same function $f_1(K_\varphi)$ is crucial for the constructions in [8] because the partial gauge fixing employed there replaces $\sqrt{f_1(K_\varphi)}$ with a fixed function on space (rather than phase space). The same gauge-fixing function is then used in both places, in the constraint $C[L]$ or its solutions and in the transformation back to unrestricted hypersurface deformations from which a line element can be constructed. However, this construction, which is equivalent to assuming $f_2(K_\varphi) = \sqrt{f_1(K_\varphi)}$ in (25), violates the condition (21) required for unrestricted hypersurface deformations to follow for the modified constraint. (For the specific $f_1(K_\varphi)$ considered by GOP, f_2 should have an additional cosine factor, or equivalently have a doubled argument of the sine function.) The constructions of [8] therefore violate hypersurface deformations.

How can GOP then claim to have performed crucial steps toward demonstrating general covariance in this setting? Unfortunately, much of the constructions are obscured by an application of incompletely defined mixtures of gauge fixings and idiosyncratic notions of observables. Here, it suffices to highlight only a few of the shortcomings found in the GOP analysis. (For more details, see [24].) Continuing with the replacement of $\sqrt{f_1(K_\varphi)}$ by a gauge-fixing function that depends only on space, GOP replace any appearance of $\sqrt{f_1(K_\varphi)}$ with gauge-fixing functions (on space) derived from the classical solutions for K_φ in two specific slicings. Implicitly, the authors simply remove the modification in this way because they indirectly equate $\sqrt{f_1(K_\varphi)}$ with K_φ, mediated by the gauge-fixing function. As a result, they do not test how non-classical $f_1(K_\varphi)$ can be consistent with covariance. It is also problematic that this step in a rather careless gauge-fixing procedure replaces a phase-space function K_φ that does not Poisson commute with the constraints with a spatial function that does obey this commutation property. The procedure turns a K_φ-dependent expression for E^φ, obtained by solving $C[L] = 0$, into a function that Poisson commutes with $C[L]$. GOP then call the result a Dirac observable, even though E^φ is not gauge invariant.

After replacing K_φ with a spatial function, the resulting expression for E^φ still does not Poisson commute with the diffeomorphism constraint and is therefore not a Dirac observable, even if K_φ could meaningfully be replaced. The same expression for E^φ also depends on E^x, which is not a spatial invariant. Indeed, unlike $C[L]$, the diffeomorphism constraint (12) depends on K_x and therefore does not Poisson commute with E^x. GOP arrive at their conclusion about E^φ being a Dirac observable by misidentifying E^x as a Dirac observable because the (loop) quantization procedure they use establishes a correspondence between an operator $\hat{E}^x$ and labels of a spherically symmetric spin network state [17,25] that are unchanged by the spatial shifts of a finite diffeomorphism. However, having a correspondence between a classical object, E^x, that is not a Dirac observable and a quantum operator, $\hat{E}^x$, that is a Dirac observable may indicate that the theory fails to have the correct classical limit. Since this way of imposing the diffeomorphism constraint is directly inherited from more general constructions in the full theory of loop quantum gravity [26,27], the issues revealed by our analysis of [8] might hint at deeper problems within the kinematics of loop quantum gravity.

4. Conclusions

Our discussion of phase-space dependent coefficients in hypersurface deformations has clarified a previously puzzling issue of partial Abelianizations in spherically symmetric models: Is it possible for partial Abelianizations to simplify the construction of quantum modifications of hypersurface deformation generators and, at the same time, retain full access to all transformations required for general covariance? We have shown that the answer is negative. A simplified construction of modified generators is based on the absence of structure functions in partially Abelianized brackets obtained for a specific choice of phase-space dependent gauge generators (lapse and shift functions). However, the partial Abelianization is maintained only if the new generators are then restricted to be phase-space independent. This condition renders the transformation from hypersurface-deformation brackets to partially Abelian brackets non-invertible. Access to unrestricted hypersurface deformations and general covariance is therefore lost in a partially Abelianized setting. Consistent modifications of the partially Abelian brackets then do not necessarily imply consistent realizations of general covariance.

A recent paper [8] by Gambini, Olmedo and Pullin has implicitly recognized this shortcoming and instead proposed to test general covariance in a tedious case-by-case study of presumed invariants, beginning with a discretized version of the line element. We have pointed out a specific place (the choice of modification functions f_1 and f_2) where hypersurface deformations are treated inconsistently in these constructions, which may perhaps lead to improved versions of the transformations considered by GOP. However, correcting this inconsistency requires an analysis of unrestricted hypersurface deformations even in the partially Abelianized setting, making sure that the transformation between these two versions of the brackets can be inverted. It is therefore impossible to analyze covariance in isolation from general hypersurface deformations, as proposed by GOP. No-go results [28] for covariance in models of loop quantum gravity, partially based on various analyses of modified hypersurface deformations, therefore cannot be evaded by the constructions of GOP.

Funding: This research was funded by NSF grant number PHY-1912168.

Institutional Review Board Statement: Not applicable.

Informed Consent Statement: Not applicable.

Acknowledgments: The author thanks Michele Schiavina for discussions and Rodolfo Gambini, Javier Olmedo and Jorge Pullin for sharing a draft of [8].

Conflicts of Interest: The author declares no conflict of interest.

References

1. Katz, J. Les crochets de Poisson des contraintes du champ gravitationne. *Comptes Rendus Acad. Sci. Paris* **1962**, *254*, 1386–1387.
2. Arnowitt, R.; Deser, S.; Misner, C.W. The Dynamics of General Relativity. In *Gravitation: An Introduction to Current Research*; Witten, L., Ed.; Wiley: New York, NY, USA, 1962. Reprinted in *Gen. Rel. Grav.* **2008**, *40*, 1997–2027.
3. Hojman, S.A.; Kuchař, K.; Teitelboim, C. Geometrodynamics Regained. *Ann. Phys.* **1976**, *96*, 88–135. [CrossRef]
4. Komar, A. Consistent Factor Ordering Of General Relativistic Constraints. *Phys. Rev. D* **1979**, *20*, 830–833. [CrossRef]
5. Blohmann, C.; Barbosa Fernandes, M.C.; Weinstein, A. Groupoid symmetry and constraints in general relativity. 1: kinematics. *Commun. Contemp. Math.* **2013**, *15*, 1250061. [CrossRef]
6. Blohmann, C.; Schiavina, M.; Weinstein, A. A Lie-Rinehart algebra in general relativity. *arXiv* **2022**, arXiv:2201.02883.
7. Gambini, R.; Pullin, J. Loop quantization of the Schwarzschild black hole. *Phys. Rev. Lett.* **2013**, *110*, 211301. [CrossRef]
8. Gambini, R.; Olmedo, J.; Pullin, J. Towards a quantum notion of covariance in spherically symmetric loop quantum gravity. *Phys. Rev. D* **2022**, *105*, 026017. [CrossRef]
9. Dirac, P.A.M. The theory of gravitation in Hamiltonian form. *Proc. Roy. Soc. A* **1958**, *246*, 333–343.
10. Stasheff, J. Constrained Poisson algebras and strong homotopy representations. *Bull. Am. Math. Soc.* **1988**, *19*, 287–290. [CrossRef]
11. Stasheff, J. Homological Reduction of Constrained Poisson Algebras. *J. Diff. Geom.* **1997**, *45*, 221–240. [CrossRef]
12. Schätz, F. Invariance of the BFV-complex. *Pac. J. Math.* **2010**, *248*, 453–474. [CrossRef]
13. Pradines, J. Théorie de Lie pour les groupoïdes différentiables. Calcul différenetiel dans la catégorie des groupoïdes infinitésimaux. *Comptes Rendus Acad. Sci. Paris Sér. A* **1967**, *264*, A245–A248.
14. Schiavina, M. BV-BFV Approach to General Relativity. Ph.D. Thesis, Eidgenössische Technische Hochschule Zürich, Zurich, Switzerland, 2015.
15. Cattaneo, A.S.; Schiavina, M. BV-BFV approach to General Relativity: Einstein-Hilbert action. *J. Math. Phys.* **2016**, *57*, 023515. [CrossRef]
16. Rinehart, G.S. Differential forms on general commutative algebras. *Trans. Amer. Math. Soc.* **1963**, *108*, 195–222. [CrossRef]
17. Bojowald, M. Spherically Symmetric Quantum Geometry: States and Basic Operators. *Class. Quantum Grav.* **2004**, *21*, 3733–3753. [CrossRef]
18. Bojowald, M.; Swiderski, R. The Volume Operator in Spherically Symmetric Quantum Geometry. *Class. Quantum Grav.* **2004**, *21*, 4881–4900. [CrossRef]
19. Bojowald, M.; Swiderski, R. Spherically Symmetric Quantum Geometry: Hamiltonian Constraint. *Class. Quantum Grav.* **2006**, *23*, 2129–2154. [CrossRef]
20. Reyes, J.D. Spherically Symmetric Loop Quantum Gravity: Connections to 2-Dimensional Models and Applications to Gravitational Collapse. Ph.D. Thesis, The Pennsylvania State University, University Park, PA, USA, 2009.
21. Bojowald, M.; Reyes, J.D.; Tibrewala, R. Non-marginal LTB-like models with inverse triad corrections from loop quantum gravity. *Phys. Rev. D* **2009**, *80*, 084002. [CrossRef]
22. Bojowald, M.; Paily, G.M.; Reyes, J.D. Discreteness corrections and higher spatial derivatives in effective canonical quantum gravity. *Phys. Rev. D* **2014**, *90*, 025025. [CrossRef]
23. Bojowald, M.; Brahma, S.; Reyes, J.D. Covariance in models of loop quantum gravity: Spherical symmetry. *Phys. Rev. D* **2015**, *92*, 045043. [CrossRef]
24. Bojowald, M. Comment on "Towards a quantum notion of covariance in spherically symmetric loop quantum gravity". *arXiv* **2022**, arXiv:2203.06049.
25. Bojowald, M.; Kastrup, H.A. Symmetry Reduction for Quantized Diffeomorphism Invariant Theories of Connections. *Class. Quantum Grav.* **2000**, *17*, 3009–3043. [CrossRef]
26. Rovelli, C.; Smolin, L. Loop Space Representation of Quantum General Relativity. *Nucl. Phys. B* **1990**, *331*, 80–152. [CrossRef]
27. Ashtekar, A.; Lewandowski, J.; Marolf, D.; Mourão, J.; Thiemann, T. Quantization of Diffeomorphism Invariant Theories of Connections with Local Degrees of Freedom. *J. Math. Phys.* **1995**, *36*, 6456–6493. [CrossRef]
28. Bojowald, M. Black-hole models in loop quantum gravity. *Universe* **2020**, *6*, 125. [CrossRef]

 MDPI

Article

Polyadic Analogs of Direct Product

Steven Duplij

Center for Information Technology (WWU IT), Universität Münster, Röntgenstrasse 7-13, 48149 Münster, Germany; douplii@uni-muenster.de

Abstract: We propose a generalization of the external direct product concept to polyadic algebraic structures which introduces novel properties in two ways: the arity of the product can differ from that of the constituents, and the elements from different multipliers can be "entangled" such that the product is no longer componentwise. The main property which we want to preserve is associativity, which is gained by using the associativity quiver technique, which was provided previously. For polyadic semigroups and groups we introduce two external products: (1) the iterated direct product, which is componentwise but can have an arity that is different from the multipliers and (2) the hetero product (power), which is noncomponentwise and constructed by analogy with the heteromorphism concept introduced earlier. We show in which cases the product of polyadic groups can itself be a polyadic group. In the same way, the external product of polyadic rings and fields is generalized. The most exotic case is the external product of polyadic fields, which can be a polyadic field (as opposed to the binary fields), in which all multipliers are zeroless fields. Many illustrative concrete examples are presented.

Keywords: direct product; direct power; polyadic semigroup; arity; polyadic ring; polyadic field

MSC: 16T25; 17A42; 20B30; 20F36; 20M17; 20N15

1. Introduction

The concept of a direct product plays a crucial role in algebraic structures in the study of their internal constitution and their representation in terms of better known/simpler structures (see, e.g., [1,2]). For instance, in elementary particle physics, the decomposition of a gauge symmetry group of the model to the direct product gives its particle content [3,4]. Furthermore, the concept of semisimplicity in representation theory is totally based on the direct product (see, e.g., [5,6]).

The general method of the construction of the external direct product is to take the Cartesian product of the underlying sets and endow it with the operations from the algebraic structures under consideration. Usually this is an identical repetition of the initial multipliers' operations componentwise [7]. In the case of polyadic algebraic structures, their arity comes into the game, such that endowing the product with operations becomes nontrivial in two aspects: the arities of all structures can be different (but "quantized" and not unique) and the elements from different multipliers can be "entangled" meaning that the product is not componentwise. The direct (componentwise) product of n-ary groups was considered in [8,9]. We propose two corresponding polyadic analogs (changing arity and "entangling") of the external direct product which preserve its associativity, and therefore allow us to analyze polyadic semigroups, groups, rings and fields.

From a mathematical viewpoint, the direct product is also important, especially because it plays the role of a product in a corresponding category (see, e.g., [10,11]). For instance, the class of all polyadic groups for objects and polyadic group homomorphisms for morphisms form a category which is well-defined, because it has the polyadic direct product [12,13] as a product.

Here we also consider polyadic rings and fields in the same way. Since there exist zeroless polyadic fields [14], the well-known statement (see, e.g., [2]) of the absence of

Citation: Duplij, S. Polyadic Analogs of Direct Product. *Universe* **2022**, *8*, 230. https://doi.org/10.3390/universe8040230

Academic Editor: Andreas Fring

Received: 25 January 2022
Accepted: 7 April 2022
Published: 8 April 2022

binary fields that are a direct product of fields does not hold in the polyadic case. We construct polyadic fields which are products of zeroless fields, which can lead to a new category (which does not exist for binary fields): the category of polyadic fields.

The proposed constructions are accompanied by concrete illustrative examples.

2. Preliminaries

In this section we briefly introduce the usual notation; for details see [15]. For a non-empty (underlying) set G the *n-tuple* (or *polyad* [16]) of elements is denoted by $(g_1, \ldots, g_n)$, $g_i \in G$, $i = 1, \ldots, n$, and the Cartesian product is denoted by $G^{\times n} \equiv \overbrace{G \times \ldots \times G}^{n}$ and consists of all such n-tuples. For all elements equal to $g \in G$, we denote n-tuple (polyad) by a power (g^n). To avoid unneeded indices we denote with one bold letter $(\mathbf{g})$ a polyad for which the number of elements in the n-tuple is clear from the context, and sometimes we will write $\left(\mathbf{g}^{(n)}\right)$. On the Cartesian product $G^{\times n}$ we define a polyadic (or n-ary) operation $\mu^{(n)} : G^{\times n} \to G$ such that $\mu^{(n)}[\mathbf{g}] \mapsto h$, where $h \in G$. The operations with $n = 1, 2, 3$ are called *unary, binary and ternary*.

Recall the definitions of some algebraic structures and their special elements (in the notation of [15]). A (one-set) *polyadic algebraic structure* $\mathcal{G}$ is a set that is G-closedwith respect to polyadic operations. In the case of one n-ary operation $\mu^{(n)} : G^{\times n} \to G$, it is called *polyadic multiplication* (or *n-ary multiplication*). A one-set *n-ary algebraic structure* $\mathcal{M}^{(n)} = \left\langle G \mid \mu^{(n)} \right\rangle$ or *polyadic magma (n-ary magma)* is a set that is G-closed with respect to one n-ary operation $\mu^{(n)}$ and without any other additional structure. In the binary case $\mathcal{M}^{(2)}$ was also called a groupoid by Hausmann and Ore [17] (and [18]). Since the term "groupoid" was widely used in category theory for a different construction, the so-called Brandt groupoid [19,20], Bourbaki [21] later introduced the term "magma".

Denote the number of iterating multiplications by ℓ_μ, and call the resulting composition an *iterated product* $\left(\mu^{(n)}\right)^{\circ \ell_\mu}$, such that

$$\mu'^{(n')} = \left(\mu^{(n)}\right)^{\circ \ell_\mu} \stackrel{def}{=} \overbrace{\mu^{(n)} \circ \left(\mu^{(n)} \circ \ldots \left(\mu^{(n)} \times \mathrm{id}^{\times(n-1)}\right) \ldots \times \mathrm{id}^{\times(n-1)}\right)}^{\ell_\mu}, \tag{1}$$

where the arities are connected by

$$n' = n_{iter} = \ell_\mu (n-1) + 1, \tag{2}$$

which gives the length of an iterated polyad $(\mathbf{g})$ in our notation $\left(\mu^{(n)}\right)^{\circ \ell_\mu}[\mathbf{g}]$.

A *polyadic zero* of a polyadic algebraic structure $\mathcal{G}^{(n)}\left\langle G \mid \mu^{(n)} \right\rangle$ is a distinguished element $z \in G$ (and the corresponding 0-ary operation $\mu_z^{(0)}$) such that for any $(n-1)$-tuple (polyad) $\mathbf{g}^{(n-1)} \in G^{\times(n-1)}$ we have

$$\mu^{(n)}\left[\mathbf{g}^{(n-1)}, z\right] = z, \tag{3}$$

where z can be in any place on the l.h.s. of (3). If its place is not fixed it can be a single zero. As in the binary case, an analog of positive powers of an element [16] should coincide with the number of multiplications ℓ_μ in the iteration (1).

A (positive) *polyadic power* of an element is

$$g^{\langle \ell_\mu \rangle} = \left(\mu^{(n)}\right)^{\circ \ell_\mu}\left[g^{\ell_\mu (n-1)+1}\right]. \tag{4}$$

We define associativity as the invariance of the composition of two n-ary multiplications. An element of a polyadic algebraic structure g is called ℓ_μ-*nilpotent* (or simply *nilpotent* for $\ell_\mu = 1$), if there exist ℓ_μ such that

$$g^{\langle \ell_\mu \rangle} = z. \tag{5}$$

A *polyadic (n-ary) identity* (or neutral element) of a polyadic algebraic structure is a distinguished element e (and the corresponding 0-ary operation $\mu_e^{(0)}$) such that for any element $g \in G$ we have

$$\mu^{(n)}\left[g, e^{n-1}\right] = g, \tag{6}$$

where g can be in any place on the l.h.s. of (6).

In polyadic algebraic structures, there exist *neutral polyads* $\mathbf{n} \in G^{\times(n-1)}$ satisfying

$$\mu^{(n)}[g, \mathbf{n}] = g, \tag{7}$$

where g can be in any of n places on the l.h.s. of (7). Obviously, the sequence of polyadic identities e^{n-1} is a neutral polyad (6).

A one-set polyadic algebraic structure $\left\langle G \mid \mu^{(n)} \right\rangle$ is called *totally associative* if

$$\left(\mu^{(n)}\right)^{\circ 2}[\mathbf{g}, \mathbf{h}, \mathbf{u}] = \mu^{(n)}\left[\mathbf{g}, \mu^{(n)}[\mathbf{h}], \mathbf{u}\right] = invariant, \tag{8}$$

with respect to the placement of the internal multiplication $\mu^{(n)}[\mathbf{h}]$ on the r.h.s. on any of n places, with a fixed order of elements in the any fixed polyad of $(2n-1)$ elements $\mathbf{t}^{(2n-1)} = (\mathbf{g}, \mathbf{h}, \mathbf{u}) \in G^{\times(2n-1)}$.

A *polyadic semigroup* $\mathcal{S}^{(n)}$ is a one-set S one-operation $\mu^{(n)}$ algebraic structure in which the n-ary multiplication is associative, $\mathcal{S}^{(n)} = \left\langle S \mid \mu^{(n)} \mid \text{associativity (8)} \right\rangle$. A polyadic algebraic structure $\mathcal{G}^{(n)} = \left\langle G \mid \mu^{(n)} \right\rangle$ is σ-*commutative*, if $\mu^{(n)} = \mu^{(n)} \circ \sigma$, or

$$\mu^{(n)}[\mathbf{g}] = \mu^{(n)}[\sigma \circ \mathbf{g}], \quad \mathbf{g} \in G^{\times n}, \tag{9}$$

where $\sigma \circ \mathbf{g} = \left(g_{\sigma(1)}, \ldots, g_{\sigma(n)}\right)$ is a permutated polyad and σ is a fixed element of S_n, the permutation group on n elements. If (9) holds for all $\sigma \in S_n$, then a polyadic algebraic structure is *commutative*. A special type of the σ-commutativity

$$\mu^{(n)}\left[g, \mathbf{t}^{(n-2)}, h\right] = \mu^{(n)}\left[h, \mathbf{t}^{(n-2)}, g\right], \tag{10}$$

where $\mathbf{t}^{(n-2)} \in G^{\times(n-2)}$ is any fixed $(n-2)$-polyad, is referred to as *semicommutativity*. If an n-ary semigroup $\mathcal{S}^{(n)}$ is iterated from a commutative binary semigroup with identity, then $\mathcal{S}^{(n)}$ is semicommutative. A polyadic algebraic structure is called (uniquely) *i-solvable*, if for all polyads $\mathbf{t}$, $\mathbf{u}$ and element h, one can (uniquely) resolve the equation (with respect to h) for the fundamental operation

$$\mu^{(n)}[\mathbf{u}, h, \mathbf{t}] = g \tag{11}$$

where h can be on any place, and $\mathbf{u}, \mathbf{t}$ are polyads of the needed length.

A polyadic algebraic structure which is uniquely i-solvable for all places $i = 1, \ldots, n$ is called a *n-ary* (or *polyadic*) *quasigroup* $\mathcal{Q}^{(n)} = \left\langle Q \mid \mu^{(n)} \mid \text{solvability} \right\rangle$. An associative polyadic quasigroup is called an *n-ary* (or *polyadic*) *group*. In an n-ary group $\mathcal{G}^{(n)} = \left\langle G \mid \mu^{(n)} \right\rangle$ the only solution of (11) is called a *querelement* of g and is denoted by $\bar{g}$ [22], such that

$$\mu^{(n)}[\mathbf{h}, \bar{g}] = g, \quad g, \bar{g} \in G, \tag{12}$$

where $\bar{g}$ can be on any place. Any idempotent g coincides with its querelement $\bar{g} = g$. The unique solvability relation (12) in an n-ary group can be treated as a definition of the unary (multiplicative) *queroperation*

$$\bar{\mu}^{(1)}[g] = \bar{g}. \tag{13}$$

We observe from (12) and (7) that the polyad

$$\boldsymbol{n}_g = \left(g^{n-2}\bar{g}\right) \tag{14}$$

is neutral for any element of a polyadic group, where $\bar{g}$ can be on any place. If this i-th place is important, then we write $\boldsymbol{n}_{g;i}$. In a polyadic group the *Dörnte relations* [22]

$$\mu^{(n)}[g, \boldsymbol{n}_{h;i}] = \mu^{(n)}\left[\boldsymbol{n}_{h;j}, g\right] = g \tag{15}$$

hold true for any allowable i, j. In the case of a binary group, the relations (15) become $g \cdot h \cdot h^{-1} = h \cdot h^{-1} \cdot g = g$.

Using the queroperation (13) one can give a *diagrammatic definition* of a polyadic group [23]: an *n-ary group* is a one-set algebraic structure (universal algebra)

$$\mathcal{G}^{(n)} = \left\langle G \mid \mu^{(n)}, \bar{\mu}^{(1)} \mid \text{associativity (8), Dörnte relations (15)} \right\rangle, \tag{16}$$

where $\mu^{(n)}$ is an n-ary associative multiplication and $\bar{\mu}^{(1)}$ is the queroperation (13).

3. Polyadic Products of Semigroups and Groups

We start from the standard external direct product construction for semigroups. Then we show that consistent "polyadization" of the semigroup direct product, which preserves associativity, can lead to additional properties:

(1) The arities of the polyadic direct product and power can differ from that of the initial semigroups.

(2) The components of the polyadic power can contain elements from different multipliers.

We use here a vector-like notation for clarity and convenience in passing to higher arity generalizations. Begin from the direct product of two (binary) semigroups $\mathcal{G}_{1,2} \equiv \mathcal{G}_{1,2}^{(2)} = \left\langle G_{1,2} \mid \mu_{1,2}^{(2)} \equiv (\cdot_{1,2}) \mid assoc \right\rangle$, where $G_{1,2}$ are underlying sets, whereas $\mu_{1,2}^{(2)}$ are multiplications in $\mathcal{G}_{1,2}$. On the Cartesian product of the underlying sets $G' = G_1 \times G_2$ we define a *direct product* $\mathcal{G}_1 \times \mathcal{G}_2 = \mathcal{G}' = \left\langle G' \mid \boldsymbol{\mu}'^{(2)} \equiv (\bullet') \right\rangle$ of the semigroups $\mathcal{G}_{1,2}$ via the componentwise multiplication of the doubles $\boldsymbol{G} = \begin{pmatrix} g_1 \\ g_2 \end{pmatrix} \in G_1 \times G_2$ (being the Kronecker product of doubles in our notation) , as

$$\boldsymbol{G}^{(1)} \bullet' \boldsymbol{G}^{(2)} = \begin{pmatrix} g_1 \\ g_2 \end{pmatrix}^{(1)} \bullet' \begin{pmatrix} g_1 \\ g_2 \end{pmatrix}^{(2)} = \begin{pmatrix} g_1^{(1)} \cdot_1 g_1^{(2)} \\ g_2^{(1)} \cdot_2 g_2^{(2)} \end{pmatrix}, \tag{17}$$

and in the "polyadic" notation

$$\boldsymbol{\mu}'^{(2)}\left[\boldsymbol{G}^{(1)}, \boldsymbol{G}^{(2)}\right] = \begin{pmatrix} \mu_1^{(2)}\left[g_1^{(1)}, g_1^{(2)}\right] \\ \mu_2^{(2)}\left[g_2^{(1)}, g_2^{(2)}\right] \end{pmatrix}. \tag{18}$$

Obviously, the associativity of $\boldsymbol{\mu}'^{(2)}$ follows immediately from that of $\mu_{1,2}^{(2)}$, because of the componentwise multiplication in (18). If $\mathcal{G}_{1,2}$ are groups with the identities $e_{1,2} \in G_{1,2}$, then the identity of the direct product is the double $\boldsymbol{E} = \begin{pmatrix} e_1 \\ e_2 \end{pmatrix}$, such that $\boldsymbol{\mu}'^{(2)}[\boldsymbol{E}, \boldsymbol{G}] = \boldsymbol{\mu}'^{(2)}[\boldsymbol{G}, \boldsymbol{E}] = \boldsymbol{G} \in \mathcal{G}$.

3.1. Full Polyadic External Product

The "polyadization" of (18) is straightforward

Definition 1. *An n'-ary full direct product semigroup $\mathcal{G}'^{(n')} = \mathcal{G}_1^{(n)} \times \mathcal{G}_2^{(n)}$ consists of (two or k) n-ary semigroups (of the same arity $n' = n$)*

$$\boldsymbol{\mu}'^{(n)}\left[\mathbf{G}^{(1)}, \mathbf{G}^{(2)}, \dots, \mathbf{G}^{(n)}\right] = \begin{pmatrix} \mu_1^{(n)}\left[g_1^{(1)}, g_1^{(2)}, \dots, g_1^{(n)}\right] \\ \mu_2^{(n)}\left[g_2^{(1)}, g_2^{(2)}, \dots, g_2^{(n)}\right] \end{pmatrix}, \tag{19}$$

where the (total) polyadic associativity (8) of $\mu'^{(n')}$ is governed by those of the constituent semigroups $\mathcal{G}_1^{(n)}$ and $\mathcal{G}_2^{(n)}$ (or $\mathcal{G}_1^{(n)} \dots \mathcal{G}_k^{(n)}$) and the componentwise construction (19).

If $\mathcal{G}_{1,2}^{(n)} = \left\langle G_{1,2} \mid \mu_{1,2}^{(n)}, \bar{\mu}_{1,2}^{(1)} \right\rangle$ are n-ary groups (where $\bar{\mu}_{1,2}^{(1)}$ are the unary multiplicative queroperations (13)), then the queroperation $\bar{\boldsymbol{\mu}}'^{(1)}$ of the full direct product group $\mathcal{G}'^{(n')} = \left\langle G' \equiv G_1 \times G_2 \mid \boldsymbol{\mu}'^{(n')}, \bar{\boldsymbol{\mu}}'^{(1)} \right\rangle$ $(n' = n)$ is defined componentwise as follows:

$$\bar{\mathbf{G}} \equiv \bar{\boldsymbol{\mu}}'^{(1)}[\mathbf{G}] = \begin{pmatrix} \bar{\mu}_1^{(1)}[g_1] \\ \bar{\mu}_2^{(1)}[g_2] \end{pmatrix}, \quad \text{or} \quad \bar{\mathbf{G}} = \begin{pmatrix} \bar{g}_1 \\ \bar{g}_2 \end{pmatrix}, \tag{20}$$

which satisfies $\mu'^{(n)}[\mathbf{G}, \mathbf{G}, \dots, \bar{\mathbf{G}}] = \mathbf{G}$ with $\bar{\mathbf{G}}$ on any place (cf. (12)).

Definition 2. *A full polyadic direct product $\mathcal{G}'^{(n)} = \mathcal{G}_1^{(n)} \times \mathcal{G}_2^{(n)}$ is called derived if its constituents $\mathcal{G}_1^{(n)}$ and $\mathcal{G}_2^{(n)}$ are derived, such that the operations $\mu_{1,2}^{(n)}$ are compositions of the binary operations $\mu_{1,2}^{(2)}$, correspondingly.*

In the derived case, all the operations in (19) have the form (see (1) and (2))

$$\mu_{1,2}^{(n)} = \left(\mu_{1,2}^{(2)}\right)^{\circ(n-1)}, \quad \boldsymbol{\mu}^{(n)} = \left(\boldsymbol{\mu}^{(2)}\right)^{\circ(n-1)}. \tag{21}$$

The operations of the derived polyadic semigroup can be written as (cf., the binary direct product (17) and (18))

$$\boldsymbol{\mu}'^{(n)}\left[\mathbf{G}^{(1)}, \mathbf{G}^{(2)}, \dots, \mathbf{G}^{(n)}\right] = \mathbf{G}^{(1)} \bullet' \mathbf{G}^{(2)} \bullet' \dots \bullet' \mathbf{G}^{(n)} = \begin{pmatrix} g_1^{(1)} \cdot_1 g_1^{(2)} \cdot_1 \dots \cdot_1 g_1^{(n)} \\ g_2^{(1)} \cdot_2 g_2^{(2)} \cdot_2 \dots \cdot_2 g_2^{(n)} \end{pmatrix}. \tag{22}$$

We will be more interested in nonderived polyadic analogs of the direct product.

Example 1. *Let us have two ternary groups: the unitless nonderived group $\mathcal{G}_1^{(3)} = \left\langle \mathrm{i}\mathbb{R} \mid \mu_1^{(3)} \right\rangle$, where $\mathrm{i}^2 = -1$, $\mu_1^{(3)}\left[g_1^{(1)}, g_1^{(2)}, g_1^{(3)}\right] = g_1^{(1)} g_1^{(2)} g_1^{(3)}$ is a triple product in $\mathbb{C}$, the querelement is $\bar{\mu}_1^{(1)}[g_1] = 1/g_1$, and $\mathcal{G}_2^{(3)} = \left\langle \mathbb{R} \mid \mu_2^{(3)} \right\rangle$ with $\mu_2^{(3)}\left[g_2^{(1)}, g_2^{(2)}, g_2^{(3)}\right] = g_2^{(1)} \left(g_2^{(2)}\right)^{-1} g_2^{(3)}$, the querelement $\bar{\mu}_2^{(1)}[g_2] = g_2$. Then, the ternary nonderived full direct product group becomes $\mathcal{G}'^{(3)} = \left\langle \mathrm{i}\mathbb{R} \times \mathbb{R} \mid \boldsymbol{\mu}'^{(3)}, \bar{\boldsymbol{\mu}}'^{(1)} \right\rangle$, where*

$$\boldsymbol{\mu}'^{(3)}\left[\mathbf{G}^{(1)}, \mathbf{G}^{(2)}, \mathbf{G}^{(3)}\right] = \begin{pmatrix} g_1^{(1)} g_1^{(2)} g_1^{(3)} \\ g_2^{(1)} \left(g_2^{(2)}\right)^{-1} g_2^{(3)} \end{pmatrix}, \quad \bar{\mathbf{G}} \equiv \bar{\boldsymbol{\mu}}'^{(1)}[\mathbf{G}] = \begin{pmatrix} 1/g_1 \\ g_2 \end{pmatrix}, \tag{23}$$

which contains no identity, because $\mathcal{G}_1^{(3)}$ is unitless and nonderived.

3.2. Mixed-Arity Iterated Product

In the polyadic case, the following question arises, which cannot even be stated in the binary case: is it possible to build a version of the associative direct product such that it can be nonderived and have different arity than the constituent semigroup arities? The answer is yes, which leads to two arity-changing constructions: componentwise and noncomponentwise.

(1) *Iterated direct product* (⊛). In each of the constituent polyadic semigroups we use the iterating (1) componentwise, but with different numbers of compositions, because the same number of compositions evidently leads to the iterated polyadic direct product. In this case the arity of the direct product is greater than or equal to the arities of the constituents $n' \geq n_1, n_2$.

(2) *Hetero product* (⊠). The polyadic product of k copies of the same n-ary semigroup is constructed using the associativity quiver technique, which mixes ("entangles") elements from different multipliers, it is noncomponentwise (by analogy with heteromorphisms in [15]), and so it can be called a *hetero product* or *hetero power* (for coinciding multipliers, i.e., constituent polyadic semigroups or groups). This gives the arity of the hetero product which is less than or equal to the arities of the equal multipliers $n' \leq n$.

In the first componentwise case **1)**, the constituent multiplications (19) are composed from the lower-arity ones in the componentwise manner, but the initial arities of up and down components can be different (as opposed to the binary derived case (21))

$$\mu_1^{(n)} = \left(\mu_1^{(n_1)}\right)^{\circ \ell_{\mu 1}}, \quad \mu_2^{(n)} = \left(\mu_2^{(n_2)}\right)^{\circ \ell_{\mu 2}}, \quad 3 \leq n_{1,2} \leq n-1, \tag{24}$$

where we exclude the limits: the derived case $n_{1,2} = 2$ (21) and the undecomposed case $n_{1,2} = n$ (19). Since the total size of the up and down polyads is the same and coincides with the arity of the double $\mathbf{G}$ multiplication n', using (2) we obtain the *arity compatibility* relations

$$n' = \ell_{\mu 1}(n_1 - 1) + 1 = \ell_{\mu 2}(n_2 - 1) + 1. \tag{25}$$

Definition 3. *A mixed-arity polyadic iterated direct product semigroup* $\mathcal{G}'^{(n')} = \mathcal{G}_1^{(n_1)} \circledast \mathcal{G}_2^{(n_2)}$ *consists of (two) polyadic semigroups* $\mathcal{G}_1^{(n_1)}$ *and* $\mathcal{G}_2^{(n_2)}$ *of the different arity shapes* n_1 *and* n_2

$$\boldsymbol{\mu}'^{(n')}\left[\mathbf{G}^{(1)}, \mathbf{G}^{(2)}, \dots, \mathbf{G}^{(n')}\right] = \begin{pmatrix} \left(\mu_1^{(n_1)}\right)^{\circ \ell_{\mu 1}}\left[g_1^{(1)}, g_1^{(2)}, \dots, g_1^{(n)}\right] \\ \left(\mu_2^{(n_2)}\right)^{\circ \ell_{\mu 2}}\left[g_2^{(1)}, g_2^{(2)}, \dots, g_2^{(n)}\right] \end{pmatrix}, \tag{26}$$

and the arity compatibility relations (25) hold.

Observe that it is not the case that any two polyadic semigroups can be composed in the mixed-arity polyadic direct product.

Assertion 1. *If the arity shapes of two polyadic semigroups* $\mathcal{G}_1^{(n_1)}$ *and* $\mathcal{G}_2^{(n_2)}$ *satisfy the compatibility condition*

$$a(n_1 - 1) = b(n_2 - 1) = c, \qquad a, b, c \in \mathbb{N}, \tag{27}$$

then they can form a mixed-arity direct product $\mathcal{G}'^{(n')} = \mathcal{G}_1^{(n_1)} \circledast \mathcal{G}_2^{(n_2)}$, *where* $n' = c + 1$ *(25).*

Example 2. *In the case of 4-ary and 5-ary semigroups $\mathcal{G}_1^{(4)}$ and $\mathcal{G}_2^{(5)}$ the direct product arity of $\mathcal{G}'^{(n')}$ is "quantized" $n' = 3\ell_{\mu 1} + 1 = 4\ell_{\mu 2} + 1$, such that*

$$n' = 12k + 1 = 13, 25, 37, \ldots, \tag{28}$$

$$\ell_{\mu 1} = 4k = 4, 8, 12, \ldots, \tag{29}$$

$$\ell_{\mu 2} = 3k = 3, 6, 9, \ldots, \quad k \in \mathbb{N}, \tag{30}$$

and only the first mixed-arity 13-ary direct product semigroup $\mathcal{G}'^{(13)}$ is nonderived. If $\mathcal{G}_1^{(4)}$ and $\mathcal{G}_2^{(5)}$ are polyadic groups with the queroperations $\bar{\mu}_1^{(1)}$ and $\bar{\mu}_2^{(1)}$ correspondingly, then the iterated direct $\mathcal{G}'^{(n')}$ is a polyadic group with the queroperation $\bar{\boldsymbol{\mu}}'^{(1)}$ given in (20).

In the same way one can consider the iterated direct product of any number of polyadic semigroups.

3.3. Polyadic Hetero Product

In the second noncomponentwise case **2)** we allow multiplying elements from different components, and therefore we should consider the Cartesian k-power of sets $G' = G^{\times k}$ and endow the corresponding k-tuple with a polyadic operation in such a way that the associativity of $\mathcal{G}^{(n)}$ will govern the associativity of the product $\mathcal{G}'^{(n)}$. In other words we construct a k-power of the polyadic semigroup $\mathcal{G}^{(n)}$ such that the result $\mathcal{G}'^{(n')}$ is an n'-ary semigroup.

The general structure of the hetero product formally coincides "reversely" with the main heteromorphism equation [15]. The additional parameter which determines the arity n' of the hetero power of the initial n-ary semigroup is the number of intact elements ℓ_{id}. Thus, we arrive at

Definition 4. *The hetero ("entangled") k-power of the n-ary semigroup $\mathcal{G}^{(n)} = \left\langle G \mid \mu^{(n)} \right\rangle$ is the n'-ary semigroup defined on the k-th Cartesian power $G' = G^{\times k}$, such that $\mathcal{G}'^{(n')} = \left\langle G' \mid \boldsymbol{\mu}'^{(n')} \right\rangle$,*

$$\mathcal{G}'^{(n')} = \left(\mathcal{G}^{(n)}\right)^{\boxtimes k} \equiv \overbrace{\mathcal{G}^{(n)} \boxtimes \ldots \boxtimes \mathcal{G}^{(n)}}^{k}, \tag{31}$$

and the n'-ary multiplication of k-tuples $\mathbf{G}^T = (g_1, g_2, \ldots, g_k) \in G^{\times k}$ is given (informally) by

$$\boldsymbol{\mu}'^{(n')}\left[\begin{pmatrix} g_1 \\ \vdots \\ g_k \end{pmatrix}, \ldots, \begin{pmatrix} g_{k(n'-1)} \\ \vdots \\ g_{kn'} \end{pmatrix}\right] = \begin{pmatrix} \left.\begin{matrix} \mu^{(n)}[g_1, \ldots, g_n], \\ \vdots \\ \mu^{(n)}\left[g_{n(\ell_\mu - 1)}, \ldots, g_{n\ell_\mu}\right] \end{matrix}\right\} \ell_\mu \\ \left.\begin{matrix} g_{n\ell_\mu + 1}, \\ \vdots \\ g_{n\ell_\mu + \ell_{\text{id}}} \end{matrix}\right\} \ell_{\text{id}} \end{pmatrix}, \quad g_i \in G, \tag{32}$$

where ℓ_{id} is the number of intact elements on the r.h.s., and $\ell_\mu = k - \ell_{\text{id}}$ is the number of multiplications in the resulting k-tuple of the direct product. The hetero power parameters are connected by the arity-changing formula [15]

$$n' = n - \frac{n-1}{k}\ell_{\text{id}}, \tag{33}$$

with the integer $\dfrac{n-1}{k}\ell_{\text{id}} \geq 1$.

The concrete placement of elements and multiplications in (32) to obtain the associative $\mu'^{(n')}$ is governed by the associativity quiver technique [15].

There exist important general numerical relations between the parameters of the twisted direct power $n', n, k, \ell_{\mathrm{id}}$, which follow from (32) and (33). First, there are non-strict inequalities for them

$$0 \leq \ell_{\mathrm{id}} \leq k-1, \tag{34}$$
$$\ell_\mu \leq k \leq (n-1)\ell_\mu, \tag{35}$$
$$2 \leq n' \leq n. \tag{36}$$

Second, the initial and final arities n and n' are not arbitrary, but "quantized" such that the fraction in (33) has to be an integer (see Table 1).

Table 1. Hetero power "quantization".

k	ℓ_μ	ℓ_{id}	n/n'
2	1	1	$n = 3, 5, 7, \ldots$ $n' = 2, 3, 4, \ldots$
3	1	2	$n = 4, 7, 10, \ldots$ $n' = 2, 3, 4, \ldots$
3	2	1	$n = 4, 7, 10, \ldots$ $n' = 3, 5, 7, \ldots$
4	1	3	$n = 5, 9, 13, \ldots$ $n' = 2, 3, 4, \ldots$
4	2	2	$n = 3, 5, 7, \ldots$ $n' = 2, 3, 4, \ldots$
4	3	1	$n = 5, 9, 13, \ldots$ $n' = 4, 7, 10, \ldots$

Assertion 2. *The hetero power is not unique in both directions, if we do not fix the initial n and final n' arities of $\mathcal{G}^{(n)}$ and $\mathcal{G}'^{(n')}$.*

Proof. This follows from (32) and the hetero power "quantization" shown in Table 1. □

The classification of the hetero powers consists of two limiting cases.

(1) *Intactless power*: there are no intact elements $\ell_{\mathrm{id}} = 0$. The arity of the hetero power reaches its maximum and coincides with the arity of the initial semigroup $n' = n$ (see *Example* 5).

(2) *Binary power*: the final semigroup is of lowest arity, i.e., binary $n' = 2$. The number of intact elements is (see *Example* 4)

$$\ell_{\mathrm{id}} = k\frac{n-2}{n-1}. \tag{37}$$

Example 3. *Consider the cubic power of a 4-ary semigroup $\mathcal{G}'^{(3)} = \left(\mathcal{G}^{(4)}\right)^{\boxtimes 3}$ with the identity e, then the ternary identity triple in $\mathcal{G}'^{(3)}$ is $\mathbf{E}^T = (e, e, e)$, and therefore this cubic power is a ternary semigroup with identity.*

Proposition 1. *If the initial n-ary semigroup $\mathcal{G}^{(n)}$ contains an identity, then the hetero power $\mathcal{G}'^{(n')} = \left(\mathcal{G}^{(n)}\right)^{\boxtimes k}$ can contain an identity in the intactless case and the Post-like quiver [15]. For the binary power $k = 2$ only the one-sided identity is possible.*

Let us consider some concrete examples.

Example 4. *Let* $\mathcal{G}^{(3)} = \left\langle G \mid \mu^{(3)} \right\rangle$ *be a ternary semigroup, then we can construct its power* $k = 2$ *(square) of the doubles* $\mathbf{G} = \begin{pmatrix} g_1 \\ g_2 \end{pmatrix} \in G \times G = G'$ *in two ways to obtain the associative hetero power*

$$\boldsymbol{\mu}'^{(2)}\left[\mathbf{G}^{(1)}, \mathbf{G}^{(2)}\right] = \begin{cases} \begin{pmatrix} \mu^{(3)}\left[g_1^{(1)}, g_2^{(1)}, g_1^{(2)}\right] \\ g_2^{(2)} \end{pmatrix}, \\ \begin{pmatrix} \mu^{(3)}\left[g_1^{(1)}, g_2^{(2)}, g_1^{(2)}\right] \\ g_2^{(1)} \end{pmatrix}, \end{cases} \quad g_i^{(j)} \in G. \tag{38}$$

This means that the Cartesian square can be endowed with the associative multiplication $\boldsymbol{\mu}'^{(2)}$*, and therefore* $\mathcal{G}'^{(2)} = \left\langle G' \mid \boldsymbol{\mu}'^{(2)} \right\rangle$ *is a binary semigroup, being the hetero product* $\mathcal{G}'^{(2)} = \mathcal{G}^{(3)} \boxtimes \mathcal{G}^{(3)}$*. If* $\mathcal{G}^{(3)}$ *has a ternary identity* $e \in G$*, then* $\mathcal{G}'^{(2)}$ *has only the left (right) identity* $\mathbf{E} = \begin{pmatrix} e \\ e \end{pmatrix} \in G'$*, since* $\boldsymbol{\mu}'^{(2)}[\mathbf{E}, \mathbf{G}] = \mathbf{G}$ $(\boldsymbol{\mu}'^{(2)}[\mathbf{G}, \mathbf{E}] = \mathbf{G})$*, but not the right (left) identity. Thus,* $\mathcal{G}'^{(2)}$ *can be a semigroup only, even if* $\mathcal{G}^{(3)}$ *is a ternary group.*

Example 5. *Take* $\mathcal{G}^{(3)} = \left\langle G \mid \mu^{(3)} \right\rangle$ *a ternary semigroup, then the multiplication on the double* $\mathbf{G} = \begin{pmatrix} g_1 \\ g_2 \end{pmatrix} \in G \times G = G'$ *is ternary and noncomponentwise*

$$\boldsymbol{\mu}'^{(3)}\left[\mathbf{G}^{(1)}, \mathbf{G}^{(2)}, \mathbf{G}^{(3)}\right] = \begin{pmatrix} \mu^{(3)}\left[g_1^{(1)}, g_2^{(2)}, g_1^{(3)}\right] \\ \mu^{(3)}\left[g_2^{(1)}, g_1^{(2)}, g_2^{(3)}\right] \end{pmatrix}, \quad g_i^{(j)} \in G, \tag{39}$$

and $\mu'^{(3)}$ *is associative (and described by the Post-like associative quiver [15]), and therefore the cubic hetero power is the ternary semigroup* $\mathcal{G}'^{(3)} = \left\langle G \times G \mid \boldsymbol{\mu}'^{(3)} \right\rangle$*, such that* $\mathcal{G}'^{(3)} = \mathcal{G}^{(3)} \boxtimes \mathcal{G}^{(3)}$*. In this case, as opposed to the previous example, the existence of a ternary identity in* $\mathcal{G}^{(3)}$ *implies the ternary identity in the direct cube* $\mathcal{G}'^{(3)}$ *by* $\mathbf{E} = \begin{pmatrix} e \\ e \end{pmatrix}$*. If* $\mathcal{G}^{(3)}$ *is a ternary group with the unary queroperation* $\bar{\mu}^{(1)}$*, then the cubic hetero power* $\mathcal{G}'^{(3)}$ *is also a ternary group of the special class [24]: all querelements coincide (cf., (20)), such that* $\bar{\mathbf{G}}^T = (g_{quer}, g_{quer})$*, where* $\bar{\mu}^{(1)}[g] = g_{quer}$*,* $\forall g \in G$*. This is because in (12) the querelement can be foundon any place.*

Theorem 1. *If* $\mathcal{G}^{(n)}$ *is an n-ary group, then the hetero k-power* $\mathcal{G}'^{(n')} = \left(\mathcal{G}^{(n)}\right)^{\boxtimes k}$ *can contain queroperations in the intactless case only.*

Corollary 1. *If the power multiplication (32) contains no intact elements* $\ell_{\text{id}} = 0$ *and does not change arity* $n' = n$*, a hetero power can be a polyadic group which has only one querelement.*

Next we consider more complicated hetero power ("entangled") constructions with and without intact elements, as well as Post-like and non-Post associative quivers [15].

Example 6. *Let* $\mathcal{G}^{(4)} = \left\langle G \mid \mu^{(4)} \right\rangle$ *be a 4-ary semigroup, then we can construct its 4-ary associative cubic hetero power* $\mathcal{G}'^{(4)} = \left\langle G' \mid \boldsymbol{\mu}'^{(4)} \right\rangle$ *using the Post-like and non-Post-associative quivers without intact elements. Taking in (32)* $n' = n$*,* $k = 3$*,* $\ell_{\text{id}} = 0$*, we obtain two possibilities for the multiplication of the triples* $\mathbf{G}^T = (g_1, g_2, g_3) \in G \times G \times G = G'$

(1) Post-like associative quiver. The multiplication of the hetero cubic power case takes the form

$$\boldsymbol{\mu}'^{(4)}\left[\mathbf{G}^{(1)},\mathbf{G}^{(2)},\mathbf{G}^{(3)},\mathbf{G}^{(4)}\right]=\begin{pmatrix}\mu^{(4)}\left[g_1^{(1)},g_2^{(2)},g_3^{(3)},g_1^{(4)}\right]\\ \mu^{(4)}\left[g_2^{(1)},g_3^{(2)},g_1^{(3)},g_2^{(4)}\right]\\ \mu^{(4)}\left[g_3^{(1)},g_1^{(2)},g_2^{(3)},g_3^{(4)}\right]\end{pmatrix},\quad g_i^{(j)}\in G, \tag{40}$$

and it can be shown that $\boldsymbol{\mu}'^{(4)}$ is totally associative; therefore, $\mathcal{G}'^{(4)}=\left\langle G'\mid\boldsymbol{\mu}'^{(4)}\right\rangle$ is a 4 ary semigroup.

(2) Non-Post associative quiver. The multiplication of the hetero cubic power differs from (40)

$$\boldsymbol{\mu}'^{(4)}\left[\mathbf{G}^{(1)},\mathbf{G}^{(2)},\mathbf{G}^{(3)},\mathbf{G}^{(4)}\right]=\begin{pmatrix}\mu^{(4)}\left[g_1^{(1)},g_3^{(2)},g_2^{(3)},g_1^{(4)}\right]\\ \mu^{(4)}\left[g_2^{(1)},g_1^{(2)},g_3^{(3)},g_2^{(4)}\right]\\ \mu^{(4)}\left[g_3^{(1)},g_2^{(2)},g_1^{(3)},g_3^{(4)}\right]\end{pmatrix},\quad g_i^{(j)}\in G, \tag{41}$$

and it can be shown that $\boldsymbol{\mu}'^{(4)}$ is totally associative; therefore, $\mathcal{G}'^{(4)}=\left\langle G'\mid\boldsymbol{\mu}'^{(4)}\right\rangle$ is a 4-ary semigroup.

The following is valid for both the above cases. If $\mathcal{G}^{(4)}$ has the 4-ary identity satisfying

$$\mu^{(4)}[e,e,e,g]=\mu^{(4)}[e,e,g,e]=\mu^{(4)}[e,g,e,e]=\mu^{(4)}[g,e,e,e]=g,\quad\forall g\in G, \tag{42}$$

then the hetero power $\mathcal{G}'^{(4)}$ has the 4-ary identity

$$\mathbf{E}=\begin{pmatrix}e\\ e\\ e\end{pmatrix},\quad e\in G. \tag{43}$$

In the case where $\mathcal{G}^{(3)}$ is a ternary group with the unary queroperation $\bar{\mu}^{(1)}$, then the cubic hetero power $\mathcal{G}'^{(4)}$ is also a ternary group with one querelement (cf., Example 5)

$$\bar{\mathbf{G}}=\overline{\begin{pmatrix}g_1\\ g_2\\ g_3\end{pmatrix}}=\begin{pmatrix}g_{quer}\\ g_{quer}\\ g_{quer}\end{pmatrix},\quad g_{quer}\in G,\;\; g_i\in G, \tag{44}$$

where $g_{quer}=\bar{\mu}^{(1)}[g]$, $\forall g\in G$.

A more nontrivial example is a cubic hetero power which has different arity to the initial semigroup.

Example 7. *Let $\mathcal{G}^{(4)}=\left\langle G\mid\mu^{(4)}\right\rangle$ be a 4-ary semigroup, then we can construct its ternary associative cubic hetero power $\mathcal{G}'^{(3)}=\left\langle G'\mid\boldsymbol{\mu}'^{(3)}\right\rangle$ using the associative quivers with one intact element and two multiplications [15]. Taking in (32) the parameters $n'=3$, $n=4$, $k=3$, $\ell_{id}=1$ (see third line of Table 1), we obtain for the ternary multiplication $\boldsymbol{\mu}'^{(3)}$ for the triples $\mathbf{G}^T=(g_1,g_2,g_3)\in G\times G\times G=G'$ of the hetero cubic power case the form*

$$\boldsymbol{\mu}'^{(3)}\left[\mathbf{G}^{(1)},\mathbf{G}^{(2)},\mathbf{G}^{(3)}\right]=\begin{pmatrix}\mu^{(4)}\left[g_1^{(1)},g_2^{(2)},g_3^{(3)},g_1^{(3)}\right]\\ \mu^{(4)}\left[g_2^{(1)},g_3^{(2)},g_1^{(2)},g_2^{(3)}\right]\\ g_3^{(1)}\end{pmatrix},\quad g_i^{(j)}\in G, \tag{45}$$

which is totally associative, and therefore the hetero cubic power of 4-ary semigroup $\mathcal{G}^{(4)} = \left\langle G \mid \mu^{(4)} \right\rangle$ is a ternary semigroup $\mathcal{G}'^{(3)} = \left\langle G' \mid \boldsymbol{\mu}'^{(3)} \right\rangle$, such that $\mathcal{G}'^{(3)} = \left(\mathcal{G}^{(4)}\right)^{\boxtimes 3}$. If the initial 4-ary semigroup $\mathcal{G}^{(4)}$ has the identity satisfying (42), then the ternary hetero power $\mathcal{G}'^{(3)}$ has only the right ternary identity (43) satisfying one relation

$$\boldsymbol{\mu}'^{(3)}[\mathbf{G}, \mathbf{E}, \mathbf{E}] = \mathbf{G}, \quad \forall \mathbf{G} \in G^{\times 3}, \tag{46}$$

and therefore $\mathcal{G}'^{(3)}$ is a ternary semigroup with a right identity. If $\mathcal{G}^{(4)}$ is a 4-ary group with the queroperation $\bar{\mu}^{(1)}$, then the hetero power $\mathcal{G}'^{(3)}$ can only be a ternary semigroup , because in $\left\langle G' \mid \boldsymbol{\mu}'^{(3)} \right\rangle$ we cannot define the standard queroperation [16].

4. Polyadic Products of Rings and Fields

Now we show that the thorough "polyadization" of operations can lead to some unexpected new properties of ring and field external direct products. Recall that in the binary case the external direct product of fields does not exist at all (see, e.g., [2]). The main new peculiarities of the polyadic case are:

(1) The arity shape of the external product ring and its constituent rings can be different.
(2) The external product of polyadic fields can be a polyadic field.

4.1. External Direct Product of Binary Rings

First, we recall the ordinary (binary) direct product of rings in notation which would be convenient to generalize to higher-arity structures [14]. Let us have two binary rings $\mathcal{R}_{1,2} \equiv \mathcal{R}_{1,2}^{(2,2)} = \left\langle R_{1,2} \mid \nu_{1,2}^{(2)} \equiv (+_{1,2}), \mu_{1,2}^{(2)} \equiv (\cdot_{1,2}) \right\rangle$, where $R_{1,2}$ are underlying sets, whereas $\nu_{1,2}^{(2)}$ and $\mu_{1,2}^{(2)}$ are additions and multiplications (satisfying distributivity) in $\mathcal{R}_{1,2}$, correspondingly. On the Cartesian product of the underlying sets $R' = R_1 \times R_2$ one defines the *external direct product ring* $\mathcal{R}_1 \times \mathcal{R}_2 = \mathcal{R}' = \left\langle R' \mid \boldsymbol{\nu}'^{(2)} \equiv (+'), \boldsymbol{\mu}'^{(2)} \equiv (\bullet') \right\rangle$ by the componentwise operations (addition and multiplication) on the doubles $\mathbf{X} = \begin{pmatrix} x_1 \\ x_2 \end{pmatrix} \in R_1 \times R_2$ as follows:

$$\mathbf{X}^{(1)} +' \mathbf{X}^{(2)} = \begin{pmatrix} x_1 \\ x_2 \end{pmatrix}^{(1)} +' \begin{pmatrix} x_1 \\ x_2 \end{pmatrix}^{(2)} \equiv \begin{pmatrix} x_1^{(1)} \\ x_2^{(1)} \end{pmatrix} +' \begin{pmatrix} x_1^{(2)} \\ x_2^{(2)} \end{pmatrix} = \begin{pmatrix} x_1^{(1)} +_1 x_1^{(2)} \\ x_2^{(1)} +_2 x_2^{(2)} \end{pmatrix}, \tag{47}$$

$$\mathbf{X}^{(1)} \bullet' \mathbf{X}^{(2)} = \begin{pmatrix} x_1 \\ x_2 \end{pmatrix}^{(1)} \bullet' \begin{pmatrix} x_1 \\ x_2 \end{pmatrix}^{(2)} = \begin{pmatrix} x_1^{(1)} \cdot_1 x_1^{(2)} \\ x_2^{(1)} \cdot_2 x_2^{(2)} \end{pmatrix}, \tag{48}$$

or in the polyadic notation (with manifest operations)

$$\boldsymbol{\nu}'^{(2)}\left[\mathbf{X}^{(1)}, \mathbf{X}^{(2)}\right] = \begin{pmatrix} \nu_1^{(2)}\left[x_1^{(1)}, x_1^{(2)}\right] \\ \nu_2^{(2)}\left[x_2^{(1)}, x_2^{(2)}\right] \end{pmatrix}, \tag{49}$$

$$\boldsymbol{\mu}'^{(2)}\left[\mathbf{X}^{(1)}, \mathbf{X}^{(2)}\right] = \begin{pmatrix} \mu_1^{(2)}\left[x_1^{(1)}, x_1^{(2)}\right] \\ \mu_2^{(2)}\left[x_2^{(1)}, x_2^{(2)}\right] \end{pmatrix}. \tag{50}$$

The associativity and distributivity of the binary direct product operations $\boldsymbol{\nu}'^{(2)}$ and $\boldsymbol{\mu}'^{(2)}$ are obviously governed by those of the constituent binary rings $\mathcal{R}_1$ and $\mathcal{R}_2$, because of the componentwise construction on the r.h.s. of (49) and (50). In the polyadic case, the construction of the direct product is not so straightforward and can have additional unusual peculiarities.

4.2. Polyadic Rings

Here we recall definitions of polyadic rings [25–27] in our notation [14,15]. Consider a polyadic structure $\left\langle R \mid \mu^{(n)}, \nu^{(m)} \right\rangle$ with two operations on the same set R: the m-ary addition $\nu^{(m)} : R^{\times m} \to R$ and the n-ary multiplication $\mu^{(n)} : R^{\times n} \to R$. The "interaction" between operations can be defined using the polyadic analog of distributivity.

Definition 5. *The polyadic distributivity for $\mu^{(n)}$ and $\nu^{(m)}$ consists of n relations*

$$\mu^{(n)}\left[\nu^{(m)}[x_1,\dots x_m], y_2, y_3, \dots y_n\right]$$
$$= \nu^{(m)}\left[\mu^{(n)}[x_1, y_2, y_3, \dots y_n], \mu^{(n)}[x_2, y_2, y_3, \dots y_n], \dots \mu^{(n)}[x_m, y_2, y_3, \dots y_n]\right] \tag{51}$$
$$\mu^{(n)}\left[y_1, \nu^{(m)}[x_1,\dots x_m], y_3, \dots y_n\right]$$
$$= \nu^{(m)}\left[\mu^{(n)}[y_1, x_1, y_3, \dots y_n], \mu^{(n)}[y_1, x_2, y_3, \dots y_n], \dots \mu^{(n)}[y_1, x_m, y_3, \dots y_n]\right] \tag{52}$$
$$\vdots$$
$$\mu^{(n)}\left[y_1, y_2, \dots y_{n-1}, \nu^{(m)}[x_1,\dots x_m]\right]$$
$$= \nu^{(m)}\left[\mu^{(n)}[y_1, y_2, \dots y_{n-1}, x_1], \mu^{(n)}[y_1, y_2, \dots y_{n-1}, x_2], \dots \mu^{(n)}[y_1, y_2, \dots y_{n-1}, x_m]\right], \tag{53}$$

where $x_i, y_j \in R$.

The operations $\mu^{(n)}$ and $\nu^{(m)}$ are totally *associative*, if (in the invariance definition [14,15])

$$\nu^{(m)}\left[\mathbf{u}, \nu^{(m)}[\mathbf{v}], \mathbf{w}\right] = invariant, \tag{54}$$
$$\mu^{(n)}\left[\mathbf{x}, \mu^{(n)}[\mathbf{y}], \mathbf{t}\right] = invariant, \tag{55}$$

where the internal products can be on any place, and $\mathbf{y} \in R^{\times n}$, $\mathbf{v} \in R^{\times m}$, and the polyads $\mathbf{x}$, $\mathbf{t}$, $\mathbf{u}$, $\mathbf{w}$ are of the needed lengths. In this way both algebraic structures $\left\langle R \mid \mu^{(n)} \mid assoc \right\rangle$ and $\left\langle R \mid \nu^{(m)} \mid assoc \right\rangle$ are *polyadic semigroups* $\mathcal{S}^{(n)}$ and $\mathcal{S}^{(m)}$.

Definition 6. *A polyadic (m,n)-ring $\mathcal{R}^{(m,n)}$ is a set R with two operations $\mu^{(n)} : R^{\times n} \to R$ and $\nu^{(m)} : R^{\times m} \to R$, such that:*

(1) *they are distributive (51)–(53);*

(2) *$\left\langle R \mid \mu^{(n)} \mid assoc \right\rangle$ is a polyadic semigroup;*

(3) *$\left\langle R \mid \nu^{(m)} \mid assoc, comm, solv \right\rangle$ is a commutative polyadic group.*

In case the multiplicative semigroup $\left\langle R \mid \mu^{(n)} \mid assoc \right\rangle$ of $\mathcal{R}^{(m,n)}$ is commutative, $\mu^{(n)}[\mathbf{x}] = \mu^{(n)}[\sigma \circ \mathbf{x}]$, for all $\sigma \in S_n$, then $\mathcal{R}^{(m,n)}$ is called a *commutative polyadic ring*, and if it contains the identity, then $\mathcal{R}^{(m,n)}$ is a (m,n)-*semiring*. A polyadic ring $\mathcal{R}^{(m,n)}$ is called *derived*, if $\gneqq^{(m)}$ and $\mu^{(n)}$ are repetitions of the binary addition $(+)$ and multiplication $(\cdot)$, whereas $\langle R \mid (+) \rangle$ and $\langle R \mid (\cdot) \rangle$ are commutative (binary) group and semigroup, respectively.

4.3. Full Polyadic External Direct Product of (m,n)-Rings

Let us consider the following task: for a given polyadic (m,n)-ring $\mathcal{R}'^{(m,n)} = \left\langle R' \mid \boldsymbol{\nu}'^{(m)}, \boldsymbol{\mu}'^{(n)} \right\rangle$ to construct a product of all possible (in arity shape) constituent rings $\mathcal{R}_1^{(m_1,n_1)}$ and $\mathcal{R}_2^{(m_2,n_2)}$. The first-hand "polyadization" of (49) and (50) leads to

Definition 7. *A full polyadic direct product ring $\mathcal{R}'^{(m,n)} = \mathcal{R}_1^{(m,n)} \times \mathcal{R}_2^{(m,n)}$ consists of (two) polyadic rings of the same arity shape, such that*

$$\boldsymbol{\nu}'^{(m)}\left[\mathbf{X}^{(1)}, \mathbf{X}^{(2)}, \ldots, \mathbf{X}^{(m)}\right] = \begin{pmatrix} \nu_1^{(m)}\left[x_1^{(1)}, x_1^{(2)}, \ldots, x_1^{(m)}\right] \\ \nu_2^{(m)}\left[x_2^{(1)}, x_2^{(2)}, \ldots, x_2^{(m)}\right] \end{pmatrix}, \tag{56}$$

$$\boldsymbol{\mu}'^{(n)}\left[\mathbf{X}^{(1)}, \mathbf{X}^{(2)}, \ldots, \mathbf{X}^{(n)}\right] = \begin{pmatrix} \mu_1^{(n)}\left[x_1^{(1)}, x_1^{(2)}, \ldots, x_1^{(n)}\right] \\ \mu_2^{(n)}\left[x_2^{(1)}, x_2^{(2)}, \ldots, x_2^{(n)}\right] \end{pmatrix}, \tag{57}$$

where the polyadic associativity (8) and polyadic distributivity (51)–(53) of the direct product operations $\nu^{(m)}$ and $\mu^{(n)}$ follow from those of the constituent rings and the componentwise operations in (56) and (57).

Example 8. *Consider two $(2,3)$-rings $\mathcal{R}_1^{(2,3)} = \left\langle \{\mathrm{i}x\} \mid \nu_1^{(2)} = (+), \mu_1^{(3)} = (\cdot), x \in \mathbb{Z}, \mathrm{i}^2 = -1 \right\rangle$ and $\mathcal{R}_2^{(2,3)} = \left\langle \left\{ \begin{pmatrix} 0 & a \\ b & 0 \end{pmatrix} \right\} \mid \nu_2^{(2)} = (+), \mu_2^{(3)} = (\cdot), a, b \in \mathbb{Z} \right\rangle$, where $(+)$ and $(\cdot)$ are operations in $\mathbb{Z}$, then their polyadic direct product on the doubles $\mathbf{X}^T = \left(\mathrm{i}x, \begin{pmatrix} 0 & a \\ b & 0 \end{pmatrix}\right) \in (\mathrm{i}\mathbb{Z}, GL^{adiag}(2,\mathbb{Z}))$ is defined by*

$$\boldsymbol{\nu}'^{(2)}\left[\mathbf{X}^{(1)}, \mathbf{X}^{(2)}\right] = \begin{pmatrix} \mathrm{i}x^{(1)} + \mathrm{i}x^{(2)} \\ \begin{pmatrix} 0 & a^{(1)} + a^{(2)} \\ b^{(1)} + b^{(2)} & 0 \end{pmatrix} \end{pmatrix}, \tag{58}$$

$$\boldsymbol{\mu}'^{(3)}\left[\mathbf{X}^{(1)}, \mathbf{X}^{(2)}, \mathbf{X}^{(3)}\right] = \begin{pmatrix} \mathrm{i}x^{(1)}x^{(2)}x^{(3)} \\ \begin{pmatrix} 0 & a^{(1)}b^{(2)}a^{(3)} \\ b^{(1)}a^{(2)}b^{(3)} & 0 \end{pmatrix} \end{pmatrix}. \tag{59}$$

The polyadic associativity and distributivity of the direct product operations $\boldsymbol{\nu}'^{(2)}$ and $\boldsymbol{\mu}'^{(3)}$ are evident, and therefore $\mathcal{R}^{(2,3)} = \left\langle \left\{ \left(ix, \begin{pmatrix} 0 & a \\ b & 0 \end{pmatrix}\right) \right\} \mid \boldsymbol{\nu}'^{(2)}, \boldsymbol{\mu}'^{(3)} \right\rangle$ is a $(2,3)$-ring $\mathcal{R}^{(2,3)} = \mathcal{R}_1^{(2,3)} \times \mathcal{R}_2^{(2,3)}$.

Definition 8. *A polyadic direct product $\mathcal{R}^{(m,n)}$ is called derived if both constituent rings $\mathcal{R}_1^{(m,n)}$ and $\mathcal{R}_2^{(m,n)}$ are derived, such that the operations $\nu_{1,2}^{(m)}$ and $\mu_{1,2}^{(n)}$ are compositions of the binary operations $\nu_{1,2}^{(2)}$ and $\mu_{1,2}^{(2)}$, correspondingly.*

So, in the derived case (see (1) all the operations in (56) and (57) have the form (cf., (21))

$$\nu_{1,2}^{(m)} = \left(\nu_{1,2}^{(2)}\right)^{\circ(m-1)}, \quad \mu_{1,2}^{(n)} = \left(\nu_{1,2}^{(2)}\right)^{\circ(n-1)}, \tag{60}$$

$$\nu^{(m)} = \left(\nu^{(2)}\right)^{\circ(m-1)}, \quad \mu^{(n)} = \left(\nu^{(2)}\right)^{\circ(n-1)}. \tag{61}$$

Thus, the operations of the derived polyadic ring can be written as (cf., the binary direct product (47) and (48))

$$\boldsymbol{\nu}'^{(m)}\left[\mathbf{X}^{(1)},\mathbf{X}^{(2)},\ldots,\mathbf{X}^{(m)}\right]=\begin{pmatrix} x_1^{(1)}+_1 x_1^{(2)}+_1\ldots+_1 x_1^{(m)} \\ x_2^{(1)}+_2 x_2^{(2)}+_2\ldots+_2 x_2^{(m)} \end{pmatrix}, \tag{62}$$

$$\boldsymbol{\mu}'^{(n)}\left[\mathbf{X}^{(1)},\mathbf{X}^{(2)},\ldots,\mathbf{X}^{(n)}\right]=\begin{pmatrix} x_1^{(1)}\cdot_1 x_1^{(2)}\cdot_1\ldots\cdot_1 x_1^{(n)} \\ x_2^{(1)}\cdot_2 x_2^{(2)}\cdot_2\ldots\cdot_2 x_2^{(n)} \end{pmatrix}, \tag{63}$$

The external direct product $(2,3)$-ring $\mathcal{R}^{(2,3)}$ from *Example* 8 is not derived, because both multiplications $\mu_1^{(3)}$ and $\mu_2^{(3)}$ there are nonderived.

4.4. Mixed-Arity Iterated Product of (m,n)-Rings

Recall that some polyadic multiplications can be iterated, i.e., composed (1) from those of lower arity (2), as well as those larger than 2, and so being nonderived, in general. The nontrivial "polyadization" of (49) and (50) can arise, when the composition of the separate (up and down) components on the r.h.s. of (56) and (57) will be different, and therefore the external product operations on the doubles $\mathbf{X}\in R_1\times R_2$ cannot be presented in the iterated form (1).

Let the constituent operations in (56) and (57) be composed from lower-arity corresponding operations, but in different ways for the up and down components, such that

$$\nu_1^{(m)}=\left(\nu_1^{(m_1)}\right)^{\circ\ell_{\nu1}},\quad \nu_2^{(m)}=\left(\nu_2^{(m_2)}\right)^{\circ\ell_{\nu2}},\quad 3\leq m_{1,2}\leq m-1, \tag{64}$$

$$\mu_1^{(n)}=\left(\mu_1^{(n_1)}\right)^{\circ\ell_{\mu1}},\quad \mu_2^{(n)}=\left(\mu_2^{(n_2)}\right)^{\circ\ell_{\mu2}},\quad 3\leq n_{1,2}\leq n-1, \tag{65}$$

where we exclude the limits: the derived case $m_{1,2}=n_{1,2}=2$ (60) and (61) and the uncomposed case $m_{1,2}=m$, $n_{1,2}=n$ (56) and (57). Since the total size of the up and down polyads is the same and coincides with the arities of the double addition m and multiplication n, using (2) we obtain the *arity compatibility* relations

$$m=\ell_{\nu1}(m_1-1)+1=\ell_{\nu2}(m_2-1)+1, \tag{66}$$

$$n=\ell_{\mu1}(n_1-1)+1=\ell_{\mu2}(n_2-1)+1. \tag{67}$$

Definition 9. *A mixed-arity polyadic direct product ring* $\mathcal{R}^{(m,n)}=\mathcal{R}_1^{(m_1,n_1)}\circledast\mathcal{R}_2^{(m_2,n_2)}$ *consists of two polyadic rings of the different arity shape, such that*

$$\boldsymbol{\nu}'^{(m)}\left[\mathbf{X}^{(1)},\mathbf{X}^{(2)},\ldots,\mathbf{X}^{(m)}\right]=\begin{pmatrix} \left(\nu_1^{(m_1)}\right)^{\circ\ell_{\nu1}}\left[x_1^{(1)},x_1^{(2)},\ldots,x_1^{(m)}\right] \\ \left(\nu_2^{(m_2)}\right)^{\circ\ell_{\nu2}}\left[x_2^{(1)},x_2^{(2)},\ldots,x_2^{(m)}\right] \end{pmatrix}, \tag{68}$$

$$\boldsymbol{\mu}'^{(n)}\left[\mathbf{X}^{(1)},\mathbf{X}^{(2)},\ldots,\mathbf{X}^{(n)}\right]=\begin{pmatrix} \left(\mu_1^{(n_1)}\right)^{\circ\ell_{\mu1}}\left[x_1^{(1)},x_1^{(2)},\ldots,x_1^{(n)}\right] \\ \left(\mu_2^{(n_2)}\right)^{\circ\ell_{\mu2}}\left[x_2^{(1)},x_2^{(2)},\ldots,x_2^{(n)}\right] \end{pmatrix}, \tag{69}$$

and the arity compatibility relations (66) and (67) hold valid.

Thus, two polyadic rings cannot always be composed in the mixed-arity polyadic direct product.

Assertion 3. *If the arity shapes of two polyadic rings $\mathcal{R}_1^{(m_1,n_1)}$ and $\mathcal{R}_2^{(m_2,n_2)}$ satisfy the compatibility conditions*

$$a(m_1-1)=b(m_2-1), \tag{70}$$

$$c(n_1-1)=d(n_2-1), \qquad a,b,c,d\in\mathbb{N}, \tag{71}$$

then they can form a mixed-arity direct product.

The limiting cases, undecomposed (56) and (57) and derived (62) and (63), satisfy the compatibility conditions (70) and (71) as well.

Example 9. *Let us consider two (nonderived) polyadic rings*

$$\mathcal{R}_1^{(9,3)}=\left\langle\{8l+7\}\mid\nu_1^{(9)},\mu_1^{(3)},l\in\mathbb{Z}\right\rangle, \tag{72}$$

$$\mathcal{R}_2^{(5,5)}=\left\langle\{M\}\mid\nu_2^{(5)},\mu_2^{(5)}\right\rangle, \tag{73}$$

where

$$M=\begin{pmatrix}0&4k_1+3&0&0\\0&0&4k_2+3&0\\0&0&0&4k_3+3\\4k_4+3&0&0&0\end{pmatrix},\quad k_i\in\mathbb{Z}, \tag{74}$$

and $\nu_2^{(5)}$ and $\mu_2^{(5)}$ are the ordinary sum and product of 5 *matrices. Using (66) and (67) we obtain $m=9$, $n=5$, if we choose the smallest "numbers of multiplications" $\ell_{\nu1}=1$, $\ell_{\nu2}=2$, $\ell_{\mu1}=2$, $\ell_{\mu2}=1$, and therefore the mixed-arity direct product nonderived* $(9,5)$*-ring becomes*

$$\mathcal{R}^{(9,5)}=\left\langle\{\mathbf{X}\}\mid\boldsymbol{\nu}'^{(9)},\boldsymbol{\mu}'^{(5)}\right\rangle, \tag{75}$$

where the doubles are $\mathbf{X}=\begin{pmatrix}8l+7\\M\end{pmatrix}$ and the nonderived direct product operations are

$$\boldsymbol{\nu}'^{(9)}\left[\mathbf{X}^{(1)},\mathbf{X}^{(2)},\ldots,\mathbf{X}^{(9)}\right]$$
$$=\left(\begin{array}{c}8\left(l^{(1)}+l^{(2)}+l^{(3)}+l^{(4)}+l^{(5)}+l^{(6)}+l^{(7)}+l^{(8)}+l^{(9)}+7\right)+7\\\begin{pmatrix}0&4K_1+3&0&0\\0&0&4K_2+3&0\\0&0&0&4K_3+3\\4K_4+3&0&0&0\end{pmatrix}\end{array}\right), \tag{76}$$

$$\boldsymbol{\mu}'^{(5)}\left[\mathbf{X}^{(1)},\mathbf{X}^{(2)},\mathbf{X}^{(3)},\mathbf{X}^{(4)},\mathbf{X}^{(5)}\right]$$
$$=\left(\begin{array}{c}(8l_\mu+7)\\\begin{pmatrix}0&4K_{\mu,1}+3&0&0\\0&0&4K_{\mu,2}+3&0\\0&0&0&4K_{\mu,3}+3\\4K_{\mu,4}+3&0&0&0\end{pmatrix}\end{array}\right), \tag{77}$$

where, in the first line, $K_i=k_i^{(1)}+k_i^{(2)}+k_i^{(3)}+k_i^{(4)}+k_i^{(5)}+k_i^{(6)}+k_i^{(7)}+k_i^{(8)}+k_i^{(9)}+6\in\mathbb{Z}$, $l_\mu\in\mathbb{Z}$ is a cumbersome integer function of $l^{(j)}\in\mathbb{Z}$, $j=1,\ldots,9$, and in the second line $K_{\mu,i}\in\mathbb{Z}$ are cumbersome integer functions of $k_i^{(s)}$, $i=1,\ldots,4$, $s=1,\ldots,5$. Therefore, the polyadic ring (75) is the nonderived mixed arity polyadic external product $\mathcal{R}^{(9,5)}=\mathcal{R}_1^{(9,3)}\circledast\mathcal{R}_2^{(5,5)}$ (see Definition 9).

Theorem 2. *The category of polyadic rings* `PolRing` *can exist (having the class of all polyadic rings for objects and ring homomorphisms for morphisms) and can be well-defined, because it has a product as the polyadic external product of rings.*

In the same way one can construct the iterated full and mixed-arity products of any number k of polyadic rings, merely by passing from the doubles $\mathbf{X}$ to k-tuples $\mathbf{X}_k^T = (x_1, \dots, x_k)$.

4.5. Polyadic Hetero Product of (m,n)-Fields

The most crucial difference between the binary direct products and the polyadic ones arises for fields, because a direct product's two binary fields are not a field [2].The reason for this lies in the fact that each binary field $\mathcal{F}^{(2,2)}$ necessarily contains 0 and 1, by definition. As follows from (48), a binary direct product contains nonzero idempotent doubles $\begin{pmatrix} 1 \\ 0 \end{pmatrix}$ and $\begin{pmatrix} 0 \\ 1 \end{pmatrix}$ which are noninvertible, and therefore the external direct product of fields $\mathcal{F}_1^{(2,2)} \times \mathcal{F}_2^{(2,2)}$ can never be a field. In the opposite case,polyadic fields (see Definition 10) can be zeroless (we denote them by $\widehat{\mathcal{F}}$),and the above arguments do not hold for them.

Recall the definitions of (m,n)-fields (see [27,28]). Denote $R^* = R \setminus \{z\}$, if the zero z exists (3). Observe that (in distinction to binary rings) $\left\langle R^* \mid \mu^{(n)} \mid assoc \right\rangle$ is not a polyadic group, in general. If $\left\langle R^* \mid \mu^{(n)} \right\rangle$ is the n-ary group, then $\mathcal{R}^{(m,n)}$ is called a (m,n)-*division ring* $\mathcal{D}^{(m,n)}$.

Definition 10. *A (totally) commutative (m,n)-division ring $\mathcal{R}^{(m,n)}$ is called a (m,n)-field $\mathcal{F}^{(m,n)}$.*

In n-ary groups there exists an "intermediate" commutativity, known as semicommutativity (10).

Definition 11. *A semicommutative (m,n)-division ring $\mathcal{R}^{(m,n)}$ is called a semicommutative (m,n)-field $\mathcal{F}^{(m,n)}$.*

The definition of a polyadic field can be expressed in a diagrammatic form, analogous to (16). We introduce the *double Dörnte relations*: for n-ary multiplication $\mu^{(n)}$ (15) and for m-ary addition $\nu^{(m)}$, as follows

$$\nu^{(m)}\left[\mathbf{m}_y, x\right] = x, \tag{78}$$

where the (additive) neutral sequence is $\mathbf{m}_y = \left(y^{m-2}, \tilde{y}\right)$, and $\tilde{y}$ is the additive querelement for $y \in R$ (see (14)). In distinction with (15), we have only one (additive) Dörnte relation (78) and one diagram from (16) only, because of the commutativity of $\nu^{(m)}$.

By analogy with the multiplicative queroperation $\bar{\mu}^{(1)}$ (13), introduce the *additive unary queroperation* by

$$\tilde{\nu}^{(1)}(x) = \tilde{x}, \quad \forall x \in R, \tag{79}$$

where $\tilde{x}$ is the additive querelement (13). Thus, we have

Definition 12 (Diagrammatic definition of (m,n)-field)**.** *A (polyadic) (m,n)-field is a one-set algebraic structure with 4 operations and 3 relations*

$$\left\langle R \mid \nu^{(m)}, \tilde{\nu}^{(1)}, \mu^{(n)}, \bar{\mu}^{(1)} \mid associativity, distributivity, double\ Dörnte\ relations \right\rangle, \tag{80}$$

where $\nu^{(m)}$ and $\mu^{(n)}$ are commutative associative m-ary addition and n-ary associative multiplication connected by polyadic distributivity (51)–(53), $\tilde{\nu}^{(1)}$ and $\bar{\mu}^{(1)}$ are unary additive queroperation (79) and multiplicative queroperation (13).

There is no initial relation between $\tilde{\nu}^{(1)}$ and $\bar{\mu}^{(1)}$; nevertheless the possibility of their "interaction" can lead to further thorough classification of polyadic fields.

Definition 13. *A polyadic field $\mathcal{F}^{(m,n)}$ is called quer-symmetric if its unary queroperations commute*

$$\tilde{\nu}^{(1)} \circ \bar{\mu}^{(1)} = \bar{\mu}^{(1)} \circ \tilde{\nu}^{(1)}, \tag{81}$$

$$\tilde{\bar{x}} = \bar{\tilde{x}}, \quad \forall x \in R, \tag{82}$$

in the other case $\mathcal{F}^{(m,n)}$ is called quer-nonsymmetric.

Example 10. *Consider the nonunital zeroless (denoted by $\widehat{\mathcal{F}}$)polyadic field $\widehat{\mathcal{F}}^{(3,3)} = \left\langle \{ia/b\} \mid \nu^{(3)}, \mu^{(3)} \right\rangle$, $\mathrm{i}^2 = -1$, $a, b \in \mathbb{Z}^{odd}$. The ternary addition $\nu^{(3)}[x,y,t] = x + y + t$ and the ternary multiplication $\mu^{(3)}[x,y,t] = xyt$ are nonderived, ternary associative and distributive (operations are in $\mathbb{C}$). For each $x = ia/b$ $(a, b \in \mathbb{Z}^{odd})$ the additive querelement is $\tilde{x} = -ia/b$, and the multiplicative querelement is $\bar{x} = -ib/a$ (see (12)). Therefore, both $\left\langle \{ia/b\} \mid \mu^{(3)} \right\rangle$ and $\left\langle \{ia/b\} \mid \nu^{(3)} \right\rangle$ are ternary groups, but they both contain no neutral elements (no unit, no zero). The nonunital zeroless $(3,3)$-field $\widehat{\mathcal{F}}^{(3,3)}$ is quer-symmetric, because (see (82))*

$$\tilde{\bar{x}} = \bar{\tilde{x}} = i\frac{b}{a}. \tag{83}$$

Finding quer-nonsymmetric polyadic fields is not a simple task.

Example 11. *Consider the set of real 4×4 matrices over the fractions $\frac{4k+3}{4l+3}$, $k, l \in \mathbb{Z}$, of the form*

$$M = \begin{pmatrix} 0 & \frac{4k_1+3}{4l_1+3} & 0 & 0 \\ 0 & 0 & \frac{4k_2+3}{4l_2+3} & 0 \\ 0 & 0 & 0 & \frac{4k_3+3}{4l_3+3} \\ \frac{4k_4+3}{4l_4+3} & 0 & 0 & 0 \end{pmatrix}, \quad k_i, l_i \in \mathbb{Z}. \tag{84}$$

The set $\{M\}$ is closed with respect to the ordinary addition of $m \geq 5$ matrices, because the sum of fewer of the fractions $\frac{4k+3}{4l+3}$ does not give a fraction of the same form [14], and with respect to the ordinary multiplication of $n \geq 5$ matrices, since the product of fewer matrices (84) does not have the same shape [29]. The polyadic associativity and polyadic distributivity follow from the binary ones of the ordinary matrices over $\mathbb{R}$, and the product of 5 matrices is semicommutative (see 10). Taking the minimal values $m = 5$, $n = 5$, we define the semicommutative zeroless $(5,5)$-field (see (11))

$$\mathcal{F}_M^{(5,5)} = \left\langle \{M\} \mid \nu^{(5)}, \mu^{(5)}, \tilde{\nu}^{(1)}, \bar{\mu}^{(1)} \right\rangle, \tag{85}$$

where $\nu^{(5)}$ and $\mu^{(5)}$ are the ordinary sum and product of 5 matrices, whereas $\tilde{\nu}^{(1)}$ and $\bar{\mu}^{(1)}$ are additive and multiplicative queroperations

$$\tilde{\nu}^{(1)}[M] \equiv \tilde{M} = -3M, \quad \bar{\mu}^{(1)}[M] \equiv \bar{M} = \frac{4l_1+3}{4k_1+3}\frac{4l_2+3}{4k_2+3}\frac{4l_3+3}{4k_3+3}\frac{4l_4+3}{4k_4+3}M. \tag{86}$$

The division ring $\mathcal{D}_M^{(5,5)}$ is zeroless, because the fraction $\frac{4k+3}{4l+3}$, is never zero for $k,l \in \mathbb{Z}$, and it is unital with the unit

$$M_e = \begin{pmatrix} 0 & 1 & 0 & 0 \\ 0 & 0 & 1 & 0 \\ 0 & 0 & 0 & 1 \\ 1 & 0 & 0 & 0 \end{pmatrix}. \tag{87}$$

Using (84) and (86), we obtain

$$\tilde{\nu}^{(1)}\left[\bar{\mu}^{(1)}[M]\right] = -3\frac{4l_1+3}{4k_1+3}\frac{4l_2+3}{4k_2+3}\frac{4l_3+3}{4k_3+3}\frac{4l_4+3}{4k_4+3}M, \tag{88}$$

$$\bar{\mu}^{(1)}\left[\tilde{\nu}^{(1)}[M]\right] = -\frac{1}{27}\frac{4l_1+3}{4k_1+3}\frac{4l_2+3}{4k_2+3}\frac{4l_3+3}{4k_3+3}\frac{4l_4+3}{4k_4+3}M, \tag{89}$$

or

$$\widetilde{\overline{M}} = 81\overline{\widetilde{M}}, \tag{90}$$

and therefore the additive and multiplicative queroperations do not commute independently of the field parameters. Thus, the matrix $(5,5)$-division ring $\mathcal{D}_M^{(5,5)}$ (85) is a quer-nonsymmetric division ring.

Definition 14. *The polyadic zeroless direct product field $\widehat{\mathcal{F}}'^{(m,n)} = \left\langle R' \mid \boldsymbol{\nu}'^{(m)}, \boldsymbol{\mu}'^{(n)} \right\rangle$ consists of (two) zeroless polyadic fields $\widehat{\mathcal{F}}_1^{(m,n)} = \left\langle R_1 \mid \nu_1^{(m)}, \mu_1^{(n)} \right\rangle$ and $\widehat{\mathcal{F}}_2^{(m,n)} = \left\langle R_2 \mid \nu_2^{(m)}, \mu_2^{(n)} \right\rangle$ of the same arity shape, whereas the componentwise operations on the doubles $\mathbf{X} \in R_1 \times R_2$ in (56) and (57) still remain valid, and $\left\langle R_1 \mid \mu_1^{(n)} \right\rangle$, $\left\langle R_2 \mid \mu_2^{(n)} \right\rangle$, $\left\langle R' = \{\mathbf{X}\} \mid \boldsymbol{\mu}'^{(n)} \right\rangle$ are n-ary groups.*

Following Definition 11, we have

Corollary 2. *If at least one of the constituent fields is semicommutative, and another one is totally commutative, then the polyadic product will be a semicommutative (m,n)-field.*

The additive and multiplicative unary queroperations (13) and (79) for the direct product field $\widehat{\mathcal{F}}^{(m,n)}$ are defined componentwise on the doubles $\mathbf{X}$ as follows

$$\tilde{\boldsymbol{\nu}}'^{(1)}[\mathbf{X}] = \begin{pmatrix} \tilde{\nu}_1^{(1)}[x_1] \\ \tilde{\nu}_2^{(1)}[x_2] \end{pmatrix}, \tag{91}$$

$$\bar{\boldsymbol{\mu}}'^{(1)}[\mathbf{X}] = \begin{pmatrix} \bar{\mu}_1^{(1)}[x_1] \\ \bar{\mu}_2^{(1)}[x_2] \end{pmatrix}, \quad x_1 \in R_1,\ x_2 \in R_2. \tag{92}$$

Definition 15. *A polyadic direct product field $\widehat{\mathcal{F}}'^{(m,n)} = \left\langle R' \mid \boldsymbol{\nu}'^{(m)}, \tilde{\boldsymbol{\nu}}'^{(1)}, \boldsymbol{\mu}'^{(n)}, \bar{\boldsymbol{\mu}}'^{(1)} \right\rangle$ is called quer-symmetric if its unary queroperations (91) and (92) commute*

$$\tilde{\boldsymbol{\nu}}'^{(1)} \circ \bar{\boldsymbol{\mu}}'^{(1)} = \bar{\boldsymbol{\mu}}'^{(1)} \circ \tilde{\boldsymbol{\nu}}'^{(1)}, \tag{93}$$

$$\widetilde{\overline{\mathbf{X}}} = \overline{\widetilde{\mathbf{X}}}, \quad \forall \mathbf{X} \in R', \tag{94}$$

in the other case, $\widehat{\mathcal{F}}'^{(m,n)}$ is called a quer-nonsymmetric direct product (m,n)-field.

Example 12. *Consider two nonunital zeroless $(3,3)$-fields*

$$\widehat{\mathcal{F}}_{1,2}^{(3,3)} = \left\langle \left\{ \mathrm{i}\frac{a_{1,2}}{b_{1,2}} \right\} \mid \nu_{1,2}^{(3)}, \mu_{1,2}^{(3)}, \tilde{\nu}_{1,2}^{(1)}, \bar{\mu}_{1,2}^{(1)} \right\rangle, \quad \mathrm{i}^2 = -1, \quad a_{1,2}, b_{1,2} \in \mathbb{Z}^{odd}, \tag{95}$$

where ternary additions $\nu^{(3)}_{1,2}$ *and ternary multiplications* $\mu^{(3)}_{1,2}$ *are the sum and product in* $\mathbb{Z}^{odd}$, *correspondingly, and the unary additive and multiplicative queroperations are* $\tilde{\nu}^{(1)}_{1,2}[\mathrm{i}a_{1,2}/b_{1,2}] = -\mathrm{i}a_{1,2}/b_{1,2}$ *and* $\bar{\mu}^{(1)}_{1,2}[\mathrm{i}a_{1,2}/b_{1,2}] = -\mathrm{i}b_{1,2}/a_{1,2}$ *(see Example 10). Using (56) and (57) we build the operations of the polyadic nonderived nonunital zeroless product* $(3,3)$*-field* $\widehat{\mathcal{F}}'^{(3,3)} = \widehat{\mathcal{F}}^{(3,3)}_1 \times \widehat{\mathcal{F}}^{(3,3)}_2$ *on the doubles* $\mathbf{X}^T = (\mathrm{i}a_1/b_1, \mathrm{i}a_2/b_2)$ *as follows*

$$\boldsymbol{\nu}'^{(3)}\left[\mathbf{X}^{(1)},\mathbf{X}^{(2)},\mathbf{X}^{(3)}\right] = \begin{pmatrix} \mathrm{i}\dfrac{a_1^{(1)}b_1^{(2)}b_1^{(3)} + b_1^{(1)}a_1^{(2)}b_1^{(3)} + b_1^{(1)}b_1^{(2)}a_1^{(3)}}{b_1^{(1)}b_1^{(2)}b_1^{(3)}} \\ \mathrm{i}\dfrac{a_2^{(1)}b_2^{(2)}b_2^{(3)} + b_2^{(1)}a_2^{(2)}b_2^{(3)} + b_2^{(1)}b_2^{(2)}a_2^{(3)}}{b_2^{(1)}b_2^{(2)}b_2^{(3)}} \end{pmatrix}, \tag{96}$$

$$\boldsymbol{\mu}'^{(3)}\left[\mathbf{X}^{(1)},\mathbf{X}^{(2)},\mathbf{X}^{(3)}\right] = \begin{pmatrix} -\mathrm{i}\dfrac{a_1^{(1)}a_1^{(2)}a_1^{(3)}}{b_1^{(1)}b_1^{(2)}b_1^{(3)}} \\ -\mathrm{i}\dfrac{a_2^{(1)}a_2^{(2)}a_2^{(3)}}{b_2^{(1)}b_2^{(2)}b_2^{(3)}} \end{pmatrix}, \quad a_i^{(j)}, b_i^{(j)} \in \mathbb{Z}^{odd}, \tag{97}$$

and the unary additive and multiplicative queroperations (91) and (92) of the direct product $\widehat{\mathcal{F}}'^{(3,3)}$ *are*

$$\tilde{\boldsymbol{\nu}}'^{(1)}[\mathbf{X}] = \begin{pmatrix} -\mathrm{i}\dfrac{a_1}{b_1} \\ -\mathrm{i}\dfrac{a_2}{b_2} \end{pmatrix}, \tag{98}$$

$$\bar{\boldsymbol{\mu}}'^{(1)}[\mathbf{X}] = \begin{pmatrix} -\mathrm{i}\dfrac{b_1}{a_1} \\ -\mathrm{i}\dfrac{b_2}{a_2} \end{pmatrix}, \quad a_i, b_i \in \mathbb{Z}^{odd}. \tag{99}$$

Therefore, both $\left\langle \{\mathbf{X}\} \mid \boldsymbol{\nu}'^{(3)}, \tilde{\boldsymbol{\nu}}'^{(1)} \right\rangle$ *and* $\left\langle \{\mathbf{X}\} \mid \boldsymbol{\mu}'^{(3)}, \bar{\boldsymbol{\mu}}'^{(1)} \right\rangle$ *are commutative ternary groups, which means that the polyadic direct product* $\widehat{\mathcal{F}}'^{(3,3)} = \widehat{\mathcal{F}}^{(3,3)}_1 \times \widehat{\mathcal{F}}^{(3,3)}_2$ *is the* nonunital zeroless *polyadic field. Moreover,* $\widehat{\mathcal{F}}'^{(3,3)}$ *is quer-symmetric, because (93) and (94) remain valid*

$$\bar{\boldsymbol{\mu}}'^{(1)} \circ \tilde{\boldsymbol{\nu}}'^{(1)}[\mathbf{X}] = \tilde{\boldsymbol{\nu}}'^{(1)} \circ \bar{\boldsymbol{\mu}}'^{(1)}[\mathbf{X}] = \begin{pmatrix} \mathrm{i}\dfrac{b_1}{a_1} \\ \mathrm{i}\dfrac{b_2}{a_2} \end{pmatrix}, \quad a_i, b_i \in \mathbb{Z}^{odd}. \tag{100}$$

Example 13. *Let us consider the polyadic direct product of two zeroless fields, one of them being the semicommutative* $(5,5)$*-field* $\widehat{\mathcal{F}}^{(5,5)}_1 = \mathcal{F}^{(5,5)}_M$ *from (85), and the other one being the nonderived nonunital zeroless* $(5,5)$*-field of fractions* $\widehat{\mathcal{F}}^{(5,5)}_2 = \left\langle \left\{ \sqrt{\mathrm{i}\frac{4r+1}{4s+1}} \right\} \mid \nu^{(5)}_2, \mu^{(5)}_2 \right\rangle$, $r,s \in \mathbb{Z}$, $\mathrm{i}^2 = -1$. *The double is* $\mathbf{X}^T = \left(\sqrt{\mathrm{i}\frac{4r+1}{4s+1}}, M\right)$, *where* M *is in (84). The polyadic nonunital zeroless direct* *product field* $\widehat{\mathcal{F}}'^{(5,5)} = \widehat{\mathcal{F}}_1^{(5,5)} \times \widehat{\mathcal{F}}_2^{(5,5)}$ *is nonderived and semicommutative, and is defined by* $\widehat{\mathcal{F}}^{(5,5)} = \left\langle \mathbf{X} \mid \boldsymbol{\nu}'^{(5)}, \boldsymbol{\mu}'^{(5)}, \tilde{\boldsymbol{\nu}}'^{(1)}, \bar{\boldsymbol{\mu}}^{(1)} \right\rangle$*, where its addition and multiplication are*

$$\boldsymbol{\nu}'^{(5)}\left[\mathbf{X}^{(1)}, \mathbf{X}^{(2)}, \mathbf{X}^{(3)}, \mathbf{X}^{(4)}, \mathbf{X}^{(5)}\right] = \begin{pmatrix} \sqrt{\mathrm{i}}\dfrac{4R_\nu+1}{4S_\nu+1} \\ \begin{pmatrix} 0 & \dfrac{4K_{\nu,1}+3}{4L_{\nu,1}+3} & 0 & 0 \\ 0 & 0 & \dfrac{4K_{\nu,2}+3}{4L_{\nu,2}+3} & 0 \\ 0 & 0 & 0 & \dfrac{4K_{\nu,3}+3}{4L_{\nu,3}+3} \\ \dfrac{4K_{\nu,4}+3}{4L_{\nu,4}+3} & 0 & 0 & 0 \end{pmatrix} \end{pmatrix}, \tag{101}$$

$$\boldsymbol{\mu}'^{(5)}\left[\mathbf{X}^{(1)}, \mathbf{X}^{(2)}, \mathbf{X}^{(3)}, \mathbf{X}^{(4)}, \mathbf{X}^{(5)}\right] = \begin{pmatrix} \sqrt{\mathrm{i}}\dfrac{4R_\mu+1}{4S_\mu+1} \\ \begin{pmatrix} 0 & \dfrac{4K_{\mu,1}+3}{4L_{\mu,1}+3} & 0 & 0 \\ 0 & 0 & \dfrac{4K_{\mu,2}+3}{4L_{\mu,2}+3} & 0 \\ 0 & 0 & 0 & \dfrac{4K_{\mu,3}+3}{4L_{\mu,3}+3} \\ \dfrac{4K_{\mu,4}+3}{4L_{\mu,4}+3} & 0 & 0 & 0 \end{pmatrix} \end{pmatrix}, \tag{102}$$

where $R_{\nu,\mu}, S_{\nu,\mu} \in \mathbb{Z}$ *are cumbersome integer functions of* $r^{(i)}, s^{(i)} \in \mathbb{Z}$, $i = 1, \ldots, 5$, *and* $K_{\nu,i}, K_{\mu,i}, L_{\nu,i}, L_{\mu,i} \in \mathbb{Z}$ *are cumbersome integer functions of* $k_j^{(i)}, l_j^{(i)} \in \mathbb{Z}$, $j = 1, \ldots, 4$, $i = 1, \ldots, 5$ *(see (84)). The unary queroperations (91) and (92) of the direct product* $\widehat{\mathcal{F}}^{(5,5)}$ *are*

$$\tilde{\boldsymbol{\nu}}'^{(1)}[\mathbf{X}] = \begin{pmatrix} -3\sqrt{\mathrm{i}}\dfrac{4r+1}{4s+1} \\ -3M \end{pmatrix}, \tag{103}$$

$$\bar{\boldsymbol{\mu}}'^{(1)}[\mathbf{X}] = \begin{pmatrix} -\sqrt{\mathrm{i}}\left(\dfrac{4s+1}{4r+1}\right)^3 \\ \dfrac{4l_1+3}{4k_1+3}\dfrac{4l_2+3}{4k_2+3}\dfrac{4l_3+3}{4k_3+3}\dfrac{4l_4+3}{4k_4+3}M \end{pmatrix}, \quad r, s, k_i, l_i \in \mathbb{Z}, \tag{104}$$

where M *is in (84). Therefore,* $\left\langle \{\mathbf{X}\} \mid \boldsymbol{\nu}'^{(5)}, \tilde{\boldsymbol{\nu}}'^{(1)} \right\rangle$ *is a commutative* 5*-ary group, and* $\left\langle \{\mathbf{X}\} \mid \boldsymbol{\mu}'^{(5)}, \bar{\boldsymbol{\mu}}'^{(1)} \right\rangle$ *is a semicommutative* 5*-ary group, which means that the polyadic direct product* $\widehat{\mathcal{F}}'^{(5,5)} = \widehat{\mathcal{F}}_1^{(5,5)} \times \widehat{\mathcal{F}}_2^{(5,5)}$ *is the nonunital zeroless polyadic semicommutative* $(5,5)$*-field. Using (90) we obtain*

$$\tilde{\boldsymbol{\nu}}'^{(1)}\bar{\boldsymbol{\mu}}'^{(1)}[\mathbf{X}] = 81\bar{\boldsymbol{\mu}}'^{(1)}\tilde{\boldsymbol{\nu}}'^{(1)}[\mathbf{X}], \tag{105}$$

and therefore the direct product $(5,5)$*-field* $\widehat{\mathcal{F}}'^{(5,5)}$ *is quer-nonsymmetric (see (81)).*

Thus, we arrive at

Theorem 3. *The category of zeroless polyadic fields* `zlessPolField` *can exist (having the class of all zeroless polyadic fields for objects and field homomorphisms for morphisms) and can be well-defined, because it has a product as the polyadic field product.*

5. Conclusions

For physical applications, for instance, the particle content of any elementary particle model is connected with the direct decomposition of its gauge symmetry group. Thus, the proposed generalization of the direct product can lead to principally new physical models having unusual mathematical properties.

For mathematical applications, further analysis of the direct product constructions introduced here and their examples for polyadic rings and fields would be interesting, and could lead to new kinds of categories.

Funding: This research received no external funding.

Institutional Review Board Statement: Not applicable.

Acknowledgments: The author is deeply grateful to Vladimir Akulov, Mike Hewitt, Vladimir Tkach, Raimund Vogl and Wend Werner for numerous fruitful discussions, important help and valuable support.

Conflicts of Interest: The authors declare no conflict of interest.

References

1. Lang, S. *Algebra;* Addison-Wesley: Boston, MA, USA, 1965.
2. Lambek, J. *Lectures on Rings and Modules;* Blaisdell: Providence, RI, USA, 1966; p. 184.
3. Barnes, K.J. *Group Theory for the Standard Model of Particle Physics and Beyond;* CRC Press: Boca Raton, FL, USA, 2010; p. 502.
4. Burgess, C.P.; Moore, G.D. *The Standard Model: A primer;* Cambridge University Press: Cambridge, UK, 2007; p. 542.
5. Knapp, A.W. *Representation Theory of Semisimple Groups;* Princeton University Press: Princeton, RI, USA, 1986.
6. Fulton, W.; Harris, J. *Representation Theory: A First Course;* Springer: New York, NY, USA, 2004; p. 551.
7. Hungerford, T.W. *Algebra;* Springer: New York, NY, USA, 1974; p. 502.
8. Michalski, J. Free products of n-groups. *Fund. Math.* **1984**, *123*, 11–20. [CrossRef]
9. Shchuchkin, N.A. Direct product of n-ary groups. *Chebysh. Sb.* **2014**, *15*, 101–121.
10. Borceux, F. *Handbook of Categorical Algebra 1. Basic Category Theory;* Cambridge University Press: Cambridge, UK, 1994; Volume 50, pp. xvi+345.
11. Mac Lane, S. *Categories for the Working Mathematician;* Springer: Berlin/Heidelberg, Germany, 1971; p. 189.
12. Michalski, J. On the category of n-groups. *Fund. Math.* **1984**, *122*, 187–197. [CrossRef]
13. Iancu, L. On the category of n-groups. *Buletinul științific al Universitatii Baia Mare Seria B Fascicola matematică-informatică* **1991**, *7*, 9–14.
14. Duplij, S. Polyadic integer numbers and finite (m,n)-fields. *p-Adic Numbers Ultrametric Anal. Appl.* **2017**, *9*, 257–281. [CrossRef]
15. Duplij, S. Polyadic Algebraic Structures And Their Representations. In *Exotic Algebraic and Geometric Structures in Theoretical Physics;* Duplij, S., Ed.; Nova Publishers: New York, NY, USA, 2018; pp. 251–308.
16. Post, E.L. Polyadic groups. *Trans. Amer. Math. Soc.* **1940**, *48*, 208–350. [CrossRef]
17. Hausmann, B.A.; Ore, Ø. Theory of quasigroups. *Amer. J. Math.* **1937**, *59*, 983–1004. [CrossRef]
18. Clifford, A.H.; Preston, G.B. *The Algebraic Theory of Semigroups;* Amer. Math. Soc.: Providence, RI, USA, 1961; Volume 1,
19. Brandt, H. Über eine Verallgemeinerung des Gruppenbegriffes. *Math. Annalen* **1927**, *96*, 360–367. [CrossRef]
20. Bruck, R.H. *A Survey on Binary Systems;* Springer: New York, NY, USA, 1966.
21. Bourbaki, N. *Elements of Mathematics: Algebra I;* Springer: Berlin/Heidelberg, Germany, 1998.
22. Dörnte, W. Unterschungen über einen verallgemeinerten Gruppenbegriff. *Math. Z.* **1929**, *29*, 1–19. [CrossRef]
23. Gleichgewicht, B.; Głazek, K. Remarks on n-groups as abstract algebras. *Colloq. Math.* **1967**, *17*, 209–219. [CrossRef]
24. Dudek, W. On n-ary groups with only one skew elements. *Radovi Mat. (Sarajevo)* **1990**, *6*, 171–175.
25. Čupona, G. On $[m,n]$-rings. *Bull. Soc. Math. Phys. Macedoine* **1965**, *16*, 5–9.
26. Crombez, G. On (n,m)-rings. *Abh. Math. Semin. Univ. Hamb.* **1972**, *37*, 180–199. [CrossRef]
27. Leeson, J.J.; Butson, A.T. On the general theory of (m,n) rings. *Algebra Univers.* **1980**, *11*, 42–76. [CrossRef]
28. Iancu, L.; Pop, M.S. A Post type theorem for (m,n) fields. In *Proceedings of the Scientific Communications Meeting of "Aurel Vlaicu" University, Arad, Romania, 16–17 May 1996*, 3rd ed.; Halic, G.; Cristescu, G., Eds.; "Aurel Vlaicu" University of Arad Publishing Centre: Arad, Romania, 1997; Volume 14A, pp. 13–18.
29. Duplij, S. Higher braid groups and regular semigroups from polyadic-binary correspondence. *Mathematics* **2021**, *9*, 972. [CrossRef]

 universe

MDPI

Article

Semiheaps and Ternary Algebras in Quantum Mechanics Revisited

Andrew James Bruce

Department of Mathematics, The Computational Foundry, Swansea University Bay Campus, Fabian Way, Swansea SA1 8EN, UK; andrewjamesbruce@googlemail.com

Abstract: We re-examine the appearance of semiheaps and (para-associative) ternary algebras in quantum mechanics. In particular, we review the construction of a semiheap on a Hilbert space and the set of bounded operators on a Hilbert space. The new aspect of this work is a discussion of how symmetries of a quantum system induce homomorphisms of the relevant semiheaps and ternary algebras.

Keywords: semiheaps; ternary algebras; para-associativity; quantum mechanics

1. Introduction

Heaps were introduced by Prüfer [1] and Baer [2] as a set equipped with a ternary operation satisfying simple axioms. One can think of a heap as a group in which the identity element has been forgotten. Indeed, these axioms are satisfied in a group if we define the ternary operation as $(a, b, c) \mapsto ab^{-1}c$. For example, given a vector space, or more generally an affine space, we can construct a heap operation as $(u, v, w) \mapsto u - v + w$. Conversely, by selecting any element in a heap, one can reduce the ternary operation to a group operation, such that the chosen element is the identity element.

There is a slightly weaker notion of a *semiheap*.A semiheap is a non-empty set H, equipped with a ternary operation $[a, b, c] \in H$ that satisfies the *para-associative law*

$$[[a, b, c], d, e] = [a, [d, c, b], e] = [a, b, [c, d, e]],$$

for all a, b, c, d and $e \in H$. A semiheap is a *heap* when all its elements are *biunitary*, i.e., $[a, b, b] = a$ and $[b, b, a] = a$, for all a and $b \in H$. This condition is also referred to as the Mal'cev identities. A (semi)heap is said to be *abelian* if $[a, b, c] = [c, b, a]$ for all a, b and $c \in H$. A *homomorphism of semiheaps* $\phi : (H, [-,-,-]) \to (H', [-,-,-]')$ is a map $\phi : H \to H'$ such that $\phi([a, b, c]) = [\phi(a), \phi(b), \phi(c)]'$. For more details about heaps and related structures the reader may consult Hollings & Lawson [3] and/or Brzeziński [4].

In this paper, we re-examine the natural occurrences of semiheaps in the formalism of standard non-relativistic quantum mechanics. The semiheaps explored here have appeared scattered in the mathematics literature under different names. However, there seems to be almost nothing written with physicists and quantum mechanics in mind. The exception here is Kerner (see [5]), who refers to "2nd type associativity" or "B-associativity", this is precisely the above para-associativity law.

In the setting of quantum mechanics, we do not just have a semiheap but also a vector space structure. A $\mathbb{C}$-vector space with a ternary product that is linear in the first and third arguments, and conjugate linear in the second argument, we will refer to as a *ternary algebra* (see [6–9]). If, in addition, the ternary product is para-associative, so defines a semiheap on the underlying set, then we speak of a *para-associative ternary algebra*. We will only deal with the para-associative case past this point. A homomorphism of para-associative ternary algebras is a linear map that is simultaneously a homomorphism of semiheaps.

Citation: Bruce, A.J. Semiheaps and Ternary Algebras in Quantum Mechanics Revisited. *Universe* **2022**, *8*, 56. https://doi.org/10.3390/universe8010056

Academic Editor: Andreas Fring

Received: 6 December 2021
Accepted: 14 January 2022
Published: 17 January 2022

We review the construction of semiheaps and ternary algebras on a Hilbert space and on the $*$-algebra of bounded operators on the said Hilbert space. While these constructions are not new, they are not well-known within the context of quantum mechanics. The new aspect of this work is a discussion of symmetries of quantum systems and how they induce semiheap and, in turn, ternary algebra homomorphisms. Generalised derivations of the ternary algebras are also discussed. We will focus on algebraic aspects of the theory and not address topological issues.

Rather generally, ternary operations and relations have a long history in physics. As examples, we have Nambu brackets (1973; [10]), the Yang-Baxter Equation (1967, 1972; [11,12]) and the BLG model of M2-branes (2007, 2009; [13,14]). A review of n-ary generalisations of Lie algebras and their physical applications can be found in [15]. We also mention that L_∞-algebras (cf. [16]) have found a wealth of applications in physics, notably through the BV-formalism of gauge theories. Returning to the Yang-Baxter equation, it has found applications in a diverse range of mathematics such as quantum groups, knot theory, braided geometry, integrable systems and noncommutative geometry. The classification of solutions to the Yang-Baxter equation is, at the time of writing, an unsolved problem. This challenge, first posed by Drinfeld in 1992, has inspired the development of various algebraic structures such as Rump's braces (see [17]) and Brzeziński's trusses (see [4,18]). For more information about current trends related to the Yang-Baxter equation the reader may consult [19,20].

It is also curious to note that, within the standard model, the number three constantly appears. Specifically, there are three generations of quarks, three generations of leptons, three fundamental forces (gravity is not included and is different), and three quarks are needed to make a baryon. Alongside this, there are three spatial dimensions and three fundamental inversions-charge (C), parity (P) and time (T). It is only the combination of CPT that is respected in all interactions. It is not known how, or indeed if, these threes are related.

2. Semiheaps Associated with Hilbert Spaces

2.1. The Semiheap and Ternary Algebra of a Hilbert Space

Given a vector space, there is no obvious way to multiply two vectors together and obtain another vector in the same space. However, if the vector space comes equipped with an inner product, then we can multiply three vectors together in a canonical way to obtain another vector. For the case at hand, we will restrict attention to (complex) Hilbert spaces as found in quantum mechanics. Typically, the Hilbert spaces in question are isomorphic to $L^2(\mathbb{R}^n) \otimes \mathbb{C}^2$, with $n = 1, 2, 3$. We will employ Dirac's notation throughout this paper. We will denote by $\mathcal{H}$ both a Hilbert space and its underlying set, the context should be clear. To emphasise the linear structure we will write $(\mathcal{H}, +)$.

Definition 1. *Let $\mathcal{H}$ be a Hilbert space, the* vector ternary product $[-,-,-] : \mathcal{H} \times \mathcal{H} \times \mathcal{H} \longrightarrow \mathcal{H}$ *is defined as*

$$[|\psi_1\rangle, |\psi_2\rangle, |\psi_3\rangle] := |\psi_1\rangle\langle\psi_2|\psi_3\rangle\,.$$

Recall that the norm of a vector is defined as $||\,|\psi\rangle\,|| := \sqrt{\langle\psi|\psi\rangle}$. It is then immediately clear that $||\,[|\psi_1\rangle, |\psi_2\rangle, |\psi_3\rangle]\,|| = |\langle\psi_2|\psi_3\rangle|\,||\,|\psi_1\rangle\,|| \leq ||\,|\psi_3\rangle\,||\,||\,|\psi_2\rangle\,||\,||\,|\psi_1\rangle\,||$ via the Cauchy–Schwarz inequality.

The following proposition is evident.

Proposition 1. *Let $\mathcal{H}$ be a Hilbert space. Then the vector ternary product, see Definition 1, is linear with respect to the first and third arguments, and conjugate linear with respect to the second entry, i.e.,*

$$\begin{aligned}
[|\psi_1\rangle + c_1\,|\psi_1'\rangle, |\psi_2\rangle, |\psi_3\rangle] &= [|\psi_1\rangle, |\psi_2\rangle, |\psi_3\rangle] + c_1\,[|\psi_1'\rangle, |\psi_2\rangle, |\psi_3\rangle]\,,\\
[|\psi_1\rangle, |\psi_2\rangle + c_2\,|\psi_2'\rangle, |\psi_3\rangle] &= [|\psi_1\rangle, |\psi_2\rangle, |\psi_3\rangle] + c_2^*\,[|\psi_1\rangle, |\psi_2'\rangle, |\psi_3\rangle]\,,\\
[|\psi_1\rangle, |\psi_2\rangle, |\psi_3\rangle + c_3\,|\psi_3'\rangle] &= [|\psi_1\rangle, |\psi_2\rangle, |\psi_3\rangle] + c_3\,[|\psi_1\rangle, |\psi_2\rangle, |\psi_3'\rangle]\,,
\end{aligned}$$

for all $|\psi_1\rangle, |\psi_2\rangle, |\psi_3\rangle \in \mathcal{H}$ *and* $c_1, c_2, c_3 \in \mathbb{C}$.

Thus, the linear structure and the vector ternary product are compatible in the above sense. Moving on to the generalised notion of associativity we have the following theorem.

Theorem 1. *The vector ternary product on a Hilbert space* $\mathcal{H}$*, see Definition 1, satisfies the para-associative law*

$$\begin{aligned}[[|\psi_1\rangle, |\psi_2\rangle, |\psi_3\rangle], |\psi_4\rangle, |\psi_5\rangle] &= [|\psi_1\rangle, [|\psi_4\rangle, |\psi_3\rangle, |\psi_2\rangle], |\psi_5\rangle] \\ &= [|\psi_1\rangle, |\psi_2\rangle, [|\psi_3\rangle, |\psi_4\rangle, |\psi_5\rangle]],\end{aligned}$$

for all $|\psi_1\rangle, |\psi_2\rangle, |\psi_3\rangle \in \mathcal{H}$*. In other words,* $(\mathcal{H}, [-,-,-])$ *is a semiheap.*

Proof. This follows via direct computation.

(i) $[[|\psi_1\rangle, |\psi_2\rangle, |\psi_3\rangle], |\psi_4\rangle, |\psi_5\rangle] = [|\psi_1\rangle\langle\psi_2|\psi_3\rangle, |\psi_4\rangle, |\psi_5\rangle] = |\psi_1\rangle\langle\psi_2|\psi_3\rangle\langle\psi_4|\psi_5\rangle$.

(ii) $[|\psi_1\rangle, [|\psi_4\rangle, |\psi_3\rangle, |\psi_2\rangle], |\psi_5\rangle] = [|\psi_1\rangle, |\psi_4\rangle\langle\psi_3|\psi_2\rangle, |\psi_5\rangle] = |\psi_1\rangle\langle\psi_2|\psi_3\rangle\langle\psi_4|\psi_5\rangle$.

(iii) $[|\psi_1\rangle, |\psi_2\rangle, [|\psi_3\rangle, |\psi_4\rangle, |\psi_5\rangle]] = [|\psi_1\rangle, |\psi_2\rangle, |\psi_3\rangle\langle\psi_4|\psi_5\rangle] = |\psi_1\rangle\langle\psi_2|\psi_3\rangle\langle\psi_4|\psi_5\rangle$.

Clearly, (i)=(ii) = (iii). □

Corollary 1. *By fixing a vector* $|\phi\rangle \in \mathcal{H}$ *we have an associated binary product* $\cdot_{|\phi\rangle} : \mathcal{H} \times \mathcal{H} \to \mathcal{H}$*, given by* $|\psi_1\rangle \cdot_{|\phi\rangle} |\psi_2\rangle := [|\psi_1\rangle, |\phi\rangle, |\psi_2\rangle]$*, i.e., the binary product satisfies*

$$\left(|\psi_1\rangle \cdot_{|\phi\rangle} |\psi_2\rangle\right) \cdot_{|\phi\rangle} |\psi_3\rangle = |\psi_1\rangle \cdot_{|\phi\rangle} \left(|\psi_2\rangle \cdot_{|\phi\rangle} |\psi_3\rangle\right).$$

If the quantum system under consideration has a non-degenerate ground state $|0\rangle$ (normalised, i.e., $\langle 0|0\rangle = 1$), then we have a canonical associative binary product.

Note that we do not have a heap, and so any associated binary product does not lead to a group structure. Specifically, the Mal'cev identities

$$[a, b, b] = a, \qquad \text{and} \qquad [b, b, a] = a\,,$$

are not, in general, satisfied. Explicitly, we see that

$$[|\psi_1\rangle, |\psi\rangle, |\psi\rangle] = |\psi_1\rangle\langle\psi|\psi\rangle\,.$$

Thus, if $|\psi\rangle$ is normalised, i.e., $\langle\psi|\psi\rangle = 1$, then $|\psi\rangle$ is *right unitary*. That is

$$[|\psi_1\rangle, |\psi\rangle, |\psi\rangle] = |\psi_1\rangle\,.$$

Again, assuming that $|\psi\rangle$ is normalised, $\mathrm{P}_{|\psi\rangle} := |\psi\rangle\langle\psi|$ projects an arbitrary vector onto $|\psi\rangle$. Thus,

$$[|\psi\rangle, |\psi\rangle, |\psi_3\rangle] = \mathrm{P}_{|\psi\rangle}(|\psi_3\rangle)\,.$$

Thus, the binary product defined in Corollary 1, defines a semigroup, i.e., a set with an associative binary product. In analogy with the situation for heaps, we refer to this semigroup as the *semigroup retract (with respect to* $|\phi\rangle$*)* of the semiheap $(\mathcal{H}, [-,-,-])$.

Proposition 2. *Let* $|\phi\rangle$ *and* $|\psi_1\rangle \in \mathcal{H}$ *be non-orthogonal vectors, i.e.,* $\langle\phi|\psi_1\rangle \neq 0$*. Then,* $|\psi_1\rangle$ *is a regular point of the semigroup* $(\mathcal{H}, \cdot_{|\phi\rangle})$*, i.e., there exists a vector* $|\psi\rangle \in \mathcal{H}$ *(pesudoinverse) such that*

$$|\psi_1\rangle \cdot_{|\phi\rangle} |\psi\rangle \cdot_{|\phi\rangle} |\psi_1\rangle = |\psi_1\rangle$$

Proof. Setting $|\psi\rangle := \frac{|\phi\rangle}{\langle\phi|\psi_1\rangle}$ provides the required vector. Explicitly,

$$|\psi_1\rangle \cdot_{|\phi\rangle} |\psi\rangle \cdot_{|\phi\rangle} |\psi_1\rangle = |\psi_1\rangle \frac{|\phi\rangle}{\langle\phi|\psi_1\rangle} \langle\phi|\psi_1\rangle = |\psi_1\rangle\,.$$

□

It is clear from the definition of the vector ternary product that

$$[\mathbf{0}, |\psi_2\rangle, |\psi_3\rangle] = [|\psi_1\rangle, \mathbf{0}, |\psi_3\rangle] = [|\psi_1\rangle, |\psi_2\rangle, \mathbf{0}] = \mathbf{0}\,,$$

where $\mathbf{0} \in \mathcal{H}$ is the zero vector. In other words, the zero vector is an absorbing element for the vector ternary product. Thus, $\mathbf{0}$ is also an absorbing element in any semigroup $(\mathcal{H}, \cdot_{|\phi\rangle})$, that is, multiplication of any vector by the zero vector on the left or right, yields the zero vector. Similarly, the semigroup $(\mathcal{H}, \cdot_{\mathbf{0}})$ is a null semigroup, i.e., $|\psi_1\rangle \cdot_{\mathbf{0}} |\psi_2\rangle = \mathbf{0}$, for all vectors $|\psi_1\rangle$ and $|\psi_2\rangle \in \mathcal{H}$.

From Proposition 1, Theorem 1 and the above discussion we see that a Hilbert space naturally comes with the structure of a ternary algebra in which the ternary product defines a semiheap (see [7] for further generalities on ternary algebras). Note that we have conjugate linearity in the second argument of the product rather than linearity.

Definition 2. *Let $\mathcal{H}$ be a Hilbert space. Then the ternary algebra $(\mathcal{H}, +, [-,-,-])$ defined via Proposition 1 and Theorem 1 is referred to as the* vector ternary algebra.

Example 1. *Consider the complex line $\mathbb{C}$ and define the inner product as $\langle z_1, z_2\rangle = \bar{z}_1 z_2$ for arbitrary complex numbers z_1 and z_2. Then the vector ternary product is given by*

$$[z_1, z_2, z_2] = z_1 \bar{z}_2 z_3\,.$$

Thus, the complex line is a ternary algebra over itself.

Example 2. *The Hilbert space we consider is finite-dimensional and given by the span of two orthonormal vectors "spin up" and "spin down"*

$$\mathcal{H} = \mathrm{Span}_{\mathbb{C}}\{|\uparrow\rangle, |\downarrow\rangle\} \cong \mathbb{C}^2\,.$$

The non-zero vector ternary products of the basis elements are

$$[|\uparrow\rangle, |\uparrow\rangle, |\uparrow\rangle] = |\uparrow\rangle\,, \qquad [|\uparrow\rangle, |\downarrow\rangle, |\downarrow\rangle] = |\uparrow\rangle\,,$$
$$[|\downarrow\rangle, |\downarrow\rangle, |\downarrow\rangle] = |\downarrow\rangle\,, \qquad [|\downarrow\rangle, |\uparrow\rangle, |\uparrow\rangle] = |\downarrow\rangle\,.$$

All other vector ternary products are equal to the zero vector $\mathbf{0} \in \mathcal{H}$. Note that there are 8 possible vector ternary products to consider. Using the linearity and conjugate linearity one can deduce the vector ternary product for arbitrary vectors (not necessarily normalised). For example

$$[a|\uparrow\rangle, b|\uparrow\rangle, c|\uparrow\rangle + d|\downarrow\rangle] = a\bar{b}c|\uparrow\rangle\,,$$

with a, b, c and $d \in \mathbb{C}$.

Example 3. *The orthonormal basis of states for the one-dimensional harmonic oscillator is countably infinite as each basis vector is labelled by $n \in \mathbb{N}$ (including zero). The Hilbert space here is, of course, $L^2(\mathbb{R})$. The vector ternary product can be written in this natural basis (and then using linearity and conjugate linearity to deduce the product of arbitrary vectors) as*

$$[|n_1\rangle, |n_2\rangle, |n_3\rangle] = |n_1\rangle\, \delta_{n_2 n_3}\,.$$

Remark 1. *All quantum systems with a finite or countably infinite number of states, e.g., the hydrogen atom, have a vector ternary product that can easily be expressed in a similar way to the previous example.*

Recall that a linear map $\varphi : \mathcal{H} \to \mathcal{H}'$ between Hilbert spaces is said to be *bounded* if there exists some $r > 0$ such that $|| \, \varphi|\psi\rangle \, ||' = r \, || \, |\psi\rangle \, ||$. It is a well-known result that boundedness implies continuity of a linear map and vice versa. A *bounded linear isometry* is a bounded linear map $\varphi : \mathcal{H} \to \mathcal{H}'$ such that $\varphi^\dagger \varphi = 1_{\mathcal{H}}$.

Proposition 3. *Let $\mathcal{H}$ and $\mathcal{H}'$ be Hilbert spaces and let $\varphi : \mathcal{H} \to \mathcal{H}'$ be a bounded linear isometry. Then φ is morphism of semiheaps*

$$\varphi : (\mathcal{H}, [-,-,-]) \longrightarrow (\mathcal{H}', [-,-,-]') \, .$$

Proof. Directly, using $\mathbb{C}$-linearity and the condition that the bounded linear map be an isometry, we observe that

$$\begin{aligned} \varphi[|\psi_1\rangle, |\psi_2\rangle, |\psi_3\rangle] &= \varphi(|\psi_1\rangle\langle\psi_2|\psi_3\rangle) = \varphi(|\psi_1\rangle)\langle\psi_2|\psi_3\rangle = \varphi(|\psi_1\rangle)\langle\psi_2|\varphi^\dagger\varphi|\psi_3\rangle \\ &= [\varphi|\psi_1\rangle, \varphi|\psi_2\rangle, \varphi|\psi_3\rangle]' \, . \end{aligned}$$

□

Remark 2. *If we consider bounded linear maps that are not isometries, then we will not, in general, have a homomorphism of the relevant semiheaps.*

As we are considering linear maps, it is clear that bounded linear isometries are also ternary algebra homomorphisms.

Unitary operators, i.e., bounded operators such that $U^\dagger U = UU^\dagger = 1_{\mathcal{H}}$, form a group, $\mathcal{U}(\mathcal{H})$, and their action on $\mathcal{H}$ are isometries. In particular, the action $\rho_U : \mathcal{H} \to \mathcal{H}$ is $|\psi\rangle \mapsto U|\psi\rangle$ for arbitrary $U \in \mathcal{U}(\mathcal{H})$. We then have the following corollary.

Corollary 2. *Let $\mathcal{U}(\mathcal{H})$ be the group of unitary operators on a Hilbert space $\mathcal{H}$. Furthermore, let $(\mathcal{H}, [-,-,-])$ be the associated semiheap. Then the action on $\mathcal{U}(\mathcal{H})$ on $\mathcal{H}$ is a semiheap isomorphism and so an isomorphism of ternary algebras.*

Symmetries in quantum mechanics are usually understood as *projective representations* of some group G. That is, we have a map

$$U : G \longrightarrow \mathcal{U}(\mathcal{H}) \, ,$$

such that $U(g_1)U(g_2) = \omega(g_1, g_2) \, U(g_1, g_2)$, with $\omega : G \times G \to U(1) := \{z \in \mathbb{C}, \; | \, |z| = 1\}$, being referred to as the *Schur factor*. Associativity implies that $\omega(g_1, g_2)\omega(g_1 g_2, g_3) = \omega(g_1, g_2 g_3)\omega(g_2, g_3)$. Assuming that $U(e) = 1_{\mathcal{H}}$ (as standard) implies that $\omega(e, e) = 1$. One can also deduce that $\omega(g, e) = \omega(e, g) = 1$ and $\omega(g, g^{-1}) = \omega(g^{-1}, g)$. If $\omega(g_1, g_2) = 1$ for all $g_1, g_2 \in G$, then we have a *unitary representation*. Wigner's theorem (see [21]) tells us that symmetries in quantum mechanics act via either projective or unitary representations. We thus, in general, have an "action up to a factor" $\rho_U(-) : G \times \mathcal{H} \to \mathcal{H}$ given by $(g, |\psi\rangle) \mapsto U(g)|\psi\rangle$.

Corollary 3. *Let $\mathcal{U}(\mathcal{H})$ be the group of unitary operators on a Hilbert space $\mathcal{H}$ and let $U : G \longrightarrow \mathcal{U}(\mathcal{H})$ be a projective representation. Furthermore, let $(\mathcal{H}, [-,-,-])$ be the associated semiheap. Then, for any $g \in G$, $\rho_U(g) : \mathcal{H} \to \mathcal{H}$ is a semiheap homomorphism and so a homomorphism of ternary algebras.*

Remark 3. *The dual of a Hilbert space also comes with the canonical structure of a semiheap and ternary algebra by defining* $[\langle\psi_3|, \langle\psi_2|, \langle\psi_1|] := \langle\psi_3|\psi_2\rangle\langle\psi_1|$. *By construction we have* $[|\psi_1\rangle, |\psi_2\rangle, |\psi_3\rangle]^\dagger = [\langle\psi_3|, \langle\psi_2|, \langle\psi_1|]$. *Note that although we can canonically identify a Hilbert space and its dual, we consider them as distinct spaces.*

The vector ternary product can be extended to direct sums of Hilbert spaces as follows. Recall that the (orthogonal) direct sum $\mathcal{H} = \mathcal{H}_1 \oplus \mathcal{H}_2$ comes equipped with an inner product given by

$$(|\psi_1\rangle + |\phi_1\rangle, |\psi_2\rangle + |\phi_2\rangle) \longmapsto \langle\psi_1|\psi_2\rangle + \langle\phi_1|\phi_2\rangle\,.$$

Then, the vector ternary product is given by

$$\begin{aligned}[|\psi_1\rangle + |\phi_1\rangle, |\psi_2\rangle + |\phi_2\rangle, |\psi_3\rangle + |\phi_3\rangle] &:= |\psi_1\rangle\langle\psi_2|\psi_3\rangle + |\phi_1\rangle\langle\phi_2|\phi_3\rangle \\ &= [|\psi_1\rangle, |\psi_2\rangle, |\psi_3\rangle] + [|\phi_1\rangle, |\phi_2\rangle, |\phi_3\rangle]\,.\end{aligned}$$

This construction extends to the orthogonal direct sum of any finite number of Hilbert spaces.

Example 4. *In supersymmetric quantum mechanics, the relevant Hilbert space is the (orthogonal) direct sum on the bosonic sector* $\mathcal{H}_0$ *and the fermionic sector* $\mathcal{H}_1$, *i.e.,* $\mathcal{H} = \mathcal{H}_0 \oplus \mathcal{H}_1$. *Of course, being orthogonal implies that linear combinations of bosonic and fermionic states cannot be physically realised. Nonetheless, we can still consider the vector ternary product on the direct sum as the sum of two vector ternary products on each sector.*

Remark 4. *Note that, as vector spaces,* $\mathbb{C} \oplus \mathbb{C} \simeq \mathbb{C}^2$, *and more over, they are isomorphic as metric spaces. Specifically, the induced metric on* $\mathbb{C} \oplus \mathbb{C}$ *is given by* $\langle z_1 + z_1', z_2 + z_2'\rangle = \bar{z}_1 z_2 + \bar{z}_1' z_2'$. *Similarly, on* $\mathbb{C}^2$ *the standard metric is given by* $\langle Z_1, Z_2\rangle = Z_1^\dagger Z_2 = \bar{z}_1 z_2 + \bar{z}_1' z_2'$, *where* $Z_i = (z_i, z_i')^T$. *However, the associated semiheaps are not identical, and so the associated ternary algebras are distinct. In particular,* $[z_1 + z_1', z_2 + z_2', z_3 + z_3'] = z_1\bar{z}_2 z_3 + z_1'\bar{z}_2' z_3' \in \mathbb{C} \oplus \mathbb{C}$, *while*

$$[Z_1, Z_2, Z_3] = Z_1\langle Z_2, Z_3\rangle = \begin{pmatrix} z_1\bar{z}_2 z_3 + z_1'\bar{z}_2\, z_3 \\ z_1\bar{z}_2' z_3' + z_1'\bar{z}_2' z_3' \end{pmatrix} \in \mathbb{C}^2\,.$$

Considering Example 2, and picking the natural representation $|\uparrow\rangle = (1,0)^T$ *and* $|\downarrow\rangle = (0,1)^T$, *we see that we are using the natural metric on* $\mathbb{C}^2$ *and its associated semiheap structure. If we used the induced semiheap structure on* $\mathbb{C} \oplus \mathbb{C}$, *then of the 8 possible ternary products (using the natural basis), the only non-zero ones are* $[|\uparrow\rangle, |\uparrow\rangle, |\uparrow\rangle] = |\uparrow\rangle$ *and* $[|\downarrow\rangle, |\downarrow\rangle, |\downarrow\rangle] = |\downarrow\rangle$. *In particular, we note that there are at least two natural semiheap structures on* $\mathbb{C}^2$ *induced by the same underlying metric structure.*

Similarly, the vector ternary product can be extended to the tensor product of Hilbert spaces. We denote the (completed) tensor product as $\mathcal{H} = \mathcal{H}_1 \otimes \mathcal{H}_2$. We remark that composite quantum systems are always described via the tensor products of their components. Basic elements of $\mathcal{H}$ are pairs which, as standard, we write as $|\psi\rangle \otimes |\phi\rangle$. The inner product (used for the completion) is, on basic elements, given by

$$(|\psi_1\rangle \otimes |\phi_1\rangle, |\psi_2\rangle \otimes |\phi_2\rangle) \longmapsto \langle\psi_1|\psi_2\rangle\,\langle\phi_1|\phi_2\rangle\,,$$

which is then extended via linearity. The vector ternary product (on basic elements) is given by

$$\begin{aligned}[|\psi_1\rangle \otimes |\phi_1\rangle, |\psi_2\rangle \otimes |\phi_2\rangle, |\psi_3\rangle \otimes |\phi_3\rangle] &:= (|\psi_1\rangle \otimes |\phi_1\rangle)\langle\psi_2|\psi_3\rangle\langle\phi_2|\phi_3\rangle \\ &= |\psi_1\rangle\langle\psi_2|\psi_3\rangle \otimes |\phi_1\rangle\langle\phi_2|\phi_3\rangle \\ &= [|\psi_1\rangle, |\psi_2\rangle, |\psi_3\rangle] \otimes [|\phi_1\rangle, |\phi_2\rangle, |\phi_3\rangle]\,.\end{aligned}$$

We observe that quite as expected, the vector ternary product on a tensor product of Hilbert spaces is the tensor product of the vector ternary products. This construction then generalises to any finite tensor product of Hilbert spaces.

2.2. Bounded Linear Operators and Their Ternary Algebra

We will denote the $*$-algebra of bounded (so, continuous) operators on $\mathcal{H}$ by $\mathcal{B}(\mathcal{H})$. Following our previous notation, we may also mean by $\mathcal{B}(\mathcal{H})$ just the set of bounded linear operators, the context should be clear. If we want to consider just the vector space structure then we will write $(\mathcal{B}(\mathcal{H}), +)$.

Definition 3. *Let $\mathcal{H}$ be a Hilbert space and let $\mathcal{B}(\mathcal{H})$ be the the $*$-algebra of bounded operators on $\mathcal{H}$. The* operator ternary product $[-,-,-] : \mathcal{B}(\mathcal{H}) \times \mathcal{B}(\mathcal{H}) \times \mathcal{B}(\mathcal{H}) \longrightarrow \mathcal{B}(\mathcal{H})$ *is defined as*

$$[A_1, A_2, A_3] := A_1 A_2^\dagger A_3 .$$

Remark 5. *We focus on bounded linear operators to avoid mathematical subtleties with taking adjoints and forming algebras under composition.*

Remark 6. *The ternary product of bounded operators is closely related to the notion of a ternary ring of operators between Hilbert spaces as first introduced by Hestenes [22] and extended to the C^*-algebra case by Zettl [23].*

Proposition 4. *The operator ternary product on $\mathcal{B}(\mathcal{H})$, see Definition 3,*

1. *is linear in the first and third arguments, conjugate linear in the second argument, and*
2. *satisfies the para-associative law, or in other words, $(\mathcal{B}(\mathcal{H}), [-,-,-])$ is a semiheap.*

Proof. Part (1) is clear from the definition. Part (2) follows from a direct calculation. Specifically,

(i) $\big[[A_1, A_2, A_3], A_4, A_5\big] = A_1 A_2^\dagger A_3 A_4^\dagger A_5$,

(ii) $\big[A_1, [A_4, A_3, A_2], A_5\big] = [A_1, A_4 A_3^\dagger A_2, A_5] = A_1 (A_4 A_3^\dagger A_2)^\dagger A_5 = A_1 A_2^\dagger A_3 A_4^\dagger A_5$,

(iii) $\big[A_1, A_2, [A_3, A_4, A_5]\big] = A_1 A_2^\dagger A_3 A_4^\dagger A_5$.

Clearly, (i) = (ii) = (iii). □

Definition 4. *Let $\mathcal{B}(\mathcal{H})$ be the the $*$-algebra of bounded operators on a Hilbert space $\mathcal{H}$. Then the ternary algebra $(\mathcal{B}(\mathcal{H}), +, [-,-,-])$ defined via Proposition 4 is referred to as the* operator ternary algebra.

Example 5. *Considering the complex line, it is clear that $\mathcal{B}(\mathbb{C}) = \mathrm{Mat}_{1\times 1}(\mathbb{C}) = \mathbb{C}$. Thus, the operator and vector ternary products are identical, see Example 1.*

Example 6. *Continuing Example 2, as the Hilbert space is isomorphic to $\mathbb{C}^2$, it is clear that $\mathcal{B}(\mathbb{C}^2) \cong \mathrm{Mat}_{2\times 2}(\mathbb{C})$. To set some notation, we denote the components of a matrix for the standard basis as $A_i^{\ j}$ and the components of the Hermitian conjugate as $\bar{A}^j_{\ i}$. Then the components of the operator ternary product are*

$$[A, B, C]_i^{\ j} = A_i^{\ k} \bar{B}^l_{\ k} C_l^{\ j} .$$

The operator ternary product for $\mathbb{C}^n$ $(n \in \mathbb{N})$ is of the above from.

As mentioned earlier, unitary operators, i.e., bounded operators such that $U^\dagger U = U U^\dagger = \mathbb{1}_{\mathcal{H}}$, form a group. Because we have the structure of a group and $U^{-1} = U^\dagger$, we have the following corollary. Alternatively, one needs only check the Mal'cev identities, and in this case, it is obvious they hold.

Corollary 4. *The group of unitary operators $\mathcal{U}(\mathcal{H})$ on a Hilbert space $\mathcal{H}$ is a heap under the operator ternary product.*

As standard, we will denote the commutator of bounded operators as $[A_1, A_2] := A_1A_2 - A_2A_1$, for arbitrary A_1 and $A_2 \in \mathcal{B}(\mathcal{H})$. We remind the reader that $[A_1, A_2]^\dagger = -[A_1^\dagger, A_2^\dagger]$, and that we can cast the Jacobi identity into the Jacobi–Leibniz form

$$[A_1, [A_2, A_3]] = [[A_1, A_2], A_3] + [A_2, [A_1, A_3]] . \tag{1}$$

Proposition 5. *The following identity holds for the operator ternary product on $\mathcal{B}(\mathcal{H})$, see Definition 3,*

$$\big[A_1, [A_2, A_3, A_4]\big] = \big[[A_1, A_2], A_3, A_4\big] - \big[A_2, [A_1^\dagger, A_3], A_4\big] + \big[A_2, A_3, [A_1, A_4]\big] ,$$

for all A_1, A_2, A_3 and $A_4 \in \mathcal{B}(\mathcal{H})$.

Proof. Directly we observe that

$$\begin{aligned}\big[A_1, [A_2, A_3, A_4]\big] &= A_1A_2A_3^\dagger A_4 - A_2A_3^\dagger A_4\, A_1 \\ &= A_1A_2A_3^\dagger A_4 - A_2A_3^\dagger A_4A_1 - A_2A_1A_3^\dagger\, A_4 \\ &+ A_2A_1A_3^\dagger A_4 - A_2A_3^\dagger A_1A_4 + A_2A_3^\dagger A_1\, A_4 \\ &= \big[[A_1, A_2], A_3, A_4\big] - \big[A_2, [A_1^\dagger, A_3], A_4\big] + \big[A_2, A_3, [A_1, A_4]\big] .\end{aligned}$$

□

We interpret Proposition 5 as a generalised version of the Leibniz rule for the commutator over the ternary product, and this should be compared with (1). We make the following observation.

Corollary 5. *If $A_1 \in \mathcal{B}(\mathcal{H})$ is self-adjoint, i.e., $A_1^\dagger = A_1$, then $[\mathrm{i}A_1, -]$ is a derivation over the operator ternary product on $\mathcal{B}(\mathcal{H})$, i.e.,*

$$\big[\mathrm{i}A_1, [A_2, A_3, A_4]\big] = \big[[\mathrm{i}A_1, A_2], A_3, A_4\big] + \big[A_2, [\mathrm{i}A_1, A_3], A_4\big] + \big[A_2, A_3, [\mathrm{i}A_1, A_4]\big] .$$

The unitary group $\mathcal{U}(\mathcal{H})$ acts on $\mathcal{B}(\mathcal{H})$ via similarity transformations. That is, $\rho_U : \mathcal{B}(\mathcal{H}) \to \mathcal{B}(\mathcal{H})$ is given by $A \mapsto U^\dagger A U$, for arbitrary $U \in \mathcal{U}(\mathcal{H})$. We then have the following proposition.

Proposition 6. *Let $\big(\mathcal{B}(\mathcal{H}), [-,-,-]\big)$ be the semiheap associated with bounded linear operators on a Hilbert space $\mathcal{H}$. Then, the action of the unitary group $\mathcal{U}(\mathcal{H})$ on $\mathcal{B}(\mathcal{H})$ is a semiheap homomorphism.*

Proof. The proposition is proved via direct calculation. Specifically,

$$\begin{aligned}\rho_U([A_1, A_2, A_3]) &= U^\dagger[A_1, A_2, A_3]U = U^\dagger A_1A_2^\dagger A_3U = U^\dagger A_1U(U^\dagger A_2^\dagger U)U^\dagger A_3\, U \\ &= [U^\dagger A_1U, U^\dagger A_2U, U^\dagger A_3U] = [\rho_U(A_1), \rho_U(A_2), \rho_U(A_3)] .\end{aligned}$$

□

Corollary 6. *Let $\mathcal{U}(\mathcal{H})$ be the group of unitary operators on a Hilbert space $\mathcal{H}$ and let $U : G \longrightarrow \mathcal{U}(\mathcal{H})$ be a projective representation. Furthermore, let $\big(\mathcal{B}(\mathcal{H}), [-,-,-]\big)$ be the semiheap associated with bounded linear operators. Then, for any $g \in G$, $\rho_U(g) : \mathcal{H} \to \mathcal{H}$ is a semiheap homomorphism and so a homomorphism of ternary algebras.*

Note that $[A_1, A_2, A_3]^\dagger = [A_3^\dagger, A_2^\dagger, A_1^\dagger]$ and so the operator ternary product is well-behaved with respect to taking adjoints. We denote the set of bounded self-adjoint operators, so the bounded observables, as $\mathcal{B}_s(\mathcal{H})$. Two operators A and $B \in \mathcal{B}_s(\mathcal{H})$ are said to be *compatible bounded observables* if they commute, i.e., $AB = BA$. A *compatible set of bounded observables* is a subset of $\mathcal{B}_s(\mathcal{H})$ such that all elements are pairwise compatible, that is, they pairwise commute. Naturally, a sub-semiheap of a semiheap is a subset that is closed with respect to the semiheap operation.

Proposition 7. *Let $\mathcal{B}_s(\mathcal{H})$ be the set of bounded observables on a Hilbert space $\mathcal{H}$. Then any compatible set of bounded observables is closed with respect to the operator ternary product. In other words, any set of compatible bounded observables forms a sub-semiheap of $(\mathcal{B}(\mathcal{H}), [-,-,-])$.*

Proof. Consider three arbitrary (not necessarily distinct) bounded observables A, B and $C \in \mathcal{B}_s(\mathcal{H})$. Then directly

$$[A,B,C]^\dagger = C^\dagger B A^\dagger = CBA = [C,B,A].$$

Upon the assumption these bounded observables pairwise commute we see that $CBA = ABC$ and so $[A,B,C]^\dagger = [A,B,C]$ as required. □

2.3. Distributivity of Operators and Derivations

From the definition of the vector ternary product on a Hilbert space $\mathcal{H}$, see Definition 1, we have the following "distributive law",

$$A[|\psi_1\rangle, |\psi_2\rangle, |\psi_3\rangle] = [A|\psi_1\rangle, |\psi_2\rangle, |\psi_3\rangle], \tag{2}$$

for all $|\psi_1\rangle, |\psi_2\rangle$ and $|\psi_3\rangle \in \mathcal{H}$, and all $A \in \mathcal{B}(\mathcal{H})$. The following was first, to our knowledge, uncovered by Kerner [5]. Let us suppose the Hilbert space in question is finite or countable infinite. Furthermore, let us fix an orthonormal basis $\{|n\rangle\}_{n\in\mathbb{N}}$. With respect to this fixed basis, any vector and operator can be written as

$$|\psi\rangle = \sum_{n=1}^{\infty} c_n |n\rangle, \qquad A = \sum_{l,m}^{\infty} a_{ml} |l\rangle\langle m|.$$

Then, combining the two above expressions

$$A|\psi\rangle = \sum_{n,m,l=1}^{\infty} c_n a_{ml} |l\rangle\langle m|n\rangle = \sum_{n,m,l=1}^{\infty} c_n a_{ml} [|l\rangle, |m\rangle, |n\rangle]. \tag{3}$$

By employing semiheaps and para-associative ternary algebras, we have a unification scheme in which vectors (states) and operators (observables) are treated as the same. It is linear combinations of triplets of vectors that are central to the theory rather than separately vectors and operators.

The distributivity law (2) can be written in the form of a generalised Leibniz rule, and this should directly be compared with Proposition 5.

Proposition 8. *Let $\mathcal{H}$ be a Hilbert space and let $[-,-,-]$ be the associated vector ternary product. Then any bounded linear operator $A \in \mathcal{B}(\mathcal{H})$ satisfies a generalised ternary Leibniz rule*

$$A[|\psi_1\rangle, |\psi_2\rangle, |\psi_3\rangle] = [A|\psi_1\rangle, |\psi_2\rangle, |\psi_3\rangle] - [|\psi_1\rangle, A^\dagger|\psi_2\rangle, |\psi_3\rangle] + [|\psi_1\rangle, |\psi_2\rangle, A|\psi_3\rangle],$$

for all $|\psi_1\rangle, |\psi_2\rangle$ and $|\psi_3\rangle \in \mathcal{H}$.

Proof. In light of (2), we require that $-[|\psi_1\rangle, A^\dagger|\psi_2\rangle, |\psi_3\rangle] + [|\psi_1\rangle, |\psi_2\rangle, A|\psi_3\rangle] = 0$. However, this is the case for any bounded operator A as, directly from Definition 3,

$$[|\psi_1\rangle, A^\dagger|\psi_2\rangle, |\psi_3\rangle] = |\psi_1\rangle\langle\psi_2|A|\psi_3\rangle = [|\psi_1\rangle, |\psi_2\rangle, A|\psi_3\rangle].$$

□

Definition 5. *Let $\mathcal{H}$ be a Hilbert space and let $[-,-,-]$ be its associated vector ternary product. A bounded linear operator $D \in \mathcal{B}(\mathcal{H})$ is said to be a* derivation of the vector ternary product *on $\mathcal{H}$ if it satisfies the ternary Leibniz rule*

$$D[|\psi_1\rangle, |\psi_2\rangle, |\psi_3\rangle] = [D|\psi_1\rangle, |\psi_2\rangle, |\psi_3\rangle] + [|\psi_1\rangle, D|\psi_2\rangle, |\psi_3\rangle] + [|\psi_1\rangle, |\psi_2\rangle, D|\psi_3\rangle],$$

for all $|\psi_1\rangle, |\psi_2\rangle$ and $|\psi_3\rangle \in \mathcal{H}$.

There is a one-to-one correspondence between anti-self-adjoint and self-adjoint operators given by multiplication by $\mathrm{i} = \sqrt{-1}$. Specifically, if A is anti-self-adjoint, then $\mathrm{i}A$ is self-adjoint, i.e., $(\mathrm{i}A)^\dagger = \mathrm{i}A$. Conversely, if B is self-adjoint, then $\mathrm{i}B$ is anti-self-adjoint, i.e., $(\mathrm{i}B)^\dagger = -\mathrm{i}B$. The following proposition appears in ([5] Section 6).

Proposition 9. *There is a one-to-one correspondence between the set of derivations of the vector ternary product on $\mathcal{H}$ and the set of bounded observables $\mathcal{B}_s(\mathcal{H})$.*

Proof. In light of (2), it is clear that $\langle\psi_2|D^\dagger|\psi_3\rangle + \langle\psi_2|D|\psi_3\rangle = 0$ if a bounded linear operator is a derivation. Thus, as the vectors in $\mathcal{H}$ are arbitrary, $D^\dagger = -D$. That is, D must be anti-self-adjoint. We can always find a unique self-adjoint operator $A \in \mathcal{B}(\mathcal{H})$ such that $D = \mathrm{i}A$. Conversely, any self-adjoint operator A corresponds to an anti-self-adjoint operator $\mathrm{i}A = D$. □

Proposition 10. *Derivations of the vector ternary product on a Hilbert space $\mathcal{H}$ are closed under the commutator.*

Proof. If D_1 and D_2 are anti-self-adjoint operators, then $[D_1, D_2]^\dagger = -[D_1, D_2]$, i.e., the commutator is also anti-self-adjoint. □

It is clear that the linear combination $a\,D_1 + b\,D_2$ is also anti-self-adjoint for a and $b \in \mathbb{R}$. Note, rather obviously, this is not the case for linear combinations with complex coefficients with non-zero imaginary parts. We then have the following observation.

Corollary 7. *Derivations of the vector ternary product on a Hilbert space $\mathcal{H}$ form a real Lie algebra with respect to the commutator bracket.*

2.4. The Heapification of Addition of Vectors

Note that Propositions 5 and 8 suggest that for a para-associative ternary product, the generalisation of Leibniz rule should be of the form $D[a,b,c] = [Da,b,c] - [a,D^\dagger b,c] + [a,b,Dc]$ for all elements a, b and c. This is in contrast to the obvious direct generalisation of the Leibniz rule. In particular, we note that there is a linear combination of objects of the form "$+ - +$" and that this is a sign that a heap operation is at play here.

From the Abelian group structure of addition of elements of a Hilbert space we can construct an Abelian heap operation as

$$\{|\psi_1\rangle, |\psi_2\rangle, |\psi_3\rangle\} = |\psi_1\rangle - |\psi_2\rangle + |\psi_3\rangle\,. \tag{4}$$

This Abelian heap is then viewed as replacing the operation of the addition of vectors. The generalised Leibniz rule (see Proposition 8) can then be cast into the form

$$A[|\psi_1\rangle, |\psi_2\rangle, |\psi_3\rangle] = \{[A|\psi_1\rangle, |\psi_2\rangle, |\psi_3\rangle], [|\psi_1\rangle, A^\dagger|\psi_2\rangle, |\psi_3\rangle], [|\psi_1\rangle, |\psi_2\rangle, A|\psi_3\rangle]\}\,.$$

The natural question is what replaces the ring distributive laws of multiplication over the addition.

Proposition 11. *Let $\mathcal{H}$ be a Hilbert space, let $[-,-,-]$ be its associated vector ternary product and let $\{-,-,-\}$ be the associated Abelian heap operation given by (4). We then have the following distributive laws.*

(i) $[\{|\psi_1\rangle,|\psi_2\rangle,|\psi_3\rangle\},|\psi_4\rangle,|\psi_5\rangle] = \{[|\psi_1\rangle,|\psi_4\rangle,|\psi_5\rangle],[|\psi_2\rangle,|\psi_4\rangle,|\psi_5\rangle],[|\psi_3\rangle,|\psi_4\rangle,|\psi_5\rangle]\}$,

(ii) $[|\psi_1\rangle,\{|\psi_2\rangle,|\psi_3\rangle,|\psi_4\rangle\},|\psi_5\rangle] = \{[|\psi_1\rangle,|\psi_2\rangle,|\psi_5\rangle],[|\psi_1\rangle,|\psi_3\rangle,|\psi_5\rangle],[|\psi_1\rangle,|\psi_4\rangle,|\psi_5\rangle]\}$,

(iii) $[|\psi_1\rangle,|\psi_2\rangle,\{|\psi_3\rangle,|\psi_4\rangle,|\psi_5\rangle\}] = \{[|\psi_1\rangle,|\psi_2\rangle,|\psi_3\rangle],[|\psi_1\rangle,|\psi_2\rangle,|\psi_4\rangle],[|\psi_1\rangle,|\psi_2\rangle,|\psi_5\rangle]\}$.

Proof. We will only prove (i) as the other two follow in the same way. We note that via the linearity of vector ternary product that

$$[\{|\psi_1\rangle,|\psi_2\rangle,|\psi_3\rangle\},|\psi_4\rangle,|\psi_5\rangle] = [|\psi_1\rangle,|\psi_4\rangle,|\psi_5\rangle] - [|\psi_2\rangle,|\psi_4\rangle,|\psi_5\rangle] + [|\psi_3\rangle,|\psi_4\rangle,|\psi_5\rangle]\,.$$

□

The above considerations suggest that one can define and study non-empty sets equipped with an Abelian heap and a semiheap structure that satisfy the preceding distributive laws. Informally, such algebraic systems are rings in which both the addition and multiplication are now para-associative ternary operations. This should be compared with Brzeziński's *trusses* (see [4,18]) in which the addition is replaced by an Abelian heap and the multiplication remains a binary operation.

3. Concluding Remarks

In this paper, we have re-examined the semiheaps and associated para-associative algebras that are naturally present in the mathematical setup of quantum mechanics. In particular, their symmetries and generalised derivations have been studied, this has, to the author's knowledge, not been explored before. Interestingly, semiheaps allow one to treat vectors and operators as non-distinct objects (see (3)). The action of an operator on a state is replaced by a linear combination of triplets of states fed into the vector ternary product. As far as we know, this observation, first made by Kerner, has not been exploited in quantum mechanics.

In conclusion, quantum mechanics has provided much inspiration for the study of operator algebras and noncommutative structures. Similarly, quantum mechanics provides the impetus for the investigation of ternary and non-associative structures.

Funding: This research received no external funding.

Data Availability Statement: Not applicable

Acknowledgments: The author thanks Steven Duplij for his encouragement to complete this work. A special thank you goes to Tomasz Brzeziński for introducing the author to heaps and related structures. The author is very grateful to the anonymous referees for there helpful comments and suggestions.

Conflicts of Interest: The author declares no conflict of interest.

References

1. Prüfer, H. Theorie der Abelschen Gruppen. *Math. Z.* **1924**, *20*, 165–187. [CrossRef]
2. Baer, R. Zur Einführung des Scharbegriffs. *J. Reine Angew. Math.* **1929**, *160*, 199–207. [CrossRef]
3. Hollings, C.D.; Lawson, M.V. *Wagner's Theory of Generalised Heaps*; Springer: Cham, Switzerland, 2017; p. xv+189. ISBN 978-3-319-63620-7. [CrossRef]
4. Brzeziński, T. Trusses: Paragons, ideals and modules. *J. Pure Appl. Algebr.* **2020**, *224*, 106258. [CrossRef]
5. Kerner, R. Ternary and non-associative structures. *Int. J. Geom. Methods Mod. Phys.* **2008**, *5*, 1265–1294. [CrossRef]
6. Abramov, V.; Kerner, R.; Liivapuu, O.; Shitov, S. Algebras with ternary law of composition and their realization by cubic matrices. *J. Gen. Lie Theory Appl.* **2009**, *3*, 77–94. [CrossRef]
7. Bazunova, N.; Borowiec, A.; Kerner, R. Universal differential calculus on ternary algebras. *Lett. Math. Phys.* **2004**, *67*, 195–206. [CrossRef]

8. Kerner, R. Ternary generalizations of graded algebras with some physical applications. *Rev. Roum. Math. Pures Appl.* **2018**, *63*, 107–141. Available online: http://imar.ro/journals/Revue_Mathematique/pdfs/2018/2/4.pdf (accessed on 13 January 2022).
9. Michor, P.W.; Vinogradov, A.M. n-ary Lie and associative algebras. *Rend. Sem. Mat. Univ. Pol. Torino* **1996**, *54*, 373–392.
10. Nambu, Y. Generalized Hamiltonian dynamics. *Phys. Rev. D* **1973**, *7*, 2405–2412. [CrossRef]
11. Baxter, R.J. Partition function of the eight-vertex lattice model. *Ann. Phys.* **1972**, *70*, 193–228. [CrossRef]
12. Yang, C. Some exact results for the many-body problem in one dimension with repulsive delta-function interaction. *Phys. Rev. Lett.* **1967**, *19*, 1312–1315. [CrossRef]
13. Bagger, J.; Lambert, N. Modeling multiple M2-branes. *Phys. Rev. D* **2007**, *75*, 045020. [CrossRef]
14. Gustavsson, A. Algebraic structures on parallel M2-branes. *Nucl. Phys. B* **2009**, *811*, 66–76. [CrossRef]
15. Azcárraga, J.A.D.; Izquierdo, J.M. n-ary algebras: A review with applications. *J. Phys. A Math. Theor.* **2010**, *43*, 293001. [CrossRef]
16. Lada, T.; Stasheff, J. Introduction to SH Lie algebras for physicists. *Internat. J. Theoret. Phys.* **1993**, *32*, 1087–1103. [CrossRef]
17. Rump, W. Braces, radical rings, and the quantum Yang-Baxter equation. *J. Algebr.* **2007**, *307*, 153–170. [CrossRef]
18. Brzeziński, T. Trusses: Between braces and rings. *Trans. Am. Math. Soc.* **2019**, *372*, 4149–4176. [CrossRef]
19. Nichita, F.F. (Ed.) *Hopf Algebras, Quantum Groups and Yang-Baxter Equations*; MDPI: Basel, Switzerland, 2019.
20. Nichita, F.F. (Ed.) *Non-Associative Structures and Other Related Structures*; MDPI: Basel, Switzerland, 2020.
21. Wigner, E.P. Group theory and its application to the quantum mechanics of atomic spectra. In *Pure and Applied Physics*; Expanded and Improved Edition; Academic Press: New York, NY, USA; London, UK, 1959; Volume 5, p. xi+372.
22. Hestenes, M.R. A ternary algebra with applications to matrices and linear transformations. *Arch. Ration. Mech. Anal.* **1962**, *11*, 138–194. [CrossRef]
23. Zettl, H. A characterization of ternary rings of operators. *Adv. Math.* **1983**, *48*, 117–143. [CrossRef]

MDPI

Article

Maxwell's Equations in Homogeneous Spaces for Admissible Electromagnetic Fields

Valery V. Obukhov [1,2]

[1] Institute of Scientific Research and Development, Tomsk State Pedagogical University, Tomsk 634041, Russia; obukhov@tspu.edu.ru

[2] Laboratory for Theoretical Cosmology, International Center of Gravity and Cosmos, Tomsk State University of Control Systems and Radio Electronics, Tomsk 634050, Russia

Abstract: Maxwell's vacuum equations are integrated for admissible electromagnetic fields in homogeneous spaces. Admissible electromagnetic fields are those for which the space group generates an algebra of symmetry operators (integrals of motion) that is isomorphic to the algebra of group operators. Two frames associated with the group of motions are used to obtain systems of ordinary differential equations to which Maxwell's equations reduce. The solutions are obtained in quadratures. The potentials of the admissible electromagnetic fields and the metrics of the spaces contained in the obtained solutions depend on six arbitrary time functions, so it is possible to use them to integrate field equations in the theory of gravity.

Keywords: Maxwell's vacuum equations; Hamilton–Jacobi equation; Klein–Gordon–Fock equation; algebra of symmetry operators; separation of variables; linear partial differential equations

Citation: Obukhov, V.V. Maxwell's Equations in Homogeneous Spaces for Admissible Electromagnetic Fields. *Universe* **2022**, *8*, 245. https://doi.org/10.3390/universe8040245

Academic Editors: Steven Duplij and Michael L. Walker

Received: 31 March 2022
Accepted: 12 April 2022
Published: 15 April 2022

1. Introduction

A special place in mathematical physics is occupied by the problem of exact integration of the equations of motion of a classical or quantum test particle in external electromagnetic and gravitational fields. This problem is closely related to the study of the symmetry of gravitational and electromagnetic fields in which a given particle moves. A necessary condition for the existence of such symmetry is the admissibility of the algebra of symmetry operators, given by vector and tensor Killing fields, for spacetime and the external electromagnetic field. The algebras of these operators are isomorphic to the algebras of the symmetry operators of the equations of motion of a test particle—Hamilton–Jacobi, Klein–Gordon–Fock, or Dirac–Fock. At present, two methods are known for the exact integration of the equations of motion of a test particle. These are the methods of commutative and noncommutative integration. The first method is based on the use of commutative algebra of symmetry operators (integrals of motion) that form a complete set. The complete set includes linear operators of first and second degree in momentum formed by vector and tensor Killing fields of complete sets of geometric objects of V_4. The method is known as the method of complete separation of variables (in the Hamilton–Jacobi, Klein–Gordon–Fock, or Dirac–Fock equations). The spaces in which the method of complete separation of variables is applicable are called Stackel spaces. The theory of Stackel spaces was developed in [1–12]. A description of the theory and a detailed bibliography can be found in [13–16]. The most frequently used exact solutions of the gravitational field equations in the theory of gravity were constructed on the basis of Stackel spaces (see, e.g., [17–19]). These solutions are still widely used in the study of various effects in gravitational fields (see, e.g., [20–27]).

The second method (noncommutative integration) was developed in [28]. This method is based on the use of algebra of symmetry operators, which are linear in momenta and constructed using Killing vector fields forming noncommutative groups of motion of spacetime G_3 and G_4. The algebras of the symmetry operators of the Klein–Gordon–Fock equation constructed using the algebras of the operators of the noncommutative motion group of space V_4 are complemented to a commutative algebra by the operators of differentiation of the first order in 4 essential parameters. Among these spacetime manifolds, the homogeneous spaces are of greatest interest for the theory of gravity (see, e.g., [29–36]).

Thus, these two methods complement each other to a considerable extent and have similar classification problems (by solving the classification problem, we mean enumerating all metrics and electromagnetic potentials that are not equivalent in terms of admissible transformations). Among these classification problems, the most important are the following.

Classification of all metrics of homogeneous and Stackel spaces in privileged coordinate systems. For Stackel spaces, this problem was solved in building the theory of complete separation of variables in the papers cited above. For homogeneous spaces, this problem was solved in the work of Petrov (see [37]).

Classification of all (admissible) electromagnetic fields applicable to these methods. For the Hamilton–Jacobi and Klein–Gordon–Fock equations, this problem is completely solved in homogeneous spaces (see [38–43]). In Stackel spaces, it is completely solved for the Hamilton–Jacobi equation and partially solved for the Klein–Gordon–Fock equation (see [14–16]).

Classification of all vacuum and electrovacuum solutions of the Einstein equations with metrics of Stackel and homogeneous spaces in admissible electromagnetic fields. This problem has been completely solved for the Stackel metric (see [17–20]). However, this classification problem has not yet been studied for homogeneous spaces.

The solutions to these problems can be viewed as stages of the solution of a single classification problem. In the first two stages, we find all relevant gravitational and electromagnetic fields that are not connected by field equations. In the third stage, using the results of the first two stages, we find metrics and electromagnetic potentials that satisfy the Einstein–Maxwell vacuum equations and have physical meaning.

Thus, for the complete solution to the problem of uniform classification, the Einstein–Maxwell vacuum equations must be integrated using the previously found potentials of admissible electromagnetic fields and the known metrics of homogeneous spaces in privileged (canonical) coordinate systems. This problem can also be divided into two stages. In the first stage, all solutions of Maxwell's vacuum equations for the potentials of admissible electromagnetic fields should be found. The present work is devoted to this stage. In the next stage, the plan is to use the obtained results for the integration of the Einstein–Maxwell equations. This will be the subject of further research. The present work is organized as follows.

Section 2 contains information from the theory of homogeneous spaces, which will be used later, and definitions and conditions for the potentials of admissible electromagnetic fields, written in canonical frames associated with motion groups of a homogeneous space.

In the Section 3 Maxwell's vacuum equations are written in canonical frames.

The Section 4 contains all solutions of Maxwell's vacuum equations for homogeneous spaces admitting groups of motions $G_3(I) - G_3(VI)$.

2. Homogeneous Spaces

By the accepted definition, a spacetime manifold V_4 is a homogeneous space—if a three-parameter group of motions acts on it—whose transitivity hypersurface V_3 is endowed with the Euclidean space signature. Let us introduce a semi-geodesic coordinate system $[u^i]$, in which the metric V_4 has the form:

$$ds^2 = g_{ij}du^i du^j = -du^{0^2} + g_{\alpha\beta}du^\alpha du^\beta, \quad det|g_{\alpha\beta}| > 0. \tag{1}$$

The coordinate indices of the variables of the semi-geodesic coordinate system are denoted by the lower-case Latin letters: $i, j, k, l = 0, 1, \ldots, 3$. The coordinate indices of the variables of the local coordinate system on the hypersurface V_3 are denoted by the lower-case Greek letters: $\alpha, \beta, \gamma, \sigma = 1, \ldots, 3$. A 0 index denotes the temporary variable. Group indices and indices of nonholonomic frames are denoted by $a, d, c = 1, \ldots, 3$. Summation is performed over repeated upper and lower indices within the index range.

There is another (equivalent) definition of a homogeneous space, according to which the spacetime V_4 is homogeneous if its subspace V_3, endowed with the Euclidean space signature, admits a set of coordinate transformations (the group G_3 of motions spaces V_4) that allow the connection of any two points in V_3. (see, e.g., [44]). This definition directly implies that the metric tensor of the V_3 space can be represented as follows:

$$g_{\alpha\beta} = e^a_\alpha e^b_\beta \eta_{ab}, \quad ||\eta_{ab}|| = ||a_{ab}(u^0)||, \quad e^a_{\alpha,0} = 0, \quad det||a_{ab}|| = {l_0}^2, \tag{2}$$

while the form:

$$\omega^a = e^a_\alpha du^\alpha$$

is invariant under the transformation group G_3. The vectors of the frame e^a_α (we call them canonical) define a nonholonomic coordinate system in V_3, and their dual triplet of vectors:

$$e^\alpha_a, \quad e^\alpha_a e^b_\alpha = \delta^b_a, \quad e^\alpha_a e^a_\beta = \delta^\alpha_\beta$$

define the operators of the G_3 algebra group:

$$\hat{Y}_a = e^\alpha_a \partial_a, \quad [\hat{Y}_a, \hat{Y}_b] = C^c_{ab} \hat{Y}_c.$$

The Killing vector fields ξ^α_a and their dual vector fields ξ^a_α form another frame in the space V_3 (we will call it the Killing frame) and another representation of the algebra of group G_3. In the dual frame, the metric of the space V_3 has the form:

$$g_{\alpha\beta} = \xi^a_\alpha \xi^b_\beta G_{ab}, \quad \xi^\alpha_a \xi^b_\alpha = \delta^b_a, \quad \xi^\alpha_a \xi^a_\beta = \delta^\alpha_\beta, \tag{3}$$

where G_{ab} are the nonholonomic components of the $g_{\alpha\beta}$ tensor in this framework. The vector fields ξ^α_a satisfy the Killing equations:

$$g^{\alpha\beta}_{,\gamma} \xi^\gamma_a = g^{\alpha\gamma} \xi^\beta_{a,\gamma} + g^{\beta\gamma} \xi^\alpha_{a,\gamma} \tag{4}$$

and form the infinitesimal group operators of the algebra G_3:

$$\hat{X}_a = \xi^\alpha_a \partial_\alpha, \quad [\hat{X}_a, \hat{X}_b] = C^c_{ab} \hat{X}_c. \tag{5}$$

The Killing equation in the ξ^α_a frame has the following form:

$$G^{ab}_{|c} = G^{ad} C^b_{dc} + G^{bd} C^a_{dc} \quad (|a = \xi^\alpha_a \partial_\alpha). \tag{6}$$

Indeed, substituting the expression:

$$g^{\alpha\beta} = \xi^\alpha_a \xi^\beta_b G^{ab}$$

into Equation (4), we get

$$G^{ab}((\xi^\alpha_{a|c} \xi^\beta_b - \xi^\alpha_a \xi^\beta_{c|b}) + (\xi^\alpha_a \xi^\beta_{b|c} - \xi^\beta_a \xi^\alpha_{c|b})) + \xi^\alpha_a \xi^\beta_b G^{ab}_{|c} = 0.$$

Substituting here the commutation relations (5), we get:

$$(G^{ab}_{|c} - G^{ad} C^b_{dc} - G^{bd} C^a_{dc}) \xi^\alpha_a \xi^\beta_b = 0.$$

The Hamilton–Jacobi equation for a charged test-particle in an external electromagnetic field with potential A_i is:

$$H = g^{ij} P_i P_j = m, \quad P_i = p_i + A_i, \quad p_i = \partial_i \varphi. \tag{7}$$

The integrals of motion of the free Hamilton–Jacobi equation are given using Killing vector fields as follows:

$$X_a = \xi^i_a p_i, \tag{8}$$

Thus, the symmetry of the space given by the Killing vector fields is directly related to the symmetry of the equations of the geodesics given by the integrals of motion. The Hamilton–Jacobi method makes it possible to find these integrals and use them to integrate the geodesic equations. Therefore, the study of the behavior of geodesics is necessary for the study of the geometry of space.

The linear momentum integral of Equation (7) has the following form:

$$X_a = \xi^i_a P_i + \gamma_a, \tag{9}$$

where γ_α are some functions of u^i. Equation (7) admits a motion integral of the form (8) if H and $\hat{X}_a$ commute under Poisson brackets:

$$[H, \hat{X}_a]_P = \frac{\partial H}{\partial p_i}\frac{\partial \hat{X}_a}{\partial x^i} - \frac{\partial H}{\partial x^i}\frac{\partial \hat{X}_a}{\partial p_i} = 0 \rightarrow g^{i\sigma}(\xi^j_a F_{ji} + \gamma_{a,i})P_\sigma = 0. \tag{10}$$

Hence:

$$\gamma_{a,i} = \xi^j_a F_{ij}, \quad F_{ji} = A_{i,j} - A_{j,i}. \tag{11}$$

Thus, the admissible electromagnetic field is determined from Equation (11) (see [41]). In [39,40] it was proved that in the case of a homogeneous space, the conditions of (11) can be represented as follows:

$$\mathbf{A}_{a|b} = C^c_{ba}\mathbf{A}_c, \tag{12}$$

at the same time:

$$\gamma_a = -\mathbf{A}_a \rightarrow \hat{X}_a = \xi^\alpha_a \partial_\alpha.$$

Here, it is denoted that:

$$\mathbf{A}_a = \xi^i_a A_i,$$

It can be shown that Equation (12) forms a completely integrable system. This system specifies the necessary and sufficient conditions for the existence of algebra of integrals of motion that are linear in momenta for Equation (7). Note that in admissible electromagnetic fields given by the conditions (12), the Klein–Gordon–Fock equation:

$$\hat{H}\varphi = (g^{ij}\hat{P}_i\hat{P}_j)\varphi = m^2\varphi, \quad \hat{P}_k = \hat{p}_k + A_k, \quad \hat{p}_k = -\imath\hat{\nabla}_k$$

also admits an algebra of symmetry operators of the form (see [39,41]):

$$\hat{X}_a = \xi^i_a \hat{\nabla}_i$$

$\hat{\nabla}_i$ is the covariant derivative operator corresponding to the partial derivative operator—$\hat{\partial}_i = \imath\hat{p}_i$ in the coordinate field u^i. Function φ is a scalar field, $m = const$. All admissible electromagnetic fields for the homogeneous spacetime are found in [39]. We will use the results of A.Z. Petrov [37]. We follow the notation used in this book with minor exceptions. For example, the nonignorable variable x^4 will be denoted u^0, etc.

3. Maxwell's Equations for an Admissible Electromagnetic Field in Homogeneous Spacetime

Consider Maxwell's equations with zero electromagnetic field sources in homogeneous spacetime in the presence of an admissible electromagnetic field:

$$\frac{1}{\sqrt{-g}}(\sqrt{-g}F^{ij})_{,j} = 0, \quad g = det|g_{\alpha\beta}|. \tag{13}$$

when $i = 0$ from the system (13), the equation follows:

$$\frac{1}{\sqrt{-g}}(\sqrt{-g}g^{\alpha\beta}A_{\beta,0})_{,\alpha} = 0. \tag{14}$$

Using the Killing Equations (4) and (5), we can obtain:

$$\frac{g_{|a}}{g} = 2\xi^{\alpha}_{a,\alpha}.$$

Indeed,

$$-\frac{g_{|a}}{g} = g^{\alpha\beta}_{|a}g_{\alpha\beta} = G^{bc}_{|a}G_{bc} + 2\xi^{\alpha}_{a,\alpha} + 2C_a = 2\xi^{\alpha}_{a,\alpha} \quad (C_a = C^{b}_{ab}).$$

Substituting this expression and the relation (12) into Equation (14), we get:

$$G^{ab}C_b\mathbf{A}_{a,0} = 0. \tag{15}$$

In the case of spaces with groups $G_3(I), G_3(II), G_3(VIII), G_3(IX) C_a = 0$. That is why Equation (15) is satisfied. In the case of the groups $G_3(III), -G_3(VII)$ $C_a = const\delta_{a3}$, and from (15) it follows:

$$\eta^{3a}\tilde{\mathbf{A}}_{a,0} = 0, \quad \tilde{\mathbf{A}}_a = A_\alpha e^{\alpha}_{a}. \tag{16}$$

For $i = \alpha$ we have:

$$\frac{1}{\sqrt{g}}(\sqrt{g}g^{\alpha\beta}F_{\beta 0})_{,0} + \frac{1}{\sqrt{g}}(\sqrt{g}g^{\alpha\beta}g^{\gamma\sigma}F_{\beta\sigma})_{,\gamma} = 0. \tag{17}$$

We transform Equation (17) using the (2) frame. The first term then has the form:

$$\frac{1}{\sqrt{-g}}(\sqrt{-g}g^{\alpha\beta}F_{\beta 0})_{,0} = -\frac{1}{l_0}(l_0\eta^{ab}\tilde{\mathbf{A}}_{a,0})_{,0}e^{\alpha}_{b}, \quad (l_0)^2 = det|\eta_{ab}|.$$

The second term using the (3) frame, the relations (12), and the commutation relations between the operators of the group can be reduced to the following form:

$$\frac{1}{\sqrt{g}}(\sqrt{g}g^{\alpha\beta}g^{\gamma\sigma}F_{\beta\sigma})_{,\gamma} = \frac{1}{2}G^{a_2b_1}C^{a}_{a_2b_2}(2C_{b_1}G^{bb_2} + C^{b}_{a_1b_1}G^{a_1b_2})\xi^{\alpha}_{b}\xi^{\beta}_{a}e^{c}_{\beta}\tilde{\mathbf{A}}_c.$$

So Equation (17) can be written as follows:

$$\frac{1}{l_0}(l_0\eta^{ab}\tilde{\mathbf{A}}_{b,0})_{,0} = \tilde{W}^{ba}\tilde{\mathbf{A}}_b, \tag{18}$$

where:

$$\tilde{W}^{ab} = (e^{a}_{\beta}\xi^{\beta}_{a_1})(e^{a}_{\alpha}\xi^{\alpha}_{b_1})W^{a_1b_1}, \quad W^{ab} = \frac{1}{2}G^{a_2b_1}C^{a}_{a_2b_2}(2C_{b_1}G^{bb_2} + C^{b}_{a_1b_1}G^{a_1b_2}). \tag{19}$$

Then, Maxwell's equations can be represented as follows:

$$\beta^{a}_{,0} = l_0\tilde{W}^{ba}\tilde{\mathbf{A}}_b, \tag{20}$$

$$\tilde{\mathbf{A}}_{a,0} = \frac{1}{l_0}\beta^{b}\eta_{ab}. \tag{21}$$

4. Maxwell's Equations for Spaces Type I–VI According to Bianchi Classification

The group operators in the canonical coordinate set of homogeneous spaces type I–VI according to the Bianchi classification can be represented as follows (see [37]):

$$X_1 = p_1, \quad X_2 = p_2, \quad X_3 = (ru^1 + \varepsilon u^2)p_1 + nu^2p_2 - p_3. \tag{22}$$

The values k ε, n for each group take the following values.)

$$G(I): \ k=0, \quad \varepsilon=0, \quad n=0.$$

$$G(II): k=0, \quad \varepsilon=1, \quad n=0.$$

$$G(III)\ k=1, \quad \varepsilon=0, \quad n=0.$$

$$G(IV)\ k=1, \quad \varepsilon=1, \quad n=1.$$

$$G(V): k=1, \quad \varepsilon=0, \quad n=1.$$

$$G(VI)\ k=1, \quad \varepsilon=0, \quad n=2.$$

Structural constants can be represented as follows:

$$C^c_{ab} = k\delta^c_1(\delta^1_a\delta^3_b - \delta^3_a\delta^1_b) + (\varepsilon\delta^c_1 + n\delta^c_2)(\delta^2_a\delta^3_b - \delta^3_a\delta^2_b) \to C_a = -(k+n)\delta^3_a \tag{23}$$

Find the frame vectors $[\xi^\alpha_a]$, $[e^\alpha_a]$ and their dual vectors $[\xi^a_\alpha]$, $[e^a_\alpha]$.

$$\xi^\alpha_a\xi^b_\alpha = e^\alpha_a e^b_\alpha = \delta^b_a, \quad \xi^\alpha_a\xi^a_\beta = e^\alpha_a e^a_\beta = \delta^\alpha_\beta.$$

For this, we use the metrics of homogeneous spaces and the group operators given in [37].

$$\xi^\alpha_a = \delta^1_a\delta^\alpha_1 + \delta^2_a\delta^\alpha_2 + \delta^3_a(\delta^\alpha_1(ku^1 + \varepsilon u^2) + \delta^\alpha_2 nu^2 - \delta^\alpha_3), \tag{24}$$

$$\xi^a_\alpha = \delta^a_1\delta^1_\alpha + \delta^a_2\delta^2_\alpha + \delta^a_3(\delta^1_\alpha(ku^1 + \varepsilon u^2) + \delta^2_\alpha nu^2 - \delta^3_\alpha),$$

$$e^\alpha_a = \delta^1_a\delta^\alpha_1\exp(-ku^3) + \delta^2_a(-\delta^\alpha_1\varepsilon u^3\exp(-ku^3) + \delta^\alpha_2\exp(-nu^2)) + \delta^\alpha_3\delta^3_a, \tag{25}$$

$$e^\alpha_a = \delta^a_1\delta^1_\alpha\exp(ku^3) + \delta^2_a(\delta^\alpha_1\varepsilon u^3\exp nu^3 + \delta^\alpha_2\exp nu^2)) + \delta^3_\alpha\delta^3_a.$$

With these expressions, we find the matrix $\tilde{W}^{ab}$ (19).

$$\tilde{W}^{ab} = \frac{1}{{l_0}^2}[\delta^a_1\delta^b_1(\varepsilon g_{11} + \varepsilon(n-k)g_{12} - kng_{22})\exp(-2nu^3) \tag{26}$$

$$-(\delta^a_1\varepsilon u^3 + \delta^a_2)(\delta^b_1\varepsilon u^3 + \delta^b_2)kng_{11}\exp(-2ku^3) +$$

$$[\delta^b_1(\delta^a_1\varepsilon u^3 + \delta^a_2)n(g_{12} + \varepsilon g_{11})) + \delta^a_1(\delta^b_1\varepsilon u^3 + \delta^b_2)k(g_{12} - \varepsilon g_{11})].$$

Here (see [37]):

$$g_{11} = a_{11}\exp 2ku^3, \quad g_{12} = (\varepsilon u^3 a_{11} + a_{12})\exp(n+k)u^3, \quad g_{22} = (\varepsilon {u^3}^2 a_{11} + 2\varepsilon a_{12} + a_{22})\exp 2nu^3,$$

Maxwell's Equations (20) and (21) become:

$$\dot{\beta}^b = \frac{1}{l_0}[\delta^a_1\delta^b_1(\varepsilon g_{11} + \varepsilon(n-k)g_{12} - kng_{22})\exp(-2nu^3) \tag{27}$$

$$\{-(\delta^a_1\varepsilon u^3 + \delta^a_2)(\delta^b_1\varepsilon u^3 + \delta^b_2)kng_{11}\exp(-2ku^3) +$$

$$[\delta^b_1(\delta^a_1\varepsilon u^3 + \delta^a_2)n(g_{12} + \varepsilon g_{11})) + \delta^a_1(\delta^b_1\varepsilon u^3 + \delta^b_2)k(g_{12} - \varepsilon g_{11})]\tilde{\mathbf{A}}_a,$$

$$\beta^a = l_0\eta^{ab}\tilde{\mathbf{A}}_{b,0}. \tag{28}$$

The dots denote the time derivatives. The components $\tilde{\mathbf{A}}_a$ are defined by the solutions of the (12) $\mathbf{A}_b$ system of equations using the formulas:

$$\tilde{\mathbf{A}}_a = e^{\alpha}_{a}\xi^{b}_{\alpha}\mathbf{A}_b \tag{29}$$

Further solutions of the system of Equation (27) for homogeneous spaces with groups of motions $G_3(I - VI)$ are given. Spatial metrics are given in the book [37]. Solutions for the system (12) can be found in [38],

$$\alpha_a = \alpha_a(u^0).$$

4.1. Group $G_3(I)$

As the parameters k, n, ε and C^a_{bc} equal zero, $G_3(I)$ is an Abelian group. The components of the vector electromagnetic potential have the form:

$$\mathbf{A}_a = \tilde{\mathbf{A}}_a = A_a = \alpha_a,$$

Substituting these expressions into the system of Equations (27) and (28), we obtain the following system of ordinary differential equations:

$$\dot{\beta}^a = 0 \rightarrow \beta^a = c^a = const;$$

$$l_0\dot{\alpha}_a = a_{ba}c^b \quad \rightarrow \alpha_q = \int \frac{a_{ab}c^b}{l_0}du^0, \quad {l_0}^2 = det|a_{ab}|.$$

All components of a_{ab} are arbitrary functions of u^0.

4.2. Group $G_3(II)$

For the group $G_3(II)$ the parameters k, n, ε have the following values: $k = n = 0, \varepsilon = 1$.

The components of the vector electromagnetic potential in the frames $[\xi^{\alpha}_{a}]$ and $[e^{\alpha}_{a}]$ have the form:

$$\mathbf{A}_1 = \alpha_1, \quad \mathbf{A}_2 = \alpha_2 + \alpha_1 u^3, \quad \mathbf{A}_3 = \alpha_1 u^3 - \alpha_3; \quad \tilde{\mathbf{A}}_a = \alpha_a.$$

Substituting these expressions into the system of Equations (27) and (28), we obtain the following system of ordinary differential equations:

$$l_0\dot{\beta}_a = \alpha_1 a_{11}\delta_{1a} \rightarrow l_0\dot{\beta}_1 = \alpha_1 a_{11}, \quad \beta_2 = c_2, \quad \beta_3 = c_3 \quad (\beta_a = \delta_{ab}\beta^b); \tag{30}$$

$$l_0\dot{\alpha}_a = a_{1a}\beta_1 + a_{2a}c_2 + a_{3a}c_3, \quad {l_0}^2 = det|a_{ab}| \quad (c_a = const,). \tag{31}$$

Set of equations(30) and (31) contains five equations for 11 functions:

$$l_0, \quad a_{ab}, \quad \alpha_a, \quad \beta_1.$$

We should consider separately the variants $\alpha_1 = 0$ and $\alpha_1 \neq 0$.

1. $\alpha_1 = 0 \rightarrow \beta_1 = c_1 = const$. Then the set of Equations (30) and (31) has a quadrature solution:

$$\alpha_q = \int \frac{a_{qb}c_{b_1}\delta^{bb_1}}{l_0}du^0 \quad (q = 2, 3).$$

For $a = 0$, Equation (31) implies a linear dependence of the components a_{1q} :

$$c_1 a_{11} + c_2 a_{12} + c_3 a_{13} = 0.$$

All independent components of a_{ab} are arbitrary functions of u^0.

2. $\alpha_1 \neq 0$. Consider the following Equations (30) and (31) from the system:

$$l_0\dot{\alpha}_1 = (a_{11}\beta_1 + c_2a_{12} + c_3a_{13}), \quad l_0\dot{\beta}_1 = a_{11}\alpha_1. \tag{32}$$

Let us take the function a_{11} out of (32). As a result, we obtain:

$$({\alpha_1}^2 - {\beta_1}^2)_{,0} = \frac{2\alpha_1}{l_0}(c_2a_{12} + c_3a_{13}).$$

Hence:

$$\beta_1 = \zeta\sqrt{{\alpha_1}^2 - 2\int\frac{\alpha_1}{l_0}(c_2a_{12} + c_3a_{13})du^0} \quad (\zeta^2 = 1).$$

>From the remaining equations of the system, we get:

$$\alpha_q = \int\frac{(a_{1q}\beta_1 + a_{2q}c_2 + a_{3q}c_3)}{l_0}du^0 \quad (q = 2,3); \quad a_{11} = \frac{l_0\dot{\beta}_1}{\alpha_1}.$$

The functions l_0, α_1, and all components of a_{ab}, except a_{11}, a_{33}, are arbitrary functions of u_0. The component a_{33} results from the equation ${l_0}^2 = det|a_{ab}|$:

$$a_{33} = \frac{{l_0}^2 + a_{11}{a_{23}}^2 + a_{22}{a_{13}}^2 - 2a_{12}a_{13}a_{23}}{a_{11}a_{22} - {a_{12}}^2} \tag{33}$$

4.3. Group $G_3(III)$

For the group $G_3(III)$ the parameters k, n, ε have the following values: $k = 1$, $n = \varepsilon = 0$.

The components of the vector electromagnetic potential in the frames $[\zeta^\alpha_a]$ and $[e^\alpha_a]$ have the form:

$$\mathbf{A}_1 = \alpha_1 \exp u^3, \quad \mathbf{A}_2 = \alpha_2, \quad \mathbf{A}_3 = \alpha_1 \exp u^3 - \alpha_3.$$

Substituting these expressions into the system of Equations (27) and (28), we obtain the following system of ordinary differential equations:

$$l_0\dot{\beta}_a = \alpha_1a_{12}\delta_{2a} \to l_0\dot{\beta}_2 = \alpha_1a_{12}, \quad \beta_1 = c_1, \quad \beta_3 = 0; \tag{34}$$

$$l_0\dot{\alpha}_a = a_{2a}\beta_2 + a_{1a}c. \tag{35}$$

Here and further, Equation (16) is used, according to which $\beta_3 = 0$. The system of Equations (30) and (31) contains five equations for 11 functions:

$$l_0, \quad a_{ab}, \quad \alpha_a, \quad \beta_2.$$

We should separately consider the variants $\alpha_1 = 0$ and $\alpha_1 \neq 0$.

1. $\alpha_1 = 0 \to \beta_2 = c_2 = const$. In this case the
   Then set of equations (30) and (31) has a solution in quadratures:

$$\alpha_q = \int\frac{a_{qb}c_{b_1}\delta^{bb_1}}{l_0}du^0 \quad (q = 2,3).$$

   >From (31) it follows a linear dependence of the components a_{1q} :

$$c_1a_{13} + c_2a_{23} = 0 \to a_{12} = ba_{11}, \quad \beta_1 = b, \quad \beta_2 = 1.$$

   l_0 and all independent components of a_{ab} are arbitrary functions of u^0. The component a_{33} is found from Equation (33).
2. Let $\alpha_1 \neq 0$. Consider the following equations from system (30) and (31):

$$l_0\dot{\alpha}_1 = a_{12}\beta_2 + c_1 a_{11}, \quad l_0\dot{\beta}_2 = a_{12}\alpha_1. \tag{36}$$

from system (36), it follows:

$$({\alpha_1}^2 - {\beta_2}^2)_{,0} = \frac{2\alpha_1}{l_0} c_1 a_{11}.$$

Hence:

$$\beta_2 = \xi\sqrt{{\alpha_1}^2 - 2\int \frac{\alpha_1}{l_0}(c_1 a_{11} + c_3 a_{13}) du^0} (\xi^2 = 1).$$

>From the remaining equations of the system, we get:

$$\alpha_q = \int \frac{(a_{2q}\beta_2 + a_{1q}c_1 + a_{3q}c_3)}{l_0} du^0 \quad (q = 2,3); \quad a_{11} = \frac{l_0\dot{\beta}_2}{\alpha_1}.$$

The functions l_0, α_1 and all components of a_{ab}, except a_{11}, a_{33}, are arbitrary functions of u_0. The component a_{33} results from Equation (33).

4.4. *Group* $G_3(IV)$

For the group $G_3(IV)$ the parameters k, n, ε have the values: $k = n = \varepsilon = 1$.

The components of the vector electromagnetic potential in the frames $[\xi^\alpha_a]$ and $[e^\alpha_a]$ have the form:

$$\mathbf{A}_1 = \alpha_1 \exp u^3, \quad \mathbf{A}_2 = (\alpha_2 + \alpha_1 u^3)\exp u^3,$$
$$\mathbf{A}_3 = (\alpha_1(u^1 + u^2 + u^2u^3) + \alpha_2 u^2)\exp u^3 - \alpha_3;$$
$$\tilde{\mathbf{A}}_a = \alpha_a.$$

Maxwell's Equations (20) and (21) reduce to the following system:

$$l_0\dot{\beta}_a = \delta_{1a}(a_{11}(\alpha_1 + \alpha_2) - \alpha_1 a_{22} + \alpha_2 a_{12}) + \delta_{2a}(\alpha_1 a_{12} - a_{11}(\alpha_1 + \alpha_2)). \tag{37}$$

$$l_0\dot{\alpha}_a = \beta_2 a_{a2} + \beta_1 a_{a1}, \quad \beta_3 = 0. \tag{38}$$

from the system (38) it follows:

$$\dot{\alpha}_3 = \int \frac{\beta_2 a_{32} + \beta_1 a_{31}}{l_0} du^0. \tag{39}$$

Let us now consider the remaining equations.

(A) $\beta_1 \neq 0$.

>From the system (37) it follows:

$$a_{12} = \frac{1}{\beta_1}(l_0\dot{\alpha}_2 - \beta_2 a_{22}) \quad a_{11} = \frac{1}{{\beta_1}^2}(l_0(\dot{\alpha}_1\beta_1 - \dot{\alpha}_2\beta_2) + {\beta_2}^2 a_{22}), \tag{40}$$

Using these relations, we obtain a consequence from the remaining equations of the system (37) and (38):

$$\beta_1\dot{\beta}_2 - \beta_2(\dot{\beta}_1 + \dot{\beta}_2) = \alpha_1\dot{\alpha}_2 - (\alpha_1 + \alpha_2)\dot{\alpha}_1. \tag{41}$$

With Equation (41), the dependent functions α_a, β_a can be expressed in terms of the independent functions. Let us write down the solutions.

1. $(\alpha_1\beta_1 + \beta_2(\alpha_1 + \alpha_2))\beta_2 \neq 0$.

$$\beta_1 = \beta_2(b - \ln\beta_2 - \int \frac{\alpha_1\dot{\alpha}_2 - (\alpha_1 + \alpha_2)\dot{\alpha}_1}{{\beta_2}^2} du^0);$$

$$a_{22} = \frac{l_0(\dot{\alpha}_2(\alpha_1 + \alpha_2) - \beta_1(\dot{\beta}_1 + \dot{\beta}_2))}{\alpha_1\beta_1 + \beta_2(\alpha_1 + \alpha_2)}.$$

$l_0, a_{13}, a_{23}, \varphi$ are arbitrary functions of time. The function a_{33} is expressed in terms of these functions using the relation (33)

2. $\alpha_1\beta_1 + \beta_2(\alpha_1 + \alpha_2) = 0, a_{22}$, is an arbitrary function, depending on u^0.

$$\alpha_1 = a\exp\varphi + b\exp\varphi, \quad \alpha_2 = (1+e)\alpha_1 \quad \beta_2 = a\exp\varphi - b\exp\varphi,$$

$$\beta_1 = e\beta_2 \quad (e = const).$$

$l_0, a_{13}, a_{23}, \varphi$ are arbitrary functions of time. The function a_{33} is expressed in terms of these functions using the relation (33).

3. $\beta_2 = 0$.

$$\alpha_2 = \alpha_1(a + \ln\alpha_1), \quad a_{12} = \frac{l_0\dot{\alpha}_2}{\beta_1}, \quad a_{11} = \frac{l_0\dot{\alpha}_1}{\beta_1}, \quad a_{22} = \frac{l_0(\dot{\alpha}_2(\alpha_1 + \alpha_2) - \dot{\beta}_1\beta_1)}{\alpha_1\beta_1}$$

$l_0, a_{13}, a_{23}, \alpha_1, \beta_1$ are arbitrary functions of time. The function a_{33} is expressed in terms of these functions using the relation (33).

(B) $\beta_1 = 0$. Maxwell's equations take the form:

$$l_0\dot{\beta}_2 = \alpha_1 a_{12} - (\alpha_1 + \alpha_2)a_{11}, \quad l_0\dot{\beta}_2 = -\alpha_1 a_{22} + (\alpha_1 + \alpha_2)a_{12};$$

$$l_0\dot{\alpha}_1 = \beta_2 a_{12}, \quad l_0\dot{\alpha}_2 = \beta_2 a_{22}.$$

The set of equations has the following

(a) $(\alpha_1 + \alpha_2) \neq 0$.

$$\beta_2 = \xi\sqrt{b + 2\int\frac{1}{l_0}(\dot{\alpha}_1(\alpha_1 + \alpha_2) - \alpha_1\dot{\alpha}_2)du_0}. \quad a_{12} = \frac{l_0\dot{\alpha}_1}{\beta_2}, \quad a_{22} = \frac{l_0\dot{\alpha}_2}{\beta_2}.$$

$$a_{11} = \frac{l_0(\alpha_1\dot{\alpha}_1 - \beta_2\dot{\beta}_2)}{\beta_2(\alpha_1 + \alpha_2)}$$

$l_0, a_{13}, a_{23}, \alpha_1, \alpha_2$ are arbitrary functions of time. The function a_{33} is expressed in terms of these functions using relation (33).

(b) $\alpha_2 = -\alpha_1 \rightarrow \alpha_1 = a\exp\varphi - b\exp\varphi \quad \beta_2 = a\exp\varphi + b\exp\varphi, \quad a_{12} = \frac{l_0\dot{\alpha}_1}{\beta_2}$ $a_{22} = \frac{l_0\dot{\alpha}_2}{\beta_2}$.
$l_0, a_{11}, a_{13}, a_{23}, \varphi, \beta_1$ are arbitrary functions of time. The function a_{33} is expressed in terms of these functions using the relation (33).

4.5. *Group $G_3(V)$*

For the group $G_3(V)$ the parameters k, n, ε have the values: $k = n = 1, \varepsilon = 0$. The components of the vector electromagnetic potential in the frames $[\xi^\alpha_a]$ and $[e^\alpha_a]$ have the form:

$$\mathbf{A}_1 = \alpha_1\exp u^3, \quad \mathbf{A}_2 = \alpha_2\exp u^3, \quad \mathbf{A}_3 = (\alpha_1 u^1 + \alpha_2 u^2)\exp u^3 - \alpha_3;$$

$$\tilde{\mathbf{A}}_a = \alpha_a.$$

Maxwell's Equation (18) reduces to the following system of equations:

$$l_0\dot{\alpha}_a = \beta_2 a_{a2} + \beta_1 a_{a1}, \quad \beta_3 = 0. \tag{42}$$

$$l_0\dot{\beta}_a = \delta_{1a}(a_{12}\alpha_2 - \alpha_1 a_{22}) + \delta_{2a}(a_{12}\alpha_1 - a_{11}\alpha_2), \quad (43)$$

Hence:

$$\dot{\alpha}_3 = \int \frac{\beta_2 a_{32} + \beta_1 a_{31}}{l_0} du^0,$$

$$l_0\dot{\alpha}_1 = (a_{11}\beta_1 + a_{12}\beta_2), \quad l_0\dot{\alpha}_2 = (a_{12}\beta_1 + a_{22}\beta_2). \quad (44)$$

1. $\alpha_1 \neq 0$. From the set of equations (43) it follows:

$$a_{12} = \frac{1}{\alpha_1}(l_0\dot{\beta}_2 + \alpha_2 a_{11}), \quad a_{22} = \frac{1}{{\alpha_1}^2}(l_0(\dot{\beta}_2\alpha_2 - \dot{\beta}_1\alpha_1) + a_{11}{\alpha_2}^2). \quad (45)$$

Substituting (45) into (44) , we get the corollary:

$$\beta_1\dot{\beta}_2 - \beta_2\dot{\beta}_1 = \alpha_1\dot{\alpha}_2 - \alpha_2\dot{\alpha}_1. \quad (46)$$

$$a_{11}(\alpha_1\beta_1 + \alpha_2\beta_2) = l_0(\dot{\alpha}_1\alpha_1 - \dot{\beta}_2\beta_2). \quad (47)$$

>From (46), it follows:

$$\alpha_2 = \alpha_1(b + \int \frac{\beta_1\dot{\beta}_2 - \beta_2\dot{\beta}_1}{{\alpha_1}^2} du^0),$$

Let us consider (48).

(a) $\alpha_1\beta_1 + \alpha_2\beta_2 \neq 0$. Then, we have:

$$a_{11} = \frac{l_0(\alpha_1\dot{\alpha}_2 - \alpha_2\dot{\alpha}_1)}{\alpha_1\beta_1 + \alpha_2\beta_2};$$

$l_0, a_{13}, a_{23}, \alpha_1, \beta_a$ are arbitrary functions of time. The function a_{33} is expressed in terms of these functions using the relation (33).

(b) $\alpha_1\beta_1 + \alpha_2\beta_2 = 0 \quad \rightarrow \alpha_1\dot{\alpha}_1 - \beta_1\dot{\beta}_1 = 0, \quad \alpha_1\dot{\alpha}_2 + \beta_2\dot{\beta}_1 = 0.$
>From this, it follows:

$$\alpha_1 = a\exp\varphi + b\exp\varphi, \quad \beta_2 = a\exp\varphi - b\exp\varphi, \quad \alpha_2 = -l\alpha_1, \quad \beta_1 = l\beta_2,$$

where $a, b, l = const, \varphi = \varphi(u^0)$.
$l_0, a_{11}, a_{13}, a_{23}$ are arbitrary functions of time. The function a_{33} is expressed in terms of these functions using the relation (33).

2. $\alpha_1 = 0$. From the system (43), it follows:

$$a_{12} = \frac{l_0\dot{\beta}_1}{\alpha_2}, \quad a_{11} = -\frac{l_0\dot{\beta}_2}{\alpha_2}, \quad a_{22} = \frac{l_0(\dot{\alpha}_2\alpha_2 - \dot{\beta}_1\beta_1)}{\alpha_2\beta_2}, \quad \beta_1 = a\beta_2,$$

here $a = const, l_0, a_{13}, a_{23}, \alpha_2, \beta_2$ are arbitrary functions of time. The function a_{33} is expressed in terms of these functions using the relation (33).

4.6. *Group $G_3(VI)$*

For the group $G_3(VI)$, the parameters k, n, ε have the following values: $k = 1$ $n = 2, \varepsilon = 0$. The components of the vector electromagnetic potential in the frames $[\xi^\alpha_a]$ and $[e^\alpha_a]$ have the form:

$$\mathbf{A}_1 = \alpha_1 \exp u^3, \quad \mathbf{A}_2 = \alpha_2 \exp 2u^3, \quad \mathbf{A}_3 = \alpha_1 u^1 \exp u^3 + 2\alpha_2 u^2 \exp 2u^3 - \alpha_3;$$

$$\tilde{\mathbf{A}}_a = \alpha_a.$$

Maxwell's Equation (18) has the form:

$$l_0\dot{\alpha}_a = \beta_2 a_{a2} + \beta_1 a_{a1}. \tag{48}$$

$$l_0\dot{\beta}_a = \delta_{1a}(a_{12}\alpha_2 - 2\alpha_1 a_{22}) + \delta_{2a}(a_{12}\alpha_1 - 2a_{11}\alpha_2), \quad \beta_3 = 0, \tag{49}$$

and from the system (48), it follows:

$$\dot{\alpha}_3 = \int \frac{\beta_2 a_{32} + \beta_1 a_{31}}{l_0} du^0.$$

$$l_0\dot{\alpha}_1 = (a_{11}\beta_1 + a_{12}\beta_2), \quad l_0\dot{\alpha}_2 = (a_{12}\beta_1 + a_{22}\beta_2). \tag{50}$$

I $\beta_1 \neq 0$, from system (48), it follows:

$$a_{12} = \frac{1}{\beta_1}(l_0\dot{\alpha}_2 - \beta_2 a_{22}), \quad a_{11} = \frac{1}{{\beta_1}^2}(l_0(\dot{\alpha}_1\beta_1 - \dot{\alpha}_2\beta_2) + a_{22}{\beta_2}^2). \tag{51}$$

Substituting (51) into (48), we get:

$$a_{22}(\alpha_1\beta_1 + 2\alpha_2\beta_2) = l_0(\alpha_2\dot{\alpha}_2 - \dot{\beta}_1\beta_1), \tag{52}$$

$$(2\alpha_1\beta_1 + \alpha_2\beta_2)(2\alpha_2\dot{\alpha}_1 + \dot{\beta}_2\beta_1) = (\dot{\beta}_1\beta_2 + 2\dot{\alpha}_2\alpha_1)(\alpha_1\beta_1 + 2\alpha_2\beta_2) = 0 \tag{53}$$

Using this relation, we get the following solutions:

(1) $\alpha_1\beta_1 + 2\alpha_2\beta_2 \neq 0$. From (52) it follows:

$$a_{22} = \frac{l_0(\dot{\alpha}_2\alpha_2 - \dot{\beta}_1\beta_1)}{(\alpha_1\beta_1 + 2\alpha_2\beta_2)}.$$

Denote:

$$\alpha_q = a_q \exp\varphi. \quad \beta_q = b_q \exp\varphi \quad (q = 1, 2),$$

where a_q, b_q, φ are functions of u^0. From Equation (53), we get:

$$\dot{\varphi} = \frac{(\dot{b}_1 b_2 + 2\dot{a}_2 a_1)(a_1 b_1 + 2a_2 b_2) - (2a_1 b_1 + a_2 b_2)(2a_2\dot{a}_1 + \dot{b}_2 b_1)}{(2a_1 a_2 + b_1 b_2)(a_1 b_1 - a_2 b_2)};$$

$$a_{12} = \frac{l_0(\dot{\varphi}a_2 + a_2) - b_2 a_{22}}{b_1}; \quad a_{11} = \frac{l_0((a_1 b_1 - a_2 b_2)\dot{\varphi} + \dot{a}_1 b_1 - \dot{a}_2 b_2) + {b_2}^2 a_{22}}{{b_1}^2};$$

$$a_{22} = \frac{l_0((a_2^2 - b_1^2)\dot{\varphi} + \dot{a}_2 a_2 - \dot{b}_1 b_1)}{2a_1 b_1 + a_2 b_2}.$$

$l_0, a_{13}, a_{23}, a_q, b_q$ are arbitrary functions dependent on time. The function a_{33} is expressed by these functions using the relation (33)

(2) $\dot{\alpha}_2\alpha_2 - \dot{\beta}_1\beta_1 = 0 \quad \rightarrow \quad \alpha_1\beta_1 + 2\alpha_2\beta_2 = 0.$ a_{22}—is an arbitrary function from u^0;

$$\alpha_2 = a\exp\varphi - b\exp(-\varphi), \quad \beta_1 = a\exp\varphi + b\exp(-\varphi).$$

>From this, it follows:

(a)

$$\alpha_1 = -\frac{\beta_2}{2}(\frac{a\exp\varphi - b\exp(-\varphi)}{a\exp\varphi + b\exp(-\varphi)});$$

$$a_{12} = l_0\dot{\varphi} - \frac{\beta_2 a_{22}}{\beta_1}, \quad a_{11} = \frac{l_0(\dot{\alpha}_1\beta_1 - \dot{\alpha}_2\beta_2) + \beta_2^2 a_{22}}{\beta_1^2}$$

(b) $\dot{\varphi} = 0$

$$\beta_1 = 1, \quad \alpha_2 = -2b, \quad \alpha_1 = -b\beta_2, \quad a_{12} = -\beta_2 a_{22}, \quad a_{11} = -bl_0\dot{\beta}_2 + {\beta_2}^2 a_{22}.$$

where $l_0, a.b = const a_{22}, a_{13}, a_{23}, \beta_2, \varphi$ are arbitrary functions dependent on time.

II $\beta_1 = 0$.
>From (48) and (49) it follows:

$$a_{12} = \frac{2l_0\dot{\alpha}_2\alpha_2}{\beta_2}, \quad a_{22} = \frac{l_0\dot{\alpha}_2}{\beta_2}, \quad a_{11} = \frac{l_0(2b^2\dot{\alpha}_2{\alpha_2}^3 - \beta_2\dot{\beta}_2)}{2\alpha_2\beta_2}, \quad \alpha_1 = b{\alpha_2}^2. \quad (54)$$

$l_0, a_{22}, a_{13}, a_{23}, \alpha_2\beta_2$ depends arbitrarily on time functions. The function a_{33} is expressed in terms of these functions using the relation (33).

5. Conclusions

The performed classification of admissible electromagnetic fields will be used in the search for electrovacuum solutions of the Einstein–Maxwell equations. As is already known, the components of the Ricci tensor of a homogeneous space in the frame (2) depend only on time. In order for Einstein's equations with matter to be proven as an integrable system of ordinary differential equations, the equations of motion of matter must be subordinated to the conditions of space symmetry. These conditions were fulfilled first by the potentials of the electromagnetic fields determined in this work.

Funding: This research received no external funding.

Institutional Review Board Statement: Not applicable.

Informed Consent Statement: Not applicable.

Data Availability Statement: The data that support the findings of this study are available within the article.

Acknowledgments: The work is partially supported by the Ministry of Education of the Russian Federation, Project No. FEWF-2020-003.

Conflicts of Interest: The author declares no conflict of interest.

References

1. Stackel, P. Uber die intagration der Hamiltonschen differentialechung mittels separation der variablen. *Math. Ann.* **1897**, *49*, 145–147. [CrossRef]
2. Stackel, P. Ueber Die Bewegung Eines Punktes In Einer N-Fachen Mannigfaltigkeit. *Math. Ann.* **1893**, *42*, 537–563. [CrossRef]
3. Jarov-Jrovoy, M.S. Integration of Hamilton-Jacobi equation by complete separation of variables method. *J. Appl. Math. Mech.* **1963**, *27*, 173–219.
4. Eisenhart, L.P. Separable systems of stackel. *Math. Ann.* **1934**, *35*, 284–305. [CrossRef]
5. Levi-Civita, T. Sulla Integraziome Della Equazione Di Hamilton-Jacobi Per Separazione Di Variabili. *Math. Ann.* **1904**, *59*, 383–397. [CrossRef]
6. Shapovalov, V.N. Symmetry of motion equations of free particle in riemannian space. *Russ. Phys. J.* **1975**, *18*, 1650–1654. [CrossRef]
7. Shapovalov, V.N.; Eckle, G.G. Separation of Variables in the Dirac Equation. *Russ. Phys. J.* **1973**, *16*, 818–823. [CrossRef]
8. Shapovalov, V.N. Symmetry and separation of variables in a linear second-order differential equation. I, II. *Russ. Phys. J.* **1978**, *21*, 645–695.
9. Bagrov, V.G.; Meshkov, A.G.; Shapovalov, V.N.; Shapovalov, A.V. Separation of variables in the Klein-Gordon equations I. *Russ. Phys. J.* **1973**, *16*, 1533–1538. [CrossRef]
10. Bagrov, V.G.; Meshkov, A.G.; Shapovalov, V.N.; Shapovalov, A.V. Separation of variables in the Klein-Gordon equations II. *Russ. Phys. J.* **1973**, *16*, 1659–1665. [CrossRef]
11. Bagrov, V.G.; Meshkov, A.G.; Shapovalov, V.N.; Shapovalov, A.V. Separation of variables in the Klein-Gordon equations III. *Russ. Phys. J.* **1974**, *17*, 812–815. [CrossRef]
12. Shapovalov, V.N. Stäckel spaces. *Sib. Math. J.* **1979**, *20*, 1117–1130. [CrossRef]

13. Miller, W. *Symmetry and Separation of Variables*; Cambridge University Press: Cambridge, UK, 1984; 318p.
14. Obukhov, V.V. Hamilton-Jacobi equation for a charged test particle in the Stackel space of type (2.0). *Symmetry* **2020**, *12*, 12891291. [CrossRef]
15. Obukhov, V.V. Hamilton-Jacobi equation for a charged test particle in the Stackel space of type (2.1). *Int. J. Geom. Methods Mod. Phys.* **2020**, *17*, 2050186. [CrossRef]
16. Obukhov, V.V. Separation of variables in Hamilton-Jacobi and Klein-Gordon-Fock equations for a charged test particle in the stackel spaces of type (1.1). *Int. J. Geom. Methods Mod. Phys.* **2021**, *18*, 2150036. [CrossRef]
17. Carter, B. New family of Einstein spaces. *Phys. Lett.* **1968**, *25*, 399–400. [CrossRef]
18. Carter, B. Separability of the Killing-Maxwell system underlying the generalized angular momentum constant in the Kerr-Newman black hole metrics. *J. Math. Phys.* **1987**, *28*, 1535. [CrossRef]
19. Bagrov, V.G.; Obukhov, V.V. Classes of exact solutions of the Einstein-Maxwell equations. *Ann. Der Phys.* **1983**, *40*, 181–188. [CrossRef]
20. Mitsopoulos, A.; Mitsopoulos, A.; Tsamparlis, M.; Leon, G.; Paliathanasis, A.; Paliathanasis, A. New conservation laws and exact cosmological solutions in Brans-Dicke cosmology with an extra scalar field. *Symmetry* **2021**, *13*, 1364. [CrossRef]
21. Rajaratnam, K.; Mclenaghan, R.G. Classification of Hamilton-Jacobi separation In orthogonal coordinates with diagonal curvature. *J. Math. Phys.***2014**, *55*, 083521. [CrossRef]
22. Chong, Z.W.; Gibbons, G.W.; Pope, C.N. Separability and Killing tensors in Kerr-Taub-Nut-De Sitter metrics in higher dimensions. *Phys. Lett.* **2005**, *609*, 124–132. [CrossRef]
23. Vasudevan, M.; Stevens, K.A.; Page, D.N. Separability of The Hamilton-Jacobi And Klein-Gordon Equations In Kerr-De Sitter Metrics. *Class. Quantum Gravity* **2005**, *22*, 339–352. [CrossRef]
24. Nojiri, S.; Odintsov, S.D.; Oikonomou, V.K. Modified gravity theories on a nutshell: Inflation, bounce and late-time evolution. *Phys. Rep.* **2017**, 692, 1–104. [CrossRef]
25. Bamba, K.S.; Capozziello, S.; Nojiri, S.; Odintsov, S.D. Dark energy cosmology: The equivalent description via different theoretical models and cosmography tests. *Astrophys. Space Sci.* **2012**, *342*, 155. [CrossRef]
26. Capozziello, S.; De Laurentis, M.; Odintsov, S.D. Hamiltonian dynamics and Noether symmetries in extended gravity cosmology. *Uropean Phys. J.* **2012**, *72*, 2068. [CrossRef]
27. McLenaghan, R.G.; Rastelli, G.; Valero, C. Complete separability of the Hamilton-Jacobi equation for the charged particle orbits in a Lienard-Wiehert field. *J. Math. Phys.* **2020**, *61*, 122903. [CrossRef]
28. Shapovalov, A.V.; Shirokov, I.V. Noncommutative integration method for linear partial differential equations. Functional algebras and dimensional reduction. *Theor. Math. Phys.* **1996**, *106*, 3–15. [CrossRef]
29. Osetrin, E.K.; Osetrin, K.E.; Filippov, A.E. Stationary homogeneous models of Stackel spaces of type (2.1). *Russ. Phys. J.* **2020**, *63*, 57–65. [CrossRef]
30. Osetrin, E.; Osetrin, K.; Filippov, A. Spatially Homogeneous Conformally Stackel Spaces of Type (3.1). *Russ. Phys. J.* **2020**, *63*, 403–409. [CrossRef]
31. Osetrin, E.; Osetrin, K.; Filippov, A. Plane Gravitational Waves in Spatially-Homogeneous Models of type-(3.1) Stackel Spaces. *Russ. Phys. J.* **2019**, *64*, 292–301. [CrossRef]
32. Mozhey, N.P. Affine connections on three-dimensional pseudo-Riemannian homogeneous spaces. I. *Russ. Math. J.* **2013**, *57*, 44–62. [CrossRef]
33. Garcia, A.; Hehl, F.W.; Heinicke, C.; Macias, A. The Cotton tensor in Riemannian spacetimes. *Class. Quantum Gravity* **2004**, *21*, 1099–1118. [CrossRef]
34. Marchesiello, A.; Snobl, L.; Winternitz, P. Three-dimensional superintegrable systems in a static electromagnetic field. *J. Phys. Math. Gen.* **2015**, *48*, 395206. [CrossRef]
35. Breev, A.I.; Shapovalov, A.V. Noncommutative integration of the Dirac equation in homogeneous spaces. *Symmetry* **2020**, *12*, 1867. [CrossRef]
36. Breev, A.I.; Shapovalov, A.V. Vacuum quantum effects on Lie groups with bi-invariant metrics. *Int. J. Geom. Methods Mod. Phys.* **2019**, *16*, 1950122. [CrossRef]
37. Petrov, A.Z. *Einstein Spaces*; Pergamon Press: Oxford, UK, 1969.
38. Obukhov, V.V. Algebra of symmetry operators for Klein-Gordon-Fock Equation. *Symmetry* **2021**, *13*, 727. [CrossRef]
39. Obukhov, V.V. Algebra of the symmetry operators of the Klein-Gordon-Fock equation for the case when groups of motions G_3 act transitively on null subsurfaces of spacetime. *Symmetry* **2022**, *14*, 346. [CrossRef]
40. Obukhov, V.V. Algebras of integrals of motion for the Hamilton-Jacobi and Klein-Gordon-Fock equations in spacetime with a four-parameter groups of motions in the presence of an external electromagnetic field. *J. Math. Phys.* **2022**, *63*, 023505. [CrossRef]
41. Magazev, A.A. Integrating Klein-Gordon-Fock equations in an extremal electromagnetic field on Lie groups. *Theor. Math. Phys.* **2012**, *173*, 1654–1667. [CrossRef]
42. Magazev, A.A. Constructing a complete integral of the Hamilton-Jacobi equation on pseudo-riemannian spaces with simply transitive groups of motions. *Math. Physics, Anal. Geom.* **2021**, *24*, 11.

[CrossRef]
43. Magazev, A.A.; Shirokov, I.V.; Yurevich, Y.A. Integrable magnetic geodesic flows on Lie groups. *Theor. Math. Phys.* **2008**, **156**, 1127–1140. [CrossRef]
44. Landau, L.D.; Lifshits, E.M. *Theoretical Physics. Field Theory*, 7th ed.; Butterworth-Heinemann: Oxford, UK, 1988; Volume II, 512p, ISBN 5-02-014420-7.

MDPI
St. Alban-Anlage 66
4052 Basel
Switzerland
Tel. +41 61 683 77 34
Fax +41 61 302 89 18
www.mdpi.com

Universe Editorial Office
E-mail: universe@mdpi.com
www.mdpi.com/journal/universe

www.ingramcontent.com/pod-product-compliance
Lightning Source LLC
LaVergne TN
LVHW071549170726
843515LV00008B/2243